THE REAL GOODS

NINTH EDITION

SOLAR LIVING
SOURCEBOOK

THE REAL GOODS
INDEPENDENT LIVING BOOKS

Paul Gipe, *Wind Power for Home & Business: Renewable Energy for the 1990s and Beyond*

Athena Swentzell Steen, Bill Steen, David Bainbridge, *The Straw Bale House*

Michael Potts, *The Independent Home: Living Well with Power from the Sun, Wind, and Water*

The Union of Concerned Scientists, *Renewables Are Ready*

Gene Logsdon, *The Contrary Farmer*

David Easton, *The Rammed Earth House*

Edward Harland, *Eco-Renovation: The Ecological Home Improvement Guide*

Leandre Poisson and Gretchen Vogel Poisson, *Solar Gardening: Growing Vegetables Year-Round the American Intensive Way*

Real Goods Solar Living Sourcebook: The Complete Guide to Renewable Energy Technologies and Sustainable Living, Ninth Edition, edited by John Schaeffer

Real Goods Trading Company in Ukiah, California, was founded in 1978 to make available new tools to help people live self-sufficiently and sustainably. Through seasonal catalogs, a thrice-annual newspaper (*The Real Goods News*), a periodic *Solar Living Sourcebook*, as well as a book catalog and retail outlets, Real Goods provides a broad range of renewable-energy and resource-efficient products for independent living.

"Knowledge is our most important product" is the Real Goods motto. To further its mission, Real Goods has joined with Chelsea Green Publishing Company to co-create and co-publish the Real Goods Independent Living Book series. The titles in this series are written by pioneering individuals who have firsthand experience in using innovative technology to live lightly on the planet. Chelsea Green books are both practical and inspirational, and they enlarge our view of what is possible as we enter the next millennium.

Ian Baldwin, Jr.
President, Chelsea Green

John Schaeffer
President, Real Goods

NINTH EDITION

THE REAL GOODS
SOLAR LIVING
SOURCEBOOK

The Complete Guide To Renewable Energy Technologies and Sustainable Living

EXECUTIVE EDITOR JOHN SCHAEFFER

Edited by Doug Pratt and the Real Goods Staff

A REAL GOODS INDEPENDENT LIVING BOOK

DISTRIBUTED BY:
CHELSEA GREEN PUBLISHING COMPANY
WHITE RIVER JUNCTION, VERMONT

Printed in the United States of America with soy-based inks on recycled paper
1 2 3 4 5 6 7 8 9 10

ISBN 0–930031–82–2
The Real Goods Solar Living Sourcebook is the ninth edition of the book previously published as the
Alternative Energy Sourcebook, with over 250,000 in print.

DISTRIBUTED BY:
CHELSEA GREEN PUBLISHING COMPANY
P.O. Box 428, White River Junction, Vermont 05001

REAL GOODS TRADING CORPORATION
555 Leslie St., Ukiah, California 95482-5507
Business office: (707) 468-9292; fax (707) 468-9394
To order: call 1-800-762-7325 or fax (707) 468-9486
For technical information: (707) 468-9292 ext. 2210
For renewable energy orders: (800) 919-2400 0r fax (707) 462-4807
Renewable energy E-mail: techs@realgoods.com
E-mail: realgood@realgoods.com
Home page: http://www.realgoods.com/

FOR THE EARTH

May we preserve and nurture it in our every action

———————————————————————

CONTENTS

ACKNOWLEDGMENTS

This ninth edition *Sourcebook* has truly been a team effort from start to finish. Thanks for all the time and energy from our excellent technical writing team of Doug Pratt, Jeff Oldham, Douglas Bath, and Gary Beckwith. Much appreciation for all the inspiration and vision for this project from Dave Smith and Stephen Morris. Thanks to Doug Pratt and Mike Leon for their great editing and graphics work on the Ninth edition.

— *John Schaeffer*

INTRODUCTION

ACCORDING TO Congress' Office of Technology Assessment, at present consumption rates, the world's known oil reserves will be exhausted in 41 years. If that estimate is true, we will have no more oil in 2037, which will relegate oil, as an energy source, to a mere blip of 175 years in the five-billion-year history of the earth. The stranglehold that the oil companies have exerted over every aspect of American society began when 23-year-old John D. Rockefeller started his first oil business back in 1862. For the next 125 years, the oil companies (aided and abetted by the United States Government) have aggressively and systematically fought the development of any competitive energy source, including renewables. Monopolistic utilities were just as protective of their turf. Solar energy first got a toehold near the turn of the century, when Clarence Kemp invented the first solar water heater in 1891. Within ten years, one million homes and 300,000 pools in Southern California were powered by solar water heaters. Fearful utilities quickly and effectively crushed this burgeoning new technology by offering free natural gas installations. Renewable energy went into a tailspin for nearly half a century, until the advent of photovoltaic cells in the 1950s, which were originally developed for the space program. The utilities simply couldn't figure out how to extend their wires into space. NASA knew then what consumers are discovering now - that all energy derives from the sun.

The utilities and the oil companies long ago grasped the fact that, when confronted with the choice between long-range environmental consequences and short-term economic impacts, the consumer will almost always be governed by the latter. Today, as we struggle to achieve responsible energy independence, we find ourselves constantly ensnared in the task of maintaining balance between our short-term interests and the long-range consequences of our actions.

It wasn't that long ago that independent homes were in the mainstream. In 1912, only one-sixth of the homes in America were plugged into the utility electric grid. By 1927 that figure had grown to two-thirds. By 1960, the transition to grid power was complete. But now the pendulum has finally begun to swing back. Today, solar power has come of age. Each year, the amount of solar energy falling on the earth is ten times that of all fossil fuel reserves and uranium combined. All of America's power needs could be met with solar plants on 59,000 square kilometers, or less than one-third the land area that is now occupied by U.S. military facilities. Solar has become, for many homeowners, just another invisible source of electricity, albeit one that doesn't pollute, make noise, or rack up monthly bills, and one that works year-in and year-out. There are now well over 100,000 homes

in the U.S. and another several hundred thousand homes in the developing world that derive their electricity from photovoltaics. In Indonesia alone there are one million PV systems gradually migrating from the drawing board to service at the rate of 100,000 a year.

Yet the world's energy consciousness still exists within a vacuum. While renewable energy technology has clearly come of age, the governments of the world have been painfully slow to catch up. As we hurtle toward our greatest environmental crisis since the dawn of human life, people are finally beginning to consider the worldwide impact of environmental degradation. Pollution is out of control, our oxygen-producing forest cover is being eliminated, habitats of the remaining species of plants and animals are disappearing, and our food supply is in danger of diminishing. In the 18 short years that Real Goods has been around we have witnessed major ecological disasters in places like Bhopal, India; Valdez, Alaska; Love Canal; Three Mile Island; Chernobyl , not to mention a vicious war for oil in the Persian Gulf. (Ironically, we also watched our sales spike with each of these catastrophes.) We are experiencing a planetary breakdown of the earth's life-support systems: air and water contamination, holes in the protective ozone layer, and global warming. The thread that ties most, if not all, of our environmental problems together is our insidious dependence on, and our insatiable appetite for, cheap oil.

The beginnings of Real Goods and the renewable energy movement for me personally can be traced back to one simple but significant event that happened to me in 1977. I had graduated from U.C. Berkeley in 1971 and left campus with a used tent and a sleeping bag in tow for the wilds of Mendocino County, and also with a taste for living communally on 290 acres of primitive earth with 20 other naive but optimistic dreamers. I suspected that life would never be the same again. We built houses out of a hand-crafted sawmill that consisted of a large saw blade mounted on a pulley that was crudely affixed to the wheel of a 1962 Chevy. Second-growth redwoods would screech through this mill, soon to become the posts and floor beams for our houses. We dug springs by hand, hauled countless sacks of gravel up and down the hills, and mounted gas pumps to deliver water to our ridge top. We grew our own organic food, milled our own flour, dug pit privies, and debated mercilessly at "Sunday morning meetings" about the politics of community, individualism vs. collectivism, and what constituted "appropriate technology." There was, of course, no electricity, and the kerosene lamps' dimming glow got harder and harder to read by night. I began to long for the creature comforts that I had been raised with. Helping to build a community literally from the ground up had been a rich and rewarding experience, but, after several years, living communally with the problems of 20 other people just got old. Soon I took a job 35 miles away in Ukiah at the County Computer Center.

The year was 1977. After several hundred 80-mile commutes to work, I discovered some old lightbulbs marked "12-volt" gathering dust on the shelf in a local hardware store in Ukiah. My curiosity piqued, I brought the bulbs home to my 12' x 18', single-walled redwood cabin. What was about to happen would change my life forever. Armed with some crude 12-gauge copper wire, a porcelain Edison-based socket, a pair of pliers, and my Swiss Army knife, to my joy and amazement I discovered that the 12-volt lightbulbs could be operated with the juice from my car battery , and in one fell swoop the twentieth century arrived to my humble dwelling. This experience must not have been unlike what happened to Thomas Edison in New York 95 years earlier on September 6, 1882, in a dingy Wall Street warehouse, when he connected a crude, coal-fired boiler to a steam engine and dynamo, linked the plant by underground wire to a block of nearby office buildings, and flipped the switch illuminating 158 lightbulbs for J.P. Morgan and the New York Times.

Electricity is empowering. Electricity is addictive. In 1977, from the vantage point of my small cabin in the wilderness, I experienced in microcosm what the technological age had brought to the twentieth century. As a child growing up in the fifties we all took for granted the General Electric ad campaigns so eloquently championed by Ronald Reagan on the radio, and the dominant belief that the "American way of life" equated how much you consumed with how American you were. Here I was on a commune unlearning my upbringing, reading with kerosene lamps, chopping wood for my handmade woodstove, digging with hand tools , and now, suddenly, this light was shining, powered by my car battery! At the time I didn't foresee the potential. Before long I was reentering the consumer age with every appliance the RV industry could churn out in 12-volt, from blenders to 12-inch TVs. I became the techno-radical in a community of neo-Luddites. Little did I realize that I was tinkering with the rudiments of what was to become the "independent home movement."

So, now that my home was slowly becoming powered with 12-volt electricity, I began to feel better about my polluting 80-mile-per-day commute, because my car had suddenly been transformed from a mere gas-guzzler into a battery charger for my house. I opened up the first Real Goods in Willits in 1978 to serve all the other urban refugees who, like me, were experimenting with independent homes and power systems. I kept thinking that there just had to be a better way to charge batteries than driving around in circles or firing up smelly diesel generators. Then, one day, a guy walked into the store and offered to sell me a large quantity of the first photovoltaic modules I'd ever seen, for about $100 per watt. They were outrageously expensive, but irresistible, promising a new, nonpolluting, clean and quiet way to electrify the wilderness. So I bought the first photovoltaic modules to be sold at retail in America. The dominant thinking at the time was that solar was the pipe dream of environmentalists and hippies, and that this complicated and costly new technology would never amount to much more than an inventor's plaything. Utility-generated electricity was still dirt cheap, and power companies were just drooling about how much money they could make with nuclear power plants. (Remember, this was still a year before Three Mile Island, and nearly 12 years before utility companies discovered that saving energy could be cheaper than building new power plants.)

Northern California quickly became the petri dish for the study of residential solar-electric energy, as one oil company executive after another Lear-jetted into town to observe this bizarre phenomenon: PVs were selling as fast as we could supply them to the latest influx of ex-urbanite back-to-the-land settlers. In the 19 years since that first 12-volt lightbulb lit up in my remote cabin, nothing short of a

renewable-energy revolution has occurred, spreading from the hills of Mendocino County all the way to the center of power in Washington, D.C., where Real Goods has been involved as an integral part of a design team that will hopefully soon bring solar power and energy-efficiency to the White House, with President Clinton's blessing.

But what about the cost of solar energy? Is independence affordable? In the long view, the answer is clearly "yes." For rural electrification PV has always been the least-cost choice. Even in 1978, when we sold PV modules for $100 per watt, they were still cheaper than utility line extensions for distances greater than one-quarter mile. With many utilities currently doubling and tripling the cost of line extensions, this trend is likely to continue, making PV more and more cost effective. But what of massive-scale "urban photovoltaics?" When we deal with energy economics, we must first resolve our uncertainty about the true costs of nonrenewable resources. Do we use the cost at the gas pump or the electric meter, which are based upon short-term extraction and distribution costs, or should we take a longer view, to assess costs based upon the time it takes to reduce biomass to petroleum (500,000+ years) and to abate the pollution effects of imperfect combustion. These "externalities," plus countless others, including health effects, societal effects, and military costs, are extremely difficult to quantify. As Amory Lovins puts it, "I have trouble believing we'd have put half a million troops in the Gulf if Kuwait just grew broccoli." Should we account for government subsidies to the fossil fuel industry? This is a huge topic and beyond the scope of this introduction, but one that is absolutely central to Real Goods. In the crudest of business realities, if something is perceived not to pay for itself in a reasonable amount of time, we cannot sell it.

Even in the short view, where the cost issue is cloudy, renewables are catching up quickly. Historical trends are encouraging and tend to confirm the conventional wisdom that PV will be cost-effective by the end of the decade if not earlier. In 1992 dollars, utility company electricity sold for $4/kWh in 1892, $0.60/kWh in 1930, and then dropped to a low of $0.07/kWh in 1970. Now, in 1996, it is beginning to approach $0.10/kWh on average and appears to be on the increase. Examining the costs of PV-generated electricity, the U.S. Department of Energy (DOE) estimates it to be $0.25/kWh now, and predicts a price of $0.12 to $0.20/kWh as we approach the year 2000; then DOE estimates a lowering to $0.05/kWh in the period from 2010 to 2030. PV prices have dropped from $100 per watt in 1978 to around $5 per watt today. Over a similar period, PV production has grown from a scant 3.3 megawatts in 1980 to over 80 megawatts in 1995. So you be the judge , basically, cost-effectiveness is in the eye of the rate-payer and depends upon how much you value independent living.

As we go to press on our 9th edition of the Sourcebook, I can say with confidence that we, as a company, are doing everything possible to walk our talk. As of summer-1995, my personal fifteen year conversion to solar energy was completed with the installation of a 2760-watt solar electric system at my house. As with most homeowners, my first question was economic. How long would it take to pay back the investment? With the capable assistance of the Real Goods technical team, we calculated that my system, a "line-intertie" interfaced with my local utility for both a "buy" and a "sell" meter, would be paid off, in full, through utility savings in less than 20 years. By age 65 my system will have totally paid for itself and electric bills will never again darken my door. When the sun is out, we sell power to PG&E; when it's not, we buy it back. Moreover, this is something every American can potentially do. With net metering (utility companies must pay consumers the same rate for electricity as they charge them) now law in California and many eastern states PV becomes still more cost effective.

In April 1996 our long awaited Solar Living Center opened. Powered 100% by PV and wind generators, our headquarters is a living testament to the viability of renewable energy sources and sustainable technologies. With the most efficient day-lighting system in the country and constructed of 800 rice straw bales with 500 tons of thermal mass eliminating the need for active heating and cooling systems, our Solar Living Center will soon become the world's premier educational and demonstration center for the new millennium.

As we go to print with our 9th Edition of the Sourcebook, I am more optimistic than ever about the tide turning toward a sustainable future. I have witnessed Real Goods blossom from that first 12-volt lightbulb in the woods in 1977 into a vital company that stands at the forefront of the renewable-energy movement in the mid-1990s. As the sustainability revolution continues, we must maintain our commitment to bringing energy sensibility to the world by embracing renewable-energy solutions, decreasing pollution, and encouraging alternatives to the internal combustion engine and the consumption of fossil fuel. In everything we do, we must keep a sustainable society clearly within our vision. We must practice material moderation despite our habits and a culture that rewards overconsumption. We must employ efficient products in an economy that values instant gratification and tolerates waste. Sustainability must become the air we breathe, the water we drink, and the way we move through this awesomely beautiful world. Real Goods is a small, but very proud part of this revolution. Thank you for walking beside us.

— John Schaeffer, President, April, 1996
Real Goods Trading Company
Ukiah, California

CHAPTER 1

INDEPENDENT LIVING

THIS NEW NINTH EDITION of the *Sourcebook* comes at a turning of the tide in favor of sensible planetary energy management. We have begun again to build sustainability into our projects as did our predecessors before the twentieth century's energy glut took hold. The excesses of the petroleum era are beginning to ebb. Two or three decades ago, a brave and stubborn few of us began swimming against the exploitative current by rejecting mass-produced housing, insisting instead on more fitting homes, and learning how to build and power them ourselves. We proved the concept that we can live comfortably by improving energy use in existing houses, and have thereby hastened the time when we can all live within the energy budget allotted to us by our planet and its sun.

For almost a decade the *Sourcebook* has been at the center of this renewable energy renaissance.

Our cautious joy results from the happy conjunction of technological wonders and a growing global commitment to reform wasteful habits. During the past hundred years, mainstream energy behavior became increasingly inefficient as consumption intensified. Through overpopulation and careless exploitation of the earth's resources we have overshot our planet's carrying capacity. Nevertheless, we can continue to live as well as we do now on only one-fifth of the energy we presently use, simply by displacing inefficiencies.

By conserving our resources and using them much more efficiently, we can gain the time needed to solve other pressing problems. Unluckily, our reforms may have come too late, and we may well face a time of scarcity. Many global resources, including the carbon-based fossil fuels buried within our planet, are nearly depleted, and some critical resources like oil will vanish even if our usage is decreased to a fifth of its current level. We, and our children, are left to repent for our

forebearers' greed. We must accelerate efforts to reject wastefulness, and to use abundant, self-renewing energy sources in ever more efficient ways.

Much of this resource madness results from an honest mistake: the notion, widely advertised in the 1950s, that technology, particularly nuclear power, was on the brink of providing a clean, simple, inexhaustible source of perpetual energy. Electricity, it was said, would become too cheap to be metered. Many of us, eager to believe in triumphant science, remember this vision of effortless plenty. Prudent but tiresome habits of looking into the future were cast aside in favor of energy hedonism. We became a culture devoted to short-term gain.

Seduced by such visions, we are now reluctant to give up our cars and electric toothbrushes. A pattern of consumption based on the infinite beneficence of technology dominates commerce. How could we have failed to consider the consequences of our unbridled consumption? Even our nuclear dreams became nightmares once we tallied up the costs of safe containment of dangerous waste.

Some folks turned their backs on the easy life. Stubborn, independent homesteaders saw the short-sightedness of consumerism and were suspicious of salesmen who wooed a whole culture with "scienterrific" wonders. Others experienced the loss of family to a workaholic world, alienation from nature and the elements, the mystification of everyday activities, and the inability to carry a task to completion using only hands and wits. Increasingly, these people sought to reconnect themselves with a more authentic existence.

Still, we were intrigued by our own ingenuity, the scientific rapture of the age. Was Big Science really necessary? Or could we use technology in appropriate measures within our homes?

As soon as the industrial age had dawned, reaction against industrialization began. Until recently, Luddites have been isolated from society's mainstream. The new wave of starry-eyed scientists, beginning with back-to-the-landers in the late 1960s and 1970s, have redefined independent living to include the best of old and new, consolidating the real gains of materialism with a broadened sense of planetary stewardship. Technology is not rejected, but rather turned to better purpose. We are here not to dominate the Earth's gifts, but rather to make things better for all life. This requires us to lengthen our sense of time. We must learn to plan for 2995 as well as for 2020. Simultaneously, we must extend our concept of family to include all life on this lovely little planet.

WHAT IS INDEPENDENT LIVING?

Independent living does not require subscription to this creed, but awareness and acceptance of these precepts seems to be universal among people who try to live sustainably. In every other way — politically, economically, stylistically, nutritionally — people living in independent homes comprise the most diverse group imaginable. Only one other trait defines them: they love their homes and plan to live in them forever. This is a profound departure from a cultural norm in which a family moves, on average, every four to seven years.

To live sustainably and independently, we shrink the boundaries of our basic needs. While we willingly pay high prices for luxuries like oranges, chocolate, and

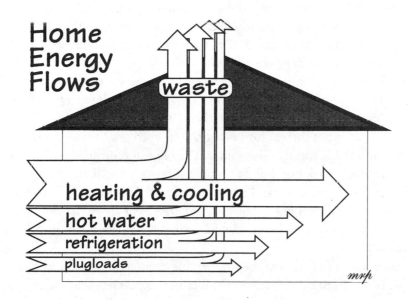

Home Energy Flows

waste

heating & cooling

hot water

refrigeration

plugloads

mrp

winter lettuce, we need not pay premiums for necessities. We should look in our own backyards before gazing over horizons. In most industrialized countries, electricity is generated by fossil fuels burned dirtily far away. It is transmitted through "the grid," a far-flung and inefficient artifact of a time when visions of free power unfettered by consequence warped our plans.

We can harvest enough electricity for our needs from the roofs of our houses, from wind on the ridges, and from water flowing down their slopes. Energy also comes to us from surplus vegetation, from the power in waves, and from deep within the earth. These sustainable, clean, local sources of power are the mainspring of an independent lifestyle, and the theme of this book.

Independent living comes in two basic flavors: the full-bodied romance of life off the grid, and the tamer but equally beneficial undertaking of energy-aware life on the grid. Both choices reject unthinking dependence on unknown resources. On or off the grid, we learn to preface each act of consumption with a thought for the resource and the whole-life costs associated with the act. We practice "ecologic," developing a keen sense of place and always striving for the long view, with attention paid to the invisible, the indigenous, and the fragile.

Ever since early humans lived in caves, home has been the realm of the heart and hearth, the place of family, joy, and fulfillment. Gradually, almost unnoticeably, over this Century of Progress, our culture has come to accept factory-built commodity housing where we cower before the flickering of a cathode-ray tube, a paltry fire that originates in some faraway generating station. Many of us would rather return to the home-centered hearth of an earlier age by reclaiming the home design process from developers. Nevertheless, a quarter century from now, four out of five of us will still be living in existing housing that was built when the promise of free power held sway. Ironically, free energy is at hand right now, but it comes not from the outlet, but from the sun, the original source of earthly life.

Most of our energy is spent on heating and cooling the spaces we inhabit. By designing homes so they work with, rather than in spite of, indigenous energy

elements, we can build energy-sensible dwellings under any weather regime. Even today, houses, neighborhoods, and whole cities are being built with their backs turned to abundant solar resources. Many architects and most builders consider the environment a problem rather than an opportunity.

Michael Reynolds, an innovative designer whose cavelike earthships, built from indigenous adobe and recycled materials, dot the high desert landscape around Taos, New Mexico, thinks differently. He says, "What you need to understand is, everything we need is out there. Now, an earthship needs just a subtle amount of heat to stay comfortable through the winter. We've been putting in little $300 gas stoves that burn fossil fuel, but that doesn't make much sense. I started trying to think, Where is heat? and how can I make friends with it, and get it?" (His ingenious solution can be found in my book, *The Independent Home: Living Well with Power from the Sun, Wind, and Water*, which makes a good companion volume to the *Sourcebook*.)

How can we turn our faces to the sun? In this chapter, you will find books and tools to enable and justify the independent life. In the chapters that follow, you will find tools for making a house into a sustainable home, and directions and hints for committing to live independently.

— *Michael Potts*

INDEPENDENT LIVING

The Independent Home

Michael Potts. There is a movement by forward-thinking individuals creating homes that are self-contained from an energy perspective and designed with a sharp eye towards planetary impact. These people sacrifice none of the conveniences of modern life, but they refuse to accept that environmental pollution is the inevitable by-product. This is the governing concept of *The Independent Home.* These are true homes, where you accomplish convenience, aesthetics, and even luxury without the massive input of fossil fuels or nuclear power. These are the homes examined by Real Goods' own Michael Potts in his book, *The Independent Home.*

Subtitled *Living Well with Power from the Sun, Wind, and Water,* this is not a book about technical wizardry or philosophical rigidity, but of ultimate common sensibility. It is about balance in lifestyle. Potts travelled extensively to find the pioneers who are using renewable technologies to reverse the power dependency that has imbued our culture in the last century. The range of backgrounds, philosophies, and personalities is as diverse as the range of home styles. What they have in common is that they sacrifice little, while gaining the deeper satisfaction that comes from living lightly on the planet. The image comes through loud and clear — the independent home is a key that unlocks the independent mind.

The Independent Home is a handsome book with 60 photographs of homes and homesteaders, and 35 illustrations that demonstrate the technologies. This is the first book to document both the how-to and the why-to of the energy independence movement in the 1990s. 300 pages, paperback, 1993.

80-220   The Independent Home $20⁰⁰

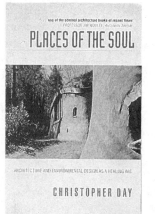

Places Of The Soul

Christopher Day. One of the great architectural thinkers of our time, Christopher Day demonstrates in this book how building design can start with people and place and how buildings can develop organically from these foundations. Our surroundings affect us physically and spiritually. Day helps us understand how different qualities in our structures can either nurture or poison us, body and soul. 192 pages, paperback, 1990.

80-214   Places of the Soul $19⁰⁰

Homing Instinct:
Using Your Lifestyle To Design & Build Your Home

John Connell. This book takes the curriculum of the progressive Yestermorrow School and puts it into print. Yestermorrow, located in Warren, Vermont, advocates interactive methodologies where the homeowner becomes integrally involved in every aspect of the design/build experience. The end product may not be traditional but usually results in an affordable, environmentally sound, durable structure that creates a profound sense of owner satisfaction. This comprehensive volume may not be a substitute for taking the Yestermorrow Design/Build course, but it is the next best thing. Author (and Yestermorrow founder) John Connell has done a great job of delving into the myriad decisions that need to be made in constructing even the simplest home. 404 pages, hardback, 1993.

80-228   Homing Instinct $35⁰⁰

In the Absence Of The Sacred:
The Failure Of Technology & The Survival Of The Indian Nations

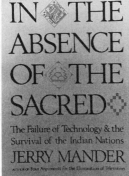

Jerry Mander. A brilliant and provocative book on the effects of technology on society with emphasis on its effects on indigenous peoples and the healing potential of their alternative ways of life. "A skewering critique of modern technology, in which cars, telephones, computers, banks, biogenetics and television . . . all are shown to be part of a mad 'megatechnology' that is destroying the world's resources and robotizing its peoples." —Kirkpatrick Sale. "Inspiring, sometimes gripping . . . Through Mander's eyes, native peoples are not quaint relics; they become sources of precisely the practical wisdom our species needs not only for survival but for renewal." —Francis Moore Lappe

Mander is the author of *Four Arguments for the Elimination of Television.* 446 pages, paperback, 1991.

82-252   In the Absence of the Sacred $14⁰⁰

Skills For Simple Living

Betty Tillotson. This attractive book, a collection of best tips from 20 years of *The Smallholder* magazine, is so packed with information that even if only half of it applies to your life, you'll still be getting more than your money's worth. You'll find recipes for ginger beer and skin salves, tips for seasoning hardwoods, designs for a meat-and-fish dryer and a clay oven, organic deterrents to garden pests, instructions for building rock-and-cement walls, handicrafts for children and adults, and the list goes on and on. The index is excellent. 218 pages, paperback, 1991.

82-166   Skills for Simple Living $15⁰⁰

Feng Shui:
A Layman's Guide To Chinese Geomancy

When age-old traditions reinforce our prejudices, we are glad. The ancient art of *feng shui*, or geomancy, brings such a tradition to the siting and orientation of a house (or grave). Much of the advice is self-evident: geomancers recommend that we build on high ground instead of in the valley, and that trees to the northwest of a house bring good fortune, but on the south they do not. Upon reflection, this is common sense advice. Unlike our own careless house siting practices, this tradition has withstood the test of millennia and remains to this day an important architectural consideration in the Orient. The wisdom and experience encapsulated in this art are at times humorous to moderns — three, four, or eight bedrooms symbolize ill fortune — but are intriguing because they transforms the act of siting and orienting a house from accidental to meditative. The best book we have found is *Feng Shui: A Layman's Guide to Chinese Geomancy*. Literate, clear, beautifully illustrated, including wonderful color pictures of *luopans* or geomancer's compasses. 123 pages, paperback, 1992.

80-234   Feng Shui $10⁰⁰

Sustainable Cities:
Concepts And Strategies For Eco-City Development

B. Walter, L. Arkin, R. Crenshaw. This book shows how urban development and the environment can coexist in a successful partnership that will result in better air quality, cleaner water, more nutritious food, less commuting time, quieter and more nature-oriented surroundings, a richer community life, less urban stress, and improved health for the people who live and work in cities. Strategies include ecological design, resource management, sustainable technologies and construction techniques, economic techniques, and more. "If I had to choose one book whose information could create a unity of purpose around an environmental agenda among citizen, development, and government groups, *Sustainable Cities* would be it." —Lester Brown, *Worldwatch*. 354 pages, paperback, 1992.

80-235   Sustainable Cities $20⁰⁰

Real Goods Independent Living Book
Renewables Are Ready

Energy use in the world has reached a critical crossroads. To prevent exhaustion of the last of our fossil fuels, we must commit ourselves to two goals: reduction of present levels of energy consumption, and widespread development and implementation of practical, affordable, renewable energy technologies. In *Renewables Are Ready*, authors Nancy Cole and P.J. Skerrett tell the inspirational stories of Americans who have made renewable energy a reality in their own communities, including:

• Debby Tewa, who with the Solar Electric Enterprise is helping to bring low-cost photovoltaic power to residents of the Hopi reservation in Arizona
• Frank Rucker, who saved his Vermont school district more than one million dollars by convincing citizens to heat the regional high school with an efficient, clean-burning system that uses locally plentiful biomass — wood chips
• the people of Waverly, Iowa, are pioneering the use of wind turbines to generate electricity

All of the ideas presented in *Renewables Are Ready* can be implemented in a city, town, or neighborhood through individual and group initiatives. The authors explain the variety of new technologies now available, and identify five key strategies that will help overcome typical economic and social barriers to bringing renewable energy solutions into the mainstream. 256 pages, paperback, 1995.

80-259 Renewables Are Ready $20⁰⁰

Nature, to be commanded, best be obeyed.
—Francis Bacon

WHOLE-LIFE ANALYSIS: TAKING THE LONG VIEW

There is no such thing as a free lunch, and we expect to pay at least the cost of anything we buy. As the distances from source to manufacturer to consumer have lengthened, real costs of consumption have been shifted from the foreground into the infrastructure of modern life. For example, transportation costs are hidden in "background" charges — highway, fuel, property, and income taxes and tax-funded subsidies. Other costs drift in the wind — the exhaust from our transport and power generation infrastructure — to be paid by ourselves or our descendants, at some distant time or place. Sometimes the hidden costs are greater than the apparent price. For example, how much would the people of Ukraine need to add to their electric bills to pay for the cleanup of Chernobyl? Several energy-cost studies undertaken to "monetize the externalities" tell us that the real cost of the electricity we buy will finally be reckoned at three times what we pay. In other words, our 10-cent kilowatt-hours honestly cost the planet 30 cents.

There are two key factors in assessing the whole cost of any product: the cost to buy the product, and the cost to operate and maintain it. To account for the product's whole-life cost, we must add two more factors: the true cost to extract and create the product, and the true cost of disposing of it after use. Portions of these costs are usually borne by society in hidden ways.

LIGHT SAVINGS

	Phillips SLS 20W	Incandescent 75-watt bulb
Cost of bulb	$29	$0.50
Product life	4.5 years	167 days
Watts used annually	20	75
Energy cost	$4.38	$16.42
Bulbs replaced in 4.5 years	0	10
Total cost	$48.71	$78.89
Savings over life	$30.18	

Compact fluorescent lightbulb.

Let me propose a contest, a whole-life cost analysis, to illuminate this idea.

Ladies and gentlemen, in this corner, the champion and historically most-favored light-source, the incandescent bulb. In the opposite corner, the challenger, a compact fluorescent (CF) light. Costs of creation and disposal are roughly equal: glass, bits of strategic metal. In the first round, considering only equipment cost, the champion wins easily: who would buy 60 watts of light for $25 when you can buy a package of four champs for under three bucks? Even though the cheaper bulbs last only a tenth as long, and it takes ten incandescents to compare fairly with a single compact fluorescent, the challenger's equipment cost appears to be more than three times the champion's. Round one to the champ!

Unfortunately, many consumers tune out after the first round. For industrial users with thousands of bulbs, the economy of changing a bulb once instead of ten times decides the contest in favor of the compact fluorescent even without their operating advantage, which becomes apparent in the next round . . .

Round Two: the cost to operate the bulbs. On my CF's package the makers have thoughtfully printed a cost comparison which shows their product in a most favorable light. This newfangled lightbulb consumes only a quarter of the electricity drawn by the conventional bulb to produce the same amount of light. Over the whole life of the bulbs, which is cheaper? Assuming that electricity and lightbulbs will not increase in price — an unlikely assumption! — incandescent bulbs (remember, it will take nine or ten) will use so much more energy over the lifespan of the compact fluorescent that the overall equipment plus operating cost will be almost twice (188%) the challenger's, or an average of $35 more per bulb! This TKO ends the contest decisively after two rounds. A smart consumer will unquestionably use as many CFs as possible, because, in the long view, that is the cheaper way.

This same contest played out between other historical champions and modern challengers generally leads to the same result. New technology is often more costly to buy, but more efficient to operate. In the days of cheap power, equipment cost was so dominant and energy so cheap that little thought was given to operating costs. Manufacturing decisions and aesthetic preferences (for example, that a refrigerator's compressor "looks better" on the bottom) still persist. Yet in today's energy market, with everything from high-mileage cars to high-tech windows, from super-efficient refrigerators to super-insulated houses, the best long-term bet is to pay almost anything to get better operating costs.

The old wisdom, waste not, want not, once again prevails.

— *Michael Potts*

CHAPTER 2

LAND

"Buy land, son; the Lord isn't making more of it."

—Attributed by Mark Twain to his father

THERE IS NOTHING MORE REAL in this life than land. Earth gives us each thing we have, and at the end we return those gifts to it. "Owning land" is a relatively modern notion and by no means a universal practice. In fact, to many of the world's peoples who live intimately with the land, the idea of ownership is incomprehensible. We who seek an alternative way of living also find dissonance between the accident of "ownership" and our visceral fealty to the land which owns us all. We resolve this contradiction by becoming the land's stewards, assuming responsibility to and for the land which gives us life. By looking about ourselves at a globe marred by exploitation, overcrowding, and short-sightedness, we have discovered that only through honoring the land, with all its creatures and qualities, can we truly honor ourselves.

The search for good land is like the search for a mate. We certainly hope to find sustenance, partnership, comfort, and stability. In the end, if we truly wish to settle peacefully and productively, we must find love and passion for the land as well. This nation began in a wave of western migration motivated by a hunger for land and independence, but somehow the habit of migration stuck, while the regard for land went astray. Our national failure to settle well has led to our tragic pattern of exploitation, insensitivity, restlessness, and estrangement. Real-estate sales and the vertiginous overvaluation of land depends on the predictability of this pattern: the average American family uproots and moves every four to seven years, and our whole culture has come to thrive on this lonely rootlessness. The "back to the land" movement which began in the late 1960s, and which fostered the alternative energy industry during a prolonged petroleum winter, differs from prevailing American culture precisely in its love for the land.

Most of us thrill to the magnificence of mountain vistas or the glittering promise of tropical islands in the sun. Such prospects, even as photographs in

16

magazines or on the flickering television screen, make our spirits soar. But those images depict land on a grand scale; for most of us, the land we can own is a much smaller plot, and the emotions we feel for it are subtler. Coming as we do from a time when holdings grew ever smaller and more urban, and where mates and property were described as chattels, a word derived from the same root as cattle, it is hardly surprising that we in the back-to-the-land generation have had to look deeply into our hearts and our past to rediscover love, whether for another human or for a piece of land. Even our words undercut the strong but subtle ties that grow between ourselves, our human partners, and our land.

As we begin to rediscover our connection with the earth, it is only to be expected that we should have lost the art of choosing land well. Almost a century of easy energy has led to habits of use and patterns of settlement completely divorced from the inherent qualities of the land. A conventional house is situated not to take best advantage of the sun and the special features of a site, but to fit within arbitrary boundaries and to accommodate the whims of the planner and catskinner who bring in the road and powerlines; such persons are not responsible to the land and the families that will live on it, but to the developers who expect a quick profit. With horrible uniformity, buildings accumulate in dark gulches, cheek-by-jowl along narrow streets, or like pimples on a smooth mountain brow, without regard for what is best for land, dwelling, or family. Occasionally, more sensitive developers proudly leave some of the more imposing trees, or employ building designs and materials evolved over centuries, but seldom as well as possible. We who are third- and fourth-generation inhabitants of these cold boxes have neither the tools nor the tradition to meaningfully evaluate the suitability of these structures for the lives we wish to lead.

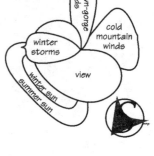

Site rose

In a triumphant throwing-out of the baby with the bath water, many back-to-the-land pioneers sold their luxuries, packed up their necessities, and headed for the boonies. We admire their spirit, and acknowledge that there was wisdom as well as desperation in their movement away from the urban center. Since the beginning of land stewardship, back when agriculture was being invented and humans took responsibility for the abiding fertility of the land, it has been wise counsel to seek land that no one else wants, and make it home. In a few brutal years of inadequate shelter — flapping plastic windows, leaking roofs, flickering lamplight, harsh mornings on splintery floors — we regained the immediate sense of the elements from which our energy-rich tradition had sheltered us, and a renewed understanding of the value of a well-conceived, well-built, and well-provisioned home. For many of us now, a pilgrimage to the denatured houses of our past, of our parents and our unawakened peers, is fraught with discomforts and puzzlements we are loathe to share with them: how could we have lived this way, so heedlessly, so wastefully, so uncomfortably? What great energy, what denial, are invested in ignoring the unsuitability of these cheerless domiciles, whose inhabitants are usually awaiting a chance to move to newer, larger, better-situated discomfort!

A new tradition, a quarter of a century old, has grown up now, and a new wave of settlers is moving back to the land without suffering the privations of those early pioneers. These lucky souls have as their guide and inspiration all of the attempts, failures, renewed trials, and successes of the original back-to-the-landers. Today our

RECLAIMING SPOILED LAND

Most of the best land on our continent has already been used, and often we must start our tenure and stewardship by healing prior injuries. Many of the original tenants on this land religiously avoided excesses, but European settlers over the past four centuries have done untold harm. The wounds are various, ranging from the inadvertent introduction of unexpectedly vigorous transplants, which kill native species and dominate the landscape, to the deliberate and perhaps even criminal contempt for the land which lays acreage to waste in poison dumps, slag heaps, and mine tailings. We must believe that, with few exceptions (nuclear contamination comes to mind, along with some particularly nasty aromatic chemical dumps like Love Canal) all land is salvageable.

We will never be able to restore the ecosystem to its pristine, presettlement state. The most complex systems — the American wetland, now filled in for parking lots and development, and the global rainforest, still being sacrificed for cellulose and beef — are now all but lost. These majestic habitats, together with great plains and our ancestral forests, take millennia to heal, and so we must start their convalescence immediately.

Fertile and arable lands that we have claimed for homes and farms, as well as the low hills, plains, and bottomlands that have made this continent such an idyllic home and breadbasket, recover more quickly, and we can hope to heal them through sensitive stewardship within our lifetimes. Independent agriculture, like independent living, finds the same fault with factory farms as we found with factory housing: site is battered to maximize immediate product, and succeeds only temporarily or through heroic infusions of imported energy in the form of pesticides, fertilizers, and intensive fossil-fueled tilth.

Many of us are coming to prefer a sustainable, Earth-attuned and life-enhancing, organic agriculture. A first step, then, is to return to natural measures. We discover, after more than a century of tinkering, that many exotics, plants, and animals imported from other continents, carry with them "hitchhikers" and lack the biological or environmental controls which keep them in balance in their native ecosystems. Again and again we find that kudzu, telapia, mango, eucalyptus, mongoose, and feral cats, goats, and pigs are scourges on a fragile land where they were never meant to range. While abating past errors, we must be careful not to commit new ones in a rush for quick miracles. Monocropping — covering an expanse with a single life-form — is abhorrent to nature, and nature's backlash often comes with surprising vigor. Only by restoring land to its happy, indigenous state, in which the aboriginal biota regain their original diversity, integrity, and interrelatedness, may we prevent environmental catastrophe.

— *Michael Potts*

vision of what works, and what does not, is clearing, and the tools and techniques found in this *Sourcebook* are better than ever before. By contrast, the crowded suburban single-family dwellings distill the worst of transient Americana, wind whistling coldly between ticky-tacky unshared walls through a space too small to garden. Cohousing, be it intentional or from necessity, where families huddle warmly together within shared walls, is a better way to crowd people. Most of us spend much of our lives crowded thus, and many prefer it; the point is, it should be done well, so families may thrive. A hardy few will wish to live stoutly independent beyond the end of pavement and powerlines, and for them the solutions of factory housing — stud and mud walls, aluminum sash windows, all-electric heating — are also unworkable. For them, the stuff of survival is found in simple solutions: indigenous materials, inspirations from the region's original inhabitants, and self-built homes which look to elemental forces (the sun, wind, falling water, and unfailing bounty of the soil) for sustenance.

The sun works. It is the original source of all earthly energy. All land bends first to the sun, and to its minions, wind and weather. Even the flattest plain and craggiest hillslope have clement spots sheltered from the tempest, often protected by no more than a slight fold in the earth, or a line of trees. Geomancers (diviners of earth signs) claim to be able to find these spots, as dowsers find buried water, by sensing the feng shui, the flow of earth, air, fire, and water. Much is made, in the Orient, of the meeting of feng shui master and architect when a skyscraper is planned. On a homelier scale but with equal solemnity, in the woods and mesas, hill slopes, prairies, and rich bottomlands of this continent, a generation of conscious homebuilders deliberates the best accommodation to seasonal sunlight, prevailing wind and water flows, available local materials, and appropriate technologies for making a home.

It takes time to learn a site. Too often, we come as urbanized new settlers from afar, and the secrets of storms, heat and cold, and successful regional designs are mysterious to us. Under time pressure, we act hastily, when the sense of the land abides and so must be divined. A site may never be known in less than a full cycle of seasons — a year — and with less effort than by living with it for every hour of every day through every season. Perhaps where a longer cycle, like the proverbial hundred-year winter pertains, we may not learn the land for seven years, and may still be surprised after seven times seven. Traditionally, at the end of the road, old-time residents born of native parents hold that only by growing up on the land can it be known. To them, a half century is far too short a perspective. Newcomers are wise to solicit and hearken to their quiet wisdom.

In my view, the oldest and most honorable inhabitants are the aboriginal trees, grasses, and the whole community of native biota. Because this community has endured, many of its lessons can be learned in an afternoon, if we are still and pay attention: Those wind-sculpted trees bending southeast have taken root on the south slope below the rim of a little ridge. In their lee on the western side, moss and an orchid grow. From this modest observation we can see how to build a home that will be proof against a chill prevailing northwest wind, and below a berm facing just east of south. These gross considerations can (and should) be incorporated in our plans. When there is an acre of land or more, we must devote many

The Stoumen farm in
Humboldt County in
Northern California

pleasant afternoons to judging the many sites and orientations. Stories abound of
how, finally, in conducting this gentle homage of learning the land, the proper site
calls out to the settler.

Wise are the settlers who first build modestly and live with the land for the
statutory year or even longer. By building once in best accord with our precepts of
a site and the simple principles of sun and experience, we will not go far wrong,
and will learn much. Gradually, as we learn the rhythms and particularities of our
land, unexpected blessings and hazards become apparent. When we come at last to
build our Home, if we do so as passionate advocates of the land, aware of all its
favors, we are most likely to settle well within our community and endure.

By cunning use of weight-bearing, sheathing, sealing, retaining, and insulating
members, with mass and glass, rock, wood, metal, and plastic, we fashion a living
machine. A dwelling's first responsibility, and where the most money is spent, is in
sheltering comfortable space. Cheap energy led to a building style that ignored
natural forces like sun, wind, and water and achieved comfort through powerful
technology: heating, ventilating, and air conditioning, or HVAC, as it is called by
the building industry. This machinery is rated in tons and consumes prodigious
amounts of energy. As energy becomes more precious, we find that appropriate
building techniques and clever adaptations to our site can nearly always accomplish

the same effect while consuming much less energy. To heat our homes passively, we employ the "greenhouse effect" of mass and glass, whereby sunlight is converted to warmth and stored in masonry or earthen masses behind glass, which hold the heat within the envelope. To cool our homes, we encourage natural ventilation and prevailing winds to pass through a shaded interior. In both cases, we use as much insulation as we can. Whenever possible, we use thermal mass, a substantial pile of material with a high specific heat, like rock or water, which tends to maintain a constant, preferably comfortable, temperature. When passive means are insufficient, we seek nonpolluting, sustainable, and generally indigenous sources of energy, such as biomass or surplus wind or hydroelectricity.

As we said in Chapter 1, most of us live, and will continue to live, in existing houses. By making the same commitment to the land, even to a tiny city lot, apartment roof, or balcony, we find that economies and unexpected opportunities abound. By bringing house and site into accord, we may discover (for example) that a disused guest room or study is a "morning room," one blessed with warming morning sun, and this space may become our home's solar "furnace room" by improving the exposure, adding thermal glass, a greenhouse, or a Trombe wall. At the very least, we should commit to the planet, and resolve to be as efficient in our dwelling as we may be.

In this chapter we have assembled the finest array available of tools for finding, restoring, and assuming stewardship; locating and orienting a home; and beginning the process of homebuilding.

— *Michael Potts*

LEARNING THE LAND

David and Mary Val Palumbo live in a large independent home on a forested site in upstate Vermont. Their super-insulated wood frame home uses biomass for heat and hot water, and a hybrid solar, micro-hydro, and propane gen-set system. David and his family practiced the classic two-dwelling approach to their land, building an efficient solar cabin, living on and learning the land, then building the big house.

David Palumbo at home

Homecoming: David Palumbo's Story

We knew we were looking for a good sized piece, with woods; I guess we just assumed we'd be connected to utilities, but when it became clear that the properties that were right for us didn't have power, I set about getting myself educated. We were glad not to add to the need for more nuclear power, and we're individualists who pride ourselves on adventure, so relying on ourselves for our own power fit right in.

We looked at a lot of properties. I often have to hold people back when they move away from the city; few people have the necessary discipline. You don't want to buy land without studying it carefully: you want to make sure you understand about rights of way for utilities, about neighbors, about how the seasons and weather extremes treat your property before you commit. We went slowly, and ended up with a better site than we knew.

We wanted a place with privacy, but close to commerce and Mary Val's work. We wanted to be at the end of our own road, so we could choose our neighbors. We wanted woods, at least 15 acres, we thought. After we got over the surprise of going off the grid, we knew we needed a good balance of energy sources.

This land needed some bushwhacking before we could appreciate it. We knew from checking the soil and drainage that there were two good house sites. We lived in a tent on the best site while we built a small, efficient, quick house, 24' x 24', facing exactly south on the second best site, so we could start living on the land. We did a lot right with it, and it's still a very workable place. The PVs are within easy broom reach for clearing off the snow. We learned what we needed to know to start planning this house.

This is a cloudy place, and PV is not enough by itself this far north. Also, we use some big power tools, so we knew we would need a good-sized generator for back-up. Our original concept was to use as much PV as we could, and make up the difference with a propane generator. There are two houses and a big shop on our grid, so we manage it like a small utility.

We got the idea for hydro while prospecting for sites for a pond. We found three streams on the land, and one of them runs year-round. Getting the hydro system working was more hit-and-miss than it had to be; I know how to do it now with much less fuss, and I'm looking forward to doing this one over, but it works well, so we only use the propane generator when we run the bigger tools, like the planer.

We didn't know how many children we would have when we built this place, so we made it big. All the framing and trim lumber came from the land, so we used as much indigenous material as we could. We used a lot of wire, because we wired for 12, 24, and 110 volts. During the winter, it takes some time to manage all the systems, the electricity, the gasifier, the children. The kids go to school and childcare, and Mary Val commutes to her work. During the week I run the house and do my work helping people put their technology together properly.

The big Essex wood gasifier runs all year long. It burns at about 1800°–2000° F; when it's cold out, I fill the firebox two or three times a day, but when it's warm, I fill it once every two or three days. We thought a lot about the way the systems would interact as we planned and built the house, so there's a wood chute in the garage for getting the wood to the basement and the gasifier. And my workshop, which is above the garage, has a trap door and stair so we can move equipment up and down easily. The root cellar turned out well, because I can work the variables. Many people forget that you have to control humidity as well as temperature to keep things from rotting. Domestic water is another problem: we have plenty of pressure, but if everyone starts a shower at the same time, the gravity feed line just can't supply enough, so I've put a pressure tank into the system for just such times, and now we can hardly tell there's anyone else on the system.

I studied the way Native Americans, especially the ones who lived around here, treated the land. Flatlanders get emotional about cutting trees, but the regenerative capacity of the northern woods is staggering. You don't need to replant, and you can't keep the woods back. There are 250 wild apple trees on our land, planted, I suppose, by the original farmer who cleared it. The forest is crowding them out, but they are important for the wildlife, for bear, deer, and grouse. We've worked with the Department of Agriculture to release those trees from competition to the south, and now we're investigating some edge and patch cutting. Patch cutting (clear-cutting a small area, no more than a half-acre, defined by the forester) can make flatlanders really howl, but the edge of the regenerating forest, where the poplars start, is a crucial part of the woods habitat, where most of the animals thrive. The young poplars, for instance, are necessary for grouse reproduction. We're doing what the original inhabitants did, bringing sunlight into the forest.

Finding And Buying Your Place In The Country

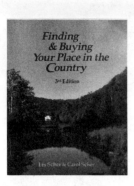

By Les and Carol Scher. For practical dreamers, here's a list of all the things you should consider before you buy rural land…including several we wish we had not forgotten. Much of the information can be found in miscellaneous sources, but here it is in one authoritative, convenient, well-organized book. Plenty of anecdotes and evaluations to help the reader make an informed decision. The cautions and checklist are a strong foundation for a lifetime project. 411 pages, paperback, 1992.

80-184   Finding and Buying Your Place in the Country $25⁰⁰

How To Buy Land Cheap

Edward Preston. A man from Sweden used this book to buy a house and land in the United States for $750! The author describes numerous ways to find amazing deals on land, including how to buy property seized from drug dealers, how to get a free house in most big cities, and how to get in on the biggest bonanza of all — land sold by bankrupt S&Ls at fire sale prices. The text contains hundreds of complete, current addresses, carefully worded form letters that get results, and even a glossary of terms to guide you through the arcane language of land transactions. 141 pages, paperback, 1991.

82-183   How to Buy Land Cheap $15⁰⁰

CHAPTER 3

SHELTER

AT REAL GOODS WE BELIEVE that sustainability is the key to the future. To live sustainably, we must strive to find ways in which our own species and all others can continue to exist together indefinitely. To accomplish this goal, we must savor every drop of our limited resources, so they can be used to their maximum potential. We must also curb the rate at which we are putting pollution into the atmosphere, landfills, and surface water, before we reach the point at which our planet's natural systems can no longer handle the contamination.

All too often, we rely on scientists and technology to pull us out of our crises. "Oh, they'll figure out a way to fix it before it gets too bad," we say. But if we listen to the most prominent scientists of today, they are the ones telling us that we have to change to survive; they have no trick up their sleeves. We believe that the time is now to be making these necessary changes.

One of the things that we humans do here on Earth is live in and build houses: shelter is one of the necessities of life. We should strive to minimize construction waste, avoid the use of nonrenewable materials, and, most importantly, design structures that require minimal energy input to maintain indoor comfort. The common building and design practices of today are far from sustainable, as forests dwindle and lumber prices continue to skyrocket. Most houses are built with little regard for their natural sites and microclimates, so that we have to outfit them with huge air conditioners and heaters. Then we pour thousands of dollars and tons of fossil fuels into these machines, at untold cost to the environment and to our heirs. This practice clearly is not sustainable, and the only question is whether the inevitable change will occur sooner or later. If we keep waiting for later, it will one day be too late. Clearly, now is the time to investigate and utilize sustainable building practices. Many of the best methods are thousands of years old, and by combining these traditional techniques with modern technology, we can help open the door to

a truly sustainable future. In this chapter on shelter, we provide information on some of the best and most innovative construction and design methods, ones that will carry us into the 21st century and beyond.

PASSIVE SOLAR DESIGN AND CONSTRUCTION

Before plunging into specific design details, let's review some of the general strategies that lead to sustainability in building. In striving to minimize the amount of energy we use (and also the money we spend) on keeping our buildings cozy, we find ourselves adhering to a versatile, innovative, yet ancient, set of principles — passive solar design.

The primary goal of passive solar design is to maximize the usefulness of the geography and weather conditions surrounding the structure. For the last hundred years, we have been building houses that are completely cut off from the environment. In most areas, the sun beats down on our houses during the winter, but we ignore this valuable solar energy heating scheme and crank up our heaters instead. Conversely, even though it may be quite cool outside at night during the summer, we keep our doors and windows closed and run our air conditioning. Even for those who do open up their houses at night, there's not enough thermal mass inside the house to keep it cool throughout the following day. The principles of passive solar heating and passive cooling, on the other hand, use these and other environmental conditions to their maximum potential.

There are a myriad of ways to design passive solar houses. A passive solar house can look virtually identical to a conventional house, or it can be very different and innovative. What binds all passive solar houses together as a group is their use of three principles — Gain, Mass, and Containment (it's easy to remember this by the acronym GMC). Any house that incorporates these three principles to a moderate degree can be considered a passive solar house, and will reap the benefits of this ancient but ingenious technology.

Gain refers to the practice of allowing the sun's heat to come into the building envelope. The trick is to allow the right amount of sun in, and then only when it's wanted. There are two types of gain, direct and indirect. Direct gain is achieved by installing the right amount of glass, or glazing, on the south side of the structure, which will heat up the house during sunny days. Once the heat enters the house, Mass and Containment make sure it stays there. Most areas in the U.S. receive at least some sunshine during the winter, and consequently have something to gain from southern glazing.

It is important not to have too much southern glazing, as this can cause overheating. This is the primary complaint of people who live in passive solar houses — too much heat! The Solar Energy Research Institute (SERI) and the Department of Energy (DOE) have created a rule of thumb to prevent overheating: determine the square footage of the room directly exposed to the glazing, and make sure that there is not more than 12% of this area in glass. Staying below this figure will generally prevent overheating. Just as important is to build properly sized overhangs, which will allow low winter sun in, but keep high summer sun out.

Indirect gain, on the other hand, is when the sun's heat is absorbed in one place and then transferred to the living space. Because there is a buffer between the intense sun and the heated space, there is rarely a problem with overheating. Examples of indirect gain are thermosiphoning hot air collectors, Trombe walls, water walls, and active solar heating.

The second principle of passive solar is thermal mass. It is important to use high-mass materials, such as concrete floors or walls, in the construction of a passive solar house. The mass acts as a storage medium and a buffer for the heat; if the mass gets heated up during the day (either by direct or indirect gain) it will slowly dissipate this heat and keep the house warm at night. The thermal mass in a passive solar house is analogous to the batteries in a photovoltaic system.

Last, and most important, is containment. None of the other passive solar principles work without containment. When builders refer to containment, they are talking about "buttoning up the house," or "sealing up the building envelope." We have to keep all of this heat that we have collected from the sun inside our house, or it isn't going to do us any good at all. There are two components to containment, insulation and air infiltration reduction, and both are essential. Many people use plenty of insulation, but they neglect to have adequate air infiltration reduction. An example is a house with R-50 insulation in the walls, but poor window and door seals. It doesn't matter how much insulation you have; it will be ineffective unless you pay equal attention to all of the nooks and crannies where cold air can get in. When you build a house that you plan to live in for a long time, it is better to err on the side of overemphasis when it comes to containment.

It is important to keep the thermal mass materials mentioned above inside the building envelope (insulate the outer shell) so that they act as a storage medium instead of a sieve.

By following the GMC rule when designing your house, you will find that your need to use fuel for heating will diminish and sometimes vanish, depending on your location. The best places for passive solar are, of course, places that have the most sun in the winter, like Colorado, Arizona, California, Nevada, and New Mexico. Although these states often experience cold winter air temperatures, many existing passive solar houses there have little need for a back-up heating source. Yet even in areas where the winter sun isn't so abundant, using passive solar principles can only help. And, having a home with high thermal mass and a good containment system will increase the efficiency of any heating system.

Another advantage of a passive solar house is its ability to stay cool in the summertime. Deciduous plantings and properly sized overhangs above south-facing glazing prevent unwanted solar gain in summer, while the thermal mass and containment properties reverse their winter roles and act as "nature's air conditioner." The trick is to take advantage of cool nights. By opening up all the windows of the house at night, the high mass materials are cooled down. In the morning, close the windows. The containment system keeps out the unwanted heat, while the cool high-mass materials stay that way throughout the day, just like in a cave. A radiant barrier in the roof is another essential and very effective tool for keeping cool in the summer.

The principles of passive solar heating and passive cooling design are very flexible. With creativity, they can be incorporated into just about any kind of

building, from a log cabin to a split-level suburban with a two-car garage. The choice of materials to use and the appearance of the house is wide open.

In this chapter, we mention several types of building techniques. Most use all or some of the passive solar principles, or could easily be adapted to do so. Whether you plan to build a log home, a straw bale home, or whatever your choice, we strongly encourage you to use all three principles — Gain, Mass, and Containment — when designing your home.

— Gary Beckwith

ARCHITECTURE

When considering building your own home, please remember the tale of the three little pigs. One chose a house of straw, another of sticks, and the third chose to build with brick. That simple tale illustrates the relationship of shelter to context. This article is also about shelter and context. In the case of humans, each of us responds to numerous and various Big Bad Wolves which will blow at the walls of the home we imagine for ourselves: fire, flood, and earthquake, to name just a few. Recognizing these environmental and other contextual forces in the planning stages can help prepare our homes to meet and shrug off the stresses to which they may be subjected in the future.

Depending upon where we are in the world we will find a different climate, as well as different resources and methods of construction. Working with the tools, materials, and techniques most appropriate to the local area can save both money and energy. In most cases there may be a variety of appropriate responses to environmental and personal economic situations, so it is helpful to consider, research, and prioritize our options. Let's look at several available alternatives and their advantages and disadvantages, in order to help select what might be appropriate in your particular endeavor.

PASSIVE SOLAR CHECKLIST	
	❏ East-West Axis
	❏ South-facing glazing
	❏ Overhangs
	❏ North side earth-berm
	❏ Thermal mass inside building envelope
	❏ Extra insulation in walls and roof
	❏ Radiant Barrier in roof
	❏ Open airways inside to promote circulation
	❏ Air infiltration reduction
	❏ Double-pane windows or better
	❏ Little or no glazing on north side
	❏ Day lighting
	❏ Air-to-air heat exchanger

As we consider alternatives, remember that the specific nature of a region and site will exert a great deal of influence in your choice. If, for example, you live near a quarry of nice building stone, you might be less likely to select adobe as your chief construction material.

In all cases, consider first the extreme environmental forces which prevail at your site. These might include some of the following: high winds, hurricanes, severe thunderstorms (including hailstorms and tornadoes), extreme heat (tropical or arid), or extreme cold (with either little snow or a great deal of snow). Consider whether your area experiences wet summers with little fire potential or dry summers with a high fire danger; also take into account the possibility of flooding, erosion, earthquakes, landslides, or ground movement. Determine how much sunshine and shade the site receives. Check for stagnant air, or for materials containing radon or other contaminants in indigenous materials or the site itself.

Of great importance, but more subtle, is the geological nature of the site, its stability, and load-carrying capacity. The nature of these factors is usually hard to detect without a great deal of experience and local knowledge, and is something best evaluated with the help of a geotechnical engineer. Consult with this engineer as soon as you start considering basic siting issues.

The environmental stresses unique to the region and the specific site (the Big Bad Wolves) should influence your choice of construction methods as much as the availability of labor and materials. Also important are issues of timing, financial constraints, and your own personal capabilities as a builder or those of builders in your area.

Before you select any building material, be sure to familiarize yourself with the indigenous homes in your area. Such homes best exemplify the traditional and successful uses of materials and building strategies.

Stone

Let us begin with a look at stone. If we find a good source of sound (i.e., nonhazardous), workable material, and we look to the indigenous homes of the area, we will probably find examples of its use which will help us with our task.

Two ways to work with this material come immediately to mind: formed wall construction, in which stones and mortar are built up in courses within reusable slipforms, and laid stone, wherein each stone is nestled and mortared atop the stones below. The skill required for the latter method is much greater, as is the time and cost of the finished work. In temperate climates a stone building's outer skin insulates poorly, and in most cases an interior insulating layer is either attached to the stone, using foam, or built as a frame wall within, using loose fill or batts. Stone walls are heavy, require solid footings, and must be reinforced with steel to resist earthquakes and high wind forces, which need to be accounted for in almost all parts of the planet. More skill is required to reinforce laid stone than formed stone work. Care is needed in the sizing of the stones, especially at the corners and around and above openings. Pointing and fastidious cleanup can, however, produce a beautiful, almost ageless structure. For guidance and inspiration, look to indigenous buildings in your area and also research the stone-building techniques of Helen and Scott Nearing, Frank Lloyd Wright, Ernest Flagg, and Ken Kern.

Building with stone has several advantages. Stone is, of course, fire-resistant, and can be made earthquake-resistant too if it is well engineered, reinforced, shaped solidly, and proportioned properly with respect to its thickness-to-height ratio and frequency of bracing. Very demanding, yet rewarding, in terms of physical labor, stonework buildings take a long time to construct, but can be very cost-effective if owner-built. They are durable, beautiful, and improve with age. If well-designed, their thermal performance, due to their large mass, can be very efficient.

Wood

Especially in the historic communities of the eastern United States, there are a wealth of buildings made using post-and-beam techniques. The structural frame of the building is typically prefabricated and can be worked either on- or off-site during the non-building season, then brought to the site and erected efficiently.

This method is rather wood-intensive, as it utilizes mostly high-quality, free-of-heart, large, and sometimes long timbers. In most cases the timbers are milled into rectangular sections, although I have used round poles for this work as well. If you have a good source of indigenous wood that fits these criteria, this can be an excellent way to build.

It is important to detail the frame properly with rigid geometry, lateral knee-bracing, and integration of frame and exterior sheathing in a way that makes a strong, resilient, and secure unit to resist earthquakes and high winds. Another issue with post-and-beam construction is the vulnerability of the frame to rot or the presence of insects. Old frames found in the East were often made of chestnut or oak, which are very strong woods. Softwood species have different characteristics and are susceptible to damage, so care must be taken in detailing, especially with regard to the vapor barrier and insulation. When utilizing this construction method, care needs to be taken if the danger of forest or brush fire is a factor at the site. This concern would be addressed in siting, landscaping, and the choice of skin and roof materials; for example, stucco, tile, and metal.

The design of the joints is of primary importance in this type of construction; my favorite joints are those found in the New World Dutch barns of the Mohawk region of New York State, where freestanding posts have the mortise notched entirely through the timber, with the tenon passing through and extending beyond by a foot, and back-doweled.

Interior finish can be either paneling or gypsum board. If paneling is used, apply over gypsum board for mass and some fire resistivity. I prefer plaster interiors to gypsum board with timber frames. If wood timbers are planed, sanded, and finished with a pickling stain, the wood tends to stay lighter in color, and brighten the interior of the home. Chamfering the edges of the timbers softens the corners and is friendlier to the touch and eye. It is tempting to hurry, but taking care to finish exposed timbers before they go into place is well worth the time and effort.

The exterior skin of a post-and-beam home must be carefully selected for it to be fire-resistant. Careful engineering and construction is required for it to be earthquake- or hurricane-resistant. Heavy timbers require heavy lifting and team-work to erect, and skill in constructing joints that fit. Special tools and huge saws are required for production work, but these structures can also be built using only hand tools. This building style's low thermal mass can be compensated for by using super-insulation techniques and huge centrally located fireplaces or interior thermal mass.

Domes

Domes are especially dear to my heart, since my wife, O'Malley, and I built and live in one at our farm. They can be fabricated off-site, indoors, and without a great deal of space, as the pieces are small and light. The structure is comprised of a few standardized and repeated strut and hub sizes of small dimension. Extreme accuracy is required when fabricating, which can be achieved by using jigs. Domes are easy to erect, because all pieces are light and the time it takes to erect them is short. Sheathing the dome can be fast as well, since standardization is the name of the game. As Buckminster Fuller observed, the dome uses a minimum of materials to enclose a maximum of volume. There is, however, no free lunch: the equalizer comes in the detailing of windows and doors, papering, roofing, interior finishing, and cabinetry. These structures are extremely strong and can withstand almost any force acting upon them if they are firmly anchored to a strong foundation. They are difficult to insulate to any great degree as they have such a thin frame (four to six inches), and they usually have no attic space. Loose foam insulation could be used but I suspect that infiltration losses would be quite high.

When placing windows and doors, the integrity of the dome shape can be lost, resulting in a slow sagging of the shape and, ultimately, failure through fatigue of the adjacent members, and collapse. Therefore, the removal of any struts is discouraged. If it is necessary to remove a strut, try to mitigate the effect with structural engineering, stiffening, and the addition of other members. Aesthetically pleasing weathertight roofing is difficult to achieve, particularly if it also needs to be resistant to forest or brush fires.

Low thermal mass is a problem with domes, but the efficient shape can be easily vented and heated. Interiors are slow to build and tedious to finish as all shapes are unusual and either round or triangular. Heavy lifting is not required, material use can be efficient, and materials are affordable. The shelter goes up and is sheathed quickly, but work proceeds very slowly after that. For more information, take a look at Buckminster Fuller's works and *The Dome Book*.

Log Cabins

Logs are a classic pioneer building material. Is there is a good source of this wood near your building site? In the Northwest and Southeast logs are more plentiful than in other areas of the United States. In the national forests there are often standing dead trees, killed by fire or disease, which are not good for much else than use in log homes or structures. This method of construction lends itself to prefabrication and off-site construction. If logs of a large diameter are used, they are so heavy to handle they can only be placed and fitted by using a crane to move them about. Nevertheless, this work can be done on-site with the proper equipment. Work can be done during the off-season and prepared so that, when the weather is

right, the logs, already cut, notched, kerfed, and fitted, can be assembled. As with timber frames, it is important to choose logs that are not already the home for hungry wood-eating insects and creatures.

Room sizes and proportions need to be oversized due to the often massive thickness of the logs. The foundation must be built to support the enormous weight of the structure. The sealant used between the logs and the detailing at corners, windows, and doors is of great importance in providing a tightly sealed interior. Without proper detailing, cold air infiltration and the resulting heat loss can be quite high. Detailing at doors and window openings also needs to allow for a great deal of settlement, since the logs shrink as they dry. If the walls get too high, especially if they are unbraced, earthquake and wind load concerns may result. Log homes do not seem to lend themselves to hot, dry climates, and seem to work best in cold, dry, snowy winters and mild, wet summers. The quality of the fit and the nature and quality of the logs are of greatest concern. The scale, proportion, and elegance of the corners and openings is a source of great aesthetic beauty, as are the use of small logs for railings, floor beams, and so forth. Wood has little specific heat and indifferent insulating ability (approximately 1 R unit per inch of thickness) but a house built of thick logs, carefully fitted and sealed against infiltration, and with a large central masonry fireplace that provides thermal mass as well as heat, can be wonderfully cozy.

Prefabricated log structures — made with smaller dimension softwood that has been totally milled and prepackaged — seldom make satisfactory homes. Their walls are too thin, the wood quality is often poor, and their thermal properties are marginal; these countrified, cookie-cutter structures are very expensive to heat and resource-inefficient.

Cordwood construction, an unusual but promising log-building technique, uses smaller dimensioned wood to good effect. It involves laying-up short pieces of logs perpendicular to the wall like bricks.

Log construction in the appropriate location can be very environmentally responsive. Using fire-resistant roofing materials and appropriate landscaping and siting, it does not pose an unacceptable fire hazard in regions with temperate summers. With proper hold-downs and engineering, this kind of construction can also provide adequate seismic resistance and high wind resistance. Log structures have fair insulative qualities and can be energy efficient so long as infiltration losses are eliminated through detailing, material selection, timber dryness, and fastidious construction. Increasing solar aperture, by using larger expanses of southern windows for heat gain, requires special engineering to ensure the continuity of the structure's lateral resistance.

New loglike materials made from laminated scrap wood now make it possible to follow these construction techniques anywhere.

Adobe

Adobe is one of the oldest and most widely used building methods. In the U.S. it is also primarily found in the Southwest and California. It is also appropriate in the western high desert of the Sierran-Cascade rainshadow, which extends all the way up into southeastern British Columbia. Its use is most economical in regions with prevalent adobe (clay) soils and a relatively dry climate, especially in areas of the Southwest near Santa Fe, and thence down into northern Mexico. This very labor-intensive building technique requires sunny, dry weather to cure the bricks if they are made on-site. If they are purchased from outside the area, trucking costs will be significant. In regions that have hot, dry summers and cold winters, adobe can be utilized without building an insulated wall, since its mass provides a "thermal flywheel" that moderates temperature swings and ensures a comfortable home. Adobe homes work particularly well in New Mexico, one of the sunniest spots on the continent. Other areas are not so lucky, and so, in most cases, it is advisable to provide insulation as well. Foamboard insulation is best applied to the exterior of the walls, then plastered over with stucco cement or adobe mud plaster.

The major concern with adobe is its seismic resistance. Classic adobe buildings have withstood centuries of temblors because their buttressed walls are relatively thick for their height, and are topped with sturdy wood bond-beams. Adobe requires bracing of some sort. Modern, thin-walled adobe constructions have proven to be very unsafe in an earthquake unless they are of simple rectangular geometry carefully reinforced with steel and concrete, braced frequently, tied together with a bond-beam at the top, and securely linked to their foundations. One method of bracing the adobe wall utilizes an integral concrete frame, which acts to hold the walls together.

Building with adobe is laborious. However, if the house is owner-built, using materials found on site, it can be inexpensive. The mix for the bricks (sand/clay content and waterproofing compounds used to keep the blocks from deteriorating in the rain), the detailing at the foundation and roof, and the integration of reinforcing frames are all quite important to the shelter's solidity and longevity. Adobe is one of the most inspiring materials to work with, and it is certainly one of the oldest building materials found in North America. If executed properly in the appropriate locale, adobe can provide a wonderful home; constructed poorly and in the wrong place, it can be a deathtrap.

Rammed Earth

Rammed earth combines many of adobe's best qualities with the advantages of slipformed walls used in stone construction. Dirt from the site, which must be of favorable constitution, is mixed with proper proportions of clay, sand, water, and cement. The earth is then tamped into reusable forms to infill walls between concrete frames, foundations, and bond beams, which tie the structure into an earthquake-resistant frame. Engineering and good detailing are required to provide safety and durability. If attention is paid to detailing at window and door openings, the rammed earth dwelling can have the same gracious and solid feeling as adobe with less labor to produce. Given the right site, rammed earth construction can be economical for the owner-builder. If the walls are made sufficiently thick, they are thermally massive and highly insulative, often precluding the need for further insulation. In severe climates, however, the walls should be further insulated.

It is possible to apply this material using a gunnite spray rig as Pneumatically-Installed Stabilized Earth (PISE); the material is sprayed against a plywood form from the outside to a thickness of 18 to 24 inches. In mild climates, such as parts

of California, Arizona, and New Mexico, this thickness may be sufficient without the need for additional insulation. However, the formula for the wall mix can be varied, or the wall thickness reduced greatly, if insulation will also be installed. The walls need to be engineered and reinforced with steel to make a seismically safe home. Rammed earth is also fire-resistant.

Earthships

Earthships are an interesting footnote in architectural history. A sort of cross between rammed earth and adobe, they employ opportunistic resources, used tires, and aluminum cans in a clever passive-solar direct-gain strategy and are often sunk into a hillslope, or "earth-integrated." First, a hole is excavated such that the front wall is at or below counter height, then the tires are laid in a bricklike pattern and laboriously filled with soil and compacted. The tires swell and interlock under the pressure of manually rammed earth, and become very thick and resilient. As in an adobe wall, integrity is further secured by a wooden or cement bond-beam atop the wall. Chinks between tires are stuffed with used and partially crushed aluminum cans. (Needless to say, prodigious amounts of liquid are consumed during the tamping process.) Roofing consists of the classic *vegas* (large wooden girders) and *latillas,* or modern laminated beams, along with plywood and foam sheathings.

Exterior walls and rounded, sculpted interior surfaces are plastered and painted to look like adobe and rammed earth homes. The use of tires and soda cans as a substrate for plaster is an interesting refuse-disposal concept for which Michael Reynolds, a Taos architect, deserves credit. Incorporating aspects of the earthship concept would be possible with many of the other construction techniques already mentioned. In fact the earthship plan and section is the classic earth-integrated concept first built in the Danube River valley some 7,500 years ago. Reynolds' earthships have a sloping glass front wall, which is usually oriented east of south to reduce late afternoon overheating.

Earthships often have integral power and plumbing systems and are designed to be completely self-sufficient: water from roof catchment, photovoltaic electricity,

Dug into the ground, the earthship's thermal mass keeps it cool in summer and warm in winter.

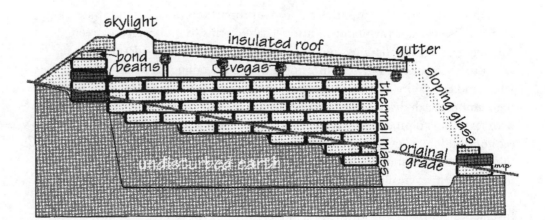

and innovative indoor waste disposal are all common features. Interior "finger walls," often painted in dark colors, provide enclosed thermal mass. The low winter sun shines in during the day and heats these interior walls; during the cold desert nights they reradiate their heat, so a well-balanced earthship hovers around 65° F with no expenditure of energy. In the summer, the shaded finger walls are cooled by evening cross-ventilation, then provide "radiant cool" to keep the interior at a comfortable temperature even during scorching southwestern summer days.

We can judge earthship longevity only by their short (ten-year) history, but they incorporate the benefits and share the risks of rammed earth and adobe construction: they are fireproof, seismically earthquake-resistant (if well-integrated), thermally massive, made of appropriate materials, inexpensive, and indigenous. Earthships can be, and have been, built for very little money by owner-builders. With straw bale and modern rammed earth techniques, they represent a refreshing and optimistically innovative approach to shelter.

Building with Earth and Sun

All these ways of building can elegantly transform different materials and methods into appropriate, environmentally integrated homes. These transformations can be in harmony with their context and our lifestyle; they can be functional, economical, and attuned to our needs. Responsible and responsive ways of building have evolved for all the thousands of years that people have sought to protect themselves from the elements and provide for their comforts. Successful strategies have been developed, refined, and adapted down through the ages as part of our cultural heritage.

In preparing to build, take into account which of your choices will add to our planetary wealth or borrow from it, and choose wisely. For example, in the Northern Hemisphere, solar buildings face south in order to capture the sun and minimize their exposure to the frigid north; build accordingly, and your home will cost less to build and maintain. It is also possible to build unwisely, and for more than a century we have, but always at great cost. Whether you are building your home with stone, earth, timbers, tires, or straw, learn reverence for your site, climate, and materials; look with insight and wisdom at your ergonomic and spiritual needs; and make homes that are safe, durable, and inspiring. Add to the wealth of this planet by treating tasks and materials with respect and care. Live peacefully in a home that sets an example of environmental adaptivity and offers a resilient response to the awesome forces of nature.

I like to think that, when we build, Eric Sloane, Frank Lloyd Wright, Ernest Flagg, Helen and Scott Nearing, Thomas Jefferson, and Buckminster Fuller join us as we put our homes together to ward off the forces of the Big Bad Wolves and set a course to save the planet. This is what we are doing in our own small way when we build something clean, strong, healthy, and lasting. We want to leave a legacy that will inspire and nurture our families and those that follow.

Look to the indigenous buildings in your region and remember the story of the Three Little Pigs. Consider also the layout of your home and try to provide sun-filled warmth and light to spaces you use during the day. Focus your sunny living areas toward the calming, healing, and ever-changing nature in your garden.

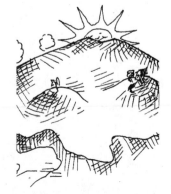

Architecture begins when we go beyond shelter. Architecture isn't the result of necessity, it's what takes place after the necessities are taken care of. To make inspiring homes, we must aspire to build something more than just shelter.

The most appealing thing about owner-built solar architecture is that it seeks to make affordable, buildable homes possible for people of modest means, so that they might become independent of those two great American goads: the mortgage and the energy bill. Seeing earthships dotting a steep forested hillside without spoiling it called to my mind Buckminster Fuller's idea that we are all together in a sort of lifeboat, floating on a universal sea, needing to balance our use of resources and energy in order to continue on our way in harmony. By conceiving and building our homes so they may be of the earth and directly powered by the sun, we certainly set architecture on a better and more sustainable course.

— *Jonathan Stoumen, AIA*

Solar Design Associates is a pioneer in the field of sustainable building design. Over the past 15 years, they have completed many passive solar structures. We asked Steven Strong, Solar Design's president, to select one of their passive solar houses and take it apart to show us what it does and how it works.

ENVIRONMENTALLY RESPONSIVE HOUSE DESIGN

There are many cost-effective options you can put to use now to reduce or eliminate your dependence on nonrenewable fuels. For true energy efficiency, your home and its energy system should be designed together to work together. The house featured here is a good example of this approach, which begins with an energy-efficient building envelope.

By carefully choosing the materials, insulation, infiltration barriers, and finishes for the exterior structural system, and incorporating state-of-the-art innovations such as low-emissivity (low-E) glazing, we create a living environment with high thermal integrity. It is an environment that requires a minimum amount of energy to maintain warmth and comfort, and whose geometry is configured to take maximum advantage of passive solar gain.

Passive Solar Design

Passive solar design involves the direct utilization of solar energy (that is, without the energy having been processed by an active, or energy-consuming, mechanical system). Through proper orientation, north wall sheltering, south-facing glass, sun-controlling architecture, and the use of the structure as thermal mass, the sun can play a direct and significant role in your home and its energy support.

By designing to optimize this free solar gain, and utilizing proper insulation, high-performance glazing, internal mass, and other energy-conserving elements, the

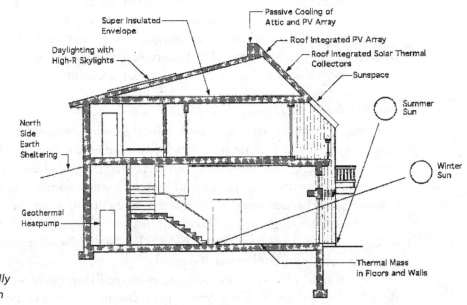

Labels in figure:
- Super Insulated Envelope
- Passive Cooling of Attic and PV Array
- Roof Integrated PV Array
- Roof Integrated Solar Thermal Collectors
- Daylighting with High-R Skylights
- Sunspace
- Summer Sun
- North Side Earth Sheltering
- Winter Sun
- Geothermal Heatpump
- Thermal Mass in Floors and Walls

An environmentally responsible design

energy requirements for your home — and hence the operating costs — will be substantially reduced, even before you consider any alternative-energy hardware.

These concepts are illustrated in the house in the accompanying figure which incorporates direct passive solar gain through generous south-facing glass, daylighting through high-R skylights and sunspace sloped glass, and internal thermal mass in walls and floors. Passive solar thermosyphon cooling is employed to cool the attic space while also keeping the PV array cool.

With supplemental space-heating requirements reduced to a minimum by the high-performance building envelope, there are many options to supply the remaining thermal energy that your new house will need. In fact, the exterior envelope of your new home can be made so energy-efficient that, in many areas of the country, your new home may require no conventional "back-up" heating system at all.

Solar Domestic Water Heating

All homes require domestic hot water, and we almost always recommend the inclusion of an active solar domestic water-heating system. The system is sized to satisfy the domestic hot-water requirements calculated for your family, and the solar collector array may be designed as an integrated part of the structure.

The center section of the south roof on the solar house in the photograph features a 150–square-foot array of roof-integrated solar thermal collectors for domestic water heating. The collector loop is a drainback configuration for maximum efficiency.

This integrated collector design results in greatly enhanced aesthetics while reducing the loss of thermal energy from the back and sides of the solar collectors. The solar array is assembled using the highest-quality selective-surface, black nickel-chrome–plated, all-copper solar absorber plates with a tempered, high-transmission glass cover that can serve as the finished weathering skin of the roof.

Of course, if the aesthetics of an integral roof are not your primary concern, or if suitable roof area is not available, factory-built collectors can be mounted over a finished roof or on a rack structure to heat your domestic water.

Active Solar Space Heating

For residences whose size and/or geographical location necessitate an additional supply of thermal energy for space heating to supplement that contributed by passive solar gain, we often recommend increasing the size of the active solar collector array beyond that required to heat the home's domestic hot water. This option makes thermal energy collected and stored during sunny periods available to provide space heating when the weather turns bad. During the warmer months, when your home's space heating demand is low, the active solar system can provide heat to a swimming pool or spa, if desired.

For even distribution of thermal energy to the living space, we favor a radiant floor-heating system. Such a system is quiet, consistent, takes up no space, and delivers an even, upwardly-radiated warmth throughout the house. It helps to distribute direct passive solar gain and does not interfere with furniture placement. It is an elegant and most efficient method of heating a space.

Space Cooling

Even when energy-conscious design practices and cooling-load avoidance strategies are employed, buildings in some areas of the country may require mechanical cooling to maintain comfort levels during some portion of the year.

Typical air-conditioning systems use a compressor-driven refrigeration loop to remove heat from a space and then force it outside into an even warmer ambient air mass. This requires a large amount of electrical energy.

In those climates where ambient humidity is low, evaporative cooling is often used successfully to create a comfortable indoor environment with much less energy than conventional compressor-driven refrigeration systems. These systems bring in fresh outdoor air and humidify it before delivering it to the space. The humidification process dramatically lowers the perceived air temperature, providing comfort.

In areas with high humidity, earth-coupled geothermal heat pumps provide air conditioning with a much greater operating efficiency than common air-source systems. Geothermal, or water-source, heat pumps use water from a well, river, lake, or pond as the thermal source and sink as opposed to the outdoor air. Heat removed from a space flows much more willingly into a colder water mass than a hotter outdoor air mass, so less energy is required.

In addition, these units can be configured to supply domestic hot water essentially free during operation. Water-source heat pumps also run in reverse, extracting usable thermal energy from the water source to provide space heating at a

Environmentally responsive residence features earth-sheltering, passive solar heating and cooling, super-insulation, internal thermal mass, air-to-air heat-recovery ventilation, and an earth-coupled, geothermal heat pump system to provide supplemental space conditioning and domestic hot-water heating. The Brookline, Massachusetts, home also features a roof-integrated solar array which provides thermal energy from active solar collectors, electricity from photovoltaic modules, and daylighting from special heat-rejecting sloped glass over the two-story central sunspace.

coefficient of operating performance (COP) of four to one. For every unit of electrical energy consumed, four times as much heat energy is delivered to the space.

Electricity

Other residential energy-system configurations are available to attain higher degrees of energy efficiency or to make a house fully energy-independent. For example, you can generate all of your own electricity with a properly designed solar photovoltaic (PV) system.

Photovoltaic is the ideal source of power for an environmentally responsive home, whether it is your principal residence or a vacation retreat. With the right design, the sunlight that falls on your homesite will power your home. Your PV power system fan can also be easily configured to provide on-site solar recharging of solar electric vehicles — now, or in the future.

Solar Design Associates has designed and built many photovoltaic-powered residences across the United States. Some of these homes are utility-interactive — in other words, residents sell surplus power generated by their PV system to the local power company during sunny days and then "buy back" power at night. Others are completely stand-alone systems with no connection to the utility grid; they store electrical energy in a battery bank for use at night and during cloudy weather.

Whenever possible, we try to integrate the photovoltaic array with the roof of

the structure. This produces superior aesthetic results and can actually improve the performance of both the PV system and the house as a whole.

The south roof of the house in the photograph is a single glass plane incorporating photovoltaic, solar thermal collectors, and daylighting. The PV array uses large-area PV modules to form the roof surface and finished weathering skin.

Experience has shown that this configuration produces the highest efficiency by providing a thermosyphon of cooling airflow behind the modules entering at the south eave and exiting at the north ridge. The PV modules are cooled while, simultaneously, the attic space is cleared of unwanted solar heat. Depending on where the project is located, this strategy can help reduce or eliminate cooling requirements.

We also have design experience with wind- and water-power systems, although these are heavily site-dependent sources and thus are less frequently employed.

— *Steven J. Strong*

Steven J. Strong is president of Solar Design Associates in Harvard, Massachusetts. The firm has designed energy-independent homes across the country from Maine to Hawaii. Steven's book, The Solar Electric House, *is described on page 95.*

BUILD YOUR OWN

Building you own house is a courageous undertaking, but it can also be extremely gratifying, and can save you a lot of money. Those of you who have experience in carpentry, construction, electrical work, or plumbing may be able to do some, most, or all of the work yourselves, with a little help from your friends.

Whatever kind of structure you are going to build, there are certain things you are going to have to deal with. Many of us here at Real Goods built our own houses, and got some good down-to-earth advice from veterans of the process. The following tips are a distillation of lessons we learned along the way. Hopefully some of the mistakes we made can help you not make the same ones. (New mistakes are always more interesting.)

— *Gary Beckwith*

On Where to Build
"Stay away from north-facing slopes. They're very cold."

— *Terry Hamor*

"I would have chosen a building site nearer to the main road, and with better gravity-flow for water instead of building a water tower."

— *Debbie Robertson*

"Don't buy too far out in the boonies. Elbow room is great, but there are practical limits. My property was five miles out a steep, rough dirt road. That's fine if you're independently wealthy, and don't have kids or friends. The difficult commute was my primary reason for selling the property."

— *Doug Pratt*

On Being Realistic with Time and Money Plans

"Learn patience. There is always so much to do. Be patient and it all gets done eventually."

— *Doug Pratt*

"If I had to do it over again, I'd realize that it takes three times as long and costs three times as much as expected. I'd have my finances together so it could be built within a year's time instead of . . . 15 years!"

— *Debbie Robertson*

On When to Move In

"Get a cheap portable living space initially. I had a refurbished school bus that allowed me to move onto the property with a minimum of development. This saved rent and allowed me to check out solar access and weather patterns before choosing a building site and designing a house. A house trailer can do the same thing.

Don't move in until it's finished. It's real tempting to move in once the walls and roof are up. Resist if at all possible."

— *Doug Pratt*

"I wouldn't have moved into an unfinished structure."

— *Debbie Robertson*

"I lived in a school bus until it was finished. [It was good to be able to] move into the shade in the summer and into the sun in the winter. Buses are cheaper than trailers, and come with a motor and charging system. One of my biggest mistakes was moving into the house before it was 100% completed. Most plugs and switches are still not done 13 years later! Finish it before you move in, or you never will."

— *Jeff Oldham*

"Good advice, if you're that kind of person. I've got to live with something awhile before I see the best solution, so I say, identify essential living segments — a dry bed and a functional kitchen at least — and move right in. Keep projects contained with dropcloths and clean up after, and you end up with a functional, well-tuned house."

— *Michael Potts*

"Try to complete each task or area before starting another."

— *Robert Klayman*

General Building Plans

"Use passive solar design. This was one of the things I did right. My house was cool in the summer and warm in the winter, and, with an intelligent design, yours can be too. Avoid the temptation to overglaze on the south side. You'll end up with a house that's too warm in the winter and cools off too quickly at night.

Buy quality stuff. Cheap equipment will drive you crazy with frustration and eat up your valuable time. Stay away from cheap poly pipe. The connections always leak. Get quality PVC."

— *Doug Pratt*

"Hire a carpenter and become his apprentice."

— *Terry Hamor*

"Take an honest assessment of your skills. Decide up front which tasks you can accomplish through your own efforts and which will require assistance. Assign a dollar amount and time required for each. Establish a working budget, which you should update throughout the project."

— *Robert Klayman*

"Money is no object when it comes to insulation! Don't use aluminum door or window frames. Aluminum is too good a conductor of heat. Use wood or fiberglass instead.

Pay attention to little details. It's not a matter of which roofing or siding system you choose; it's the small details like flashing, corner joints, drainage runoffs that really count."

— *Jeff Oldham*

"Design your wood storage with the ability to load from the outside and retrieve from the inside. Build a laundry chute and a dumbwaiter. Have a small utility bathroom accessible from a back door. Plan for, or install, a dishwasher."

— *Debbie Robertson*

We strongly suggest that you get in touch with an organization that is set up to assist people who are designing and building their own homes. Here are a couple that we recommend:

East Coast:
Yestermorrow School
RR1 Box 97-5
Warren, VT 05674
(802) 496-5545

West Coast:
The Owner-Builder Center
1250 Addison St. #209
Berkeley, CA 94702
(510) 848-6860

We also rounded up some of the best books available to help you on your self-building adventure.

Straw bale construction

STRAW BALE CONSTRUCTION

Rediscovery of a century-old building technique may prove to be the answer to our current search for an affordable substitute for lumber. Inexpensive, easy to use, fire-resistant, and environmentally harmonious, bales of straw could revolutionize home construction methods in the 1990s.

Matts Myhrman and Judy Knox of Out On Bale in Tucson, Arizona, are considered by many to be the father and mother of the present straw bale renaissance. They have traveled around the country educating builders and local building department officials about the safety, utility, and elegance of straw bale building methods. They've also interviewed the inhabitants and photographed and examined straw bale houses dating back to the turn of the century. They have found plastered straw bale houses in the Sandhills area of Nebraska dating back to 1903, some of which are as sound today as the day they were built. These old buildings were constructed with straw from native rangeland grasses and plastered with mud. Straw or hay was used because it was the only building material these early homesteaders could find. The Sandhills had no trees and the sandy soil was not good for the sod construction methods used by other homesteaders of that era.

Today, because of bad forestry management practices and the pressure of too much building, lumber prices are rising and quality is going down. This has forced builders to look to less expensive alternative building materials. At the same time, plagued by poor air quality, California and other states are in the process of banning agricultural burning of straw because it creates a major air pollution problem. If straw is baled rather than burned, this waste cellulose can be turned into one of the most promising alternative building materials: straw bales.

California's North Central Valley has roughly 400,000 acres dedicated to rice cultivation. After the rice is harvested the straw remains in the field and creates a major disposal problem for farmers. Other types of straw decay quite rapidly when exposed to moisture, but rice straw is high in silica and decays very slowly. The high silica content of rice straw also gives it extra strength and longevity. In other regions, local agriculture often provides potential building materials in the form of grain straws like wheat, oat, rye, and barley, which are generally regarded as waste products after the grain has been removed and can be used for straw bale construction.

Since straw matures in a matter of months, millions of tons of these environmentally sound building materials are produced *annually*. Compared to the decades it takes for trees to grow large enough for the sawmill, the advantages offered by straw bales look even better. Straw bale construction is the perfect win/win solution for farmers and builders.

Modern load-bearing plastered straw-bale wall construction goes like this: A concrete slab, or floor, is poured with short pieces of rebar (metal concrete-reinforcement rods) sticking up out of the concrete about 12 inches high on two-foot centers around the slab perimeter. After waterproofing the top of the foundation, the bottom bales of the wall are wrapped with a polyethylene sheet (to prevent moisture migration from below) and impaled on the rebar. Subsequent bales are stacked on this bottom layer like bricks, with each new layer of bales offset by half over the bales below. Rebar pieces are driven into selected bale layers pinning them to the bales below. Plumbing is generally placed in interior frame walls. Electrical wiring is done on or recessed into the bales after the wall is up. Electrical outlet boxes are attached to wooden stakes driven into the bales. Preassembled window and door frames are set in place as the walls are constructed.

Straw bale wall

The frames are then pinned to the surrounding bales with dowels to hold them in place before the stucco is applied.

When the walls are finished and have completely compressed under the weight of the finished roof and ceiling for six weeks or so, the bales are wrapped with stucco mesh or chicken wire. Stucco is then troweled onto the wire, coating the walls inside and out. This forms a bug-proof, fire-resistant envelope around the straw and securely attaches the electrical fixtures to the wall.

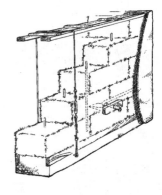

A large roof overhang and gutters help keep rain off the exterior walls and lessen the possibility of moisture migration. Straw bales can take some moisture on the exposed bale-ends, since the gaps between the straws are too big for capillary action to carry moisture into the bale. Of course, painting the stucco can go a long way toward keeping the bales dry. Water entering from the top of the wall would definitely cause rot problems since it has no way to get out. If the bales are kept dry (less than 14% moisture) before and during installation, the building will last for many decades. The Nebraska straw bale buildings still standing today have outlasted many of their wood-frame counterparts.

In load-bearing straw bale construction, a rigid collar or plate is built on top of the straw bale walls and tied to the foundation. The roof rafters or trusses are then attached to this collar. The non–load-bearing option for straw bale construction (currently more acceptable to conservative building code officials), involves supporting the roof on a framework of vertical posts and horizontal beams (wood, masonry, or metal) and using the bales strictly as a thick, insulative infill to complete the wall.

A straw bale wall built with three-string bales (tight construction grade bales are typically 16" x 23" x 46") will produce a super-insulated wall with an R-50 insulation rating. That's two to three times better insulation than most kinds of new construction, without the environmental hazards of formaldehyde-laced fiberglass. Straw bale walls are highly fire-resistant because, with stucco on each side of the wall, there is too little oxygen left inside to support combustion. Two-string bales can be used for construction if they are sufficiently compacted. They are smaller in size, have a lower insulation rating, and do not have the structural strength of three-string bales. Therefore, good, tight, three-string construction-grade bales are preferable.

Straw bale construction is quick and inexpensive, and results in a highly energy-efficient structure. The technique is easy to learn, and the wall system for a whole building can be stacked in just a day or two. This construction method also allows for some interesting building shapes. To change the shape of a wall or put a friendly looking radius on a sharp corner, use a weed whacker to trim the bales to the desired shape. Bales can also be bent in the center to form curved walls, and interesting insets and artwork can easily be carved into the straw before the stucco is applied. The two-foot-thick walls leave room for attractive window seats like those found in old stone castles in Europe. A straw knife (or hay saw) is another handy tool for shaping bales. This antique farm tool can sometimes be found at auctions or secondhand stores. Just look for something that looks like a very large machete with oversized saw teeth cut into the blade.

Engineering testing, already completed in Arizona and currently underway in

New Mexico, coupled with historic evidence in other parts of the U.S., as well as in Mexico, Canada, Finland, and France, provides building officials with affirming information regarding the safety and structural integrity of straw bale buildings.

If the current trend continues, it won't be long before construction-grade straw bales appear at your local lumberyard. If you want to save money, reduce your heating and cooling costs, and help prevent the current overcutting that is destroying our nation's forests and watersheds, straw bale construction is definitely the way to go.

— Ross Burkhardt

NONTOXIC BUILDING MATERIALS

Once you have decided what type of home you are going to build, you still have to decide on what materials to use for your wall coverings, floor coverings, insulation, framing members, and more. It is important to make choices that promote the health of you and your family.

In the past we have used many materials that have proven harmful, leading to maladies collectively known as "toxic house syndrome" or "sick building syndrome." Some products, such as lead paint, lead piping, and asbestos, have been taken off the market and removed from our houses (at great expense in terms of human suffering as well as money). Many more still being used today are on a growing list of materials associated with potential health risks. Some examples include fiberglass, as well as the formaldehyde-based adhesives used in certain kinds of carpeting, paints, and floor finishes. We are told that the potential health risks of these synthetic materials are being studied, but we tend to believe that "studying" is often only a time-delay tactic that allows manufacturers to dump their inventories on unsuspecting buyers before their products' true risks are acknowledged.

A growing movement in reaction to synthetics of all kinds may, however, be painting this complex subject with too broad a brush; energy costs and other issues must also be considered in selecting materials. Environmentally correct paints imported from Europe may not be a good energy bargain because of the environmental cost of shipping them overseas. In striving for better, safer materials to use in our homes, we must still retain some sense of perspective, and realize that the rush to market of a faddish natural product may do as much damage as a well-conceived artificial product.

It is time for us to look into the future and choose materials that provide a lifetime's worth of service without introducing a potential health risk. Most of these materials also have less of an effect on the environment, as many are recycled or natural products. Even if the cost is a little higher, investments made in our health must be of the highest priority.

We have rounded up some books and resource guides that will help you make healthy choices when selecting your building materials.

— Real Goods tech staff

PASSIVE SOLAR

Efficient House Sourcebook

Robert Sardinsky & the Rocky Mountain Institute. This directory lists and critiques, complete with addresses, prices, and contact names, the periodicals, books, schools, organizations, and agencies that deal with all aspects of resource-efficient house design, construction, retrofit, and much more. "Most of the entries are excerpted deftly enough to be considered information sources themselves. You'd have to subscribe to a truckload of periodicals to keep up with what's presented here in one book."—J. Baldwin, *Whole Earth Review*. 165 pages, paperback, 1992.

80-108   Efficient House Sourcebook $17⁰⁰

Passive Solar Energy

Bruce Anderson & Malcolm Wells. This easy-to-read sourcebook, written by two of the most knowledgeable experts on passive solar design, provides many practical ideas for energy conservation and for adapting existing houses to passive solar heating. There hasn't been a book in print on passive solar energy for almost five years. This valuable resource will help you design your home to minimize heating and cooling loads. The book discusses siting for passive solar, how solar chimneys work, and how the sun can also cool. Building costs are covered and a resource list is included. Excellent for anyone planning their new efficient home or addition. 160 pages, paperback, 1994.

80-196   Passive Solar Energy $25⁰⁰

Guide to Resource Efficient Building Elements

Published by the non-profit Center for Resourceful Building Technology in Missoula, Montana, this is **the** reference on earth-friendly building products. The listings are logically grouped by building element, such as foundations, framing, sheathing, roofing, doors and windows, interior finishes, etc. Over 300 product types are covered from manufacturers throughout the US and Canada. Company contact information is given, plus a description of the product, a brief explanation of applications, and pertinent information on available sizes, colors, fire ratings, and environmental features. The Guide is particularly strong on engineered wood products, use of reconstituted paper and agricultural fibers, recycled-content materials, and innovative new materials that are produced a resource-efficient manner.

This is a must for builder, owner-builders, and designers interested in low-impact building. This is the newest 5th Edition, published in Sept. 1995. 114 pages, paperback.

80-325   Guide to Resource Efficient Building $25⁰⁰

The Sunshine Revolution

Harald Rostvik. Norwegian architect Harald Rostvik is one of Europe's foremost experts on passive solar design. His new book provides an excellent overview of solar technology today and examines world politics for solar energy policy. He discusses solar architecture, solar cars, boats and planes, solar cooling and heating, solar cooking, desalination, and crop drying around the world. This is a valuable resource for environmentalists, consumers, investors, professional designers, architects, engineers, and educators.

"This is the most fascinating, fact-filled, colorfully illustrated book on solar energy that I've seen in my entire life." —John Schaeffer, President, Real Goods. "This is the most comprehensive introduction to solar energy I've ever encountered. The handsome book is from Norway, giving a refreshing international viewpoint to U.S. readers numbed by the images and half-truths tiresomely repeated in so many of our publications. Any fool can make solar schemes work in Arizona; Scandinavians have to be sharp. The presentation is thorough and easily understood. It should be especially useful to school teachers and others who need to be confidently knowledgeable in basic solar energy matters—politics and all." —J. Baldwin.

300 colorful, incredible illustrations. 188 pages, hardback, 1992.

80-190	The Sunshine Revolution	$39⁰⁰

ECCO House

The ECCO House plans deliver dramatic energy efficiency in an elegant 480– square-foot home. This compact two bedroom, one bath home feels a lot larger than its square footage suggests. It is small but not spartan. Since perception of space expands as lines of sight lengthen, ECCO House achieves spaciousness via disappearing walls, high light sources and an 18-foot ceiling. The living room opens onto a private courtyard with a slide-wire canopy that shades from the summer sun or lets in winter sun. The master suite includes a whirlpool bath. Nestled beneath the staircase is a single-drum washer/dryer. The energy and space-efficient ECCO House is ideal for adding a second unit, or for starting out small, or for downsizing your lifestyle when the kids are grown. The price of the plans includes a telephone design consultation followed by sketches for customizing your home that result. The house can be built on any type of foundation and can be built from kit panels or from scratch. Designed by a leading passive solar innovator, this custom home includes the amenities of an elegant lifestyle but with all the benefits of a low-cost, low maintenance living space. The video describes the evolution of the ECCO House. Designer Tim Maloney explains the details of the construction and points out the space saving features. If you are considering an ECCO House, this is an excellent first-hand tour. 30 minutes.

92-211	ECCO House Plans (Set of 3 plans) 24" x 36"	$395⁰⁰
92-212	ECCO House Video	$34⁰⁰

Building With Straw, Vol. I: A Straw Bale Workshop

Learn all the basics of straw bale construction in this video of a weekend workshop. Follow along as a group of novices gains hands-on experience constructing a two-story post and beam and bale addition to a century old lodge. As the weekend unfolds you'll see and hear how to design to maximize the solar aspects of your site, build a straw bale wall from the foundation up, then finish off your walls with a simple stucco coating. Includes slide presentation showcasing a variety of straw bale structures, and a printed insert detailing cost factors. 73 minutes.

80-237	Building with Straw, Vol. I	$30⁰⁰

Building With Straw, Vol. II: A Straw Bale Home Tour

What does a straw bale house look like? See for yourself in this video, which takes you on a tour of ten straw bale structures, ranging from a simple owner-built home costing $7.50 per square foot to a custom bank-financed home costing $100.00 per square foot. Straw bale construction, a technique developed nearly a century ago in Nebraska, is currently experiencing a world-wide renaissance centered in the American Southwest. In this video you'll hear the personal insights and hindsights of ten modern-day pioneers in New Mexico and Arizona who have built their dream homes out of straw. Learn from their experiences and get ideas for your own straw bale home. 60 minutes.

80-238	Building with Straw, Vol. II	$30⁰⁰

STRAW BALE

Real Goods Independent Living Book
The Straw Bale House

This fascinating and useful book describes the exceptionally durable and inexpensive method of plastered straw bale construction. Whether building an entire house or a more modest space such as a home office, building with straw bales is easy to learn and can be more time, cost, and energy-efficient than traditional construction methods. Benefits include:

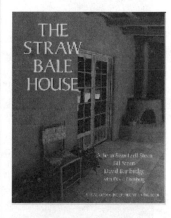

• Super-insulation, with R-values as high as R-57.

• Good indoor air quality and noise reduction.

• Speedy construction processes (walls can be erected in a single weekend).

• Low construction costs, as little as $10 per square foot (depending on owner involvement).

• Use of a natural and abundant renewable resource that can be grown sustainably in one season.

• Reduction of air pollutants created by burning agricultural waste straw.

This valuable book describes these benefits in an understandable and interesting way and is a beneficial resource for both those ready to build quiet, comfortable spaces with straw and those just exploring the idea. 336 pages, paperback, 1994.

80-248 The Straw Bale House $30⁰⁰

How To Build Your Elegant Home With Straw Bales – Video & Manual Package

Here is the best straw bale building how-to package available at any price. The 90-minute video is professionally done. It's informative and entertaining. The narrator has a great ability to clarify unfamiliar techniques and simplify the complex. A single building is carefully documented from site layout through move-in. Many innovative and labor-saving building tricks are revealed. The folks who put this package together have built over 20 straw bale structures over the past 7 years, and have distilled much of their hard-won knowledge into a product anyone can enjoy. All types of bale construction are covered, as well as intelligent site orientation, climate-specific design considerations, flooring options, utilities, and even a bit on solar systems, composting toilets, and graywater use. The 62-page companion manual is a great idea that improves the tape utility. Everything in the tape is also covered in the manual for quick on-site review. This ensures you don't forget anything important as you're building. Includes plenty of simple drawings. For simple, inexpensive, elegant, do-it-yourself building technology you can't beat straw bale, or the presentation of this inspired package.

80-095 How-To Video/Manual $79⁹⁵

Straw Knife

An easy way to notch bales to fit between post-and-beam construction is with this tool. A great tool for a wide variety of bale forming and fitting. Less expensive than a noisy, dust-producing chainsaw.

54-315 Straw Knife $79⁹⁵

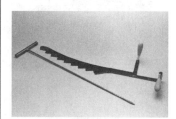

Straw Needle & Knife

Straw Bale Needle

This is the tool you need to make your own non-standard bales. Use it to make half bales at the ends of every other row or those odd-length bales needed when you set a window into your straw wall. Since the needle is 3 feet long it can be used to penetrate bales horizontally for pulling ties through to anchor chicken wire to each side of the wall.

54-316 Straw Bale Needle $19⁹⁵

YURTS

Pacific Yurts

The yurt is an architectural wonder, a legendary dwelling that has been in continuous use for centuries. Invented by the nomadic Mongols of Siberia, this circular, domed abode is as near-perfect a blend of beauty, simplicity, and functionality as any human habitation throughout history.

The Pacific Yurt is ideal as either a recreation retreat or a year-round residence for "living lightly on the Earth." It may be kept simple or may feature modern amenities such as plumbing, electricity, and multilevel deck systems. Other uses for the yurt include a hot tub/spa enclosure, a workshop/studio, a resort/conference center, a ski hut, a remote base camp, or temporary housing for an owner-builder.

The easily ventilated yurt is well adapted to warmer climates or the Alaskan North Slope. It can be insulated and equipped with a heater or woodstove. Even in extreme cold, it heats efficiently and comfortably. Naturally strong, it can be reinforced to withstand high winds and heavy snow.

The low cost per square foot makes the Pacific Yurt an outstanding value and an economical alternative to higher-priced standard frame structures. You can easily transport your yurt in a small pickup, then set it up quickly, virtually anywhere, for a comfortable stay. You can take it home, or leave it up permanently. These yurts are surprisingly easy to erect. The 30-foot yurt takes two people less than a day to set up, and the 12-foot model only a few hours. Materials are of the finest quality available, including a center ring of cross-laminated, kiln-dried fir, select fir rafters, galvanized steel tension cable, electronically bonded, vinyl-laminated polyester top cover, and large clear vinyl windows with screens. All top covers come with a 5-year pro-rated warranty, and the custom DuroLast top covers come with a 15-year warranty. Many custom options are available, including extra windows, solar skylight arc, insulation, and extra doors.

Once you've tasted yurt life, you may conclude that the human race made a mistake in abandoning the old nomadic ways.

Pacific Yurts

•12'D Yurt	115 Sq. Ft. 8'H 350# wt.	$2,395
•14'D Yurt	155 Sq. Ft. 8'9"H 450# wt.	$3,140
•16'D Yurt	200 Sq. Ft. 9'3"H 550# wt.	$3,595
•20'D Yurt	314 Sq. Ft. 10'H 700# wt.	$4,690
•24'D Yurt	452 Sq. Ft. 11'6"H 900# wt.	$5,450
•30'D Yurt	706 Sq. Ft. 13'H 1200# wt.	$6,990

Shipping and packing are extra. Allow 4 to 6 weeks for delivery. All yurts are shipped freight collect FOB Oregon. Do not send money to us; we will refer you directly to the manufacturer.
Send SASE for color brochure and prices on options.

Fast Framer

Build A Frame — In A Flash

Here's the easiest way to build a storage shed, vehicle cover, playhouse, or greenhouse. Just slip straight-cut lumber into the framing connectors; they shape the structure for you and eliminate the need for difficult angled cuts. All you'll need is a hammer, saw, screws, and nails. (Once your frame is up, you can use plastic, fiberglass, plywood, metal roofing, or canvas to cover the frame). The plastic E-Z UP 2' x 2' kit contains 15-pieces (9 angles and 6 feet) and will frame a structure up to 6' L x 6' H x 7' W. The galvanized steel Fast Framer framing 2' x 4' kit contains 36 pieces (24 angles and 12 feet) and will make a 7' L x 8' H x 8' W structure. Illustrated instructions included. Kits can be joined end-to-end for longer structures.

E-Z UP Framer

54-514 Fast Framer	$59⁹⁵
54-515 EZ-UP Framing Kit	$29⁹⁵

TIPIS

Original American Dream Home

A durable tipi provides a warm and dry home even in damp coastal rain forests, heavy winds, snow, and sub-zero temperatures. Handcrafted in Oregon, this modern alternative shelter is the earliest in a long line of historic North American dwellings. Based on traditional Sioux designs, this updated version is large enough to house up to three full-time occupants or ten to twelve short-term campers. Tipi cover and liner, essential for insulation and creating draft for the interior campfire, are made of tightly-woven, weather-resistant, 100% cotton duck canvas. Special attention to every detail means it's constructed to last: water-repellent fabric is treated to resist mold, mildew, and fungi; stitched with rot-proof, industrial-strength Dacron thread; lacing pin holes are reinforced with sturdy canvas webbing; no metal grommets to rip fabric. Complete bundles of hand-peeled, 24-foot tipi poles are available. All poles are carefully thinned from forests to allow Ponderosa pine to flourish. Set-up requires 17 tipi poles. Complete your tipi package with door

Nomadic Tipi

cover and ground cover. Includes 25-page set-up instruction booklet. The Nomadic Tipi provides plenty of comfortable living space. 18 ft. diameter, 243 square feet of floor area, and over 14 feet of head room in the center. Makes a great backyard guest house or weekend getaway lodging. Allow 2-3 weeks for delivery. Set-up package includes lacing pins, stakes, and all the rope you need to set up your tipi. Call before ordering and we'll send you a color sheet with six different handpainted designs

- •92-205   Nomadic Tipi, 18',
  Cover of 13 oz. marine finish cotton duck $549⁰⁰
- •92-206   Nomadic Tipi, 6', Liner of 10 oz. marine finish
  cotton duck $229⁰⁰
- •92-207   Door Cover $29⁹⁵
- •92-208   Ground Cover, 20 oz. Vinyl/Nylon $179⁰⁰
- •92-209   Set of 17 Poles, 24 ft. long (shipped freight collect) $229⁰⁰
- •92-213   Set-up Package $59⁰⁰

SHELTER SYSTEMS

Options for LightHouses

Mosquito Net Doors: No-see-um netting is used in addition to fabric.

Floor: Made of a material similar to the blue rip-stop tarps you may be familiar with but with fire retardant and of a higher grade.

Sun Shade: 6-feet by 12-feet black and silver rip-stop with four grip clips.

Liner: A duplicate of the outer covering. It comes with fasteners and cords attached, creating a dead air space between the layers.

Porch: 5-1/2-feet by 5-1/2 feet arched square, it is tied to the main domes with two poles supporting the outer corners.

Net Wall: Side wall panels are opened out functioning as awnings, exposing the panels of netting underneath. These net panels must be installed at the time of purchase.

The LightHouse

The LightHouse sidewalls are 5-1/2 ounce polyester canvas, which is watertight, flame retardant, breathable, resistant to sun degradation, and will not rot or mildew. Its color is light tan for good reflectivity and interior brightness. The translucent skylight above the sidewalls is constructed of a woven, rip-stop, UV-resistant film. It creates a pleasing interior light and opens to provide an excellent fresh air vent that is rainproof. A sunshade of silver and black rip-stop covers the top of the dome to block and reflect the sun. The frame is PVC tubing for strength and longevity. Patented Lexan clips connect the LightHouse's covering without puncturing or weakening. Each panel of the LightHouse is shingled over the next so that the dome breathes and remains leak proof.

The LightHouse 18 and LightHouse 14, 18 and 14 feet in diameter, have four tipi-style doors spaced evenly around the dome. The LightHouse 10 is 10 feet in diameter has one door. There are no zippers to fumble with or break. Clear vinyl windows above the doors let you see out in all directions. In tropical weather the sidewalls can be rolled up to provide unsurpassed ventilation. There's always plenty of light and fresh air in the LightHouse. The LightHouse comes complete with stakes, guy lines, vent tubes, spare parts, and an instruction booklet that details floors, site selection, anchoring, cooling, winterizing, and stove installation.

•92-201	LightHouse 18	$725.00
•92-256	LightHouse 18 Floor	$99.00
•92-253	LightHouse 18 Winter Liner	$639.00
•92-259	LightHouse 18 Net Wall	$99.00
•92-202	LightHouse 14	$639.00
•92-255	LightHouse 14 Floor	$79.00
•92-210	LightHouse 14 Winter Liner	$525.00
•92-258	LightHouse 14 Net Wall	$79.00
•92-203	LightHouse 10	$359.00
•92-254	LightHouse 10 Floor	$49.00
•92-257	LightHouse 10 Winter Liner	$279.00
•92-252	LightHouse Mosquito Door	$14.95
•92-251	LightHouse Porch	$59.95
•92-204	LightHouse Sun Shade	$29.95

Shipping to Alaska & Hawaii:

Size	1st class mail	UPS 2nd day air
10'	$45.00	$60.00
14'	$60.00	$75.00
18'	$85.00	$105.00

Shelter Systems Solar-Dome

Many greenhouses are too big and too expensive for the average gardener, but not the Solar-Dome. Vegetables and flowers receive the growing conditions they need: shelter from cold, rain, wind, frost, and birds, with plenty of room for the gardener to work standing up, store tools, build up flats, hang potted plants, or care for mature plants. The Solar-Dome sets up in 20 minutes without tools. All hubs are factory attached and all poles are interchangeable. Just insert them into the connectors and it's up. Turn or move the dome into any desired position or lift to wherever it's needed most in the garden plot. If you're not using your Solar-Dome for one season or more, just take out the poles, roll the greenhouse up, and store it in a closet or on a shelf.

The Solar-Dome is built of a super-strong woven rip-stop greenhouse covering. It is treated with ultra-violet inhibitors, designed specifically for long sun exposure in greenhouse use. The filtered light that comes through is excellent for plants and will never burn leaves like glass or clear vinyl will. It transmits 90% of available light. The clips that join the PVC poles at each hub have a patented design. They're made of an especially strong plastic called Lexan. They grip the canopy, providing greater strength than sewn seams or other grommets.

The ventilation tubes ensure plenty of fresh air with the doors open or hooked closed. Also included are stakes, hooks for hanging potted plants, and an instruction manual. There are no zippers in the Solar-Dome. Simple hook closures have proved to be the best. Carts and wheelbarrows can be wheeled in and out, as the Solar-Dome's doors open completely along the ground.

•92-245	Solar-Dome 20 (20'D x 10'H)	$859⁰⁰
•92-246	Solar-Dome 18 (18'D x 9'H)	$679⁰⁰
•92-247	Solar-Dome 14 (14'D x 7'H)	$575⁰⁰
•92-248	Solar-Dome 11 (11'D x 6'4"H)	$325⁰⁰
•92-249	Solar-Dome 8 (8'D x 7'4"H)	$175⁰⁰

Shipping to Alaska & Hawaii:

Size	1st class mail	UPS 2nd day air
8', 11'	$45⁰⁰	$55⁰⁰
14'	$60⁰⁰	$75⁰⁰
18'	$85⁰⁰	$105⁰⁰
20'	$110⁰⁰	$130⁰⁰

SKYLIGHTS

SunPipe Skylights:

Efficient and easy-to-install, SunPipe adds beautiful, natural daylight to your kitchen, hallway, bathroom, office, or studio. SunPipe sends light from above your roof, through the attic, to a diffusing dome in the ceiling. Super-reflective lining inside the pipe insures maximum light transfer. Illumination on sunny days is equivalent to 900+ watts of incandescent lighting; 100 to 500+ watts on cloudy day. A SunPipe, properly installed, doesn't leak. The SunPipe transmits no direct sun rays, so there's no solar heat gain, nor color fading or bleaching. The 13-inch diameter aluminum SunPipe requires no cutting of joists or rafters, no framing, taping, drywall, or finishing. Typical skylight wells limit the spread of light into a room. The SunPipe diffuses the light uniformly, floor to ceiling; it can light up a whole room. Installation takes only 3 hours on average. The kits include a 4-foot pipe, clear top dome, white bottom dome, sealing materials, hardware, and installation guide. Two and four foot extensions may be stovepiped together to form the required height between ceiling and roof. (Measure from top of ceiling to the position on the roof where the pipe will be located, and add a minimum of 10 inches, or up to 30 inches if a longer pipe will capture more direct sunlight.) The optional SunScoop may be attached to the top of the dome, facing south, to capture up to 170% more sunlight during the winter . In choosing your SunPipe, use the 13-inch diameter kit (includes flashing) for 8 to 9 foot ceilings; the 21-inch kit (does not include flashing) for 10 to 20 foot ceilings.

•63-180	SunPipe Kit 13"	$349⁹⁵
•63-181	SunPipe 13 Extension 2'	$79⁹⁵
•63-182	SunPipe 13 Extension 4'	$149⁹⁵
•63-183	SunScoop 13"	$49⁹⁵
•63-184	SunPipe Kit 21"	$689⁹⁵
•63-185	SunPipe 21 Extension 2'	$159⁹⁵
•63-186	SunPipe 21 Extension 4'	$259⁹⁵
•63-187	SunScoop 21"	$99⁹⁵

BOOKS/VIDEOS

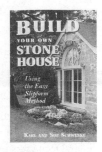

Build Your Own Stone House

Karl & Sue Schwenke. A concise explanation of slip-form construction, the easiest way for novice builders to make sound, tall stone walls. This readable book follows the authors as they build their own home, with the idiosyncracy, anecdotes, and personality that result from such a project. 164 pages, paperback, 1991.

80-147	Build Your Own Stone House	$12⁰⁰

Building Stone Walls

By John Vivian. This concise and elegant little book tells all you need to know to build a monument to yourself. The simple approach, detailed pictures and the emphasis on safety and forethought set a fitting tone for the monumental task. Wall-making is a contemplative as well as a physical endeavor, and the author captures the spirit and passes it along admirably — a classic little book. 105 pages, paperback, 1993.

80-180	Building Stone Walls	$9⁰⁰

Cordwood Masonry Video

Earthwood Building School's Rob and Jaki Roy have built enough cordwood masonry houses to determine the best use of materials, encounter the unexpected, and learn short-cuts and construction tricks. Refreshingly rough around the edges, this substantative video provides solid, clear, and simple step-by-step instruction to enable the inexperienced or professional to create a dream house of their own. The authors cover foundations, wall construction, natural insulation, window and door placement, mortar mixing, wood selection and drying techniques, material estimating, and more. Environmentally sensitive, enduring, and remarkably inexpensive, owner-built housing is within reach of millions of people thanks to solid help like this instructional video. 88 minutes.

80-096		$40⁰⁰

Building With Stone

Charles McRaven. Nothing can rival stone for its beauty and durability. *Building with Stone* is an introduction to the art and craft of creating stone structures. The book covers acquiring stone, the needed tools, mortaring, and laying stone. It explains the construction of walls, wells, arches, barbecue pits, fireplaces, root cellars, dams, and homes. An excellent guide for anyone working with stone. 192 pages, paperback, 1989.

80-213   **Building with Stone** $15⁰⁰

Low-Cost Pole Building Construction

Ralph Wolfe. Pole construction is a resource-efficient way to enclose space. After a brisk discussion of principles, from weak soils and termites to wind loading and what to buy before the pole-raising party (beer and work gloves), nearly half the book is devoted to examples of pole buildings and sample plans for a barn, a storage shed, a woodshed, a garage, a house, and three cabins. With building experience and this book, you could build your own. Clear illustrations, plans, and abundant tables. 178 pages, paperback, 1993.

80-183   **Low-Cost Pole Building Construction** $13⁰⁰

Underground Buildings

Malcolm Wells. An architect's sketchbook of 26 years of work. This is a delightful, irreverent, and very personal account of the struggle against the current of the architectural mainstream. Wells includes hundreds of sketches to illustrate the potential, the successes, and the failures of solar energy efficient, underground buildings. Handwritten and easy-to-read. 200 pages, spiralbound, 1990.

80-146   **Underground Buildings** $15⁰⁰

How To Build An Underground House

Malcolm Wells. His fourth book about underground architecture. He has followed and pioneered underground architecture since 1964, his earlier books having sold 120,000 copies. The United States only boasts 3,000-4,000 earth dwellings due to inertia. This is a very elegantly and simply written book that talks about not only the underground house, but the appliances and energy systems needed to sustain it. Wonderful graphics, drawings, and plans, all hand-done in Wells' inimicable style. A must for undergrounders! 96 pages, paperback, 1991.

80-155   **How to Build an Underground House** $12⁰⁰

Sheds

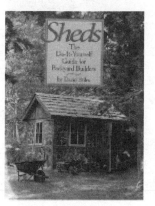

Both novice and skilled weekend carpenters will appreciate award-winning designer David Stiless' inspired and practical advice on how to build a professional-looking and aesthetically pleasing 8' x 10' tool shed out of environmentally safe materials. Includes step-by-step illustrated instructions and suggested daily labor schedules. Instructions and building designs for alternative backyard structures — like a child's play shed and a pool pavilion — are also included. Illustrated with drawings and color photos. 142 pages, paperback, 1993.

80-326   Sheds $18⁰⁰

Earthship

Professionally produced video by Dennis Weaver of his independent sustainable living space, Earthship. Narrated with the same enthusiasm and down-home twang he brings to those Great Western Savings commercials. The video is a solid prospectus for a responsible way to build, using surplus material — tires and cans. The book is a builder's guide for an innovative construction technique integrating recycled waste with conventional building materials. The primary construction material is used tires, a post-consumer disposal nightmare, which can be recycled into a building, free. A single tire filled with tamped earth weighs 300 pounds and is stable, immovable, and cheap: the houses in the book cost from $20 per square foot. Interior walls made with adobe-covered cans are strong, light, and can be sculpted to any shape — an exciting concept. Just as small problems lead to big problems, so do small solutions lead to big solutions. Weaver grinds his favorite axe — the greenhouse effect — and emphasizes the 3 Rs: Reduce, Reuse, Recycle. The video is approximately 35 minutes. The companion books, *Earthship*, Volume 1 (230 pages), Volume 2 (260 pages), and Volume 3 (257 pages), by Michael Reynolds, are comprehensive how-to manuals showing how to use these new building materials. They provide detailed information from design to completion. Vol. 1 shows how to build an Earthship, and is an excellent introduction. Vol. 2 explains interior details and integrates water, waste, lighting, heating and electrical systems, and includes a chapter about getting permits. Vol. 3 adds structural and mechanical details, and provides a vision of community and urban planning concepts.

80-117	Earthship Video	$29⁹⁵
80-118	Earthship, Vol. 1	$29⁹⁵
80-131	Earthship, Vol. 2	$29⁹⁵
80-245	Earthship, Vol. 3	$29⁹⁵

Root Cellaring

Root cellaring, as explained by Mike and Nancy Bubel, is a no-cost, low-technology, energy-saving way to keep the harvest fresh all year long. *Root Cellaring* tells you how to choose the fruits and vegetables that will store best; specific storage requirements for nearly 100 home garden crops; and how to build your own root cellar, indoors or out. Describes and illustrates actual root cellars around the country and includes excellent vegetable recipes. 281 pages, paperback. 1991

80-413	Root Cellaring	$13⁰⁰

The $50 & Up Underground House Book

Mike Oehler. How to build a comfortable underground home for as little as $50 to $15,000, written by a maverick builder. This valuable fifth edition contains many practical tips for designing energy-efficient underground houses with views, natural light, and ventilation. Instructions include excavating by hand, constructing a built-in greenhouse, and purchasing inexpensive building materials. Because earth dwellings are tucked into the landscape, these attractive homes are well suited to most natural terrains. Whatever your building plans, this lively book about low-cost construction is a great buy! 115 pages, paperback, 1992.

80-320	The $50 & Up Underground House Book	$14⁰⁰

New Compact House Designs
27 Award-Winning Plans — 1,250 Square Feet Or Less

In the future, the ideal home will be smaller, more energy-efficient and better suited to its environment. An excellent sourcebook for both current and prospective homeowners, this hands-on guide showcases the winning designs in a competition that spanned the continent. Each building plan is for a single family dwelling with at least two bedrooms and a gross floor area of 1,250 square feet or less. The designs reflect a broad spectrum of geographic and stylistic diversity, including a New England farmhouse, a Southwestern hacienda, an Elizabethan cottage and a postmodern home. Compiled by architect Don Metz, each entry features a detailed project description with judges' comments, site and floor plans, elevation and section drawings. 188 pages, paperback, 1991.

80-089	New Compact House Designs	$18⁰⁰

Alternative Housebuilding

Mike McClintock. Beautiful duotone illustrations carefully detail the planning and construction of log, timber-frame, pole, cordwood-stone-earth masonry, and earth-sheltered houses. Full of ideas, step-by-step construction advice, and elegant examples that can help raise the level of owner-built homes to architecture. The author, syndicated how-to columnist for the *Washington Post*, provides sound techniques, excellent source listings. 367 pages, paperback, 1989.

80-119	Alternative Housebuilding	$20⁰⁰

The Self-Build Book
How To Enjoy Designing And
Building Your Own Home

Architects Jon Broome and Brian
Richardson combine inspiration
and practical information on
putting theory into practice,
making environmentally friendly
choices on building materials and
energy conservation, and on
organizing the project beyond the
building process, including land, financing, permission, and
professional help—and portrays detailed insights into some
of the most useful building methods for self builders. The
revision of this British classic is particularly useful to self-
builders worldwide, because it revisits real projects to see
how the passage of time has affected the attitudes of the
original builders. The book will appeal to forward-thinking
builders (and architects) who want exposure to techniques
not yet commonly practiced in America. This book truly is
(in the words of one reviewer) "a celebration of the
unconventional." 271 pages, paperback, 1995.

80-062   The Self-Build Book $30⁰⁰

Dwelling On Earth

David Easton. This is the best book
on rammed earth we have found.
Rammed earth construction links the
purity of natural materials with the
beauty of indigenous design. Earth
walls have the thermal mass you need
for your passive solar home. David
Easton, an innovator, designer, and
builder of thick earth walls, expresses
his reverence for nature-based architecture in this practical
guide to earthbuilding techniques. Ancient methods of
compacting soil were used to construct the extraordinary
multi-storied cliff dwellings by the Anasazi Indians in the
Southwest nearly 2,000 years ago. Modern earthbuilders
combine traditional and contemporary techniques to create
stable, aesthetically pleasing earthen walls. This
comprehensive manual describes both classic and up-to-date
applications for rammed earth construction. It covers
foundation and site preparation, form building, and
compacting the earth either by hand tamping or machine.
Handsome, evocative illustrations both instruct and inspire.
Indispensable for those seriously interested in rammed earth
works. 115 pages, spiralbound, 1991.

80-416   Dwelling on Earth $25⁰⁰

Energy Conservation In
Housing

David Meinhert. Provides designers,
builders and remodelers a collection
of useful tables, clear plans, and
well-thought-out ideas about the
efficient use of energy in the living
space. The book avoids the trap of
jargon in presenting designs that
efficiently manage heat loss and gain
in tightly constructed homes while maintaining an adequate
supply of fresh air. 150 pages, paperback, 1990.

80-143   Energy Conservation in Housing $14⁰⁰

Earth To Spirit: In Search
Of Natural Architecture

With the advent of modern
technology, an increasingly
sterile variety of architecture
has come to dominate our
urban and industrial
landscapes. In response to this
deadening trend, author David
Pearson has used his
background in housing, holistic
design, and environmental issues to rediscover architecture
that creatively integrates the human habitat with the natural
environment. With camera in hand, he's traveled the globe
in search of buildings that emphasize the beauty and freedom
of nature. His journey spans the history of architecture,
revealing ancient structures, traditional dwellings, and
contemporary buildings that display a sensitivity to the
innate ecological wisdom of the natural world. An inspiring
blend of images and ideas, his book teaches us how to bring
architecture into modern lives with landscapes in such a way
that land, home, and spirit may co-exist in harmony. 160
pages, paperback, 1995.

80-091   Earth to Spirit $18⁰⁰

The Natural House Book

David Pearson. This book modestly bills itself as "a comprehensive handbook to show how to turn any house or apartment into a sanctuary for enhancing your well-being," and delivers much of what it promises. The author is an architect, planner, and London-based eco- and healthy-housing consultant. This handsome, color-illustrated book borrows historical and ethological successes in shelter and applies them to modern housing design. It considers every aspect of healthy living space by analyzing the whole house — healthy as contrasted to dangerous — and the elements that go into houses, the living systems and materials used in constructing the space, and then considering each of the kinds of rooms we use for living, sleeping, cooking, bathing, exercising, and growing plants. This book succeeds because it takes a whole-systems approach. 288 pages, paperback, 1989.

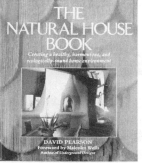

80-120   The Natural House Book $20⁰⁰

EcoRenovation

Edward Harland. The American edition of Edward Harland's wonderful *EcoRenovation* appears here for the first time. Originally published in Great Britain, this utilitarian book is perfect for homeowners who want to make their existing homes as "green" as possible. It considers the major home-related ecological concerns and issues one at a time, and makes practical recommendations about how to find the best solutions in your home. We especially like his focus on energy issues and nontoxic products. Every chapter ends with a list of priorities for action, where most of us will find valuable advice. Clear treatments of often-neglected considerations, and novel approaches like an excavated underground room, make this little book a gem. Well organized and illustrated. Co-published by Real Goods and Chelsea Green. 288 pages, paperback, 1994.

80-242   EcoRenovation $17⁰⁰

Building With Junk And Other Good Stuff

Jim Broadstreet. A master scrounger's guide to the millions of dollars of building supplies thrown away every day. For dedicated recyclers and builders on a budget this book covers every aspect of home building from floors to roofing and from bankers to building inspectors. Readable and informative. *The Whole Earth Catalog* says it well: "detailed, true, and full of wit." 162 pages, hardback, 1990.

80-142   Building with Junk $20⁰⁰

How To Plan, Contract, And Build Your Own Home

Richard M. Scutella & Dave Heberle. Building or buying a new home is likely to be the greatest single purchase in your life, and the comprehensive information in this book will help you to consider every detail. The authors cover every imaginable topic under the four major categories of what to build, how to build it, where to build it, and who should build it. Also discussed are some areas often neglected in other books, such as energy-efficient lighting, garage design, home safety and security, selecting major appliances, and current information on plumbing and wiring. 410 pages, paperback.

82-143   How to Plan, Contract & Build $20⁰⁰

Treehouses

Leave your earthbound perspective behind with this guide to the most ingenious branch of architecture. Builder-author Peter Nelson shows you how to design and build a treehouse — everything from how to pick the right tree to shingling the roof. Includes stories and full-color photographs of dozens of treehouses, ranging from kids' playhouses to artistic arboreal palaces. The centerpiece of this book is a well-illustrated photo essay of the construction of a spectacular octagonal treehouse poised 30 feet up in an old growth fir. Anyone who has ever built a treehouse, or dreamed of living out on a limb, will find this book irresistible. 128 pages, paperback, 1994.

80-331 Treehouses $20⁰⁰

A Real Goods Independent Living Book
The Rammed Earth House
Rediscovering The Most Ancient Building Material

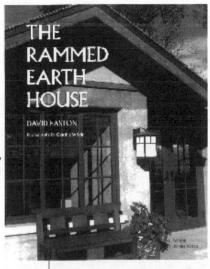

For more than ten thousand years humans have been using earth as a primary building material. Brick and adobe structures are familiar, of course, but now people in every part of the US are discovering the beauty and simplicity of an ancient earth-building technique: the use of stabilized rammed earth.

Rammed earth, as practiced today, involves mixing basic material on-site—earth, water, and a little cement—then tamping the mixture into wooden forms to create thick, sturdy masonry walls. This method of building creates a structure of timeless beauty and Old World charm, a home that offers its inhabitants a powerful sense of security and well-being. Earth-built homes have a permanence and solidity altogether lacking in so many of today's modular, pre-fab, stick-built houses.

Rammed earth houses have other advantages, too, including:

● Resource efficiency. A well-built earth house will last hundreds of years, which means that the natural materials needed for construction are used wisely and sustainably.

● Energy efficiency. Massive earth masonry walls retain solar heat in winter and cooler nighttime temperatures in summer, making rammed earth homes both comfortable and economical year-round.

● Design flexibility and aesthetics. Rammed earth walls allow for beautiful interior detailing. not only are the walls themselves attractive, but niches, nooks, doors, and windows are easy to incorporate, adding both style and functionality.

● Site-specific qualities. A rammed earth home is made of materials indigenous to the site. When sensitive design and construction are undertaken hand-in-hand with ecological landscaping of the environs, the rammed earth house achieves an interrelationship with its surroundings that only intensifies over time.

The Rammed Earth House offers clear and complete information for owner-builders, contractors, architects, and anyone interested in housing innovations that put into practice the philosophy of sustainability. Author David Easton is an experienced designer and builder of homes and commercial building, and one of the world's leading authorities on rammed earth construction. Complementing and enriching the book's text Cynthia Wright's photographs capture the sensuous character and abiding appeal of this beautiful and appropriate style of building. 224 pages, paperback, 1995.

80-063 The Rammed Earth House $30⁰⁰

CHAPTER 4

HARVESTING ENERGY

IT IS IMPOSSIBLE FOR most of us to envision life without electricity. In our daily lives we use prodigious amounts of it, wasting more than half of what we buy. In developing nations, a quarter of the average family's budget goes toward batteries and fuel for lighting and cooking, and with similar inefficiencies. Modern life thrives on electricity's cool, invisible current. As soon as we are sheltered from storms and the dog days of August, we want our tools and conveniences.

In our lifetimes, we can hope that most if not all households will optimize their ability to be energy self-sufficient. Since sunshine falls democratically on us all and can be harvested easily, our homes can be producers as well as consumers of energy.

If we choose to live beyond the fringes of powerlines without living like troglodytes, we must become our own little power companies and take responsibility for the whole electric enterprise: generating and storing it; transmitting it from source to demand; and distributing it among the loads, where it sheds light, performs work, amuses and instructs us, and powers all the myriad uses we have created in this electricity-infatuated era. Suppliers as well as consumers, off-the-gridders quickly come to grips with the finite energy budget within which we all must learn to live comfortably. Should we consume carelessly, exhausting our natural resources, we are lucky if we can resort to unpleasant measures like fossil-fueled generators or hauling batteries to town to make up the shortfall. On the self-sufficient homestead, even with careful energy management, there may be a few anxious weeks at the beginning of winter when the sun fails (the batteries are low), water is precious, and the laundry may be done less often. This is the most basic kind of demand-side management: electrical usage is matched to availability. Once the sky has burst, this same household may rejoice in so much hydroelectricity and water flowing in the nearby stream that lights and electric heaters are left burning night and day to keep from overcharging the batteries; this is also demand-side management, and we should all be so lucky!

Most home power systems use more than one source and are thus said to be *hybrid.* The homestead in the preceding paragraph used both photovoltaic (PV) and micro-hydro power, but likely combinations include two or more choices from the practical renewable source list: solar, micro-hydro, wind, and biomass. During dismal energy-producing times, most independently powered homes keep a fossil-fuel back-up generator as the unpleasant power source of last resort. Large public electricity providers similarly diversify their sources, but normally generate the bulk of their baseload power with fossil-fuel-fired generating plants. In contrast, the homestead power provider takes pride in firing up the gen-set rarely, if ever. This small miracle of conservation is accomplished by means of careful load management: buying energy-efficient appliances, using them efficiently, turning lights and phantom appliance loads off when not in use, and coordinating the use of heavy loads (washing machine, dishwasher, power tools, and well-pump) so that the system is not overwhelmed by demand. Home energy system "ratepayers" are seldom unaware of where their power comes from. We sacrifice nothing except nonchalance; most of us insist that living within our finite energy budget is a meaningful but simple challenge, one which we gladly accept.

What is found to be true for those living beyond the powerline and pavement also proves useful and true at home on the grid. In the next few years we may expect to see a pervasive movement toward domestic generation of electricity, wherein most houses can expect to harvest (using affordable, off-the-shelf technology) about as much electricity as they consume. An integral part of this change will be *net metering,* which means that the electricity we generate and that which we consume sells for the same price. If, at the end of the month when the meter is read, it has spun in the utility company's favor, we pay; if in our favor, they pay. Until conservation and independence are directly reinforced in this way, we will find it difficult to take the great and practical strides toward energy self-sufficiency that are available to us and necessary for the planet. In this scenario, the utilities (which literally become, in accordance with Amory Lovins' vision, power brokers) use the transcontinental grid and their existing transmission and distribution system to buffer supply and demand, and for this greatly valuable but technically feasible service they will be fairly compensated. This kind of integrated resource planning allows us, after nearly a century of dependence on electricity (and, inadvertently, on petrochemicals) to tap the simple, renewable, abundant, and democratically available resources first, saving fossil fuels for more urgent uses.

Whether as ratepayers of a utility company or as our own power providers, our best immediate source of energy is what has been called "the conservation powerplant." Even the best modern appliances waste nearly half the energy they consume through heat loss and poor design. Many of our smaller appliances are secret energy criminals, consuming electricity when we think they are turned off. In the self-powered home, though, electricity is too costly to waste, and so every effort is made to insulate, to stop air infiltration, to make heating and cooling appliances more efficient, to use the most thrifty form of refrigeration and domestic hot water

Over the long haul of life on this planet, it is the ecologists, and not the bookkeepers of business, who are the ultimate accountants.

—Stewart Udall

production, and to keep lights and phantom loads turned off when not in use. In conventional homes, the difference between thrifty and careless use may mean only a few dollars per month on the energy bill, and historically it has been difficult for us to be as careful as we should be. Yet, as we become more aware of the real costs of careless consumption — damaged forests and lakes, dirty air, reddened eyes, and raw throats — and as energy costs escalate, the point is more easily apprehended. Ben Franklin might as well have said, "a watt saved is a watt earned."

Sensible energy management also requires that we use appropriate power to satisfy our needs. Electricity is an exceedingly costly and inefficient way to heat things, although it is so safe, simple, and clean that it is the most practical power for such heating appliances as blow-dryers, irons, toasters, and laser printers. Heating space and water with electricity are unpardonable sins from a resource perspective. Americans are particularly offensive because they hoard large quantities of hot air and water. Our conventional water heaters maintain dozens of gallons of water at a piping hot temperature 24 hours a day, just in case someone needs a teaspoonful. This mindless indulgence costs the average household almost as much as space heating and cooling, and is particularly difficult to justify when we consider that solar domestic hot water systems are efficient and easy to install and operate. We should be getting at least four-fifths of our hot water from the sun, and the rest should come from instantaneous water heaters which heat only the water that is actually used. (Solar domestic hot water is treated more fully in Chapter 7.)

The sun is our primary and most generous source of power. Across the Sunbelt, the southern tier of states, sunshine falls at a rate of almost a kilowatt of energy per square meter. Photovoltaic modules, large transistor-like devices which first became available in the last quarter of this century, can convert up to 20% of that energy directly into electricity, and so a properly oriented square meter of roof exposed to full sun can typically harvest more than a kilowatt-hour of energy each day. The south-facing roof of an average 2,000–square-foot home can harvest up to 40 kilowatt-hours of energy each day, or twice the preconservation average home usage. Even in Alaska and along the northern tier of states, where cloudy days and lower sun intensities prevail, the average roof can power the average home, provided it is well designed and the roof is correctly oriented and unshaded (although this requires large storage capacity). Surprisingly, many of the most successful PV-powered homes are located in northern states. In the high Rockies, for instance, where winter days are often bright and where cool temperatures, altitude, and reflected light from snowfall amplify PV efficiency, independently powered homes often rely on solar electricity alone.

The sun and the Earth's seasonal precession drives the global weather engine, and at favorable homesites this engine can be harnessed and made to produce bountiful electricity. Water- and wind-generated electricity is often a perfect match for photovoltaic power, because when the sun is not shining it is often windy or wet. Hydroelectric power harvested from water falling in a mountain watercourse is the most favorable source, running reliably for months at a time. Because of its nonstop reliability, micro-hydro systems are designed with reduced battery storage, and are the least expensive form of home power available. Micro-hydro is clean, simple, and trouble-free. At many windy sites, usually where there is a steady wind regime with average wind speed exceeding ten miles per hour, wind power is attractive and practical at the homestead level. Wind's inconstancy and unpredictability often complicate its application, and so it represents a good second or third source for a hybrid system.

Other intriguing sources of energy, such as biomass, geothermal, and oceanic (including tidal, wave, and thermal), are either available under special circumstances or are currently being developed; but the bulk of renewable homestead energy is harvested from sun, water, and wind. Independent homes often take advantage of biomass, in the form of firewood, for domestic space heating. Use of this sustainable resource has become increasingly sophisticated as our knowledge of forest management and woodstove technology has improved.

This chapter is the heart of our *Sourcebook,* because harvesting energy is the single most important factor in achieving a comfortable independent lifestyle. Here and in the following chapters we tell all you need to know to inventory your energy needs, design and install your energy harvesting system, and commence living under your own power.

— *Michael Potts*

NANCY WASHBURN'S HOME ALMOND, WISCONSIN

Wisconsin winters are cold with moderate snowfall. The spring and fall seasons are cool, while the summers are quite warm with pleasantly cool nights. Our frost-free growing season is usually 90 days long. The sun shines a lot during most of the year, so solar energy is available to provide most of our energy needs. When Nancy was planning her home, she was concerned about where her energy was coming from: she did not like the fact that some of her power was coming from a nuclear power plant, and the rest was coming from fossil fuels. She did not want her new home to contribute to the proliferation of these technologies, so she chose to power her home with renewable energy. Luckily, even in 1980 there was a wealth of local expertise to help her design and build an energy-efficient home. Mark, the builder who helped Nancy design and build her house, also lived in an off-grid home, and his experience was invaluable.

The house is earth-bermed and of passive solar design, with large amounts of glazing on the south wall on both levels. The lower (main) level is almost completely bermed, with only the south wall exposed. The upper level has the roof on the north with a few windows east and west, and lots of glazing to the south. All windows are low-emissivity (low-E) glass filled with argon gas; they are also fitted with "window quilts" to reduce heat loss at night. The main floor is concrete. When the sun is out, the floor's thermal mass heats up, releasing the stored heat for use at night and during cloudy weather. Auxiliary heat is provided with an airtight woodburning stove.

Nancy moved into her new home at the end of January 1981. Her electrical system then consisted of two 6-volt EV (electric vehicle) batteries which powered

SPECIFICATIONS

System Voltage: 12 Volts DC
Charging Capacity:
  PV: 702 Watts
  Wind: 1000 Watts
Storage System:
  Batteries: Sixteen 6-Volt DC 220AH —
    Lead-Acid
Inverter: Heart 2800-12E6, 2800-Watt
Input: 12-Volt DC
Output: 240-Volt AC

Nancy Washburn on her catwalk

the home's 12-volt DC lighting and a radio. To charge the batteries, she hauled them to her sister's house miles away, where she recharged them with grid power. Lugging the heavy batteries around got old quickly, so during the first week of the following spring, she installed two Solarex 35-watt PV modules. "It was a real luxury not having to haul batteries anymore," Nancy recollected. This first system was manually controlled and poorly fused, but it worked! With only limited charging capacity, the batteries were subjected to deep discharges. It gets very cold up here, and the batteries were kept in an unheated area, so one time, when they were in a state of deep discharge, they actually froze.

Water comes from a 180-foot-deep well with a 3/4-horsepower 240-volt AC submersible pump down the hole. Three large storage tanks located on the second floor of her home provide gravity flow. At first, when the storage tanks needed filling, a generator was started manually to operate the well pump. The generator also charged the batteries during cloudy weather, but the battery charger Nancy had was small, and it took a lot of generator time to get the batteries charged.

Nancy was dissatisfied with this system, and another upgrade soon followed. Four Solarex 30-watt modules were added, along with four new 6-volt EV batteries. System controls were improved by the addition of a controller and a DC load center. And a closed loop solar water heater was also added to the home that year.

Nancy's first refrigerator was an interesting experiment: A well-insulated room was built into the northeast corner of the lower level. The room had an outside door, as well as a small door opening into the kitchen area, and was equipped with 15 55-gallon drums filled with water. The theory was that the drums of water would freeze during the winter with the outside door left open. Then, with the door closed during the warmer months, the room would remain cold all summer. Unfortunately, the room only stayed cold through May. There were also other troubles with this fridge, like barrels freezing solid, splitting open, and leaking water all over the place. In 1984 the barrels were removed and a 10–cubic-foot super-insulated box was built into the cold room using the existing kitchen access door. A 12-volt DC marine compressor keeps the box cold. This new load required more charging capacity, so four Solarex 38-watt modules were added at this time.

A couple of years later, more upgrades to the system were installed. A Heart split-phase inverter made AC available. This powered the well pump, a computer, a TV, and a vacuum cleaner. A custom-built freezer joined the fridge in the cold room. During this upgrade, a new PV Systems Company controller/load center was chosen to upgrade the system. This load center also upgraded the fusing and the metering components of the system. An automatic power transfer switch eliminated the need to manually switch loads over to the generator when it was operating. A more efficient, larger capacity battery charger was added at the same time.

In the next upgrade, four Solarex 60-watt modules joined the others already on the roof.

The early 1990s brought a larger Heart inverter, a Line Tamer, new batteries with HydroCaps, and two more Solarex 60-watt modules to accommodate a color TV, new computer, microwave, VCR, and washing machine. The HydroCaps are amazing: the batteries have required no watering since they were installed. The

The Washburn system

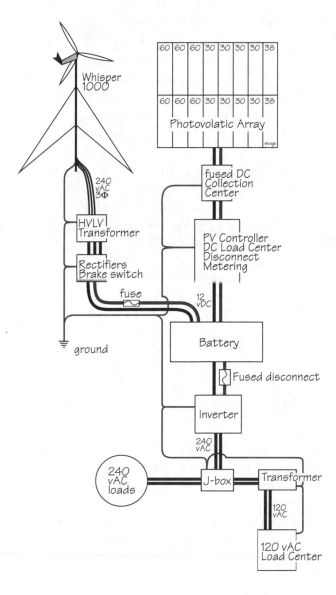

larger inverter allowed Nancy to run her AC loads simultaneously. Her original inverter would run the well pump alone, or the house loads alone, but not both at the same time.

While the gravity water system worked well, it did not create a satisfactory flow at the faucets or in the shower. Since the only water heater was solar, there was no hot water during cloudy periods, so the next upgrade was to the water system: two of the storage tanks were removed, and the other one was converted to operate as a pressure tank. An automatic pressure switch was installed on the pressure side to control the pump, making its operation completely automatic. With a pressurized water system in place, an Aquastar Model 80-LP on-demand water heater was added to the water system as a back-up to the solar heater. A space-heating loop was included in the plumbing of the water heater so that, by adjusting a thermostat, water can be pumped by a circulating pump through the Aquastar, which turns itself on, then to a fan convector located in the downstairs bathroom, which quickly heats the room. During periods when Nancy is gone and the woodstove cannot be stoked, this system keeps the whole home from freezing.

In Wisconsin, November and December are the cloudiest months of the year, and the days are short. The PV output is at its annual low, and so generator time is concentrated during these months. This time of year is also one of the windiest. To take advantage of this resource, and with thoughts of an electric car in the future, a Whisper 1000 HV wind machine is being installed on an 84-foot tilt-up tower near the home. The gasoline generator has been sold. According to our calculations, the wind system should eliminate the need for it. Nancy says, "It will be a thrill not to rely on a gasoline-powered generator anymore."

After 13 years of off-the-grid living, Nancy is very satisfied with her system. Her home is comfortable and she enjoys all the modern appliances she needs. The electrical system has grown and evolved along with the PV industry, and is now state-of-the-art. Who knows what the future holds? Whatever happens, this off-grid home will continue to run well into the foreseeable future.

— Bob Ramlow
Real Goods Snow Belt Energy Center,
Amherst, Wisconsin

THE HARLAN HOME
CASPAR, CALIFORNIA

Rob Harlan and his family — Lonnie, Kamala, Jessica, and Ben — live in an off-the-grid home about three miles from the Pacific coast of Northern California. Although it would only have cost a few thousand dollars to bring a power company hookup to the building site, Rob's commitment to renewable energy led him to power his home and heating systems independently. As a building contractor specializing in solar-home construction, Rob was also interested in living with the technologies he installs.

Rob, who grew up in Chicago and majored in Biology at Grinnell College in Iowa, was drawn to the Mendocino area by a feeling of community which he had not experienced before. "In my search for a meaningful life after college," says Rob, "I found direction and community support here, and I came to realize that if I wanted to make the world a better place, I had to start with myself." In the 1970s this belief in honest and responsible living led Rob to protest Pacific Gas & Electric Company's start-up of the Diablo Canyon Nuclear Power plant — a nuke of questionable design located near a known earthquake fault. The plant was eventually brought on-line despite massive protests, but the experience galvanized Rob's interest in positive alternatives. As a first step to living under his own power, Rob started harvesting electricity for his bedroom lights with a photovoltaic module mounted on the roof of his rental home.

Rob's first solar home was a small "mobile shack" which he built himself and powered with four PVs. Eventually, after several years of working in the building trades, Rob was able to buy his own piece of land in the pygmy forest near Caspar, California. Pygmy forests are areas where the trees have been stunted by extremely poor soil conditions. A 100-year-old tree may grow only six or eight feet tall under these conditions, and the land is inexpensive because it is considered undesirable. The area is ideal for solar living: no tall trees block the sun, and the site affords an expansive view of the Pacific Ocean.

Rob designed and built the home for himself, incorporating elements of Oriental and Southwest aesthetics with his commitment to energy-efficient independent living. Soon after completing construction on his home as it was originally conceived, Rob's family became more numerous when his partner, Lonnie Meyer, and her three teenaged children joined him. This precipitated a new construction project — the addition of bedrooms for the children — and complicated the demands on the solar electric and heating systems.

Rob's thoughtful and aesthetically pleasing design includes many features which help moderate the home's internal temperature through direct solar gain and passive storage. A small, concrete-block Trombe wall along the south side of the family room also serves as the back of a river-rock hearth for the woodburning stove. Double-glazed south-facing windows and skylights bring solar heat into the home, where it is absorbed and stored in the thermal mass of the tiled floor and plastered walls. The walls of Rob's home are constructed with two-by-six framing and R-19 insulation, while the ceilings feature exposed beams, with rigid insulation rated at R-30 applied over plywood backing and finished with bamboo matting, which is visible from below.

Builders today are faced with a dilemma concerning the lumber they use in their homes. If they want attractive, top-quality wood for trim and siding, they will most likely be forced to pay premium prices and buy lumber milled from old-growth, virgin trees. Second-growth and recycled lumber are difficult to work with and much less pleasing to look at. Rob chose to avoid this dilemma by minimizing the amount of exposed lumber used in his home. Exterior and interior walls are finished in a manner that wraps the plaster around corners and into window and door openings to eliminate the need for trim. Exterior walls are covered with stucco and stone. Besides creating a soft, flowing, Southwest adobe look, this strategy

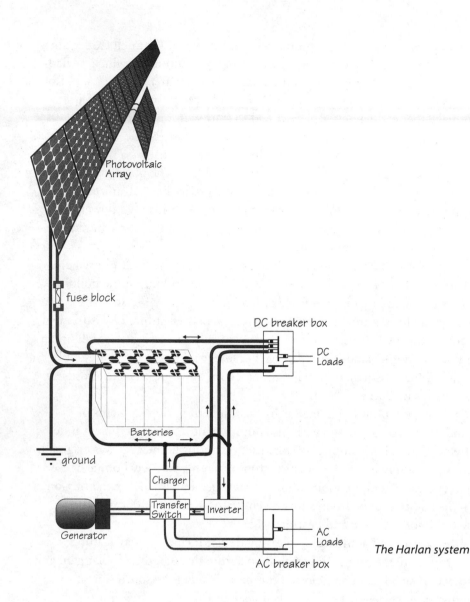

Photovoltaic Array

fuse block

DC breaker box

DC Loads

ground

Batteries

Charger

Transfer Switch

Inverter

Generator

AC Loads

AC breaker box

The Harlan system

allowed him to increase the thermal mass in the walls significantly. To add further to interior thermal storage, Rob doubled layers of gypsum board on south-facing walls that would be exposed to daytime sun. This added very little cost and practically eliminated waste, because he was able to use up all his scraps in the first layer.

The home's primary source of electricity is an array of 24 recycled Arco photovoltaic modules (a mixture of 16-2000s and M-44s), each with 48 watts of peak output. These PVs charge 20 nickel/iron batteries with a total storage capacity of 1200 amp/hours. Over the years, Rob has lived with and learned to care for lead/acid batteries, but when the opportunity to purchase these recycled nickel/iron batteries came along, he converted to them because they are less toxic, more forgiving, and with proper care may last the rest of his life (even though they were 40 years old when he got them).

One disadvantage to the nickel/iron batteries is a wider variation in the voltage output than with the commoner lead/acid types. According to Rob, his batteries often vary in output from 10.0 to 16.5 volts. While most of the components of his system can handle a wide voltage range, his inverter requires input voltage between 10.5 and 15.3 volts. "Toward the end of a bright day, the batteries will be charged up to their maximum voltage. We seldom need the inverter during the day, but if we do, we have to bring the voltage down, so we turn on a bunch of lights or redirect some energy to heating water through the Enermaxer."

Rob's electrical system delivers two kinds of electricity, 12-volt DC direct from the batteries and 110-volt AC through the inverter. The DC power generated by the photovoltaics and stored in the batteries is used to run quartz-halogen lights, a 12-cubic-foot Sunfrost refrigerator, and a 12-volt permanent magnet roller-type water pump which draws about 25 amps. The water pump is located 300 feet from the house and works well most of the time with varying voltages, although if the voltage is too low it cannot attain its shut-off pressure. For his 120-volt AC power needs, a Trace 2012 inverter provides "house current" for such things as power tools, grain mill, VCR and TV, and even the hair dryer and iron, which Rob reports are necessities in a household with teenagers.

Hot water for the home is provided by two Aquastars with solar batch preheaters. Rob hopes to reduce the amount of wood he needs to burn in the winter by retrofitting an underfloor radiant heating system in the living room. This system could be heated with hot water from solar collectors or propane. Propane is used for cooking and to power a back-up generator for charging the batteries during the occasional prolonged periods of fog which occur on the coast. Rob reports that the propane generator is quiet and reliable.

Rob is working with John Takes at Burkhardt Turbines to build a solar-powered electric work vehicle from a recycled Volkswagen Rabbit truck. When completed, the truck will sport photovoltaic charging panels on a tiltable lumber rack.

— John Birchard

THE PALUMBO HOUSE
HYDE PARK, VERMONT

David and Mary Val Palumbo did not set out to build an independent home when they started looking to move to rural northern Vermont a decade ago, but their favorite site was far enough beyond the nearest utility powerline that it saved dollars and made sense to pioneer alternative energy. They started out in a tent while building a small house on the second-best building site, so they could live with the land for a while before building the big house. They are hands-on people, so they and their friends did most of the work. Wherever possible, they used indigenous, sustainably harvested, nonpolluting materials in their construction. Four years passed from the time they set up the tent until they moved into their home. Their strategy was classic: they built a small, well-designed house on the second-best site the first summer, then built the big house as time and materials allowed.

Hyde Park is one of the coldest spots in the lower 48 states, and it takes some serious heat to keep a house warm. The large, comfortable Palumbo home relies on

wood, a renewable resource in good supply locally, which fuels a gasifier. A gasifer is an efficient high-tech woodstove which burns so hot that the fuel is literally turned into a gas before it burns, so combustion is efficient and flue gases are clear and nonpolluting. The gasifier runs year-round, and produces domestic hot water in abundance as well as hydronic heating. This strategy would be problematic in an area where ice storms and other causes take the utility power down frequently, but the Palumbos are their own power company, and their energy flow is trouble-free.

The hybrid supply side of the Palumbo's electrical system employs three sources: solar, hydro, and fossil fuel back-up. Most of the electricity comes from a micro-hydro generator and from a sizeable array of photovoltaic modules on the shop roof. The hydro source was not part of the original plan, but was discovered in the process of developing a pond. Domestic water and water for the micro-hydro come from sources up the hill on the Palumbos' land, so no energy is needed to create water pressure. In times of extra need or low productivity from the renewable sources, a propane generator can be used to recharge the batteries or power major power tools in the shop. Propane was selected as the generator fuel because it is relatively abundant, nonpolluting, and runs the generator's engine efficiently. Energy is stored in a large bank of industrial lead-acid batteries, which are kept in a

The Palumbo system

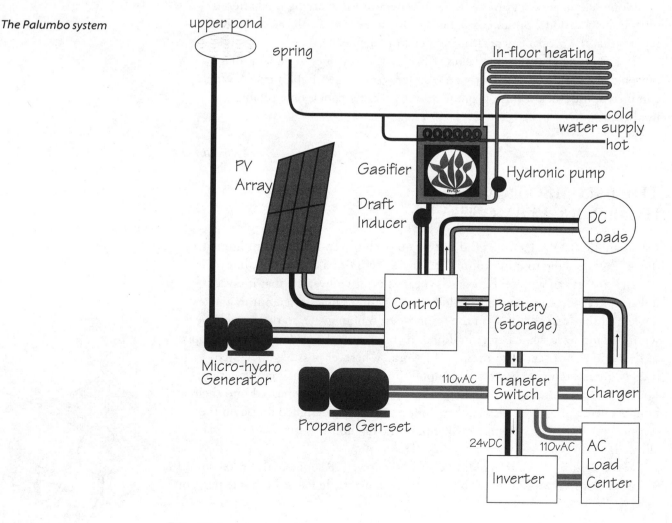

fireproof, vented battery chamber in the basement between the root cellar and the furnace room.

The Palumbo Power Company system began as a necessity, but has clearly become a work of conscience and vocation as their house has grown. The demand side consists of several different types of electricity distributed throughout the house: conventional house current powers standard appliances, 24-volt DC powers more special-purpose devices like the Sun Frost refrigerator, and 12-volt DC is used for some lighting, including the innovative LED baseboard lighting. The battery bank is wired as two 12-volt sub-banks. Originally, the gasifier's all-important draft-inducer fan was switched back and forth to keep the batteries balanced, but now the sides are hard-wired to loads carefully balanced by experience. Most loads run on house current (110-volt AC) supplied by an inverter. A transfer switch directs electrical traffic when the generator cuts in, because electricity from the generator can damage parts of the solar and hydro system. David Palumbo urges caution when switching on loads after the generator starts, because the generator produces wildly inconsistent electricity until it has warmed up and settled down. He has solved the problem by adding a time-delay relay on the generator. The system is constantly being improved; as we go to press, one of the new generation of Trace sine-wave inverters is being tested in parallel with the trusty old inverter. Dave says it has solved most of the interference problems.

— Michael Potts

JOHN & NANCY SCHAEFFER'S HOME
TALMAGE, CALIFORNIA

Nancy, John, Jesse, Sara, and Ashley Schaeffer had been living comfortably in their house in the foothills east of Ukiah for several years when the idea for Real Goods the Catalog Company burst upon them. When your family is already connected to the grid, it can be hard to justify disconnecting, and so for many years the Schaeffers have been working on their own "conservation power plant," using the most energy-efficient appliances available: a Sun frost refrigerator, a thrifty heat pump, Energy-Star computer equipment, compact fluorescent lights, super-low-flow toilets and dishwasher, and solar yard lighting. Now the Schaeffers are generating their own power, and selling some back to the utility.

Here's how the system works: Just below the brow of the east-west ridge on which the Schaeffer home sits, a tracked 3-kilowatt array of photovoltaic modules follows the bright Ukiah sun for an average of seven hours a day, generating about 20 kilowatt-hours daily, or a little more than the family consumes. Naturally, on gloomy days when the sun is absent, the modules are inactive, but electricity is still needed for refrigeration, lighting, the water pressure pump, washer, dryer, dishwasher, TV, stereo, hot tub, and other miscellaneous loads amounting to about 16 kilowatt-hours a day of demand. In a typical off-the-grid system, such a generous collection of appliances would require about 80 kilowatt-hours of battery

storage, which is a big load of batteries in anybody's book. This is the chief reason why the Schaeffers decided not to take their home completely off the grid. But with this new approach, called a *utility intertie,* battery storage is reduced to a small (20 kWh) back-up bank to cover grid outages.

The new Trace inverter allows the Schaeffers to use the grid for primary storage. In this part of California, as in most other places in the United States, the utility's peak demand coincides with PV's peak supply, in the early summer afternoons when the sun is beating down, and air conditioning and agricultural pumping is thrumming at its maximum. Pacific Gas & Electric, the local utility, is delighted to borrow all the nonpolluting kilowatts they can at this time, paying them back in the evening and morning when demand is much lower and their system has excess capacity due to hard-to-dispatch generators that must be kept on-line around the clock. So, during the afternoon, the Schaeffer system exports electricity through the Trace unit back through the meter and into the grid. In the early evening, when the PVs shut down and in-home demand increases, the Trace orchestrates the flow of electricity in the opposite direction, and the meter turns back in PG&E's favor. When the meter reader comes by on his monthly rounds, he should find, especially in the summer months, that the meter reads *less* than it did the previous month.

The new Trace inverter does its business by being smart. Of course it performs the basic service of turning DC from PV or batteries into AC, but it also monitors the quality of the AC connected to it and tries to correct distortion by damping peaks and filling valleys. If it is supplying power to the grid, and the grid goes down, the Trace disconnects in milliseconds. This is important, because the grid may have been turned off for servicing, and some human may be about to work the powerlines. (Remember that distribution lines are at high voltage, which local transformers convert to much lower-voltage house current. Transformers work both ways, so house current sent back the other way would be stepped up to a high enough voltage to give some unsuspecting lineman quite a thrill.) From a utility perspective, the Trace unit represents the future, a time when homes all over the country will be supplying part of the baseload power through what the utilities call *distributive generation.* Power companies dream of a time when these small, decentralized generators will be "dispatchable," meaning that, with a cellular phone call or a signal sent down the line, any site holding back extra power will add it to the grid to handle peak loads. At the same time, the utility system's protection engineers worry about *islanding,* the idea that, if the Schaeffers and their neighbors are all generating, their little branch of the grid may be energetic enough to stay up when the rest of the grid goes down. This is not so much a problem as an engineering challenge, because as islanding becomes more likely, the utilities will have to adapt their grid systems to accommodate an island's ability to stay alive when the rest of the system fails, while at the same time providing a method for bringing the system down for servicing. Ironically, the biggest problem with the new Trace unit is its tendency to correct distortion: the utilities define "in-spec power" as whatever they happen to be delivering, and, if it is distorted, they want that distortion left alone in a sort of modern-day interpretation of the medieval notion that might makes right.

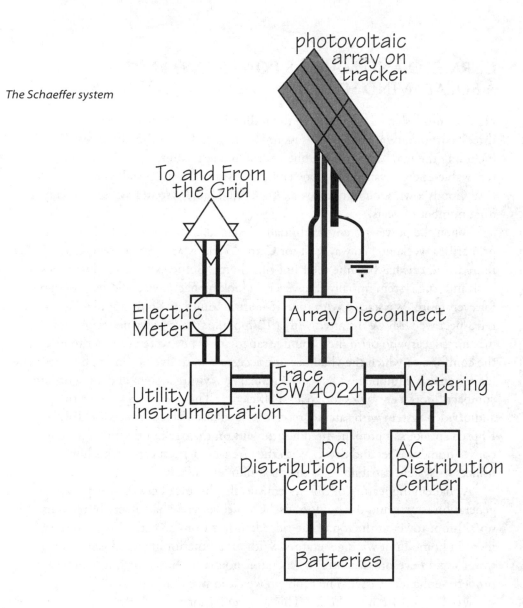

photovoltaic
array on
tracker

To and From
the Grid

Electric
Meter

Array Disconnect

Utility
Instrumentation

Trace
SW 4024

Metering

DC
Distribution
Center

AC
Distribution
Center

Batteries

The Schaeffer 3 kilowatt PV-utility intertie system cost $22,700 including enginering and installation. With net metering now in effect in California, John calculates pay back in 16.6 years.

— *Michael Potts, with help from John Schaeffer and Jeff Oldham*

CARMELO AND MARLA'S POWER SYSTEM: A SOLAR/WIND HYBRID

Carmelo looked forward to each storm so that he could sit at his kitchen table, watch the amps roll in on his monitor, and gloat over his energy wealth.

Nestled into a shaded oak grove in the rolling hills of Northern California is a three-bedroom, two-bath 2,000-square-foot home. The entire homestead, including the main house and a one-bedroom guest cabin, is run by a hybrid renewable-energy system. The power system was primarily installed in 1989–90 by Real Goods' own Doug Pratt (a.k.a., Dr. Doug). The system has been in daily use for a number of years.

When the power system was initially designed and installed, the home was primarily a weekend getaway spot for Carmelo and Marla. Since then, Carmelo has retired, and this has become their full-time home. Some expansion capabilities were built into the system initially. However, in looking back, we could have designed for even more. We started with four 48-watt modules on an 8-module tracker; in retrospect we wish we'd started with a 12-module tracker. Carmelo and Marla are urban refugees who prize their country sanctuary. They were not eager to give up the comforts to which they had become accustomed to live out in the boonies. The main house is equipped with a conventional AC wiring system and all lights are compact fluorescent types for greatest efficiency. There is also a state-of-the-art audio/video system with satellite receiver, video disc player, projection TV, large Klipsch speakers, a propane-fired hot tub out on the deck under the stars, and a conventional washer and dryer. With the often-used guest cabin, this homestead utilizes its system capabilities to the maximum efficiency.

A challenging feature of this system was the distance between the power generation source and the point of use. Carmelo already had a very heavy-duty, tilt-up, 50-foot tower at the top of the ridge left over from an earlier disastrous experiment in home-built wind generators. Without a functioning wind generator, the tower was an eyesore. But with a generator, no matter how small, it became a property-enhancing asset. The problem we faced was that the top of the ridge wind site was 500 feet from the house. This was too far for cost-effective transmission of the wind generator's low-voltage output. This site also happens to have the best solar access to take maximum benefit from a tracking mount. The solution was to put the energy-collection equipment on the ridge top and build a small power shed at the site. The shed contains the batteries in one compartment, and the control, safety, monitoring, and inverter equipment in a separate compartment. (Batteries produce hydrogen gas during charging and *must* be isolated from the rest of the gear.) The inverter converts the stored DC power into more easily transmitted 120-volt AC. The higher the voltage, the easier it is to transmit, so the house runs everything on conventional AC power. There are plenty of vents on both the battery and control sides of the shed. The batteries need to be well ventilated to prevent hydrogen buildup, and the inverter needs to be well ventilated because it produces a small amount of waste heat. If that heat is allowed to build up, the inverter's surge capacity will be diminished. Actually, there is a vent directly under the inverter on the raised floor of the shed.

The wind generator was installed initially in order to do something useful with

Carmelo and Marla's solar/ wind hybrid system

the existing tower. Carmelo was looking more at resale value than at power output. Big 1000-watt and larger wind generators were too expensive, and mid-sized 400- to 700-watt generators didn't exist at that time. So, more for property enhancement than for power generation, we settled on the early Wind Seeker 250-watt generator. Fortunately this generator didn't cost too much. But a funny thing happened the first winter. Much to everyone's surprise, when the winter storms started to blow in, Carmelo suddenly had more energy than he knew what to do with. The cumulative effects of a power source that sometimes worked 24 hours a day were astounding. Carmelo looked forward to each storm so that he could sit at his kitchen table, watch the amps roll in on his monitor, and gloat over his energy wealth. The King Midas of renewable energy had arrived!

Getting the energy from the ridge top to the house actually turned out to be the easy part of this installation. Carmelo really enjoys his gizmos and wanted

complete monitoring and some control capabilities at his house. His control panel in the kitchen allows him to check battery voltage, PV charging input, wind charging input, or inverter/battery charger input/output. There is also a flashing red low-voltage warning light on the panel. The control panel has controls to remote-start his generator and is equipped with a generator-on indicator light. Finally there is a wind-up timer that turns on the water pump(s). All this control and monitoring power required almost a dozen runs of 14–2 UF romex wire between the house and power shed. More money was spent on monitoring and control wire than on power transmission wire. This is unusual for any system, but it gave Carmelo peace of mind.

During installation we buried two extra runs of monitoring wire, which turned out to be a lifesaver. We found that Carmelo's satellite receiver would forget all its satellite position programming if it didn't receive power every two to three days. Reprogramming was a three- to four-hour nightmare, not a job to face every weekend at your relaxing country getaway. To correct the problem we used a 12-volt timer that would switch on inverter power to a dedicated satellite receiver outlet for one hour each morning. The receiver drew enough wattage to bring the inverter out of the power-saving stand-by mode and into conventional operation for that hour. The receiver recharges its internal battery, and all the programming is safe for another day. Even though he lives at the house full-time now, Carmelo still leaves the satellite receiver plugged into the dedicated outlet, so that if he leaves for a few days he doesn't have to worry about it. Solving the receiver problem used up one spare run of 14–2 wire. Then Carmelo got a 12-volt answering machine. However, we had never run any direct 12-volt power to the house because of the

CARMELO AND MARLA'S SYSTEM

8   Siemens M-75 PV Modules
1   Zomeworks 8-Module Tracker
1   Wind Seeker Wind Generator
1   Custom-Made 50-Foot Tilt-Up Tower
1   Trace C30A+ Charge Controller
6   Trojan L-16 Deep-Cycle Batteries
1   Trace 2012SB Inverter/Battery Charger
1   Todd 15.5V/75A Battery Charger
1   SCI Battery Saver
1   8-Position Fuse Box
1   Customized Photron Energy Monitor

Approximate cost: $10,000 including installation

voltage drop over 500 feet. Fortunately, the answering machine only needed a few milliamps to operate (an incredibly tiny amount of power), so we were able to supply it with the other spare 14–2 wire run.

It was sometime during that first winter, wallowing in excess energy, that the idea to convert the hot tub was hatched. Heat for the tub was supplied originally with a 1500-watt electric heater, which took almost three full days of generator run time to bring the tub up to temperature. Converting to a small, instantaneous propane heater brought the heating time down to about three hours. The hot-tub pump was a two-speed unit that was operated as a 1/8 horsepower (hp) pump in heating/ circulation mode, or a 3/4 hp pump when operated for jets and foam. In 1/8 hp mode it only drew about 70 watts. With some careful plumbing and a bit of modification to the controls of the tankless heater, the circulation pump provided sufficient pressure differential to trigger the propane heater. Wintertime surplus power was now being used to run the hot tub pump. Voilà! Hot-tubbing in the dead of winter without generator noise!

The power system consists of eight Siemens M-75 photovoltaic modules on a Zomeworks tracking mount, a Wind Seeker 250-watt wind generator, and an ancient 3000-watt propane-powered generator for back-up. Six 350–amp-hour L-16 deep-cycle batteries provide over 1000 amp-hours of storage capacity in this 12-volt system. Charge control is handled by a Trace C30A+. A Trace 2012SB provides the AC power that is used for everything except the telephone answering machine. The monitoring/control panel was custom-made on site, but we basically started with a digital volt/amp meter like the SCI Mark III Battery Monitor and added multiple inputs to the meter, and additional controls as needed.

Available technology has changed considerably in the years since Carmelo's system went in. Some components of his system have been replaced over the years for better performance; others have been changed because we learned as we went along. His 3-kilowatt propane generator was never able to take advantage of the Trace SB battery charger's full power. For that you need 4 to 5 kw. So a more efficient, but smaller potential output, Todd 75-amp/15.5-volt battery charger was added last year. The original AC-powered water pumps, a 1/3 hp submersible in the shallow well, and a 1/3 hp booster pump at the pond, drew too much power for the quantity of water they delivered. The well pump also seemed to suffer frequent failures, probably due to browning-out during start-up surges, when it must pull power through almost 1,000 feet of wire. This summer, the pumps were replaced with more efficient Shurflo 115-volt AC diaphragm pumps. The Shurflos deliver 50% less gallons per minute, but they're so efficient that they require less than a quarter the energy per gallon delivered. This simple conversion has freed up considerable energy for other uses in the electrical system. The low-voltage alarm light has never worked well. The alarm that was available at the time has little or no time delay, so it will trip whenever the washing machine or any other momentary surge load starts. As a low-voltage warning, it has been more an annoyance than a useful tool. The charge controller was changed from an SCI Charger Model I to a Trace C30A+ last summer. The SCI wasn't pushing the batteries hard enough during the day to achieve full charging, and the charge control set points are not adjustable. The Trace controller cured this performance problem.

If this system were to be installed today, the first-generation Wind Seeker 250-watt unit that we installed would be replaced by the current generation Wind Baron NEO+ at 750 watts. Price for the 750-watt NEO+ is only a little more than Carmelo paid for the early 250-watt unit. This is too good a deal for anyone with even intermittent wind power to pass up.

If this system were to be installed today, Dr. Doug would recommend an active Wattsun tracker rather than the passive Zomeworks. At the time of installation there was no choice in trackers. The Zomeworks goes to sleep at the end of the day facing west. The rising sun warms it up and the tracker rolls over in the morning. When cold, the Zomeworks is slow to wake up and roll over to face the sun. An hour and a half of lost charging time is significant in the winter. Also, on the ridge top site the Zomeworks sometimes gets blown off course. In Carmelo's case, if it's windy enough to blow the Zomeworks off track, the Wind Seeker is more than making up the difference, but with a Wattsun he could have had both power sources working simultaneously.

How does a long-time renewable energy user feel about his investment? Carmelo says, "I've been extremely pleased with the freedom, power, and lack of maintenance from this system. For the money spent, nothing else could have given me the reliability, not to mention the free, quiet operation. Having a high-quality energy system significantly increased my property value, and made it possible to get conventional bank financing. The only complaint I've got is that the inverter and controls are so far away. One big mistake I made was buying a used generator. Also, stake all buried wire and plumbing runs so you can find them, and make diagrams and maps of everything as you go.

"My favorite experience was the day my 70-year-old dad from Italy showed up while we were pouring concrete footings for the wind tower. (When the original homebuilt turbine failed, it tore out a couple of the guy-wire footings.) The PV/battery/inverter system was already up and running, and we were using it to run the cement mixer. I showed my Dad how we were using sunshine to mix concrete. This whole photovoltaic thing was new to him. I'd told him about what I was doing already, but he thought I was pulling his leg. He walked around the rest of the day grinning, shaking his head, and muttering about sunshine to mix concrete. After a lifetime of professional masonry work, he was utterly amazed."

— *Doug Pratt*

YOLANDA AND DANTE
PETALUMA, CALIFORNIA

Between the rolling ranchland of Sonoma County and the southern edge of the great forests of the Pacific Northwest there is a refuge of solitude for two very special people. The roads in this neighborhood all have the familiar sentries of power poles and overhead wires interconnecting the dwellings with distant power plants. A careful look reveals that one particular house also has a fleet of solar modules facing squarely toward the sun.

I first met Yolanda in the Real Goods retail store in Ukiah. She was very eager to learn how to be more energy-independent, and was equally determined to put into practice the required technology. A few days later, she had placed an order for a Sun Frost refrigerator, and we began to work on a renewable energy system which could handle a good portion of the house's demands.

Dante and Yolanda did not have a clear idea of how much power they would have to generate to satisfy their needs. We decided that a system which required some supervision would offer the best learning environment. They would need to monitor the system, and, if the batteries became excessively depleted, the utility company hookup could be switched back on as the primary power source, allowing the solar array some time to replenish the batteries. We could have chosen a fully automatic system to do this, but this would have isolated them somewhat from their goal of reducing power consumption.

Yolanda and Dante had lived for many years in the golden hills above Oakland, California. When the devastating fire of 1992 swept through those hills and destroyed their home, Yolanda and Dante were ready for a change in scenery and attitude. The hustle and danger of the San Francisco Bay area, coupled with the likelihood of yet another fire in a decade or two, prompted their exit. "The recent fire was not an accident," says Dante. "Big fires have been sweeping over those hills for generations; it's the natural scheme of things for that bioregion. We didn't want to go through the rebuilding process only to see it happen all over again."

Moving north to a small ranch in Sonoma County offered the best in relaxation while still having reasonable access to the Bay Area, where they maintain business activities. The property they purchased had been degraded by prior inhabitants. The house, the proverbial "fixer-upper," was clearly in need of renovation, and the stagnant, polluted pond would have discouraged many people with less vision. Yet Dante and Yolanda could feel the warmth of possibilities here. They proceeded to take on this challenge as an opportunity.

As a strong advocate for libertarian ideals, Dante is inspired to invest in his own independence, be it political or otherwise. In spite of the availability of utility power on-site, he wanted to be able to live with his own power, not rented power. His priorities were to incorporate solar space heating into the remodeling of the home and to use the appropriate technology to clean up the pond.

"Rancho de Libertad" is a home remodeled in a southwestern style. The adobe tile floor incorporates a radiant floor hydronic heating system, which makes use of an active solar thermal system to provide the heat. A tremendous leap in energy efficiency was made when they purchased a 16-cubic-foot Sun Frost refrigerator/ freezer. This dramatically reduced the electricity required in the home, and also brought the possibility of an independent energy system within economic reality.

Still, electrical power consumption remained high, as the filtration pumps for the pond were running many hours per day. Clearly, an alternative to the grid-powered pumps would need to be found if Dante and Yolanda were to achieve energy self-reliance.

The Systems

All normal house loads, the pumps that circulate the solar heated water in the radiant floor system, the jet pump at the well, and a spa/jacuzzi are all powered by a photovoltaic system designed by me, Douglas Bath, Real Goods Application Engineer. The system was originally specified to consist of 12 photovoltaic modules mounted on a Wattsun active tracker. The power is processed by an Ananda Powercenter, and the energy is stored in a 750-amp-hour 24-volt Chloride battery bank. The solar array has since been amended by the addition of another four solar modules which are nontracking. A nice feature of the Ananda Powercenter is that a second controller can be installed in the field, which allows for considerable future expansion. A Cruising Equipment Amp-Hour Plus meter helps determine the depth of battery discharge, as well as displaying system voltage and power consumption/production.

Seven years of California drought and abuse had left the pond on the property a scummy, polluted, anaerobic mess. Dante installed a 1 hp AC pump to provide both filtration through three large sand filters and aeration via a central fountain discharge into the pond. The many hours of pump run time resulted in an unacceptably high utility power bill.

With the experience of one independent energy system under his belt, Dante installed a 90-volt DC, 1 hp pump, which runs off a tracking photovoltaic array. This powers the pump directly without battery support during sunny days. The restoration of the pond is further aided by four Floatron solar water purifiers. These help to slowly diminish the algae buildup of many years. Dante cruises out in his rowboat daily to remove the minerals that accumulate on the Floatron electrodes. He enjoys his regular maintenance regime, as it gives him the chance to enjoy his pond and witness the remediation. The clarity of the pond continues to improve, and the bullrushes and newly stocked fish are flourishing.

Dante and Yolanda tell me about the relaxation their move to the country has brought and the profoundly healthful quality of life away from the bustle of the

The purifying pond

metropolis they left behind. The renovation of the buildings and rejuvenation of the pond helped soothe their urban-battered psyches. Yolanda now has time to devote to gardening, saving stray cats, and keeping the chickens away from both. Whenever I call, I always ask how Carmen and Miranda (two of the chickens) are doing. I think that they are being looked after by some Real Good people.

— Douglas Bath

SYSTEM PROFILE

Customer: Yolanda and Dante
Site: Petaluma, California

System Description:
Photovoltaic array with utility backup
12 Siemens M-75 (tracking)
4   Solarex MSX-60 (fixed)
1   Chloride 750-amp-hour 24-volt industrial battery (12 ea. 2V cells)
1   APT 3–200 Powercenter
1   Cruising Equipment Amp-Hour Plus Meter
1   Trace 2624SB Inverter
Approximate Cost: $14,000

BADGERSETT RESEARCH FARM GREENHOUSE
CANTON, MINNESOTA

Badgersett is a family-owned business, developed by Philip and Mary Rutter with the help of their two sons, Brandon and Perry. Former university researchers, the Rutters transformed their research into their lifework: the development of woody plants as producers of food, or "woody agriculture."

The Research Center has had a long-term partnership with the Hubei Academy of Agricultural Sciences in China and the Hubei Provincial Forestry Department. Their work is to develop chestnut woody agriculture for hand-labor systems. In 1993, the Rutters began a cooperative project with the University of Minnesota Center for Alternative Plant and Animal Products to pursue research in hybrid hazelnuts. Philip has been President of both the American Chestnut Foundation and the Northern Nut Growers Association.

The farm is located in the extreme southeastern corner of Minnesota, on hilly land about 60 miles west of the Mississippi River. Summers are hot and winters are cold. The Rutters planted their first trees more than 15 years ago and began large plantings in 1981. This spring they had about 6,000 chestnuts and about 5,000 hazels.

The heart of the Badgersett Research Farm is the solar greenhouse. The energy system for the greenhouse was developed by Real Goods. The greenhouse is semi-earth-sheltered, because the farm is in a very cold growing area, Zone 4B, which is colder than other locations within Hardiness Zone 4. The building was codesigned with architect Roald Gundersen, who also worked on Biosphere 2 in Arizona. The front is bermed four feet in the ground, and in the back is six to seven feet in the

BADGERSETT SYSTEM PROFILE

Component	Number and Manufacturer
Solar Panels	6 Siemens M-75
Mounting Structure	Zomeworks Top-of-Pole
Charge Disconnect	30-amp/two-pole safety disconnect
Charge Controller	Trace C30 (*Not* the C30A)
System Monitor	Cruising Equip. Amp-Hour Plus
Battery Bank	24 6V/220A Golf Cart Batteries
Battery Interconnects	2/0 cables with crimped and soldered ring lugs
Load Disconnect	400-amp Fused Disconnect
DC Load Center	6-Circuit Load Center
Inverter	Trace 2624SB
Inverter Cables	BC10, 10-Foot, 4/0 Cables
Other	Wind Baron NEO wind turbine with 45-amp inline fuse
	2 16-inch Intake/Exhaust Fans
	Shurflo Medium Flow Pump
	Shurflo Submersible Pump
System Cost	Approximately $5,600

At this time the Wind Baron has not yet been installed (the extra power has not been needed), but is expected to be in place by the time this reaches print.

Dr. Doug's Notes on Bagersett's System
There are a couple of unusual uses of equipment in this system. The submersible well pump and the Wind Baron NEO are both using the same wire run. If the NEO is making more power than the pump needs, then excess power flows to the battery. If the NEO is making less power than the pump needs, then power flows from the batteries to the pump. Wires can be two-way streets! Electricity will always flow from the higher voltage point to the lower voltage point.

There is no charge controller on this system. The Trace C30 Load Controller is used to turn on or off alternative loads such as wind fans or grow lights when battery voltage hits the user-adjustable high and low points.

ground (so the back wall is almost totally earth-sheltered). The roof is super-insulated, with glass facing only south. This is a building that is designed as a serious business, not as a hobby.

According to Philip Rutter, "It works! So far this year we've grown about 20,000 plants in here, which is actually a small fraction of what we expect the building to be able to produce once we have some logistics worked out on how to handle the plant material as it goes through the production process — from planting, to growth, to getting it sold. When you are dealing with so many plants, just the sheer detail of how many times you have to pick something up and move it gets to be terribly important, and we figure it will take at least two years to work out some of the kinks there. But, it is working. One of the things that strikes me, having seen other commercial greenhouses, is that anybody who is in business to make money is out of their mind not to earth-shelter buildings like this.

"This is not a far-out alternative kind of building," Rutter relates. "This is a very straightforward money-making operation at this point. In fact, from an efficiency standpoint, this building makes a tremendous amount of sense. The cost of construction is higher than for standard greenhouses, but in the long run, over a ten-year time span, we really expect that this building is going to more than pay for the additional construction cost by saving energy. In addition to that, the energy is substantially more reliable. In a traditional, above-ground greenhouse, if you have a heating system failure in the winter, you can lose your crop in a hurry. Unless the roof blows off, which is really not possible, this building can't freeze in the wintertime because of its earth-sheltered and super-insulated aspects. That security may make the difference between whether our business survives in the long term. You only need one catastrophe to put you out of business.

"We are not plugged into the grid at all. There are aspects to financing the power that many businesses really don't consider. In order to run a powerline in here we would have to dedicate a substantial amount of land to the power company, we would have to clear trees, and we would have to take land out of production to maintain the right of way for the powerlines, which in turn would mean we would lose the productive potential of that land in perpetuity. And then we would get a bill every month regardless of how much power we use. In addition, if we have a month where something requires a lot of extra power, all of a sudden we have another great big fat bill, regardless of whether we have income or not at the time. That is one of the things that kills new businesses: standing expenses when cash flow is irregular.

"In this system, though, the power is built into the cost of construction. We own the power equipment; it is business equipment. We can depreciate the equipment on our taxes, which helps out, and there is no monthly bill. The ability to put the power cost into the cost of construction is actually a tremendous benefit, because it means you are paying for your power as part of the mortgage, and there are tax advantages to mortgages that you simply don't get when you pay a power bill to the local utility every month, month-in and month-out. In spite of the fact that the initial up-front cost sometimes looks steep, all you have to do is draw out the ten-year cost for it to make tremendous sense all of a sudden from a strictly financial standpoint.

"The building is designed to run with both a wind generator and solar modules, which makes excellent sense for us here in Minnesota. November, for example, is likely to be very cloudy and stormy, a time in which we get a lot of wind and not that much sun. January, on the other hand, is likely to be extremely bright and very still, although an occasional storm will make a good wind for a couple of days. But most of the time it is brilliantly sunny here, and the fact that the solar modules are more efficient when cold to some extent compensates for the shorter sunlight hours. In addition, the building was designed so that it has a large dished snow and ice field right in front of it, which bounces a tremendous amount of solar radiation both into the building for heat and onto the solar modules. We discovered the best position for our solar modules last February by watching the amp-hour meter while we moved the panels around until we got maximum output. If we pointed the array just perpendicular to the sun, we found that we were getting a couple of amps *less* power than if we pointed it straight up and down so that it was getting both sun and snow bounce. The difference really is substantial. At the moment the wind system is not yet operational, partly because we've been able to run the building satisfactorily by hand-tracking the solar modules. We adjust their mount about three times a day to take advantage of the movement of the sun, which gives us enough power to keep us going. We could use more power to run our circulation fan and some auxiliary lights inside, but in fact we can get along without them, so we have been dealing with the various other emergencies of how to get this business started."

This is one of the more unusual power systems Real Goods has been asked to design. Badgersett uses their greenhouse from February 15 through December 15. There is *no* back-up heating system. Warm daytime air is fan-forced below the 2-foot-deep gravel floor. In case of overheating the fans will vent the greenhouse. The power system provides energy for lighting, fans, water pumping, and, when they are installed later this year, opening and closing the large insulating shutters.

— Doug Pratt

PLANNING DOMESTIC POWER: FELICIA AND CHARLIE COWDEN'S STORY

Charlie and Felicia Cowden live off-the-grid in an house they built on Kauai's North Shore. In the aftermath of Hurricane Iniki, which bludgeoned Kauai in September 1992, they helped their neighbors get their power back. In sunnier times, the Cowdens run the Hanalei Surf Company. As you will see from their story, Felicia and Charlie's move off the grid involved a process of learning and adaptation, but the conclusions they have reached ring true for anyone who uses electricity.

Felicia: The biggest stumbling block on the way to power independence is our culture's custom of wasting incredible amounts of energy. We waste more than we

Felicia and Charlie Cowden

use. You can't simply buy energy independence: you have to make a brain investment, and learn some things.

We started moving toward conservation when we were living on the grid in Princeville, by replacing incandescent lights with compact fluorescents.

Charlie: When we were deciding to buy our property and build, whether to pay the money to trench and connect to the powerlines at the highway, or go solar, I didn't know anything about alternative energy sources. So I asked my contractor, "Should I do it?" He didn't know much about it either, but he said, "It's the future, man, you gotta do it."

Felicia: We throttled our energy use way back when we moved to our new house. We were limited financially, and could only get six panels. We figured we would need more, but during the sunshine months, we're full by noon, and we haven't given anything up. We just changed our energy habits.

When we first moved in, it was like living in a science experiment. It soon became very apparent that there were things we hadn't thought of before. Living in my house has made me much more conscious of consumption, so when I visit other people, I see incandescent lights, and can't help but notice all the things being done inefficiently. It feels sinful! Solar power's biggest gift to the environment is showing people that it's possible to live well without being wasteful. Since moving in, I've become an energy evangelist, always trying to get more people into conservation.

There are so many things we wish we'd known! Nobody told us about phantom loads, things that use electricity whenever they're plugged in, like the clock on a microwave . . . The real energy criminals of the small appliance world are the remote products that require chargers, like electric toothbrushes and cordless razors. They draw substantial current to charge their batteries, then when the batteries are full, they continue to trickle energy. That's a lot of burnt dead dinosaurs for a few minutes of minor convenience, and we all have to breathe the smoke . . . If push comes to shove, you can always brush your teeth yourself.

Charlie: In our house we put our phantom loads in power strips, but it would be better to put switched outlets in the wall.

Felicia: A perfect example: a recessed microwave with the outlet behind it, where you can't reach it to unplug its phantom load. Lots of times, people want to install their microwaves that way. If you put a wall switch on it, you can turn it off when it's not needed.

If I leave all our phantom loads — the VCR, stereo, TV, microwave, and adding machine — plugged in, even if they are turned off, they consume more electricity than my panels produce. If I can turn their outlets off, I produce more power than I need. So you don't have to do austerity things, you just have to be careful not to waste.

We learned a lot since we built our house, most of it what the power company calls "load management" or "demand-side management." I can't run the stereo when I'm doing a wash, because I chose a power-hog double-agitator washing machine. Now, I'd buy a front loader.

Charlie: You've got to make small amounts of electricity do your work. I saw a front-loading washer in the store, and I thought they'd made a mistake, because the energy tag said it used a tenth the power of the top loaders.

Felicia: It's all in the torque. You've got to understand the difference between inductive loads, like motors and little transformers, and resistive loads, like lights. Inductive loads are real power-hungry, especially when they start. For example, you learn not to run the vacuum cleaner when the washer's running. The garbage disposal puts a big strain on our system . . .

Charlie: . . . and the television picture gets real small.

Felicia: If I wake up in the morning and see the batteries aren't depleted, I look for high-demand things to do, like clothes-washing and vacuuming. Load management depends on weather: if it's cloudy, and the batteries are depleted, I think about conservation.

Charlie: People think they've got to make sacrifices, give up TV, and live in the dark. That's not necessary, they just have to pay attention . . .

Felicia: When guests come to our house or people come up to work, they are fascinated by our system, the fact that we have all the electrical things we need. When their friends come by, they like to play with the meter, show what the system is doing. It's instructive and fun.

It helps if the system is shipshape, like it was meant to be part of the house. When people see batteries in a corner and all sorts of cables, it makes an image of alternative energy as a subculture lifestyle. That's not necessary or even true. It works better if it's clean . . .

When we talk about energy self-sufficiency, we try to get people looking away from the money issues. Maybe you save money going off the grid, but there are more important things. When you go to sleep at night, there's a zero electromagnetic field. And think of the tons of stuff you aren't putting into the atmosphere. It's a bonus if you save money.

— *Michael Potts*

Excerpted by permission from The Independent Home: Living Well with Power from the Sun, Wind, and Water *(Chelsea Green Publishing, 1993)*

DESIGNING A LARGE OFF-THE-GRID POWER SYSTEM

Real Goods used to package Remote Home Kits — prepackaged solar energy systems for generating power. After countless hours of making substitutions, adjusting the component lists, and revising system prices, we decided to change our approach. To provide better service, we now custom-design each energy system to the specific needs of the user.

People used to call up after seeing our Remote Home Kits, and say, "I think I need a Kit #3," and it often took a long time to get to the conclusion that it was not the right system for them. Now we begin at the right place: What are the electrical needs of this particular user, and how can they best be met?

Ellen's Story

Everyone has different needs when they come to us for help in designing an energy system. Certain issues are universal and have to be addressed with every system we design. We'll let our conversations with Ellen Brodsky (not her real name) show you some of the issues that we frequently work with, and what her system looks like. You may have several of the same concerns, and your system may end up looking very much like Ellen's.

Ring ring . . . ring ring . . .

Dr. Doug: Hi, Dr. Doug here. How can I help you today?

Ellen: This is Ellen Brodsky calling. I'm considering buying a really beautiful piece of land in Colorado, and there's no utility power there. A friend gave me your phone number and told me that you would be able to help. Am I calling the right place?

The Doctor: You certainly are! We design remote energy systems all the time. Tell me a bit about the land and what you want to do there.

Ellen: Well, it's on a south-facing slope, so I guess it would be good for solar. We have to heat the house, of course, but there is little need for air conditioning. It's just my husband and me, and our electricity use is pretty average. We turn the lights off when we're not using them. My husband runs a joke-writing business from home on his computer, which he uses a lot.

The Doctor: It's great that you have southern exposure. Now, the first step in designing your electrical system is to determine your energy demand. We do this by listing all the appliances that you want to run in your house and estimating how many hours each day you want to run them. I have a handy chart for you to fill out which will help us do this figuring. After we finish talking today, I can send the chart to you. Or, if you want you can get a copy of the *Solar Living Sourcebook,* our guide to energy conservation and renewable energy; the chart is on pages 152. Another good book that tells you what it takes to live under your own power is *The Independent Home: Living Well with Power from the Sun, Wind, and Water.* You're thinking about taking a big step, seceding from the grid, and you'll want to learn more about what you're getting into.

Now, let me give you a little background that will help you fill out the chart when you get it.

Ellen: Okay. Please talk slowly; I'm taking notes.

The Doctor (slowly): The fundamental rule is that you need to be as efficient as possible with your energy use, because you are creating (and paying for) your own power. For this reason, we have to find the best way to meet your needs with the least amount of energy possible. This will keep your system affordable.

Some appliances consume so much electric energy that it's not cost-effective to run them off your PV system. This includes anything that uses electricity to produce heat, such as electric stoves, water heaters, and baseboard space heaters. Toasters are okay if you don't make a lot of toast. All of these heating needs can be satisfied with propane, which most off-the-gridders use.

With some appliances, the models that are in stores today are not very efficient, and it's worth investing in a more efficient model. This way, you'll need fewer solar modules, and your system will cost less. This applies primarily to lighting and refrigeration. I highly recommend compact fluorescent lighting and either a Sun Frost refrigerator, which uses a fraction of the energy that typical models use, or a propane refrigerator, which uses no electricity at all.

The space heating and cooling needs of your house are best satisfied with passive solar architecture. I highly recommend the book *Passive Solar Energy,* which shows ways in which you can design your home to heat and cool itself without electricity, propane, or any other energy source. You'll probably want to have a woodstove for back-up.

Efficient lighting and refrigeration; entertainment equipment like televisions, VCRs, and stereos; kitchen appliances; water pumping; computers, microwave ovens, and many other appliances are all reasonable loads for your PV system. Our System Demand Planning Chart asks you to list all your appliances, and, when you have filled it out, you can call me or send a copy back to me and we'll design your system together.

Ellen: That sounds great! I think I'd like to order those books right now, because we want to get started learning everything about going off-the-grid.

Two weeks after Ellen got her off-the-grid library, Dr. Doug received a copy of the System Demand Planning Chart, all filled out. Ellen did a great job — she didn't list any heaters or incandescent lights, and she remembered not to use a standard refrigerator. Here's what the chart looked like:

SYSTEM DEMAND PLANNING CHART (ELLEN BRODSKY)

Appliance	Quantity	Wattage	Hours/Day	Days/Week
Kitchen lights	2	40	4	6
Living-room lights	4	18	2.5	6
Computer	1	55	6	6
Office lights	2	18	8	6
Laser printer	1	1000	.3	4
Blender	1	300	.1	2
Coffee grinder	1	100	.02	6
Sewing machine	1	75	1	1
Bedroom lights	1	18	1	7
TV	1	60	2	6

Appliance	Quantity	Wattage	Hours/Day	Days/Week
VCR	1	15	2	6
Stereo	1	40	3	7
Microwave	1	1000	.16	5
Vacuum	1	750	.25	1
Washing machine	1	1500	1	1
Water pump	1	60	1	7
Ceiling fan	3	20	8	7

The Doctor used the figures on the Planning Chart to design the following system for Ellen. He sent her a copy of his recommendations and called her on the phone to discuss it.

ELLEN'S SYSTEM

Component	Dr. Doug's Recommendation
Solar Panels	6 Siemens PC-4 modules
Mounting Structure	Wattsun Tracker
Charge Disconnect	Ananda Powercenter
Charge Controller	Ananda Powercenter
System Monitor	Cruising Equipment AmpHour Plus
Battery Bank	8 L-16 batteries
Battery Interconnects	2/0 cables with crimped and soldered ring lugs
Load Disconnect	Ananda Powercenter
Load Center	Ananda Powercenter
Inverter	Trace 2624 SB with battery charger
Inverter Cables	4/0 cables with crimped and soldered ring terminals
Other	Low Battery Alarm, Lightning Arrestor
Approximate Cost	$7,930 (minus Hard Corps member's 5% discount for a total of $7,533)

Ring ring . . . Ring ring . . .

Ellen: Hello?

The Doctor: Hi, this is Dr. Doug calling. I wanted to make sure that you received the system design I sent to you and see if you had any questions about it.

Ellen: Well, yes I did receive it and, actually, I was just looking it over and making a list of questions to ask you. First of all, what is this Ananda Powercenter thing that's listed for several of the components?

The Doctor: The APT Powercenter is a new advance in PV technology. It takes several of the essential monitoring components in a PV system, and offers them prewired together in an enclosure. The Powercenter will save you a lot of time and money when it's time to install your system. It has the added advantage of being UL-approved, which helps meet most local building codes. You could buy all the components individually, but you would only save a few hundred dollars, which you'll spend quickly on an electrician and on time spent for the installation. I have some more literature on the Powercenter which I can send you if you'd like.

Ellen: That would be great. What is that battery charger listed with the inverter?

The Doctor: I listed a battery charger in your system because I always recommend some kind of a back-up energy source. The system I designed for you will provide up to 100% of your energy needs for about eight months of the year. For the rest of the year, I recommend you use a gas or propane generator to top off your batteries once a week or so, unless you have a good wind or hydro source you haven't told me about. Most folks only have to run the generator for a few hours a week. I could design the system to accommodate 100% of your needs all year round, but then it would be much more expensive and greatly oversized during the spring, summer, and fall. It's also nice to have a generator there so you can run big appliances like power tools every once in a while.

Ellen: Actually, we already have a 6000-watt generator. Can we just use that one?

The Doctor: That's perfect. All you have to do is connect it to your inverter. Whenever you turn on the generator, you'll be charging up your batteries. I've included a low battery alarm on your system, which will tell you if your batteries get low in the winter. That's when it's time to click on your generator for a few hours.

Something we've learned about off-the-grid systems is that the power system should be installed first, before any other construction starts. That way, you can build with quiet, free solar power most of the time, and fire up the generator only when you need to run a big power tool.

Ellen: So, you really think that this system will allow me to do all these things, without utility power?

The Doctor: There are tens of thousands of people in this country with systems very similar to yours. Such systems are very reliable — actually more reliable than utility power. I've heard lots of stories of the utility going down in a storm, and the off-the-gridders smiling all the way to the refrigerator. Even in periods of no sun, the battery storage keeps your system working and providing power.

Ellen: That sounds great. My husband and I have decided to go ahead and buy the land. I've been reading *The Independent Home* while my husband reads *Passive Solar Energy* and we're both getting excited about this move. I know we'll be calling you back soon to order this system. Thank you for all your help!

One month later, Ellen called up to order the system. Within a year after that, her passive solar home was built and the PV system installed. Ellen has been running all the appliances listed on her Planning Chart, and she's never had a power outage. Her husband's joke-writing business blossomed recently after he appeared on "America's Funniest Home Videos," and he's thinking about adding some more office space and additional computer equipment. The modular nature of this system will allow them to upgrade easily in the future.

— *Gary Beckwith*

PHOTOVOLTAICS

Photovoltaic modules are the devices that have made reliable power beyond the powerlines possible. PV is the most widely used alternative power source. Even those lucky enough to have a viable hydro or wind site often choose to have a few PV modules for back-up or seasonal use. A PV module produces electrical current when it is exposed to sunlight. The technology is closely related to and was a spinoff of 1950s transistor technology. PVs provide clean, uncomplicated power whenever and wherever the sun shines on them, so, not surprisingly, space applications funded most of the early research and development.

A Brief Technical Explanation

A single PV cell is a sandwich consisting of two very thin wafers generally made of very pure silicon. These wafers have been doped with elements that produce a surplus of electrons in one layer (called the n-layer) and a deficit of electrons in the other layer (called the p-layer). When bombarded by the photons in sunlight, some of these electrons are liberated and start to flow. Electricity results not from the heat of the sunlight, but from millions of these liberated electrons flowing away from the n-layer. Some of these wandering electrons make their way to metallic conductors on the silicon surface, then flow on through the electrical circuit. The PV cell acts like an electron pump. A single silicon cell produces just under half a volt, while amperage is dependent on cell size and efficiency. A module consists of many individual cells that are arranged and wired in series and parallel to provide the required voltage and amperage output. The module is encapsulated with tempered glass or some other transparent material on the front surface and some kind of protective and waterproof material on the back. The edges are sealed for weatherproofing, and there is often an aluminum frame holding everything together in a mountable unit. A junction box or wire leads providing electrical connections can usually be found on the module's backplane. Although truly weatherproof encapsulation was a problem with the early modules assembled 15 years ago, we have not seen any problems with glass-faced modules in years.

Construction Types

There are currently three commercial production technologies for PV cells.

Single Crystal: This is the oldest and most expensive technique, but it is still the most efficient sunlight conversion technology available. *Boules* (large cylindrical loafs) of pure single-crystal silicon are grown in an oven, then sliced into wafers, doped, and assembled. This is the same process used in manufacturing transistors and integrated circuits, and so is very well-developed, efficient, and clean. Silicon crystals are characteristically blue (because they absorb all other colors), and single crystalline cells look like deep blue glass.

Multicrystalline: In this technique, which is also called polycrystalline, less monolithic loaves are grown or cast, then sliced into wafers off a large block of multicrystalline silicon. It is slightly lower in conversion efficiency compared to single crystal, but the process is less exacting and so manufacturing costs are lower. Crystals are usually on the order of a centimeter (two-fifths of an inch) and can usually be seen in the cell's deep blueness.

Amorphous: The silicon material is vaporized and deposited on glass or stainless steel. This production technology costs less than the other methods, but the cells are also less efficient. Early production methods produced a product that faded up to 50% in output over the first few years before stabilizing. Present day technology claims to have dramatically reduced this fading problem. These cells are often almost black in color.

The industry has standardized on a module output which fits into a 12-volt DC regime. This means they usually produce 14 to 18 volts output, as the source voltage must be higher for battery charging. Twelve volts is a relatively safe voltage, and familiar as the standard voltage for automotive electrical systems. Dr. Doug advises, "While not impossible, it's pretty difficult to hurt yourself on such a low voltage. Keep your tongue out of the sockets and you will be okay." For more advice, be sure to read the chapter on batteries and safety starting on page 176. Batteries can be very dangerous if mishandled.

Multiple modules can be wired in parallel or series, just as their individual cells are, to achieve any desired output. The modular design of PV panels allows systems to grow and change as system needs change. Modules of different manufacture and age can be intermixed with no problems, so long as all modules have rated voltage output within about 1.0 volt of each other. Buy what you can afford now, then add to it in a few years when you can afford to expand.

Efficiency?

On average, the sun delivers 1000 watts (1 kilowatt) per square meter at noon on a clear day. This is defined as a "full sun" and is the benchmark by which modules are rated and compared. That is certainly a nice round figure, but it is not what most of us actually see. Dust, water vapor, air pollution, seasonal variations, altitude, and module temperature all affect how much power your modules actually receive. (For instance, the 1991 eruption of Mt. Pinatubo in the Philippines reduced available sunlight worldwide by 10% to 20% for a couple of years.) It is reasonable to assume that most sites will actually average about 85% of full sun, unless they are over 7,000 feet in elevation, in which case they may receive more than 100% of what is considered full sun.

PV modules do not convert 100% of the energy that strikes them into electricity. Current commercial technology averages about 13% efficiency for single- and multicrystalline cells, and 6% for amorphous cells. Much higher conversion rates are achievable in the laboratory by using experimental cells made with esoteric and rare elements, which are currently far too expensive to see commercial production.

How Long Do PV Modules Last?

PV modules last a long, long time. How long we honestly do not yet know, as the oldest terrestrial modules are barely 30 years old. Ask us again in 30 more years and we will have a better answer. All full-size modules carry 10- to 20-year warranties, reflecting their manufacturers' faith in the durability of these products. PV technology is closely related to transistor technology. Based on our experience with transistors, which just fade away after twenty years of constant use, most manufacturers have been confidently predicting 20-year or longer lifespans. However, keep in mind that PV modules are only seeing six to eight hours of active use per day, so we may find that lifespans of 60 to 80 years are normal. Cells that were put into the truly nasty environment of space in the early 1960s are still functioning perfectly.

Payback Time

In the early years of the PV industry there was a nasty rumor circulating that said PV modules would never produce as much power over their lifetimes as it took to manufacture them. During the very early years of development, when PV cells were being used exclusively for spacecraft, this was true. We often grumble that this rumor is perpetuated by the glow-in-the-dark nuke-loving types who have had a controlling grip on this country's energy policy for far too long. The truth is that PV modules pay back their manufacturing energy investment in 1.4 to 10 years time, depending on module type, installation climate, and other conditions. In fact, a preponderance of the embodied energy (energy required to extract and manufacture) is contained in the aluminum frame (on the order of 0.8 years for an average module).

Maintenance

It's almost laughable how easy maintenance is for PV modules. Having no moving parts makes them practically maintenance-free. Basically, you keep them clean. If it rains irregularly or if the birds visit often, you would be wise to wipe the modules down. Do not hose them off when they're hot, since uneven thermal shock could theoretically break the glass. Wash them in the morning or evening.

Control Systems

Conventional controls for PV systems are usually simple. When the battery reaches a full-charge voltage, the charging current can be directed elsewhere, or the circuit opened. Without a load, module voltage rises 5 to 10 volts and stabilizes harmlessly. When the battery voltage drops to a certain set-point, the charging circuit is closed and the modules go back to charging. Most controllers offer a few other whistles and bells like nighttime disconnect, LED indicator lights, etc. See the Controls and Monitors section in Chapter 5 for a complete discussion of controllers.

PV Performance in the Real World

Skeptics and pessimists knew it all along: PV modules could not possibly be all that simple. Even the most elegant technology is never perfect, so here are a few things to watch out for.

Wattage ratings on PV modules are given under ideal laboratory conditions at room temperatures. There are two factors which directly affect module performance out in the real world: percentage of full sun, and temperature.

Full Sun: As already discussed in the Efficiency section, most of us seldom see 100% of full sun conditions in our locations. If you are not getting full, bright, shadow-free sun, then your PV output will be reduced. If you are not getting bright enough sun to cast fairly sharp-edged shadows, then you do not have enough sun to harvest much useful electricity.

Temperature: The current from all modules fades somewhat at higher temperatures. This is not an important consideration until ambient temperatures climb above 80° F, which is not uncommon in the sun. The backs of modules should be as well-ventilated as possible. In very hot installations where surplus water is available, sprinklers are sometimes employed to cool the modules. On the positive side of this same issue, all modules increase output at colder temperatures, as in the wintertime when we need all the help we can get. We have seen cases when the modules were producing 30% to 40% over specs on a clear, cold winter morning

with a fresh reflective snow cover and hungry batteries. What this all boils down to is this: Derate your module output by 15% as a general rule of thumb. If you are designing a panel-direct system (where the modules are connected directly to the pump or fan without charge controllers or batteries), derate by 20%, or by 30% for direct applications in a hot climate.

Shadows: Even a tiny amount of shading dramatically affects module output. Electron flow is like water flow. It flows from high voltage to low voltage. Normally the module is high and the battery or load is lower. A shaded portion of the module drops to very low voltage. Electrons from other portions of the module and even from other modules in the array will find it easier to flow into the low-voltage shaded area than into the battery.

Module Mounting

Modules will catch the maximum sunlight, and therefore have the maximum output, when they are oriented perpendicular (at right angles) to the sun. This means that tracking the sun across the sky from east to west will give you more power output. Tracking mounts are expensive. Due to economies of scale, they're usually only worthwhile on larger PV systems, generally ones with eight or more modules. All systems are most productive if the modules are perpendicular (within five degrees) to the sun at noon, the most productive time of the day. In the winter, modules should be at the angle of your latitude plus 23 degrees, which is the *precessional angle* of the sun. In the summer, your latitude minus the same 23-degree angle is ideal. (On a practical level, many residential systems will have power to burn in the summer, and seasonal adjustment may be unnecessary.) As noted above, modules should have some air space behind them to promote air flow and better cooling.

Generally speaking, PV arrays that consist of eight or more modules are better off on a tracking mount, and smaller arrays are usually better placed on fixed mounts. This rule of thumb is far from ironclad, however, and there are good reasons to use either kind of mounting. See the PV Mounting/Trackers section for a more through discussion and a large selection of mounting technologies.

Photovoltaic Summary

Advantages	Disadvantages
1. No moving parts	1. High initial cost
2. Ultra-low maintenance	2. Only works in direct sunlight
3. Extremely long life	3. Sensitive to shading
4. Noncorrosive parts	4. Lowest output during shortest day length
5. Easy installation	
6. Modular design	5. Low-voltage output difficult to transmit
7. Universal application	
8. Safe low-voltage output	
9. Simple controls	

— Doug Pratt

SYSTEM DESIGN

The New Solar Electric Home

Joel Davidson. Gives you all the information you need to set up a first-time PV system, whether it's a remote site, grid connect, marine, or mobile, stand-alone, or auxiliary. Good photos, charts, graphics, and tables. Written by one of the pioneers. A good all-around book for getting started with alternative energy. 408 pages, paperback, 1990.

80-101   The New Solar Electric Home $19⁰⁰

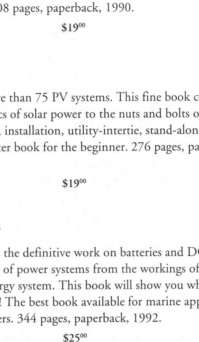

The Solar Electric House

Steven Strong. The author has designed more than 75 PV systems. This fine book covers all aspects of PV from the history and economics of solar power to the nuts and bolts of panels, balance of systems equipment, system sizing, installation, utility-intertie, stand-alone PV systems, and wiring instruction. A great starter book for the beginner. 276 pages, paperback, 1993.

80-800   The Solar Electric House $19⁰⁰

Living On 12 Volts With Ample Power

David Smead & Ruth Ishihara. This book is the definitive work on batteries and DC refrigeration systems. It thoroughly covers all aspects of power systems from the workings of solar panels to the optimization of a balanced energy system. This book will show you why your battery charger may be killing your batteries! The best book available for marine applications. Highly recommended for serious 12-volt users. 344 pages, paperback, 1992.

80-103   Living on 12 Volts with Ample Power $25⁰⁰

The Solar Electric Independent Home Book

Paul Jeffrey Fowler. A good, very basic primer for getting started with PV, written for the lay person. Lots of good charts, clear graphics, a solid glossary, and appendix. Includes 75 detailed CAD diagrams, and the up-to-date text includes recent changes in PV technology. This is one of the best all-round books on wiring your PV system. 174 pages, paperback, 1993.

80-102   The Solar Electric Independent Home Book $17⁰⁰

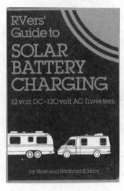

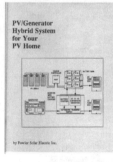

SYSTEM DESIGN

Practical Photovoltaics Electricity From Solar Cells

Richard J. Komp, Ph.D. This is the newly updated third edition of this classic, popular reference on solar electricity. Author/physicist Komp combines a thorough technical discussion of manufacturing technology with practical how-to advice for wiring, mounting, etc. Includes how PV cells and modules are manufactured, including new develoments and technologies. Also offers well-illustrated instructions for building your own module from bare cells. The easiest, most complete education on photovoltaics available. Paperback, 200 pp., 1995.

80-107   Practical Photovoltaics $19⁰⁰

RVers Guide To Solar Battery Charging

Noel and Barbara Kirkby. The authors have been RVing for over 20 years and have applied photovoltaics to their independent lifestyle. This book includes numerous example systems, illustrations, and easy-to-understand instructions. The *Whole Earth Catalog* calls it, "A finely detailed guide to installing PV systems in your motorhome, trailer, boat, or cabin." 176 pages, paperback, 1993.

80-105   RVers Guide to Solar Battery Charging $13⁰⁰

PV/Generator Hybrid System For Your PV Home

A great little basic primer and operating manual for the typical residential PV system with generator back-up. This booklet explains what each component is, and how it works in the system. It gives pointers on generator selection, battery sizing, and lots of diagrams detailing wiring options between the generator and inverter. However, it doesn't deal with the PV system. This booklet just explains how the generator and battery charger fit into the typical system, and gives a wealth of information on smart system operation. Paperback, 8-1/2" x 11", 1991, 25 pages.

80-337 PV/Generator Hybrid System for Your PV Home $8⁰⁰

The Solar Worksheet System Design Software

This custom system design software from Maximum RGB Ltd. is designed and produced entirely on alternative energy. It is based on the same worksheets we use here in the *Sourcebook* with the advantage of automatic computations and pull-down menues for common appliances and loads. Allows you to quickly and convieniently try out different appliances, batteries, and modules. Currently available for the Mac.

81-103 The Solar Worksheet for the Mac $15⁰⁰

The Solar Simulator System Design Software for Windows

This photovoltaic system simulator allows you to quickly and easily try such variables as location (within the continental US), season, tracking or fixed mounts, inverter sizing, and PV module or battery type and quantity. Weather data is based on actual seasonal averages for 38 cities scattered around the country. A simplified computer worksheet page will figure your electrical loads, or more precise figures from the printed worksheets here in the Sourcebook can be entered directly. Wire sizing is figured automatically based on distances you input, plus schematic diagrams and shopping lists with estimated total costs are generated and can be printed. A random generator simulates your designed system performance under variable weather and appliance load conditions. An extensive Help directory is included.

Runs on Microsoft Windows version 3.1 or later, requires 1.7MB of hard disk space, VGA monitor, available on 3.5" disc only.

81-050   Solar Simulator for Windows $30 ⁰⁰

WHO YA GONNA CALL?

Energy issues can be confusing. Fortunately, Real Goods has a quiver of qualified personnel who collectively have had hundreds of years of experience designing, living with, and teaching about energy efficiency and production. Our two dimensional figurehead, "Dr. Doug" (who is neither a real Doug nor a Doctor), represents our ideal of an energy technician who has the bumps and bruises of the school of hard knocks, and is willing to share the lessons and point out the speed bumps. We do have some real people with real names who might have the expertise you are looking for, so when the question "Who ya gonna call?" comes up, you might want to think about directing your inquiry to one of these folks:

Independent Home Energy Systems
Terry Hamor, **Bill Simmons** and **Steve Rogers** have helped hundreds of people realize the dream of living without the utility grid. Need a solar pumping system or want a small system for that Borneo expedition? Call them at **800/919-2400**.

Solar Hot Water
If you want to take advantage of the sun's abundant energy for pools, spas, domestic hot water, or space heating, you should call **Bob Ramlow** at **715-824-3982**. Bob has been installing solar thermal systems for 14 years and is your source for the best solar thermal information and products.

Water Quality Products   Everybody talks about the water but nobody does anything about it; well, almost nobody! "Our Water Guy" **Rodger Breslin** has been helping people get the right filter for years. Call Rodger at **800-762-7325 x 2229**.

Regional Retail Technicians
If you are fortunate enough to be in proximity to one of our Regional Retail Centers, call your local technician, or simply drop by, they'll help you realize your dreams... just a little closer to home. **Gary Beckwith** in Eugene, OR, **541-334-6960**; **Doug Livingston** in Hopland, CA, **707-744-2100**, or **B.J. Welling**, Amherst, WI, **715-824-3982**

Real Goods in Canada
For a current product list, technical advice, or to be added to the Real Goods Canadian mailing list contact: *Marc Dupuis and Gillian Browning* at Sunfire Energy **604-352-2001** or write to **Box 848, 5569 Taghum, Nelson, BC V1L 6A5**

Solar Architecture & Engineering
Steven Strong of Solar Design Associates (and author of The Solar Electric Home) has designed some of the most spectacular solar buildings in the world. From custom solar homes to entire photovoltaic powered communities, Steve and his team have become the leaders in creating beautiful, sustainable, solar powered buildings, and are the right architects to hire for very special projects. Solar Design Associates can be contacted through the Real Goods Renewable Energy Division at **707-468-9292 x 2210**.

Real Goods Power Systems Packages

We Can Design the Perfect Renewable Energy System for You

• Friendly, knowledgeable, large technical staff
• Systems designed for all needs and budgets
• Largest selection & inventory available
• Best customer service in the business
• Highly adaptable and expandable systems
• More reliable than utility power
• Often cheaper than utility power

The typical Fulltimer system takes only 2' x 5' inside your utility room.

The Weekender Power Kit

This modest kit for intermittent cabin use has all the major wiring and components pre-assembled. You simply mount the big pieces, (control board, PV module(s), batteries, & lamp) and connect the included wires. A Siemens M55 PV module with roof mounts, all controls, monitoring, safety equipment, wiring, 15-watt fluorescent lamp, and detailed instructions are included. Provides 12V DC power for lights and entertainment equipment. Kit will accept up to five PV modules for future expansion. An optional inverter to run AC appliances can be added at any time. Requires two golf-cart, or RV/Marine deep-cycle batteries, not included due to shipping costs. No substitutions with this kit. Optional pole mount recommended if you don't have a south-facing, shade-free roof surface. Provides an easy way to get started on PV with plenty of room to grow.

92-261	Weekender Power Kit	$895⁰⁰
13-401	Optional 2-mod Top-Pole Mount	$65⁰⁰

The Homesteader

Our smallest Power System Package is designed to handle all the lighting, entertainment, and small kitchen appliances for a modest, energy-conserving household of one to four people in a full-time home. With a basic Package of Powercenter, options, inverter, minimal PV system, mounting structure, and batteries, this System starts out at about $3,800.

Example: The Hellers, Big Island, Hawaii (the rainy side, no less!)

5	Quad-Lam PV module sets
1	Homebuilt mounting system
1	Trace 2012SB (an older model, superseded by the 2512)
1	APT-3 Powercenter w/options
6	Batteries

$3,900 Total System Package (approx.)

**Example:
The Campbells, North-central Washington, 35 mi. from the Canadian border. A simpler, pared-down Fulltimer system for this part-time home, that can easily be upgraded in the future.**

6	48-watt PV modules
1	8-module pole-top mount
1	Trace DR2424 inverter
8	Batteries

Charge Control, Safety Equipment $4,400 Current Cost to Date (Eventual cost, approximately $6500)

The Solar Sultan

Our largest Power System Package is usually designed around the state-of-the-art Trace 4,000-watt sine-wave inverter. This powerful inverter produces a clean, precise, multi-stepped approximation of a sine wave that can run any household appliance, plus problem loads, like sensitive audio equipment, variable-speed ceiling fans, or laser printers. This package can be designed for any size household, with a moderate degree of energy efficiency. With a basic Package of Powercenter, options, inverter, minimal PV system, mounting structure, and batteries, this System starts out at about $12,000.

Example: The Glanzs, northern-central Colorado, just east of the frontal range.

12	55-watt PV modules
1	Active tracker mount
1	750-watt wind turbine
1	Trace SW4024 inverter
1	APT Powercenter
1	Large Pacific Chloride battery bank

$13,000 Total System Package (approx.)

The Hellers are retired and built their system over several years as creature comfort demands and fixed incomes allowed. Lew says, "...the Electric company line goes right past our home. I wish I could get it removed. It's ugly!" Solar costs more up front, but the savings in the long run are tremendous, particularly in Hawaii, which has sky high electric rates. Hot water comes from a home-built solar heater, and the refrigerator is propane powered.

The Solar Fulltimer

Our middle-sized Power System Package can handle all lighting, entertainment, and kitchen appliance loads, plus heavy surge loads like power tools and washing machines. This Package is for moderately energy-conserving households of two or more people in a full-time home. With a basic Package of Powercenter, options, inverter, minimal PV system, mounting structure, and batteries, this System starts out at about $6,400.

The Campbells are using their new house part-time for now. Space is available on the pole-top mount for two more modules when they start living full-time in this beautiful house. All household electrical needs are run through the inverter. There are a propane powered refrigerator, stove, hot water heater, and generator. A solar hot water system is planned. They'll never pay a power bill again.

All Packages include the following Powercenter options:
- Lightning arrestor
- Vista 3 digital monitor
- Smartlite for remote monitoring
- Two 20-amp DC load breakers
- Appropriately sized battery & inverter cables
- 2" & 3/4" Romex connectors
- Spare main fuse

We have found these options to be almost universally necessary. Countless other options are available which your technician will recommend.

The Glanz's new passive solar, straw bale house features a state-of-the-art hybrid alternative energy that uses both PV and wind energy to supply all of their electrical needs for this stylish, almost-free-to-operate home.

Siemens Photovoltaic Modules

Siemens is a German multi-national company with photovoltaic production facilities in Vancouver, Washington and Camarillo, California. These are the former Arco modules which have been the industry bench- mark for years. Arco Solar was purchased by Siemens several years ago. Same American employees, same facilities, same high quality.

Unless noted otherwise all Siemens modules use the most efficient single crystal silicon, conventional low-iron tempered glass, EVA encapsulation, tedlar back sheets, anodized aluminum framing, and weather-tight junction boxes.

Siemens employs the least sensible module designation system of any manufacturer in the business. The numbers in the module designation have nothing to do with wattage output. So we have a situation where the M-75 module produces 48 watts, while the PC-4 module produces 75 watts. Con-fused? Good, you're not alone. (Siemens, are you listening?) The moral of this story is read the specifications before ordering the module.

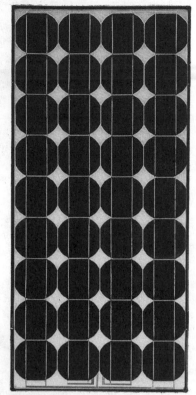

Siemens PC-4

Siemens PC-4

Siemens has created the ProCharger line of photovoltaic panels with much lower prices. There's no sacrifice in quality and reliability with these 75-watt single crystalline panels. The secret to bringing the price down is that the cells are larger, and they're grown in the same shape that they're used — something like an octagon. Less trimming, less waste, and much lower prices! Cells are wired 36 in series for higher voltage output. The JF module has a conventional anodized aluminum frame and junction box. It can be used with any system voltage up to 600 volts open circuit. Made in the USA.

Rated Watts: 75 watts @ 25°C
Rated Power: 17 volts, 4.4 amps
Open Circuit Volts: 22 volts @ 25°C
Short Circuit Amps: 4.8 amps @ 25°C
L x W x D: 47.3" x 20.8" x 2.5"
Construction: single crystal, tempered glass
Warranty: within 10% of rated output for ten years
Weight: 16.5 lb/7.5 kg

11-106   Siemens PC-4-JF 75-Watt $495⁰⁰

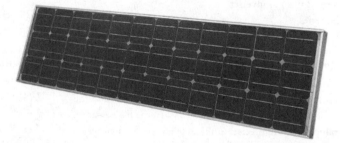

Siemens M75

The Siemens M75 is efficient, attractive, easy to install, and comes with a wired-in bypass diode in each junction cover. The M75 consists of 33 cells in series. Each module includes an easy-to-understand instruction manual. The M75 has been the industry standard battery charging module for many years. Not for hot climates (above 80°F). UL-Listed. Made in USA.

Rated Watts: 48 watts @ 25°C
Rated Power: 15.9 volts, 3.02 amps
Open Circuit Volts: 19.8 volts @ 25°C
Short Circuit Amps: 3.4 amps @ 25°C
L x W x D: 48" x 13" x 1.4"
Construction: single crystal, tempered glass
Warranty: within 10% of rated output for ten years
Weight: 11.6 lb/5.2 kg

11-101   Siemens M75 $349⁰⁰

Siemens M55

The M55 is a higher voltage standard module, consisting of 36 cells in series. It is ideal for water pumping applications, hotter climates, or rooftop RV applications where higher voltage is required. It is the best module to use in extremely hot climates as high-temperature voltage drop is kept tolerable. UL-listed. Made in USA.

Rated Watts: 55 watts @ 25°C
Rated Power: 17.4 volts, 3.15 amps
Open Circuit Volts: 21.7 volts @ 25°C
Short Circuit Amps: 3.4 amps @ 25°C
L x W x D: 50.9" x 13" x 1.4"
Construction: single crystal, tempered glass
Warranty: within 10% of rated output for ten years
Weight: 12.6 lb/5.7 kg

11-105   Siemens M55 $379⁰⁰

Siemens M55

Siemens M65

The M65 is designed primarily for RV, marine, or remote home usage, when only one module is employed. The M65 contains 30 cells wired in series. It is a self-regulating module that decreases its current output from 3 amps to less than 0.25 amp when the battery approaches full charge, eliminating the need for a charge controller. Battery must be 83 amp-hours or larger or a charge controller will be necessary. Not recommended for hot climates where temperatures exceed 90° F, or for multi-module systems. UL-listed. Made in USA.

Rated Watts: 43 watts @ 25°C
Rated Power: 14.6 volts, 2.95 amps
Open Circuit Volts: 18 volts @ 25°C
Short Circuit Amps: 3.3 amps @ 25°C
L x W x D: 42.6" x 13" x 1.4"
Construction: single crystal, tempered glass
Warranty: within 10% of rated output for ten years
Weight: 10.5 lb/4.8 kg

11-103   Siemens M65 $399⁰⁰

Siemens M20

The M20 is a compact, self-regulating module ideal for RVs, boats, and remote homes where needs are minimal, use is intermittent, or space is limited. As the battery approaches full charge, the M20 decreases the current output from 1.37 amps to less than 0.25 amp, eliminating the need for a charge controller. Siemens recommends at least 70 amp-hours of battery storage for each M20 module. Not recommended for hot climates. Made in USA.

Rated Watts: 22 watts @ 25°C
Rated Power: 14.5 volts, 1.38 amps
Open Circuit Volts: 18.0 volts @ 25°C
Short Circuit Amps: 1.60 amps @ 25°C
L x W x D: 22.4" x 13" x 1.4"
Construction: single crystal, tempered glass
Warranty: within 10% of rated output for ten years
Weight: 5.6 lb/2.5 kg

11-107   Siemens M20 $285⁰⁰

Siemens M20

Solarex Photovoltaic Modules

Solarex is an American owned company with photovoltaic assisted production facilities in Maryland. The Solarex line offers exceptional performance in high temperature climates, as all the modules offer 16.8 volts or higher output. Solarex uses polycrystalline cells coated with a titanium dioxide anti-reflective material, low-iron tempered glass, EVA encapsulation, tedlar back sheets, bronzed aluminum framing, and large weather-tight junction boxes for all their modules, unless noted otherwise.

MSX-60

Solarex MSX-60

The MSX-60 is Solarex's production solar panel. Now with a 20-year warranty on the 60-watt modules! Made in USA.

Rated Watts: 60 watts @ 25°C
Rated Power: 17.1 volts, 3.5 amps
Open Circuit Volts: 21.1 volts @ 25°C
Short Circuit Amps: 3.8 amps @ 25°C
L x W x D: 43.63" x 19.75" x 1.97"
Construction: polycrystalline, tempered glass
Warranty: within 10% of rated output for 20 years
Weight: 15.9 lb/7.2 kg

11-501   Solarex MSX-60 $425⁰⁰

Solarex SA-5

The Solarex SA-5 is nearly identical to the old Arco G-100. This is an amorphous silicon module with a lexan frame. It is attractively priced and ideal for small applications and science projects. Made in USA.

Rated Watts: 5 watts @ 25°C
Rated Power: 15 volts, 300 milliamps
Open Circuit Volts: 19 volts @ 25°C
Short Circuit Amps: 330 milliamps @ 25°C
L x W x D: 13 5/8" x 13 5/8" x .89"
Construction: amorphous silicon, lexan frame
Warranty: ten years
Weight: 2.4 lb/1.1 kg

11-521   Solarex SA-5 $89⁰⁰

SA-5

Solarex SX-30 Unbreakable Lite

This module is a good choice for sailboats or scientific expeditions. Will bend to mount on boat cabin tops, can be stepped on or have things dropped on it without damage.

Rated Watts: 30 watts @ 25°C
Rated Power: 17.8 volts, 1.68 amps
Open Circuit Volts: 21.3 volts @ 25°C
Short Circuit Amps: 1.82 amps @ 25°C
L x W x D: 24.25" x 19.5" x 0.38"
Construction: polycrystalline, tedlar coating
Warranty: one year
Weight: 4.84 lb/2.2 kg

11-511   Solarex SX-30 $299⁰⁰

Solarex SX-20 Unbreakable Lite

The 20-watt module in this series is a good choice for recharging video camera batteries, some small water pumps, or heavy laptop computer users.

Rated Watts: 18.5 watts @ 25°C
Rated Power: 17.8 volts, 1.12-amps
Open Circuit Volts: 21.0 volts @ 25°C
Short Circuit Amps: 1.16 amps @ 25°C
L x W x D: 17.5" x 19.5" x 0.38"
Construction: polycrystalline, tedlar coating
Warranty: one year
Weight: 3.28 lb/1.49 kg

11-512   Solarex SX-20 $269⁰⁰

Solarex SX-10 Unbreakable Lite

The 10-watt module is an excellent choice for backpackers, charging laptop computers, or recharging nicad batteries.

Rated Watts: 10 watts @ 25°C
Rated Power: 17.5 volts, 0.57 amps
Open Circuit Volts: 21.0 volts @ 25°C
Short Circuit Amps: 0.6 amps @ 25°C
L x W x D: 17.5" x 10.5" x 0.38"
Construction: polycrystalline, tedlar coating
Warranty: one year
Weight: 1.8 lb/0.82 kg

11-513   Solarex SX-10 $139⁰⁰

The Solarex Lite Series

The Lite series of Solarex modules are made without glass and are very thin, lightweight, and highly portable. Typically they're used for boating, camping, RVs, scientific expeditions, or mobile communications. This series of panels is great for recharging laptop computers and video cameras in remote locations. Consists of a layer of aluminum bonded to fiberglass mat with the silicon cells on top. Environmentally sealed with Tedlar (a non-yellowing plastic glazing material). These modules will flex slightly, but are extremely tough. Has 3-meter lead wire attached. Made in the USA.

Solarex Lite

See page 110 for more information on running laptop computers and other appliances in remote locations.

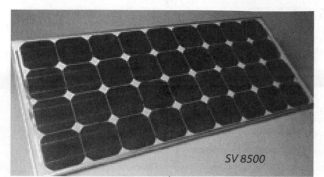

SV 8500

Solavolt Photovoltaic Modules

Solavolt is a joint U.S./ Japanese endeavor. The bare PV laminates are manufactured by Sharp in Japan, one of the oldest, most experienced PV manufacturers in the world. Laminates are tested, rated, framed, and have junction boxes installed by Photocomm in Arizona. Solavolt modules are top-quality with single-crystal silicon, low-iron tempered glass covers, tedlar backsheets, anodized aluminum frames, and roomy weather-tight junction boxes. They are the first single-crystal modules with partial shade tolerance.

Solavolt SV8500 PV Modules
The Highest Efficiency, Ever for Production Modules, Plus Shade – Tolerance!

The Solavolt single – crystal PV cells post a record – breaking 16.5% efficiency for production cells. We've seen higher efficiencies in the laboratory using exotic materials, but never in production modules before. Higher efficiency means more of the available sunlight is turned into electricity; thus you get more power from a smaller, and less expensive, module.

Each individual cell (the dark round or square bits inside the module) is manufactured with power bypass ability. This prevents massive power drop and spot overheating when a portion of the module is shaded. This is another first for single crystal modules. Previously, a small fist sized shade patch would put the entire module out of production. The Solavolt modules can tolerate small amounts of shade, with only a proportionate power drop.

Has a full 10 year warranty, low iron tempered glass, EVA encapsulation, anodized aluminum frames, large weather tight junction boxes, and tedlar back sheets. Standard output voltage modules; can be intermixed with other similar voltage modules.

> *Rated Watts: 85.5 watts @ 25° C*
> *Rated Power: 17.4 volts, 4.91 amps*
> *Open Circuit Volts: 22.0 volts @ 25° C*
> *Short Circuit Amps: 5.5 amps @ 25° C*
> *L x W x D: 47.2" x 20.9" x 1.4"*
> *Weight: 18.7 lb/8.5 kg*

11–151 Solavolt SV8500 $539⁰⁰

Single Interconnect Cables

These ready-made 30" single-wire 10-gauge cables can be used for interconnection on all modules (the Siemens M-series modules require snipping off the forked connectors). Wire is USE-2 rated and is sunlight resistant. 1/4" forked connectors are crimped on both ends.

11-156 Single Interconnect Cable $4⁰⁰

Series/Parallel Interconnect Cables

These ready-made 30" 10-gauge cables have three color-coded wires, and can be used for either series or parallel interconnects with all modules (the Siemens M-series modules require snipping off the forked connectors). Cable jacket is sunlight resistant, wires are class-THHN. 1/4" forked connectors are crimped on both ends of each wire.

11-157 Series/Parallel Interconnect Cable $6⁰⁰

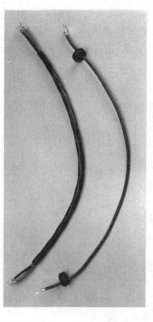

Duravolt 5000
Vandal-resistant, shade-tolerant, and fade-proof at high temperatures

The DV5000 is an excellent choice for RVs or vandal-prone sites. Rocks bounce harmlessly off this 47-watt unbreakable module, even bullets only reduce output by the percentage of surface area removed. DuPont's made-for-PV Tefzel® glazing, instead of glass, makes them incredibly tough. Built-in bypass diodes between individual cells make them able to continue useful charging during partial shade. Amorphous silicon construction makes them fade-proof at high temperatures (like on the roof of an RV). You can't find a tougher, harder working, more durable photovoltaic module at a more reasonable price. Not strongly recommended for large power arrays or tracking mounts, where the large size of this module is a liability. Made in USA.

Duravolt 5000

> *Rated Watts: 47 watts @ 25°C*
>
> *Rated Power: 16.8 volts, 2.8 amps*
>
> *Open Circuit Volts: 23.7 volts @ 25°C*
>
> *Short Circuit Amps: 3.6 amps @ 25°C*
>
> *L x W x D: 52.0" x 27.0" x 1.5"*
>
> *Construction: amorphous, Tefzel® glazing*
>
> *Warranty: within 10% of rated output for ten years*
>
> *Weight: 21.0 lb./9.5 kg*

11–149   Duravolt 5000 $299⁰⁰

Kyocera G51

The G51 is Kyocera's 36-series-connected-cells standard building block module. It is compatible with any other 36-cell modules.

Rated Watts: 51.0 watts @ 25° C

Rated Power: 16.9 volts, 3.02 amps

Open Circuit Volts: 21.2 volts @ 25° C

Short Circuit Amps: 3.25 amps @ 25° C

L x W x D: 38.9" x 17.6" x 2.0"

Construction: polycrystalline, tempered glass

Warranty: within 10% of rated output for 12 years

Weight: 13.0 lb/5.9 kg

11-201   Kyocera G51 $339⁰⁰

Kyocera G51

Kyocera Photovoltaic Modules
Kyocera is a Japanese company with all manufacturing facilities in Japan. They make an excellent polycrystalline module with the industry's tightest range of allowed performance deviation (3%), and an outstanding 12, year warranty. Kyocera modules use low-iron tempered glass covers, EVA encapsulation, aluminum foil back sheets, anodized aluminum frames, and weather-tight junction boxes.

Uni-Solar UPM-880

This Uni-Solar module uses a rigid frame. No glass glazing and a flexible base material make this module close to vandal-proof. The black anodized frame allows the module to withstand the severe twists that can occur with earthquakes or high winds without being damaged, yet provides the structural rigidity for mounting on a tracker or to roof or pole mounts. Made in the USA. Ten-year warranty.

> *Rated Watts: 22 watts @ 25° C*
>
> *Rated Power: 15.6 volts, 1.4 amps*
>
> *Open Circuit Volts: 22 volts @ 25° C*
>
> *Short Circuit Amps: 1.4 amps @ 25° C*
>
> *L x W x D: 47" x 13.5" x 1.4"*
>
> *Construction: amorphous, Tefzel glazing*
>
> *Warranty: within 10% of rated output for ten years*
>
> *Weight: 8 lb/3.6 kg*

UPM-880

11-223   Uni-Solar UPM-880 $199⁰⁰

Thanks to a benevolent arrangement of things, the greater part of life is sunshine.

—Thomas Jefferson

Uni-Solar MBC-525

The MBC series are flexible modules with nickel-plated brass grommets at the corners. These lightweight modules even float! While not "foldable" they may be rolled up to a 6-inch diameter for maximum packability without damage. Good for marine applications. Made in the USA.

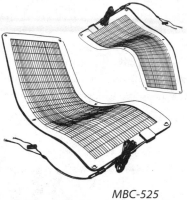

MBC-525

> *Rated Watts: 22 watts @ 25°C*
> *Rated Power: 15.6 volts, 1.4 amps*
> *Open Circuit Volts: 22 volts @ 25°C*
> *Short Circuit Amps: 1.8 amps @ 25°C*
> *L x W x D: 50.5" x 15.5" x 0.75"*
> *Construction: amorphous, Tefzel glazing*
> *Warranty: within 10% of rated output for three years*
> *Weight: 4 lb/1.8 kg*

11-220   Uni-Solar MBC-525 $289⁰⁰

Uni-Solar MBC-262

A smaller version of marine or backpacking charger. Good for battery maintenance or laptop computers. Made in USA.

> *Rated Watts: 11 watts @ 25°C*
> *Rated Power: 15.6 volts, 0.7 amps*
> *Open Circuit Volts: 22 volts @ 25°C*
> *Short Circuit Amps: 0.85 amps @ 25°C*
> *L x W x D: 27.3" x 15.5" x 0.25"*
> *Construction: amorphous, Tefzel glazing*
> *Warranty: within 10% of rated output for three years*
> *Weight: 2.5 lb/1.14 kg*

11-221   Uni-Solar MBC-262 $169⁰⁰

Uni-Solar MBC-131

An outstanding battery trickle charger unit for boats, RVs, planes, camping lanterns, or any application that requires battery charging. Made in USA.

> *Rated Watts: 5.5 watts @ 25°C*
> *Rated Power: 15.6 volts, 0.35 amps*
> *Open Circuit Volts: 22 volts @ 25°C*
> *Short Circuit Amps: 0.45 amps @ 25°C*
> *L x W x D: 27.3" x 7.8" x 0.25"*
> *Construction: amorphous, Tefzel glazing*
> *Warranty: within 10% of rated output for three years*
> *Weight: 1.5 lb/0.68 kg*

11-222   Uni-Solar MBC-131 $109⁰⁰

PV BATTERY CHARGERS

Charge Your Camcorder With The Sun

Use this 5.4-watt folding panel to charge 6 to 9-volt video and camcorder batteries. It includes a charging tray designed for most handycam-type batteries and a disconnect cable with an adapter that fits many electronic devices. Weighing only 22 oz. this solar charger also comes with Velcro® mountings and harness to attach it to your pack for easy, on-the-trail charging. Includes a rugged nylon stuff sack. For use with nicad batteries only. (Not applicable for Canon.)

> *Specifications: Power 5.4W, Voltage 9.66V, Current 561mA*
> *Dimensions: Charging unit measures 5-1/2" x 7-1/2" x 3/4"*
> *Solar panels each measure 4-1/2" x 6"*
>
> *Typical Charge Times (6V, 1200mAh): 2 hrs. 10 min.*

11-112 Camcorder Charger $269⁰⁰

Car Battery Maintainer

Our Solar Car Maintainer is a self-contained, self-regulated, portable battery charger. When exposed to the sun, it provides a continuous trickle charge to your battery to deliver optimal power and performance. Batteries typically leak or discharge power at a rate of approximately 10% per month. Parking a car for a long period always carries the risk of finding a dead battery when you return. Use our Solar Car Maintainer to keep your batteries fully charged. Place it on your dashboard and plug it into your cigarette lighter. If your car's cigarette lighter doesn't work when the ignition is off, instructions and an adapter are included to connect it directly to your fuse box. It comes with a 4-foot cord. Size:12-3/4" L x 3-1/4" W x 1/2" D. Produces 40 milliamps at 16.5 volts.

11-117   Car Battery Maintainer $39⁰⁰

Improved Solar Super Charger

Our new high-powered Solar Super Charger will charge any two nicad batteries at a time. The holder fits AAA, AA, C, or D sizes. The improved single-crystal solar module measures 3 1/2" x 10", is waterproof, nearly unbreakable, and has a 16" wire lead. Output is 4.5 volts at 250 mA. Will easily charge a pair of our high-capacity AA nicads in 6 hours or less.

50-212 Improved Solar Super Charger $25⁰⁰

Clip-On Solar Charger

Employ this 50-milliamp 16-volt solar module to keep your boat, RV, tractor, golf cart, or motorcycle battery topped off when the vehicle is standing unused. No more unpleasant surprises, no spoiled outings or missed appointments when the battery proves too weak to start the engine. You can also use it to run a radio, 12-volt fan, etc. The 6.3 by 6.9 by 0.7-inch module is water- weather- and rust-proof, corrosion-resistant, and extremely strong. The stand allows the panel to be adjusted to the proper angle to receive maximum sunlight. Comes with connecting cable and large alligator clips.

11-119   Solar Charger $29⁰⁰

PV POWER KITS

The beauty of solar electric power systems lies in their modularity. Solar panel installations are like building block constructions that can always be added to and enhanced in the future. Unlike most consumer items that have obsolescence built in, you can use your solar panels virtually forever and add on as power needs and budget capabilities expand.

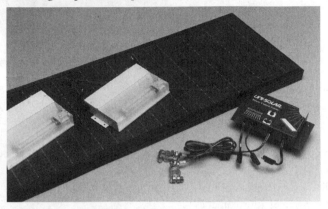

Simple 22-Watt Solar Starter System

The UniKit-2 gives you a simple solar starter system. It includes a 22-watt PV module, two 9-watt compact fluorescent light fixtures, 60 feet of plug-together wiring, and a complete Power Control Center. Just add a 60 amp-hour or larger deep-cycle battery and you're in the lighting business! The PV module is an unbreakable 22-watt Uni-Solar UPM-880. The lights are switched with pull chains. The Power Control Center protects your battery from overcharge and deep discharge, and LED's indicate PV module operation, battery conditions and low-voltage disconnect. The Power Control Center is expandable to 140 watts (up to six UPM-880 PV modules) and has a spare pin jack and an automotive connector for wiring further 12-volt accessories, such as cassette players, radios, TVs, or compressors. The wiring kit contains 25 feet of wire to run from the PV module to the Power Control Center, 10 feet of wire for each light, plus 15 feet of spare system wire. This kit is ideal if you need just a couple of lights now and plan to expand later. Made in the USA.

11-219	UniKit-2 Power System	$349⁰⁰

The Weekender Power Kit

This is as simple as we can make it! The Weekender is a completely prewired kit which needs only mounting and connection to the PV panel and the battery on site. All the fusing, safety disconnect, and charge control components are pre-mounted and wired together before shipping. You don't have to become a PV expert to put this kit into operation. Complete and well-illustrated installation instructions are included. This is a simple and cost-effective way to provide electrical power for any small outbuilding, studio, or remote cabin. The Weekender Power Kit comes complete with a full-size Siemens M55 55-watt solar module, a Smartcharger 20-amp controller with graphic metering fused safety disconnect, fuse panel, 12-volt outlet, 15-watt fluorescent lamp, and all necessary mounting, wiring, and complete installation instructions. This photovoltaic-power kit is designed to provide power for general room lighting, desk lighting, fans, other 12-volt equipment; or, with the appropriate inverter, (optional) you can power stereos, computers, printers, TVs, and other small 120-volt appliances. There are five extra slots in the fuse panel for future expansion. Electrical production is approximately 240 watt-hours per day with five hours of direct sun. The power is stored in batteries for use at night or on weekends. You can expand the photovoltaic array on this system with up to a total of five modules, for a daily production of 1250 watt-hours per day.

The Weekender Power Kit is supplied with a single module and rooftop mounts. It is shipped without batteries as the cost of shipping batteries is prohibitive. A pair of common electric golf-cart batteries are recommended. These will cost $60 to $90 each. A pole top mounting option is available if the building will be in the shade or not oriented south. Kit is complete, no substitutions are available.

92-261	Weekender Power Kit	$895⁰⁰
13-401	2-Panel Top-of-Pole Mount	$65⁰⁰

A Mini Solar Power Station
For Solar-Powered Fountains and More...

We designed our Solar Power Station to run a backyard fountain, but then realized it could do many other small jobs. Such as energy efficient yard lights, recharging laptop computers, or any 120-volt AC appliance up to 50-watts. The Solar Power Station makes power from direct sunlight with a photovoltaic panel, and stores it in a maintenance-free sealed battery. The 50-watt inverter converts the stored power to conventional 120-volt AC as needed. A low voltage disconnect prolongs battery life by automatically shutting the inverter off when power falls too low, and keeps it off until the battery is partially replenished. High and Low indicator lights show the battery state of charge. You can manually control use with the on/off switch to run your fountain (or other appliance) when *you* choose, or the low voltage disconnect will decide when to run it for you. Will run our 10-watt Fountain Pump for approximately three hours when fully charged, or about two hours daily with full sun. All stainless steel mounting tilts or swivels in any direction and will screw down to a 4x4 post, or a 2x4 railing. PV panel output is approximately 5-watts, a built-in charge controller prevents overcharging. Battery capacity is 7-amphours, has a two to three year life expectancy, and is easily replaceable. Size is 18.5"L x 12.5"W x 3.5"D. Weighs approximately 12.5 lbs. One-year mfg. warranty.

27-124   **Solar Power Station** $239⁰⁰

Highly recommended for use with our 4 and 10-watt Fountain Pumps, see page 299.

Detail of back

TAKE YOUR COMPUTER ANYWHERE!

Laptop and notebook technology enables us to go anywhere and bring our computer along. However, most of these portable computers have battery packs that allow you to use the computer for a maximum of a couple hours. This puts a limitation on these otherwise extremely versatile computers. Here's a solution: use our Solarex Unbreakable Module to generate electricity that is stored in the Power Pack. You can then convert from 12-volt DC to 120-volt AC with a Powerstar inverter. Think a bit about what you would like to do with your computer and give us a call. We'll make sure you get the right system. Here's a checklist of information that is good to know before you call:

- ❐ Type of computer and its power consumption
- ❐ Find out if 12-volt DC charger is available from either another manufacturer or another supplier
- ❐ Weather conditions where you are going and your usage patterns while there
- ❐ Have your power cord available when you call, if possible.

Solar Charging For Your Laptop Computer
Mercury Solar System

For eco road warriors who can't be tied down to battery chargers and AC plugs, the SunRunner Universal offers laptop battery charging anyplace the sun shines. Simply place the hi-output single-crystal panels in sunlight and plug the six-foot cord into your laptop's power port. Solar power will either assist the internal battery (if the computer is running), or recharge the battery (if the computer is off or asleep). The panels fold into the size of a standard 3-ring binder (9"x 12-1/2"). With a stylish black nylon cordova cover, it weighs only 40 oz. The panels have heavy-gauge aluminum backing for lifelong durability on the road. Output can be up to 26 volts or 1.5 amps depending on computer model. An adapter is required. Please call for the list of currently supported models. One year mfg. warranty.

To ensure you receive the correct adaptor we need your laptop brand and model #, plus a daytime phone #. Please have this information ready when you call.

• 11-115 Mercury System $489⁰⁰

Shipped from manufacturer

*KIS Mercury Solar System
Note: Nice picture, but for best performance don't allow partial shading on your module like our photographer did here.*

—Dr. Doug

Neptune Solar Power For Your PowerBook

The unbreakable, flexible Neptune is built specifically for use with Apple's PowerBook 100-series portable computers. With good sunlight you can run your PowerBook all day without having to plug into AC power. Or, you can recharge your PowerBook battery by leaving the panel in the sun and simply plugging into the PowerBook. Neptune output is about half the AC adapter output. The weatherproof nylon cordova-bound 1.75 lb. panel folds in half to fit inside a briefcase or computer carrying case. The Neptune has nylon carrying straps and two outside pockets, one for the 6' power cord and a larger one for loose disks or papers. Folded size is 8.5" by 15.5". It comes with a well-written Owner's Manual detailing use, battery conservation measures, and free software that tracks your battery performance and gives graphic indicators of battery charge. *Not for use with the 500-series PowerBook; use the Mercury model for this higher-powered laptop.*

11-123 Neptune $289⁰⁰

PV MOUNTING AND TRACKING STRUCTURES

PV modules are usually supplied with frames, but some kind of mounting structure is required to hold several modules together or to get the modules off the roof and allow some cooling air behind them. PV modules produce the most power when perpendicular (at right angles) to the sun. The perfect mounting structure would follow the sun across the sky every day. Tracking mounts do this, but their mechanical complexity drives the cost up significantly. For smaller systems it's usually more cost-effective to use fixed mounts. In the following pages you'll find numerous mounting structures, each with its own particular place in an independent energy system. We'll try to explain the advantages and disadvantages of each style to help you decide if a particular mount belongs in your system.

In order of complexity, your choices are:

- RV Mounts
- Home-Built Mounts
- Fixed Mounts (with or without adjustable tilt)
- Top-of-Pole Fixed Mounts
- Passive Tracker
- Active Tracker

RV Mounts

Because of wind resistance and never knowing which direction the RV will be facing next, most RV owners simply attach the modules flat on the roof. RV mounts raise the module an inch or two off the roof for cooling. They can be used for simple one- or two-module home systems as well. Simple and cheap, most of them are made of aluminum for corrosion resistance. Disadvantages are their fixed tilt and orientation.

Home-Built Mounts

Want to do-it-yourself? No problem. Anodized aluminum is the preferred material due to its corrosion resistance, but mild steel can be used just as well, so long as you're willing to touch up the paint occasionally. Sometimes slotted steel angle stock is available in galvanized form. Wood is not recommended because PV modules can last for over 40 years, and even treated wood won't hold up well when exposed to the weather for that long. Make sure that no mounting parts will ever cast shadows on the modules. Make them adjustable-tilt if possible.

Fixed Mounts

This is probably the most common mounting structure style we sell. The Ground or Roof Mounts, or the Side-of-Pole Mounts from Zomeworks are examples. The ground/roof racks have telescoping back legs for seasonal tilt and variable roof pitches. This mounting style can be used for roof, ground, or even flipped over and used on south-facing walls. Use concrete footings for ground

mounts. The racks are designed to withstand wind speeds of up to 120 mph. They don't track the sun though, and getting snow off of them is sometimes troublesome. Ground mounting leaves the modules vulnerable to grass growing up in front, or to rocks kicked up by mowers. Roof or wall mounting is preferred.

Top-of-Pole Fixed Mounts

Made by Zomeworks and designed to withstand winds up to 120 mph, this mounting style is the best choice for snowy climates. With nothing underneath it, snow tends to slide right off. For small or remote systems, pole-top mounts are the least expensive and simplest choice. We almost always use these for one- or two-module pumping systems. Tilt and direction can be easily adjusted; in fact you can track the sun manually with this mounting style. The pole is standard iron pipe and is not included (pick it up locally to save on freight). Make sure that your pole is tall enough to give sufficient burial depth (see the charts in this chapter) and still clear livestock, snow, weeds, etc. Ten feet is usually sufficient. Taller poles are sometimes used for theft deterrence. Pole size depends on the specific mount (check the chart in this chapter).

Passive Trackers

Made by Zomeworks, the Track Rack follows the sun from east to west using just the heat of the sun for energy. No extra source of electricity is needed at all; a simple, effective, and brilliant design solution. The north-south axis is seasonally adjustable manually. Maintenance consists of two squirts with a grease gun once every year. Tracking will help substantially in the summer and somewhat in the winter. The two major problems with this technology are wind disturbances, and slow "wake-up" when cold. The tracker will go to "sleep" facing west. On a cold morning it may take more than an hour for the tracker to warm up and roll over toward the east. In winds over 25 mph, the passive tracker will probably be blown off course. This tracker can withstand winds of up to 85 mph (provided you follow the manufacturer's recommendations for burying the pipe mount), but will not track at high wind speeds.

Active Trackers

The Wattsun Tracker uses linear actuators like those on satellite TV dishes to track the sun. A small controller at the top of the array is programmed to keep equal illumination on the photocells at the base of an obelisk. Power comes from tapping off a single PV module or from the optional 3-watt driver module. Power use averages a minuscule 0.5 watt per hour. There is a small nicad battery pack in the control box which drives the array back to the east when no power is sensed from the PV driver panel, or about half an hour after sunset. Wattsun Trackers are available in single-axis, east-west tracking with manual north-south adjustment, or optionally in dual-axis, which automatically tracks both east-west and north-south. Active trackers average 10% to 15% more collection per day than a passive tracker in the same location. A disadvantage of this mounting type is its initial cost. Due to fixed costs for controllers and linear actuators, Wattsun Trackers are more economical for larger PV arrays. Other disadvantages are reliance on electronic gizmos and mechanical parts. Controller and actuator failures are becoming rarer as Wattsun continues to upgrade the quality, but they still do happen occasionally.

Special note for anyone living south of the equator: Wattsun Trackers for the Southern Hemisphere require special ordering due to mirror-image operation!

To Track or Not to Track, That Is the Question

Photovoltaic modules produce the most energy when situated perpendicular to the sun. A tracker is a mounting device that follows the sun and keeps the modules in the optimum position for maximum power output. At the right time of year, and in the right location, tracking can increase daily output by as much as 50%. But beware of the qualifiers: not every site or every system is a good candidate for tracking mounts.

Tracking is an option only if there is clear access to the sun from early in the morning until late in the afternoon. A solar window from 9 A.M. to 4 P.M. is workable; if you have greater access, more power to you (literally). Tracking will add the most power in summertime when the sun is making a high arc overhead. Trackers will add 35% to 50% to your incoming power during the summer. In wintertime tracking will only add 10% to 20%, depending on your site and latitude. If you live north of 45° latitude, winter tracking isn't going to help much; the sun makes such a low arc across the sky that fixed mounts could catch it just as well. In general, the closer to the equator you are, the more sense wintertime tracking makes.

Tracking mounts are expensive. If you've got inexpensive modules, like used Quad-Lams, you're better off just buying more modules and putting them on fixed mounts. Economies of scale make trackers less costly per module for larger systems. Trackers are typically used on small systems dedicated to water pumping. (It's the only way to get more gallons per day out of a pumping system.) Residential systems usually need to be larger than six modules to make tracking cost-effective.

Tracking is most effective at temperate latitudes if your power needs peak in the summer. Pumping and cooling applications are the most common uses. Residential power needs, on the other hand, usually peak in the winter, in which case tracking is only effective on systems of six modules or larger. In tropical latitudes, below 30°, tracking makes sense for almost any system of four modules or larger.

— *Doug Pratt*

MOUNTING

Wattsun Active Trackers

Tracking mounts are most beneficial for larger arrays, or applications with heavy summertime power use, such as pumping or air movement. The closer you are to the equator, the more year-round benefit you will receive from tracking mounts.

Wattsun has upgraded all their mid-sized trackers to a new style actuation design. This is an improvement from the old style "tilt and roll" to a new stronger style of azimuth tracker that rotates around the pole mount first, and then tilts (if the Dual Axis option is used) for elevation position. This tracking style requires less ground clearance, as the array is always parallel to the ground or roof-top. There are no large array corners pointing down toward the ground, or up into the air to catch the wind. This new azimuth tracker provides a full 180^0 of east to west rotation, and up to 80^0 of tilt. The tilt angle is adjusted automatically with the Dual Axis option, and is strongly recommended for all applications. The tilt angle on the standard single axis tracker is manually adjustable.

The controller will accept power directly from the PV array, up to 55 volts. Approximate power use for the tracker is a minscule 12 watt-hours per day. Outputs to the tracking motors are short circuit protected. The electronics are completely potted, waterproof, and over-temperature protected.

Trackers, depending on model, mount on 3", 4", or 6" schedule 40 steel pipe (not included). They can withstand winds up to 100 mph. Because they break down into smaller component parts, most Wattsuns can be shipped by UPS. Stainless steel hardware is available for severe marine environments at additional cost. See the Tracker/Pole chart at the end of this product section for recomendations on pipe size and hole depth. Mfg. warranty is ten years.

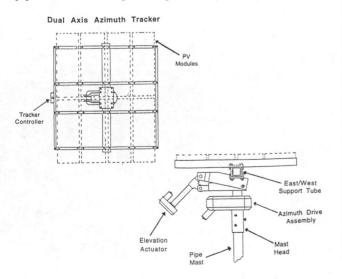

Dual Axis Azimuth Tracker

PV Modules

Tracker Controller

East/West Support Tube

Azimuth Drive Assembly

Elevation Actuator

Pipe Mast

Mast Head

•13-250	Wattsun TR 4 Siemens M75, M55	$745 00
•13-251	Wattsun TR 6 Siemens M75, M55	$795 00
•13-252	Wattsun AZ 8 Siemens M75, M55	$995 00
•13-160	Wattsun AZ 10 Siemens M75, M55	$1,095
•13-244	Wattsun AZ 12 Siemens M75, M55	$1,195
•13-253	Wattsun AZ 16 Siemens M75, M55	$1,465
•13-254*	Wattsun AZ 18 Siemens M75, M55	$1,795
•13-255*	Wattsun TR 24 Siemens M75, M55	$1,995
•13-285	Wattsun TR 4 Siemens PC-4	$825 00
•13-284	Wattsun AZ 6 Siemens PC-4	$1025
•13-283	Wattsun AZ 8 Siemens PC-4	$1,245
•13-282*	Wattsun AZ 10 Siemens PC-4	$1,495
•13-281*	Wattsun AZ 12 Siemens PC-4	$1,775
•13-286*	Wattsun TR 18 Siemens PC-4	$2,195
•13-256	Wattsun TR 4 Solarex MSX-60	$795 00
•13-257	Wattsun AZ 6 Solarex MSX-60	$945 00
•13-258	Wattsun AZ 8 Solarex MSX-60	$1,195
•13-164	Wattsun AZ 10 MSX-60	$1,295
•13-259*	Wattsun AZ 12 Solarex MSX-60	$1,395
•13-261*	Wattsun TR 18 Solarex MSX-60	$2,095
•13-262	Wattsun TR 4 Kyocera G51	$775 00
•13-263	Wattsun TR 6 Kyocera G51	$845 00
•13-264	Wattsun AZ 8 Kyocera G51	$1,045
•13-165	Wattsun AZ 10 Kyocera G51	$1,145
•13-265	Wattsun AZ 12 Kyocera G51	$1,245
•13-267	Wattsun AZ 18 Kyocera G51	$2,095
•13-166	Wattsun TR Solavolt 4 SV8500	$895 00
•13-167	Wattsun AZ Solavolt 6 SV8500	$1,165
•13-168	Wattsun AZ Solavolt 8 SV8500	$1,325
•13-169	Wattsun AZ Solavolt 10 SV8500	$1,625
•13-170	Wattsun AZ Solavolt 12 SV8500	$1,965
•13-274	Wattsun Stainless Steel Hardware, 4 Modules	$40 00
•13-275	Wattsun Stainless Steel Hardware, 6 Modules	$50 00
•13-276	Wattsun Stainless Steel Hardware, 8 Modules	$60 00
•13-277	Wattsun Stainless Steel Hardware, 12 Modules	$80 00
•13-278	Wattsun Stainless Steel Hardware, 16 Modules	$90 00
•13-279	Wattsun Stainless Steel Hardware, 18 Modules	$110 00
•13-280	Wattsun Stainless Steel Hardware, 24 Modules	$120 00
•13-218	Wattsun Optional PV Drive Panel	$79 00

Note: TR denotes a Tilt & Roll type of tracker

AZ denotes a Azimuth type of tracker

* *Must be shipped by common carrier; too large for UPS. Call our technicans for prices on larger quantities, or for more information.*

Wattsun TR-type trackers for south of the equator require special ordering due to mirror-image operation.

Wattsun Dual Axis Options

With the new style of azimuth tracker, the Dual Axis option is strongly recommended for all trackers. The tilt range is approximately 80°.

- •13-207 Wattsun Dual Axis Option for 4-Module Trackers $225.00
- •13-208 Wattsun Dual Axis Option for 6-Module & Larger $365.00

Zomeworks Track Racks

The Zomeworks is a passive solar tracker. It operates with gravity and the heat of the sun. The east and west tubes are hollow, have a small connecting tube at the bottom, and a shadow-plate partially covering them. The drive mechanism is freon warmed by the sun and forced from one tube to the other. So long as both tubes see equal sunlight by pointing directly at the sun, pressures are equalized. As the sun moves, the rack follows. The east-west range is 110°. The rack will go to "sleep" facing west at the end of the day. The rising sun in the morning will warm it and the rack rolls over to start again. In cold weather wake-up can take a couple hours. Easy to install and seasonally adjustable, the Track Rack comes with a ten-year warranty.

The north-south axis is seasonally adjustable over a 40° range for top performance all year long. All refrigerant joints are silver soldered, and there are no plastic drive components. The Zomeworks Track Rack is welded together and constructed of painted mild steel. Stainless-steel racks are available for extra cost. The racks are designed to withstand winds up to 85 mph, although tracking is iffy above 25 mph. Custom designed shock absorbers dampen motion in high winds. See specification page for mounting pole size and hole recommendations. Racks are custom-made for specific modules; always specify which modules you're using and your approximate latitude.

Passive Trackers available for all modules.
Please call for pricing.

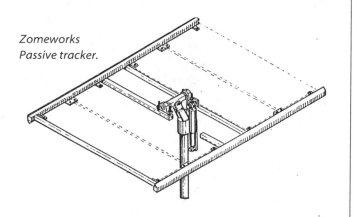

Zomeworks
Passive tracker.

•13-101	Zomeworks 1-2 Panel Siemens M55, 65, 75	$375.00
•13-102	Zomeworks 4 Panel Siemens M55, M75	$525.00
•13-103*	Zomeworks 6 Panel Siemens M55, M75	$735.00
•13-104*	Zomeworks 8 Panel Siemens M55, 65, 75	$885.00
•13-105*	Zomeworks 12 Panel Siemens M55, 65, 75	$1,085
•13-106*	Zomeworks 14 Panel Siemens M55, 65, 75	$1,295
•13-131	Zomeworks 1-2 Panel Solarex MSX-60	$385.00
•13-132*	Zomeworks 4 Panel Solarex MSX-60	$690.00
•13-133*	Zomeworks 6 Panel Solarex MSX-60	$850.00
•13-134*	Zomeworks 8 Panel Solarex MSX-60	$1,040
•13-135*	Zomeworks 12 Panel Solarex MSX-60	$1,300
•13-171	Zomeworks Marine Bearings	$20.00

* Too large for UPS; must be shipped common carrier.
All Track Racks shipped freight collect from New Mexico.
Marine bearings must be installed at time of purchase.
Stainless-steel track racks are available; call for quote.

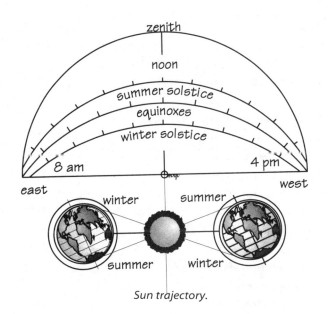

Sun trajectory.

Rapid Response West Canister Shadow Plate

This shadow plate from Zomeworks speeds morning tracker wake-up. It's available as an upgrade for installed Track Racks, and as an option for new Track Racks. The wide west-side shade is made from polished aluminum to reflect more sunlight onto the canister. This special shadow plate will cut wake-up time by approximately 40%. Specify the mount and type of module you have.

•13-161	Zomeworks Shadow Plate 8/12/14 module	$115⁰⁰
•13-162	Zomeworks Shadow Plate 4-6 module	$100⁰⁰
•13-163	Zomeworks Shadow Plate for 2 module	$70⁰⁰

Shipped freight collect from New Mexico

•13-178	1 SV8500	$112⁰⁰
•13-179	2 SV8500	$138⁰⁰
•13-180	3 SV8500	$155⁰⁰
•13-181	4 SV8500	$160⁰⁰
•13-182	5 SV8500	$280⁰⁰

Ground Or Roof Mount Fixed Racks

This line of mounting structures from Zomeworks are sturdy, economical racks made of 6061-T-6 structural aluminum angle in standard mill finish. Their telescoping rear struts offer quick and easy seasonal adjustment from 15 to 65 degrees. They're designed to withstand winds up to 125 mph. Stainless-steel mounting hardware for the modules is provided with each rack. The customer or installer provides the anchor bolts. Designed for ground mounting, roof mounting, or flip them over for south wall mounting.

The series of mounts uses what we call the "ladder style" of mounting. These modules are positioned with their length horizontal, and are stacked one above the other. The number and brand of modules each rack will hold is listed.

•13-342	2 Siemens M-series	$120⁰⁰
•13-343	3 Siemens M-series	$130⁰⁰
•13-344	4 Siemens M-series	$144⁰⁰
•13-345	6 Siemens M-series	$157⁰⁰
•13-346	8 Siemens M-series	$281⁰⁰
•13-347	2 Siemens PC-4JF	$138⁰⁰
•13-348	3 Siemens PC-4JF	$155⁰⁰
•13-349	4 Siemens PC-4JF	$160⁰⁰
•13-350	5 Siemens PC-4JF	$280⁰⁰
•13-351	2 Solarex MSX60	$140⁰⁰
•13-352	3 Solarex MSX60	$155⁰⁰
•13-353	4 Solarex MSX60	$163⁰⁰
•13-354	5 Solarex MSX60	$199⁰⁰
•13-355	6 Solarex MSX60	$330⁰⁰
•13-183	1 DV5000	$120⁰⁰
•13-184	2 DV5000	$144⁰⁰
•13-185	3 DV5000	$157⁰⁰
•13-186	4 DV5000	$281⁰⁰

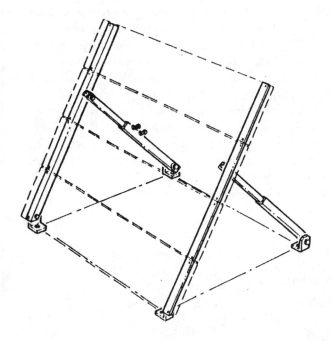

Zomeworks Top-of-Pole Fixed Racks

Zomeworks offers a standard pole-mounted fixed rack, made from heavy-duty steel pipe gimbals with mill-finish aluminum mounting rails. Horizontal and vertical axes of rotation are locked in place by hex-head set-bolts. These racks are easier to install than lean-to style racks because careful alignment is not necessary. They are also better in snowy climates because the snow slides off easily. You can manually track the sun by rotating the rack on the mounting pole. Stainless-steel module mounting hardware is provided with each rack. Poles are not included. Item numbers below show quantity and module brand. Please be sure to specify which type of panel you are using. See specification page for pole size and hole recommendations. Racks are custom-made for specific modules, always specify which modules you're using!

•13-400	Zomeworks 1 Siemens M-series	$60⁰⁰
•13-401	Zomeworks 2 Siemens M-series	$63⁰⁰
•13-408	Zomeworks 3 Siemens M-series	$95⁰⁰
•13-402	Zomeworks 4 Siemens M-series	$163⁰⁰
•13-403	Zomeworks 6 Siemens M-series	$195⁰⁰
•13-404	Zomeworks 8 Siemens M-series	$285⁰⁰
•13-430	Zomeworks 10 Siemens M-series	$410⁰⁰
•13-405*	Zomeworks 12 Siemens M-series	$440⁰⁰
•13-419*	Zomeworks 14 Siemens M-series	$490⁰⁰
•13-399	Zomeworks 1 Siemens PC-4	$65⁰⁰
•13-421	Zomeworks 2 Siemens PC-4	$74⁰⁰
•13-425	Zomeworks 3 Siemens PC-4	$165⁰⁰
•13-422	Zomeworks 4 Siemens PC-4	$182⁰⁰
•13-423	Zomeworks 6 Siemens PC-4	$275⁰⁰
•13-431	Zomeworks 8 Siemens PC-4	$340⁰⁰
•13-432*	Zomeworks 10 Siemens PC-4	$475⁰⁰
•13-433*	Zomeworks 12 Siemens PC-4	$500⁰⁰
•13-407	Zomeworks 2 MSX-60	$74⁰⁰
•13-410	Zomeworks 3 MSX-60	$165⁰⁰
•13-412	Zomeworks 4 MSX-60	$195⁰⁰
•13-414	Zomeworks 6 MSX-60	$275⁰⁰
•13-416	Zomeworks 8 MSX-60	$340⁰⁰
•13-434*	Zomeworks 10 MSX-60	$475⁰⁰
•13-418*	Zomeworks 12 MSX-60	$500⁰⁰
•13-195	Top-Pole Mount 1 SV8500	$65⁰⁰
•13-196	Top-Pole Mount 2 SV8500	$74⁰⁰
•13-197	Top-Pole Mount 3 SV8500	$165⁰⁰
•13-198	Top-Pole Mount 4 SV8500	$182⁰⁰
•13-199	Top-Pole Mount 6 SV8500	$275⁰⁰
•13-200	Top-Pole Mount 8 SV8500	$340⁰⁰
•13-295*	Top-Pole Mount 10 SV8500	$475⁰⁰
•13-296*	Top-Pole Mount 12 SV8500	$500⁰⁰

Too large for UPS; must ship via common carrier.
All shipped from New Mexico.

•13-297	Pole-Top Mount 1 DV5000	$63⁰⁰
•13-298	Pole-Top Mount 2 DV5000	$163⁰⁰
•13-299	Pole-Top Mount 3 DV5000	$195⁰⁰
•13-300	Pole-Top Mount 4 DV5000	$285⁰⁰

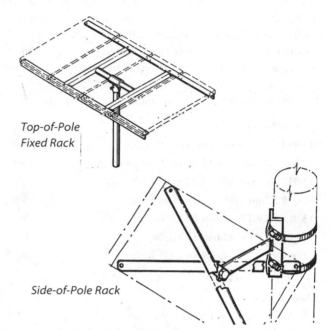

Top-of-Pole Fixed Rack

Side-of-Pole Rack

Zomeworks Side-Of-Pole Racks

These racks are adjustable from zero to ninety degrees. Two hose clamps secure the rack to the side of a mounting pole, which is not included. Mounting poles, typically 1-1/2 to 3-inch OD pipe, should be purchased locally. Racks have mill finish aluminum module mounting rails, and mild steel structural parts. The item codes below with the prices specify which modules each rack will hold. Only the most common modules are listed; others are available. Call our technical department. Be sure to specify which panel you will be using.

•13-301	Zomeworks 1 Siemens M20	$47⁰⁰
•13-311	Zomeworks 1 Siemens M65-M75-M55	$62⁰⁰
•13-321	Zomeworks 2 Siemens M65-M75-M55	$120⁰⁰
•13-331	Zomeworks 3 Siemens M65-M75-M55	$160⁰⁰
•13-341	Zomeworks 4 Siemens M65-M75-M55	$190⁰⁰
•13-316	Zomeworks 1 MSX-60	$75⁰⁰
•13-323	Zomeworks 2 MSX-60	$135⁰⁰
•13-464	Zomeworks 1 PC4JF	$75⁰⁰
•13-465	Zomeworks 2 PC4JF	$150⁰⁰
•13-466	Zomeworks 3 PC4JF	$200⁰⁰
•13-467	Zomeworks 4 PC4JF	$243⁰⁰

RV Mounting Structures

Zomeworks offers two lines of rooftop RV mounting structures, simple flush mounts or more efficient tilt-up mounts. You may not always park in a direction to make use of the tilt-up feature, but when you can, it increases output by 30% to 40%. Always drive with panels in the flat position. If you're not interested in making the adjustment, we also offer fixed flush mounts. All units are constructed of aluminum and available for one or two panel systems. If you have more than two panels, use more than one mount.

Tilting Mounts

•13-915	Zomeworks 1 Siemens M-series	$45⁰⁰
•13-918	Zomeworks 2 Siemens M-series	$65⁰⁰
•13-917	Zomeworks 1 Solarex, or PC4JF	$55⁰⁰
•13-920	Zomeworks 2 Solarex, or PC4JF	$75⁰⁰
•13-307	RV Tilt Mount 1 SV8500	$55⁰⁰
•13-308	RV Tilt Mount 2 SV8500	$75⁰⁰
•13-309	RV Tilt Mount for 1 DV5000	$60⁰⁰

Flush Mounts

•13-909	Zomeworks 1 Siemens M-series	$29⁰⁰
•13-912	Zomeworks 2 Siemens M-series	$35⁰⁰
•13-911	Zomeworks 1 Solarex, or PC4JF	$25⁰⁰
•13-914	Zomeworks 2 Solarex, or PC4JF	$39⁰⁰
•13-313	RV Flush Mount 1 SV8500	$25⁰⁰
•13-314	RV Flush Mount 2 SV8500	$39⁰⁰
•13-317	RV Flush Mount 1 DV5000	$35⁰⁰

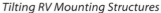

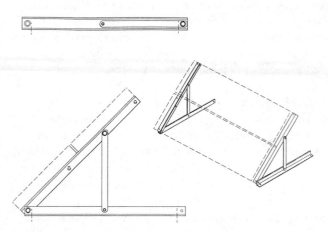

Tilting RV Mounting Structures

Aluminum Universal Solar Mounts

For simple installation it's hard to beat the economy of our aluminum PV mounting structures. Each mount will hold up to four Siemens (Arco) M-Series modules, three Siemens PC-4 modules, three Kyocera modules, 3 Solarex modules (MSX-53, 56, 60). Aluminum angle rails adjust to three seasonal tilt angles. It can mount on a deck, wall, or roof. All stainless-steel hardware is included to secure your solar panels to the mounts, but anchor bolts must be provided by the installer. Instructions are included. Specify module type.

13-702	Solar Mount, 3 or 4 PV Modules	$109⁰⁰

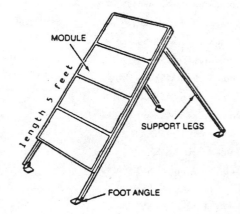

Tracker & Top-of-Pole Mounting Chart

Pipe Size Sch. 40 steel (actual O.D.)	Zomeworks Track Racks	Zomeworks Top-of-Pole fixed mounts	Wattsun Trackers	Minimum pipe height above ground	Recommend hole depth	Recommended hole diameter (fill w/concrete)
2" (2.38")	1 or 2 module SI, SX, G51, PC-4	1 module PC-4 1 or 2 module SI, SX, G51 3 module SI		48"	30"	10"
2-1/2" (2.88")	4 module SI, G51	2 module PC-4 3 module SX, PC-4, G51 4 module SI, G51		48"	32"	12"
3" (3.5")	4 module SX, PC-4 6 module SI, G51	4 module SX, PC-4 6 module SI, G51	4 module SI, PC-4, SV85, G51, SX 6 module SI, PC-4, SV85, G51	52"	35"	12"
4" (4.5")	6 module SX, PC-4 8 module SI, G51	6 module SX, PC-4 8 module SI, G51	6 module SX 8 module SI, PC-4, SV85, G51, SX 10 module SI, PV-4, SV85, G51, SX 12 module SI, G51, SX 16 module SI	58"	44"	16"
5" (5.56")	8 module SX, PC-4 10 module SI, G51 12 module SI, G51	8 module SX, PC-4 10 module SI, G51 12 module SI, G51		66"	50"	18"
6" (6.63")	10 module SX, PC-4 12 module SX, PC-4 14 module SI 16 module SI	10 module SX, PC-4 12 module SX, PC-4 14 module SI,	12 module PC-4, SV85 18 module SI, PC-4, SV85, G51, SX 24 module SI	96"	52"	20"

Note: Minimum Pipe Height above ground allows approximately 24" for snow and weed clearance. Some models are less, call our tech staff for specific tracker figures.
PC-4=Siemens ProCharger 4
SI=Siemens M55, M65, M75
SX=Solarex MSX60, MSX64
G51=Kyocera G51
SV85=Solavolt 8500

FOSSIL-FUELED GENERATORS

In an ideal world we would never have need for noisy, pollution-spewing, greenhouse-gas–producing, resource-gobbling generators. But then, our ideal world would need to be equipped with no seasonal variations, clouds that only appear at night, and 50¢/watt PV modules. That might be pretty boring, since human beings seem to thrive on adversity. At least for the time being, fossil-fueled generators have a limited but important niche to fill in many independent energy systems.

If you live outside the tropics, and intend to power your operation through the winter primarily with photovoltaics, it is probable that your expensive modules will spend the majority of their time idle. Adding enough renewable PV capacity to satisfy that last 10% of your energy needs can easily account for 50% to 60% of total system cost. If you strongly oppose the use of fossil-fueled generators all of us here at Real Goods commend your strong environmental stance, but we also feel obliged to explain that this kind of investment might not make sense for you economically. We will happily help you put your commitment to total renewables into practice, especially because we all secretly wish we could do so, too. The problem is, most of us can't afford to.

But before you cave in and buy that infernal combustion beast (the one that leaks dead-dinosaur drool), you should first take a long hard look at your building site and the feasibility of wind or hydro power. These are our renewable secondary sources of choice, both good wintertime energy sources that are covered elsewhere in this *Sourcebook* (see page 123 and page 133).

Great! Now that we've sent the rabid environmentalists scurrying off to other parts of the *Sourcebook,* the motorheads among us can peel off our disguises, pop open a six-pack, and start talking machinery! Remember the horsepower races of the 1960s before pollution controls ruined everything? Remember the original Ford Cobra 427 that could do 0 to 100 to 0 in 10 seconds (if the driver could stay conscious)? Betcha never suspected that Dr. Doug actually spent 18 years as a professional auto mechanic before starting his present job. To tell the truth, modern clean-burning auto technology makes more power from smaller engines that run longer with far less maintenance, so he had to find another job!

The fact is, most folks are going to buy a generator as one of their first investments when they start developing a homestead. The original generator can build the house and then get itself semiretired once the alternative energy power system becomes operational. At Real Goods, though, we do things a little differently. We like to get the renewable energy system operational first and use it to build the house, with the generator playing a back-up role from the beginning. The carpenters, and you, will appreciate working without a noisy generator. Using the generator to run large power tools and to cover the occasional wintertime energy shortfall will be the cheapest way to get these needs met. Most independent system owners are proud to put fewer than 100 hours per year on their generators. Maintenance costs are negligible, and the life expectancy is measured in decades for generators that see only limited service like this.

When purchasing a generator, you'll have the choice of models that will use one of the three types of fuel: gasoline, propane, or diesel. Let's examine the pros and cons of each.

Gasoline

Gasoline generators are the most common type available, but they do have disadvantages. Gasoline as a fuel is dangerous to carry and sloppy to transfer. The safest way to transport it is inside your vehicle's fuel tank, or with an outboard motor gas can with a twist-lock hose and in-line pump. Gasoline leaves a lot of carbon behind when it is burned as a fuel. This reduces engine life expectancy, and, unless the engine does some full-output duty regularly, the carbon will build up inside the cylinders, requiring periodic tear-downs for cleaning.

Propane

Most gasoline generators can be converted to run on propane. As a resource, propane is more abundant than gasoline, and is also a much cleaner fuel from a carbon standpoint. Engines generally last about 50% longer when run on propane. Most independent homesteads are already going to run the water heater, stove, and maybe the refrigerator with propane, so most likely propane delivery and handling will be neat and convenient. For most homesteads propane is the best fuel choice. If your generator is a converted gasoline model, derate its output by 10% to allow for propane's slightly lower energy content.

Diesel

Diesel engines have a reputation for low-cost operation and longevity. As a fuel, diesel *is* one of the most efficient energy sources available. Diesels run a long time on surprisingly little fuel. We have found that engine life depends more on the quality of manufacture than on the fact that it's a diesel. (GM disabused us of any longevity fantasies with their ill-starred line of diesel cars in the 1980s). On the downside, diesels are *very* noisy to operate, which might actually be considered a desirable trait, since it will discourage frequent use of the generator. Recent air-quality statistics indicate that diesels are uniquely blameworthy in their production of some of the most intractable pollutants, the sulfur oxides (SO_2) and the nitrous oxides (NO_2). Generally, diesel will be the least desirable generator fuel option, unless you live in a little valley with a microclimate prone to an inversion layer and you really pine for smog. Few independent installations will require enough generator hours to justify diesel's higher cost, fuel supply hassles, and noisy operation.

Features to Look for in a Generator

Most homesteads will need a generator of 4 kW to 6 kW capacity. This is large enough to run the washing machine and catch some battery charging at the same time. If you're considering one of the full-sized Trace inverters with the battery-charging option (the most convenient way to add generator charging capacity), these can manage as much as 4 kW when the batteries are really hungry. Generators under 2.5 kW aren't large enough to run many power tools or washers, so avoid them unless you are trying to tread very lightly.

Remote starting capability is a mighty handy feature to have. It can save many a cold, dark, wet slog out to the generator. (Heed the voice of experience!) It also makes it possible to use some of the auto-start units that sense low battery voltage and automatically start the generator, charge the batteries to a user-set voltage, and then shut the thing down for you. You'll read about it elsewhere, but it tickles us to note that the newest Trace unit can be programmed for a "quiet time" during which it will not fire the generator, even when the batteries have dropped below the gen-start set-point. This feature is *better* than downtown.

Quality

If your generator is only going to log one hundred hours of use per year, it obviously need not be the highest quality available. However, we like the higher-quality units for their starting reliability, easier maintenance, more quiet operation, and remote starting capabilities. Cheapo generators use cheapo throwaway (i.e., nonserviceable) engines. They are generally unpleasant to be around and fail to meet expectations.

— Doug Pratt

HYDROELECTRICITY

If you could choose any renewable energy source to use, hydro is the one. If you don't want to worry about a conservation-based lifestyle; always nagging kids to turn off the lights, watching the voltmeter, basing every appliance decision on energy efficiency; then you had better settle next to a nice year-round stream! Hydropower, given the right site, can cost as little as a tenth that of a PV system of comparable output. Hydropower users are often able to run energy-consumptive appliances that would bankrupt a PV system owner, like large side-by-side refrigerators and electric space heaters. Hydro may require more effort to install, but even a modest hydro output over 24 hours a day, rain or shine, will add up to a large cumulative output. Hydro systems get by with smaller battery banks because they only need to cover the occasional heavy power surge rather than four days of cloudy weather. Hydro turbines can be used in conjunction with any other renewable energy source, such as PV or wind, to charge a common battery bank. Especially in the West there are seasonal creeks with substantial drops that flow only in the winter. This is when power needs are at their highest and PV input is at its lowest. Small hydro systems are well worth developing, even if used only a few months out of the year, if those months coincide with our highest power needs. So, now that we have you all enthusiastic, what makes a good hydro site and what else do you need to know?

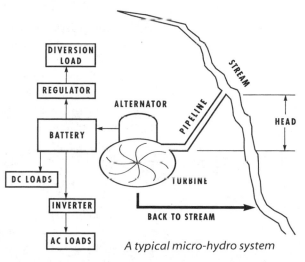

A typical micro-hydro system

What is a "Good" Hydro Site?

The Columbia River in the Pacific Northwest has some really great hydro sites, but they aren't exactly homestead-scale (or low-cost). Within the hydro industry the kind of home-scale site and system we deal with are called micro-hydro. The most cost-effective hydro sites are located in the mountains. Hydropower output is determined by water's volume times its vertical fall (jargon for the vertical fall is head). We can get approximately the same power output by running 1,000 gallons per minute through a 2-foot drop as by running 2 gallon per minute through a 1,000-foot drop. In the former scenario, where lots of water flows over a little drop, we are dealing with a low-head/high-flow situation, which is not truly a micro-hydro site; turbines that can handle thousands of gallons are large, bulky, expensive, and extremely site-specific. At the homestead level, low-flow/high-head systems are perfect for a small Harris turbine which can handle a maximum of 120 gallons per minute, and requires a minimum 20-foot fall in order to make any useful amount of power. In general, any site with more than 100 feet of fall will make an excellent micro-hydro site, but many sites with less fall, down to 20-feet, can be very productive also. The more head, the less volume will be necessary to produce a given amount of power. Check the output charts at the end of this section for a rough estimate of what your site can deliver.

Our Hydro Site Evaluation service will estimate output for any site, plus it will size the piping and wiring, and factor in any losses from pipe friction or wire resistance. See the example.

A hydro system's fall doesn't need to happen all in one place. You can build a small collection dam at one end of your property and pipe the water to a lower point, collecting fall as you go. It's not unusual to have several thousand feet of pipe in a hydro system with only a hundred feet of head.

What If I Have a High-Volume/Low-Head Site?

You are in the catbird seat for sure, but the variables are beyond the scope of this book. Apart from very significant engineering issues, there are also likely to be regulatory hurdles. Contact the DOE's Renewable Energy Clearinghouse at (800) 363-3732, or use Internet access at: http://www.ern.doe.gov for more free information on low-head hydro than you ever thought was possible.

How Do Micro-Hydro Systems Work?

The basic parts of a micro-hydro system are the penstock, to deliver the water, the turbine, that turns water pressure and volume into rotational energy, the alternator or generator, that turns the rotational energy into electricity, the regulator, to control the generator or dump excess energy depending on regulator style, and the wiring to deliver the electricity. Our micro-hydro systems also use batteries, to store energy, and usually an inverter to convert the low voltage direct current that can be stored in batteries into the higher voltage alternating current that most of our appliances and lights use (120 volt AC).

Most micro-turbine systems use a small DC alternator or generator that delivers a small but steady energy flow that is accumulated in a battery bank. This gives us a couple of important advantages. First, the battery system allows the user to store up energy, and expend it, if needed, in short, powerful bursts (like a washing machine starting the spin cycle), so we can start and run larger appliances, motors, and tools than the turbine could manage by itself. The batteries allow us to use substantially more energy for short periods than our turbine is producing, so long as the battery and inverter are designed to handle the load. Second, DC charging means that precise control of alternator speed is not needed, as is required for 60 Hz AC output. This saves thousands of dollars on control equipment.

Harris Turbines

The Harris turbine is the simplest and most appropriate micro-hydro generator we have encountered. A cast silicone-bronze Pelton turbine wheel is mated with a low-voltage DC alternator, which produces power for charging batteries and running household appliances.

The alternators used for the Harris are the world's most common models: vintage 1960s and 1970s Delco or Motorcraft automotive units with windings that are customized for each individual application. (For us to customize the windings correctly, it is very important for you to fill out one of the Hydro Site Survey forms when ordering.) Parts that wear, brushes and bearings, are commonly available at any auto parts store, making maintenance for these units easy anywhere in the world. Bearing and brushes require replacement at intervals anywhere from one to five years depending on how hard the unit is working.

Depending on the volume and fall at your micro-hydro site, Harris turbines will produce from 1 kWh (1000 watt-hours) to 30 kWh per day. For comparison, the typical American home consumes 10 to 15 kWh per day with no particular energy conservation, so with a good site, it is fairly easy to live a conventional lifestyle.

How Many Nozzles?

The Harris turbine can be supplied with one, two, or four nozzles. The maximum flow rate for any single nozzle is 30 gallons per minute (gpm), so if you have 40 gpm to work with at peak flow, and want to use it all, you will need two nozzles. Many users buy two or four nozzle turbines, so that individual nozzles can be turned on and off to meet variable power needs or water availability. The nozzles are replaceable because they eventually wear out, especially if there is grit in the water. They come in 1/16" increments, from 1/16" thru 1/2". (This is another reason you must fill out the Hydro Site Survey form, so that your turbine can be fitted with the proper size nozzles.) The first nozzle on any turbine doesn't have a shutoff valve, while all nozzles beyond the first one are supplied with ball valves for easy, visible operation.

Power Transmission

A disadvantage of this system can be difficulty of power transmission from the turbine to the batteries. This is more of a problem with high output sites, and less of a problem with modest output sites. Low-voltage power is difficult to transmit if large quantities or long distances are involved. The turbine should be as close to the batteries as is practical. Transmission distances of more than 500 feet often require expensive large-gauge wire or technical gimmickry. A typical installation has the batteries at the house on top of the hill, where the good view is, and the turbine at the bottom of the hill, where the water ends its maximum drop. With longer distances we sometimes recommend installing the batteries and inverter at the turbine site, as this allows us to transmit 120 volt power. Please consult with Real Goods Technical Staff about this option. Longer transmission distances are possible, but this drives system prices up.

Controllers

Hydro generators require special controllers or regulators. Controllers designed for photovoltaics may damage the hydro generator, and will very likely be crispy-crittered themselves if used. We can't simply open the circuit when the batteries get full like we can with PV. So long as the generator is spinning, we have to find someplace for the energy to go. Controllers have been developed especially for hydro systems that take surplus power beyond what is needed to keep the batteries charged, and divert it to a secondary load, usually a water- or space-heating element. So extra energy heats either domestic hot water or the house. These diversion controllers are also used with some wind generators, and can be used for PV control as well if this is a hybrid system. Examples are the Enermaxer control, or Ananda's Tapering Diversion Regulator (TDR).

Site Evaluation

Okay, you have a fair amount of drop across your property, you have enough water flow, and you think micro-hydro is a definite possibility. What happens next? Time to get outside and take some measurements, then fill in the necessary information on the Hydro Site Survey form. By looking at your completed form, the techies here at Real Goods can calculate which turbine and options will best fill your needs, as well as what size pipe and wire and which balance-of-system components you require. Then we can calculate specific output and costs so that you can decide if it is worth the installation.

Distance Measurements. Keep the turbine and the batteries as close together as practical. As discussed earlier, distances greater than 500 feet can get expensive. The more power you are trying to move, the more important distance becomes.

We need to know the distance from the proposed turbine site to the batteries (how many feet of wire), and the distance from the turbine site to the water collection point (how many feet of pipe). These distances are fairly easy to determine.

Fall Measurement. Next, we need to know the vertical fall from the collection point to the turbine site. This measurement is a little tougher. If there is a pipeline in place already, or if you can run one temporarily and fill it with water, this part is easy. Simply install a pressure gauge at the turbine site, make sure the pipe is full of water, and turn off the water. Read the static pressure (which means no water movement in the pipe), and multiply your reading in pounds per square inch (psi) by 2.31 to obtain the vertical drop in feet. If the water pipe method isn't practical, you'll have to survey the drop or use a very accurate altimeter.

A two-nozzle and four-nozzle (shown upside-down) Harris hydro turbine.

The following instruction represent the classic method of surveying. You've seen survey parties doing this and have doubtless always wanted to attend a survey party, so this is your big chance to get in on the action. You'll need a carpenter's level, a straight sturdy stick about six feet tall, a brightly colored target that you will be able to see a few hundred feet away, and a friend to carry the target and to make the procedure go faster and more accurately. (It's hard to party alone.)

Stand the stick upright and mark it at eye level. Five feet even is a handy mark that simplifies the mathematics, if that's close to eye level for you. Measure and note the length of your stick from ground level to your mark. Starting at the turbine site, stand the stick upright, hold the carpenter's level at your mark, make sure it is level, then sight across it uphill toward the water source. With hand motions and body English, guide your friend until the target is placed on the ground at the same level as your sightline, then have your friend wait for you to catch up. Repeat the process, carefully keeping track of how many times you repeat. It is a good idea to draw a map to remind you of landmarks and important details along the way. If you have a target and your friend has a stick (marked at the same height, please) and level, you can leapfrog each other which makes for a shorter party. Multiply the number of repeats between the turbine site and the water source by the length of your stick(s) and you have surveyed the vertical fall. People actually get paid to have this much fun!

Flow Measurement. Finally, we need to know the flow rate. If you can, block the stream and use a length of pipe. Time how long it takes to fill a 5-gallon bucket. Dividing the time by five yields seconds per gallon. Divide by 60 to get gallons per minute. If the flow is more than you can dam up or get into a 4-inch pipe, or if the force of the water sweeps the bucket out of your hands, forget measuring; you've got plenty!

Conclusion

Now you have all the information needed to guesstimate how much electricity your proposed system will generate based on the Projected Hydro Turbine Output Charts at the end of this section. This gives a rough indication as to whether your hydro site is worth developing, and which alternator option is best. If you think you have a real site, fill out the Hydro Site Survey form, and send it along with $10 to the Technical Dept. at Real Goods. We will run your figures through our computer sizing program, which gives us projected power output and allows us to size plumbing and wiring for the least power loss at the lowest cost, plus a myriad of other calculations necessary to design a working system. You can find an example at the end of this section.

Because Harris turbines are customized for each installation, it is absolutely essential that you either fill out the Hydro Site Survey form that follows and mail it in, or give the data to one of our technical staff by phone before ordering a turbine.

These charts give a rough idea of how much output to expect from the Harris turbines at specific head and flow rates. The High Output option can increase output for some sites, particularly low head, low speed sites. The precise output curves aren't easily graphable, so if you have trouble figuring it out from the charts below, give our technical staff a call and we'll talk it over.

Projected Standard Hydro Turbine Output (in Watts)

GPM	Feet of Head						
	25	50	75	100	200	300	600
3	-	-	-	-	25	60	125
6	-	-	-	10	80	110	250
10	-	-	35	65	40	225	450
15	-	40	60	105	225	350	600
20	25	65	100	150	300	450	-
30	45	100	175	230	450	550	-
100	140	350	500	650	-	-	-

Wattage output figures based on actual measurement with Standard Output alternator. A fan option must be used on all systems producing over 20 amps. (280 watts @ 12V, 560 watts @ 24V)

Projected High Output Hydro Turbine Production (in Watts)

GPM	Feet of Head						
	25	50	75	100	200	300	600
3	-	-	-	-	40	70	150
6	-	-	10	20	100	150	300
10	-	15	45	75	180	275	550
15	-	50	85	120	260	400	800
20	25	75	125	190	375	550	1100
30	50	125	200	230	580	800	1500
100	200	425	625	850	1500	-	-

Wattage output figures based on actual measurement with High Output alternator option. A fan option must be used on all systems producing over 30 amps. (420 watts @ 12V, 840 watts @ 24V)

These output charts are based on actual output measurements from a working site. Multiply this number by 24 to get daily watt-hours generated. For comparison, the average American house uses 10,000 to 20,000 watt-hours daily, though with careful lamp and appliance selection, plus some conservation activity, we can pare this down to 4,000-5,000 watt-hours without sacrificing basic lifestyle.

REAL GOODS
HYDROELECTRIC SITE SURVEY FORM

Name_____

Address_____

Phone #_____Date_____

Pipe Length:_____ (from water intake to turbine site)

Pipe Diameter:_____(only if using existing pipe)

Available Water Flow:_____(in gallons per minute)

Vertical Fall:_____(from water intake to turbine site)

Turbine to Battery Distance:_____(one way, in feet)

Transmission Wire Size:_____(only if existing wire)

House Battery Voltage:_____(12, 24, ??)

Alternate estimate (if you want to try different variables)

Pipe Length:_____ (from water intake to turbine site)

Pipe Diameter:_____(only if using existing pipe)

Available Water Flow:_____(in gallons per minute)

Vertical Fall:_____(from water intake to turbine site)

Turbine to Battery Distance:_____(one way, in feet)

Transmission Wire Size:_____(only if existing wire)

House Battery Voltage:_____(12, 24, ??)

For a complete computer printout of your hydroelectric potential, including sizing for wiring and piping, please fill in the above information and send to Real Goods along with $10. Refundable with system purchase.

17-001   HydroElectric Evaluation $10

CALCULATION OF HYDROELECTRIC POWER POTENTIAL

ENTER HYDRO SYSTEM DATA HERE:	Customer: Meg A. Power
PIPELINE LENGTH:	1300.00 Ft.
PIPE DIAMETER:	4.00 INCHES
AVAILABLE WATER FLOW:	100.00 G.P.M.
VERTICAL FALL:	200.00 Ft.
HYDRO TO BATTERY DISTANCE:	50.00 Ft. (1 way)
TRANSMISSION WIRE SIZE:	2.00 AWG #
HOUSE BATTERY VOLTAGE:	24.00 VOLTS
HYDRO GENERATION VOLTAGE:	29.00 VOLTS

Power produced at hydro:	*Power delivered to house:*
49.78 AMPS	49.78 Amps
29.00 VOLTS	28.20 VOLTS
1443.53 WATTS	1403.59 WATTS

Four nozzle, 24v, high output w/cooling turbine required

PIPE CALCULATIONS

HEAD LOST TO PIPE FRICTION:	7.61 Ft.
PRESSURE LOST TO PIPE FRICTION:	3.29 PSI
STATIC WATER PRESSURE:	86.62 PSI
DYNAMIC WATER PRESSURE:	83.33 PSI
STATIC HEAD:	200.01 Ft.
DYNAMIC HEAD:	192.40 Ft.

HYDRO POWER CALCULATIONS

OPERATING PRESSURE:	83.33 PSI
AVAILABLE FLOW:	100.00 GPM
WATTS PRODUCED:	1443.53 WATTS
AMPERAGE PRODUCED:	49.78 AMPS
AMP-HOURS PER DAY:	1194.65 AMP HOURS
WATT-HOURS PER DAY:	34644.83 WATT HOURS
WATTS PER YEAR:	12645362.71 WATT HOURS

LINE LOSS (USING COPPER)

TRANS. LINE ONE WAY LENGTH:	50.00 FEET
VOLTAGE:	29.00 VOLTS
AMPERAGE:	49.78 AMPS
WIRE SIZE #:	2.00 AMERICAN WIRE GAUGE
VOLTAGE DROP:	0.80 VOLTS
POWER LOST:	39.95 WATTS
TRANSMISSION EFFICIENCY:	97.23 PERCENT
PELTON WHEEL RPM WILL BE:	2969.85 AT OPTIMUM WHEEL EFFICIENCY

This is an estimate only! Due to factors beyond our control (construction, installation, incorrect data, etc.) we cannnot guarantee that your output will match this estimate. We have been conservative with the formulas used here and most customers call to report more output than estimated. However, be forewarned! We've done our best to estimate conservatively and accurately, but there is no guarantee that your unit will actually produce as estimated.

MEASURING HYDROPOWER

Fulminations from a Real Goods Applications Engineer

Why is it that we in the United States are so steadfast in our refusal to adopt the metric system? Our obstinacy leaves us nearly alone in the world and brings us nothing but grief as we attempt to participate in the new global economy. One would think that the young and iconoclastic renewable-energy industry would have jumped headfirst into the warm and rational clutches of metrification years ago, but noooooo . . . Instead, the Powers That Be cling to the archaic terminology of a era that has proven to belong on the ash heap of history.

This is particularly true when the subject is small-scale hydroelectric power. Converting feet and gallons per minute to watts requires tremendously complicated conversions between units inherently at odds. Trying to figure out the logic of the fudge factors used in these unit conversions can be very frustrating when all you really want to figure out is how much power is available. To this end, I present the following to those curious few who would like to be able to estimate power output from a hydroelectric site. I submit that the metric system allows us to easily and accurately make use of simple math and physics concepts to arrive at a conclusion which is both reasonable and comprehensible.

Analysis of Hydropower Potential

The potential energy available from falling water can be derived from the general equation describing the potential energy of an elevated object:

Potential Energy = m x g x h

where: m = mass

                 g = gravitational acceleration

                 h = height

A given mass of water at rest a given height above the ground will embody a potential energy expressed in *joules,* the metric unit of energy measurement.

We are not interested so much in this absolute potential under static (not moving) conditions, but rather in the *rate* of power production with a given flow rate of water.

Therefore, our rate of power production is based upon the *mass per second* which can fall this distance, which yields the rate of power production expressed as *joules/second,* otherwise known as *watts.*

In other words:

Wattage (maximum theoretical) = (mass/sec) x g x h

or

Wattage = kilograms/second x (9.8 meters/second/second) x h (in meters)

Example: What is the potential wattage available from 5 liters/sec falling 100 meters?

Water has a density of 1 gram/cubic centimeter, so 1 liter has a mass of 1 kilogram. Therefore:

Wattage = 5 kg/sec x (9.8 m/sec/sec) x 100 m = 4750 watts

To relate this to a practical hydroelectric system, certain inefficiencies must be accounted for. The turbine will not convert all of the available potential energy into mechanical energy. The electric generator or alternator will not convert all of the mechanical energy into electrical power. Our Pelton wheel alternator units have an overall efficiency of about 40%, and so, to estimate the power produced, multiply theoretical yield by .40. Using the example cited above, multiplying 4750 watts by our efficiency factor of 40% results in a theoretical maximum output of 1900 watts (1.9 kilowatts). There will also be friction losses in the supply pipe which reduce the effective "head," but ignoring them does not compromise our reasonable baseline estimate for the site.

The 5 liters/second of water processed would be equal to (for those who must have English units) 79.25 gallons per minute (1 liter/second = 15.85 gallons/minute), and 100 meters is equal to 328.1 feet.

— *Douglas Bath*

HYDRO-ELECTRIC

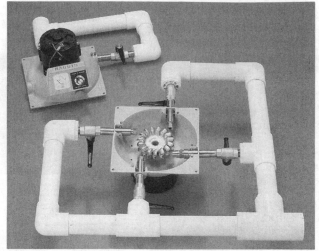

*A two-nozzle and four-nozzle
(shown upside-down) Harris hydro turbine.*

Harris Hydroelectric Turbines

When calculated on a cost-per-watt basis, a hydroelectric generator can cost as little as one-tenth as much as a photovoltaic (solar) system of equivalent power, and sometimes can be cheaper than grid power. Solar only generates power when the sun is shining; hydro generates power 24 hours a day. The generating component of the Harris turbines is an automotive alternator (Delco or Motorcraft, depending on system requirements) equipped with custom-wound coils appropriate for each installation. The rugged turbine wheel is a one-piece Harris casting made of tough silicon bronze. There are hundreds of these wheels in service, with no failures to date. The aluminum wheel housing serves as a mounting for the alternator and up to four nozzle holders. It also acts as a spray shield, redirecting the expelled water into the collection box. Harris Hydroelectric Turbines are available in several different nozzle configurations to maximize the output of the unit. The particular number of nozzles that you need is a function of the available flow in gpm and the existing pipe diameter.

Here is a chart with some general rules for sizing the number of nozzles on the system you will need, but bear in mind that we need to size your system exactly. *All turbines have custom windings; we need the Hydro Site Survey information to properly assemble a turbine for your site!*

GPM	Number of Nozzles
5 to 30	1
30 to 60	2
60 to 120	4

Here is a chart with the output limits for Standard and optional High Output alternators. Please note that if your site will exceed the wattage limits for the Standard alternator, you will need to purchase the High Output option.

Alternator output limits	12 volt	24 volt
Standard:	375 watts	750 watts
High Output:	750 watts	1500 watts

You may also wish to install the High Output alternator to produce more wattage from your site. We have approximate output charts a few pages back for both alternators. If your site is marginal the High Output option may make the difference you need.

- •17-101   1 Nozzle Turbine $750⁰⁰
- •17-102   2 Nozzle Turbine $850⁰⁰
- •17-103   4 Nozzle Turbine $995⁰⁰
- •17-131   High Output Alternator   add $200⁰⁰
- •17-132   24-Volt Option   add $50⁰⁰
- •17-133   Low Head Option (less than 60')   add $40⁰⁰
- •17-134   Extra Nozzles $5⁰⁰
- •17-135   Fan Unit   add $70⁰⁰
- •17-136   48 Volt Option   add $150⁰⁰

Pelton Wheels

For the small hydroelectric do-it-yourselfer, we offer the same reliable and economical Pelton wheel used on the complete turbines above. Harris silicon bronze Pelton wheels resist abrasion and corrosion far longer than polyurethane or cast aluminum wheels. These are 5" diameter high quality castings that can accommodate nozzle sizes of 1/16" through 1/2". Designed with threads for Delco or Motorcraft alternators.

- •17-202   Silicon Bronze Pelton Wheel $269⁰⁰

WIND ENERGY

Small Wind Turbines Come of Age

Ed Wulf was building his dream home in Southern California's Tehachapi Mountains. But when he learned that the local utility would charge him $50,000 to bring in power, he said, "I can do better than that." And he did.

Wulf didn't have far to look for his solution. His picture window opens onto one of the world's largest wind power plants. The 5,000 wind turbines across from Ed Wulf are the world's largest producers of wind-generated electricity. Each year they churn out enough electricity to serve the residential needs of 500,000 energy-hungry Californians.

For regulatory and logistical reasons, the area's wind power plants couldn't help Wulf. He had to go it alone. But their very existence convinced him that wind energy would work for him. Wulf set out to install his own stand-alone power system, a hybrid that uses the area's abundant wind and solar energy. Now Wulf's single Bergey wind turbine attracts nearly as much attention from awed tourists as the big machines on the nearby hillsides.

Wind Works

To Ed Wulf and the others like him across the country, wind energy works. It works economically and reliably. Whether it's a single wind turbine serving a remote homestead or thousands of machines supplying bulk power to an electric utility, wind energy has finally come of age after more than a decade of development. Though once bedeviled by poor performance and unreliability, wind turbines in the United States have now operated for nearly one billion hours and generated more than 15 billion kilowatt-hours of electricity. Today thousands of wind machines, both big and small, work dependably day in, day out. Some small wind turbines have even proven more dependable in remote power systems than the conventional engine generators they were originally designed to supplement.

Off-the-Grid

Outside of mechanically pumping water, wind turbines are best known for their ability to generate power at remote sites. They've distinguished themselves in this role for decades. During the 1930s, when only 10% of the nation's farms were served by electricity, literally thousands of small wind turbines were in use, primarily on the Great Plains. These "home light plants" provided the only source of electricity to homesteaders in the days before the Rural Electrification Administration brought electricity to all.

Today, three-fourths of all small wind turbines built are destined for stand-alone power systems at remote sites. Some find their way to homesteads in Canada and Alaska far from the nearest village. Others serve mountaintop telecommunications sites where utility power could seldom be justified.

An increasing number are being put to use in the lower 48 states by homeowners determined to produce their own power, even though they could just as easily buy their electricity from the local utility.

Data compiled by Pacific Gas & Electric Company (PG&E) in a 1990 study of its service area, found the number of stand-alone power systems mushrooming at a rate of 29% per year. They expect the market to continue expanding as urbanites increasingly move to rural areas not currently served by the Northern California utility. The business prospects looked so enticing that PG&E toyed with the idea of providing the stand-alone home power systems

itself, instead of building additional power lines. Other utilities are doing just that.

The government of New South Wales now subsidizes stand-alone power systems for remote cattle stations in Australia's outback in lieu of extending a power line from the provincial utility. Even a utility as conservative as Electricité de France has found that it makes economic sense for them to install wind turbines in rural areas of France and its overseas territories rather than extend their lines to any and all.

These utilities have learned what many living off-the-grid have discovered for themselves, anyone more than one-half mile from the utility line who has an average wind speed of 9 mph will find wind energy more economic than other alternatives.

The Electric Power Research Institute has gone so far as to suggest that in some cases it may make more economic sense to remove some under-used transmission lines in the United States and serve the loads with hybrid stand-alone power systems rather than continue maintaining the line.

Hybrids

A decade ago if you wanted a wind system for a remote site the dealer would happily oblige, sizing the wind turbine and the batteries to carry your entire load. That's not the case today. Both Bergey Windpower and Northern Power Systems, two small wind turbine pioneers, have successfully demonstrated wind and solar hybrids at remote telecommunications stations. These hybrids capitalize on each technology's assets.

In many areas wind and solar resources complement each other: winter's winds are balanced by summer sun, thus enabling designers to reduce the size — and cost — of each component. They've found that these hybrids perform even better when coupled with small backup

generators to reduce the battery storage needed. Many of those living off-the-grid, like Tehachapi's Ed Wulf, reached the same conclusion intuitively.

Typically a micro turbine, such as Southwes Windpower's Windseeker, or a small wind turbine, such as a Bergey 1500, a modest array of PV modules, and a small generator will suffice for most domestic uses. Though PG&E found that most Californians living off-the-grid had backup generators, they seldom used them. In a properly matched-source off-the-grid system, the backup generator provided peace of mind, but little electricity.

Village Electrification

It's only a short conceptual hop from designing a hybrid system for one American family to dozens in a village in a developing nation. Extending utility service from the cities to remote villages in developing countries is a seldom affordable luxury. More and more developing countries are turning to wind and solar energy as a less expensive, more reliable, and quicker way to meet the electrical needs of rural areas.

Low per capital consumption magnifies a hybrid system's benefits because so little electricity is needed to raise the quality of life. Two 10-kilowatt turbines, which would supply two homes with electric heat in the United States, can pump safe drinking water for 4,000 people in Morocco.

When one turbine isn't enough, the modularity of small wind turbines enables tailoring the off-the-grid system to the village's needs. Mexico's first wind farm consisted of six Bergey Excels. At the village of Xcalac on the Yucatan peninsula, Bergey Windpower installed an array of their 10-kilowatt wind turbines, 12 kilowatts of PV, and a 35-kilowatt diesel generator. The hybrid power system offset the construction of a proposed $3.2 million power line that may never be built to the poor province. Such mini wind farms are not limited solely to developing countries. In Laredo, Texas a community college installed four 10-kilowatt wind turbines to pump water in a stand-alone irrigation system. And a popular winery in Southern France, Chateau Lastours, operates seven 10-kilowatt French Aerowatt turbines so they can disconnect from the grid of the nuclear utility.

Advanced Small Wind Turbines

The growing popularity of small wind turbines is due largely to their greatly improved reliability. In contrast to the designs of a decade ago with complicated drive trains and mechanical governors, these machines are the height of simplicity. They typically use direct drive permanent-magnet alternators and automatically furl the rotor in high winds. No simpler means for controlling wind turbines has ever been devised.

These advanced small wind turbines are represented by Southwest Windpower's Windseeker and Air series, World Power Technologies' Whisper series, and Bergey Windpower's 850, 1500, and Excel models.

Manufacturers expect further refinements of these designs in the years ahead as incremental improvements continue to boost performance and result in lower costs. No one expects any earth-shaking new breakthrough will revolutionize small wind turbines. As Ed Wulf found near Tehachapi, wind energy is here today. Wind works.

The Bergey 1500 Advanced Small Wind Turbine

What You Need to Make Wind Work

First and most important of all you need a place to put the wind turbine. The site should be well exposed to the wind and free of any obstructions within 200 feet. If there are any nearby trees, the turbine must be mounted on a tower at least 30 feet above the tallest tree. And keep in mind that trees often grow taller, particularly softwoods.

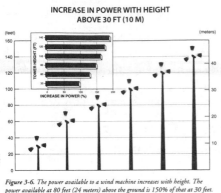

Figure 3-6. *The power available to a wind machine increases with height. The power available at 80 feet (24 meters) above the ground is 150% of that at 30 feet.*

Next, determine if you have enough wind. Hybrid power systems for living off-the-grid require less wind than those that have to compete directly with utility power. If possible, measure the wind at your site with a recording anemometer over several seasons. We offer the very moderately priced (for a recording anemometer) Totalizer 2100 in the Product Section in our section on Wind Accessories . Anemometers are many times cheaper than even a modest wind generator.

Next you need a reliable wind turbine. There are several on the market and we offer the ones we've found to be most reliable in our Product Section. The appendix of *Wind Power for Home & Business* (item 80-192, $35) contains an extensive and up-to-date list of wind turbine manufacturers worldwide. The appendix lists most available wind turbines, including micro turbines, small wind turbines, and medium-sized wind turbines like those used in California wind farms.

Then determine the height of tower you need. "The taller the tower, the greater the power" is an adage that has been proven time and again. For micro turbines, towers of 40-60 feet tall may be sufficient. Small wind turbines typically justify towers 100 feet tall or more depending upon the terrain.

And don't forget the paperwork. Check if there are any regulations governing wind turbines in your area and apply for any necessary permits.

Above all don't cut corners. If wind power systems are designed and installed with care, they will last a lifetime. For more on siting wind turbines, and how to install them safely see *Wind Power for Home & Business.*

Estimating Output

If you know the average annual wind speed at your site, you can quickly size up its potential output by using tables of Annual Energy Output provided by most wind turbine manufacturers. The accompanying table estimates the annual generation from wind turbines of different sizes based on the diameter of their rotors. For example, at a site with a 10 mph average wind speed at the top of the tower, a Marlec micro turbine with a rotor 1 meter (3.3 feet) in diameter will generate 200 kWh per year; a Windseeker with a rotor 1.5 meter (4.9 feet) in diameter will produce 450 kWh per year; a Bergey 1500, which uses a rotor slightly larger than 3 meters (9.8 feet) in diameter, will generate 1,800 kWh per year; its big brother, the 7-meter (23-foot) Bergey Excel, will generate 10,000 kWh per year or more than five times as much. Medium-sized turbines will generate considerably more: a 80 kW turbine 18 meters (60 feet) in diameter will produce 60,000 kWh per year, enough for a small business, and the biggest commercial machines available today, 500 kW machines with rotors 40 meters (130 feet) in diameter can produce nearly 300,000 kWh per year under the same conditions. Real Goods limits its sales and support to turbines designed for residential applications.

Typical Costs

The cost of a wind power system includes the cost of the wind turbine itself, the tower, and its installation. The total cost of micro turbines can be as little as $500-$1,500 depending upon the tower used and its height. Bigger machines are more costly, but also more cost-effective. Whether wind energy is a good investment at your site depends on a host of factors, including the average wind speed, the installed cost, inflation, utility buy back rates, taxes, and so on. (For more on how to determine cost effectiveness see *Wind Power for Home & Business*, especially the chapter, "Economics: Does Wind Pay?".) If the wind turbine will be part of an off-the-grid power system and there is at least a 9 mph average wind speed, a wind turbine will nearly always make economic sense.

Paul Gipe is the author of Wind Power for Home & Business. *He has written and lectured extensively about wind energy.*

— *Paul Gipe*

ESTIMATED ANNUAL ENERGY OUTPUT AT HUB HEIGHT
(IN THOUSAND KWH/YR)

Average Wind Speed	Rotor Diameter (m)	1	1.5	3	7	18	40
(mph)	(ft)	3.3	4.9	9.8	23	60	130
9		0.15	0.33	1.3	7	40	210
10		0.20	0.45	1.8	10	60	290
11		0.24	0.54	2.2	13	90	450

Adapted from *Wind Power for Home & Business* (Chelsea Green Publishing, 1993)

TYPICAL WIND SYSTEM INSTALLED COST

	Rotor Diameter (m)	(ft)	Swept Area (m²)	kW	Approx. Installed Cost
Micro-Turbines	1	3	0.75	0.25	$1,500
Small	3	10	7	1.5	$5,000
Turbines	7	23	40	10	$20,000
Medium-Sized	18	60	250	100	$125,000
Turbines	40	131	1250	500	$600,000

Conclusions

Real Goods offers several mini-micro to mid-sized wind turbines, all of them designed for residential use. We also carry all the balance of system equipment necessary for a completely off-the-grid power system, or for utility intertie system, plus books, and wind measuring devices. See the Product Section. Our large technical staff is highly skilled and diversified. We can help you upgrade your system, or design a complete new system, explain installation and operation procedures, and often help troubleshoot the occasional problem.

Sources of Information:

American Wind Energy Association; 122 C St. NW, 4th Floor; Washington, DC 20001; phone: (202) 408-8988; fax (202) 408-8536. National trade association that publishes a monthly *Windletter* with occasional articles on small wind turbines.

Wind Power for Home & Business published by Chelsea Green. Appendix contains maps of world wind resources; estimates of annual energy output for wind turbines of all sizes; estimates of water-pumping windmill capacities; listings of institutions working with wind energy in the United States, Canada, and Europe; manufacturers of small- and medium-sized wind turbines and farm windmills; and sources for towers, recording anemometers, anchors and guying hardware, and do-it-yourself plans.

Wind turbines in California generate enough electricity to offset nearly 3 billion pounds of global warming gases each year. Newer, more cost-effective wind turbines will soon enable construction of wind farms elsewhere in the United States. Developers are currently planning projects for Minnesota, Iowa, Maine, Wyoming, and the Pacific Northwest.

INTERCONNECTED WITH THE GRID

While the bulk of wind turbines interconnected with the grid in the United States are found on wind farms, that's not the case in Denmark. Two-thirds of the 3,500 wind turbines in Denmark are used by homeowners, farmers, and small businesses scattered across the country. The remainder are clustered into small wind farms, some owned by cooperatives.

While Americans were busy erecting 1-kilowatt wind turbines in their backyards during the early 1980s, the Danes were installing 55-kilowatt machines in theirs. Today the size of the average wind turbine installed by individuals in Denmark is nearing 200 kilowatts, and it is capable of generating enough electricity for 50 to 75 Danish families.

Wind development has followed different tracks in the two countries because of different regulatory policies. Danish law requires utilities to pay 85% of the retail rate for any wind-generated electricity. (German utilities pay 90% of the retail rate.) Wind-generated electricity is also exempt from Denmark's stiff energy taxes. Because medium-sized wind turbines are more economical than small turbines, the Danish system encourages users to buy the most cost-effective wind turbine available and profitably sell any excess electricity to the utility for about 10 cents per kilowatt-hour.

In contrast, American utilities pay small-scale windpower producers only 35% to 40% of the retail rate, or 3 to 4 cents per kilowatt-hour. This effectively discourages the sale of excess electricity to the utility, even though this is legally permitted. Under these conditions, the wind turbine must be sized to meet only domestic consumption, limiting homeowners in the United States to small wind turbines.

The higher buy-back rate in Denmark also encourages Danes to form cooperatives in which they buy two or more turbines, site them to best advantage, and then share in the revenues from the sale of surplus electricity to the utility.

The Danish model for wind development has spread to Germany and the Netherlands, where wind energy is growing apace. By the year 2,000 analysts expect these northern European countries to have installed nearly three times the wind-generating capacity now operating in the United States.

— Paul Gipe

What Others Say About Paul Gipe's Wind Power For Home & Business

"... the amateur will find it to be a bible of wind turbines... all that's missing is the tool box."
— Yves-Bruno Civel, editor of *Systemes Solaires* (France)

"It collects and organizes decades of experience — good and bad, funky and slick. Names are named. Facts, tips, legends, physics, ... all the book learning a wind-harnesser needs to know ... Gipe has gathered years of field experience, added lots of photographs and diagrams to an easily understood text, and represented it without a trace of forked tongue. You'll find the basics of siting, choosing, installing, and running your own system, as well as useful knowledge of the commercial machines festooning hillsides around the world ... This is the wind power book to get."

— J. Baldwin,
editor and contributor to the *Whole Earth Review*

"A new U.S. book on wind power takes an unprecedented step by listing medium-sized wind turbines, which are a good deal when they are bought secondhand by homeowners, farmers, or small businesses — and it lists those to stay away from ... Also of note in an American book, it advocates the use of cooperative ownership to increase the use of wind turbines in the Midwest."

— Ros Davidson, *Windpower Monthly* (Denmark)

"Over a period of almost two decades, Paul Gipe has been the single individual most responsible for helping to carry the wind energy message to the general public. In Wind Power for Home & Business Paul has captured a great deal of what he has learned and produced a practical how-to guide to small wind turbines. For anyone considering the purchase of a small wind system, this book will be invaluable."

— Randall Swisher, executive director,
American Wind Energy Association.

Wind Power For Home and Business

Paul Gipe. A Real Goods Independent Living Book. Real Goods and Chelsea Green joined together to publish the most complete reference on all aspects of modern (post-1970s) wind energy machines for homes and businesses. Wind energy technology has changed a lot in the last twenty years. No longer characterized by a do-it-yourself contraption in the backyard that could barely power a water pump, today's state-of-the-art wind generators are efficient, powerful and inexpensive in the long run. New turbine designs can now generate all the power you need for home use if you live in a good wind area. Whether you're committed to using a wind power system or just want to learn about it, this book will tell you all you need to know. Author Paul Gipe has had a part in nearly every aspect of wind energy's development since the mid-1970s — from measuring wind regimes, to siting and installing wind turbines. Gipe is currently president of the American Wind Energy Association, and works at Tehachapi Wind Farm, the world's largest producer of wind-generated electricity. He has written and lectured extensively about wind power. Extensive appendixes and references. 414 pages, paperback, 1993.

80-192   Wind Power for Home and Business $35⁰⁰

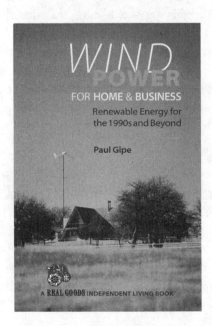

Southwest Windpower Air 303
The First "Wind Module"

The nearly silent-running 300-watt output Air 303 is designed for mounting on a simple 1-1/2" pipe tower. The manufacturer claims that rooftop mounting is okay, (where there is some power boost due to wind compression passing over the building) but we urge caution. No wind turbine, not even the Air 303 which has been engineered for a minimum of noise, is quiet under all conditions. Speed regulation is accomplished with the flexible carbon-fiber composite blades which change shape according to windspeed. The blades have an aggressive foil shape allowing low speed startup, and will reach a maximum speed at approximately 35 mph wind speed. Above this speed, the blades start acting as brakes, and simply maintain a steady rpm. (Although things get increasingly noisy at higher wind speeds in our own tests.)

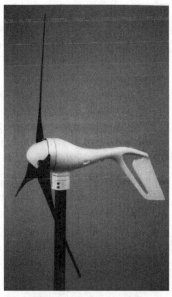

Air 303 Marine

Dr. Doug gives the Air 303 wind turbine an enthusiastic two thumbs up. Not only is it unreasonably cute for a wind generator, but is is low-cost, lightweight, maintenance-free, nearly silent, as easy to mount as a TV antenna, and can be treated, and wired, like a PV module. Everything you've ever wanted in a wind turbine with a 3-year warranty to boot.

The swoopy body is a non-corrosive cast aluminum alloy. It is unpainted in the standard version, powder-coated white in the marine version. The brushless permanent-magnet alternator uses neodyminum magnets for the greatest power in the least space. An LED indicator light gives visual assurance when generating power. The Air 303 uses a sophisticated internal regulator unlike any other wind turbine. It can stand alone, and is externally adjustable for any battery type, or it can be adjusted to its maximum, 18 volts, and be wired through your existing PV regulator. Any number of these turbines may be wired together and run through a single large regulator, just like PV modules.

Demand for these exceptional turbines has far outstripped supply at times. Please check for availability before ordering. Three-year mfg. warranty.

16-130	Air 303 Marine 12-volt	$795[00]
16-131	Air 303 Marine 24-volt	$795[00]
16-132	Air 303 Std. 12-volt	$550[00]
16-133	Air 303 Std. 24-volt	$550[00]

Specifications for Southwest Windpower Air 303

Rated Wattage Output:	300 @ 28 mph wind speed
Available Voltages:	12 or 24 volts regulated DC
Maximum Design Wind Speed:	120 mph
Cut-in Wind Speed:	5 mph
Number of Blades:	3
Rotor Diameter:	45 inches
Tower Top Weight:	13 lbs.
Generator Type:	Brushless Permanent Magnet Alternator
Tower Type:	1-1/2" Sch. 40 steel pipe
Warranty:	3 years

Windseeker 502

Windseeker 500-Watt Turbines – Wind Generators Made Simple

Windseeker was the original developer of simple, homestead-scale wind turbines almost ten years ago. They use a brushless design with high-powered neodymium permanent magnets. Start-up speed is about 5 mph, with peak output at 32 mph. No routine maintenance is required. The 20 lb. lightweight design is largely cast aluminum, with stainless steel hardware for corrosion-free longevity. All Windseeker 500 models mount on standard 2" steel pipe. Easy to erect, low-cost pipe tower designs are included in the Owner's Manual. An elegantly simple tilt-back governor is used to prevent overspeed in high winds, yet allows useful charging to continue. Wind speeds up to 120 mph can be survived unattended.

The quieter, smoother, longer-lasting 3-blade model is recommended for all but the most ideal sites. The judder of a two-bladed machine trying to track a new wind direction while under load is hard on bearings and in extreme cases has caused blades to break. (Repairable, but expensive.) Three-bladed machines don't have this problem.

All Windseeker models use basswood blades with a tough urethane two-coat finish, and polyurethane UV tape to protect the leading edge. Magnets are nickel-plated for corrosion resistance, and the entire assembly is powder-coated white. Marine versions have sealed alternators, and higher-grade stainless components. Voltage must be specified at time of order, 12- and 24-volt are standard. Two year manufacturer's warranty.

- 16-137 Windseeker 502, 2-blade $875.00
- 16-138 Windseeker 502, Marine 2-blade $890.00
- 16-139 Windseeker 503, 3-blade $1075
- 16-140 Windseeker 503, Marine 3-blade $1099

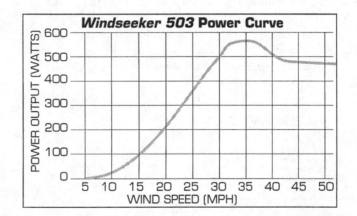

Specifications for Southwest Windpower 500 series

Rated Wattage Output:	500 @ 32 mph wind speed
Available Voltages:	12 or 24 volts regulated DC
Maximum Design Wind Speed:	120 mph
Cut-in Wind Speed:	5 mph
Number of Blades:	2 or 3 depending on model
Rotor Diameter:	60 inches
Tower Top Weight:	20 lbs.
Generator Type:	Brushless Permanent Magnet Alternator
Tower Type:	2" Sch. 40 steel pipe
Warranty:	2 years

Windseeker 503

Whisper Wind Generators

Whisper produces a series of simple, cost-effective wind generators. Designed and built in the USA, the Whisper series is engineered to last, with no scheduled lubrication or maintenance, only an annual inspection. Proprietary magnetic field configuration in the brushless alternator design delivers low cog start-up, which means you get more power in lighter winds. Start-up speed is approximately 7 mph for all models. The counter-balance governor eliminates springs and self activates to assure generator survival in hurricane force winds (up to 120 mph). Standard 600 and 1000 generators are equipped with a 2-bladed rotor made of epoxy-coated wood. The 600 is one-piece rotor, the 1000 is two-piece with spring steel mounts. An optional 3-bladed epoxy-coated wood rotor is available. The standard 3000 model is equipped with a 2-bladed rotor of carbon-fiberglass with spring steel mounts. A 3-bladed carbon-fiberglass rotor is optional. The low voltage models offered below can be easily user-configured for battery systems from 12- to 48-volt (except the Whisper 3000, which is not available in 12-volt). High voltage models are also available and can be user-configured for 64- to 240-volt output for water pumping or electric heating. Call our tech staff for pricing and specs on high-volt models. The Whisper controller includes the rectifier and shut off switch that will act as a brake, but no voltage control. For unattended automatic operation we recommend an Enermaxer controller and heating elements (either air or water) equal to wind machine wattage. Complete fold-over tower plans are included in owner's manual. Two year mfg. warranty.

Whisper 1000

Whisper 3000

- 16-117   Whisper 600 $980⁰⁰
- 16-120   3-blade wooden rotor for 600   add $100⁰⁰
- 16-119   Whisper 1000 $1590
- 16-120   3-blade wooden rotor for 1000   add $120⁰⁰
- 16-109   Whisper 3000 Low Volt $3880
- 16-120   3-blade carbon-fiber rotor for 3000   add $380⁰⁰
- 25-107   Enermaxer 60 amp Controller (for 600 & 1000) $285⁰⁰
- 25-153   Enermaxer 120 amp/24 volt Diversion Controller (for 3000) $375⁰⁰
- 25-149   12V/25 amp water heater element $117⁰⁰
- 25-148   24V/25 amp water heater element $117⁰⁰
- 25-155   48V/20 amp water heater element $105⁰⁰
- 25-109   12V/20 amp air heater element (may be wired in series &/or parallel) $25⁰⁰

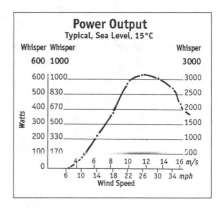

Power Output
Typical, Sea Level, 15°C

Model	600	1000	3000
Rated Wattage Output:	600	1000	3000
Wind Speed @ Rated Power:	26 mph	25 mph	25 mph
Rotor Diameter:	7 feet	9 feet	14.8 feet
Tower Top Weight:	40 lbs.	55 lbs.	130 lbs.
Tower Type:	2.5" steel pipe	2.5" steel pipe	5" steel pipe

Bergey Wind Generators

Bergey is the absolute Rolls-Royce of the home wind turbines. Yes, they're expensive initially, but if simple direct drive, passive controls, extensive corrosion protection, and freedom from routine maintenance counts as a plus to you then this is the turbine you want. Recommended maintenance is this: once a year on a windy day, walk out to the tower and look up. If the blades are turning, everything is okay. The Bergey Windpower line features models at 850 watts, 1500 watts, and the Excel at 10,000 watts. Bergey is one of the few American windplant manufacturers that survived the post-1986 tax credit crash, and today they are still thriving and growing. Bergey wind generators are wonderful in their simplicity: no brakes, pitch changing mechanism, gearbox, or brushes. An automatic furling design forces the generator and blades partially out of the wind at high wind speeds, but still maintains maximum rotor speed. AUTOFURL™ uses aerodynamics and gravity only, there are no brakes, springs, or electromechanical devices to reduce reliability. These turbines are designed to operate unattended at wind speeds up to 120 mph and with complete loss of load. The patented POWERFLEX blades combine torsionally flexible blades with precisely located pitch weights near the tips. Aerodynamic and centrifugal forces act together to twist the blades toward the best angle for each wind speed. The fiberglass blades are exceptionally strong and fatigue resistant. Power is transmitted from the turbine to controller as wild voltage 3-phase AC making for easy transmission and much smaller wiring requirements. Both models include the most comprehensive owners' manuals of any wind turbine on the market. Plenty of site selection, anchoring, erecting, and tuning help.

Bergey 1500-Watt Wind Generator

The 1500-watt unit is designed for battery charging or water pumping only. Voltage output and use must be specified when ordering. The included control unit features a solid state regulator that tapers the final charge and provides transient (lightning) protection between the turbine and batteries. *Shipped freight collect from Oklahoma.*

- •16-201   Bergey 1500-24 Turbine (24V) $4,795
- •16-202   Bergey 1500-xx Turbine (12, 32, 48, or 120VDC) $4,995

Guyed Lattice Tower Kits for 1500 series

- •16-218   Bergey Tower Kit, 12.2 meter (40 ft.) $1,395
- •16-219   Bergey Tower Kit, 18.3 meter (60 ft.) $1,825
- •16-220   Bergey Tower Kit, 24.4 meter (80 ft.) $2,395
- •16-221   Bergey Tower Kit, 30.5 meter (100 ft.) $2,875

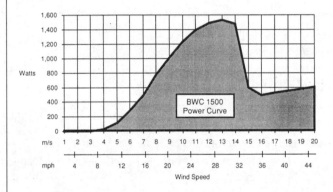

Specifications for Bergey 1500 series	
Rated Wattage Output:	1500 @ 28 mph wind speed
Available Voltages:	12 to 120 volts regulated DC
Maximum Design Wind Speed:	120 mph
Cut-in Wind Speed:	8 mph
Number of Blades:	3
Rotor Diameter:	10 feet
Tower Top Weight:	168 lbs.
Generator Type:	Brushless Permanent Magnet Alternator
Tower Type:	Conventional guyed section or freestanding
Warranty:	2 years

Bergey Excel 10,000-Watt Wind Generator

The Excel 10K unit can be used for utility intertie, battery charging (48V only), or water pumping. Use and output voltage must be specified when ordering. Control or Powersync inverter units are included in price. A 220 VAC/50 Hz option is available at no extra charge.

- •16-200   Bergey Excel-R/48 Turbine, 48VDC $17,975
- •16-203   Bergey Excel-S 240VAC/60Hz† $19,475
- •16-204   Bergey Excel-PD, for direct water pumping only $17,975

† 220VAC/50Hz option available at no charge

Guyed Lattice Tower Kits for Excel series
- •16-211   Bergey Tower Kit, 18 meter (60 ft.) $4,825
- •16-212   Bergey Tower Kit, 24 meter (80 ft.) $5,425
- •16-213   Bergey Tower Kit, 30 meter (100 ft.) $6,175
- •16-214   Bergey Tower Kit, 37 meter (120 ft.) $7,175

Shipped freight collect from Oklahoma.
Other tower types & heights also available.

Call for pricing and additional information on these and other Bergey products.

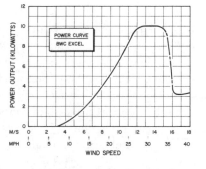

Bergey 850-Watt Turbines
Engineered, Reliable Wind Power

Bergey wind turbines are the absolute best available for homestead-scale battery charging. Routine maintenance consists of: "Once a year on a windy day walk out to the wind tower. Look up. If the blades are turning, everything is okay." Careful engineering has produced a turbine with a minimum of moving parts. Simplicity equals reliability. Start-up is at 8 mph, peak output is at 28 mph wind speed. The three POWERFLEX® non-weathering fiberglass blades present an aggressive airfoil at lower wind speeds. As speed increases, carefully engineered weights twist the blades, tuning the airfoil shape to the wind speed. Bergey's unique AUTOFURL™ system uses simple wind pressure and gravity to pull the blades at right angles as wind speeds go over 35 mph. This cuts rotor speed, but allows continued unattended output at speeds up to 120 mph.

Rotor diameter is 8 feet. Tower top weight is 86 lbs. for this direct-drive, permanent-magnet, brushless, low speed alternator. Output is 3-phase AC for ease of transmission and smaller cable sizes. The included rectifier/controller is mounted near the batteries. Mounting fits a standard 4" steel pipe. Instructions for assembling and raising the tower are included. Two-year mfg. warr.

- ● 16-215   Bergey 850 Turbine, 12v $2,095
- ● 16-216   Bergey 850 Turbine, 24v $1,995

Guyed Lattice Tower Kits for 850 series
- •16-207   Bergey Tower Kit, 13.4 meter (44 ft.) $835⁰⁰
- •16-208   Bergey Tower Kit, 19.5 meter (64 ft.) $1,210
- •16-209   Bergey Tower Kit, 25.6 meter (84 ft.) $1,695

Specifications for Bergey Excel

Rated Wattage Output:	10,000 @ 27 mph wind speed
Available Voltages:	48VDC, or 240V/50-60Hz AC
Maximum Design Wind Speed:	120 mph
Cut-in Wind Speed:	8 mph
Number of Blades:	3
Rotor Diameter:	23 feet
Tower Top Weight:	1020 lbs.
Generator Type:	Brushless Permanent Magnet Alternator
Tower Type:	Conventional guyed section or freestanding
Warranty:	2 years

Specifications Charts for Bergey 850 on next page, 146.

Specifications for Bergey 850, cont. from page 145

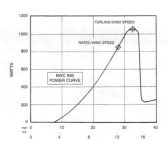

Specifications for Bergey 850	
Rated Wattage Output:	850 @ 28 mph wind speed
Available Voltages:	48VDC, or 240V/50-60Hz AC
Maximum Design Wind Speed:	120 mph
Cut-in Wind Speed:	8 mph
Number of Blades:	3
Rotor Diameter:	8.0 feet
Tower Top Weight:	86 lbs.
Generator Type:	Brushless Permanent Magnet Alternator
Tower Type:	4" Sch. 5 steel pipe
Warranty:	2 years

WIND ACCESSORIES

2100 Totalizer Anemometer

The 2100 Totalizer is a moderately-priced instrument that accurately determines your average wind speed. The Totalizer is an odometer that counts the amount of wind passing through the anemometer, and you find the average wind speed in miles per hour by dividing miles of wind over elapsed time. An excellent choice for determining the average wind speed at your site. Read it once a day or once a month, so long as you note when the last reading was taken you can determine the average. A 9-volt alkaline battery (included) provides 1-year of operation. The readout can be mounted in any protected environment. Includes a Maximum #40 Anemometer, 60-feet of sensor cable, battery, stub mast for mounting, and instructions.

63-354   The 2100 Totalizer $239⁰⁰

Hand-Held Wind Speed Meter

This is an inexpensive and accurate wind speed indicator. It features two ranges, 2 to 10 mph and 4 to 66 mph. A chart makes easy conversions to knots. Speed is indicated by a floating ball viewed through a clear tube. Many of our customers are curious about the wind potential of various locations and elevations on their property, but are reluctant to spend big money for an anemometer just to find out. Here is an economical solution. Try taping this meter to the side of a long pipe, (use a piece of tape over the finger hole for high range reading) and have a friend hold it up while you stand back and read the meter with binoculars. This is very helpful in determining wind speeds at elevations above ground level. Includes protective carrying case and cleaning kit.

63-205   Wind Speed Meter $19⁰⁰

SYSTEM SIZING WORKSHEETS

This Design Guide provides a simple and convenient method for determining total household electrical needs, and then sizing the photovoltaic and battery system to meet those needs. 90% of the renewable energy systems we design are PV-based, so these worksheets deal primarily with PV. If you are fortunate enough to have a viable wind or hydro power source you'll find output information for these sources in their respective chapters of the *Sourcebook*. Our technical staff has considerable experience with these alternate sources and will be glad to help you size a system; give us a call.

For those computer literate types, or at least computer owners, we've got a couple of programs to do all this for you in a somewhat automated fashion. Back on page 97 there's the *Solar Worksheet* (item # 81-103, $15) for Macs, and the *Solar Simulator* (item 81-050, $30) for Windows. Both programs will help you figure household electrical needs, and then help size a system for your climate. The *Simulator* goes a little farther by testing your system design against random-generated weather and use patterns, drawing a system schematic, and totaling system cost.

General Information

Conserve, conserve, conserve! As a rule of thumb, it will cost about $3 to $4 worth of equipment for every watt-hour you must supply. Trim your wattage to the bone! Don't use incandescent light bulbs or older standard refrigerators.

By all means read the section on Power System Design starting on pp. 153 of the *Sourcebook* before you tackle these worksheets (you'll save us both a lot of extra work, and you'll learn a lot in the process).

If you're intimidated by this whole process don't feel like the Lone Ranger. Our tech staff is here to hold your hand and help you through the tough parts. We do need you to fill out the first couple of charts and work down to Line 5, the Total Household Watt-Hours per Day. We can pick up the design process from there. You are in the best position to make lifestyle decisions; how late do you stay up at night, are you religious about always turning the light off when leaving a room, are you running a home business or is the house empty five days a week, do you hammer on your computer for 12 hours per day, does your pet iguana absolutely require his rock heater 24 hours a day? These are questions we can't answer for you. So figure out your watt-hours, let us know what we're shooting for.

Determine the total electrical load in watt-hours per day.

The two forms below allow you to list every appliance, both AC and DC powered, how much wattage it draws, how many hours per day it runs, and how many days per week. This gives us a daily average for the week, as some appliances, like a washing machine, may only be used occasionally.

Some appliances may only give the amperage and voltage on the nameplate. We need wattage. Multiply the amperage by the voltage to get wattage. Example: a blender nameplate says, "2.5A 120V 60hz". This tells us the appliance is rated for a maximum of 2.5 amps at 120 volts/60 cycles per second. 2.5 amps times 120 volts equals 300 watts. Beware of using nameplate amperage however. For safety reasons this must be the highest amperage the appliance is capable of drawing. Actual running amperage is often much less. This is particularly true for entertainment gear. Many conventional appliances are listed on the Appliance Watt Chart, page 152.

Line by Line Instructions (a la Internal Revenue Service, but for a worthy cause)

Line 1   Total all Average Watt-Hours/Day in the column above.

Line 2   For AC appliances multiply the watt-hours total by 1.1 to account for inverter inefficiency (typically 90%). This gives the actual DC watt-hours that will be drawn from the battery.

Line 3   DC appliances are totalled directly, no correction necessary.

Line 4   Insert the total from line 2 above.

Line 5   Add the AC and DC watt-hour totals to get the Total Household DC Watt-Hours/Day. At this point you can fax or mail the Design forms to us and after a phone consultation we'll put a system together for you.

Do you prefer rugged self-reliance? Bully for you! Forge on.

Line 6   Insert the voltage of the battery system; 12-volt or 24-volt are the most common. Talk it over with one of our tech staff before deciding on a higher voltage as control and monitoring equipment is sometimes hard to find.

Line 7   Divide the total on line 5 by the voltage on line 6.

Line 8   This is our fudge factor that accounts for losses in wiring, batteries, and allows a small safety margin. Multiply line 7 by 1.2.

Line 9   This is the total amount of energy that needs to be supplied to the battery every day on average.

Line 10   This is where guesswork rears its ugly head. How many hours of sun per day will you see? Our Solar Insolation Map gives the average daily sun hours for the worst month of the year. You probably don't want to design your system for worst possible conditions. Energy conservation during stormy weather, or a back-up power source can allow use of a higher hours per day figure on this line and reduce the initial system cost.

Line 11   Divide line 9 by line 10, this gives us the total PV current needed.

Line 12   Decide what PV module you want to use for your system. You may want to try the calculations with several different modules. It all depends on how you need to round up or down to meet your needs.

Line 13   Insert the amps of output at rated power for your chosen module.

Line 14   Divide line 11 by line 13 to get the number of modules required in parallel. You will almost certainly get a fraction left over. Since we don't sell fractional PV modules you'll need to round up or down to a whole number. We conservatively recommend any fraction from 0.3 and up be rounded upward.

If yours is a 12-volt nominal system you can stop here and transfer your line 14 answer to line 19. If your nominal system voltage is something higher than 12 volts, then forge on.

Line 15   Enter the system battery voltage. Usually this will be either 12 or 24.

Line 16   Enter the module nominal voltage. This will be 12 except for unusual special-order modules.

Line 17   Divide line 15 by line 16. This will be how many modules you must wire in series to charge your batteries.

Line 18   Insert the figure from line 14 and multiply by line 17.

Line 19   This is the total number of PV modules needed to satisfy your electrical needs. Too high? Reduce your electrical consumption, or, add a secondary charging source such as wind or hydro if possible, or, a stinking, noisy, troublesome, fossil-fuel gobbling generator. (No! We aren't biased... but we've got a better sense of humor than the IRS.)

Battery Sizing Worksheet

Line 20   Enter your Total Daily Amp-Hours from line 9

Line 21   Reserve battery capacity in days. We usually recommend about three to seven days of back-up capacity. Less reserve will have you cycling the battery excessively on a daily basis, which results in lower life expectancy. More than seven days capacity starts getting so expensive that a back-up power source should be considered.

Line 22   You can't use 100% of the battery capacity (unless you like buying new batteries). 80% is the maximum, and we usually recommend to size at 50% or 60%. This makes your batteries last longer, and leaves a little emergency reserve. Enter a figure from .5 to .8 on this line.

Line 23   Multiply line 20 times line 21, and divide by line 22. This is the minimum battery capacity you need.

Line 24   Select a battery type. The most common for household systems are golf carts @ 220 amp-hours, or L-16s @ 350 amp-hour. See the Battery Section for more details. Enter the amp-hour capacity of your chosen battery on this line.

Line 25   Divide line 23 by line 24; this is how many batteries you need in parallel.

Line 26   Your system nominal voltage from line 6.

Line 27   The voltage of your chosen battery type.

Line 28   Divide line 26 by line 27; this gives you how many batteries you must wire in series for the desired system voltage.

Line 29   Enter the number of batteries in parallel from line 25.

Line 30   Multiply line 28 times line 29. This is the total number of batteries required for your system.

AC Device	Device Watts	X	Hours of Daily Use	X	Days of Use per Week	÷	7	=	Average Watt-hrs per Day
		X		X		÷		=	
		X		X		÷		=	
		X		X		÷		=	
		X		X		÷		=	
		X		X		÷		=	
		X		X		÷		=	
		X		X		÷		=	
		X		X		÷		=	
		X		X		÷		=	
		X		X		÷		=	
		X		X		÷		=	
		X		X		÷		=	
		X		X		÷		=	
		X		X		÷		=	
		X		X		÷		=	
		X		X		÷		=	
		X		X		÷		=	
		X		X		÷		=	
		X		X		÷		=	
		X		X		÷		=	
		X		X		÷		=	

1. Total AC Watt-Hrs./Day

2. X 1.1 = Total Corrected DC Watt-Hrs./Day

DC Device	Device Watts	X	Hours of Daily Use	X	Days of Use per Week	÷	7	=	Average Watt-hrs per Day
		X		X		÷		=	
		X		X		÷		=	
		X		X		÷		=	
		X		X		÷		=	
		X		X		÷		=	
		X		X		÷		=	

3. Total DC Watt-Hrs./Day

3	(from previous page) Total DC Watt-Hrs./Day	
4	Total Corrected DC Watt-Hrs./Day from Line 2 +	
5	Total Household DC Watt-Hrs./Day =	
6	System Nominal Voltage (usually 12 or 24) ÷	
7	Total DC Amp-Hrs./Day =	
8	Battery losses, wiring losses, safety factor X 1.2	
9	Total Daily Amp-Hour Requirement =	
10	Estimated Design Insolation (hours per day of sun, see map) ÷	
11	Total PV Array Current in Amps =	
12	Select a Photovoltaic Module for Your System	
13	Module Rated Power Amps ÷	
14	Number of Modules Required in Parallel =	
15	System Nominal Voltage (from line 6 above)	
16	Module Nominal Voltage (usually 12) ÷	
17	Number of Modules Required in Series =	
18	Number of Modules Required in Parallel (from Line 14 above) X	
19	**Total Modules Required** =	

Battery Sizing Worksheet

20	Total Daily Amp-Hour Requirement (from line 9)	
21	Reserve Time in Days X	
22	Percent of Useable Battery Capacity ÷	
23	Minimum Battery Capacity in Amp-Hours =	
24	Select a Battery for Your System, Enter Amp-Hour Capacity ÷	
25	Number of Batteries in Parallel =	
26	System Nominal Voltage (from line 6)	
27	Voltage of Your Chosen Battery (6 or 12 usually) ÷	
28	Number of Batteries in Series =	
29	Number of Batteries in Parallel (from line 25 above) X	
30	**Total Number of Batteries Required**	

WATT CHART FOR TYPICAL APPLIANCES
(Figures in parentheses are additional starting wattage required when applicable.)

Use the manufacturer's specs for your particular appliance if possible, but be careful of nameplate ratings which are the highest possible electrical draw for that appliance.

Description	Watts
Refrigeration:	
22 cu. ft. auto defrost (approximate run time 7-8 hours per day)	700(2200)
12 cu. ft. SunFrost refrigerator(approximate run time 6-9 hrs. per day)	58(700)
Standard freezer (runs approximately 7-8 hrs. per day)	700(2200)
10 cu. ft. Sun Frost freezer (runs approximately 6-9 hrs. per day)	88(700)
Kitchen Appliances:	
Dishwasher cool dry	700(1400)
hot dry	1450(1400)
Trash compactor	1500(1500)
Can opener (electric)	100
Microwave (.5 cu. ft.)	1200
Microwave (.8 to 1.5 cu. ft.)	2100
Exhaust hood	144
Coffee maker	1200
Food processor	400
Toaster (2 slice)	1200
Coffee grinder	100
Blender	350
Food dehydrator	600
Mixer	120
Range, small burner	1250
Range, large burner	2100
Water Pumping:	
AC Jet Pump (1/3 hp), 300 gal. per hour, 20' well depth, 30 psi	750(1400)
AC Submersible Pump (1/2 hp) 40' well depth, 30 psi	1000(6000)
DC pump for house pressuresystem (typical use is 1-2 hrs. per day)	60

Description	Watts
DC submersible pump (typical use is 6 hrs. per day)	50
Entertainment/Telephones:	
TV (25-inch color)	170
TV (19-inch color)	80
TV (12-inch black & white)	15
Video Games (not incl. TV)	20
Satellite system, 12 ft. dish with auto orientation/remote control	45
VCR	30
Laser disk/CD player	30
AC powered Stereo (avg. volume)	55
AC Stereo, home theater	500
DC powered Stereo (avg. volume)	15
CB (receiving)	10
Cellular telephone (on standby)	20
Radio telephone (on standby)	25
Electric piano	30
Guitar amplifier (avg. volume)	40
(Jimi Hendrix)	8500

Description	Watts
General Household:	
Typical fluorescent light (60W equivalent)	15
Incandescent lights as indicated on bulb	
Electric clock	4
Clock radio	5
Electric blanket	400
Iron (electric)	1200
Clothes washer	1150
Dryer (gas)	500(1800)
Dryer (electric)	5750(1800)
Vacuum cleaner, average	900
Central vacuum	1500(1500)
Furnace fan 1/4 hp	600(1000)
1/3 hp	700(1400)
1/2 hp	875(2350)
Garage door opener 1/4 hp	550(1100)

	Watts
Alarm/security system	6
Air conditioner 1500/ton or /10,000 BTU	(2200)
Office/Den:	
Computer/Modem	55
14" color monitor	100
14" monochrome monitor	25
Ink jet printer	35
Dot matrix printer	200
Laser printer	1200
Fax machine standby	10
printing	500
Electric typewriter	200
Adding machine	8
Electric pencil sharpener	100
Hygiene:	
Hair dryer	1500
Waterpik	90
Whirlpool bath	750(1000)
Hair curler	750
Electric toothbrush (charging stand)	6
Shop:	
Worm drive 7-1/4" saw	1800(3000)
AC table saw, 10"	1800(4500)
AC grinder, 1/2 hp	1080(2500)
Hand drill, 3/8"	400
Hand drill, 1/2"	600

POWER SYSTEM DESIGN PRIMER

Here is this *Sourcebook's* meat-and-potatoes main course (or tofu and steamed veggies for you New Age types). In this section we'll discuss what you can reasonably expect from a renewable energy system, which appliances are most appropriate, and where more energy-efficient strategies should be sought.

Types of Electricity

What, you mean there are different kinds of electricity? There sure are! The most basic distinction we make is between alternating current (AC) and direct current (DC). AC is what utilities commonly supply to their customers, and is what we call "house current." AC power reverses the direction of current flow periodically. In the U.S. we have settled on 120 volts and 60 cycles per second as the standard. In other countries they've sometimes settled on other voltages and frequencies. That's why your small appliances like shavers and blow dryers may not work when you take them to Europe. AC power is easier to generate and much easier to transmit over long distances than DC power, which is why it has become the world standard; yet AC, unlike DC, cannot be stored.

Electrical storage devices, commonly called batteries, only accept and produce DC power. DC power flows in one direction only, from negative to positive. Most alternative energy power sources produce DC power, which can be saved up in batteries for later use. Batteries allow us to run much larger electrical loads for a short period than the power source alone could support.

System Voltage: 12- or 24-Volt?

DC alternative energy systems commonly use either 12 or 24 volts for their electrical collection and storage systems. Which voltage is used depends mostly on the size of the system. It is always easier to transmit energy if the voltage is higher. Wires can be smaller, a single charge controller can handle twice the wattage input, and there can be real cost savings at higher voltages. For DC systems processing less than 2000 watt-hours per day, a 12-volt system is usually the best choice, because a broader range of small appliances and systems components are available at this "automotive voltage." For systems processing more than 2000 watt-hours daily we usually recommend a 24-volt configuration. Sometimes a smaller system will be designed to allow for future expansion, and 24-volt is the prudent choice. Sometimes larger systems have specific 12-volt appliances they need to run (or the system has grown over the years). Occasionally, a very large system will use even higher voltage, but access to monitoring equipment and balance-of-system components is often a problem at higher voltages. 12 and 24 volts have become the industry standards and most common equipment is now available in either voltage.

But My House Runs on 120-Volt AC!

Most of the independent homes we design run primarily on 120-volt house current, because their owners want to take advantage of mainstream appliances and conveniences, and the mainstream flows at 120-v AC/60 Hz power. Fortunately there is a gizmo called an *inverter* which bridges the gap, changing DC energy into more generally useful AC house current. Mass-produced wiring hardware is de-

signed for AC, and there is a wide range of high-quality, mass-produced products available. DC appliances tend to be more expensive, harder (or impossible) to find, and often of lower quality. We pay a small efficiency premium for passing our power through the inverter, but for most appliances, it's worth it.

Conservation

Your single most important job when planning your power system is to ensure that no precious power is wasted. Start-up costs of renewable energy systems can be quite high, and you will not want to buy excess capacity only to throw it away. Every watt-hour wasted by an inefficient appliance, or consumed by an appliance which could be eliminated by intelligent design, costs approximately $3. We are not proposing a substantial lifestyle change, just the application of appropriate technology and (un)common sense.

In electrical jargon, we call anything that consumes electric energy a *load*. A couple pages back we have a series of worksheets to help you determine your average daily energy needs, and what equipment you'll require to meet that load. In the sections below we have divided generally appropriate from generally inappropriate loads, and spent a little time discussing each.

Please understand that in this section we are writing for the benefit of the typical PV-powered system owner, who runs most household loads with an inverter. This kind of setup represents about 90% of all the independent power systems we know of. Those lucky few with good hydro or wind potential may be able to ignore most of our rantings about conservation. In fact, many hydro system owners are in the enviable position of having to burn off excess power in water- or space-heating elements. Again, common sense should dictate how parsimonious (or creative) you will need to be in your use of power. If you can get plenty of incoming power at reasonable cost, then there is no need to be draconian in your conservation measures. On the other hand, if you are contemplating an independent PV-powered system but cannot live without central air conditioning, a 29-cubic-foot, frost-free, ice- and water-dispensing refrigerator, and dimmer-operated incandescent lighting, then be prepared to spend a load of money on your system.

APPROPRIATE ELECTRICAL LOADS

Most electrical systems just happen, pieced together over time, but an independent home's system is most cost-effective and operates best if designed as a whole. Here are some considerations that will help in that planning.

Lighting

In the typical utility-powered house, lights account for a large piece of the energy pie — up to 50% of the total electrical load is consumed by incandescent bulbs. Compact fluorescent or standard tube-type fluorescent lights, in contrast, produce as much light as incandescents, yet use only one-fifth as much power (saving 80%). In other words, incandescent bulbs have no business in your alternative energy house! Actually, they do make great little heaters, since approximately 90% of the energy you feed them is converted to heat. But this is hardly a cost-effective way to heat your house. A device which converts only 10% of the energy it consumes into useful work (in this case, visible light) is not welcome in a home with an energy budget. This applies to any bulb that makes light by means of a glowing filament, including quartz halogen bulbs, which are only 10% to 15% more efficient than incandescents. They come in as a very distant second choice when compared to fluorescents.

We generally recommend using AC power for lighting. You pay a small efficiency premium for passing power through the inverter, but AC lights are mass-produced, which makes them significantly cheaper, and they are available at a better quality and selection than DC appliances. Choosing AC for lighting also allows you to use conventional wiring techniques and mass-produced equipment. Any system with more than four or five lights will be cheaper and easier to install and maintain using AC house current.

We have been conditioned to prefer overlighted environments and generally use light poorly. Efficiency can be improved immensely by making intelligent use of the lights you have. A reading light at ceiling level needs five times the wattage of a light at your shoulder. Use lower-level lighting for general room illumination and task lighting for specific needs. Kitchens are an excellent example. The old standard practice of placing a single large fixture in the middle of the room leaves you working in your own shadow at the countertops. Instead, put a small fixture in the middle of the room, enough to keep you from bumping into the furniture or stepping on the kids, and put small, switchable task lights under the wall cabinets to light the countertop when you're working at that space. See our AC Lighting section for more information (page 377).

Refrigeration

We Americans have been spoiled by many years of cheap power and enormous side-by-side, ice- and water-dispensing refrigerators. This uniquely American monster will happily consume 5000 to 7000 watt-hours on an average day. At close to $3 per installed watt-hour, this kind of watt-sucking appliance is not affordable in an independently powered house (nor for anyone else who cares a hoot about what we leave for our children). Even a more modest-sized conventional AC-powered fridge consumes 2000 to 5000 watt-hours per day. But fear not! In later chapters you will find some reasonable solutions. For small households, propane-powered fridges make good sense, and they do not use any electrical power at all. They will burn approximately 1.5 gallons of propane per week, which you are likely to have on hand already for cooking and water heating. For larger households, a super-efficient refrigerator like the Sun Frost (which comes in 16- and 19-cubic-foot models) is the best choice. A Sun Frost will typically use about 50% of the power that a best-of-breed, "energy-saving," mass-produced fridge of similar size will gobble up to do the same job. Like most high-efficiency devices, Sun Frosts are expensive initially, but if you figure how many PV modules they save you from having to buy, the payoff is immediate!

Some new mass-produced refrigerators are using under 1000 watt-hours per day. Even though this is higher than the Sun Frost's energy use, the lower initial price may allow you to buy additional PV modules and still come out cheaper than the Sun Frost with PV modules. See our Refrigeration section in Chapter 6 for a complete discussion.

Kitchen Appliances

Microwave ovens, blenders, mixers, food processors — all the common kitchen appliances — are appropriate loads if your inverter is sized to handle them and if you practice demand-side management by using them one at a time, and conservatively when energy supplies are lower. Even though some of these appliances may draw quite a lot of power, they do so for relatively short periods of time, and their cumulative energy use is within reason. Microwave ovens in particular are highly efficient appliances. Do not try to base your power usage calculations on the microwave's wattage ratings. Commonly this will be the cooking wattage, which represents only 40% to 60% of the oven's total power use. Beware, also, of appliances with clocks: these "energy criminals" draw an unconscionable amount of power 24 hours a day, and can quickly bring an independent power system to its knees.

Heating and cooling with electricity is spectacularly inefficient. Watch out for electrical heating elements like toasters, coffee makers, or waffle irons. While your inverter may be able to drive them, they will rapidly deplete your batteries. Toasters are fine if you run only a cycle or two and only when your batteries are in a decent state of charge. If it has been cloudy for two days, it is better to get out Grandma's toasting fork and use the woodstove or stovetop burner. (Or go without toast this morning. What sacrifices we make for energy responsibility!) We discuss coffee makers further under Inappropriate Electrical Loads. (Don't panic! We will tell you some energy-efficient ways to support your java habit.)

This just in from Lawrence Berkeley Laboratory: A gas oven consumes more electricity to bake a potato than a microwave oven does. That's right, more electricity. The glow bar ignitor, which draws 350 to 400 watts during oven start-up, continues to draw power the whole time the oven is on. To bake a moderately sized potato, LBL researcher Brian Pon found that a gas oven started cold used 200 watt-hours, while a microwave used 110 watt-hours. The gas oven, of course, also consumes gas. (See Home Energy magazine, Nov./ Dec. 1993.) Pretty amazing, huh? Want to escape this kind of bizarre "energy-saving" mentality? Ask about gas stoves without a glow bar at your local appliance dealer. There are still some responsible appliance manufacturers out there who make electronic ignition or standing pilot ovens.

Ovens and Stoves

Almost all gas stoves use electronic igniters by mandate these days, because pilot lights are so wasteful and can be dangerous. Piezo electric igniters only use a few watts, and only when actually clicking to start the flame. But watch out for stoves with electric clocks (an "energy criminal" that steals power all the time and keeps the inverter on — see the Clocks section under Inappropriate Electrical Loads for more information). You may be able to disconnect the wiring going to the clock but leave the ignitor functional. And watch for glow-plug oven-ignition systems. The glow-plug chosen by many manufacturers to ignite the oven draws 300 watts for as long as the oven is turned on (whether the burner for the oven is turned on or not). Some manufacturers make energy-responsible gas ranges with electronic or standing pilot ignitions for both the stovetop and oven. If efficiency is an issue in your home, it is worth seeking out these appliances. Most will work without being plugged in, although almost all ranges have lights, clocks, and other gewgaws unrelated to the basic cooking function because we Americans are so gadget-happy. Ask the friendly salesman, and insist on a good answer. Appropriate stoves for an alternatively powered house should be designed to operate in a power failure and to be lit with a match or striker.

Entertainment Equipment

Stereos, CD players, computers, TVs, VCRs: you got 'em, we all love 'em. Fear not, there is no reason to give them up just because you move to your idyllic place in the country. But, here again, you need to exercise some common sense. Please read the section on inverters and modified sine-wave vs. pure sine-wave power starting on page 224. For some of our electronic toys, it's important.

Let's look over this general subject one item at a time to relieve confusion.

Computers: Personal computers are one of the easiest appliances to run on renewable energy. Most will happily accept modified sine-wave power with no distress. An average computer will draw about 50 watts, a monochrome monitor will draw an additional 20 to 30 watts, or the typical VGA color monitor an additional 50 to 75 watts. For those who spend a serious four hours or more in front of the computer on a daily basis, you might want to consider a laptop unit or some of the other ultra-low power-use options available. Many on-the-grid computer users already use an alternative energy system, in the form of an uninterruptible power supply (UPS). Industrial strength UPSs often use the same batteries and inverters employed in an independent home; the Real Goods tech staff can help you come up with an effective solution.

It's sometimes smart to use a small, separate inverter to run the computer system rather than the main household inverter, especially if the main inverter is subject to sudden surge loads from pumps, washing machines, or power tools. Computer power supplies are sensitive to such spikes, and will often spontaneously reboot, costing you precious data. A small separate inverter running off a common battery bank usually solves this problem. If big loads draw battery voltage down too much, a separate battery bank may be in order.

A WORD ABOUT SOLAR OVENS

In our food-in-a-hurry culture, we often lose some of life's greatest pleasures. We discovered this one at a job site where electricity was limited — the solar panels were carried in, up 2,000 feet of rocky trail, and we thought twice before using the power saw. A simple solar oven or box cooker took an hour or two to get food to the same state of piping readiness as a microwave would have achieved in a minute or two, but the food seemed to taste much better. For us, something almost mystical occurred when the burritos and artichoke dip were heated directly by the sun. We concluded, then and there, that conventional ovens — whether gas, electric, or microwave — should be reserved for days when the sun doesn't shine. At higher altitudes, even sun filtered through the mares'-tails and thin overcast brings a sun-oven up to a perfect cooking temperature surprisingly quickly. During the last hour before lunch, you get to think about the luscious morsels heating to perfection in their cozy sunroom. Try it yourself; we think you'll like it a lot.

— Michael Potts

Printers: Some computer printers can be choosier than computers about the type of power they consume. Others, laser printers in particular, are high-order power gobblers. The common printer types break down as follows:

Dot Matrix: This is the common noisy kind of printer, which will run happily off modified sine-wave power but will draw an average of 200 watts while running. If your inverter and battery system can support the load, no problem.

Jet Types: Ink Jets, Desk Jets, Bubble Jets — every manufacturer has its own brand name. They all work great on modified sine-wave power, and the best part is they only draw 25 to 30 watts on average. Jet printers are the best choice for self-powered households. Low power use and the ability to do graphics make them a good replacement for power-hungry laser printers.

Laser Printers: Not recommended for renewable energy systems. Laser printers consume large amounts of power (1100 to 1500 watts is average) keeping their fusers hot, and many are also intolerant of modified sine-wave power. If you absolutely *must* use a laser printer, turn it on only when you mean to use it, and batch your printing so it is done efficiently. Call the Real Goods technical staff for recommendations on inverters and printers that will work together as well as possible.

Stereos: Some stereos will pass an audible buzz through the speakers when operating on alternative energy systems. Aficionados call this buzz the "60-cycle hum," and it is often particularly offensive when the stereo is running on modified square-wave power. This problem plagues both AC- and DC-powered audio systems. It does not hurt the stereo, but it annoys the listener. The presence and volume of this buzz depends on the power supply inside the stereo and what kind of equipment is present in your particular alternative energy system. (Sometimes, as in a conventionally powered stereo, the buzz is caused by proximity between power and signal wires, so experiment with moving the wires around before you take the system out and shoot it.) More AC equipment with better-quality filtering is on the market, but there are too many brands and too many models to be able to point out specific good ones or bad ones. The best advice we can offer is this: plug it in and try it out. If you're buying new gear, try to find an understanding dealer who will let you try it out before committing to purchase.

There are some advantages to staying with DC for a stereo system. Power consumption of a typical AC-powered receiver/stereo will average 30 to 50 watts. The typical DC-powered boombox or automotive stereo will only draw 10 to 15 watts to do the same job. Most analog electrical systems like stereos confine noisy alternating current to a shielded box in the back of the bus, because AC hum is a bear to filter out and is best kept away from noise-prone circuitry. If you leave the radio on all day long for background noise, you will certainly want to choose DC-powered gear. Automotive systems can satisfy the most demanding audio freak, while consuming much less energy than their wasteful AC-powered cousins.

Many portable boom boxes, whether they have external 12-volt plugs or not, actually run on 12 volts. If the unit takes eight batteries, it's a 12-volt, though you may have to wire up your own 12-volt power plug. Many of these portable stereos also have output jacks for larger, better-quality home speakers. The obedient speaker cares not a peep whether it is driven by an AC- or DC-gobbling master.

TVs and VCRs

We very rarely have serious problems with TVs or VCRs running on modified sine-wave power. All inverters broadcast interference; it goes with their job. Occasionally the TV or the antenna is too close to the inverter and picks up this interference. Use coax cable for the antenna lead-in and any other signals that get amplified before you hear them (between the tape deck and the amplifier, for example) and keep the entertainment gear at a little distance from the inverter. (If you are at all worried about EMR — electromagnetic radiation — you will want to keep your inverter at a distance from any place where you habitually spend time anyway.) Twisting the inverter input cables around each other also helps cure the problem; the radio frequency interference (RFI) radiating from one cable tends to cancel the RFI from the other cable. Some TVs will display a bit of video interference, a line or multiple lines across the screen gradually creeping upward, or "sparklies" and snow in solid colors, especially red. It doesn't hurt the TV; the cure is either a sine-wave inverter or an expensive filter/line conditioner.

Other Household Appliances

Hair dryers, clothing irons, waffle irons, and any other appliances using electric heating elements can burn substantial amounts of power if used for any length of time. These appliances can be used on an independent system, but their use must be managed with a healthy dose of common sense. Most folks will have power to burn in the summer, but in the winter it is easy to burn your bridges. Keep an eye on the voltmeter and learn to understand what it tells you. If the batteries are low and you do three hours of ironing, the batteries are going to be scraping bottom, and may even have sustained irreparable damage.

Laundry Equipment/Power Tools

These appliances represent some of the largest loads we routinely run. Full-sized, 2000-watt or larger inverters will handle most any 1/2 hp washers or hand-held power tools *if* your battery bank is large enough and in a good enough state of charge. Many independent households run the washer off the inverter in the summer when there is usually power to spare, and off the back-up generator in the winter while catching a bit of battery charging simultaneously. Gas dryers only draw 300 to 400 watts to tumble the drum, and are no problem to run. Solar driers (a.k.a., clotheslines) are even easier (and much cheaper!) to run and can be used during most of the year. We have even seen houses with special clothes-drying rooms — Amory and Hunter Lovins's Rocky Mountain Institute has one — designed with good solar exposure and thermal behavior. If you want to run an electric dryer, see the Power Hogs heading under Inappropriate Electrical Loads.

Almost all hand-held and some stationary power tools can be run with a full-sized inverter. But beware! Many power tools draw large amounts of energy from the batteries, so demand-side power management (or "common sense" for the independent power producer) is again important. If you make a few cuts with a power saw, no problem; if you are cutting all day, however, you will probably need to start the generator periodically.

These appliances and tools have brutal starting surges that may take all of the inverter's power momentarily, so do not run the computer or other voltage-sensitive loads at the same time as power tools or try to run a load of wash with the same inverter. Dr. Doug reports that, since his CD player always skips to the first track and starts over every time the power saw is started, he has learned to enjoy tapes while doing carpentry.

Rechargeable Power Tools

These are generally fine to use but deserve a few words of warning. Most rechargers switch over to a trickle charge when the charge cycle is over, but, they never completely shut off. We call this phenomenon a *phantom load,* and try to eliminate it from a responsibly managed power regime. Charge your battery, remove it from the charger, and then unplug the charger. Also, be aware that with modified sine wave, some chargers never come off of full charge at all, and the expensive battery pack goes into meltdown. Black and Decker and some Sears Craftsman chargers are the ones we've heard about most often in this regard. Keep an eye on the charger the first time you use it to make sure it will turn off when the battery or charger starts to get warm. Many companies make 12-volt DC chargers (to be recharged off the cigarette-lighter plug in a vehicle on the job). This makes better sense than taking DC from your big batteries, inverting it to run the 120-volt AC charger, which rectifies it back to DC to charge the little battery. DC rechargers are *so much* more efficient that even the trickle charge is acceptable; use them in preference to AC-powered chargers if you can.

Water Pumps

This category of loads goes into the Appropriate Electrical Loads section only after a big caution! Use DC-powered pumps whenever practical. Most AC-powered water pumps are abysmally inefficient. Use them only if no viable DC-pumping system is available. AC pumps typically use *three times* as much wattage per gallon as a comparable DC pump on the same job. However, we often run into situations where only an AC submersible pump will suffice. This is particularly true when lifting more than 250 feet or in hard-freeze climates, where buried storage tanks might be impractical. See our Pumping section in the Water chapter (page 286) for further discussions of the best available solutions.

If you must use an AC-powered pump, you will need a full-sized inverter (2000 watts or larger) and a good-sized battery bank that is always kept well-charged. This arrangement will easily run a 1/3 hp submersible pump, will run a 1/2 hp submersible under most conditions, and may run a 3/4 hp submersible under lightly loaded conditions. Avoid larger-horsepower AC pumps if possible. The equipment necessary to support them gets very expensive.

For household pressure boosting, use a DC pump. We like either the cost-conscious Shurflo Medium-Flow model, or the Solar Star Booster pump for greater volume applications. Power use will be about one-third that of a comparable AC-powered booster pump. Again, see the Pumping section on page 286 for more information.

Electrical Loads to Run Directly from the Batteries

For most household appliances, the convenience, availability, and generally higher quality of AC appliances overrides any energy savings for comparable DC appliances, *with a few notable exceptions*. Motor-driven appliances like refrigerators, water pumps, and fans are significantly more efficient in DC form. This is partly because DC motors tend to be more efficient than AC motors, but mostly because these DC appliances are manufactured with energy efficiency in mind. We discuss stereos and answering machines in other parts of this chapter; if you use them a lot, go DC. Many folks run just one DC outlet into each room for emergency back-up. That way, if the inverter fails, you can still have lights while you're getting it fixed. Inverter failures run less than 1% average, but they do happen, and are worth planning for.

INAPPROPRIATE ELECTRICAL LOADS

Technically speaking, anything electrical can be powered with alternative energy; the equipment is available and reliable. Economically speaking, the cost for renewable energy power is high enough to require either very deep pockets or a willingness to conserve. (Funny how conservation always wins that race.) Listed below are appliances and loads to avoid when planning your alternative energy house.

Power Hogs (220-Volt Stuff)

Most appliances that use 220-volt AC are doing it because they use exorbitant amounts of electricity. (Submersible water pumps are sometimes the exception; they use 220-volt for ease of transmission.) Electric ranges and ovens, electric water heaters, baseboard heaters, and electric dryers are the most common offenders. Explore your solar- and gas-powered options before settling on one of these power pigs.

Air Conditioners

We accept that air conditioning makes life bearable (or possible) in many areas, but power consumption is *very* high for all current-production air conditioners. We suspect, based on research in desert regions and in areas with a high number of cooling degree-days, that much of the demand for air conditioning is simply a result of bad building design. If a building has been designed to remain cool inside despite a high outside air temperature, we believe it will require *much* less artificial cooling. We believe this investment in smart design is well worth the cost. Energy hard-liners even go so far as to say that if, after applying the best available heat-management technologies, a person still cannot live without air conditioning, then maybe humans weren't meant to be living in that particular environment.

Generators are the only way to run an air conditioner for any length of time. If yours is a dry climate, consider swamp coolers instead; they use one-third the energy of a refrigerant unit, and can be run with suitably sized, stand-alone equipment. Better still, if you haven't built your house yet, design one that doesn't need air conditioning to survive: superinsulated, earth-bermed, straw bale, and underground houses are all options (see Chapter 3 on Shelter for more information).

Heating Systems (and Pellet Stoves)

Forced-air heating systems and pellet stoves use substantial amounts of power to run the fans and combustion air blowers during the time of year (winter) when most independent systems can least afford energy drain. If you can possibly use a passive system such as a woodstove, wall-mounted gas heaters (without fans), or, better yet, good initial passive solar design, then your system will be vastly less expensive.

Coffee Makers

We're not so crazy as to suggest you give up America's favorite drug, er, beverage; Dr. Doug loves his caffeine, too! However, the typical "Mr. Caffeine" drip-type coffee maker uses 1000 to 1500 watts of power when it's actually brewing the coffee and 500 to 600 watts keeping it warm. This appliance can be a serious watt-sucker! Rather than investing megabucks in alternative energy gear to support your coffee habit, apply a little common sense.

Here are two reasonable solutions. The first totally nonelectric solution is to boil water in a teapot, pour it through a drip filter (like the common Melitta setup) and then, for best efficiency and flavor, store it in a Thermos. Ta da! You've now made coffee that's as good or better-tasting than anything Mr. Caffeine could brew without using a drop of electricity. Dr. Doug has done it this way ever since a camping trip in 1973 when some fellow campers turned him on to fresh-ground drip coffee and forever changed his life.

The second solution is to use Mr. Caffeine *just* to make the coffee, then turn it off. This won't use much more power than a cycle of the toaster. Pour the brewed coffee into a Thermos or reheat the coffee a cup at a time in the microwave. This results in much better-tasting coffee while saving plenty of power.

Conventional Mass-Produced Refrigerators

We have already discussed these dinosaurs briefly in the Appropriate Loads section (see page 155). Conventional refrigerators are big-time watt-suckers. Use propane if yours is a small family, or buy an ultra-efficient refrigerator. You will end up many dollars ahead when you figure how many PV panels the efficient fridge will save you from buying. In the next few months, the super-efficient fridges produced as a result of the government's "golden carrot" program should become available. Compare their prices and efficiencies carefully with the other alternatives before buying. Remember, mass-manufacturers are willing to spend big bucks hyping you about their righteousness in order to avoid having to be truly righteous; image is cheaper than performance, especially in the energy arena.

Food Dehydrators

These most often use our inefficient buddy, the 100-watt incandescent light bulb, in its best role: as a heat source. Energy use is very high, and efficiency is low. We would rather tell you to use free sunlight, but solar-powered dehydrator designs can be troubled by local conditions, spores, humidity, and other factors. Real Goods has plans for an excellent solar food dehydrator. See page 504.

Incandescent Light Bulbs

We discussed these also in the Appropriate Electrical Loads section. They make great little heaters, but only fair light sources. They have no business in an independently powered house. Compact fluorescent lamps are cheaper than PV modules.

Clock Radios, Plug-in Clocks, Answering Machines, Electric Toothbrushes (or Anything That Uses a "Power Cube" in the AC Socket)

These appliances use very tiny amounts of power, maybe only a watt or two, but they make the inverter stay turned on and running 24 hours a day. It might cost their manufacturers only a quarter a unit to build a meaningful, power-saving stand-by mode into the power cube, but since you, the consumer, are paying for the inefficiency, what do they care? On an inverter system, one of these loads, if it is large enough to awaken the inverter from stand-by (where it uses only 0.2 to 1.0 watt), consumes its own load *plus* the inverter's overhead, which is 5 to 10 watts. While this sounds like grasping at straws to those unfamiliar with the demands of an independent power system, losing this much power to inefficiency on a continuous, 24-hour basis can add a module or two to the size of the PV array, in order to keep from going bankrupt energy-wise.

The solution for clocks is to use battery power. A wall-mounted clock runs for nearly a year on a single AA rechargeable battery. We have found great solar-powered clock radios, good-quality battery-powered alarm clocks, and most of the other timekeepers anyone could possibly require just by looking around. Clocks on house current are ridiculously wasteful.

Most answering machines actually run on 9 or 12 volts DC, which makes international marketing easy because the manufacturer just changes the transformer "power cube" for different countries. A 12-volt DC answering machine can be run directly from a 12-volt battery, but it will probably need an isolated battery all to itself to avoid audio interference problems from the inverter. If the little power socket on the device (where the plug from the power cube goes) is marked "12

VDC" that is still not enough; these devices want pure voltage, and the hash from an inverter travels right across the batteries and through the 12-volt lines, throwing the sensitive electronics into a tizzy.

Electric toothbrushes use very little power themselves, but their charging systems are grossly inefficient. If the toothbrush has a power cube, see below. We bet you can do a better job, if you try, with elbow grease alone.

Power Cubes: Watch out for those small cube-shaped transformers that plug into the wall outlet to power a lower-voltage appliance. These villainous wastrels usually run a horrible 60% to 80% *inefficiency* (which means that, for every dime's worth of electricity they consume, they throw away six or eight cents worth). Some of these nasties, like rechargeable battery chargers for your cordless electric whatever, *always* draw power, even if there's no battery, toothbrush, razor, or cordless phone present and charging. We recommend that power cubes be kept on plug strips or switchable outlets that can be switched OFF when not in use. Use these energy criminals *only* when the appliance actually needs charging.

Phantom Loads

This one sounds interesting, doesn't it? Many modern appliances actually remain partially on when they appear to be turned off. Anything that can be powered up with a button on a remote control must remain partially on and listening to receive the "on" signal. Anything with a clock — VCRs, coffee makers, microwave ovens — uses a small amount of power all the time. In an alternatively powered home, the idea is to strive for the no-load state. That way, when nothing is drawing electricity, the inverter can go to sleep, and the batteries just sit there and store electrons. Again, plug these appliances into a power strip or switched outlet and switch them off when not in use. (We have noticed that this solution cures PMS — Perpetual Midnight Syndrome — on your VCR, too!)

Electric Vehicles

It's usually not practical to run your electric car off your remote home power system. Electric cars typically require two to three times more energy to operate than your house does (unless you drive only one or two miles a day). Furthermore, electric vehicles typically require their power at 96 to 120 volts DC, while your house only runs at 12 or 24 volts DC. Electric-vehicle charging stations are usually dedicated to vehicle charging. On the other hand, for those lucky folks who have surplus hydro or wind electricity, an EV is a great place to send the excess energy.

Now What?

Toss those energy hogs out on their ears, embrace an energy-efficient lifestyle, and gallop off into a glorious sunset. Live happily ever after.

Actually, you should turn to the worksheets (see page 147) and figure out your loads, but we liked the first answer better, for that is ultimately what we are all working toward.

— Doug Pratt and Michael Potts

CHAPTER 5

MANAGING ENERGY SYSTEMS

THE LAST CHAPTER dealt with renewable energy sources and how to harvest energy. This chapter deals with what to do with the energy once it's harvested. This is accomplished by "balance of systems" (BOS) equipment, which takes care of the safety, control, monitoring, storage, and conversion aspects of the typical renewable energy system.

In the block diagram of a typical renewable energy system (page 166), note that electricity flows from the generating source, through some safety equipment, through a charge controller, and then into the battery. Incoming power can come from photovoltaic (PV) modules, a wind generator, a micro-hydro turbine, or some other source. Different types of charging sources usually require different control technologies, but can share the same battery bank. As energy is needed, it is either drawn directly from the battery as DC current, or via an inverter as AC house current. Monitoring equipment keeps track of source activity, battery state, and power demand. This chapter explains all of the equipment "downstream" from the source.

Batteries

Batteries provide storage, accumulating surplus energy beyond what is needed to run the household while the source is productive, and doling out previously saved energy when the household needs more than charging sources can provide. Batteries have a finite capacity, like buckets: once they are full, no more energy will fit into them; when empty, no more energy will flow from them.

The stored energy in the battery can be used in the household, either directly as low-voltage DC power or processed into conventional AC house current like that supplied by the utility grid. Safety devices should be provided between any autonomous sources of power — between any energy source and the system — and

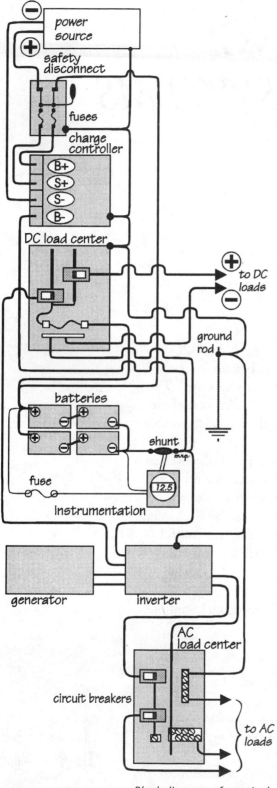

Block diagram of a typical renewable energy system

Labels in diagram:
power source
safety disconnect
fuses
charge controller
B+
S+
S-
B-
DC load center
to DC loads
ground rod
batteries
shunt
12.5
fuse
instrumentation
generator
inverter
AC load center
circuit breakers
to AC loads

batteries should be isolated from the rest of the system. This is because even small batteries are capable of releasing awesome quantities of energy suddenly or accidentally. For the sake of safety, fuses are an absolute must on any circuit that is connected to a battery. Without them, you risk burning the house down.

Controllers

Since we can fit only so much energy into a battery, it's usually necessary to have a charge controller to stop the charging process when the battery is full. Back to our bucket analogy: overfilling a bucket usually does little harm, but batteries can be damaged or even destroyed if overcharged excessively. When filling a bucket with a hose, you have to slow down as you approach full to avoid spilling. Batteries, like the bucket, can absorb very heavy charging when empty, but the energy must be metered in more slowly as the battery approaches full charge. Controllers do this job.

PV controllers have the additional duty of preventing the minute reverse current flow that happens to PV modules at nighttime. This reverse flow is very small, but with multiple modules it can become significant.

Wind and hydroelectric controllers usually have to divert any excess energy to some load that can absorb it. Typically a heating element is used. Wind and hydro sources usually cannot be turned off easily.

Monitors

If you are going to operate your own power system, you may as well become good at it — you will be repaid in longer component life and greater reliability. A system monitor allows you to "see into" the electronic world and get a handle on what's going on. Simply monitoring battery voltage gives a fairly reliable indication of the system's state of charge, and what sort of activity is occurring at the moment. Current flowing within the system can also be usefully monitored. More sophisticated monitors keep track of the quantity of energy that has passed into or out of the system, with adjustable alarm points to warn you if something is amiss.

Safety Equipment

Batteries are convenient. Their technology has been around for more than 100 years. We are all somewhat familiar with them due to the fact that every motor vehicle has one. A few words of advice, though: *Don't get complacent around batteries!* Even a small (by our standards) automotive starter battery can easily turn a carelessly placed 10-inch crescent wrench red-hot in seconds. After abuse like this, for a grand finale, a battery can even explode. (While a battery explosion is not a satisfying Hollywood-style explosion with billowing flames and slow-motion parts tumbling through the sky, it is spectacular in an understated way, with corrosive battery acid spattering around, putting eyes, skin, pets, furnishings, and structure at risk. Our advice: do not try this in your home.) We consider safety equipment designed for low-voltage DC to be *absolutely essential* on any circuit that connects to a battery. If you are unsure how to install a component or system safely, please call a member of the Real Goods technical staff. We would all rather you be safe than sorry.

When we design systems, we comply with the National Electrical Code (NEC), and use UL-listed equipment as much as possible. (UL listing for DC equipment is a new field, and consequently there is still a good deal of equipment on the market that is not UL-listed. This does not mean that it is inferior or unsafe in any way. Equipment does *not* have to be UL-approved to comply with NEC code, but a UL listing provides an additional layer of assurance, which makes building inspectors sleep much easier.)

Integrated Equipment

Someone finally got smart and put all the necessary BOS equipment into a single unit. Safety, charge control, and monitoring equipment can now be included in a single pre-engineered, prewired, pretested, and UL-approved Power Center. We think this is the direction of the future in our industry. Power Centers take up less space, take far less time to install, and easily pass building inspection. You *do* pay something extra for the clean packaging and prewiring, but if you are paying someone to install your system, their time will be reduced, and you will immediately make back part of the cost. If you are installing your own system, it depends on how much you value your time and the good looks of the finished product. There are six basic Power Center models to accommodate different system sizes, with a variety of options available to suit individual needs. Many of the options can be added in the field years after the original installation. Dr. Doug says check it out!

Inverters

Batteries store and release energy as low-voltage DC (direct current) electricity. This works fine for some appliances, but the world's power grids have settled on using higher voltage AC (alternating current) power because it transmits easily. In its AC form, electricity cannot be stored. We find it convenient to invert most of the energy we generate for our household into conventional 120-volt AC, so that

we can use common, mass-produced electrical appliances. Logically enough, the device that does this job for us is called an inverter. Modern inverters are more than 90% efficient and have failure rates well under 1%, which is more efficient and reliable than any utility company we know. You'll find much more information about these electronic marvels in the inverter section starting on page 224.

ELECTRICAL UNITS

The power consumed by most electrical appliances is stated in *watts,* the unit of energy consumption per unit of time. One watt delivered for one hour equals one *watt-hour* of energy. Wattage is the product of current (measured in amperes, or *amps* for short) times voltage. This means that one amp used at 120 volts is the same wattage as 10 amps used at 12 volts. Wattage and voltage are independent of each other: a watt at 120 volts is the same amount of energy as a watt at 12 volts. To convert a battery's amp-hour capacity to watt-hours, simply multiply the amp-hours times the voltage. To calculate how much battery capacity it will take to run an appliance for a given time, multiply appliance wattage times the number of hours it will run to get the total watt-hours, then divide by the battery voltage to get amp-hours required. For example, running a 100-watt lightbulb for 1 hour uses 100 watt-hours. If a 12-volt battery is running the light, it will consume 8.33 amp-hours (100 watt-hours ÷ 12 volts).

BATTERIES

Batteries provide energy storage. They accumulate energy as it is generated by various devices such as PV modules, wind turbines, or hydro plants. This stored energy buffer runs the household at night or during extended periods when there is no energy input — a long string of cloudy days with a PV system, for example. Batteries can be discharged rapidly to yield more current than the energy source can produce by itself, so pumps or motors can be run intermittently.

Batteries need to be treated with care, to prolong their life expectancy and for safety reasons. If commonsense caution is not used, batteries can provide enough power to cause impromptu welding accidents and even explosions.

A Little Respect

Batteries are often the most misunderstood component of an renewable energy system. Compared to all the other electronic marvels in the typical renewable energy package, the battery is a very simple electrochemical component. Energy research is an important and active field, and some promising new technologies are on the horizon. Many of these new technologies are likely to bear fruit within the next few years, but are also likely to be too expensive for terrestrial and civilian use. So, for the time being, we must make do with traditional battery design — a technology that is nearly 100 years old, but one that is tried and true and requires surprisingly little maintenance. The care, feeding, cautions, and dangers of lead-acid batteries are well documented. Safe manufacturing, distribution, and recycling systems for this technology are in place and work well. Can we say the same for a sulfur-bromine battery?

Batteries are the oldest technology in the independent power system. Improvements in the construction and packaging of batteries have been made, but their basic chemistry has not been much improved for a century or more. We often hear promises of "dramatic breakthroughs" in battery technology, but the faithful old lead-acid battery is still a serviceable workhorse which, if granted respect and minimal care, will serve us well.

Battery Capacity

A battery's capacity for storing energy is rated in ampere-hours, or *amp-hours* for short; one ampere delivered for one hour equals one amp-hour. To know how much total energy is delivered, it is necessary to know at what voltage the amp-hours are delivered. Battery capacity is listed in amp-hours at a given voltage. For instance, a typical golf-cart battery's capacity is 220 amp-hours at 6 volts.

Automotive engine-starting batteries are rated for how many amps they can deliver at a cold temperature, or *cold cranking amps* (CCA). This rating is not relevant for storage batteries. Beware of any battery that claims to be a deep-cycle storage battery and has a CCA rating.

Battery manufacturers typically rate their storage batteries at a 20-hour rate. Continuing the previous example, a golf-cart battery that is rated at 220 amp-hours

will deliver 11 amps for 20 hours. The 20-hour rate is the standard we use for all the batteries in the *Sourcebook*. This rating is designed only as a means to compare different batteries to the same standard, and is not to be taken as a performance guarantee.

Batteries are electrochemical devices, and are sensitive to temperature, charge-discharge cycle history, and age. The performance you will get from your batteries will vary with location, climate, and usage patterns. In the end, a battery rated at 200 amp-hours will provide you with twice the capacity of one rated at 100 amp-hours.

Batteries are less-than-perfect containers for storing the energy of our power systems. For every 1.0 amp-hour you remove from the batteries, it is necessary to pump about 1.25 amp-hours back in to bring the battery back to the same state of charge. This figure varies with temperature, battery type, and age, but it is a good rule of thumb by which to calculate approximate battery efficiency.

Kinds of Batteries

There are two different kinds of electrochemical batteries in common use for alternative energy systems: Lead-acid, by far the most common, and nickel-cadmium (or *nicads*). We'll look at the advantages and disadvantages of each.

Lead-Acid Batteries: The lead-acid battery cell consists of two lead plates of slightly different composition suspended in a dilute sulfuric acid solution (called the *electrolyte*) and contained within a chemically and electrically inert vessel. A battery is literally a collection of cells. A typical automotive lead-acid starting battery consists of six cells, each of which produces approximately 2 volts. These individual cells are connected in series, so their voltages are additive. Larger cells provide more storage capacity, measured in amp-hours, but the voltage output never exceeds the 2-volt peak potential of the chemical reaction which drives the cell.

Lead-acid batteries produce hydrogen gas during charging, which poses a fire or explosion risk if allowed to accumulate. The hydrogen must be vented to the outside. Lead-acid batteries will also sustain considerable damage if they are allowed to freeze. A fully charged battery can survive temperatures as low as -40° F, but as the state of charge falls, sulfur begins to bond chemically with the lead, and the electrolyte becomes closer to plain water. Electrolyte tends to stratify, with a lower concentration of sulfur ions near the top. If a battery gets cold enough and discharged enough, it will freeze. At a 50% charge level a battery will freeze at approximately 15° F. This is the lowest state of charge you should ever intentionally let your batteries reach. If freezing is a possibility, the batteries can be kept indoors (remember to vent!) as long as the house is occupied or protected against freezing. If it's an occasional-use cabin, the batteries may be buried in the ground inside an insulated box.

Lead-acid batteries age in service. Once a bank of batteries has been in service for six months to a year, it generally is not a good idea to add more batteries. A battery bank performs like a team of horses, pulling only as well as the weakest. New batteries will perform no better than the oldest cell in the bank. All lead-acid batteries in a bank should be of the same capacity, age, and manufacturer.

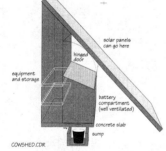

An ideal battery enclosure

The state of charge of lead-acid batteries can be monitored either the hard way, with a hydrometer, by taking up a sample of electrolyte, or the easy way with a voltmeter. Using a hydrometer is the most accurate way to monitor the battery condition, but it can be messy and hard on both your clothes and the batteries. (Wear old clothes, watch out for drips, and don't let anything drop into the battery cell.) Battery voltage is a measure of the charge level of the battery. The voltmeter is only accurate when the battery is in what is called its *at-rest state,* having been neither charged nor discharged for an hour or so. Since voltage is quite elastic and will stretch upward when a charge is applied, or stretch downward when a load is applied, the battery's internal chemistry needs time to settle down before a reading can be obtained that truly indicates the battery's state of charge. Digital meters are highly recommended for their high degree of accuracy and their ability to read fine differences. We generally recommend using a voltmeter to monitor the battery's *state of charge* on a daily level, and using a hydrometer to monitor the battery's *state of health* on a semi-annual basis. The fewer times the battery tops are opened, the better (both for your clothes and for the battery). See the Battery Care, Safety, and Maintenance section on page 176 for more information.

Batteries are built and rated for the type of "cycle" service they are likely to encounter. During a *cycle* the battery goes from its charged state through partial discharge, then is recharged. Cycles can be "shallow," yielding 10% to 15% of the battery's total capacity, or "deep," providing 50% to 80% of total capacity. No battery can withstand 100% cycling without damage, often severe. Batteries designed for shallow-cycle service, such as automotive starting batteries, will tolerate few, if any, deep cycles without sustaining internal damage. This makes them unsuitable for independent power systems. Batteries used in remote power systems must be capable of sustaining many repeated deep cycles without ill effects.

Batteries wear out, and when they do, they must not be disposed of carelessly. The active ingredients in these batteries, lead and sulfuric acid, can be highly toxic and dangerous and need to be handled with great respect. Every stream on the Hawaiian island of Maui, for example, has been found to contain certain trace poisons which can only result from the improper dumping of lead-acid batteries. In addition to posing a danger if it enters the water cycle, lead is considered a strategic metal, and has salvage value, so it should be recycled. Turn in old batteries for recycling at auto-parts stores or recycling centers. A majority of the lead in new car batteries comes from recycled batteries.

Several common types of lead-acid batteries are presented here in order of worst to best.

Car Batteries: The most common type of lead-acid battery is the automotive battery, sometimes called a "starting battery." This type of lead-acid battery has many thin lead plates and is designed to deliver hundreds of amps for a few seconds to start a car. Starting batteries are only designed to cycle about 10% of their total capacity and to recharge quickly from the alternator after cycling. They are *not* designed for the deep-cycle service demanded by remote home power systems, and will fail fairly quickly when used in a deep-cycling application. In many developing nations, installations use this type of battery simply

because nothing else is available, and in some instances premature battery failure has irrevocably harmed the cause of rational delivery of electrical systems. Shallow-cycle batteries are appropriate only for systems that will never give up more than 10% of their capacity, and will normally "float" at a fully charged state. It is a false economy to use the wrong kind of batteries in any installation.

RV or Marine Deep-Cycle Batteries: This generic category includes all of the 12-volt batteries that Sears, Montgomery Ward, K-Mart, and other large retailers sell as "deepcycle," "RV," or "marine" batteries. They are always 12-volt, and have a capacity between 80 and 160 amp-hours. These batteries represent a compromise between starting batteries and true deep-cycle batteries. Many of them are actually used as starting batteries by RV owners who simply don't know any better. They will give far better deep-cycle service than starting batteries, and may be the ideal choice for a beginning system that you plan to expand later. Life expectancy for these batteries is typically two to three years.

Gel-Cell Sealed Batteries: Gel-cell sealed batteries have the acid either gelled or put into a spongelike filler. They have the advantage of being completely sealed. They can operate in any position, even sideways or upside down, and will not leak acid or gas. Because their electrolyte moves more slowly, these batteries cannot tolerate high rates of charging or discharging for extended time periods, although their thinner plates will allow high rates for a short time. Their sealed construction, which makes them ideal for some limited applications, makes it impossible to check individual cell conditions with a hydrometer. The higher voltage charge rates commonly used in PV systems will cause gassing and eventual premature cell failure due to water loss. We recommend this type of battery only in situations where hydrogen gassing during charging cannot be tolerated, where the battery is going to be moved and handled a great deal, or where the battery needs to fit into unique, tight spaces. Boats, UPS computer power supplies, and remote scientific expeditions are the most common uses. Special lower-voltage charge controls *must* be used with these batteries. Life expectancy is two to three years for most gel cell batteries.

Lead-Calcium or "Telephone Company" Batteries: During the past ten years telephone companies have been upgrading much of their switching equipment from the old-style 48-volt relay type to newer solid-state equipment. When a telephone station is changed over, the monster battery bank that ran the old equipment is either discarded or recycled. Many of these shallow-cycle lead-calcium batteries are finding new homes in remote power systems. The typical life expectancy for these batteries is 15 to 20 years, although there are some on the market that are said to last 50 years or more. These batteries will work fine in remote power systems if you treat them carefully. These are shallow-cycle batteries that rarely experienced more than a 15% cycle in telephone service. If you are careful never to discharge them deeply, these batteries can give years of excellent service. As usual, though, there is a trade-off with using phone-company batteries. While they can sometimes be found cheap, or even free for the hauling, their sheer weight and size make them difficult to contend with. Some of these batteries weigh close to 400 pounds per 2-volt cell! Because cycle capacity is limited to 15% or 20%, you have to buy, move, and install five or six times more battery mass than is required for true deep-cycle batteries. Remember, even though the phone company battery may be rated at 1680 amp-hours, you can only use 20% of that, or 336 amp-hours, which isn't much by renewable energy standards.

True Deep-Cycle Batteries: These batteries are specifically designed for storage and deep-cycle service. They tend to have larger and thicker plates. This is the type of battery that is best suited for use with renewable energy systems. They are designed to have a majority of their capacity used before being recharged. They are available in many sizes and types, the most common being 6-volt and 2-volt configurations for ease of movement. Once in place, the multiple batteries are series- and parallel-wired for your basic system voltage. These batteries are built to survive hundreds or even thousands of 80% cycles, though for greatest life expectancy we recommend 50% as the normal maximum discharge. Save that 30% difference for emergencies, and never use the last 20% unless you enjoy buying new batteries. The less deeply you regularly cycle your batteries, the longer they will last. The three most commonly available batteries within this group are the "golf-cart" types with a three- to five-year life expectancy, the L-16 series with a seven- to ten-year life expectancy, and Industrial Chloride forklift-type batteries with a 15- to 20-year life expectancy. Deep-cycle batteries are probably your best battery investment. We usually recommend at least the golf-cart types for beginning systems.

Nicad (Nickel-Cadmium) Batteries

The nickel-cadmium battery reaction is very different from the lead-acid battery reaction. These cells use a base, potassium hydroxide, as the electrolyte instead of an acid. The electrolyte does not enter into a chemical reaction with the plate materials; it only acts as a transfer medium for the electrons. So the specific gravity of the electrolyte does not change with the state of charge. The battery's positive pole is composed of nickel compounds, and the negative pole is composed of the metallic element cadmium. Cadmium is a highly toxic element and should be treated with great respect and caution. Nicads enjoyed a tremendous surge of popularity a few years ago, with many folks jumping over to them for the increased life expectancy they offered without fully understanding their drawbacks.

Pocket-plate construction is the nicad battery type that is most likely to be found in storage-battery service. There is considerable controversy concerning the life expectancy of the common pocket-plate nicads. Life expectancy can vary from a few hundred cycles to tens of thousands of cycles depending on how deeply the cells are cycled. Contrary to popular belief, these cells do not respond well to deep cycling. Major manufacturers rate pocket-plate cells at approximately 500 cycles with 80% depth of discharge (DOD) cycling. This is not much better than a Trojan L-16 lead-acid battery, which costs a fraction of what you would pay for a nicad.

There *is* a major difference, however, in the way nicads age and die as compared to lead-acid batteries. Lead-acid batteries gradually decrease in capacity over time until they reach around 80% of new capacity. From this point on, battery

HELPFUL HINTS FOR LEAD-ACID BATTERIES

For more details, see page 176.

1. All batteries will self-discharge and can sulfate when left for long periods unattended and where no charge controller or trickle-charge device is employed. Disconnect them in this case. However, self-discharge is not prevented by disconnecting batteries alone. Keep battery tops clean and dry.
2. Keep batteries warm. At 0° F, 50% of their rated capacity is lost.
3. Batteries must be vented to prevent explosion of hydrogen and other gases.
4. Do not, under any circumstances, locate any electrical equipment in a battery compartment.
5. Wear old clothes when working with an electrolyte solution as they will soon be full of holes.
6. Be careful of metal or tools falling between battery terminals. The resulting spark can cause a battery to be destroyed or damaged.
7. Baking soda neutralizes battery acid, so keep a few boxes on hand.
8. Check the water level of your battery once a month. Excessive water loss indicates gassing and the need for a charge controller or voltage regulator. Use distilled water only.
9. Protect battery posts from corrosion. Use a professional spray or heavy-duty grease or Vaseline. If charging has ceased or been reduced and the generator checks out, chances are corrosion has built up where the cable terminal contacts the battery post, preventing the current from entering the battery. This can occur even though the terminal and battery post look clean.
10. Never smoke or light a match near a battery, especially when charging.
11. Take frequent voltage readings. The battery manufacturer will provide instructions about the voltage of a cell when their battery type is fully charged. The voltmeter is the best way to monitor a battery accurately.
12. Batteries gain a "memory" about how they are being used. Large deviations from regular use after this memory has been established can adversely affect performance. Incurred memory can be erased by discharging the battery system 95%. Recharge it to 140% of its capacity at a slow rate, then rapidly discharge the battery again completely and recharge to normal capacity. A generator/battery charger combination should be used for this procedure.
13. Use fewer, larger cells wired in series rather than lots of small cells in parallel. For example use two 6-volt batteries in series rather than two 12-volt batteries in parallel. It is safer in the case of a shorted cell, as you have half the chance of random cell failure, half the maintenance effort, and in general, thicker and more rugged plates — certainly more rugged than popular marine or RV deep-cycle batteries.

— Doug Pratt and Michael Potts

capacity starts dropping very rapidly. Within a couple months of this point, the lead-acid battery will be unable to hold any charge. This rapid loss in battery capacity is like falling off a cliff. In contrast, nicads die more gracefully. There is no sudden cliff that they drop off; their capacity just gradually diminishes. The rate at which it diminishes is relatively fast with DOD cycles in excess of 50%, but it is almost negligible with DOD cycles of less than 20%. Nicad battery manufacturers consider a battery to be "used up" when it reaches 80% of its original capacity. Many alternative energy users probably have nicad battery banks that are far below the 80% mark; yet, because these batteries still hold a charge and perform at a satisfactory level, why complain? Unless you are tracking battery capacity with an expensive cumulative amp-hour meter, you will probably never notice the slow loss of capacity in a nicad.

Nicads have some great advantages:

- They can be extremely long-lived if cycled at about 30% capacity or less. Deeper cycling will cut sharply into life expectancy.
- Nicads have flat discharge-voltage curves. The voltage on nicads remains constant until the last 10% of capacity. This characteristic is easier on inverters and other voltage-sensitive appliances.
- Batteries of different sizes, capacities, and ages can be added together, some thing that is absolutely forbidden with lead-acid batteries.
- Nicads do not lose capacity at low temperatures as quickly as lead-acid batteries do. They can tolerate freezing with no adverse effects, although they won't work while frozen.

Nicads also have some big disadvantages:

- The single biggest disadvantage is **cost.** Nicads are very expensive initially. New nicads may be three to ten times as costly as comparable lead-acid batteries. We sometimes offer used nicads, which are somewhat less expensive.
- Nicad batteries display the same voltage from 10% to 90% of charge capacity. This means that a voltmeter cannot be used to gauge the state of charge. It is necessary to use a sophisticated and expensive amp- or watt-hour accumulator to accurately judge how much energy you have in storage.
- Nicads are incompatible with lead-acid batteries. You can't mix the two battery types in a single system.
- Most inverters and charge controllers now on the market do not provide the higher voltages that nicads prefer when charging.
- Even though the batteries may live for long periods, they need periodic electrolyte changes. Potassium hydroxide is difficult to handle and highly toxic. Great caution must be used while juggling heavy batteries to pour out the old electrolyte. In fact, this is probably not a good home project.

— *Doug Pratt*

BATTERY CARE, SAFETY, AND MAINTENANCE

The following are some of the most important and commonly asked questions regarding battery safety and maintenance.

What basic safety issues are important when I'm working on my batteries?

1. Protect your eyes with goggles and your hands with rubber gloves. Battery acid is a slightly dilute sulfuric acid. It will burn your skin after a few minutes' exposure, and your eyes almost immediately. Keep a box or two of baking soda and at least a quart of clean water in the battery area at all times. Flush any battery acid that comes in contact with skin with plenty of water. If you get acid in your eyes, flush with clear water for 15 minutes and then seek medical attention.
2. Wear old clothes! No matter how careful we try to be around batteries, we always seem to end up with holes in our jeans. Wear something you can afford to lose, or at least have holes in.
3. Tape the handles of your tools or treat them with Plastic-Dip so that they can't possibly short out between battery terminals. Even small batteries are capable of awesome energy discharges when they are short-circuited. The larger batteries commonly used in alternative energy systems can *easily* turn a 10-inch crescent wrench red-hot in seconds while melting the battery terminal into a useless puddle, and, for the grand finale, possibly exploding and starting a fire. This is more excitement than most of us need in our lives.
4. Stop thinking *"Oh, none of that will happen to me; I'm careful!"* All of the stupid, avoidable catastrophes above *have* happened to us, and we're very careful and skilled, too. Don't take needless chances; it's too easy to be safe.

What should I do to get my batteries ready for winter?

1. Check the water level in your batteries and fill with distilled water. Batteries use more water when they're being fully charged every day. *Don't use tap water!* The trace minerals can poison the battery.
2. Clean the battery tops. We've been gassing the batteries at full charge all summer. The condensed fumes and dust on the tops of the batteries start to make a pretty fair conductor after a few months. Batteries will significantly discharge across the dirt between the terminals. That power is lost to you forever! Sponge off the tops with a baking soda/water solution, or use the battery-cleaner sprays you can get at the auto-parts store. Follow the baking soda/cleaner with a clear-water rinse. *Make sure that cell caps are tight and that none of the cleaning solution gets into the battery cell!* This stuff is deadly poison to the battery chemistry. Clean your battery tops once or twice a year.
3. Clean and/or tighten the battery terminals. If you were smart and coated *all* the exposed metal parts around your battery terminals with grease or Vaseline when you installed them, they'll still be corrosion-free. If not, then

take them apart, scrub or brush as much of the corrosion off as possible, then dip or brush with a baking-soda solution until all fizzing stops. Then scrape some more. Keep this up until you get all the blue-green crud off. (It's like cancer: if you don't get it all, it will return). Then carefully cover all the exposed metal parts with grease when you put it back together. If your terminals are already clean, then just gently snug up the bolts. Lead is a soft metal and will gradually "creep" away from the bolts.

4. Run a hydrometer test on all the cells. (Wait at least 48 hours after adding water before you run this test). Your voltmeter is great for gauging the battery's state of charge, while the hydrometer is for gauging its state of health. Use the good kind of hydrometer with a graduated float (not the cheapo floating-ball type). Real Goods sells one for only $6 (Product #15-702). What we're looking for in this test is not the state of charge, but the difference in points between cells. In a healthy battery bank all cells will read within 10 points of each other. Any cell that reads 20 to 25 points lower is starting to fail. You may have two or three months to get a replacement set. At 50 points difference or more, the bad cell is sucking the life out of your battery bank, and needs to get out of there now! Don't pay any attention to the color-coded good/fair/recharge markings on the float. These pertain only to automotive batteries, which use a slightly hotter acid. We're only looking for differences between cells.

What if I find a significant difference between cells on the hydrometer test?

Well, the first question is *how much* difference?

At 10 to 20 points difference, you might just need a good equalizing charge. Run your batteries up to about 15.0 volts (that's for 12-volt systems — you folks at higher voltages can figure it out) and hold them at between 15.0 and 15.5 for three to five hours. This will even out any small differences between cells that have developed over time. It's a good idea to do an equalization every two to six months; think of it as a minor tune-up for your batteries. Equalizing is more important in the winter, when batteries tend to run at lower charge levels than in the summer. At more than 20 points difference, you're probably looking at a dying cell. Replacing a single battery out of a larger bank usually isn't a good idea. We need to look at what percentage of full life expectancy the batteries are at now. If the batteries have seen 20% or more of their typical life expectancy, it probably isn't a good idea to simply replace the bad battery. For instance, a set of golf-cart batteries (three to five years typical life expectancy) with two years on them is not a good candidate for a single battery replacement. A cell in a set of Industrial Chloride batteries that's two years old, though, *is* a good candidate (though we've never had a Chloride cell fail anywhere near that early in life). A new battery installed in an older battery bank will be dragged down to the performance level of the worst cell in the bank. So don't mix old and new batteries unless you're willing to sacrifice a considerable amount of life expectancy on the new cells.

The *Battery Book for Your PV Home* (see page 181) covers in detail what we've lightly brushed over here.

I don't want to limp through another winter always being low on power and damaging my batteries. What's the best way to prevent damage?

1. Get a decent system monitor. Nothing kills off your batteries faster than repeated, excessively deep cycles. A digital voltmeter is the very minimum monitoring equipment you need. The voltmeter is comparable to the gas gauge in your car: you can operate without it, but you're running blind and may do actual damage if things quit in the wrong place. Overly discharging your batteries causes short life expectancy. Meters are cheaper than batteries. Copy the battery state-of-charge chart below, post it by your meter and pay attention to it. Note that this chart is for battery voltage measured *at rest*. Another possibility, particularly if your system is being operated by non-technical people, is to install one of the audible low-battery alarms.

2. Add an efficient battery charger. Provided you've got a 1000-watt or larger generator, the highly efficient Todd battery chargers put more watts per gallon of gas into your battery than any other charger on the market. For generator-based charging, you want the 15.5-volt models for the fastest bulk charging. The Todd will work just fine in conjunction with other battery chargers you might already have.

For larger systems the SB battery-charger option on the newer Trace inverters is the simplest way to add battery-charging capability, and it offers more power, more control, and greater reliability than the Todd chargers.

BATTERY STATE OF CHARGE AT REST	Voltage Reading	Percent of Full Charge
	12.6	100%
	12.4-12.6	75%-100%
	12.2-12.4	50%-75%
	12.0-12.2	25%-50%
	11.7-12.0	0%-25%

ASK DR. DOUG

Dear Dr. Doug:

I'm just getting started on my power system. Should I go big on the battery bank, assuming I'll grow into it?

The answer is yes and no. Yes, you should start with a somewhat larger battery bank than you absolutely need. Over time most folks find more and more things to use power for once it's available. But if this is your first venture into remote power systems and battery banks, then we usually recommend that you start with some "trainer batteries." So no, don't invest too heavily in batteries your first couple of years. The golf-cart–type, deep-cycle batteries make excellent trainers. They are modestly priced, will accept moderate abuse without harm, and are commonly available. You're bound to make some mistakes and do some horrible things with your first set of batteries. You might as well make mistakes with inexpensive batteries. In three to five years, when the golf-cart–type trainers wear out, you'll be much more knowledgeable about what you need and what quality you're willing to pay for.

Dear Dr. Doug:

The battery bank I started with two years ago just doesn't have enough capacity for us anymore. Is it okay to add some more batteries to the bank?

Lead-acid batteries age in service. Generally it isn't recommended to add new cells in series after a battery bank is six months to one year old. The new batteries will be dragged down to the performance level of the worst cell in the bank. Different batteries have different life expectancies, so we really need to consider how long the

I live in a very cold climate. How do I keep my batteries from freezing, and what happens if they do?

Freezing is usually fatal for lead-acid batteries. The expanding ice crystals bend internal plates, puncture separators, and push active lead material off the plate grids. There usually isn't much left after a hard freeze. The good news is that batteries don't freeze easily. At full charge it takes a temperature of at least -40° F, but, as the batteries discharge, the freezing point rises. At 50% depth of discharge, the lowest you ever want to take your batteries under normal conditions, they'll freeze at around 0° F.

If you're powering a full-time residence, we recommend installing the batteries indoors. Build a little lean-to enclosure against an outside wall with venting from inside to the outside. It's important that the roof be sloped to help the hydrogen rise and pass outside. You only need a little 2-inch diameter vent top and bottom to keep everything safe; plus you'll be pulling a little warm household air through the enclosure to keep the batteries toasty and happy.

If your place is a part-time recreational residence and wintertime temperature can

bank should last. For instance, it might be acceptable to add more cells to a large Industrial Chloride set at two years of age, since this set is only at 10% of life expectancy.

On the other hand, a Sears "deep-cycle" at two years of age is at 100% of life expectancy. Don't add to your battery bank once it's dipped to 10% of its life expectancy. At that point, it's better to just run them out or trade with a neighbor. All batteries in a bank should be the same age, capacity, and manufacturer's brand. Nicads are a different story. It's okay to add more capacity at any time (any time you can afford it, that is), and the cells don't have to be of the same capacity.

Dear Dr. Doug:
I keep hearing rumors about some great new battery technology that will make lead-acid batteries obsolete in the near future. Is there any truth to this, and should I wait to invest in batteries?
Ah, pie in the sky someday . . . The truth is there are at least a dozen battery technologies in the laboratory now. Some of them look very promising, but none of them are likely to give lead-acid a run for your money within the next three to five years. Lead-acid is also back in the laboratory — this old dog can still learn some new tricks. Lead-acid technology is going to be around, and is going to continue to give the best performance per dollar, for a long time to come.

Fully spent, a battery's electrolyte has turned to water, all ions having been taken up by the cathode and anode. At the peak of readiness, a battery's plates have given up the ions to the electrolyte, which is now a corrosive mixture of water and sulphuric acid.

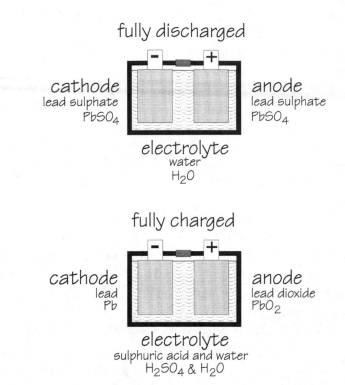

fully discharged

cathode
lead sulphate
$PbSO_4$

anode
lead sulphate
$PbSO_4$

electrolyte
water
H_2O

fully charged

cathode
lead
Pb

anode
lead dioxide
PbO_2

electrolyte
sulphuric acid and water
H_2SO_4 & H_2O

drop below -40° F, then we recommend either installing the batteries in the basement or buried in a box in the ground with 2-inch rigid foam insulation on top and sides. The bottom can be sand or finely crushed gravel. The ground will be warmer than air temperatures and we want that ground contact. This will prevent freezing in all but the most extreme climates.

In either case, *leave the charging system operational to ensure that the batteries stay charged!* If you have a charge controller with adjustable voltage, like the Trace C30A, then turn the charging voltage down to about 13.5 to 13.8 volts to reduce water usage over the winter. Whatever kind of charge controller you have, make sure the batteries are topped off with distilled water before you leave.

With a minimum amount of care and attention you can make any set of batteries last up to twice as long as "normal." A couple hours a year is a small investment considering what batteries cost to replace, and how much energy most folks spend worrying about them.

— *Doug Pratt*

BATTERIES AND STORAGE

Battery Book For Your PV Home

This booklet by Fowler Electric Inc. gives concise information on lead-acid batteries. Topics covered include battery theory, maintenance, specific gravity, voltage, wiring, and equalization. Very easy to read and provides the essential information to understand and get the most from your batteries. Highly recommended by Dr. Doug for every battery powered household. 22 pages, paperback, 1991.

80-104　Battery Book for Your PV Home　　　　8^{00}

"Golf Cart" Type Deep Cycle Batteries

These batteries are excellent "trainer" batteries for folks new to battery storage systems. First-time users are bound to make a few mistakes as they learn the capabilities and limitations of their systems. These are true deep cycle batteries and will tolerate some 80% cycles without suffering unduly. Even with all the millions of dollars spent on battery research, conventional wisdom still indicates that these lead-acid "golf cart" batteries are the most cost-effective solution to battery storage. Because of the relatively low cost these batteries are hard to beat for small to medium sized systems. Typical life expectancy is approximately three to five years.

> *Battery Voltage: 6 volts*
> *Rated Capacity: 220 amp-hours*
> *L x W x D: 10-1/4" x 7-1/4" x 10-1/4"*
> *Weight: 63 lb/28.6 kg*

15-101　Deep Cycle Battery, 6V/220A-hr　　　59^{00}

Shipped Freight Collect from Northern California

L-16 Series Deep Cycle Batteries

For larger systems or folks who are upgrading from the "golf cart" batteries, these L-16s have long been the workhorse of the alternative energy industry. The larger cell sizes offer increased ampacity and lower maintenance due to larger water reserves. L-16s typically last from seven to ten years.

> *Battery Voltage: 6 volts*
> *Rated Capacity: 350 amp-hours*
> *L x W x D: 11-3/4" x 7" x 16"*
> *Weight: 128 lb/58 kg*

15-102　Deep Cycle Battery, 6V/350A-hr　　　169^{00}

Shipped Freight Collect from Northern California

"Golf Cart" Type

L-16 Series

"RV Or Marine" Deep Cycle Batteries

These batteries are ideal for small or just-starting-out systems. They represent a big step up from ordinary automotive starting batteries, which should never be used in storage systems. These batteries have high antimony plates for good deep cycle service. Normal life expectancy is 18 to 24 months.

> *Battery Voltage: 12 volts*
> *Rated Capacity: 85 amp-hours*
> *L x W x D: 10-1/4" x 6-3/4" x 8-1/2"*
> *Weight: 46 lb/20.9 kg*

•15-103　Deep Cycle Battery, 12V/85A-hr　　　85^{00}

> *Battery Voltage: 12 volts*
> *Rated Capacity: 105 amp-hours*
> *L x W x D: 13" x 7" x 9"*
> *Weight: 53 lb/24kg*

•15-104　Deep Cycle Battery, 12V/105A-hr　　95^{00}

Shipped Freight Collect from Northern California

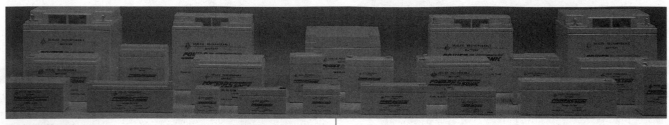

Gel Cell Batteries

Gel cell batteries can be charged and discharged more rapidly than other lead-acid batteries because the plates are very thin and the electrolyte is totally gelled. Gel cells are leakproof, sealed for life, and can operate in any position (even upside down) with an extremely low self-discharge rate. They are one of the only batteries rated for use in extreme cold environments. To summarize some of the advantages of gel cells (to justify their high cost!) over standard lead-acid batteries:

- They don't require acid checks or watering
- They can withstand shock and vibration better
- They tolerate extreme cold and extreme heat
- There is virtually no gassing or emission of corrosive acid fumes

These batteries must be charged at lower voltages than standard lead-acid batteries. Peak voltages of 13.7 to 14.1 maximum are recommended. Life expectancy is two to three years.

Amp-Hours	L x W x D	Weight
12	6" x 4" x 4"	9
31	7.75" x 5.2" x 7.25"	24
48	9.4" x 5.5" x 9.25"	39
70	10.9" x 6.75" x 9.9"	54
95	13.5" x 6.8" x 10.25"	74
160	20.75" x 8.5" x 10.0"	135
225	20.75" x 11.0" x 10.0"	168

•15-200	Gel Cell Battery, 12V/7A-hr	$30⁰⁰
•15-202	Gel Cell Battery, 12V/12A-hr	$75⁰⁰
•15-204	Gel Cell Battery, 12V/31A-hr	$85⁰⁰
•15-205	Gel Cell Battery, 12V/48A-hr	$130⁰⁰
•15-206	Gel Cell Battery, 12V/70A-hr	$160⁰⁰
•15-203	Gel Cell Battery, 12V/95A-hr	$240⁰⁰
•15-207	Gel Cell Battery, 12V/160A-hr	$385⁰⁰
•15-208	Gel Cell Battery, 12V/225A-hr	$475⁰⁰

Our Best Industrial Quality Chloride Batteries

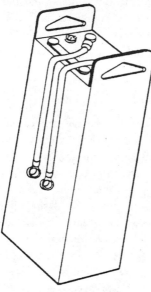

Pacific Chloride industrial batteries are the best batteries that we sell. They come with a five-year warranty, and with proper maintenance will easily last 15 to 20 years or more. Each cell is individually packaged in a steel case with two lifting handles and coated with acid-resistant paint. Moving these cells is a relatively safe and easy two-person task, even over rough terrain. The advantage of using Chloride 2-volt cells for alternative energy storage is increased performance, greater reliability, and less frequent maintenance. This is achieved by reducing:

- the number of cells needed to supply the required amp-hour capacity;
- the number of inter-cell connections subject to corrosion; and
- the number of cells that need to be watered.

Pacific Chloride cells are specifically constructed for high performance. They can withstand being brutally cycled to 80% depth of discharge for 1500 cycles; with less depth of discharge the cycle life expectancy rises rapidly. The leak-proof cell covers are thermally bonded to the cell jar. The cell terminals have leak-proof seals and the positive terminals have floating bushings designed to permit normal positive plate growth without damaging the integrity of the cell jar. The plate separators and grid structure have been designed to reduce internal resistance, increase kilowatt-hour output, and increase reliability. All inter-cell connectors with stainless steel hardware are included. The inter-cell connectors are made of premium quality, very flexible copper welding cable with plated lugs, all designed for easy installation and reduced corrosion.

The Real Goods Pacific Chloride Renewable Energy System Batteries are priced as 6-cell, 12-volt batteries. For 24-volt systems you will need to purchase two 6-cell batteries and connect all of the cells in series. These batteries are totally recyclable. Interconnects and leads are customized for each application. Specify system voltage and length of positive and negative leads when ordering.

Amp-Hours	L x W x D	Weight (for all six cells)
525	4-3/4" x 6-1/2" x 24"	480
635	5-1/2" x 6-1/2" x 24"	540
740	5-3/4" x 6-1/2" x 24"	660
845	7" x 6-1/2" x 24"	690
950	7-3/4" x 6-1/2" x 24"	750
1055	8-1/2" x 6-1/2" x 24"	780
1160	9-1/4" x 6-1/2" x 24"	840
1270	10" x 6-1/2" x 24"	930
1375	10-3/4" x 6-1/2" x 24"	990
1480	11-1/2" x 6-1/2" x 24"	1050
1585	12-1/4" x 6-1/2" x 24"	1152
1690	13" x 6-1/2" x 24"	1200

•15-421	Chloride Battery, 12V/525A-hr	$1,340
•15-422	Chloride Battery, 12V/635A-hr	$1,490
•15-423	Chloride Battery, 12V/740A-hr	$1,630
•15-424	Chloride Battery, 12V/845A-hr	$1,775
•15-425	Chloride Battery, 12V/950A-hr	$1,940
•15-426	Chloride Battery, 12V/1055A-hr	$2,135
•15-427	Chloride Battery, 12V/1160A-hr	$2,245
•15-428	Chloride Battery, 12V/1270A-hr	$2,425
•15-429	Chloride Battery, 12V/1375A-hr	$2,545
•15-430	Chloride Battery, 12V/1480A-hr	$2,755
•15-431	Chloride Battery, 12V/1585A-hr	$2,975
•15-432	Chloride Battery, 12V/1690A-hr	$3,245

Shipped freight collect from Northern California
Allow 4-6 weeks for delivery
Chloride batteries are made of lead, which is a commodity subject to price fluctuations. Check before purchasing batteries!

BATTERY CHARGERS

Todd Battery Charger

This battery charger manufactured by Todd Engineering is the lowest cost, stand-alone battery charger we've found to date. A big plus for this all solid-state charger is its amazing efficiency. The 75-amp model uses a scant 1050 watts at full output! Todd delivers close to 90% efficiency, compared to the best transformer-based chargers, which can't top 60%. This means more amps into your battery and less watts out of your generator. These are all *constant-voltage* chargers which have a steady regulated voltage output. As battery voltage rises and approaches the charger voltage, the amperage flow tapers off automatically to prevent overcharging. Our most popular model is the 75-amp/15.5-volt regulated voltage charger. It puts out 70-80% of rated output at just under 14 volts. It begins to taper only after 14 volts is achieved, allowing the fastest possible battery charging, greatly reducing generator time and fuel costs.

All Todd chargers are shipped with dual voltage output. On the front of the charger is a small terminal block with a jumper on it. With the jumper *removed* the charger operates in the lower voltage configuration, with the jumper *in place* (shorting across the two screws) the charger will operate at the higher voltage configuration. For convenience a switch can be added in place of the jumper. There is approximately a 1.0-volt difference between the two settings. This allows the lower voltage Todds to be switched from float to charge (13.6/14.6), the higher voltage Todds to be switched from charge to equalize (14.5/15.5), or the nicad Todds to be switched from 16.5 to 17.5 volts.

The 75-amp model can deliver full output, on generators as small as 2000 watts. With larger generators, this charger leaves you with enough reserve power for other chores, such as laundry, pumping, and vacuuming. It's a very versatile charger allowing series connection for 24-volt charging or unlimited paralleling to achieve as many amps as desired. It is compact (15-1/4" x 7-1/4" x 4"), lightweight (9 lbs.), can be wall mounted, has a thermostatically controlled cooling fan, and is warranted for one year. The low voltage model (13.6 volts to 14.6 volts) is designed for charging using standard AC power and in low voltage mode can be left hooked up indefinitely for unattended operation. The medium voltage model (14.5 volts to 15.5 volts) is designed for faster and more efficient charging using generator power. The high voltage model (16.5 volts to 17.5 volts) is designed for charging nicad batteries *only*. All three voltages come with an automatic temperature sensitive 2-speed fan in the 45-and 75-amp models. The 45-amp model uses less than 950 watts and the 30-amp model uses less than 650 watts.

15-631	75-amp Charger (13.6V to 14.6V)	$279⁰⁰
15-629	75-amp Charger (14.5V to 15.5V)	$279⁰⁰
15-633	75-amp Charger (16.5V to 17.5V)	$279⁰⁰
15-637	45-amp Charger (13.6V to 14.6V)	$209⁰⁰
15-628	45-amp Charger (14.5V to 15.5V)	$209⁰⁰
15-636	45-amp Charger (16.5V to 17.5V)	$209⁰⁰
15-639	30-amp Charger (13.6V to 14.6V)	$139⁰⁰
15-627	30-amp Charger (14.5V to 15.5V)	$139⁰⁰
15-638	30-amp Charger (16.5V to 17.5V)	$139⁰⁰

Too many customers are killing their Todd Chargers. Here is the scenario: The charger is permanently connected to the generator and battery bank, so that whenever the generator is turned on the batteries get charged. This KILLS chargers! The problem is that as the generator is coming up to and down from speed, the voltage is all over the place, and the charger can't take it. *Manual Solution:* Don't plug in charger until generator is up to speed, and unplug before turning off generator. *Automatic Solution:* Use Todd's 30-amp Time Delay Transfer Switch (#23-121, $55), available from Real Goods.

— *Jeff Oldham*

Schauer Chargers

These chargers have an adjustable finish voltage. They come factory preset at 14.2 volts, but are adjustable from 12.0 to 15.0 volts. This makes them ideal for float charging any type of lead-acid battery, particularly sealed or gel-cell units. The finish voltage will automatically raise or lower slightly to compensate for changes in temperature between -25° to 115° F. The charger will not operate with reversed polarity or shorted leads so it won't produce sparks near the battery. It will not drain the battery if AC power is shut off.

- 15-707 6-amp Charger (12V to 15V) $85⁰⁰
- 15-708 10-amp Charger (12V to 15V) $99⁰⁰

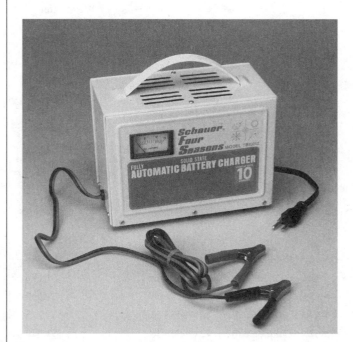

BATTERY ACCESSORIES

Battery Pal Automatic Float Charger

Leaving a car, boat, motorcycle, RV or other vehicle garaged for weeks or months almost guarantees a dead battery. The process is unstoppable. A fully charged battery, even a new one, loses approximately 1% of its charge per day. Battery Pal prevents this problem. Unlike trickle chargers, it cannot damage a battery through electrolyte boilout due to over-charging. Float circuitry senses when the battery is fully charged and automatically reduces current flow to the level that maintains battery life: 25-60 milliamps. This is an ideal unit for garages where a solar charger won't work. For use with 5 to 100 amp-hour lead-acid storage batteries.

Power Requirement: 100 to 130 VAC
Maximum Charging Output: 600 milliamps
Charge Voltage: 13.8±1 VDC

11-128　Battery Pal　39^{00}

874-L Battery Hydrometer

This full-size graduated-float style specific gravity tester is accurate and easy to use in all temperatures. Specific gravity levels are printed on the tough, see-through plastic body. It has a one-piece rubber bulb with a neoprene tip.

Use Tip: Ignore the color-coded green/yellow/red zones on the float. These are calibrated for automotive starting batteries which use a denser acid solution. Storage batteries will usually read in the red or yellow zones. Just watch for differences between individual cells.

15-702　Full-Sized Hydrometer　6^{00}

Battery Pal

2/0 Gauge Battery Cable

Full-Sized Hydrometer

Battery Service Kit

This kit has everything you need to service a round-post automotive battery. It contains an open-end wrench to remove terminal bolts and nuts, a wire brush, a battery post and terminal cleaner, a safety grip battery lifter, a terminal clamp lifter (a must if you've ever tried to remove a corroded battery terminal), an angle nose plier, a terminal clamp spreader, and a box wrench.

15-703　Battery Service Kit　52^{00}

Wing Nut Battery Terminals

This terminal converts conventional round post-type batteries to a wing nut connection adapter with a 5/16 inch stud.

15-751　Wing Nut Battery Terminal　2^{00}

Heavy-Duty Insulated Battery Cable

For interconnecting batteries in series or parallel, most systems need these insulated battery cables. These are extra heavy-duty 2/0 cables, with crimped and soldered 3/8" lug ends. The best in the industry! Don't choke your inverter's potential with too small cabling.

15-778　2/0 Gauge Battery Cable, 13 Inch　15^{00}

Light-Duty Battery Cable

These light-duty #4 gauge insulated cables are good for interconnecting smaller systems with 1000 watt or smaller inverters and modest current flows.

15-776　4 Gauge Battery Cable, 16 Inch　3^{00}
15-777　4 Gauge Battery Cable, 24 Inch　3^{50}

CONTROLLERS, MONITORS, AND SAFETY EQUIPMENT

These different functions have all been grouped together in a single section because they are interrelated, and because every renewable energy system needs one or more of each. Very often a single piece of equipment will blur the traditional boundaries and perform multiple functions for us.

Controllers

Every charging source will require a controller. The only exception is when the charging source is very small compared to the battery it charges. Only if a PV module produces 1.5% or less of the battery's ampacity (example: a PV module that produces 1.5 amps charging into a 100 amp-hour battery), will no charge control be required.

Each charging source customarily requires a different type of controller. The most basic function of a controller is to prevent battery overcharging. If batteries are routinely allowed to overcharge, it drastically reduces the life expectancy of the battery. A controller senses the battery's voltage, and reduces or stops the charging input when the voltage gets high enough. These control requirements are affected by characteristics of the source.

PV Controllers: Most PV controllers simply open the circuit between the battery and the PV array when the voltage rises to a set point. With some controllers this

DINOSAUR DIODES

When PV modules first came onto the market a number of years ago it was common practice to use a diode, preferably a very low forward-resistance Schottky, to prevent the dreaded reverse current flow at night. Early primitive charge controllers didn't deal with the problem, so installers did, with diodes. Over the years the equipment has improved and our understanding of PV operation has broadened. Even the best Schottky diodes have a 0.5-volt to 0.75-volt forward voltage drop. This means the module is operating at a slightly higher voltage than the battery. The higher the module voltage, the more electrons leak through the boundary layer between the positive and negative silicon layers in the module. These lost electrons will never come down the wire to charge the battery. The 0.5-volt higher operating voltage results in more power being lost during the day to leakage than the minuscule nighttime current loss we are trying to cure. Modern charge controllers use a relay, which has zero voltage drop, to connect the module and battery. The relay opens at night to prevent reverse flow. Diodes have become dinosaurs in the PV industry.

voltage is preset and nonadjustable, while on others it can be adjusted. A PV controller has the additional task of preventing reverse current flow at night. Most do this by opening a relay after sensing that voltage is no longer available from the modules when the sun has set. A few older designs still use diodes, a one-way valve for electricity, to accomplish this, but the relay has become the preferred method (see the sidebar on Dinosaur Diodes).

Hydro and Wind Charge Controllers: Hydro and wind controllers use different strategies to control battery voltage. A PV controller can simply open the circuit to stop the charging and no harm will come to the modules. If a hydro or wind turbine is disconnected from the battery while still spinning it will continue to generate power, but with no place to go, the voltage rises dramatically until something finally gives. With these types of rotary generators we usually use a *diverting charge controller.* A diverter will dump excess power into some kind of "dummy" load, a heater element being the most common. By fitting the right diverter, though, incandescent lights, fans, or other motors can sometimes be used.

Monitors

As investor and operator of a renewable energy system you will want to do a good job of management. The rewards will be increased system reliability, longer component life, and lower operating costs. The system monitor is the piece of equipment that allows you to peer into the electrical workings of your system and keep track of what goes on. The most basic, indispensable piece of monitoring gear is the voltmeter. Every system needs a voltmeter at the very minimum. A voltmeter measures the battery voltage, which in a lead-acid–based battery system can be used as an indicator of system activity and battery state of charge.

The basic voltmeter is a simple device, but monitors are becoming ever more sophisticated and informative as the number of independent power systems increases. Ammeters measure current flow in a circuit at the instant, and accumulating amp-hour meters give a cumulative total of amp-hours in or out of the system. New to the marketplace is the computerized "smart meter," which is even smarter than your average alternative-energy technician. Such a device can monitor a wide variety of battery and system conditions, set off adjustable alarms, divert excess power, act as a system charge controller, automatically start a back-up generator, or telephone you from a remote location in case of trouble it can't cure. Not every system or every person needs this order of sophistication in a monitor/controller, but independent system managers who are deeply interested in system performance are delighted to find these versatile devices available.

System Safety

As we mentioned in the introduction to this chapter, batteries are safer than traditional AC power in some aspects — it is difficult to shock yourself with low-voltage power — but, on the other hand, batteries can deliver much more power into a short circuit than standard AC house current is capable of. Such a high amperage flow can turn wires red-hot almost instantly, burning off the insulation

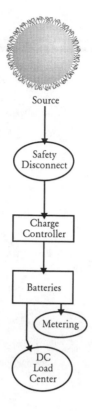

Source

Safety
Disconnect

Charge
Controller

Batteries

Metering

DC
Load
Center

Typical Power Flow

and easily starting a fire. As if that wasn't enough, once a DC arc is struck, it has much less tendency to self-snuff than an AC arc. Arc-welders prefer using DC for precisely this reason, but impromptu welding is most unwelcome in the home power system. Fuses rated for DC contain a special snuffing powder inside to prevent an arc after the fuse blows. AC fuses seldom use this extra level of protection.

The National Electrical Code wisely requires a fuse and a safety disconnect between any power source and any appliance. An independent power system has two or more power sources, the generating source (or sources — PV, wind, hydro) and the battery. To protect the charge controller, which has power sources on both sides of it, we usually use a conventional two-pole safety disconnect with DC-rated fuses, one pole wired on each side of the controller.

Inverter safety requires specialized equipment. The 1990 National Electric Code was the first to address low-voltage systems, and at that time no products meeting the needs of full-sized inverters existed. These larger, house-scale power inverters can safely handle up to 800 amps at times. Existing AC equipment either produced unacceptable voltage drop at high amperage flows, or was prohibitively expensive. Ananda Technologies has pioneered the solution to this problem. The simplest solution for existing or cost-conscious systems is the 400-Amp Fused Disconnect. This very heavy-duty, oil-filled switch is rated for 2000 amps and has been known to sustain 6000-amp surges without damage. (Kids, do not try this at home!) This switch can provide the high surge amperage that a modern inverter requires without inducing any inverter-crippling voltage drop. The integrated full-system solution is to use one of the Ananda Powercenters, which provides all the fusing, safety disconnect, monitoring, and control for the entire system in one compact, prewired, engineered, UL-approved package. We now strongly recommend the Powercenter approach for any new system.

For mid-sized, 400- to 1000-watt inverters, there is still no suitable safety disconnect on the market, unless the Powercenter package is chosen. We strongly recommend a 110-amp or 200-amp DC fuse assembly. Smaller inverters of 100 to 300 watts simply plug or hardwire into a fused outlet.

— Doug Pratt

FULLY INTEGRATED CONTROLS

APT Powercenters

The Ananda Power Technologies (APT) Powercenter combines all your renewable energy system controls, metering, and overcurrent protection in one compact, durable, attractive enclosure. They are the first products to be UL-listed for battery-based PV charge controlling and renewable energy system overcurrent protection. Powercenters are pre-wired, pre-assembled, tested, and certified to meet the strictest NEC safety standards. They can be installed in as little as 30 minutes. Nothing is safer and easier for a battery-based system than the APT Powercenter.

A Powercenter will increase the system cost by an average of about $500. However, if you are paying someone else to install your system the Powercenter cost will be recovered immediately in saved labor. If you're doing your own installation, what's your time worth, and how clean and tidy do you want your work to be? Powercenters are better looking, install faster, and provide safer installations than any other method. They allow for system adaptability and expansion. Dr. Doug gives them his heartiest seal of approval.

There are a number of UL-listed APT Powercenters models available, from a two-pole 100-amp main to a 3-pole 400-amp main. All the series 3 and 5 Powercenters come standard with a 60 amp PV control and breaker, a Smartlight battery state of charge indicator, space for further controls for PV arrays, digital monitoring, power control, utility transfer or generator start, and space for DC circuit breakers.

Options include additional solar controls, temperature compensation, low-battery alarms, low-voltage disconnects (for individual circuits), automatic generator start (two wire), AC utility transfer control, DC load control and power diversion, digital metering systems, DC load breakers, DC input breakers, lightning protection, battery and inverter cables, and much more.

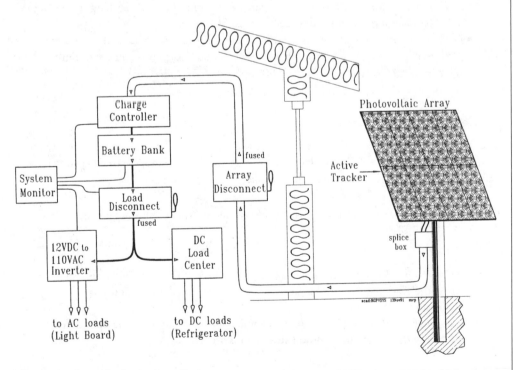

Schematic of a typical solar installation

APT photovoltaic controls which are standard with every Powercenter include the following features:

- *Equalize/Manual Switch:* Allows you to manually select Equalize, Automatic, or Off modes.
- *Field Adjustable:* Charge termination point is easily adjustable in the field, with an extremely wide adjustment range.
- *High Efficiency:* Very low resistance of only .003 ohm through the PV and load disconnect mercury contactors. No power-robbing diodes are needed with the automatic PV night-time disconnect control.
- *Mercury Contactor:* PV control with hermetically sealed construction is rated for millions of cycles. Can withstand high surges and lightning. Liquid mercury contacts provide the lowest resistance possible, even after years of use with no contacts to corrode or burn.
- *PV Disconnect:* Industrial-grade circuit breaker with #2 wire terminals and red trip-indicator.
- *LED Status Indicators:* Provide system status at a glance. GREEN is charging, YELLOW indicates charging has stopped, usually due to full batteries.
- *Factory Calibrated and Tested:* Every Powercenter is precisely calibrated (to custom specs if requested) and is thoroughly tested for several hours to allow component burn-in.
- *UL Listed:* APT is currently the only UL-listed PV controller available.

PV Controller Sizing Chart

The following table indicates the maximum wattage of PV modules that can be connected to each APT controller. Use the nameplate rated wattage of each PV module to determine how many modules can be connected to each controller.

PV Controller Size	12-volt System	24-volt System	48-volt System
60 amps	770 watts	1540 watts	3080 watts

The wattage figures above take into account the required 80% safety derating of the controller by the National Electrical Code, and the typical actual wattage output of the photovoltaic modules.

Here are the basic Powercenter models; all applications will require some options as well. **You must talk with one of our technical staff before ordering!**

Powercenter 2-4X

The 2-4X is designed primarily for systems that have only DC loads, or small DC to AC inverters (70 amps DC maximum). The weatherproof 4X enclosure allows installation in full exposure to all weather elements, as well as corrosive and dust-laden environments. The sealed relays and conformal coatings on the control boards and monitor allow the unit to operate for many years in high humidity and harsh environments.

Enclosure: 17" H x 15" W x 8" D, fiberglass, NEMA 4X type (outdoor), padlock provision, stainless steel hardware.

PV Control: One control module with equalize switch, adjustable operation, and 60 amp mercury contactor.

PV Disconnect: One 60 amp breaker with #2 wire terminals, space for up to two disconnects total.

System Monitor: Optional, order separately

Controls: Space for up to 4 control modules for PV, Gen Start, Alarms, and Power Control

DC Breakers: Space for up to 6 DC input or load breakers from 15 through 70 amps.

Maximum PV Charge Control Capacity: 120 amps.

Removable Interior: Entire interior assembly quickly removes, making conduit and wire installation easier.

Spanish Labeling: Labeling is provided in both Spanish and English for control functions (Door labeling only).

Add $50 for 48-volt Powercenter 2 orders.

- •25-181   APT Powercenter 2-4X $795⁰⁰
- •25-182   APT Powercenter 2-4X-CLD (clear-door option) $119⁰⁰

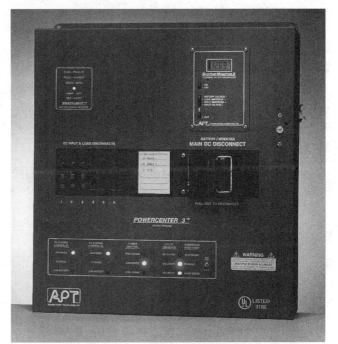

Powercenter 3 Series

Enclosure: 20" H x 18" W x 4" D, with nine 3/4" and two 2" conduit knockouts, NEMA 1 type (indoor).

PV Control: One control module with equalize switch, adjustable operation and 60-amp mercury contactor.

PV Disconnect: One 60-amp disconnect breaker with #2 wire terminals.

Charger Input: Battery charger input with #2 wire terminals. (Fused by main disconnect.)

System Monitor: APT Smartlight, Red/Amber/Green/Pulsing/ Flashing battery state of charge indicator.

Controls: Space for up to four additional control modules for PV, Gen Start, Alarms, Power Control, and Utility Transfer.

DC Breakers: Space for up to six DC input or load breakers from 15 through 100 amps.

Main Disconnect: 100 or 200 amp depending on model, 2-pole pull-out with Class T fuses.

Add $50 for 48-volt Powercenter 3 orders

•25-172	APT Powercenter 3-101	$1,095
•25-173	APT Powercenter 3-202	$1,145
•25-185	APT Powercenter 3-222	$1,295
•25-186	APT Powercenter 3-404	$1,445

Powercenter 5 Series

Enclosure: 24" H x 24" W x 6" D, with twelve 3/4" and four 2" conduit knockouts, NEMA 1 type (indoor).

PV Control: One control module with equalize switch, adjustable operation and 60 amp mercury contactor.

PV Disconnect: One 60 amp disconnect breaker with #2 wire terminals.

Charger Input: Battery charger input, for up to #2 wire. (Fused by the main disconnect.)

System Monitor: APT Smartlight, Red/Amber/Green/Pulsing/Flashing battery state of charge indicator.

Controls: Space for up to eight control modules for PV, Gen Start, Alarms, Power Control, and Utility Transfer.

DC Breakers: Space for up to 12 DC input or output breakers from 15 through 100 amps.

Main Disconnect: 200 or 400 amp, 2- or 3-pole pull-out, depending on model, with Class T fuses.

Add $50 for 48-volt Powercenter 5 orders.

200 amp 2-pole fused pull-out, with two Class T fuses

•25-175 APT Powercenter 5-202 $1,295

200 amp 3-pole fused pull-out, with three Class T fuses. For two separate inverters or systems with very high amperage charging sources.

•25-176 APT Powercenter 5-222 $1,445

400 amp 2-pole fused pull-out, with two Class T fuses

•25-177 APT Powercenter 5-404 $1,595

400 amp 3-pole fused pull-out, with three Class T fuses. For two separate inverter circuits or systems with very high amperage charging sources.

•25-178 APT Powercenter 5-444 $1,795

APT Powercenter 3 and 5 Sizing Guide

This guide is designed to help you easily determine which Powercenter model and options you will need. If you have additional questions feel free to contact the Real Goods technical staff (which you must do before ordering anyway).

1. **Size your System:** The first step will be to determine the number of solar modules and other charging sources that will be required. Then the system voltage and loads will need to be considered so the correct inverter and number of DC circuits can be properly chosen.

2. **Choose the Powercenter Model and Options:** The chart below will assist in choosing your basic Powercenter. The Options List will assist in choosing the necessary options for your system. Be sure to consider future system expansion needs.

Choosing the Correct Powercenter Model

It's important that your inverter, other DC loads, and charging sources are properly matched to the main fuses in your Powercenter. The tables below will help you to determine the proper match.

The two primary factors which determine the size of your main fused disconnect are *System Voltage* and *Inverter or DC Load*. The inverter typically creates a higher amperage flow than any other component in your system, so we tend to size the Powercenter around the inverter. At 12 volts the DC amperage is twice as much as at 24 volts for the same 120-volt AC output. Even though the inverter and DC voltage are the main considerations, all other inputs and outputs such as PV array, battery chargers, wind turbines, or other DC loads need to be considered.

The following table is based on the inverter being used to its full capacity. If you are sure that you will always be pulling much smaller than maximum loads through your inverter, then the main fuse size could be downsized accordingly. For example: a Trace 2512 normally requires an APT5-404 Powercenter. If you're only running lights, computer equipment, and entertainment equipment and never running high-surge appliances like a washing machine or power tools then you can downsize to the APT3-200 Powercenter.

If your inverter isn't shown in the table above, or you need to calculate for other high-current DC conditions, the "Time vs. Amps Fuse Chart" below can help you match the correct amount of amperage to your main fuses based on the length of time load will occur. Be sure to take into account the surge ability of the inverter or other devices.

Remember to add for the loss of efficiency through the inverter. This is typically 15% at full load. For example: If you have a 3000-watt load on the 120-volt AC side, you would actually draw about 3450 watts out of the DC battery. Divide wattage by voltage to get amperage. 3450 watts divided by 12 volts equals 288 amps. If this were a 24-volt system you would only be drawing 144 amps, a good reason for having higher-voltage storage systems when the system has to meet large loads.

Powercenter Model	12-volt System	24-volt System	48-volt System
APT2-4X (with 70 amp load breaker)	1-Heart HF600 1-Powerstar UPG400 1-Exeltech 500	1-Trace 724 1-Heart Freedom 10 1-Powerstar UPG900 1-Exeltech 1000	1-Trace 2248 (older model)
APT3-101 2 - 100-amp fuses	1-Trace 812 1-Heart Freedom 10 1-Powerstar UPG700 1-Exeltech 500	1-Trace DR1524 1-Heart EMS1800 1-Powerstar UPG1500 1-Exeltech 1000	1-Trace 2548
APT3-202/APT5-202 2 - 200-amp fuses	1-Trace DR1512 1-Heart EMS1800 1-Powerstar UPG1300 1-Exeltech 1000	2-Trace DR1524 1-Heart EMS2800 1-Trace DR2424	2-Trace 2548 1-Trace SW4048
APT3-222/APT5-222 3 - 200-amp fuses	2-Trace DR1512 2-Heart EMS1800 2-Powerstar UPG1300 2-Exeltech 1000	4-Trace DR1524 2-Heart EMS2800 2-Trace DR2424	4-Trace 2548 2-Trace SW4048
APT3-404/APT5-404 2 - 400-amp fuses	1-Trace 2512 2-Trace 2512 (stacked) 1-Heart EMS2800 1-Trace SW2512	1-Trace SW4024 2-Heart EMS2800 2-Trace DR2424	4-Trace 2548 2-Trace SW4048
APT5-444 3 - 400-amp fuses	2-Trace 2512 4-Trace 2512 (stacked) 2-Heart EMS2800 2-Trace SW2512	2-Trace SW4024 4-Heart EMS2800 4-Trace DR2424	8-Trace 2548 4-Trace SW4048

This table indicates the maximum-sized, fully loaded inverter that each Powercenter model will handle.

OPTIONS LISTS

PHOTOVOLTAIC CONTROLS

Model	Description	Price
A60P	This changes the standard 60-amp input breaker to two 30 amp parallel breakers for two PV subarrays.	$32⁰⁰
A60-ADD*	Additional 60-amp PV controller. Includes one 60-amp breaker and 60-amp mercury contactor with staged setpoints for multi-staged charge process.	$239⁰⁰
A60P-ADD*	Additional 60-amp PV controller with two 30-amp parallel input breakers. Includes two 30-amp breakers, 60-amp mercury contactor.	$269⁰⁰
2/0-TERM	Upgrades standard #2 AWG PV controller input wire terminals to #2/0 AWG.	$34⁰⁰
BTC10	Battery temperature compensation for PV controller. Automatically adjust charge setpoint according to temperature. With sensor and 10' of cable.	$29⁰⁰
BTC25	With 25' of cable.	$49⁰⁰
LA100V*	Lightning arrestor. 50,000 amp, 750 joules, 100V clamp. Use one for each input. Will rupture when capacity is reached indicating need for replacement.	$52⁰⁰
48V-ADD	Additional charge for *each* PV control module that is 48 volts.	$49⁰⁰
CONFORMAL	Circut board conformal coating for humid or harsh enviroments. Charge per board.	$16⁰⁰

* Indicates items that can be easily added as a field upgrade to Powercenters that have been shipped.
Add 10% to standard pricing when ordering as a field upgrade (covers the extra installation instructions).

TIME VS. AMPS FUSE CHART

		100 Amp Fuse	200	400
Fuse Blow Point in Amps Time	1 second	340	670	1450
	6 seconds	280	530	1100
	1 minute	195	370	800
	4 minutes	175	340	710
	8 minutes	165	310	690
	16 minutes	160	300	650
	32 minutes	145	275	600
	1 hour	130	250	550
Long Term Load Handling	1 to 3 hours	100	200	400
	3 or more hours	80	160	320

SYSTEM CONTROLS AND SIGNALS

Model	Description	Price
GEN	Generator start-stop signal module for 2-wire start type generators. Includes Start/Auto/Off and Stop/Auto/Off switches on control board, manual Start/Stop switch on Powercenter door, indicator lights, and field adjustable set points.	$239⁰⁰
GEN-RC	Remote-control switch for GEN option. Mounted on standard switch plate, fits single-gang box, or optional SE surface enclosure.	$45⁰⁰
GEN-EX	Generator exercise control. Digital, adjustable run minutes and start time up to every ten days.	$299⁰⁰
PC-AC30	AC power control. Use to turn off battery chargers or control AC loads. With 30-amp 2-pole 120/240 VAC contactor in separate enclosure, test and control switches, field adjustable set points.	$349⁰⁰
PC-DC35 NO or NC	DC power control. Use for power diversion or to control DC loads. With 35-amp mercury contactor, test and control switches, field-adjustable set points. Specify NO (normally open) or NC (normally closed) when ordering. Load breakers required, order separately.	$269⁰⁰
UT-XFER	AC utility automatic transfer switch and control module. With 30-amp 2-pole 120/240 VAC contactor in separate enclosure, test and control switches, field-adjustable set points.	$369⁰⁰
LBCO35	35-amp low battery cut-out. To disconnect specific DC loads such as pumps or electronic equipment during low battery conditions. Includes 35-amp mercury contactor, test/auto/off switch, red LED indicator, field-adjustable set point. Load breakers are needed; order separately.	$109⁰⁰
LBCO60	Same as LBCO35 but with 60-amp contactor. Order breakers separately.	$149⁰⁰
LBA	Low battery alarm signal. With 85db Piezo audible sounder mounted in Powercenter. Test/auto/off switch and red LED indicator. Adjustable setpoint.	$49⁰⁰
H/LA-2	High-low battery alarm to signal module. Provides relay closure for high and low battery conditions. Test/auto/off switch, red/yellow/green LED indicators, adjustable setpoints.	$239⁰⁰
48V-ADD	Additional charge for *each* system control module that is 48-volt.	$49⁰⁰

SYSTEM MONITORS

Monitor and shunt installation is approved by UL as part of the Powercenter system only.
These monitors and shunts are not UL-listed for use in systems without Powercenters.
Vista3, TRI-MET, E-Meter, and CM+2 normally mount in Powercenter — add RMTB option for remote monitor installation.

Mod.#	Description	Price
VISTA 3-SH	APT Vista 3 monitor and shunts installed as part of Powercenter. Reads battery voltage, input amps, and load amps on digital display.	$219^{00}
VISTA 3	Vista 3 as a stand-alone product, or as second monitor. Shunts not included.	$109^{00}
AS2	Auxiliary switch. Adds two more channels to read amps or volts. Order shunts below.	$69^{00}
SH100	100 amp/100mv shunt. Installed in Powercenter, with bussing and lugs.	$39^{00}
SH200	200 amp/200mv shunt. Installed in Powercenter, with bussing and lugs.	$59^{00}
SH500	500 amp/50mv shunt. Installed in Powercenter, with bussing and lugs.	$59^{00}
TRI-MET	Tri-Metric Amp-Hour Meter. With battery voltage, charging and charged light, net current reading and battery capacity measurement. Includes one 200 or 500 amp shunt. 12-and 24-volt systems only.	$269^{00}
E-METER	E-Meter monitor and shunt installed. Reads voltage, net amperage, cumulative amp hours (or KWH), and time remaining at average discharge. For 12-or 24-volt systems	$349^{00}
CM+2	Cruising Equip. Co. Amp-Hour Plus-2 meter. With battery voltage, two current readings, and battery capacity measurement. Includes two 500 amp shunts. For 12- and 24-volt systems only. Maximum current reading is 250 amps. Not recommended for APT5-444.	$449^{00}
RMTB	Remote meter terminal block. Wired into the Powercenter shunts and fuses, with label for easy connection. Specify monitor type.	$43^{00}
RSTB	Remote terminal block for Smartlight.	$19^{00}
SMLT-12	APT Smartlight for remote installation. Five mode battery state of charge indicator: Red/Amber/Green/Pulsing/Flashing. Just like the one in the Powercenter; install these in other locations for system monitoring. Mounted on standard switch plate. Specify system voltage, and liquid or gel battery type.	$39^{00}
SMLT-24		$39^{00}
SMLT-48		$49^{00}

For flush-mount plate that fits standard single-gang electric box, add "p" to Smart light part number, and add $6.

REMOTE MONITOR CABLE

Model	Description	Price
CBL2-22	2-conductor, 22-gauge PVC jacketed cable. For remote Smartlight.	$.20/ft
CBL7-22	7-conductor, 22-gauge PVC jacketed cable. For VISTA 3, TRI-MET, E-METER	$.55/ft
CBL10-22	10-conductor, 22-gauge PVC jacketed cable. For CM+2	$.75/ft

DC LOAD BREAKERS, INPUT BREAKERS, AND RELAYS

Model	Description	Price
LB15 *	15-amp 1-pole load breaker. With #14 to #2 wire terminal.	$32⁰⁰
LB20 *	20-amp 1-pole load breaker. With #14 to #2 wire terminal.	$32⁰⁰
LB30 *	30-amp 1-pole load breaker. With #14 to #2 wire terminal.	$32⁰⁰
LB35 *	35-amp 1-pole load breaker. With #14 to #2 wire terminal.	$36⁰⁰
LB40 *	40-amp 1-pole load breaker. With #14 to #2 wire terminal.	$36⁰⁰
LB50 *	50-amp 1-pole load breaker. With #14 to #2 wire terminal.	$39⁰⁰
LB60 *	60-amp 1-pole load breaker. With #14 to #2 wire terminal.	$39⁰⁰
LB70 *	70-amp 1-pole load breaker. With #14 to #2 wire terminal.	$65⁰⁰
LB100	100-amp 1-pole load breaker. With #4 to #2/0 wire terminal. (not for 2-4X)	$179⁰⁰

DC input breakers must have use specified, such as wind, hydro, battery charger etc.

Model	Description	Price
IB15 *	15-amp 1-pole input breaker. With #14 to #2 wire terminal. (Specify use)	$32⁰⁰
IB20 *	20-amp 1-pole input breaker. With #14 to #2 wire terminal. (Specify use)	$32⁰⁰
IB30 *	30-amp 1-pole input breaker. With #14 to #2 wire terminal. (Specify use)	$32⁰⁰
IB35 *	35-amp 1-pole input breaker. With #14 to #2 wire terminal. (Specify use)	$36⁰⁰
IB40 *	40-amp 1-pole input breaker. With #14 to #2 wire terminal. (Specify use)	$36⁰⁰
IB50 *	50-amp 1-pole input breaker. With #14 to #2 wire terminal. (Specify use)	$39⁰⁰
IB60 *	60-amp 1-pole input breaker. With #14 to #2 wire terminal. (Specify use)	$39⁰⁰
IB70 *	70-amp 1-pole input breaker. With #14 to #2 wire terminal. (Specify use)	$65⁰⁰
IB100	100-amp 1-pole input breaker. With #4 to #2/0 wire terminal. (Specify use)	$179⁰⁰
2/0 TERM	Upgrades standard input/load terminals to #2/0 terminal block.	$34⁰⁰

* These items can be easily added as field upgrades to Powercenters that have been shipped. Add 10% to list pricing if ordered as a field upgrade.

† Mercury Contactors: These are to be operated from dry contact signal such as from the Trace SW4024 inverters or other controls, with terminals, labels, and fusing included. Normally Open contacts typically are used to turn loads on when batteries are full. Normally Closed contacts are typically used to turn loads off when batteries are low or for PV control.

BATTERY AND INVERTER CABLES

Super flexible welding type cable. With crimped and soldered ring terminals, colored heat shrink and identification labels on all ends. Not UL-listed.

Model	Description	Price
BC2	100-amp cable set for battery. Two 5.5 ft #2 cables.	$39^{00}
IC2	100-amp cable set for inverter. Two 3.5 ft #2 cables.	$33^{00}
ADD2	100-amp cable extension, per cable, per foot. Extends one #2 cable. Specify + or - or both.	$4^{50}/ft.
BC2/0	200-amp cable set for battery. Two 5.5 ft #2/0 cables.	$55^{00}
IC2/0	200-amp cable set for inverter. Two 3.5 ft #2/0 cables.	$50^{00}
ADD2/0	200-amp cable extension, per cable, per foot. Extends one #2/0 cable. Specify + or - or both.	$5^{50}/ft.
BC4/0	400-amp cable set for battery. Two 5.5 ft #4/0 cables.	$85^{00}
IC4/0	400-amp cable set for inverter. Two 3.5 ft #4/0 cables.	$75^{00}
ADD4/0	400-amp cable extension, per cable, per foot. Extends one #4/0 cable. Specify + or - or both.	$8^{00}/ft.
BI-2/0	Battery interconnect cable. 14" long, #2/0 cable. Super flexible welding cable with crimped and soldered ring terminals, heat shrink both ends.	$15^{00}

SPARE FUSES, CABLE CLAMPS, AND KEYED DOOR LATCH

Model	Description	Price
F100-T	100-amp Class T spare or replacement fuse.	$9^{00}
F200-T	200-amp Class T spare or replacement fuse.	$16^{50}
F400-T	400-amp Class T spare or replacement fuse.	$36^{00}
F1/2	1/2-amp spare or replacement AGC fuse for metering and controls.	$1^{00}
KEY-PC3	Tamper-resistant key operated door latch with two keys for Powercenter 3 models.	$25^{00}
KEY-PC5	Two tamper-resistant key operated door latches with two keys for Powercenter 5 models.	$32^{00}
CC2	2" Cable clamp – every powercenter needs at least one.	$8^{95}
CC3/4	3/4" Cable clamp	$1^{39}

Trace C-30A+ Charge Controller

The Trace C30A+ Charge Controller has quickly become our best-selling basic charge controller for systems with up to 30 amps of charging current. It can be wired in parallel for arrays over 30 amps. It automatically adjusts to either 12- or 24-volt battery systems. It comes factory preset for normal wet-cell lead acid batteries, but both the high and low voltage set points are user adjustable over an approximate range of 10 to 19 volts (double for 24-volt) with a digital voltmeter. The C-30A+ features a nighttime disconnect eliminating the need for a power-robbing diode, and an equalizing switch that allows occasional equalizing charges of the battery bank. Has a 56-volt transorb lightning arrestor and 30 amp slow-blow fuse for protection. Large terminals will accept #14 thru #4 gauge wire. Reverse polarity protected. One-year full warranty.

25-103   Trace C30A+ Charge Controller $100⁰⁰

SunCycler 8-Amp PV Controller

The SunCycler is our simplest PV controller. This low cost, fully potted and sealed 12-volt controller will handle up to 8-amps of input current. The simple screw terminal connector strip will accept up to #12 gauge wire. An LED indicates charging. The high voltage disconnect point is easily adjustable from 13.8 to 16 volts with the calibrated trim pot in the upper right corner. Low voltage reconnect is set at approximately 13.2 volts. Size is 2.3" x 2.3" x 1". No night time disconnect or other bells and whistles. Just simple, adjustable, inexpensive control for small systems. *Five-year warranty!*

25-075   SunCycler 8-Amp Controller $45⁰⁰

Smartcharger 20

This is one of the slickest controller/simplified monitor packages we've ever seen. The Smartcharger 20 from APT is a 20-amp preset PV controller, with graphic multicultural LEDs that indicate when charging is occuring, approximate battery state of charge, and a warning light for extreme low battery conditions. Night time disconnect, spike and surge, and reverse polarity protection is provided. Set screw type terminals on the back of the weatherproof, fully potted controller accept up to #10 gauge wire. Size is 3" x 5" x 1.2". For modest systems this may be all the controller or monitor you'll ever need.

Two options are available. Low Voltage Disconnect (LVD), which will automatically disconnect any loads at critical low voltage (11.2v disconnect, 12.2v reconnect), and Battery Temperature Compensation (BTC), for extremely cold or hot climates, which will adjust the battery charge voltages based on battery temperature (includes 10' cable). Available in 12-or 24-volt, must specify at time of order, not field switchable. Two year mfg. warranty.

25-076   Smartcharger 20 $69⁰⁰
25-074   Smartcharger 20 w/LVD $79⁰⁰
25-073   Smartcharger 20 w/BTC $89⁰⁰
25-072   Smartcharger 20 w/LVD & BTC $99⁰⁰

M8 / M16 Sun Selector Charge Controller

If you don't need to regulate a full 30 amps of solar power, then the M8 or M16 is the way to go. Made by Sun Selector Electronics, these fine charge controllers eliminate any need for power-robbing blocking diodes, and will automatically reduce the high-voltage cut-off point as the battery approaches full charge to minimize gassing. Both the M8 and the M16 measure only 2" x 2" x 1-1/4" and are simple to install with just four wires to connect. Both units are totally encapsulated for outdoor or marine installations. The M8 will handle up to 8 amps of charging current; the M16 will handle up to 16 amps of charging current. Both units feature four LED indicators, which inform the user of charger conditions. These standard units are set up for conventional wet-cell lead-acid batteries; gel-cell units are available by special order. Five-year full warranty.

25-101	M8 Charge Controller, 12V	$65⁰⁰
25-102	M16 Charge Controller, 12V	$79⁰⁰
25-128	M8 Charge Controller, 24V	$65⁰⁰
25-104	M16 Charge Controller, 24V	$79⁰⁰

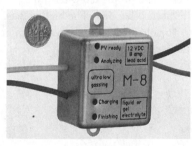

Marine Charge Controller

Sun Selector produces a special controller for marine applications with two separate battery banks. This series of controllers will charge both battery banks, with the bulk of charging going to the lowest battery, but will not allow one battery to discharge into the other. With two pairs of battery wires from controller, same as the standard M8 and M16 in all other aspects. Five-year warranty.

•25-127	M8M Dual Battery Marine Controller, 12V	$90⁰⁰
•25-129	M16M Dual Battery Marine Controller, 12V	$145⁰⁰

Enermax Universal Battery Voltage Controller

The Enermax Universal Battery Voltage Controller is an excellent voltage controller that has proven extremely reliable for hydro and wind systems. It is now available in a new re-engineered 120-amp version as well as the time-tested 60-amp version. This is the controller of choice for hydro systems, but will work with many different charging sources at once (wind, hydro, PV). It is a parallel shunt regulator that diverts any excess charging power to a dummy heating load. Typically a water- or air-heating element, though simple light bulbs could be used as well. It is made of all solid-state components so that there are no relay contacts to wear out. The desired float voltage is selected with the front-mounted rheostat. During charging, as battery voltage rises just enough power will be diverted to maintain the float voltage. The battery will float at any user-selected voltage and any excess power; up to the maximum amperage rating for the Enermax will be diverted. Water heating is the most popular diversion.

When ordering the Enermax, keep in mind that you must have an external load at least equal to your charging capacity. This is usually performed with either water-heating elements or air-heating elements. We offer 15- through 25-amp elements below. Always specify voltage when ordering. Higher voltage 100-amp units available by special order. Two-year warranty.

Specify voltage when ordering!

25-107	Enermax 60–60-amp Controller, 12, 24, 32, 36V	$285⁰⁰
25-152	Enermax 120–120-amp Controller 12V	$360⁰⁰
25-153	Enermax 120–120-amp Controller 24V	$375⁰⁰

Battery Voltage	Adjustment Range	Maximum Amperage	Idle Amperage
120-Amp Enermax Specifications			
12	13.3 - 17.0	120	0.01
24	26.6 - 34.0	120	0.01
60-Amp Enermax Specifications			
12	13.3 - 17.0	60	0.01
24	26.6 - 34.0	60	0.01
32	34.5 - 44.0	60	0.01
36	39.9 - 51.0	60	0.01

Water Heating Elements For Diversion Loads

20- and 25-amp elements are 1" MIPT, 15-amp elements are 3/4" MIPT.

25-108	12V/15A Water Heating Element	$89⁰⁰
25-149	12V/25A Water Heating Element	$117⁰⁰
25-130	24V/15A Water Heating Element	$89⁰⁰
25-148	24V/25A Water Heating Element	$117⁰⁰
25-154	32V/15A Water Heating Element	$95⁰⁰
25-155	48V/20A Water Heating Element	$105⁰⁰
25-156	2-1/2" Square Flange Adaptor for water heater	$9⁰⁰

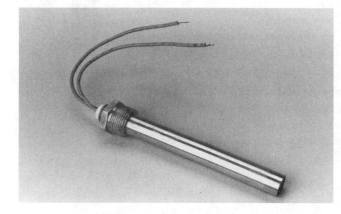

Air Heating Element For Diversion Loads

These resistors get very hot during regulation and must be mounted a safe distance from flammable surfaces. Use two or more elements in series for higher voltage systems. Air elements also tend to "sing" during regulation, so don't mount them in liveable space, they'll drive ya crazy!

25-109	12V/20A Air Heating Element	$25⁰⁰

Schottky Diode

If PV modules are used with the Enermax, then you need this 60-amp Schottky diode kit installed between the PV modules and batteries to prevent nighttime reverse current flow. (All you die-hard diode-heads, this is it! Your last big chance to use a *diode* in your PV system!)

25-704	60-amp Diode with Heat Sink	$25⁰⁰

APT PowerManager 60-Amp PV Charge Controller

The same proven quality controller used in the APT Powercenters is now available as a stand alone product. Has a 60-amp mercury contactor rated for millions of cycles with only 0.1 volt drop at full rated amperage flow. Also features automatic night time disconnect, temperature compensation (with 10' sensor cable), equalize/manual/off switching, and field adjustable voltage set points (comes factory pre-set for lead-acid wet cells unless specified otherwise). The enclosure is a durable, hinged, powder-coated white steel. Circuit boards are conformal coated to protect from moisture and corrosion. LED indicator lights provide at a glance system status. Has two 3/4" knock-outs for wiring, and accepts up to #6 gauge wire. Five-year mfg. warranty.

25-071	APT PV-60 Charge Controller	$249⁰⁰

12-or 24-volt (must specify!)

25-097	APT PV-60 Charge Controller 48 volt	$299⁰⁰

The Perfect Charge Controller for RVs or Intermittent-Use Cabins

The German-made Steca 4-amp controller is ideal for single module systems with an output of 4-amps or less. It has special gassing regulation circuitry that reduces the charging voltage from 14.1 to 13.7-volts when the batteries aren't significantly discharged overnight. This keeps your RV or cabin battery from using excess water, and increases the battery life expectancy. LED lights indicate function. 2.7" x 2.25" x 0.9" 1-yr. mfg. warranty.

25-009	Steca 4-amp Charge Control	$39⁰⁰

APT Tapering Diversion Regulators (TDR) Charge Control For Wind And Hydro

Unlike PV modules, in which the current can be simply shut off when the batteries are full, wind or hydro charging sources must have excess current diverted to a heater element to prevent battery or charger damage. APTs line of TDR controllers regulates the current to the batteries, and automatically tapers charging current to your selected finish voltage. Excess current is diverted to either built-in air heaters, or water heater elements installed in your water heater or hot tub. There is no electronic noise, these controllers will not interfere with radios or other equipment. The voltage adjustment is on the front panel, and is clearly labeled to indicate voltage and setting ranges for various battery types. An LED indicates when operating. Reverse polarity protected. 3/4" and 1-1/4" knock-outs provided. Wiring terminal strips are enclosed. TDRs can be used with any combination of charging sources, but is most commonly used with Whisper wind generators and Harris hydro turbines.

* Sizing Notes: The TDR continuous wattage rating must be equal to or greater than the wind's or hydro's rated wattage. When using the TDR as a total system regulator, just add the wattage of all charging sources, and make sure the TDR is bigger. For higher wattages multiple TDRs wired in parallel may be used.

TDR opened for detail

Internal Air Resistor TDR Models

Built-in air resistors in the ventilated enclosure minimize wiring and mounting difficulties, while protecting from burns or shock hazards.

Item #	Model	System Voltage	Continuous Wattage	Fuse Needed	Price
TDR 300-A Series Models					
25-096	TDR312-A	12	347 watts	30 amps	$395⁰⁰
25-095	TDR324-A	24	350 watts	15 amps	$395⁰⁰
25-094	TDR348-A	48	350 watts	15 amps	$445⁰⁰
TDR 600-A Series Models					
25-093	TDR612-A	12	617 watts	60 amps	$495⁰⁰
25-092	TDR624-A	24	630 watts	30 amps	$495⁰⁰
25-091	TDR648-A	48	630 watts	15 amps	$545⁰⁰
TDR 1200-A Series Models					
25-090	TDR1212-A	12	1200 watts	100 amps	$575⁰⁰
25-089	TDR1224-A	24	1260 watts	60 amps	$575⁰⁰
25-088	TDR1248-A	48	1260 watts	30 amps	$675⁰⁰

External Load Resistor TDR Models

Use with water heating elements sold below.

TDR 600-E Series Models — for one external load circuit

Item #	Model	System Voltage	Continuous Wattage	Fuse Needed	Price
25-087	TDR612-E	12	600 watts	50 amps	$345^{00}
25-086	TDR624-E	24	600 watts	30 amps	$345^{00}
25-085	TDR648-E	48	600 watts	15 amps	$395^{00}

TDR 1200-E Series Models — for two external load circuits

Item #	Model	System Voltage	Continuous Wattage	Fuse Needed	Price
25-084	TDR1212-E	12	1200 watts	100 amps	$495^{00}
25-083	TDR1224-E	24	1200 watts	50 amps	$495^{00}
25-082	TDR1248-A	48	1200 watts	30 amps	$575^{00}

TDR 1800-E Series Models — for three external load circuits

Item #	Model	System Voltage	Continuous Wattage	Fuse Needed	Price
25-081	TDR1812-E	12	1800 watts	150 amps	$595^{00}
25-080	TDR1824-E	24	1800 watts	80 amps	$595^{00}
25-079	TDR1848-E	48	1800 watts	40 amps	$645^{00}

Water Heating Elements

Industrial grade elements designed for years of trouble-free service. Screws into standard 1" water heater NPT fittings. Order one element for each external load circuit. 12/24-volt element rated at 600 watts (40-amps at 12-volt, 20-amps at 24-volt), 48-volt element rated at 500 watts (8.3 amps).

25-078 12/24v Water Heating Element $99^{00}
25-077 48v Water Heating Element $39^{00}

SPECIALTY CONTROLLERS

Trace C-30 Load Controller

This quality Trace product is very similar to the best-selling C-30A+ PV controller, but without the automatic nighttime disconnect or equalizing features. The C-30 is a battery voltage controlled relay. User-adjustable high- and low-voltage switch points allow optimum settings for differing systems. Range is approximately 10 to 19 volts (or double for 24-volt). Lightning and input protection is provided by a 56-volt transorb (a lightning arrester) and a 30-amp slow-blow fuse. Internal electronics are reverse-polarity protected and switchable for either 12- or 24-volt operation. Large terminals accept from 14-gauge through 4-gauge wire. The C-30 can be configured for either charge control (low-voltage connect/high-voltage disconnect), or load control (low-voltage disconnect/high-voltage connect). This unit can be used to protect batteries from over-discharge, to trigger low-voltage alarm systems, and to turn pumping systems on and off. The list is limited only by your imagination. One-year full warranty.

25-100   Trace C-30 Load Controller $100⁰⁰

Universal Power Controller

This is a unique, user-configurable, single-function DC set-point controller with adjustable time delay and independently adjustable high and low set points. The UPC-1 can be used with 12, 24, or 48VDC systems. The output of the UPC-1 is designed to be connected to an external relay, so controlled amperage is only limited by the size of the slave relay. It can be used for charge diversion, high or low alarms, load activation or disconnection, generator start, or other applications limited only by your imagination. Only uses 4mA to operate. Ten year warranty!

- Adjustable high and low set points with exact control voltages read by digital voltmeter.
- Selectable high or low control logic
- Selectable 12- or 120-second time delay
- Control status LED indicator
- Sealed, user replaceable, gold contact relay
- Lightning and reverse polarity protected
- Setup instruction summary right on unit

25-119   UPC-1 Controller $130⁰⁰

Universal Power Controller

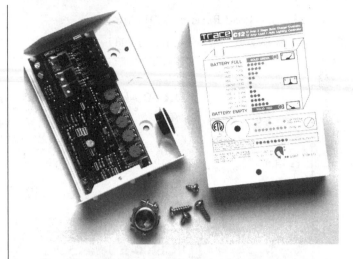

Trace C-12 Charge/Load/Lighting Controller

Welcome to the first PV controller that does it all.   It's a 3-stage, 12-amp PV charge controller with easily user-set bulk and float voltages, night-time disconnect, automatic monthly equalization, and electronic overcurrent protection at 15 amps with automatic reconnection attempts (2 attempts at 12-sec. intervals, then continuous attempts at 15-min. intervals). No more balky fuses and relays!

It's a 12-amp automatic load controller for lights, pumps, or radio/telephone equipment, with 15-amp surge allowance. Again, set points for load disconnect and reconnect are easily user-set. Loads will blink 5-minutes before disconnect, allowing the user to reduce loads, and extend run time. Even after low-voltage disconnect it allows one (only one!) 10-minute grace period by pushing an override button.

It's an automatic light controller for sign or road lighting. The light will turn on when voltage input from the PV ceases (usually a sure sign the sun has set). Run time can be set dusk to dawn, or timed any number of hours from 2 to 8, or run manually. Of course, if the battery gets too low, then low-voltage disconnect will override to protect the battery.

It even has an LED mode indicator, which shows approximate battery state of charge, or that low-voltage disconnect, overload, or equalization has occurred.

This 12-volt, 12-amp controller is compatible with any battery type including nicads. Clamping terminal strip will accept up to #10 AWG wire. Mounting screws and strain relief are included. The box is rain-tight, and the electronics are conformal coated for outdoor installations (it's still best to install in a protected location; don't tempt fate...). Certified by ETL to UL specifications 1741 (draft). Mfg. warr. is two years.

27–315   Trace C–12 Universal Controller $110⁰⁰

GenMate Auto-Start Controller

GenMate is a super-versatile computerized generator controller that can be customized to automatically start and stop nearly any electric start generator.

GenMate Auto-Start Controller (shown without case and cover)

Flipping tiny switches programs the GenMate, a one-time operation. This controller can be set to automatically start a generator at a user-selected voltage, run until another selected voltage is reached, and then turn off. Besides monitoring battery voltage, there are also two additional inputs: one line that is normally open, and one that is normally closed. If either change state then the generator will start. These can be used for manual starting, automatic pumping on a low water signal, automatic generator-driven fire pumps, a timed generator exercise cycle, practically anything you can imagine. You can also select a pre-start warning beeper, and the number of start-cycle retries to attempt before sounding a start-failure alarm. To use this controller, be sure that your generator is equipped with a low oil cutoff switch and automatic choke. 12- or 24-volt operation. Full five-year warranty. Specify generator make and model when ordering! Hondas and some others require customized controllers.

25-170   **GenMate Auto-Start Controller** $439⁰⁰

Spring Wound Timer

Our timers are the ultimate in energy conservation; they use absolutely no electricity to operate. Turning the knob to the desired timing interval winds the timer. Timing duration can be from 1 to 12 hours (or 1 to 15 minutes), with a hold feature that allows for continuous operation. This is the perfect solution for automatic shutoff of fans, lights, pumps, stereos, VCRs, and Saturday morning cartoons! It's a single pole timer, good for 10 amps at any voltage, and it mounts in a standard single gang switch box. A brushed aluminum faceplate is included. One-year warranty.

25-400   **Spring-Wound Timer, 12 hour** $34⁰⁰
25-399   **Spring-Wound Timer, 15 minute, no hold** $29⁰⁰

24-Hour 12-Volt Programmable Timer

This timer has separate dry contacts that will operate any voltage appliance up to 20 amps, including lights, pumps, fans, and appliances. It will also start and stop remote generators. Includes a 1.2-volt nicad battery, giving it a 200-hour power reserve in case your battery is dead. Trippers included allow for up to three separate on-off operations per day; more can be added if wanted. Minimum timed interval is 15 minutes. The timer is run by a 12-volt quartz movement, and best of all it draws only 0.01 amp.

25-401   **Programmable 12V Timer** $109⁰⁰

Photovoltaic Light Controller

Here's a PV-powered light controller that does it all in one affordable package. First, it's a 16-amp charge controller (the M-16 listed earlier in PV charge controls). Next it's a 20-amp light controller. When there is no voltage coming from the PV module (the sun went down), it turns on the light, or whatever. Next, it has a user-adjustable load timer. Only want your lights to run for four hours after dark? Want them to run all night? Have it your way. Finally, it's also a low voltage disconnect. If you've had ten days of cloudy weather it will protect your batteries and won't let them get run into the ground. Fully encapsulated and impervious to weather, unaffected by temperature extremes. Full five-year warranty too! What a deal!

25-157	Solar Light Controller, 12V	$139⁰⁰
25-158	Solar Light Controller, 24V	$139⁰⁰

Variable Speed Controls

These electronic variable-speed controls use less than 1/2 watt to control 12 & 24-volt applications such as lights, fans, and motors. The switches are available in 4-amp and 8-amp configurations. These work great on incandescent and halogen lamps, and also on our 12 or 24-volt fans. The 4-amp switch comes either with or without the forward-reverse switch. The 8-amp switch does not have the reverse option. The forward-reverse feature is useful for summer-winter reversal of air direction in fans.

24-101	Variable Speed Switch, 4A	$27⁰⁰
24-102	Variable Speed Switch, 4A (forward-reverse)	$31⁰⁰
24-103	Variable Speed Switch, 8A	$39⁰⁰

30 Amp General Purpose Relays

These general purpose DPDT (double-pole double-throw) power relays have either a 12-volt/169-milliamp or 24-volt/85-milliamp pull in coil depending on model choosen. They can be used remotely to switch loads larger than your DC-volt timer, load disconnect, float switch, etc. can handle alone, up to 30 amps per pole. They can also be used to switch AC circuits from DC powered timers or sensors. Because these are highly adaptable DPDT relays they can control multiple switching tasks simultaneously. Choose your system voltage.

25-112	12V/30A Relay	$34⁰⁰
25-151	24V/30A Relay	$34⁰⁰

MONITORS AND ALARMS

All ammeters operate without the need for external shunts. The meters measure 2-5/8" x 2-3/8" with a standard 2" panel mount that has a four-bolt pattern. Expanded scale voltmeters are very easy to read and extremely accurate. Rather than reading the low end of the scale (0-9 volts) which never registers in a 12-volt system (or 0-18 volts in a 24-volt system), the expanded scale voltmeters (10-16 volts or 20-32 volts) show you only what you need to see and give a more accurate reading. All voltmeters should be installed with an inline fuse. All meters can be recalibrated to assure long-term accuracy.

25-302	20 to 32V Voltmeter	$29⁰⁰	25-315	0 to 60A Ammeter	$19⁰⁰
25-311	0 to 10A Ammeter	$19⁰⁰	25-408	Inline Fuse Holder	$3⁰⁰
25-312	0 to 20A Ammeter	$19⁰⁰	24-521	AGC Fuses, 2 amp (5/box)	$2⁰⁰
25-313	0 to 30A Ammeter	$19⁰⁰			

A Larger Voltmeter

Our expanded-scale 10-16-volt voltmeters have a larger 2-1/2" x 3" face that makes them easier, and more accurate to read. Designed for panel mounting, it drops into a 2-1/4" round hole. Accuracy is ± 5%, and needle position is adjustable. Voltmeters should be installed with an inline fuse. Great for 12V systems. The most accurate we've found. Inexpensive, and can be user-calibrated.

25-301	10-16V Voltmeter	$10⁰⁰
25-408	In-line fuse holder	$3⁰⁰
24-521	AGC 2-amp fuses, 5/box	$2⁰⁰

Equus Digital Voltmeter

This Equus voltmeter shows at a glance your battery voltage with a digital LCD display to one decimal place. Digital meters are far more accurate and easier to read than conventional analog meters. It comes with a programmable bar graph that shows the state of charge and an internal light running off a separate wire that allows easy reading day or night. For a small, simple system this meter is Dr. Doug's recommended minimum. The meter mounts in a 2-1/16-inch hole. It runs on only 12 milliamps. 12-volt only.

25-341	Equus Digital Voltmeter	$49⁰⁰

SCI Mark III Battery Monitor

A rotary selector switch on this combination volt/amp meter lets you change the large LED display between three readings. 1). Incoming amperage, up to 30 amps with the built-in internal shunt, or 100 amps with the optional external shunt. 2). Battery voltage; three digits, reading to 1/10th of a volt. 3). Output amperage up to either 100 or 500 amps using one of the optional shunts. A-high voltage alarm light indicates when battery voltage exceeds 15.5 volts, and a low-voltage light turns on at 11.5 volts. (Double voltages for 24V.) Both are user adjustable. The Mark III Monitor comes mounted on a 5-1/4" x 8-1/4" flush-mount faceplate. The optional knock-out box will fully enclose the unit. Available in either 12- or 24-volt. When using the 500-amp shunt, the amp display reads 1/10 of actual value, showing 25.0 per 250 amps. Five-year full warranty.

25-361	DM III 24V Battery Monitor	$169⁰⁰
25-362	DM III 12V Battery Monitor	$169⁰⁰
25-351	100-amp Shunt	$19⁰⁰
25-364	500-amp Shunt	$29⁰⁰
25-114	Single SCI Knockout Box	$39⁰⁰

A Voltmeter For The Technically Challenged
or Anyone Who Appreciates, At-a-Glance Monitoring

For those who have never been able to figure out what the numbers on a voltmeter mean, or just didn't care until the battery died, here's the meter for you. If you can tell red from amber from green, this simple little LED indicator will show your battery capacity at a glance. Red is empty, amber is low, green is full; what could be more simple? When the battery is charging, the LED flashes amber/green and an overcharge flashes red. Don't get caught with your batteries down. Put this inexpensive, graphic meter out in the living room or kitchen where it gets noticed. Small 18 or 20 gauge wire can be run as far as 100' to the battery. An in-line fuse at the battery is needed. Even kids can understand this simple meter — no Saturday morning cartoons unless the meter is green. Time-delay logic assures accurate readings. The full-color universal battery symbols on the face makes reading the Smartlight Plus easy — simply match the LED color to the corresponding symbol. Measures 2.25" x 4". You must choose between vented, wet cells (the battery caps come off to add water), or sealed, gel-cells (the caps won't come off). One-year warranty.

25-163	Smartlight Plus, 12v wet-cell	$39⁰⁰
25-164	Smartlight Plus, 12v sealed-cell	$39⁰⁰
25-165	Smartlight Plus, 24v wet-cell	$39⁰⁰
25-166	Smartlight Plus, 24v sealed-cell	$39⁰⁰

Related Product:

25-408	In-line Fuse Holder	$3⁰⁰
24-521	2 Amp AGC fuse 5/box	$2⁰⁰

The New, Improved Tri-Metric Monitor

This low-cost, high-quality monitor was improved in Spring '96. Still has a large, bright LED display that can be read across the room, still has a "charged" light that tells you at the end of the day if your system reached full charge, and still fits conveniently in a standard double-gang electrical box. A single touch sensitive switch toggles the Tri-Metric through all the most important information, battery voltage, net amperage flow, and amp-hours removed, plus now, an accessory position which can be programmed to display any one of a wide variety of readouts. Hours since last battery full charge, total amp-hours removed since battery installation, charge efficiency, depth of latest discharge, and minimum or maximum battery voltage are choices available. The Tri-Metric will continually track all these choices, and display them whenever requested. It also has a settable "amp-hour alarm clock" that counts only discharged amp-hours (doesn't deduct charged amp-hours). This can be set to remind the user of needed battery service such as watering or equalization. It will flash "CbA" (check battery) alternately with the normal display when triggered. Works on both 12- or 24-volt systems, or 48-volt systems with the new optional lightning protector/adapter. The lightning protector works with all system voltages. Draws less than 30 milliamps. Requires 100- or 500-amp shunt, sold separately. One-year mfg. warranty.

25-363	Tri-Metric Energy Monitor	$169⁰⁰
25-303	48v Adapter/Lightning Protector	$29⁰⁰
25-351	100-amp Shunt	$19⁰⁰
25-364	500-amp Shunt	$29⁰⁰

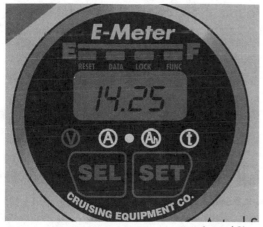

Actual Size

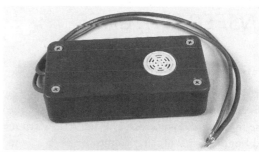

The Ultimate Battery Monitor

The E-Meter from Cruising Equipment is both the simplest, and the most complete battery monitor available. Simple? The accurate fuel gauge type meter gives at-a-glance voltage monitoring. Even flashes red when it's time to recharge! Complete? Gives volts, amps, amp-hours, kilowatt-hours, time remaining based on an adjustable rolling average of the past few minutes, historical battery information, and more. Has optional RS-232 computer output for logging of data (remotely if needed).

We found this meter to be an enormous ulcer-reliever at our free solar-powered concert series in the city park this past summer. When your renewable energy power system is pumping out 4,000-6,000 watts through the audio system, there are 3,000 happy dancing people in attendance, there's no power back-up, and the band's second set has just started, it's very satisfying to have a meter that tells you this power consumption level can be maintained for another six hours.

Mounts in a 2" (52mm) round hole, is 3" deep. Accepts 9.5-40 volts input, 500 amp shunt is included. Requires two in-line fuses, and 5 wires for hook up. Two wires must be a twisted pair. Sold below. Photo sensor on front panel adjusts display brightness according to ambient light levels. Meter goes into "sleep" mode with only bar graph fuel gauge active after 10 minutes of inactivity. Draws 28 ma in sleep mode. Tracks the number of battery recharge cycles, plus average depth of discharge, and greatest depth of discharge. Many more programing options are available. Programming is simple, and is well explained. Mfg. warranty two years.

25-346	E-Meter w/500A Shunt	$239
25-408	In-line fuse holder	$3 (needs two!)
24-521	2-Amp fuses, box/5	$2
26-524	Meter wire, 3 twisted pairs	$0.70/ft.

(specify length)

Audible Low Battery Alarm

Over-discharging your batteries can cripple your 12-volt power system. Inadvertent deep cycling of batteries drastically reduces their life. This alarm signals audibly when battery voltage falls below a preset minimum, so you can turn off lights and appliances before damage is done. The alarm is preset to come on at 11.5 volts, or can be adjusted. The alarm's dimensions are 2" x 1" x 4" inches. Voltage: nominal 12 volts.

25-342   **Low Battery Alarm** $39

Voltage Alarm

This alarm automatically detects hazardous high and low voltage fluctuations in your 12-volt system, warning you of problems with your batteries, alternator, or charging mechanism. The alarm has adjustable high and low settings, with both a visual and an audible alarm. At the factory-set low point of approximately 11.7 volts, a red LED and buzzer are activated. At the high setting of 14.6 volts, the yellow LED lights up. At 15.1 volts, the buzzer sounds. Voltage is a nominal 12 volts.

25-343   **Voltage Alarm** $75

TRANSFER SWITCHES

Automatic Transfer Switch

Transfer switches are designed as safety devices to prevent two different sources of voltage from traveling down the same line to the same appliances. This transfer switch made by Todd Engineering will safely connect an inverter and an AC generator to the same AC house wiring. If the generator is not running, the inverter is connected to the house wiring. When the generator is started, the house wiring is automatically disconnected from the inverter and connected to the generator. A time-delay feature allows the generator to start under a no-load condition and warm up for approximately one minute. Available in both 110-volt and 240-volt models. Each will handle up to a maximum of 30 amps. These switches are great for applications where utility power may be available only a few hours per day, or if frequent power outages are experienced. Housed in a metal junction box with hinged cover. Wires are clearly marked and installation schematic is included inside cover. Two-year warranty.

23-121	Transfer Switch, 30A, 110V	$55⁰⁰
23-122	Transfer Switch, 30A, 240V	$89⁰⁰

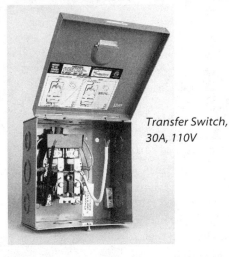

Transfer Switch, 30A, 110V

50-Amp/3-Pole Transfer Switch

For those who need to transfer more than 30 amps we offer this larger model with three heavy-duty DPDT 50-amp contactors. This contactor can be used with either 120-volt or 240 volt AC. Has a 15-second time delay to allow generator start-up under no load. Clearly marked, easy terminal strip wiring. Schematic included inside cover. 12" x 10" x 4" enclosure with multiple knockouts provided. Two year warranty.

23-118	Transfer Switch, 50A, 3 pole	$170⁰⁰

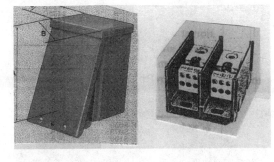

Rain-tight Junction Boxes & Power Distribution Block

To make your wiring at the PV array easier mount our two-pole power distribution block inside a rain-tight junction box. This is the easy way to join your large lead-in cables to the smaller interconnect wires from the PV array. Insert cables and tighten set screws. Primary side accepts one large cable, #6 to 350 MCM, secondary side accepts six smaller cables, #14 to #4. For use with copper or aluminum conductors.

Rain-tight boxes are 16 gauge zinc-coated steel with gray finish. Removable cover is fastened at bottom by screw. For non-corrosive environments.

24-215	Power Distribution Block, 2-pole	$29⁰⁰
24-213	10 x 8 x 4 Junction Box	$39⁰⁰
24-214	12 x 10 x 4 Junction Box	$49⁰⁰

INVERTER FUSING

Fused Disconnects

At last, a fused disconnect guaranteed to put a smile on any building inspector's face and still give your full-size inverter all the surge power it wants without voltage drop. Electric Code, and common sense, require a fuse and a disconnect between the battery and any appliance. This space-saving disconnect does both functions and can handle a 2000-amp surge for three minutes to protect your batteries and inverter. Features include a sealed, oil-filled 2000-amp switch that eliminates arcing and voltage drop to the inverter under load. These are extremely slow-blow fuses designed to tolerate the kind of high-amperage surge loads that full-size inverters produce. UL-listed Class T fuse is rated to 125 volts DC, 20,000 AIC, and wire lugs that can take up to 300 MCM wire.

24-206	Fused Disconnect Switch, 400A	$295⁰⁰
24-208	400-amp Class T Replacement Fuse	$32⁰⁰
24-216	Fused Disconnect Switch 300A	$305⁰⁰
24-217	300-amp Class T Replacement Fuse	$32⁰⁰

110-Amp Class T Fuse Block With Cover

For those who want to protect their mid-sized inverter, but don't need the big 400A switch, we also offer this 110-amp fuse with block, cover, and #2/0 lugs. This is a slow-blow fuse that will blow immediately if your battery is short circuited, but will allow several minutes of up to 150-amp use. Good for inverters up to 800 watts, it is easily installed on a wall or shelf with the provided screws. Dimensions are 2" x 2-1/2" x 5-1/2", including the cover. Rated for DC voltage up to 125 volts. Does not provide a disconnect means, and so will not satisfy NEC code.

24-507	Fuse Block, 110A Fuse and Cover	$49⁰⁰
24-508	110-amp Class T Replacement Fuse	$17⁰⁰

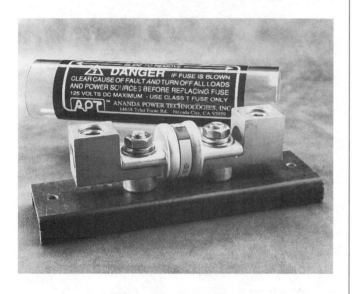

200-Amp Class T Fuse Block With Cover

A larger Class T fuse assembly for inverters up to 1500 watts. Includes 200-amp Class T fuse with block, cover, and #2/0 lugs. Does not provide a disconnect means, and so will not satisfy NEC code. Provides safe, sensible fuse protection for mid-sized inverters.

24-212	Fuse Block, 200A Fuse and Cover	$49⁰⁰
24-211	200-amp Class T Replacement Fuse	$15⁰⁰

FUSED DISCONNECTS AND DC FUSES

Safety Disconnects

Every PV system should employ a fused disconnect for safety. The function of this component is to disconnect all power-generating sources and all loads from the battery so that the system can be safely maintained and disconnected in emergency situations with the flick of a switch. We carry several different safety disconnects for different sized applications. Fuses should be sized according to the wire size you are using.

We recommend using a two-pole disconnect for the charging side of your batteries. One pole is used for the line between the panels and charge controller, and the other between the controller and the batteries. Make sure this fused disconnect can handle the maximum amperage of your array output. Another fuse or fused switch can be used to handle the discharge of your battery.

The 60-amp three-pole fused disconnect should be employed when the Enermaxer voltage regulator is used, which requires the third pole for disconnection.

24-201	30-amp 2-Pole Safety Disconnect	$35⁰⁰
24-404	30-amp 3-Pole Safety Disconnect	$99⁰⁰
24-202	60-amp 2-Pole Safety Disconnect	$55⁰⁰
24-203	60-amp 3-Pole Safety Disconnect	$99⁰⁰
24-204	100-amp 2-Pole Safety Disconnect	$140⁰⁰

Fuses sold separately (see next page)

Fuses And Fuseholders

Fuses and circuit breakers are circuit protectors. Their only function is to break an electrical circuit if the current (amps) flowing in that circuit exceeds the rating of the device. Any size fuse may be used safely as long as its rating is lower than the maximum ampacity of the smallest wire in the circuit.

Most fuses and breakers will pass three times their rated current for a few seconds, but they will open the circuit immediately in the event of a short circuit, which draws hundreds or even thousands of amps. This is necessary so that the fuse will melt or the breaker will open before the wire catches fire in the event of a short circuit. Fuses or breakers should be installed as close to the batteries as possible, and absolutely before wires pass through a flammable wall. The chart below gives the maximum ampacity for various sizes of copper wire.

Some types of wire can handle slightly more current, and all types of wire can handle much more current for very short amounts of time. For example, a 2000-watt inverter can surge to 6000 watts for a few seconds when starting large motors. During these few seconds, a 12-volt inverter is drawing 600 or more amps from the battery, but 4/0 wire and a 250-amp fuse will work fine.

Wire gauge	Maximum ampacity
14	15
12	20
10	30
8	45
6	65
4	85
2	115
0	150
2/0	175
4/0	250

If you choose not to use our highly recommended Safety Disconnects listed above, you must at a bare minimum fuse your system both between your solar array and your battery, and between your battery system and the distribution panel or load. This can be done cheaply and easily with our FRN-R Fuse Blocks, available for 30-amp fuses or for 60-amp fuses.

Don't forget to order fuses for your disconnects!

24-401	30A Fuseholder	$5⁰⁰
24-402	60A Fuseholder	$9⁰⁰
24-501	30A DC Fuse	$4⁰⁰
24-502	45A DC Fuse	$7⁰⁰
24-503	60A DC Fuse	$7⁰⁰
24-505	100A DC Fuse	$14⁰⁰

DC LOAD CENTERS AND CIRCUIT BREAKERS

Trace DC Disconnect/ Overcurrent Module

Trace DC Disconnects

Trace offers what can be a lower cost alternative for DC system overload/disconnect equipment, particularly for systems that use AC power for all or almost all loads. Available in two basic versions, 175 amps and 250 amps with a variety of options, the Trace Disconnect can provide all the necessary DC protection to meet code requirements.

Uses easy to reset circuit breakers, no fumbling for expensive replacement fuses in the dark. Breakers, all sizes, are rated for 25,000 amps interrupting capacity at 65 volts, and are UL listed for DC systems up to 125 volts. Main breaker lugs accept up to #4/0 AWG fine strand cable, no ring terminals required. 3/4", 2", and 2-1/2" knockouts are provided. Mates up with the conduit box option for DR and SW series Trace inverters. (Ohhh, matching color scheme too!) Space is available inside the disconnect for the DC current shunt(s) required by most monitors.

Options, all of which can be installed in the field, include a second main breaker for dual inverter systems, a DC bonding block for negative and ground wires, and up to four 15 or 60-amp circuit breakers for PV or other charging input, or DC power output to loads.

The Negative/Ground bonding block provides lugs for up to four 4/0, two #1/0, two #2, and four #4 cables.

A properly sized DC Disconnect with a 60-amp auxiliary breaker for the PV input gives all the protection an inverter-based system requires. Add another 15-amp auxiliary breaker if you've got a DC water pump or fridge to run and you're done.

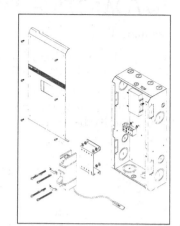

25-010	Trace DC175-amp Disconnect	$275⁰⁰
25-011	Trace DC250-amp Disconnect	$275⁰⁰
25-012	Trace 2nd 175-amp Breaker	$195⁰⁰
25-013	Trace 2nd 250-amp Breaker	$195⁰⁰
25-014	Trace 15-amp Auxiliary Breaker	$25⁰⁰
25-015	Trace 60-amp Auxiliary Breaker	$39⁰⁰
25-016	Trace Neg/Grd Bonding Block	$50⁰⁰

Trace DC Disconnect/Overcurrent Module

Recommended Sizing for Trace DC Disconnects		
Disconnect Size	Recommended With These Inverters	Minimum Cable Size
DC-175	DR-1512, 2412, 1524, & 2424. SW-3048E, & 4048	#2/0
DC-250	DR-3624, SW-2512, 4024, & 5548	#4/0

See page 246 for inverter cables.

Code Approved DC Load Centers

Being on the leading edge of the PV industry, Real Goods encourages the move toward safer, code-approved equipment. Although code-approved equipment can often be more costly, we feel when weighed together with safety factors it is a small price to pay for insurance.

The continuing move toward inverters running more and more of the household loads has reduced our selection of DC load centers. We find that a simple six-circuit center will handle the DC fusing needs for 99% of the systems. The six-circuit center meets all safety standards and utilizes only UL-listed components. All current-handling devices have been UL-approved for 12-volt through 48-volt DC applications.

All DC circuit breakers require a main fuse in-line before the breakers to provide catastrophic overload protection. If you already have a large Class T fuse for your inverter it can be used for this protection. If not, you must use the 110-Amp Class T Fuse Block on the input.

Larger enclosures are also available by special order that will handle up to 16 branch circuits with a maximum of 70 amps on an individual branch and up to 200 amps total for all branches.

Breakers must be ordered separately.
Load center does not include the 110-amp main fuse.

23-119	6-Circuit Load Center	$50⁰⁰
24-507	Class T Fuse Block, 110A Fuse and Cover	$49⁰⁰
23-131	10-amp Circuit Breaker	$16⁰⁰
23-132	15-amp Circuit Breaker	$15⁰⁰
23-133	20-amp Circuit Breaker	$18⁰⁰
23-134	30-amp Circuit Breaker	$18⁰⁰
23-136	40-amp Circuit Breaker	$20⁰⁰
23-135	50-amp Circuit Breaker	$25⁰⁰
23-137	70-amp Circuit Breaker	$40⁰⁰

AUTOMOTIVE-TYPE DC FUSE BOX & FUSES

Newmark Fuse Box

This is our bare-bones basic fuse box. It consists of an eight-position fuse block riveted into a hinged metal box with three 1/2" knockouts in place for wiring. It's made to accept AGC-type fuses up to 20 amps. It has a screw terminal for positive input from the battery and one screw terminal for each of the eight fused outputs. This fuse box only provides space for the positive wires. See our 15-Position Bus Bar below for the negative wires.

24-411   8-Position Fuse Box $15⁰⁰

8-Position Fuse Box

AGC Glass Fuses

These glass fuses go into the Newmark fuse box. The most common size for systems is the 20A fuses. They are available in the following amperages: 2A, 3A, 5A, 10A, 15A, 20A, 25A and 30A. The fuses measure 1-1/4" x 1/4" in diameter and are good up to 32V maximum. Packed five per box.

24-521	AGC Glass Fuses, 2 Amp, 5/box	$2⁰⁰	24-525	AGC Glass Fuses, 15 Amp, 5/box	$2⁰⁰
24-522	AGC Glass Fuses, 3 Amp, 5/box	$2⁰⁰	24-526	AGC Glass Fuses, 20 Amp, 5/box	$2⁰⁰
24-523	AGC Glass Fuses, 5 Amp, 5/box	$2⁰⁰	24-527	AGC Glass Fuses, 25 Amp, 5/box	$2⁰⁰
24-524	AGC Glass Fuses, 10 Amp, 5/box	$2⁰⁰	24-528	AGC Glass Fuses, 30 Amp, 5/box	$2⁰⁰

Inline Fuse Holder

There is no excuse for not fusing your equipment with a fuseholder that installs this easily. This bayonet-type fuse holder has 15 inches of 14-gauge wire in a loop that can be cut anywhere along the length for versatility. This is a "universal" type with three different springs for any length AGC style fuse. Always keep a couple of these on hand. Recommended for fusing your analog voltmeter. Fuses not included.

25-408   Inline Fuse Holder $3⁰⁰

15-Position Bus Bar

You can use this 15-position bus bar as a common negative for any fuse block.

24-431   15-Position Bus Bar $11⁰⁰

Blue Sea DC Circuit Breaker

Blue Sea DC Circuit Breakers — High Amperage Protection, Small, Inexpensive Package

Blue Sea Systems has finally created a compact DC circuit breaker at an affordable price. These sealed, surface-mounting breakers are water and vaporproof, they are thermally activated, and manually resettable, with the reset handle providing a positive trip identification. The red trip button can also be manually activated, allowing the breaker to function as a switch. They are "trip-free," and cannot be held closed against an overload current. Thermal tripping allows high starting loads, without tripping.

These simple to install, surface-mount breakers are an excellent low-cost choice for small to midsized inverters, wind generators, control and power diversion equipment, or any other application with current flows of 50 to 150 amps. For use with DC power systems up to 30 volts. For code compliance and safety Blue Seas breakers must be installed with a class T fuse between them and battery bank. See page 211 for class T fuses. Lifetime warranty!

25-167	Blue Sea 50 Amp Circuit Breaker	$35⁰⁰	25-183	Blue Sea 125 Amp Circuit Breaker	$35⁰⁰
25-168	Blue Sea 75 Amp Circuit Breaker	$35⁰⁰	25-184	Blue Sea 150 Amp Circuit Breaker	$35⁰⁰
25-169	Blue Sea 100 Amp Circuit Breaker	$35⁰⁰			

LINEAR CURRENT BOOSTERS/PUMP CONTROLS

Linear Current Booster

The Linear Current Booster (LCB) is an electronic interface between solar modules and DC motors being operated directly such as water pumps or ventilator fans. The LCB will down-convert excess voltage into amperage to provide a starting surge for motors, or to keep the motor running under marginal light conditions. DC motors don't get hot and bothered by low voltage, they simply run slower. This offers substantially improved performance for array direct pumping systems by allowing the pump to start earlier in the day and run longer through cloudy weather. We consider a current booster to be standard equipment on most pumping systems.

Sun Selector Linear Current Boosters

The LCB must be set to the proper solar module operating voltage. This can be accomplished by ordering the unit preset from the factory, or by setting the unit in the field when the unit is ordered as a "T" (tunable) model. The basic operating range is from 9 to 38 volts DC, with a maximum open circuit voltage of 50 volts DC. All LCB 20's are tuneable (no option).

An option for the LCB is the remote control (RC) feature. An extra wire is brought out of the unit which, when shorted to PV(-), will turn the LCB and the motor load off. The remote control wire can be actuated by a float switch or remote on-off switch.

The WLS-3 is an inexpensive water sensor that works with the RC option to control level in a tank. The sensor is waterproof, requires no power supply, and has an LED operation indicator. Three adjustable length probes provide a level differential adjustment. The WLS-3 can be configured to fill or empty a tank depending on which output wire is connected. All sensors can be virtually any distance from the LCB and requires only a pair of inexpensive telephone wires between them.

All Sun Selector LCBs can be configured both in parallel and in series to achieve any needed level of voltage or current. The first number in the model indicates maximum input amperage. Five-year full warranty on all LCBs and options.

25-133	LCB 3M, 12V	$69⁰⁰
25-133-RC	LCB 3M, 12V & Remote Control	$79⁰⁰
25-133-T	LCB 3M Tuneable	$79⁰⁰
25-133-T-RC	LCB 3M Tuneable & Remote Control	$95⁰⁰
25-139	LCB 3M, 24V	$69⁰⁰
25-139-RC	LCB 3M, 24V & Remote Control	$79⁰⁰
25-134	LCB 7M, 12V	$99⁰⁰
25-134-RC	LCB 7M, 12V & Remote Control	$89⁰⁰
25-134-T	LCB 7M Tuneable	$119⁰⁰
25-134-T-RC	LCB 7M Tuneable & Remote Control	$129⁰⁰
25-138	LCB 7M, 24V	$109⁰⁰
25-138-RC	LCB 7M, 24V & Remote Control	$89⁰⁰
25-159	WLS-3 (option)	$39⁰⁰
25-135	LCB 20	$375⁰⁰

Larger LCBs in 40- and 80-amp sizes are available for long-distance power transmission of PV or hydro energy. Call our tech staff for more information and pricing.

LCB 3M, 12V

LCB 20

WIRE, ADAPTERS, AND OUTLETS

Wires are freeways for electrical power. If we do a poor job designing and installing our wires, we get the same results as with poorly designed roads: traffic jams, accidents, and frustration. The proper choices of wire and wiring methods can be confusing, but with sufficient planning and thought we can create safe and durable paths for energy flow.

The National Electrical Code (NEC) provides broad guidelines for safe electrical practices. Local codes may expand upon or supersede this code. It is important to use common sense when dealing with electricity, and this might be best done by acknowledging ignorance on a particular subject and requesting advice and help from experts when you are unsure of something. The Real Goods technical staff is eager to help with particular wiring issues, although you should keep in mind that local inspectors will have the final say on what they consider the most appropriate means to the end of a properly installed electrical system; therefore, advice from them or from a local electrician familiar with both local code requirements and alternative energy systems might be more informative.

Wire

Wire comes in a tremendous variety of styles which differ in size, number, material, and type of conductors, as well as the type and temperature rating of insulation protecting the conductor.

Permanent wiring should always be run within electrical enclosures, conduit, or in side walls. Wires are prone to all sorts of threats, including but not limited to abrasion, falling objects, and children using them as a jungle gym. The insulation around the metal conductor offers minimal protection based on the assumption that the conductors will be otherwise protected. Poor installation practices that forsake the use of conduit and strain-relief fittings may lead to a breach of the insulation, which in turn could cause a short circuit leading to fire or electrocution. Electricity, even in its most apparently benign manifestations, is not a force to be managed carelessly.

A particular type and gauge of wire is rated to carry a maximum electrical current. The NEC code requires that we not exceed 80% of this current rating for continuous duty applications. With the low-voltage conditions that we run across in independent energy systems, we also need to consider the proper size of wire, that which will move the energy efficiently enough to get the job done at the other end. Using undersized wire might be acceptable from the code perspective, but DC current does not travel well, losing voltage as it goes. This problem becomes more acute with smaller conductors over longer runs. Voltage drop caused by moving the current a particular distance can result in poor performance; when voltage drops, delivered power (wattage) also drops. Some voltage drop is unavoidable when moving electrical energy from one point to another, but we can limit this to reasonable levels if we choose the right size of wire, as shown in the accompanying chart. We have an all-purpose wire sizing chart and formula in the Appendix.

WIRE SIZING CHART/FORMULA

We could give you some incomprehensible voltage drop charts (like we've done in the past), but this all-purpose formula works better.

This chart is useful for finding the correct wire size for any voltage, length, or amperage flow in any AC or DC circuit. For most DC circuits, particularly between the PV modules and the batteries, we try to keep the voltage drop to 3% or less. There's no sense using your expensive PV wattage to heat wires. You want that power in your batteries!

Note that this formula doesn't directly yield a wire gauge size, but rather a 'VDI' number which is then compared to the nearest number in the VDI column, and then read across to the wire gauge size column.

1. Calculate the Voltage Drop Index (VDI) using the following formula:

$$VDI = AMPS \times FEET \div (\% \text{ VOLT DROP} \times VOLTAGE)$$

Amps = Watts Divided by Volts   Feet = One way wire distance
% Volt Drop = Percentage of voltage drop acceptable for this circuit (typically 2% to 5%)

2. Determine the appropriate wire size from the chart below.
    A. Take your VDI number you just calculated and find the nearest number in the VDI column, then read to the left for AWG wire gauge size.
    B. Be sure that your circuit amperage does not exceed the figure in the Ampacity column for that wire size. (This is not usually a problem in low voltage circuits.)

Wire Size AWG	Copper Wire		Aluminum Wire	
	VDI	Ampacity	VDI	Ampacity
0000	99	260	62	205
000	78	225	49	175
00	62	195	39	150
0	49	170	31	135
2	31	130	20	100
4	20	95	12	75
6	12	75	•	•
8	8	55	•	•
10	5	30	•	•
12	3	20	•	•
14	2	15	•	•
16	1	•	•	•

Example: Your PV array consisting of 4 Siemens PC4 modules is 60 ft. from your 12-volt battery. This is actual wiring distance, up pole mounts, around obstacles, etc. These modules are rated at 4.4 amps, times 4 modules = 17.6 amps maximum. We'll shot for a 3% voltage drop. So our formula looks like:

$$VDI = 29.3 \quad \frac{17.6 \times 60}{(3[\%] \times 12 [V])}$$

Looking at our chart, a VDI of 29 means we'd better use #2 wire in copper, or #0 wire in aluminum. Hummm. Pretty big wire.

What if this system was 24-volt? The modules would be wired in series, so each <u>pair</u> of modules would produce 4.4 amps. Two pairs, times 4.4 amps = 8.8 amps max.

$$VDI = 7.3 \quad \frac{8.8 \times 60}{(3[\%] \times 24 [V])}$$

Wow! What a difference! At 24-volt input you could wire your array with little ol' #8 copper wire.

Adapters and Outlets

Adapters and outlets provide us with an easy way of connecting and disconnecting our loads from their power source. The roots of the independent-energy movement grew from the automobile and recreational-vehicle industry, so plugs and outlets for low-voltage applications are based on that ubiquitous creation, the cigarette lighter. As the price of high-quality inverters has dropped, so has the demand for 12-volt equipment. Still, we often run across situations where these male and female DC outlets and connectors prove to be a convenient and cost-effective method of connecting load to power source. It is important to remember that these outlets and plugs have limited current-carrying capacity, as do the cigarette lighter outlets found in cars, trucks, and recreational vehicles. Do not try to move more amperage through the wire than is specified for the circuit's fuse. Cigarette lighter plugs and outlets are rarely rated for more than 15 amps.

The conventional rules for DC outlets and plugs hold that the center of the receptacle (female outlet) and the tip of the plug (male adapter) are positive, and the outer shell of the receptacle and side contact of the plug are negative. This should be verified (with a hand-held voltmeter) rather than assumed, because the polarity can be reversed in a variety of ways. Better to check it out and fix it rather than ruin the costly DC television you just bought.

Another convention is that the power *source* is presented by means of the female outlet. This way, it is more difficult to accidentally electrocute yourself. There are some solar modules, however, that have DC male adapters on them, even though they are technically a power source. They are made this way so that you can easily charge a car battery with the solar panel through the cigarette lighter.

Some people would rather not use these cigarette lighter outlets and plugs in their houses because they look dangerous . . . and they can be. Little fingers and tools fit conveniently, and disastrously, within the supposedly "protected" live socket. Consider using an AC outlet that has the prongs perpendicular, slanted, or in some other configuration which makes it impossible to plug in a standard house-current appliance. Although it is tempting to use inexpensive standard house-current outlets, we advise against it. We know from personal experience that any savings are quickly lost when a wrong-voltage appliance is plugged in and destroyed. Such outlets have AC amperage ratings, but are unlikely to have been rated for DC applications. Use common sense, and do not try to run a large load through light-duty hardware.

— *Douglas Bath and Michael Potts*

WIRE

Type USE Direct Burial Cable

Type "USE" (Underground Service Entrance) cable is moistureproof and sunlight-resistant. It is recognized as underground feeder cable for direct earth burial in branch circuits and is the only wire you can install exposed for module interconnects. It is approved by the National Electrical Code and UL. It is resistant to acids, chemicals, lubricants, and ground water. Our USE cable is a single conductor with a sunlight-resistant jacket. It is much more durable than standard romex. 10-gauge is solid wire, 8 gauge is stranded.

		Price/Foot
26-521	USE 10-Gauge Wire	$.55
26-522	USE 8-Gauge Wire	$.55

Copper Lugs

We carry very heavy duty copper lugs for connecting to the end of your large wire from 4-gauge to 4/0 gauge (#0000). The hole in the end of the lug is 3/8-inch diameter. Wire must be soldered to copper lugs.

26-601	Copper Lug, 4-Gauge	$2.00
26-602	Copper Lug, 2-Gauge	$4.00
26-603	Copper Lug, 1-Gauge	$4.00
26-604	Copper Lug, 1/0-Gauge	$4.00
26-605	Copper Lug, 2/0-Gauge	$5.00
26-606	Copper Lug, 4/0-Gauge	$6.00

Nylon Coated Single Conductor Wire

This single conductor wire connects the components of your alternative energy system. We stock it in black only. We recommend you use red tape on the ends for positive and white tape on the ends for negative. Wire is stranded copper, with THHN jacketing. This wire should always be installed in conduit, and never in exposed situations.

The minimum order for 16-, 14-, and 10-gauge THHN is a 500-foot roll. Wire gauges of 8 and larger can be ordered in any length. Copper prices change frequently; please call to check prices before ordering.

		Price/Foot
26-534	#8 THHN Primary Wire	$0.55
26-535	#6 THHN Primary Wire	$0.65
26-536	#4 THHN Primary Wire	$0.90
26-537	#2 THHN Primary Wire	$1.35
26-538	#0 THHN Primary Wire	$2.20
26-539	#00 THHN Primary Wire	$2.75

#16, #14, #12 and #10, must be ordered in 500-foot rolls

26-531	#16 THHN Primary Wire, 500 ft.	$50.00
26-532	#14 THHN Primary Wire, 500 ft.	$60.00
26-530	#12 THHN Primary Wire, 500 ft.	$70.00
26-533	#10 THHN Primary Wire, 500 ft.	$110.00

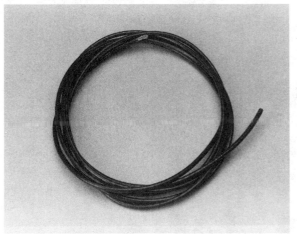

Split Bolt Kerneys

Split bolt kerneys are used to connect very large wires together or large wires to smaller wires. You must always wrap the kerney with black electrical tape to prevent corrosion and the potential for short-circuiting.

26-631	Split Bolt Kerney, #6	$6.00
26-632	Split Bolt Kerney, #4	$8.00
26-633	Split Bolt Kerney, #2	$9.00
26-634	Split Bolt Kerney, #1/0	$12.00
26-635	Split Bolt Kerney, #2/0	$15.00

Solderless Lugs

These solderless lugs are ideal for connecting large wire to small connections or to batteries.

| 26-622 | Solderless Lug (#8,6,4,2) | $2.00 |
| 26-623 | Solderless Lug (#2 to 4/0) | $7.00 |

Wiring 12 Volts For Ample Power

David Smead and Ruth Ishihara. The most comprehensive book on DC wiring to date, written by the authors of the popular book *Living on 12 Volts with Ample Power* (see page 95). This book presents system schematics, wiring details, and troubleshooting information not found in other publications. Leans slightly toward marine applications. Chapters cover the history of electricity, DC electricity, AC electricity, electric loads, electric sources, wiring practices, system components, tools, and troubleshooting. 240 pages, paperback, 1991.

| 80-111 | Wiring 12 Volts for Ample Power | $19.00 |

Electrical Wiring

The AAVIM staff. Thoroughly covers standard electrical wiring principles and procedures. This book has become an industry standard for training students, teachers, and professionals. Includes over 350 step-by-step color illustrations, covering circuits, receptacles and switches, installing service entrance equipment, and more. Revised in 1991 to include changes made in the 1990 National Electrical Code. 188 pages, paperback, 1992.

| 80-302 | Electrical Wiring | $23.00 |

ADAPTERS

Wall Plate Receptacle

Our most popular basic 12-volt receptacle is identical to an automobile cigarette lighter. It is made of a break-resistant plastic housing and all-brass metal parts and fits a standard single gang junction box. Handles up to 15-amp surge, but 7 to 8 amps is the maximum continuous current from any cigarette lighter plug. Specify brown or ivory.

26-201-B Wall Plate Receptacle, Brown $4⁰⁰

26-201-I Wall Plate Receptacle, Ivory $4⁰⁰

Replacement Plug

The SP-6 is the simplest 12-volt plug on the market. The plug is supplied with two solderless terminals for easily attaching terminals to wire. The terminals are inserted into the rear of the plug and lock into place. Can accommodate wire up to 16 gauge. Rated at 8 amps.

26-101   Replacement Plug $1⁵⁰

Heavy-Duty Replacement Plug

The SP-20 is designed for heavier-duty applications than the SP-6. It is supplied complete with three styles of end caps to accommodate coil cords, SJ-series jacketing, and other round wire. The plug is unbreakable and is rated at 10 amps.

26-102   Heavy-Duty Plug $2⁰⁰

Fused Replacement Plug

The SP-90-F5-B fused plug has a unique polarity reversing feature. It includes a 5-amp fuse and four sizes of snap-in strain reliefs which accommodate wire gauges from 24 to 16 AWG. It is rated at 8 amps continuous duty and can be fused to 15 amps for protecting any electric device.

26-103   Fused Replacement Plug $3⁰⁰

Plug Adapter

This simple, durable SP-70 plastic adapter provides an elegant conversion from any standard 110-volt fixture. Simply insert your 110-volt plug into the connector end of this adapter (no cutting!), insert a 12-volt bulb into the light socket of your lamp, and you're in 12-volt heaven for cheap!

26-104   Plug Adapter $2⁷⁵

Triple Outlet Plug

The 30-TRP triple outlet plug permits the use of three cigarette lighter plugs on the same circuit simultaneously. Rated at 12 amps. Includes heavy-duty 16-gauge wire, and a self-adhesive pad in back.

26-105   Triple Outlet Plug $9⁰⁰

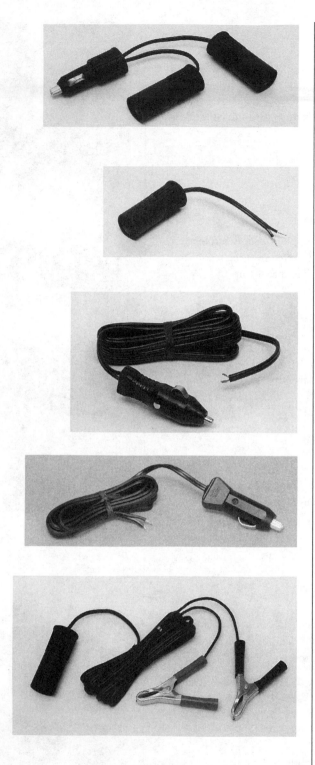

Double Outlet Adapter

The double outlet power adapter permits the use of two 12-volt products in a single outlet. Two receptacles connect to a single adapter plug with short lengths of 16-gauge wire. Rated at 10 amps.

26-107   Double Outlet Adapter $6⁰⁰

Extension Cord Receptacle

This receptacle, made of all brass parts in a break-resistant plastic housing, connects with 12-volt cigarette lighter plugs. Includes 6-inch lead wire.

26-108   Extension Cord Receptacle $3⁰⁰

Non-Fused Power Cord

This replacement power cord is handy when an existing power supply cord has been damaged or worn. It features a male cigarette lighter plug on one end attached to 8 feet of 20-gauge, polarity coded wire. The other end has the wire jacketing stripped ready for your installation. 4-amp maximum.

26-109   Non-Fused Power Cord $2⁷⁵

Fused Power Cord

The fused power cord is identical to the 26-109 listed above except that it features a fused male plug on one end. Includes replaceable 2-amp fuse.

26-110   Fused Power Cord $4⁰⁰

Extension Cords And Battery Clips

These unique cords have spring-loaded clips to attach directly to battery terminals. They increase the use and value of 12-volt products like lights, TV sets, radios, or appliances. The maximum amperage that can be put through the 1-foot cord is 10 amps and the maximum that can be put through the 10-foot cord is 4 amps.

26-111   Extension Cord and Clips, 1 ft. $6⁰⁰
26-112   Extension Cord and Clips, 10 ft. $7⁰⁰

*A house is a machine
for living in.*

—Charles Edouard
Jeanneret

Extension Cords

These low-voltage extension cords have a cigarette lighter receptacle on one end and a cigarette lighter plug on the other end. They are available in 10-foot and 25-foot lengths. The maximum recommended current is 4 amps for the 10-foot cord and 2 amps for the 25-foot cord.

26-113	Extension Cord, 10 ft.	$6⁰⁰
26-114	Extension Cord, 25 ft.	$7⁰⁰

UL-Listed Switches And Receptacles

Code approved DC rated 15-amp switches and 20-amp receptacles. All are heavy-duty commercial grade. Although expensive, these rugged components may be the only way to keep your local building inspector happy (and maybe even let you sleep better at night). They are light-years ahead of conventional RV grade plugs and switches. All are ivory colored and mount in standard electrical boxes.

26-310	Single pole 15A Switch	$6⁰⁰	26-314	Switch Plate	$1⁰⁰
26-311	3-Way 15A Switch	$8⁰⁰	26-315	Single Receptacle Plate	$1⁰⁰
26-312	Single 20A Receptacle	$7⁰⁰	26-316	Duplex Receptacle Plate	$1⁰⁰
26-313	Duplex 20A Receptacle	$16⁰⁰	26-317	Heavy Duty 15A Plug	$15⁰⁰

DC Power Converter

This converter allows you to operate a wide variety of DC powered products, including a portable stereo, cassette player, video cassette recorder, and other applications from a standard cigarette lighter receptacle. One convenient switch enables the output voltage to be set to 3, 6, 9, or 12 volts DC from a 12-volt DC input, negative ground. Output polarity may be reversed and the LED indicator shows when the adapter is in use. The complete unit, which is UL-listed, includes a 3-amp fused universal plug, 6-foot distribution cord and an assortment of four popular polarity-reversible coaxial power plugs. The fuse is replaceable.

26-121	Converter	$15⁰⁰

12-VOLT OUTLETS

12-Volt Brass Outlets

Suntronics makes an attractive line of solid brass outlets and switches designed for the 12-volt market. Constructed of .036" solid brass, they are rated at 20 amps for the outlets and 15 amps for the switches. *(Note: 20 amps is a surge rating for the outlets. Don't expect a cigarette lighter plug to sustain more than 7 to 8 amps continuously. There just isn't enough metal-to-metal contact in this plug configuration. — Dr. Doug)*

26-221	One Plug	$7⁰⁰	26-226	Switch and Plug	$11⁰⁰
26-222	Two Plugs	$10⁰⁰	26-227	Switch and Fuse	$9⁰⁰
26-223	One Switch	$7⁰⁰	26-228	Switch, Fuse, and Plug	$12⁰⁰
26-224	Two Switches	$11⁰⁰	26-229	Plug and Fuse	$10⁰⁰
26-225	Four Switches	$15⁰⁰			

INVERTERS

Alternative energy systems use batteries to store energy for later use. This is the least expensive and most universally applicable storage method available. Batteries store energy as low-voltage DC (direct current). DC is fine, in fact preferable, for some applications, particularly running motors, but most of the world operates on higher-voltage AC (alternating current). AC transmits more efficiently than DC and so has become the world standard. Therefore, it is convenient if our alternative energy system contains a device to produce conventional AC house current.

Six or seven years ago, the highly efficient, long-lived, relatively inexpensive inverters that we now have were still a pipe dream. Since that time, though, the world of solid-state equipment has advanced by extraordinary leaps and bounds, and more than 95% of the systems we put together now include an inverter.

Benefits of Inverter Use

North America and a good part of the world run on 120-volt AC. Joining the mainstream allows the use of mass-produced components, wiring hardware, and appliances. Appliances may be chosen from a wider, cheaper, and more available selection. Electricity also transmits more efficiently at higher voltages. Power distribution through the house can be done with conventional 12- and 14-gauge romex wiring using standard hardware, which electricians appreciate and inspectors understand. Anyone who has ever wrestled with heavy 10-gauge wire inside of a single gang box, as a DC system requires, will see the immediate benefit of running on AC.

An inverter system also supplies the ultimate clean, uninterruptible power supply for your computer. No spikes, no surges, no garbage! Just don't run the power saw or washing machine off the same inverter at the same time.

Modern inverters are extremely reliable. Most models have failure rates well under 1%. Efficiency averages about 90% for most models with peaks at 95% to 98%. In short, inverters make life simpler, do a better job of running your household, and ultimately save you money on appliances and lights.

Description of Inverter Operation

If we took a pair of switching transistors and set them to reversing the DC polarity (direction of current flow) 60 times per second we would have low-voltage alternating power of 60 Hz. If this power was then passed through a transformer, which can transform AC power to higher or lower voltages depending on design, we would end up with a crude 120V/60 Hz power. In fact, this was about all that early inverter designs of the 1950s did. As you might expect, the waveform was square and very crude (more about that in a moment). If the battery voltage went up or down, so did the output AC voltage, only ten times as much. Inverter design has come a long, long way from the noisy, 50% efficient, crude inverters of the 1950s. Modern inverters hold a steady voltage output regardless of battery voltage fluctuations, and efficiency is now typically in the 90% to 95% range. The waveform of the power delivered has also been dramatically improved.

ELECTRICAL TERMINOLOGY AND MECHANICS

Electricity can be supplied in a variety of voltages and waveforms. Whoa! What's that you say? Different flavors? You bet. As delivered to and stored by our battery system we've got low-voltage DC. As supplied by the utility network we've got house current, nominally 120-volts AC. What's the difference? DC electricity flows in one direction only, hence the name "direct current." It flows directly from one battery terminal to the other battery terminal. AC current alternates, switching the direction of flow periodically. The U.S. standard is 60 cycles, or alternations, per second. Other countries have settled on other standards, but it's usually either 50 or 60 cycles per second. The electrical unit that expresses how many cycles per second is *hertz*, named after Heinrich Hertz, an early electrical pioneer. So AC power is defined either as 50 Hz or 60 Hz.

Our nation-wide standard also defines the voltage. Voltage is similar to pressure in a water line. The greater the voltage, the higher the pressure. When voltage is high, it's easier to transmit a given amount of energy, but it's more difficult to contain and potentially more dangerous. House current in this country is delivered at about 120 volts for most of our household appliances, with the occasional high-consumption appliance using 240 volts. So our power is defined as 120/240 volts/60 Hz. We usually use the short form, 120V, to denote this particular voltage/cycle combination. This voltage is by no means the world standard, but is the most common one we at Real Goods deal with. Inverters are available in international voltages by special order, but not all models from all manufacturers.

Waveforms

The power supplied by the utilities is created by spinning a bundle of wires through a series of magnetic fields. As the wire moves into, through, and out of a magnetic field, the voltage gradually builds to a peak and then diminishes. The next magnetic field the wire encounters has the opposite polarity, so current flow is induced in the opposite direction. If this action was plotted against time, as an oscilloscope does, we'd get a picture of a sine wave as shown in the Wave Gallery (see page 226). Notice how smooth the curves are.

Early inverters produced square-wave alternating current. While this current alternates, it is considerably different in waveform from regular AC and causes indigestion in some appliances. Motors and heaters are fine, but solid-state equipment sometimes has a hard time with it, resulting in loud humming, overheating, or failure.

Most modern inverters produce a hybrid waveform called a quasi-, synthesized, or modified sine wave. In truth this could just as well be called a modified square wave, but manufacturers are optimists. (Is the glass half-full or half-empty?) Modified sine-wave output cures many of the problems with square wave. Most appliances will digest it and never know the difference. There are some notable exceptions to this rosy picture however, which we'll cover in detail later in the Inverter Problems section (see page 229).

Full sine-wave inverters have been available for some time, but, because of their lower efficiency and higher initial cost, they have only been used for running very specific loads. This is changing rapidly. By the time you read this, there will be a variety of high-efficiency, moderately-priced, sine-wave inverters on the market. Trace has unveiled their new 4000-watt sine-wave unit, and (pardon the pun) it looks like the wave of the future. Other manufacturers have active development programs for sine-wave equipment. We confidently expect that they will become the standard in a few years.

Inverter Output Ratings

All modern inverters are capable of briefly sustaining much higher loads than they could run continuously, because electric loads like motors require a surge to get started. This momentary overcapacity has led some manufacturers to fudge on their output ratings. A manufacturer might for instance call his unit a 200-watt inverter based on the instantaneous rating, or an 800-watt inverter based on the 30-minute rating. Happily, this is a practice that is fading into the past. Most manufacturers are taking a more honest approach as they introduce new models and are labeling inverters with their continuous wattage output rather than some fanciful number. In any case, we've been careful to list the continuous power output of all the inverters we carry. Just be aware that the manufacturer calling the inverter a 200-watt unit does not necessarily mean it will do 200 watts continuously. Read the fine print!

Wave Gallery

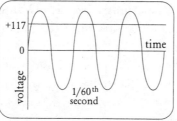

House current: 117 volt AC sine wave

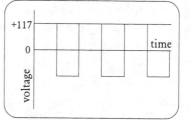

Inverter 117 volt AC square wave

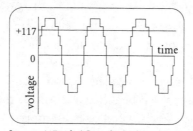

Inverter 117 volt AC synthesized sine wave

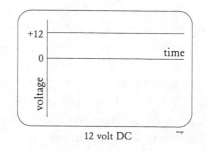

12 volt DC

READING MODIFIED SINE-WAVE OUTPUT WITH A CONVENTIONAL VOLTMETER

Most voltmeters that sell for under $100 will give you weird voltage readings if you use them to check the output of a modified sine-wave inverter. Readings ranging anywhere from 80 to 105 volts are the norm. This is because when you switch to "AC volts," the meter is expecting to see conventional utility sine-wave power. What we commonly call 110- to 120-volt power actually varies from 0 to about 175 volts through the sine-wave curve. One-hundred-twenty volts is an average called the *Root Mean Square,* or RMS for short, that is arrived at mathematically. More expensive meters are RMS-corrected; that is, they can measure out the average voltage for a complex waveform that isn't a sine wave. Less expensive meters simply assume that if you switch to "AC Volts," it's going to be a sine-wave. So don't panic when your new expensive inverter checks out at 85 volts; the inverter is fine, but your meter is being fooled. Modern inverters will hold their specified voltage output, usually about 117 volts, to plus or minus about 2%. Most utilities figure they're doing well by holding variation to 5%.

Effects of Temperature and Battery Voltage on Output Power

There are certain conditions that will reduce any inverter's ability to provide maximum output: low battery voltage, high temperature, and inadequate cables. Low battery voltage will severely limit any inverter's ability to meet a surge load, including low voltage caused by the surge. If the battery bank isn't large enough to supply the energy demand without dropping voltage, then the inverter probably won't be able to start the load. For instance, a Trace 2512 will easily start any washing machine, but not if you've only got a couple of golf-cart batteries to supply it with energy. The starting surge will be more than the battery bank can sustain, and the inverter will shut off to protect itself (and your washing-machine motor). Suppose you have eight golf-cart batteries on the 2512, but decided to save some money on the hookup cables, and used an old set of automotive jumper cables. The battery bank is capable, the inverter is capable, but enough electrons can't get through the small wire. The result is the same: low voltage at the inverter, which shuts off without starting the washer. Do not scrimp on inverter cables or on battery interconnect cables. These items are inexpensive compared to the cost of a high-quality inverter and good batteries, so do not cripple your system by trying to save a few bucks.

All inverters generate a small amount of waste heat. The harder they are working, the more heat they produce. If they get too hot, they will shut off to protect themselves or limit their output. Give the inverter plenty of ventilation. Treat it as you would a piece of stereo equipment, keeping it dry, protected, and well-ventilated.

How Big an Inverter Do I Need?

The size of your inverter will depend upon your power needs. If you only have a single appliance or two to run it does not take much. For example, say you want to run a 19-inch TV, a VCR, and a light all at once. Total up all the wattages: about 80 for the TV, 25 for the VCR, and 20 for a compact fluorescent light, for a total of 125 watts. Pick an inverter that can supply at least 125 watts continuously, and you have it.

To power a whole house full of appliances and lights requires more planning. Obviously not every appliance and light will be on at the same time. Mid-sized inverters of 600 to 1000 watts do a good job of running lights, entertainment equipment, and small kitchen appliances — in other words, most common household loads. What a mid-sized inverter will *not* do is run a mid- to full-sized microwave, a washing machine, or larger hand-held power tools. For those loads you need a full-sized 2000+-watt inverter. In truth, most households end up with one of the full-sized inverters because household loads tend to grow, and larger inverters can be equipped with very powerful battery chargers. This is the most convenient way to add battery-charging capability to a system (and the cheapest, too, if you are already buying the inverter).

How Big a Battery Do I Need?

We usually size most household battery banks to provide stored power for three to five days of autonomy during gloomy weather or while a source is being repaired. For most folks, this is a comfortable compromise between cost and function. If your battery bank is sized to provide a typical three to five days of back-up power, then it will also be large enough to handle any surge loads that the inverter is called upon to start. However, we occasionally run into situations with 3/4 hp or larger submersible well pumps or stationary power tools which require a larger battery bank simply to meet the surge load when starting. Call the Real Goods technical staff for help if you are anticipating large loads of this type.

Inverter Safety Equipment and Power Supply

In some ways batteries are safer than conventional AC power. It's very difficult to shock yourself at low voltages. But in other respects batteries are far more dangerous: they can supply many more amps into a short circuit, and once a DC arc is struck, it has little tendency to self-snuff. Starting with its 1990 edition, the National Electrical Code began addressing low-voltage DC power systems. One of the very sensible things the Code requires is fusing and a safety disconnect for any appliance connected to a battery bank. When the Code first came out, this posed some problems for large inverters. Available AC equipment either caused an unacceptable voltage drop or was prohibitively expensive and large. (We're talking about 5-foot-high switch cabinets, folks.) Fortunately, Ananda Power Technologies (APT) came to the rescue by developing first the 400-amp fused disconnect switch, and later their integrated Powercenters (see page 189).

Fusing is extremely important for any circuit connected to a battery! Without fusing you are risking burning your house down. For inverter fusing, the best choice is to use a Power Center sized for your system. For simpler systems, you generally need a 400-amp fuse for full-sized inverters and a 110-amp fuse for mid-sized inverters. Small inverters can be plugged into a 20- or 30-amp fused outlet. The size of the cables providing power to the inverter is as important as the fusing, as we noted earlier. Do not restrict the inverter's ability to meet surge loads by choking it down with undersized or lengthy cables. Ten feet is the longest practical run between the battery and the inverter.

PROBLEMS WITH INVERTER USE

We have been painting a rosy picture up to this point, but let us now have a little brutal honesty to balance things out. If you know about these problems before-hand, it is usually possible to work around them when selecting appliances. All inverters have output limits. They are rated to produce a specific wattage for a specified time. Obviously a 250-watt inverter is not going to run your power tools. So don't pick an inverter that is too small for your needs.

Waveform Problems

We have already talked about sine waves vs. modified sine waves in the section on Waveforms on page 225. Unless otherwise noted, all the inverters in the *Sourcebook* produce modified sine-wave current. This works fine for 99% of the appliances, with notable exceptions listed below. Many of these problems, and solutions for them, are also discussed in the Introduction to System Design on page 153. The problems discussed here are caused only by modified sine-wave inverters. A small sine-wave unit is often employed to power a specific appliance or two, while the rest of the household runs on the cheaper and more efficient modified sine-wave inverter. If you are looking at one of the new full-sized sine-wave units you can skip this section.

Audio Equipment: Some audio gear will pick up a 60-cycle buzz through the speakers. It doesn't hurt the equipment, but it's annoying to the listener. There are too many models and brands to say specifically which are a problem and which aren't. We've had better luck with new equipment recently. Manufacturers are starting to put better power supplies back into their gear. We can only recommend that you try it and see.

Some top-of-the-line audio gear is protected by SCRs or Triacs. These devices are installed to guard against powerline spikes, surges, and trash (which don't happen on inverter systems). They see the sharp corners on modified sine-wave as trash and will sometimes commit electrical hara-kiri to prevent that nasty power from reaching the delicate innards. Some are even smart enough to refuse to eat any of that ill-shaped power, and will not power up.

Computers: 99.9% of them run happily on modified sine wave. Most of the uninterruptible power supplies on the market have square or modified sine-wave output. For many years Real Goods ran its own 44-station PC network using a pair of Trace 2012s as an uninterruptible power supply with no problems. It has recently come to our attention that some of the newest desktop MacIntosh units (1993 models and later) won't run on modified sine wave. We presume that they've been equipped with SCRs or similar protection circuitry. This is *not* a problem with any of the laptop or older Mac models.

Laser Printers: Many laser printers are equipped with SCRs, which causes the problem detailed above. Laser printers are a poor choice for renewable energy systems anyway due to their high stand-by power use, which averages a constant 600 to 1200 watts. Inkjet printers can do almost anything a laser printer can do while only using 25 to 30 watts.

Ceiling Fans: Most variable-speed ceiling fans will buzz on modified sine-wave current. They work fine, but the noise is annoying.

Other Potential Inverter Problems

Radio Frequency Interference: All inverters broadcast radio static when operating. Most of this interference is on the AM radio band. Do not plug your radio into the inverter and expect to listen to the ball game; you'll have to use a battery-powered radio and be some distance away from the inverter. This is occasionally a problem with TV interference as well. Distance helps. Put the TV at least 15 feet from the inverter. Twisting the inverter input cables may also limit their broadcast power.

Full-Time Phantom Loads: Devices that are always using power are not a problem for the inverter to run (unless they draw so little power that they are below the turn-on threshold) but they do create system design problems. Clocks and answering machines that stay on continuously consume many times their actual wattage to operate. It may only require a watt or two to run a clock, but the inverter requires 6 to 10 watts to run. Most inverters go into a stand-by mode when no loads are sensed. In stand-by it only takes a third of a watt to keep the inverter ready. Small continuous loads like this therefore consume an enormous amount of energy. Use wind-up or battery-powered clocks instead. See the System Design introduction (page 153) for more details about answering machines and other nasty phantom loads.

— Doug Pratt

INVERTERS

PowerStar Pocket Socket 100-Watt Inverter

The Pocket Socket is smaller, lighter, and less expensive than any other inverter. It has lower radio frequency (RF) emission than any inverter in its class. We've found that the Pocket Socket will surge higher and longer than the PowerStar 200. Can be used to run desktop computers, 19-inch or smaller TVs and VCRs, or any AC appliance that draws less than 100 watts continuously. It has automatic overload and thermal shutoff protection. Uses a standard cigarette lighter plug. The PocketSocket has a replaceable 14-amp fuse for reverse polarity and catastrophic overload protection.

> *Input Voltage: 10 to 15 volts DC*
> *Output Voltage: 115 volts AC*
> *Continuous Output: 100 watts*
> *Surge Capacity: 400 watts*
> *Idle Current: 0.1A @ 12VDC (1.2 watts)*
> *World Voltages Available: none*
> *Average Efficiency: 90% at half-rated power*
> *Recommended Fusing: 30 amp*
> *Dimensions: 4.2" x 1.75" x 1.3"*
> *Weight: 8 oz.*
> *Warranty: one year*

27-101   Pocket Socket Inverter, 100W/12V $79⁰⁰

PowerStar 200-Watt Pocket Inverter

This compact inverter is one of the best mini-inverters in the industry. It is ideal for powering most 19-inch color TVs, personal computers, VCRs, video games, stereos, and lots of other small appliances directly from your car or 12-volt system. It's great for travelers to carry in their cars to power these appliances. Includes a cigarette lighter input plug. It will deliver 140 watts of 115-volts AC power continuously from your 12-volt battery. It will provide 400 peak watts and 200 watts for over two minutes. PowerStar has recently made an improvement to the 200-watt inverter. In case of overload, the unit now safely delivers as much power as it can, into that overload.

> *Input Voltage: 10 to 15 volts DC*
> *Output Voltage: 115 volts AC true RMS ± 5%*
> *Continuous Output: 140 watts*
> *Surge Capacity: 400 watts*
> *Idle Current: 0.25 amp @ 12V (3 watts)*
> *World Voltages Available: none*
> *Average Efficiency: 90% at half-rated power*
> *Recommended Fusing: 30 amp*
> *Dimensions: 5" x 2.6" x 1.7"*
> *Weight: 15 oz.*
> *Warranty: one year*

27-104   PowerStar 200 Inverter, 200W/12V $109⁰⁰

Statpower NOTEpower

The NOTEpower inverter is perfect for the on the go "hacker." Notebook computers are great as long as you only want to use them for a few hours on a trip, but extended field trips can give your little brain in a pouch that run down feeling. Plug the NOTEpower into any vehicle cigarette lighter and you can operate your computer (via its AC adapter/charger) or recharge spare computer batteries. Weighing in at just a couple of ounces, the NOTEpower can put out up to 50 watts of power that could come in handy for cellular phones, camcorders, shavers, or other rechargeable equipment. A trim 3.5" x 2.5" x 1.25", the Notepower will fit just about anywhere! If you can't get a 12 volt adapter for your product, consider the NOTEpower!

27-134 NOTEpower Inverter $69⁰⁰

PowerStar Upgradable Inverters

PowerStar's line of upgradable inverters all use the same case, and are rated in terms of their continuous true RMS power capability. The continuous power rating on the smallest inverter is 400 watts; the first upgrade takes you up to 700 watts continuous; and the final upgrade takes you up to 1300 watts continuous. The actual upgrade can be performed by the factory in a few days. The price for the upgrade is simply the difference between the two units. Therefore, there is no cost penalty for buying a smaller unit and later upgrading it, making these inverters extremely versatile!

All three units inverters feature an audible alarm low-battery warning below 10.9-volts input, a green AC power indicator lamp, one standard three-prong 115-volt plug, optional hard-wire AC output access, and a built-in connector socket for a remote-control circuit. The PowerStar design features a dual mode limiter to enable operation of appliances rated higher in power than the inverter. A moderate overload lowers the output voltage, a severe overload causes shutdown. Reset by cycling the power switch.

The small 400-watt unit is suitable for computer systems, power tools, and small appliances. Its 3000-watt surge capacity can start and run lightly loaded 1/4 hp motors.

The 700-watt unit will run a 500-watt microwave, a vacuum cleaner, a hair dryer (on medium), or a small coffee pot or hotplate. Like the smaller unit, the surge capacity is 3000 watts.

The 1300-watt unit will continuously operate a full-size microwave, a circular saw, or any 1300-watt appliance. 6000- watt surge capacity on this larger unit.

Input Voltage: 10.5 to 16.5 volts DC

Output Voltage: 115 volts AC true RMS ± 5%

Continuous Output: 380, 700, 1300 watts

Surge Capacity: 3000 watts (6000 for the 1300 watt unit)

Idle Current: 0.06 amp @ 12VDC (0.7 watt)

World Voltages Available: none

Average Efficiency: over 90% at half-rated power

Recommended Fusing: 110A Class T for 400 and 700, 200A Class T for 1300

Dimensions: 3.15" x 3.3" x 12"

Weight: 15 lbs. maximum (varies with model)

Warranty: two years

27-105	PowerStar Inverter, 400W/12V	$429⁰⁰
27-106	PowerStar Inverter, 700W/12V	$499⁰⁰
27-107	PowerStar Inverter, 1300W/12V	$799⁰⁰

PowerStar Upgradable 24 Volt Inverters

Identical to the 12-volt models listed above, except with 24-volt input. Two models available, 900 watts and 1500 watts.

Input Voltage: 21 to 33VDC

Output Voltage: 115 volts AC true RMS +/- 5%

Continuous Output: 900 or 1500 watts

Surge Capacity: 3000 or 6000 watts

Idle Current: under 0.1A @ 24VDC (2.4 watts)

World Voltages Available: none

Average Efficiency: over 90% at half-rated power

Recommended Fusing: 110A Class T fuse

Dimensions: 3.15" x 3.3" x 12"

Weight: 15 lbs. maximum (varies with model)

Warranty: two years

27-701	PowerStar Inverter, 1500W/24V	$899⁰⁰

ProWatt 250 Inverter

This Statpower inverter will surge to 500 watts, produce 300 watts for 10 minutes, 250 watts for 30 minutes, and 200 watts continuously. It has a low battery cut-out with audible alarm at 10 volts. It is voltage regulated and frequency controlled and features all of the protection features of its big brother, the ProWatt 1500. It has an 18-inch cigarette lighter cord and fits in the palm of your hand!

> *Input Voltage: 10 to 15 volts DC*
> *Output Voltage: 115 volts AC true RMS ± 5%*
> *Continuous Output: 200 watts*
> *Surge Capacity: 500 watts*
> *Idle Current: less than 0.15A @ 12V (1.8 watts)*
> *World Voltages Available: 230V/50Hz by special order*
> *Average Efficiency: 90% at half-rated power*
> *Recommended Fusing: 30 amp*
> *Dimensions: 6" x 4-1/2" x 1-1/2"*
> *Weight: 20 oz.*
> *Warranty: one year (with six-month extension for registering)*

27-109   ProWatt Inverter, 250W/12V $229⁰⁰

AC Power For Remote Locations

The reliable POCKETpower inverter from Statpower provides more wattage at lower cost than any quality inverter we've seen. The POCKET-power is rated at 400 watts surge, 200 watts for 5 minutes, or 150 watts of clean, regulated, modified sine-wave power continuously. This inverter can run TVs and VCRs, desktop and laptop computers, or fluorescent lights from any 12-volt DC source, such as your car or renewable energy system. Will shut off if overloaded or battery voltage falls below 10 volts. Has a 30" input cord, red "on" indicator lamp, and standard 3-prong outlet. One-year mfg. warranty. Weighs 21 oz.

27-114   POCKETpower Inverter $89⁰⁰

ProWatt 800 Inverter

The ProWatt 800 is only a fraction of the size and weight of competitive size inverters. This efficient inverter produces temporary power levels much higher than its 800-watt continuous rating, and will produce 1000 watts of power for ten minutes. LED bar graphs, unique to Statpower inverters, provide an excellent visual representation of the power being drawn and the state of the battery charge. These displays eliminate the guesswork that owners of other inverters have to endure when using high-powered equipment. AC hardwire capable.

Input Voltage: 10 to 15 volts DC
Output Voltage: 115 volts AC true RMS ± 5%
Continuous Output: 800 watts
Surge Capacity: 1000 watts for 10 minutes
Idle Current: less than 0.3A @ 12V (3.5 watts)
World Voltages Available: 230V/50Hz by special order
Average Efficiency: 90% at half-rated power
Recommended Fusing: 110A Class T fuse
Dimensions: 3" x 9" x 10"
Weight: 5 lbs.
Warranty: one year (with six-month extension for registering)

27-112   ProWatt Inverter, 800W/12V $495⁰⁰

ProWatt 1500 Inverter

Statpower introduced this remarkable inverter in 1992. It is rated at 1500 watts continuous output, and is capable of delivering 1800 watts for 30 minutes or 2000 watts for 10 minutes. Low battery alarm at 10.7 volts, low voltage shutdown at 10 volts. It features protection circuitry for over-temperature, overload, short circuit, and reverse polarity. Other features include AC hard-wire capability, remote on/off capability, and remote display capability. All this at a price that beats all competition.

Input Voltage: 10 to 15 volts DC
Output Voltage: 115 volts AC true RMS ± 5%
Continuous Output: 1500 watts
Surge Capacity: 2000 watts for 10 minutes
Idle Current: less than 0.6A @ 12V (7 watts)
World Voltages Available: 230V/50Hz by special order
Average Efficiency: 90% at half-rated power
Recommended Fusing: 200A Class T
Dimensions: 3" x 9" x 15"
Weight: 8 lbs.
Warranty: one year (with six-month extension for registering)

27-110   ProWatt Inverter, 1500W/12V $799⁰⁰

Exeltech Sine Wave Inverter

This reasonably priced true sine-wave inverter is tailor-made for sensitive electronic equipment that has previously suffered from the modified sine-wave inverters. Finally, one can listen to audio gear without the annoyance of a background buzz.

Exeltech models SI-250 and SI-500 feature sophisticated circuitry with output-short, output-overload, and input-reverse polarity protection. These are not for whole house applications, where more efficiency and better pricing can be had with modified sine-wave inverters. Rather, these will be used for those particular applications where the shortcomings of most other inverters are all too obvious.

Applications for the Exeltech are numerous: color monitors, stereos, electronic instruments, satellite systems, VCRs, TVs, tape players, and test equipment.

Exeltech SI-250

Input Voltage: 10.5 to 17 volts DC
Output Voltage: 117 volts AC true RMS ± 5%
Continuous Output: 280 watts
Surge Capacity: 590 watts
Idle Current: 5 watts
World Voltages Available: none
Average Efficiency: 83% at half-rated power
Recommended Fusing: 30 amp
Dimensions: 8-3/4" x 3-1/4" x 4-1/2"
Weight: 5 lbs.
Warranty: one year

27-120	Exeltech SI-250 Inverter, 280W/12V	$425.00
27-121	Exeltech SI-250 Inverter, 280W/24V	$425.00

Exeltech SI-500

Input Voltage: 10.5 to 17 volts DC
Output Voltage: 117 volts AC true RMS ± 5%
Continuous Output: 535 watts
Surge Capacity: 1180 watts
Idle Current: 5 watts
World Voltages Available: none
Average Efficiency: 83% at half-rated power
Recommended Fusing: 60 amp
Dimensions: 8-3/4" x 3-1/4" x 4-1/2" (yes, it's the same as the 250W unit)
Weight: 5 lbs.
Warranty: one year

27-122	Exeltech SI-500 Inverter, 500W/12V	$625.00
27-123	Exeltech SI-500 Inverter, 500W/24V	$625.00

Efficiency came to represent an annual energy source two-fifths bigger than the entire domestic oil industry.

—Amory Lovins

Trace 812 Inverter

The 812 is the baby member of the Trace family. It will produce a 2400-watt surge, 800 watts for 30 minutes, 650 watts for 60 minutes, or 575 watts continuously. It will power TVs, VCRs, computers, and test equipment, and it will also power most vacuum cleaners, Champion juicers, and most microwave ovens. For an extra $100 you can order an optional 25-amp battery charger in the "Standby Option." This internal 25-amp battery charger comes with a 30-amp transfer switch that switches back and forth between battery power and grid/generator power. The battery charger requires 2000 watts of generator power at a minimum to operate it. The current is not adjustable but maximum battery voltage is. Also available is the RC/3 Remote Control option allowing you to turn the unit on and off from a remote location (although with the extremely low no-load power drain you can let the unit stay on all day for around 0.4 amp-hours). The Trace 812 comes with our very highest recommendation.

We also offer a 110-amp fuse with holder for installing between the battery and the Trace 812. This will help you comply with state and local electrical codes.

Input Voltage: 10.5 to 15.8 volts DC

Output Voltage: 120 volts AC true RMS ± 3%

Continuous Output: 575 watts

Surge Capacity: 2400 watts

Idle Current: 0.017A @ 12V (0.2 watt)

World Voltages Available: 230V/50 or 60Hz (continuous output drops to 425 watts)

Average Efficiency: better than 90% from 40 to 300 watts

Recommended Fusing: 110A Class T

Dimensions: 5-3/4" x 10-1/2" x 8"

Weight: 14 lbs.

Warranty: two years

27-211	Trace 812 Inverter	$550⁰⁰
27-212	Trace 812 Inverter and Battery Charger	$650⁰⁰
27-214	Trace 812 Remote Control	$50⁰⁰
24-507	Fuse Holder, 110A Class T Fuse and Cover	$49⁰⁰

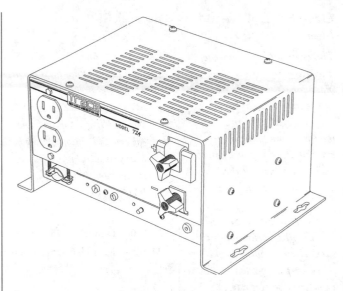

Trace 724 Inverter

The Trace 724 was created as a direct response to the demands of our customers who prefer to base their systems upon a 24-volt battery bank. The 724 will produce 1000 watts for 7 minutes, 600 watts for 55 minutes, 500 watts for 90 minutes, and 425 watts continuous. A 12-amp battery charger option is available as is a remote-control (RC/3) option that will allow remote monitoring via an LED and will also allow the user to turn the 724 on and off.

Input Voltage: 21.0 to 31.6 volts DC

Output Voltage: 120 volts AC true RMS ± 3%

Continuous Output: 425 watts

Surge Capacity: 2100 watts

Idle Current: 0.017A @ 24V (0.4 watt)

World Voltages Available: 230V/50 or 60Hz (continuous output drops to 400 watts)

Average Efficiency: better than 90% from 40 to 300 watts

Recommended Fusing: 110A Class T

Dimensions: 5-3/4" x 10-1/2" x 8"

Weight: 14 lbs.

Warranty: two years

27-221	Trace 724 Inverter	$625⁰⁰
27-222	Trace 724 Inverter and Battery Charger	$725⁰⁰
27-214	Trace 724 Remote Control	$50⁰⁰
24-507	Fuse Holder, 110A Class T Fuse and Cover	$49⁰⁰

Trace 2512 Inverter

The entire 2000 watt series of Trace inverters now offer UL approval. All residential models are Electrical Testing Labs certified to UL 1778, and RV/Marine models certified to UL 1236. Standard residential models now include the formerly optional Turbo Fan and Low Voltage Disconnect as standard equipment. Standard RV/Marine models now include the formerly optional Turbo Fan, Low Voltage Disconnect, Standby Battery Charger, and Digital Voltmeter. The external terminal block with strain relief for AC wiring will accept up to 6-gauge wire. No-load power drain is the best in the industry at 0.33 watt (0.28 amp). The first two digits of the Trace model number are the output wattage in hundreds, the second two digits are input DC voltage. (So the 2512 is 2500 watts output, 12-volt DC input). The 2512 will provide a 6000-watt surge for over a minute for starting high-surge motors.

The Standby option (the SB designation) is the cheapest and most convenient way to add generator-driven battery charging capabilities to your system. The inverter will sense the generator has started, wait ten seconds for the voltage to stabilize, then transfer all loads to the generator through the automatic built-in 30-amp transfer switch. The inverter then goes into the highly adjustable battery charging mode. When the generator shuts off everything goes back to normal inverter mode. Standby versions are equipped with an improved efficiency multi-stage battery charger. In Real Goods' own testing this new charger showed efficiencies in the high 70% range, greatly improved from the typical 50% to 60% for transformer-based chargers. Maximum charger output for the 2512 is 120 amps. You'll need a 5000- to 6000-watt generator to drive this charger to full potential. Adjustable controls allow tailoring the charge cycle to the needs of your battery bank or the capabilities of your generator. Maximum charging amperage, bulk voltage, float voltage, equalization voltage, equalization time, and more are all front-panel adjustable.

All Trace inverters are available in international voltages, 230 volts AC, in either 50 or 60 Hz (you *must* specify one) by special order. This is 230-volt AC *single phase* output, it is not splitable like 240-volt in this country is. Add three to four weeks to delivery time. If you need 240-volt output see the T-220 Transformer later in this chapter or the new DR series from Trace which is stackable for 240-volt split phase output.

All 2000 series Trace inverters may be stacked with an identical inverter to double the amperage output by using the optional stacking interface.

Input Voltage: 8.8 to 15.6 volts DC
Output Voltage: 120 volts AC RMS ±2%
Continuous Output: 2500 watts
Surge Capacity: 6000 watts
Idle Current: 0.028A @ 12V (0.33 watts)
World Voltages Available: 100, 230V/50 or 60Hz
Average Efficiency: 85% to 96% from 50 watts to rated output
Recommended Fusing: 400A Class T
Dimensions: 9-3/4" x 11-1/2" x 13-1/2"
Weight: 47 lbs.
Warranty: two years

27-224	2512 Trace Inverter	$1,345
27-225	2512SB Trace Inverter with battery charger	$1,575

Trace 2548 Inverter

The Trace 2548 will provide 2500 watts of output power continuously at 120 volts AC from a 48-volt DC input. It has a surge capacity of 6000 watts for approximately two minutes. All other features of this unit are identical to the 12-volt unit described in detail above. The battery-charger option has a maximum charge rate of 30 amps.

Input Voltage: 22.4 to 62.6 volts DC
Output Voltage: 120 volts AC true RMS ± 2%
Continuous Output: 2500 watts
Surge Capacity: 6000 watts
Idle Current: 0.012A @ 48V (0.57 watt)
World Voltages Available: 100, 230V/50 or 60Hz
Average Efficiency: 85% to 96% from 50 watts to rated output
Recommended Fusing: 200A Class T fuse
Dimensions: 9-3/4" x 11-1/2" x 13-1/2"
Weight: 47 lbs.
Warranty: two years

27-228	Trace 2548 Inverter	$1,775
27-229	Trace 2548-SB Inverter and Battery Charger	$1,995

Trace 32 And 36 Volt Inverters

We still find the occasional older wind system operating at 32- or 36-volts DC. These are voltages that most modern manufacturers have bypassed so don't build a new system around one of these, but if you've got one, Trace still supports you with a first-quality product.

The 32-volt charger option has a maximum charge rate of 45 amps, the 36-volt version 40 amps. All other features of these units are identical to the 12-volt unit described in detail above.

Trace 2232 Inverter

Input Voltage: 19.9 to 42.9 volts DC
Output Voltage: 120 volts AC true RMS ± 2%
Continuous Output: 2200 watts
Surge Capacity: 6000 watts
Idle Current: 0.016A @ 32V (0.51 watt)
World Voltages Available: 230V/50 or 60Hz
Average Efficiency: 85% to 96% from 50 watts to rated output
Recommended Fusing: 200A Class T
Dimensions: 9-3/4" x 11-1/2" x 13-1/2"
Weight: 47 lbs.
Warranty: two years

27-230	Trace 2232 Inverter	$1,675
27-231	Trace 2232-SB Inverter and Battery Charger	$1,900

Trace 2536 Inverter

Input Voltage: 19.9 to 47.0 volts DC
Output Voltage: 120 volts AC true RMS ± 2%
Continuous Output: 2500 watts
Surge Capacity: 6000 watts
Idle Current: 0.014A @ 36V (0.5 watt)
World Voltages Available: 230V/50 or 60Hz
Average Efficiency: 85% to 96% from 50 watts to rated output
Recommended Fusing: 200A Class T
Dimensions: 9-3/4" x 11-1/2" x 13-1/2"
Weight: 47 lbs.
Warranty: two years

27-232	Trace 2536 Inverter	$1,720
27-233	Trace 2536-SB Inverter and Battery Charger	$1,950

Trace 2000 Series Options
Digital Voltmeter For Trace 2000 Series

The Digital Voltmeter is a handy option for purchasers of the Standby charger. This meter monitors four conditions: battery voltage to tenths of a volt, the charge rate of the battery charger, the frequency of the generator in hertz (Hz), and the peak voltage of the charging source. The Digital Voltmeter is easily installed in five minutes by the customer.

27-301-12	Digital Voltmeter, 12V	$120[00]

Stacking Interface Module For Trace 2000 Series

The stacking interface module allows two inverters to be paralleled for double output power at 120 volts AC. If both units have the battery-charger option, the charging capability is also doubled. This is the option that Trace pioneered to give its units maximum flexibility. Both inverters must be identical models. If one has battery charger, then the other one must also. User installed.

27-308	Stacking Interface	$275[00]

Remote Control For Trace 2000 Series

The remote-control option is for use in installations where the inverter is not easily accessible. The RC-2000 comes complete with the Digital Voltmeter option listed above and 20' cable. It provides a duplicate set of control switches and indicator lamps and is only available in models with the standby battery charger option. Option RC2 is a more basic remote with on-off control and an LED that indicates on-off and search mode states.

27-306	RC2 Remote Control	$75[00]
27-309	RC-2000 with DVM for 2512	$250[00]
27-316	RC-2000 with DVM for 2232	$250[00]
27-317	RC-2000 with DVM for 2536	$250[00]
27-318	RC-2000 with DVM for 2548	$250[00]

The Trace DR-Series

The DR is the latest (and still growing) series of inverters from Trace. They are all lower cost and higher performing than the models they replace, and all DR inverters come equipped with high performance battery chargers as standard equipment. They are available in a variety of 12-or 24-volt wattages.

DR-series inverters are conventional, high-efficiency, modified sine-wave inverters. All models are Electrical Testing Lab approved to UL specifications. These inverters will automatically protect themselves from overload, high temperature, high- or low-battery voltages, all the things we've come to expect from modern inverters. You just can't hurt them externally. Fan cooling is temperature regulated, low battery cut-out is standard and is adjustable for battery bank size and level of battery protection desired. All models come with a powerful three-stage adjustable battery charger and all models have a 30-amp transfer switch built-in. Charge amperage is adjustable to suit your batteries or generator. Charging voltages are tailored by a front panel knob. Pick your battery type, eight common choices given plus two equalization selections, and you've automatically picked the correct bulk and float voltage for your batteries.

The optional battery temperature probe will allow the inverter to factor all voltage set points by actual battery temperature. This is a good feature to have when your batteries live outside and go through temperature extremes. Comes with a 15' cord.

The DR-series is stackable in series (must be identical models) for three-wire splitable 240-volt output. This is something new from Trace. Formerly, when you stacked you got more wattage, but still at just 120 volts. With the DR series, when you stack you get more wattage, and it's at 240 volts. Stacking is now easier and less expensive too; just plug in the inexpensive stacking cord and it's done.

All models are designed for wall mounting with conventional 16-inch stud. spacing.

Trace DR-Series, 12-Volt Models

Trace DR1512 Volt Inverter/Charger

Input Voltage: 10.8 to 15.5 volts DC
Output Voltage: 120 volts AC true RMS ± 5%
Continuous Output: 1500 watts
Surge Capacity: 3200 watts
Idle Current: 0.045\A @ 12V (0.54 watt)
Charging Rate: 0 to 70 amps
World Voltages Available: yes (call for details)
Average Efficiency: 94% at half rated power
Recommended Fusing: 200A Class T
Dimensions: 8-1/2" x 7-1/4" x 21"
Weight: 35 lbs.

Warranty: 2 years

27-237	Trace DR1512 Inverter/Charger, 12V	$950⁰⁰
27-313	Stacking Intertie Kit	$75⁰⁰
27-208	Optional Battery Temperature Probe, 15'	$20⁰⁰
27-324	Conduit Box option-DR Series	$60⁰⁰

Trace DR2412 Inverter/Charger

Input Voltage:10.8 to 15.5 volts DC
Output Voltage:120 volts AC true RMS ± 5%
Continuous Output: 2400 watts
Surge Capacity: 6000 watts
Idle Current: 0.045\A @ 12V (0.54 watt)
Charging Rate: 0 to 120 amps
World Voltages Available: no
Average Efficiency: 94% at half rated power
Recommended Fusing: 400A Class T
Dimensions: 8-1/2" x 7-1/4" x 21"
Weight: 45 lbs.

Warranty: 2 years

27-245	Trace DR2412 Inverter/Charger, 12V	$1285
27-313	Stacking Intertie Kit	$75⁰⁰
27-208	Optional Battery Temperature Probe, 15'	$20⁰⁰
27-324	Conduit Box option-DR Series	$60⁰⁰

See page 246 for DR-inverter cables.

Trace DR-Series 24-Volt Models
Trace DR1524 Inverter/Charger

Input Voltage: 21.6 to 31.0 volts DC

Output Voltage: 120 volts AC true RMS ± 5%

Continuous Output: 1500 watts

Surge Capacity: 4200 watts

Idle Current: 0.030\A @ 24V (0.72 watt)

Charging Rate: 0 to 35 amps

World Voltages Available: yes, call for details

Average Efficiency: 94% at half rated power

Recommended Fusing: 110A Class T

Dimensions: 8-1/2" x 7-1/4" x 21"

Weight: 35 lbs.

Warranty: 2 years

27-238	Trace DR1524 Inverter/Charger, 24V	$895⁰⁰
27-313	Stacking Intertie Kit	$75⁰⁰
27-208	Optional Battery Temperature Probe, 15'	$20⁰⁰
27-324	Conduit Box option-DR Series	$60⁰⁰

Trace DR2424 Inverter/Charger

Input Voltage: 21.6 to 31.0 volts DC

Output Voltage: 120 volts AC true RMS ± 5%

Continuous Output: 2400 watts

Surge Capacity: 7000 watts

Idle Current: 0.030\A @ 24V (0.72 watt)

Charging Rate: 0 to 70 amps

World Voltages Available: yes, call for details

Average Efficiency: 95% at half rated power

Recommended Fusing: 200A Class T

Dimensions: 8-1/2" x 7-1/4" x 21"

Weight: 40 lbs.

Warranty: 2 years

27-242	Trace DR2424 Inverter/Charger, 24V	$1225
27-313	Stacking Intertie Kit	$75⁰⁰
27-208	Optional Battery Temperature Probe, 15'	$20⁰⁰
27-324	Conduit Box option-DR Series	$60⁰⁰

Certified by ETL to meet
UL 1741 and CSA #107.1 standards

Trace DR3624 Inverter/Charger

Input Voltage: 21.6 to 31.0 volts DC

Output Voltage: 120 volts AC true RMS ± 5%

Continuous Output: 3600 watts

Surge Capacity: 10,000 watts

Idle Current: 0.030\A @ 24V (0.72 watt)

Charging Rate: 0 to 70 amps

World Voltages Available: no

Average Efficiency: 95% at half rated power

Recommended Fusing: 300A Class T

Dimensions: 8-1/2" x 7-1/4" x 21"

Weight: 45 lbs.

Warranty: 2 years

27-246	Trace DR3624 Inverter/Charger, 24V	$1475
27-313	Stacking Intertie Kit	$75⁰⁰
27-208	Optional Battery Temperature Probe, 15'	$20⁰⁰
27-324	Conduit Box option-DR Series	$60⁰⁰

Trace 12, 24, And 48-Volt Sine Wave Inverters

Trace revolutionized the inverter industry back in 1984 with superior efficiency and a product that just wouldn't break. Now, in a class by themselves, they are introducing an even more revolutionary technology that will bring inverters into the new millennium. The new series of sine wave inverter/ chargers do it all. Conventional remote inverter, uninterruptible power supply, generator management, or utility intertie, it can do any, or sometimes all of these simultaneously!

These microprocessor-controlled inverters produce a multi-stepped approximation to a sine wave that meets all utility requirements for line intertie and will run any AC appliance

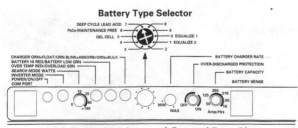

DR-Series Monitor and Control Face Plate

within its wattage capabilities. And they do it at up to 94% efficiency, a level that was previously unknown to sine wave inverters. There are 34 to 52 steps per cycle depending on battery voltage. The more heavily loaded the inverter, the less distortion in the output. Total waveform distortion is 3% to 5%. *This is cleaner power than your utility can offer you!* All models in the SW-series are Electrical Testing Lab approved to UL specifications. There are no "problem" appliances with this inverter line. If the appliance runs plugged into the wall, it will run on the SW-series. Laser printers, recording studios, ceiling fans, water pumps, sensitive electronic equipment, they all run happily.

There are three primary modes of operation for these amazing units:

Inverter Mode

In this mode the SW-series acts like a typical remote power inverter with battery charger. It has adjustable threshold watts to trigger turn-on, and adjustable search spacing to save power. It will produce rated surge power for 2 minutes. These inverters will automatically protect themselves from overload, high temperature, high or low battery voltage, all the things we've come to expect from modern inverters. You just can't hurt them externally. Fan cooling is temperature regulated, low voltage battery disconnect is adjustable. Automatic generator start can be programmed to occur at a specific battery voltage or at a specific time every day. The number of generator start attempts, the cranking time for each attempt, and the warm-up time before loading are all adjustable. Once the generator starts, the inverter will match phases and shift the household loads to the generator. (Note that there is no transfer relay, and therefore no "hiccup" in the power as this transfer occurs.) There is a very powerful three-stage DC battery charger on board which will go to work recharging the batteries while the inverter monitors the generator input amps. When amperage draw approaches the programmed generator maximum, the charger will back off, and if necessary the inverter will even add its output to the generator to start a heavy load, then go back to charging. Amperage draw from the generator will never exceed the maximum you programmed in. When the generator shuts off, either at the end of a timed run, from reaching the programmed charging voltage, or from lack of fuel, the inverter will disconnect at a programmed low voltage point and pick up all loads, again without a "hiccup" in the delivered power. Generator start can be "locked out" during certain hours. For instance you might not want the generator to start in the middle of the night. This lock-out has a "must start" override voltage that is also programmable. All of the automated generator management functions are available in all the operation modes.

Standby Mode

As an uninterruptable power supply this one can't be touched for the price. Grid power is fed through and conditioned by the SW-series inverter. No filtering or line conditioning is needed, but a surge arrestor is still recommended as high frequency voltage spikes could pass through. Worst case transfer time in case of sudden grid failure at full rated output is 16 milliseconds. Computer technicians have demonstrated that 100 milliseconds is typically fast enough for today's personal computers. In addition the SW-series inverter will support the grid if it "browns out" or if unintended loads threaten to trip the AC breaker. If the amount of AC power demanded is greater than the breaker size you've programmed into the SW inverter, it will contribute power to the system. If the grid fails and the SW inverter is running all the loads, then all the automatic generator management, and battery charging discussed above will occur. Once utility power is restored the inverter will match phases and shift the AC loads back to the utility, then shut off the generator if it was running.

Utility Interactive Mode

This mode will allow you to sell power to the utility! This is the big breakthrough that we've all been waiting for. It is simple to set up any of the SW inverters up to accomplish this, and the unit was designed to meet all intertie requirements. However, this type of installation is so new, that not all utility companies have formalized their regulations for acceptable installation. Regulations will vary from one utility to another. The utility companies have a right and a need to be careful about how power is fed into their system. We cannot guarantee that your utility will welcome your power with open arms, or even a positive attitude. Federal law requires them to buy any power from a reliable renewable energy producer but it doesn't require them to make it easy or profitable. Because of the thousands of local utility companies, each with their own codes and regulations, we may only be able to offer limited assistance obtaining approval from your local utility. *Talk to your local utility first; possession of the equipment and the ability to do so does not constitute a right to sell them power.*

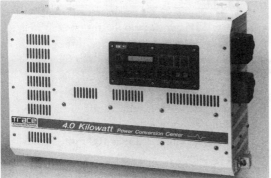
Sine-Wave Inverter

A small battery bank is still required in the Utility Interactive Mode. A larger battery bank will protect you from blackouts for as long as it lasts, or until the generator starts and takes over. (All the generator management features are still active in this mode too.) The inverter will allow the batteries to float at a programmed voltage, any power input beyond what is needed for this will be converted to AC and fed back into the grid. The SW inverter will protect itself from grid open (no connection to the grid), grid shorted (when the grid fails and it's trying to run the whole neighborhood), and islanding (when the grid fails but the neighborhood happens to be drawing exactly as much power as your renewable energy source is supplying). If the utility fails in the Utility Interactive Mode the SW inverter will go into the normal Inverter Mode to run your house without a hiccup.

In addition to all the features discussed above all the SW inverters are stackable in series for 240 volts AC three-wire splitable output at double the rated wattage continuous. Stacking only requires plugging in an inexpensive stacking cord.

There are three, adjustable, voltage-controlled SPDT relays that can be used to automatically turn external loads or charging sources on and off.

The generator start function consists of two relays. One that stays closed as long as the generator is running, and one that only closes as long as you program the starter to crank. This set-up should allow the SW series inverter to work with any remote start capable generator.

The battery charger is a sophisticated three-stage unit. During the initial "Bulk Charge" stage it will charge at a constant programmable current causing the battery voltage to rise.

A constant voltage "Absorption" stage begins after battery voltage reaches the programmed bulk charge voltage. The charge rate is gradually reduced, holding the battery voltage constant at the bulk charge voltage. The "Float" stage begins when the current required to hold the batteries at the bulk

charge voltage has tapered to a programmed low level. At this point, the battery voltage is allowed to fall to the programmed float voltage, where it is maintained until another charge cycle is initiated. A new charge cycle is triggered when the battery voltage falls to 2 volts below the float voltage for 90 seconds. A battery temperature sensor probe is standard equipment, and all programmed battery voltage set points are temperature factored. (Even the correction factor is adjustable if you want.)

Designed for wall mounting with conventional 16 inch stud spacing.

Trace SW-Series, 12-Volt Model
SW2512 Sine-Wave Inverter/Charger Specs

Full specifications for this brand-new model were not available at press time. Please call for details.

Input Voltage: 10.8 to 15.5 volts DC
Output Voltage: 120 volts AC true RMS ± 2%
Continuous Output: 2500 watts
Surge Capacity: ?? call
Idle Current: ?? call
Charging Rate: 0 to150 amps
Transfer Amps: 60 amps
World Voltages Available: ?? call
Average Efficiency: 94% peak, 85%-90% average
Recommended Fusing: 400A Class T
Dimensions: 22.5" x 15..25" x 9.0"
Weight: 90 lbs.
Warranty: 2 years

| 27-247 | SW2512 Trace Sine Wave, 12v, 2.5 kW | $2,485 |

Trace SW-Series, 24-Volt Models
SW4024 Sine-Wave Inverter/Charger Specs

Input Voltage: 20 to 34 volts DC
Output Voltage: 120 volts AC true RMS ± 2%
Continuous Output: 4000 watts
Surge Capacity: 10,000 watts for 2 min.
Idle Current: adjustable - 1 to 16.8 watts
Charging Rate: 0 to 120 amps
Transfer Amps: 60 amps
World Voltages Available: yes, call for details
Average Efficiency: 94% peak, 85%-90% average
Recommended Fusing: 400A Class T
Dimensions: 22.5" x 15..25" x 9.0"
Weight: 105 lbs.
Warranty: 2 years

| 27-236 | SW4024 Trace Sine Wave, 24v, 4 kW | $3,185 |
| 27-240 | SW3024E Export Model, various volts/hz | $3,185 |

Trace SW-Series, 48-Volt Models
SW4048 Sine-Wave Inverter/Charger Specs

Input Voltage: 40 to 68 volts DC

Output Voltage: 120 volts AC true RMS ± 2%

Continuous Output: 4000 watts

Surge Capacity: 10,000 watts for 2 min.

Idle Current: adjustable - 1 to 16.8 watts

Charging Rate: 0 to 60 amps

Transfer Amps: 60 amps

World Voltages Available: yes, call for details

Average Efficiency: 94% peak, 85%-90% average

Recommended Fusing: 200A Class T

Dimensions: 22.5" x 15..25" x 9.0"

Weight: 105 lbs.

Warranty: 2 years

| 27-241 | SW4048 Trace Sine Wave, 48v, 4 kW | $3,185 |
| 27-248 | SW3048E Export Model, various volts/hz | $3,185 |

SW5548 Sine-Wave Inverter/Charger Specs

Full specifications for this brand-new model were not available at press time. Trace also has *PV* and *Utility* versions of this inverter for utility sell-back use only, with no battery pack required. Please call for details.

Input Voltage: 40 to 68 volts DC

Output Voltage: 120 volts AC true RMS ± 2%

Continuous Output: 5500 watts

Surge Capacity: ?? call

Idle Current: ?? call

Charging Rate: 0 to 70 amps

Transfer Amps: 60 amps

World Voltages Available: ?? call

Average Efficiency: 94% peak, 85%-90% average

Recommended Fusing: 400A Class T

Dimensions: 22.5" x 15..25" x 9.0"

Weight: 136 lbs.

Warranty: 2 years

| 27-249 | SW5548 Trace Sine Wave, 48v, 5.5 kW | $3,680 |
| 27-250 | SW5548 PV Trace Sine Wave, 48v, 5.5 kW | $3,990 |

(This model for utility intertie only, please call for specifications and details).

Options For Trace SW Series
Battery Cables For SW Inverter Hook-Up

The wire sizing between the battery and the inverter is of critical importance. For this reason, Trace has specified a very heavy duty 4,000-strand, 4/0 (that's four-ought!) welding cable. They're available in either a 5-foot or a 10-foot pair of cables with mechanically crimped ring-terminals and color coded shrink wrap ends. Heavyduty cables like these are an absolute necessity with the SW inverters.

(Note, if you're using one of the APT Powercenters with your installation, they have their own inverter cable options.)

27-311	5 foot 4/0 cables, 2 each	$65⁰⁰
27-312	10 foot 4/0 cables, 2 each	$115⁰⁰

300 And 400 Amp Fused Disconnects

This fused disconnect was specifically designed for the high amperage, low voltage, ultra-low resistance needs of large inverters. The large contact, oil-filled switch is rated to handle 2,000 amps. The Class T fuse is a UL-listed slow-blow DC rated fuse that easily handles the surge requirements of even the largest inverters. Rated to 125 volts DC, 20,000AIC. Has 300 MCM wire lugs. (That's 2 sizes larger than the 4/0 cable sold above.)

(Note, if you're using one of the APT Powercenters, fusing is already provided.)

24-206	400A Fused Disconnect	$295⁰⁰
24-216	300A Fused Disconnect	$305⁰⁰

Conduit Box Option

The optional Conduit Box allows the very heavy input wiring from the batteries to be run inside conduit to comply with NEC. Has three (one on each side) 1/2", 3/4", 2", & 2-1/2" knockouts and plenty of elbow room inside for bending large wire.

27-209	Optional Conduit Box, SW-Series	$80⁰⁰

Remote Control Panel Option

Offers a complete duplicate control panel with all controls, indicator lights, and LCD display. Includes 25 feet of connection cable. (Uses standard 25 pin computer cable and connectors.)

27-210	Remote Control Panel for SW series (specify voltage)	$250⁰⁰

Stacking Interface Cord

To stack a pair of SW inverters this is all you need. Simple plug-in installation.

27-314	Stacking Intertie Cord	$35⁰⁰

Heart Inverters
Freedom Series

The Freedom series of inverters from Heart Interface offer great value. Conveniently sized in 1000, 2000, and 2500 watts of continuous operation, all Freedom series inverters come standard with a sophisticated three-stage battery charger. Also, these inverters feature automatic power-source transfer switching. This allows the inverter to recognize the availability of AC power from another source, such as a generator or the utility grid. When this power is available, the inverter will have these auxiliary power sources drive the electrical loads. The inverter draws power from the batteries only when no other source of power is available.

Freedom series inverters have electronic, thermal, and circuit breaker overload protection, as well as a unique AC reverse polarity protection. Automatic Load Detection allows these inverters to drop into a power saving idle mode when no power is needed.

An optional Remote on/off and LED bar graph indicate battery voltage, current, and charge rate, and includes status indicators for overload, reverse polarity, and low battery. The Remote panel also makes adjustment of the initial, maintenance, and equalization charge voltages possible.

Heart Freedom 10 Inverter

The UL-listed Freedom 10 is available in either 12-or 24-volts DC input. International voltage output is available by special order. Specify 50 or 60 Hz. Maximum charging amperage for the 12-volt model is 50 amps, for the 24-volt model it is 25 amps. Six-foot #2-gauge battery cables are included.

Input Voltage: 10.0 to 15.5 (or 20.0 to 31.0) volts DC
Output Voltage: 120 volts AC true RMS ± 5%
Continuous Output: 1000 watts
Surge Capacity: 3000 watts
Idle Current: 0.12A @ 12V (1.44 watts)
World Voltages Available: 230V/50 or 60Hz
Average Efficiency: 92% peak, 85% full load
Recommended Fusing: 110 Amp Class T
Dimensions: 12" x 9-3/4" x 7"
Weight: 31 lbs.
Warranty: 30 months

27-405	Heart Freedom 10 (specify input voltage)	$1,085
27-406	Heart Remote Panel	$138⁰⁰

Heart Remote Panel

Heart Freedom 20 Inverter

The UL-listed Freedom 20 is available in 12- or 24-volt DC input. International voltage output is available by special order. Specify 50 or 60 Hz. Maximum charging amperage for the 12-volt model is 100 amps, for the 24-volt model it is 50 amps. Six-foot #2/0-gauge battery cables are included.

Input Voltage: 10.0 to 15.5 (or 20.0 to 31.0) volts DC
Output Voltage: 120 volts AC true RMS ± 5%
Continuous Output: 2000 watts
Surge Capacity: 4500 watts
Idle Current: 0.12A @ 12V (1.44 watts)
World Voltages Available: 230V/50 or 60Hz
Average Efficiency: 93% peak, 84% full load
Recommended Fusing: 400 Amp Class T
Dimensions: 12" x 11-1/2" x 8-3/4"
Weight: 52 lbs.
Warranty: 30 months

27-408	Heart Freedom 20 (specify input voltage)	$1,650
27-406	Heart Remote Panel	$138⁰⁰

Heart Freedom Model 10

Heart Freedom 25 Inverter

The UL-listed Freedom 25 is also available in 12- or 24-volt DC input. International voltage output is available by special order. Specify 50 or 60 Hz. Maximum charging amperage for the 12-volt unit is 130 amps, for the 24-volt it is 65 amps. Six-foot #2/0-gauge battery cables are included.

Input Voltage: 10.2 to 15.5 (or 20.0 to 31.0) volts DC
Output Voltage: 120 volts AC true RMS ± 5%
Continuous Output: 2500 watts
Surge Capacity: 5200 watts
Idle Current: 0.12A @ 12V (1.44 watts)
World Voltages Available: 230V/50 or 60Hz
Average Efficiency: 94% peak, 86% full load
Recommended Fusing: 400 Amp Class T
Dimensions: 12" x 11-1/2" x 8-3/4"
Weight: 56 lbs.
Warranty: 30 months

27-409	Heart Freedom 25 (specify input voltage)	$1,890
27-406	Heart Remote Panel	$138⁰⁰

Heart HF 12-600 Inverter

Heart Interface's smallest inverter will run 600 watts for 25 minutes and reaches 90% efficiency at only 50 watts. It is an ideal unit for computers, printers, TVs, VCRs, and test equipment. Modified sine-wave output.

Input Voltage: 10.2 to 15.0 volts DC
Output Voltage: 120 volts AC true RMS ± 5%
Continuous Output: 300 watts
Surge Capacity: 1500 watts
Idle Current: 0.07A @ 12V (0.84 watt)
World Voltages Available: 100, 230V/60Hz
Average Efficiency: 90% at half-rated power
Recommended Fusing: 110A Class T
Dimensions: 7.45" x 7.15" x 5"
Weight: 15 lbs.
Warranty: 30 months

27-404   Heart HF 12-600X 600-Watt $600⁰⁰

Heart HF 12-600 Inverter

Vanner Voltmaster Battery Equalizer

Vanner Voltmaster Battery Equalizer

The Vanner Voltmaster allows you to upgrade to a larger 24-volt inverter system, but still use all your 12-volt appliances. The Voltmaster was designed to eliminate the overcharging of one battery in a split 24/12-volt system. All batteries will charge and discharge equally even when there is an unequal load. The device electronically monitors voltages of both batteries and transfers current whenever one battery discharges at a rate different than the other. By maintaining equalization down to 0.01 volt, the Voltmaster will extend battery life by preventing both over- and undercharging. Voltmasters will temporarily deliver greater than rated amperage into a 12-volt load; it will simply draw the excess from one 12-volt battery and make up for it later. All units are guaranteed for one year.

27-801   #60-10A, 10 amp $340⁰⁰
27-802   #60-20A, 20 amp $365⁰⁰
27-803   #60-50A, 50 amp $495⁰⁰

INVERTER OPTIONS

Trace 120V-to-240V Transformer

The Trace T-220 is a 3000-watt one-hour transformer that may be configured by the user to function as a step-up or step-down autoformer, an isolation transformer or a generator balancing transformer. Many alternative energy homesteads have everything on 12-volt or 120-volt systems, with the exception of an occasional 240-volt submersible pump. By installing the T-220 between the submersible pump and the pressure switch, the transformer will only come on when the pump is activated. 5000-watt maximum surge. Circuit-breaker protected. Can be used with any brand of 120-volt inverter.

27-307   Trace T-220 Transformer $295⁰⁰

Battery Cables For Inverter Hook-Ups

The wire sizing between the battery and the inverter is of critical importance. For this reason, we offer both 2/0 and 4/0 cables. They're available in either a 5-foot or 10-foot pair of cables with mechanically crimped 3/8" ring-terminals and color-coded shrink-wrap ends. Highly recommended to complete your inverter installation. See chart on page 213 for recommended cable sizing.

(Note, if you're using one of the APT Powercenters with your installation, they have their own inverter cable options.)

27-311   5 ft. Battery Cables, 4/0, 2 each $65⁰⁰
27-312   10 ft. Battery Cables, 4/0, 2 each $115⁰⁰
27-099   5 ft. Battery Cables, 2/0, 2 each $50⁰⁰
27-098   10 ft. Battery Cables, 2/0, 2 each $89⁰⁰

400-Amp Fused Disconnect

This fused disconnect was specifically designed for the high amperage, low voltage, ultra-low resistance needs of large inverters. The large-contact, oil-filled switch is rated to handle 2000 amps. The 400 Amp Class T fuse is a UL-listed slow-blow DC rated fuse that easily handles the surge requirements of even the largest inverters. Will handle an 800-amp surge for three minutes. Rated to 125 volts DC, 20,000AIC. Has 300 MCM wire lugs, which is two sizes larger than the 4/0 cable sold above. (Note, if you're using one of the APT Powercenters, fusing is already provided.)

24-206   400A Fused Disconnect $295⁰⁰

FUTURE TECHNOLOGIES

These are exciting times! I feel like, in the 1990s, we are finally beginning to approach life with a little more sanity. Yet, in true American form, this new attitude of ours is born mostly from a sense of desperation. We continue to foul our nest until it is about to rot through, then we scramble frantically at the last minute to clean it up. We're leaving behind the era of "want" as we realize this new era of "need." We *need* a sustainable future; our nest is a mess!

Technologies that can sustain a well-balanced future are advancing at a rapid rate. Conservation technologies for energy, water, and natural resources are maturing. Utilities are beginning to understand that it's more profitable to save energy then to produce and sell more of it.

Lighting technologies are advancing too, with improvements in fixture cooling and reflector design. There is ongoing work to reduce the use of mercury in fluorescent lamps, which poses a danger in the landfill. Ballast power quality and efficiency are making big gains. Dimmable compact fluorescent (CF) lights appeared on the European market in 1994. CF lamps just keep getting smaller and smaller, with a few manufacturers offering "triple" twin-tube configurations. Incredibly efficient white sodium vapor lights with instant restart capabilities are becoming available, pushing lighting efficiency to the edge of the envelope. A recycling infrastructure for fluorescent lamps and ballasts is being created on many overcrowded, too-toxic municipal dump sites. In some cases, it is becoming more profitable to mine metals from dumps than it is to process virgin ore! This has created a new exploratory job, commonly referred to as "dump-diving."

Myriad energy conservation systems and devices crowd the market, with lots more coming along. Energy conservation and pollution control are among the fastest-growing world markets. In the European Community nations alone, 60% of environmental problems are attributed to energy production and its use. America could run with the global leaders in this marketplace, or, as we have largely done with cars and electronics in recent decades, just kick back on the sofa and snooze. The fact that Real Goods has grown from $29,000 in sales in 1986 to over $16 million in 1995 attests to the potential growth in this sector of the economy.

New ways to deal with air conditioning are emerging, starting at the most basic level with rational, functional building design. Buildings can be designed to minimize or eliminate the need for cooling or heating. Failing this, we will soon see evaporative coolers with dehumidification systems that use a quarter of the energy of a conventional air conditioner. Solar absorption and non-CFC compressed systems with improved motors and insulation will also make it to market. Motor improvements alone will have enormous impact. Currently, half of the world's energy consumption goes to run motors and their support equipment.

We're at the dawn of the biggest explosion of building alternatives since humans moved out of the cave. Renewed interest in straw bale, earth berm, rammed earth, and adobe construction techniques is spreading like the slash-and-burn fires they displace. This revolution in shelter technology will bring with it increased structural and seismic testing, and wide acceptance in building departments across the nation.

Building materials are now being made out of garbage and waste. There are framing systems made out of recycled Styrofoam cups and steel, recycled plastic lumber, and joists made of finger-joint mill ends and pressed scrap wood beams. Siding and shingles made from composites like shredded newspaper, cardboard, plastic, tires, bottles, and cement often have much better thermal and maintenance performance than traditional materials, and frequently cost less, too. You can find carpet and floor coverings made from recycled plastic jugs, and tiles made from scrap brick pieces, automobile glass, and bottles. There is a plywood made of straw (this would be another great use for hemp fiber), and building panels made with pleated, recycled paper cores of extraordinary strength and lightness. These materials will soon be common instead of rare. I can even see the day when the Uniform Building Code will be rewritten to mandate new building construction that contains a minimum recycled content, with sustainability written into the code. Architecture must come out from behind its 200-year-old blinders and rise to the needs of a humanity living in harmony with the planet.

Breakthroughs on the renewable energy front are brewing as well. Everyone wants to know when photovoltaics are going to be cheap and affordable to all. Right now we see the price of PV becoming competitive with existing forms of utility-scale power generation before the turn of the century. Our desperate need to find nonpolluting and decentralized methods of energy production is driving this race. Yet energy efficiency remains the area where improvements can be most dramatic. Efficiencies of up to 80% may be possible with technologies that somewhat mimic the process of photosynthesis found in plants. A product is being developed now called Lumeloid™ that utilizes this technology. It is a very thin (.0003 inch) polymeric film prepared by a method similar to that now employed commercially in the manufacture of polarized film, and using much the same equipment, but with a different chemistry, and with the electrodes imbedded in the film to gather electric power. It will probably be three to five years before it's on the market. We believe that research should be focused on cell materials that are in great abundance and environmentally benign. While supporting a strong infrastructure of independent home designers, builders, installers, and support industry, we should also support existing utilities in their nervous forays into the world of PV. They need all the encouragement they can get, for it is ultimately their mega-buying power that will drive the market for cheap and abundant PV. Utilities also will lend credibility to the technology, which will result in widespread acceptance, standardization, and, most importantly for our readers, financing! I think this financing issue will be resolved in the near future: lenders will incorporate fair valuation for energy features along with other marketable aspects of a home. Real Goods now offers financing on PVs, and this may be the first step in whole-system financing!

Battery technology has been stuck in a time capsule for nearly a century. At last, new developments are upon us, with a lot of exotic formulations being investigated. As with photovoltaics, research should focus on widely available, nontoxic materials. Flywheel storage systems offer exciting promise along with fascinating engineering challenges. This idea has been resurrected with the use of safe composites that turn to dust upon a mechanical failure, rather than sending shrapnel flying in all directions. Advantages can be expected in terms of safety, high power density, and amazing cycle life. In essence, a flywheel storage system consists of a high-mass flywheel coupled to a motor generator that is suspended in a vacuum by magnetic bearings. When power is applied to the "battery," the flywheel either starts spinning or spins faster, converting the electrical energy into mechanical energy stored in the spinning flywheel for later use. The mechanical energy is converted back to electrical energy when an electrical load is placed on the motor generator, thus slowing down the flywheel. One American manufacturer has integrated a flywheel storage system into an electric vehicle and claims a 350-mile range! A little further down the road, we could see a superconductor the size and shape of a doughnut in our basements. Mass-market room-temperature superconductors are getting closer! Even chemical ("cold") fusion and Zero Point Energy may be on the horizon. Microprocessors are finding their way into the rest of the balance-of-system components. Smarter inverters are capable of amazing feats of system control and interface. Products new to this *Sourcebook* include a user-programmable inverter that produces clean sine-wave power and also serves as a charge controller, system monitor, utility intertie, battery charger, load controller, system manager, and more, all while performing its primary function in a superior manner. Monitors and metering are getting the "chip" also, and we are seeing computer-interface system monitors popping up all over. Soon all components will be in communication with one another, controlling our energy systems and homes by turning fans off and on, darkening our windows as the house heats up, and shifting loads around as conditions change. In the past we have all been limited by availability, but in the next few years we will only be limited by our imaginations.

Doesn't all of this remind you of reading a 1930s *Popular Science* magazine? Automatic atomic vacuum cleaners, commuting by airplane and parachute, cars that drive themselves — remember all that crazy stuff? Well, a surprising amount of it really has come true. As I said, we live in thrilling times!

— Jeff Oldham

FUTURE TECHNOLOGIES BOOKS

Hydrogen Fuel Cells: Key To The Future

A video presentation made by Dr. Roger Billings of the American Academy of Science, on the hydrogen fuel cell. This detailed video outlines the design and construction of a fuel cell and its uses, in particular its application to electric vehicles. Dr. Billings discusses vehicle range and what kind of infrastructure it would take to incorporate this exciting technology into our world today. Cleaner, safer, and more efficient alternatives to our present power grid distribution system are also offered. Includes footage of the Laser I electric vehicle, and a demonstration of a fuel cell. A must-see for anyone interested in hydrogen. 105 minutes.

80-303   Hydrogen Fuel Cells $35⁰⁰

Fuel From Water: Energy Independence With Hydrogen

Michael Peavey. An in-depth and practical book on hydrogen fuel technology which details specific ways to generate, store, and safely use hydrogen. Hundreds of diagrams and illustrations make it easy to understand, very informative, and practical. If you're interested in using hydrogen in any way, you will find this book indispensable. 244 pages, paperback, 1993.

80-210   Fuel from Water $20⁰⁰

CHAPTER 6

HEATING AND COOLING

AN IMPORTANT CONSIDERATION for any dwelling is providing an adequate climate-control system. Many of us live where the outdoor air temperature fluctuates considerably from the "ideal" 25° C (77° F), and, in spite of our best attempts to use passive means to control our microclimate (see Chapter 3), we may need to supplement with mechanical systems to meet our heating and cooling needs. Choosing among these systems can be difficult, and the final choice can have a big impact on the energy budget and on the pocketbook.

Whenever the mercury rises or falls outdoors, our first line of defense is to keep the heat out or in. A profound lesson learned by early solar home designers was that an abundance of south-facing windows can create massive overheating of the living space. The general rule of thumb now is to limit southern glass area to 7% of the living space's square footage. If existing windows are proving to be a problem, consider using window films to block out summer heat and help keep in winter warmth. An uninsulated drape can reduce heat loss through a window by one-third, while an insulated drape can cut the loss in half.

Generously insulated exterior walls, floors, and ceilings are an essential ingredient for stabilizing indoor temperatures. If you live in a hot climate and lack the advantage of trees shading your house, radiant barriers installed in the roof can be very effective at keeping heat out. For summer cooling, try to take advantage of cool evening breezes by having windows located so that they allow cross-ventilation. An exhaust fan connected directly to a PV module will pull hot air out of the highest spot in the living space, with make-up air coming from the cool north side or underside of your home; this can cool down the air and the furniture during hot weather, and may keep the house cool all day long if the house is well insulated and sealed up. Of course, sealing up the house is becoming more of a concern as we become more aware of the health threats posed by chemical outgassing from building materials, as well as oxygen depletion due to combustion. These concerns can easily be addressed by proper ventilation, but then, opening that window could let in the hot air we have been trying to keep out (or in). An air-to-air heat exchanger, though initially expensive, can ensure adequate ventilation while at the same time keeping the hot air where we want it to be and saving money in the long term. In low-humidity climates, consider evaporative coolers as a good way to beat the heat when these measures fall short.

At this point, it should be obvious that we at Real Goods have an aversion to air conditioning. Air conditioning represents an enormous electrical load, and probably limits the number of independent energy systems sold in the Southeast. There is research being conducted on solar air-conditioning systems, and perhaps sometime soon Real Goods will be able to recommend them. One of our creative customers designed a cooling and heating system for his 3200-square-foot home by pumping water through PVC piping under his slab foundation, and through a number of solar thermal collectors on his roof during the day in winter (heating), and through the same collectors in the summer at night. The unglazed collectors radiate the slab energy at night to the dark (or black) sky, cooling the slab down for another day. Located in the Central Valley of California, the owner pays no more than $5 in any given month for the conventional back-up heating (electric heaters) or cooling (air conditioning) that he uses as a secondary system.

You have numerous choices when considering a home heating system. If your interest is in energy independence using renewable energy sources, you would be wise to steer clear of electric heaters. Although extremely efficient at turning electricity into heat, the quantity of electricity required is usually very great, and comes at a time of year when energy supply is typically limited and demand for other electric loads is high. Forced air systems (furnaces which use blowers to move heated air to various rooms) also use extra power for inefficient fans, which may operate for several hours every day. Forced air systems often run lightly insulated ducts through unheated spaces, which cools off the delivered air somewhat, and these ducts can leak, heating up spaces best left unheated. One advantage of a forced air system is that you can close off some of the registers. This gives you some "zone-heating" control (why heat rooms that you are not in all day?), but closing down too many of the registers can make the fan system work harder than it should and increase the electrical consumption.

The *ideal* independent-energy home would require no heating or cooling; however, if your independent home falls a little short of this ideal, then you should make use of appliances that burn fuels to provide heat and require no electricity to operate. These would include woodstoves, as well as natural gas, propane, or fuel oil furnaces. Many models require little or no power to operate. Pellet stoves, which burn wood by-products or agricultural wastes, are gaining supporters, but they do require power for sensitive electronics and small blowers.

We are often asked about the feasibility of **solar space heating**. Our position is that a properly engineered solar greenhouse or **passive solar system** (a system which makes no use of pumps, fans, or controllers) may well be an excellent way to augment traditional heating systems. Care must be taken to design the system so as to provide satisfactory performance during both winter and summer. **Active**

systems, on the other hand, usually make use of solar collectors; a heat exchange medium, such as a fluid or air; an energy-storage device, such as an insulated tank or slab of concrete; and a heat exchanger, like a baseboard heater or radiator. Active systems can be complex and a poor investment. We feel that the value of such a system is debatable because, in many locations, the system provides little benefit for many months of the year, and those months requiring heating have poor solar conditions. If your weather conditions are cold with clear skies during most of the year, then an active system still might be an option to consider. **Radiant floor heating** is becoming very popular and can usually be adapted to accept solar input without a great deal of difficulty. This type of system combines the storage mechanism with the radiator in a masonry floor slab, thus reducing the number of components required in the system. Be sure to take the time to find an experienced solar contractor who can provide referrals in your area.

— *Douglas Bath*

REFRIGERATION

According to the U.S. Department of Energy, the typical household devotes 15% of its total energy budget to refrigerators and freezers. In the ultraconservative homes we strive for, this fraction can become as large as 50% to 75% of our total demand for electricity. If we wished to power a typical 1990-era residential refrigerator with a solar photovoltaic array, we would need anywhere from 10 to 30 modules for this one appliance. Clearly, refrigeration systems should be carefully chosen in order to minimize their long-term impact on our energy budget and pocketbook.

A refrigerator/freezer is nothing more than a small, insulated room with an air conditioner which pulls the heat out of the room and anything in it (the food) and pumps the heat into the kitchen. Most conventional models have fans and heating elements in the walls to eliminate condensation and to melt ice. We pay for this "frost-free" convenience with greater energy consumption. The makers of refrigerators have evolved their products through years of declining electrical costs. Competition in the marketplace has focused on convenience features (cold-water dispensers, icemakers, "biggest on the inside," etc.), forsaking energy-efficient engineering. Sacrificing energy efficiency can result in more noise in operation and prolonged operating time. The convenience of frost-free operation results in low humidity, which promotes the rapid wilting of unprotected fruits and vegetables. This is why refrigerators have "crispers," which provide a small space isolated from the arid surroundings.

Anyone with a passing grade in high school physics would recognize that the condenser, which dissipates the heat removed from inside the refrigerator, should be located on top of the refrigerator. Heat does rise, and we would like it to get away from the space we are trying to cool. Some of us have seen older refrigerators designed this way. The refrigeration cycle is driven by a compressor, which also generates significant heat. This compressor, as well, would ideally be located remotely or on top of our cool space. But a few generations of "cheap" energy have made these logical, efficient practices secondary considerations: a smooth horizon-

tal surface supporting wine racks or crayon buckets is now the norm. If we want to make the refrigerator "bigger on the inside," then why not just make the insulation in the walls that much thinner? Is it just a coincidence that some of the largest manufacturers of refrigerators and freezers also build electrical power plants? Perhaps, but we doubt it.

Cold without Kilowatts

The energy-efficient home can overcome the energy gluttony of conventional refrigerators and freezers in one of two ways: remove the electrical demand for refrigeration, or reduce the electrical demand as much as possible. Eliminating electrical demand can be accomplished either by eliminating refrigeration (generally an unacceptable solution) or by storing perishable items through nonelectric techniques. This can be done with root cellars or with fuel-driven refrigerators. Fuel-driven refrigerators and freezers use the circulation of ammonia in what is called the "absorption cycle," a technology that has been used for many years to a limited degree. Propane is usually used as the fuel to drive this cycle, with a flame only slightly more intense than a pilot light. There are other models available that burn kerosene, natural gas, and, very likely, diesel and fuel oil. Refrigerating by burning fuel offers convenience, but, in absolute terms, the efficiency is somewhat less than the most efficient compressor-driven models. Still, there are a number of places where fuel is easier to come by than electricity, and in these situations it is hard to beat a propane refrigerator.

King of the Hill

The current, and still reigning champion of energy-efficient refrigerator/freezers (going on ten years now) is the Sun Frost, and we at Real Goods are not at all ashamed to be its ardent supporters. These Sun Frost units take a commonsense approach to the goal of keeping things cold. This results in refrigerators and freezers which use 70% to 80% less energy than that required for conventional electric models of equal size. Sun Frost makes no bones about the energy required to operate their units. This energy can usually be provided by two to four solar modules with reasonable solar exposure. If you live in an area that has long sunless periods during the winter months, you may need an electrical back-up to your solar array to support the refrigerator and other electric loads during gloomy times.

Sun Frost models use compressors to circulate refrigerant just as conventional units do. But the Sun Frost compressors are smaller, quieter, and less energy-hungry than conventional refrigerator compressors. By generously insulating the storage space, placing the condenser and compressor atop the unit, and offering only "semi-automatic defrost" models, the run time on the compressor is reduced to the minimum. The compartmentalized design of Sun Frost models eliminates the transfer of moisture from stored food to the refrigerator/freezer coils. This saves energy, reduces frost buildup, and preserves food longer. Automatic-defrost models have to use energy to first evaporate the moisture from your food, then melt the ice from the cooling coils, and then recool these coils.

Sun Frost is a leader in the independent-energy industry, offering high-quality and socially responsible products, including ultra-efficient vaccine refrigerators which can operate off of a single module. These units can help bring health care that we take for granted to remote parts of the world.

The Golden Carrot

A U.S. utilities consortium has decided to take on refrigeration as an important technology to improve during the next decade. Their primary goals were a 25% reduction in power compared to comparably sized units now available, and to stop using chloroflourocarbons (CFCs) as a refrigerant and as a blowing agent in the manufacture of the insulation. It was decided that the best way to accomplish these goals was to have companies that were already producing many thousands of refrigerator units per year submit proposals outlining the methods they felt would yield the desired results, and then award the best of these proposals with a $30 million research and development grant — the so-called "Golden Carrot." Many in the renewable energy industry believe the program was carefully designed to eliminate Sun Frost from the competition. This Golden Carrot grant has been awarded to Whirlpool, who offered their 22-cubic-foot, side-by-side model for sale in 1994.

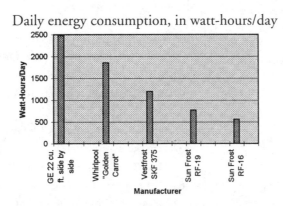

Daily energy consumption, in watt-hours/day

It is difficult to make comparisons between refrigerators of different sizes and features (auto-defrost, icemakers, and so forth). Our experience with independent-energy users is that most who have Sun Frost units are pleased they made the purchase. They find the space provided by the 16- and 19-cubic–foot models to be adequate, in spite of being less spacious than the average refrigerator/freezer. Those who use propane refrigerators often complain about the limited storage space, but relish the electricity-free operation. As industry becomes aware of an increasing demand for energy-efficient appliances, energy-conscious consumers may have more options from which to choose. The fact is that 19 nuclear power plants could be shut down today if we all were to miraculously have our existing refrigerators replaced with Sun Frost models.

— *Douglas Bath*

REFRIGERATION

Sun Frost Refrigerator/Freezers

The Sun Frost ultra-efficient design makes it practical to power a full-sized refrigerator/freezer with a renewable energy system. Refrigeration is typically the largest consumer of electrical energy in an energy-efficient home. All Sun Frosts are cooled by highly efficient top-mounted hermetically sealed compressors. The larger 16- and 19-cubic–foot models use a pair of compressors. The compressors and condenser are top mounted so they cool efficiently without having the waste heat re-enter the cabinet. The walls of the refrigerator contain 3 inches of insulation and the freezer section up to 4-1/2 inches of polyurethane foam. Frost buildup in the freezer is very slow because there is no air circulating between the refrigerator and freezer. The Sun Frost runs extremely quietly. It's usually difficult to tell if the compressor is running or not.

The single compressor RF-12, a 12-cubic–foot refrigerator/freezer, draws 336 watt-hours per day with a 70° F ambient temperature, which means that the entire refrigeration system can typically be powered from only two 48-watt PV modules! All Sun Frosts are guaranteed for two years. Sun Frosts are available from 4-to 19-cubic–feet. Refrigerator-only and freezer-only models are also available.

Have you noticed broccoli, carrots, romaine lettuce, and other vegetables turning limp after a few days in the refrigerator? This occurs because conventional refrigerators take moisture out of your food (and the air in the refrigerator) and convert it to frost in the freezer. The usual solution for this problem is airtight containers for every article of food. Sun Frost has conquered this problem and maintains high humidity inside, with no air transfer between compartments. The results are much less frost and wilting, so you don't have to use airtight containers and your food will store longer.

The Sun Frost is not completely frost-free, although it creates much less frost than conventional models. The amount is somewhat similar to that of a partially automatic defrosting refrigerator, and the unique defrost system requires only a few minutes of labor.

Sun Frost standard models are off-white Navamar (a Formica-like material), but almost any color of Formica or Navamar is available. For any color other than white there is a $100 charge. Sun Frosts are also available with a wood veneer. Red Oak, Cherry, Birch, Maple, Teak, Black Walnut, White Ash, Knotty Pine, and Honduras Mahogany are available for an extra $150. All veneered units are shipped unfinished so the customer can chose the stain. For an exact match to custom cabinets Sun Frosts can be ordered without any laminate or veneer. Call for more details.

Without the usual compressor and plumbing underneath, the bottom shelf of a Sun Frost is only 5 inches off the floor. Either use the optional Sun Frost Stand to raise the refrigerator about 13 inches, or plan to build something custom. The optional stand has two drawers on heavy-duty hardware.

In the list below, R stands for refrigerator, F for freezer. The RF-19 has freezer and refrigerator compartments of equal size, other models have conventional one-third/two-thirds division. All models are 34-1/2 inches wide and 27-1/2 inches deep with varying heights. Allow an additional 3-3/4 inches width if door will open against a wall (standard 24-inch depth countertops are okay, door will clear). All models are available in 12- or 24-volts DC, or 120- or 220-volts AC. Voltage is not field-switchable. All models use a brushless compressor design. The entire cooling system contains only one moving part. The wattage is virtually the same for the various voltages.

All Sun Frosts are made to order. Allow approximately eight weeks for delivery. *Be sure to specify if you want the hinge on the right or the left, color if desired, and the voltage!*

•62-134	Sun Frost RF-16 (DC)	$2,595
•62-134-AC	Sun Frost RF-16, 120VAC	$2,449
•62-135	Sun Frost RF-19 (DC)	$2,749
•62-135-AC	Sun Frost RF-19, 120VAC	$2,589
•62-125	Sun Frost R-19 (DC)	$2,489
•62-125-AC	Sun Frost R-19, 120VAC	$2,399
•62-115	Sun Frost F-19 (DC)	$2,789
•62-115-AC	Sun Frost F–19, 120VAC	$2,649
•62-133	Sun Frost RF-12	$1,949
•62-133-AC	Sun Frost RF-12 120VAC	$1,875
•62-122	Sun Frost R-10 (DC)	$1,575
•62-122-AC	Sun Frost RF-10 120VAC	$1,475
•62-112	Sun Frost F-10 (DC)	$1,675
•62-112-AC	Sun Frost RF-10 120VAC	$1,589
•62-131	Sun Frost RF-4	$1,349
•62-121	Sun Frost R-4	$1,349
•62-111	Sun Frost F-4	$1,349
•62-116	Sun Frost Stand, White	$255.00
•62-117	Sun Frost Stand, Color	$295.00
•62-120	Sun Frost Color Option	$100.00
•62-119	Sun Frost Wood Grain Option	$150.00

Shipped freight collect from Northern California

SUN FROST TECHNICAL DATA

Model	Power Usage @ 70° F*	Power Usage @ 90° F*	Weight Crated	Volume Crated	Exterior Dimensions
RF-19	770 W-H/Day	1030 W-H/Day	320 lbs.	46 cu. ft.	66" x 34.5" x 27.75"
R-19	380	630	310	46	66" x 34.5" x 27.75"
F-19	1250	1630	320	46	66" x 34.5" x 27.75"
RF-16	560	810	300	44	62.5" x 34.5" x 27.75"
RF-12	350	560	230	36	49.3" x 34.5" x 27.75"
R-10	190	310	215	32	43.5" x 34.5" x 27.75"
F-10	690	880	215	32	43.5" x 34.5" x 27.75"
R-4	110	160	160	23	31.5" x 34.5" x 27.75"
F-4	350	450	160	23	31.5" x 34.5" x 27.75"
RF-4	160	240	160	23	31.5" x 34.5" x 27.75"
RFV-4	160	230	160	23	31.5" x 34.5" x 27.75"

24-hour closed-door test with stated ambient outside temperature

INTERIOR DIMENSIONS OF SUN FROST REFRIGERATORS AND FREEZERS

Model	Freezer Section Height	Depth	Width	Volume	Refrigerator Section Height	Depth	Width	Volume
R,F,RF-19	24"	20-3/4"	28"	8.1 cu. ft.	24"	20-3/4"	28"	8.1 cu. ft.
RF-16	13"	20"	26"	3.91 cu. ft.	31"	20-3/4"	28"	10.4 cu. ft.
RF-12	6-1/2"	21"	26"	2.05 cu. ft.	24"	20-3/4"	28"	8.1 cu. ft.
R-10	—	—	—	—	28"	20-1/2"	27-1/2"	9.3 cu. ft.
F-10	28"	20-1/2"	27-1/2"	9.13 cu. ft.	—	—	—	—
R-4	—	—	—	—	13"	20"	26"	3.9 cu. ft.
F-4	13"	20"	26"	3.91 cu. ft.	—	—	—	—
RF-4	2-1/2"	20"	26"	.68 cu. ft.	10-1/2"	20"	26"	3.2 cu. ft.
RFV-4	4"	20"	26"	1.2 cu. ft.	6"	20"	26"	1.8 cu. ft.

Sun Frost RF-16

An Energy-Efficient And Ozone-Friendly Refrigerator

The U.S. Dept. of Energy estimates that the typical household devotes 15% of its energy budget to the refrigeration and freezing of food. The new Vestfrost offers high efficiency, environmental design at reasonable price. Using only .88 kilowatt-hour per day @ 75⁰ ambient temperature, the Vestfrost SKF 375 is an energy miser unmatched by domestic manufacturers. It uses newly formulated foam insulation and refrigerants which contain absolutely no CFC products. This unit also meets the strict European standards for recyclability.

Standing just over 6'5" tall, the refrigerator over freezer design allows easy access to the most commonly used items without bending over. Europeans have adapted to the higher upper shelf by using a handy kitchen stool. The Vestfrost has separate compressors and controls for each of the refrigerator and freezer sections. The condenser and cooling tubes are built into the walls of the unit, and are sealed so no dust collects on the working parts of the unit, which would lower the efficiency. This "sealed" design also insures that the Vestfrost will operate very quietly and never requires cleaning.

Internal volume is essentially the same as the Sun Frost RF-16. Although the power consumption is somewhat higher, this refrigerator is a reasonable alternative to Sun Frost for independent energy homes and for the environmentally-conscious homeowner. Two units can be placed side-by-side (with left and right opening doors) to create a truly magnificent refrigeration space. *Since the Vestfrost is taller and narrower than American made refrigerators, it will generally not fit into the standard refrigerator spaces with cabinets above.*

Features:

- Built in condenser gives lower energy consumption: no dust, low noise
- Separate compressors for fridge and freezer
- Can be built in
- Left or Right hand opening, field-switchable
- Fridge section has:
  - Interior light
  - Automatic Defrost
  - Flexible door shelf arrangement
- Freezer section has:
  - Defrost water outlet (fold-out) for manual defrost
  - Combination of pull-out drawers and fast freeze shelves
- Control panel at top of cabinet with:
  - Thermometer for fridge or freezer section
  - Thermostats for fridge and freezer sections
  - Fast Freeze button and lamp

Specifications:
Net Capacity:
—Fridge: 6.9 cu. ft.
—Freezer: 4.2 cu.ft.
Height: 78.7"
Width: 23.4"
Depth: 23.4"
Freezing Capacity (24 hrs): 39.5 lbs.
Electrical consumption (24 hrs): .88 kwh @ 77⁰ F ambient temp.
Electrical consumption EPA specs @ 90⁰ F : 1.2 Kwh

•62-299   Vestfrost Refrigerator $995⁰⁰

Norcold Propane Refrigerator/Freezer
Made in the USA — Highest Quality!

Norcold, America's best known name in RV and marine refrigeration has come out with this new unit for the remote home. Like all gas fridges, there are no CFCs used for coolant or insulation. It features quicker cooling than the foreign competition, and a patented "moisture control system" which keeps condensation to a bare minimum. Offers 5.6 cu. ft. in the refrigerator, and 1.9 cu. ft. in the freezer, for a total of 7.5 cu. ft. Has twin crisper bins, powder-coated shelf racks, flip-up shelf sections for tall bottles, and adjustable door and freezer shelving. Also features a stainless steel heat exchanger for longest possible life, adjustable feet, up–front controls, positive flame indication meter, and the door is easily reversible (yea!). Exterior dimensions are 25.25" W x 23.8" D x 61.5" H. Color is white only. Maximum gas consumption is 1550 BTU/hr. or .39 gal./day. Actual use will depend on ambient temperature, and will be less at any temperatures under 80°F. Can also run on 120VAC, but isn't particularly energy efficient in this mode.

•62–302 Norcold Propane Fridge/Freezer $1,265

Shipped freight collect from Scottsdale, AZ.

Koolatron 12-Volt Refrigerators

Koolatron of Canada is the world's largest manufacturer of 12-volt coolers. The secret of their cooler is a miniature thermoelectric module that effectively replaces bulky piping coils, compressors, and loud motors used in conventional refrigeration units. For cooling, voltage passing through the metal and silicon module draws heat from the cooler's interior and forces it to flow to the exterior, where it is fanned through the outer grill. The Koolatron is thermostatically controlled. Amp draw is 4.0 amps when on.

The coolers will maintain approximately 40° to 50° F below outside temperature. They can be run as either a cooler or a warmer and can be used on 12-volt DC or adapted to 120-volt AC (by ordering the optional adaptor). Constructed of high-impact plastic and super urethane foam insulation, the coolers include a 10-foot detachable 12-volt cord.

Traveller

Dimensions: 10" x 12-1/2" x 15"
Weight: 8 lbs.
Capacity: 0.25 cu. ft.

•62-521 Traveller 12V Cooler $99⁰⁰
•62-525 Koolatron AC Adapter $59⁰⁰

Voyager

Dimensions: 16-1/2" x 16-1/2" x 20"
Weight: 14 lbs.
Capacity: 0.9 cu. ft.

•62-526 Voyager 12V Cooler $149⁰⁰
•62-525 Koolatron AC Adapter $59⁰⁰

Companion

Dimensions: 13-1/2" x 16" x 19-1/2"
Weight: 13 lbs.
Capacity: 0.6 cu. ft.

•62-523 Companion 12V Cooler $169⁰⁰
•62-525 Koolatron AC Adapter $59⁰⁰

See photos on page 258.

Koolatron Traveller

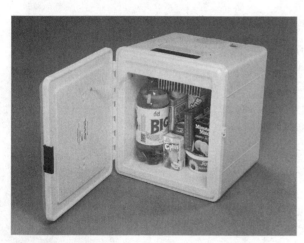

Koolatron Voyager

Koolatron Companion

Super-Efficient, Totally CFC-Free Freezer Costs Only 5¢ A Day To Run

This 7.5 cubic foot capacity, manual defrost, chest freezer uses absolutely no CFC products for insulation or coolant, is constructed of all-recyclable components, and uses a minuscule 540 watt-hours per day (at 90°F ambient temp.). That's about 50% less energy than comparable chest freezers. You can have an affordable home freezer, without high energy bills or guilt about the ozone layer.

The Vestfrost freezer was developed three years ago to satisfy the German and Swedish "green" markets. Vestfrost is a Danish manufacturer with 35 years in the refrigeration business. It is equipped with a lock and interior light on the PVC-lined, counter-balanced lid. There is an internal partition, which doubles as a drain pan during defrosting, and a pair of sliding, lift-out, wire baskets. The temperature control knob is a child-proof design, plus there are external "Power On" and "Temperature Warning" lights. A fast freeze switch is provided for sudden heavy loads. Freezing capacity is 67 lbs. per 24 hour period. A drainhole and plug are provided for defrosting. Operation is almost completely silent. Size is 33.5"H x 44.4"W x 25.6"D, Weight is approximately 150 lbs., Color is white. Warranty is one year.

• 62–301 Vestfrost Freezer $695⁰⁰

Shipped freight collect from Scottsdale, AZ

Servel Propane Refrigerators

Servel has been a household word in gas refrigeration for over 50 years. It was first marketed by the Swedish company Electrolux in 1925. In 1956, the rights to Servel were acquired by Whirlpool, which was unsuccessful with the unit, and it disappeared for 30 years. Now Servel is back with a state-of-the-art household propane refrigerator.

We've sold more than 800 of the new Servels since their reintroduction in 1989, and have one in our showroom. All continue to perform flawlessly and appear to be extremely well built. They average less than 1.5 gallons of propane per week! The body is all-white and the door is hinged on the right, like the Model-T, with no choices available. The unit comes with rust-proof racks, four in the refrigerator and two in the freezer. The spacious interior has two vegetable bins, egg and dairy racks, frozen-juice rack, ice-cube trays, and an ice bucket. An optional four-year warranty can be purchased for $59.95 from the manufacturer. The total volume is 7.7 cubic feet (6 for the refrigerator, 1.7 for the freezer). Total shelf space is 9.9 square feet. The overall dimensions are 57-3/4" H x 24-3/4" W x 24-3/4" D. Net weight is 181 lbs. It is operable on either propane or 120-volt AC (draws 275 watts almost continuously on 120 volts).

•62-300   Servel Propane Refrigerator $1,295

Shipped freight collect from California or Ohio

Frostek Propane Freezer

A freezer that works without electricity or CFCs! The Canadian produced Frostek is the largest (and perhaps only) propane-powered freezer on the market. Its 8.5 cubic foot capacity offers reliable long-term food storage, and can accommodate an entire side of beef, or quarts upon quarts of frozen fruits and vegetables.

In carefully monitored laboratory tests the powerful cooling unit maintained 8^0 F interior temperatures or less all the way up to an outside ambient temperature of 80^0 F. Even at 90^0 F it still maintains 0^0 F inside. Fuel use is approximately 1.8 to 2.5 gallons per week, depending on ambient temperature. Or 7.5 to 11 lbs. if you buy propane by the pound. An intelligent exterior thermometer shows interior conditions without opening the lid. Has easy-access bottom drain, piezo ignitor, thermostat on/off control, and runs silently.

The Frostek stands 38" high, 44.5" long, 31" deep. Inside dimensions are 24" x 37" x 15.5". There are no awkward cooling fins jutting into the food storage area. Cabinet is rustproof steel with baked enamel finish. Color is gray and white. Approved by both the American Gas Association and Canadian Gas Association. Weight is 220 lbs. Shipping weight is 275 lbs. Call for availability, production quantities are limited.

•62-308   Frostek Propane Freezer $1,785

Shipped freight collect from Ohio

ColdMachine

For those who wish to create their own custom refrigeration system, the Adler/Barbour ColdMachine system offers high efficiency, exceptional reliability, and environmental sensitivity. Over 50,000 systems are installed in private yachts of all sizes. These pre-charged systems offer easy installation for refrigerators up to 15 cubic feet, freezers up to 7 cubic feet, and refrigerator/freezers with up to 5 and 3 cubic feet, respectively. A minimum of 4 inches of insulation is required for refrigerators, and 6 inches or more for freezers. For refrigerator/freezers, make a partition of 1-inch foam core with fiberglass or formica faces, with a 1-inch opening across the top for convection air-flow return.

Units are available allowing either top or front access to the freezer compartment. Specify the vertical freezer for top access, or the horizontal freezer for front access.

All Adler/Barbour units are shipped with HFC-134a refrigerant in accordance with the Montreal Protocol Rulings.

Voltage: 12 volts DC
Current: 5-6 amp
Condenser Dimensions: 10" x 14" x 8-1/2"
Condenser Weight: 26 lbs.
Btu Rating: 300/hr. maximum
Evaporator Dimensions: 15" x 6" x 12"
Evaporator Weight: 5 lbs.

•62-534   ColdMachine PB110LV (vertical freezer) $1,175
•62-535   ColdMachine PB111H (horizontal freezer) $1,175

Isaac Solar Ice Maker

A *solar* ice maker? No we haven't been out in the sun too long, this is a real product with a number of working installations. Using only the heat of the sun, an Isaac will produce 10 to 1000 pounds of ice per day depending on model and available sunlight. The Isaac uses the Solar Ammonia Absorption Cycle that was discovered in the 1850s, and is still used on a smaller scale by all gas refrigerators. This refrigeration technology fell out of fashion in the 1930s with the arrival of cheap electricity and the discovery of the miracle refrigerant freon. The Isaac uses no electricity, gas, or freon.

During the day the Isaac stores energy in the receiving tank as high-pressure, distilled, pure ammonia. At night the user checks the sight glass to judge how much ammonia was produced, adds an appropriate amount of water to the ice compartment, and switches the valves from Day to Night positions. The ammonia is allowed to evaporate back into the collector while providing refrigeration for the ice compartment. The refrigerant cycle is sealed. In the morning the valves are switched again and the process starts all over while the ice production from the night before is harvested.

The Isaac is constructed of stainless steel for maintenance-free outdoor installation in oceanside sites (not to mention some pure industrial beauty). Many of the current installations are for fishing villages in remote sites. Maintenance consists of re-aiming the collector every four weeks to track the sun, and an occasional bucket of water to wash away any dust.

The Isaac is built-to-order in a variety of models. Delivery is 30 to 90 days. An experienced technician is required on site for installation and training; this is not included in the price. Discounts available for quantity orders. Call for more information.

•62-532	Isaac Standard Solar Ice Maker	$9,895
•05-218	Export Crating for Standard Isaac	$595.00
•62-533	Isaac Double Solar Ice Maker	$13,995
•05-219	Export Crating for Double Isaac	$995.00

Shipped freight collect from Maryland

COOLING EQUIPMENT

The marketplace offers a variety of high-efficiency fans and evaporative coolers. Most of the ones we have found useful are low-voltage units, whose DC motors are intrinsically more efficient than those of comparable AC fans. There are many inexpensive AC ceiling fans, but the future renewable-energy homeowner should carefully consider the demands which operating these fans for many hours a day may place on the home energy system. AC ceiling fans have historically represented one of the more problematic loads for an inverter to operate. The fan motors can buzz annoyingly or overheat when operated by non–sine-wave inverters.

Evaporative coolers can be a very effective cooling technology if the relative humidity is moderate to low. Fountains and porous jugs of water have been used for centuries as climate-control devices. Many independent energy users find their power production outpacing their normal demands during times of summer heat, and cooling can be an excellent load with which to take advantage of a power surplus.

There are a number of areas, though, where evaporative coolers are of no benefit. High-humidity regions, for instance, which also suffer from intense heat may require air conditioning. The air conditioner's run time can be minimized (which reduces fuel or electrical energy consumption) by super-insulating the home. Air conditioning not only reduces the temperature of the building; it also removes moisture from the air, which in turn aids our body's evaporative cooling system, called perspiration. For this reason, dehumidifiers may be somewhat less energy-intensive than air conditioners, yet both require significant energy input, which generally means that they must be powered by some source other than a battery-based system.

— *Douglas Bath*

COOLING

Solar Evaporative Air Conditioner

The Recair 18A2 turns hot, dry summer air into a refreshing indoor climate, while filtering out dust and pollutants. It can cool up to 400 square feet, using only one 60-watt solar panel and 2 to 4 gallons of water per day to cool and clean the air. Unlike cellulose cooler filters, it will not promote bacterial growth or foul odors. It will move 750 cubic feet per minute (cfm), drawing 3 amps at low speed and 4.6 amps on high speed at 12 volts DC from one 60-watt solar module or battery. It is installed with simple hand tools and connected to a garden hose or the separately available pressure tank.

Recair is the most efficient cooler of its type on the market today. Conventional air conditioners recirculate the same stale air within your house. Recair uses outside air and filters out any impurities such as pollen or dust, while adding moisture to the air, giving you a more healthful, cool indoor air environment.

Installed with a solar panel, Recair will cool when it is needed most. As the morning sun strikes the solar panel, Recair goes to work cooling your house. As the dry air is forced through the wet filter by the fan, the water is evaporated, which cools the outgoing air. The drier the air, the greater the temperature drop. The only drawback is that it will not work efficiently where relative humidity exceeds 40%. Dimensions: 17-1/2" x 22-1/4" x 20".

Note: PV module not included.

•64-201   **Solar Evaporative Cooler** $495⁰⁰

RV Evaporative Air Cooler

Hot, dirty air becomes clean, cool air as highly effective spin-spray action washes out dust, pollen, and impurities. The unit fits all standard 14" x 14" RV roof vents. Its streamlined styling and low-profile contour offer reduced wind resistance. For longer life, the motor is cooled by dry air. Operation is exceptionally quiet. It produces 450 cfm, drawing only 2.1 amps on low speed, and producing 750 cfm on high speed drawing 4.6 amps. The nonorganic industrial foam filter provides superior filtration and cooling, without bacterial growth and foul odors. The discharge grille, only 1-1/2 inches deep, has individually adjustable louvers. The built-in reservoir fills automatically from the RV system or with a garden hose, and can be hand-filled from inside the vehicle. The cooler weighs 16 pounds; dimensions are 35" x 22" x 10-1/2". There's a one-year warranty on all parts; three years on the water pump. Not recommended where average relative humidity exceeds 40%.

•64-204   **RV Evaporative Air Cooler** $429⁰⁰

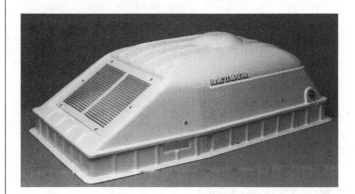

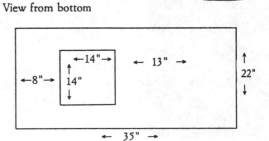

View from bottom

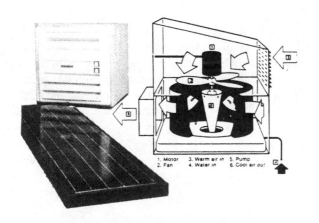

12" And 16" DC Fans

These are the same fans used in our complete Attic Fan Kits. Both models use heavy-duty 36-volt DC motors that can be run with either 12-or 24-volt systems. Frames have three rubber grommeted mounts. A nonadjustable thermostat switch that closes at 110⁰ F and opens at 90⁰ F (approx. temp.) is included. Input power and output in CFM are given in the chart below. Specs at 14 and 28 volts simulate PV-direct applications. Specs at 12 and 24 volts are battery-driven.

12" DC Fan		
Voltage Input	Amps Draw	CFM Output
12 volts	1.17 amps	740
14 volts	1.47 amps	1050
24 volts	3.20 amps	1335
28 volts	4.20 amps	1430

16" DC Fan		
Voltage Input	Amps Draw	CFM Output
12 volts	1.47 amps	810
14 volts	1.84 amps	1100
24 volts	4.20 amps	2000
28 volts	5.30 amps	2275

64-216 12" DC Fan $100⁰⁰
64-217 16" DC Fan $125⁰⁰

A Quiet, Energy-Efficient AC Fan

Your woodstove works hard to generate all that blissful heat. This powerful, quiet fan helps establish the air flow needed to distribute heat evenly throughout your house. To make the most of warm air simply set the fan on your mantelpiece, near your woodstove, or on the floor. Swivel bracket mounting hardware is included for a more permanent installation. Made of sturdy cast metal with finger guards, has a 12-foot cord with in-line switch. Draws only 13-watts of 120VAC power. Moves 50 CFM air. 4-3/4" x 4-3/4" x 1-3/4".

64-215 Super Quiet AC Fan $39⁰⁰

Solar-Powered Attic Fan Kits

Moving stagnant hot air out of your attic, greenhouse, or barn is the least energy-intensive and most effective way to keep your space cool. Using solar power to do it makes even more sense. The brighter and hotter the sun shines, the faster the fan spins. Our Solar Fan Kits give you everything you need. High-efficiency fan, matched PV module(s), mounting brackets, wiring, a thermoswitch to prevent running in cold weather, and instructions.

We offer two sizes. The 12" Fan Kit has a single 10-watt PV module, and moves 700-800 CFM. The 16" Fan Kit has a pair of 10-watt PV modules, and moves 800-1,000 CFM. Modules are 12" x 36" amorphous silicon, fan frames are rubber grommet mounted for minimal noise.

64-219	12" Attic Fan Kit	$195.00
64-220	16" Attic Fan Kit	$295.00

12" Attic Fan Kit

16" Attic Fan Kit

12-Volt Fans

These 12-volt DC axial fans are ideal for moving woodstove heat throughout the house. The brushless motor design minimizes electromagnetic interference and radio frequency (RF) interference. All motors are polarity protected. The voltage range for the nominal 12-volt DC fan is 6 to 16 volts DC. Fan blades are PBT plastic, flame rated by UL to a maximum ambient temperature of 149° F. The two smaller fans have a PBT plastic housing with a depth of 1 inch. The larger fan has an aluminum housing with a depth of 1-1/2 inches. Size shown is square.

64-211	12V Fan, 15 cfm, 0.24-amps, 2-3/8"	$45.00
64-212	12V Fan, 32 cfm, 0.25-amps, 3-1/8"	$45.00
64-213	12V Fan, 105 cfm, 0.55-amps, 4-11/16"	$45.00

24-Volt Ceiling Fan, 42" Diameter

For those of you with higher-powered 24-volt systems we
have a great higher-powered ceiling fan for you. In our tests
we found this four-bladed, pendant-mounted fan to be the
most powerful DC ceiling fan we've ever seen. Draws a
maximum of 1.5 amps at 24 volt (36 watt). Includes
balancing kit. Pendant mounting allows use on either flat or
sloped ceilings. Blades painted black (not as pictured)

64-248	24V Ceiling Fan, 42"	$199⁰⁰
24-102	Variable Speed sw.-reversing	$31⁰⁰

12-Volt Ceiling Fan

We have researched the low-voltage ceiling-fan market over
the years and carried several different brands. This ceiling fan
is still the best. It can be operated on 12 volts only and draws
1/2 to 3/4 of an amp. It has three 20-inch wooden blades
and a 10 x 8-1/2-inch base which fastens to the ceiling with
regular butterfly fasteners, or may be screwed into wood
beams. Pendants are not available.

64-231	12V Ceiling Fan, Plastic Base	$195⁰⁰

Thermofor Ventilator Opener

The Thermofor is a compact, heat-powered device which
regulates window, skylight, or greenhouse ventilation
according to temperature using no electricity and requiring
no wiring. You can set the temperature at which the window
starts to open between 55° and 85° F. It will open any
hinged window up to 15 pounds a full 15 inches, and can be
fitted in multiples on long or heavy vents. These units are
ideal for greenhouses, animal houses, solar collectors, cold
frames, and skylights.

64-302	Thermofor Solar Vent Opener	$59⁰⁰

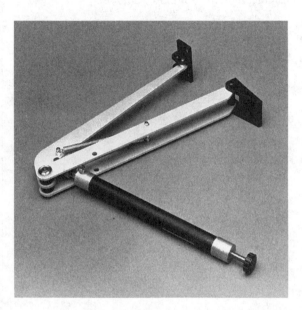

PROPANE SPACE HEATING

A super-insulated house can have remarkably small heating demands. Often, a high-efficiency propane space heater is all that is required for a home of up to 2000 square feet. All the models we offer are approved for mobile homes or trailers. The Ecotherm models draw combustion air from outside, and their direct venting requires no chimney installation. The Platinum Cat models allow great design flexibility, because the exhaust is actively vented by a small fan through plastic pipe, which can be run to any convenient exit. Carefully determine the size of your space to be heated and compare this with the heater specifications.

— Doug Pratt

Eco-Therm Direct Vent Heaters

Direct venting makes the Eco-Therm the ideal gas heater for bedrooms, bathrooms, and all other closed rooms, as well as super airtight homes. It draws in outside air for combustion and exhausts it through the same wall penatration (pipe within a pipe). It doesn't deplete oxygen in the room and doesn't waste heated air.

The sleek, slim Eco-Therm installs easily almost anywhere on an outside wall a minimum of 6 inches thick and a maximum of 12 inches thick. It should not be located under windows which can be opened. Just cut a 4-3/4-inch hole in the wall, then trim the direct vent tube to the thickness of the wall. Hang the heater on the wall and bring a 1/2-inch gas line to the valve. Since it heats so well without fans or blowers, the Eco-Therm requires no electricity. This unit is thermostat-controlled.

The special design insures that the front panel stays at a safe temperature. All components are of the highest quality. The steel heat exchanger is enamelled inside and out. The vent system is aluminum for corrosion resistance. Its strong outside terminal needs no additional protective grid, and the offset dual throat makes it waterproof. Eco-Therm heaters at altitudes above 5000 feet must have the fuel orifice replaced. This is a no-cost conversion at the factory if you let us know at the time of order placement.

Please specify either propane or natural gas. Warranty is 1 year, with a 5 year warranty on the heat exchanger. Minimum clearances: floor 4-3/4"; side wall 4"; rear 0"; ceiling 36".

Eco-Therm MV-130

> Output: 10,690 Btu/hr
> Dimensions: 27-1/8" x 24-3/8" x 6-5/8"
> Net Weight: 46 lbs.
> Shipping Weight: 55 lbs.
> Warranty: 1 year parts, 5 year heat exchanger

65-203   Eco-Therm Heater, MV-130 $439⁰⁰

Eco-Therm MV-120

> Output: 6400 Btu/hr
> Dimensions: 21-1/8" x 24-3/8" x 6-5/8"
> Net Weight: 34 lbs.
> Shipping Weight: 42 lbs.
> Warranty: 1 year parts, 5 years heat exchanger

65-204   Eco-Therm Heater, MV-120 $399⁰⁰

Eco-Therm MV-112

> Output: 4200 Btu/hr
> Dimensions: 15-3/8" x 24-3/8" x 6-5/8"
> Net Weight: 26 lbs.
> Shipping Weight: 33 lbs.
> Warranty: 1 year parts, 5 year heat exchanger

65-205   Eco-Therm Heater MV-112 $379⁰⁰

Eco-Therm MV-130

CAT 6P-12A

A 6000 Btu per hour heater with 12-volt automatic ignition. Draws less than 0.5 watt on stand-by, 8 watts when running. Equipped with On/Off/Auto switch. Propane consumption is only 1/4 pound per hour. It will warm an area up to 175 square feet. Mounting can be either surface or 2-3/4" recessed. Two-year full warranty.

> Output: 6000 Btu/hr
> Dimensions: 21" x 12" x 5-3/8"
> Net Weight: 15 lbs.
> Shipping Weight: 20 lbs.

•65-201   Catalytic Heater, 12V Ignition $489⁰⁰

CAT 1500-XL

A 5200 Btu per hour propane catalytic heater designed for use in homes. It has 120-volt automatic ignition. It is vented, thermostatically controlled, and fully approved by the AGA. It will warm an area up to 175 square feet. Draws 13 watts on stand-by, 37 watts when running. Surface mounting. Two-year full warranty.

> Output: 5200 Btu/hr
> Dimensions: 28" x 12-3/8" x 5-3/8"
> Net Weight: 17 lbs.
> Shipping Weight: 22 lbs.

•65-202   Catalytic Heater, 110V Ignition $529⁰⁰

Build Your Own Solar Hot Air Collector

There's a free winter heat source just waiting to be taken. Do you have a more or less south-facing roof? Do you have cold sunny days in the winter? Then you could cut your winter heating costs by 50% to 80% depending on your roof area and climate. In this 52-minute video a homeowner in Massachusetts takes us carefully step by step through refurbishing his large hot air collector. All the details are shown, and a complete materials and resources list is included. This is a simple, but amazingly effective homemade design that can be adapted to any size, shape, or slope of sun-facing roof. You can retrofit almost any house for inexpensive solar heating, even when no consideration was given to solar gain initially. Care is taken to insure a maintenance-free installation for many, many years of ultra low-cost heating. Only a tiny fractional horsepower blower motor automatically controlled by a temperature switch uses power. Typically this fan draws 120 watts or less, a fraction of what a furnace fan uses. It's simple, it's homemade, it saves resources, Dr. Doug highly recommends it.

80-890   Solar Hot Air Video $29⁰⁰

RESIDENTIAL HEAT-LOSS CALCULATIONS

To help determine the appropriate size for a home space heating system, we present the following "quick and dirty" calculation method. You will need to know the square footage of exterior surfaces of the dwelling (list windows and exterior doors separately from walls), the R-value of these exterior surfaces, and a reasonable estimate of the difference in temperature between the outside and inside during the winter.

Surface	Area (sq. ft.)		U-Value*		Delta-T**		Heat Loss	
Outside walls	920	x	.053	x	60° F	=	2925	Btu/hr
Windows	160	x	.667	x	60° F	=	6400	Btu/hr
Exterior doors	30	x	.067	x	60° F	=	120	Btu/hr
Cold Ceiling	1600	x	.033	x	60° F	=	3168	Btu/hr
Cold Floor	1600	x	.033	x	55° F	=	2904	Btu/hr
Volume (H x W x L) 8x32x50		x	1.0 air changes/hr				=12,800	Btu/hr
Total heat loss:							=28,317	Btu/hr

*U-VALUE = 1 divided by R-value
**DELTA-T = Inside temperature minus outside temperature

Conclusion: Purchase a 30-35,000 Btu/hour space heating system.

FUEL FOR THE HEARTH

Ever wonder why it is called the "solar" system?

We begin with a piece of inert rock. Call it Earth. The rock, similar to many other pieces of debris throughout the universe, may repose in cold, dark silence for eternity. This piece of rock, however, is fortunate enough to have a strong, stable relationship with the sun. The sunlight creates temperature differences which, under the right conditions, stir movement, enabling the chance combination of elements. Over a few million years the elements are reconfigured, so that sunlight can be converted into energy. Life happens. And life begets life and more life and more life until there is a pyramid of beings, plant and animal, interdependent upon each other.

Trace back the life forces, strands thin and intertwined, and each one extends a tiny umbilical cord to the sun.

The human animal's direct use of sunlight is vestigial. We cannot photosynthesize like green plants, so we consume vast amounts of stored sunlight in the form of food. Within our bodies the sunlight is combusted in a process remarkably similar to what occurs to a stick of maple in a woodstove. And the result of this conversion process is warmth and energy from within.

Externally, we are equally dependent on the sun. Our protection comes from the sun. Our clothing — whether derived from plants, animals, or fossil fuels — comes from the sun. Our control of our interior environment comes from the sun. We are, alas, a slash-and-burn species (not simply a slash-and-burn "society").

Whether we are living in the Stone Age in remote Philippine hills or in luxurious decadence in Southern California, we slash and burn. Of course, as we sip our crisp Chablis and watch the sun set over the Pacific, we are mercifully insulated from the consequences of our actions. With fossil fuels we have simply learned to slash and burn vertically rather than horizontally.

Follow the chain. To provide the energy that keeps us warm, that powers our computers, and produces the images that flicker on our TV screens like the embers of an ancient fire, we burn the stored energy from the sun. There is recently stored energy (wood), antique crystals (coal), and rich ambergris (oil and gas), the aged brandy of fossil fuels.

On an intellectual level we know that the earth's wine cellar is finite, but this knowledge does not appear to affect our consumption. Faced with the option of consuming or not consuming, of saving something for another day, we — the species — always choose to bring up another bottle from the cellar. Recently, though, some of the more "enlightened" members of our slash-and-burn species have begun to rediscover an elegance in consuming lower on the energy chain, closer to the sun.

Ah, but there is always the exception — nuclear power, our surrogate sun. We can simulate the sun's power, but we cannot control it. The more we learn, the more frightening are the consequences of our experiments. Here is a force powerful enough to mutate the building blocks of life, the components that evolved to live compatibly with sunlight. We say, let the scientists spend a few hundred years learning about the consequences of nuclear power before displacing millions of years of evolution.

We'll take the sun.

Modern science is only now "discovering" what all animals already know: that our well-being must have a daily dose of sunshine to maintain good physical and mental health. Even much of our spiritual sustenance comes from the sun. Our modern celebrations still commemorate natural evolutions of the solar year. Nary a civilization does not mark the winter and summer solstices with festivals of light (Christmas, the Fourth of July). Spring brings rebirth, and the recommencement of the photosynthetic process (Earth Day, May Day). Autumn celebrates the harvest (Thanksgiving), the completion of our solar collection.

Sunrises and sunsets provide inspiration to artists and beach walkers. At the moment of sunset, we instinctively become still. Here is a moment we can actually look at the sun, connect with it directly, and share the universe. Words are unnecessary. We are renewed.

All because of the sun. It all comes from sun.

At Real Goods we are sun worshippers, in the sense that we recognize our sustenance — literally and figuratively — comes from the sun. The sun, for our practical human purposes, is infinite. The stored solar energy that exists on our planet, however, is finite and must be used judiciously, lest we deplete our supply, fouling our nests in the process.

Real Goods sells "everything under the sun," which means solar modules, 12-volt Christmas tree lights, water-conserving showerheads, and books on constructing your own pond. Look deeply enough into the lineage of every product we sell, and you will see that each one, in its own way, comes from the sun. That's what makes something Real Goods.

Just as the sun is at the center of the planetary energy system, there is a center to the body's energy system — the heart. This organ pumps the sustaining life force to the extremities; without it, there is no life.

Once the simplicity of this principle has been grasped, the logic of eating lower on the food chain becomes elegantly understandable. Who needs middlemen? The lower we eat on the food chain, the closer we are to consuming sunlight.

Our homes, as well, have an energy system that is tied to a central energy source. This center is called the hearth. In the old days the hearth was the source of heat, the supplier of energy for cooking, the protection from wild animals, the place where the family gathered.

The concept of "hearth" has been all but obliterated by the age of cheap oil. Until the last half century, family life still revolved around that dreadnought of home energy, the cookstove, that marvelous appliance that performed most of the home's energy functions. With the abundance of oil and electricity, however, the separate functions of the hearth were sliced off and specialized in ways that were not conducive to the maintenance of strong family bonds.

The function of providing heat, for instance, was relegated to the basement, where a powerhouse of a furnace could climate-control every nook and cranny of a suburban split-level. Multizone heating capability compensated for abominations of design. If a heat distribution problem arose, it could be solved by simply adding more power.

The old cookstove could simmer, sizzle, bake, warm, and poach, all at the same time. The modern kitchen, by contrast, has segmented the cooking function into an infinite number of specialized functions, each with a dedicated appliance. To make coffee alone, we choose between instant in a microwaved cup of hot water, a pot run through Mr. Coffee, or a steaming cup from the espresso machine. Overkill, American-style.

The protective function of the hearth has been taken over by the built-in security system, but not very well. The more prevalent, and disturbing, trend is to achieve security by being better armed than your assailant.

As for the hearth's role in unifying the family, the need still exists. It has found a surprising path to fulfillment, as the function of achieving human connection has been surrendered to the television. We watch the mesmerizing blink of the lights as our ancestors did the flicker of the open flames. We grow attached to our electronic friends on the sitcoms as we once did to neighbors. Witness the outpouring of sentiment when a show like "Cheers" leaves the screen. If something important in the way we relate to other human beings has been lost in the process of "videoization," well, that's progress.

The hearth has been one of the casualties of the Oil Age, but it has experienced rebirth in the independent home. All proper homes have functioning hearths, from both an energy and an emotional perspective.

It seems entirely natural that Real Goods should offer a wide range of products that enhance the sense of hearth in a home. We have even set ourselves a corporate goal to expand our offerings in this area in the coming months. How best to utilize

energy for heating and cooking are two of the most important decisions a homeowner makes. We want to assist in the process by including hearth products in our definition of "Real Goods."

Critical to the choice of hearth products is understanding the fuel options, ranging from passive solar to anthracite coal, and their resulting impact on the environment. There are important decisions to be made on the type of heat generated — radiant, like the sun, or convective, like the air — and on the means of distribution. Will your home be centrally heated or space-heated? From an energy consumption perspective, these decisions will have the greatest overall impact on your energy "footprint."

Moreover, these decisions are one area where the grid-dweller and the off-the-gridder are on equal terms in terms of impact. The issues are complex, and the wide variety of appliances and fuels can make the decision a difficult one.

At the moment, our offerings are limited to items that enhance the operation, convenience, and safety of traditional hearths. In the coming months, however, it is our plan to add products and services that will enable energy-conscious consumers to achieve the same degree of control over this component of energy consumption as they do in their generation and use of electricity.

The fuels we burn in our hearths derive from the sun. If we desire energy independence, then, in many cases, our first priority will be the creation of a proper, if not perfect, hearth.

To Burn or Not to Burn

I'm totally confused.

The subject is wood, a material I have used to heat my home for the past 15 years. I know it's more work to burn wood, but I like the exercise. It makes more sense to burn calories splitting and hauling hardwood than running in front of a television set on a treadmill at the health spa. I enjoy the fresh air; I even enjoy bugging the kids to get them to help stack. Once they get past the whining about the disruption to their electronic entertainment, it is surprising how quickly they become human beings.

I like the warmth. I like the flicker and glow of embers. I like the freedom from the whir of an electric motor that blows hot air (complete with all manner of dust, mites, and hairballs) from a hole in the floor.

Sometimes, when I am stacking wood so that a summer's worth of sunshine can make it a cleaner, hotter fuel, I think of those supertankers carrying immense cargoes of black gold from the ancient forests of the Middle East. I think of the wells that Saddam Hussein set afire in Kuwait, and the noxious roar of the jet engines of the warplanes we used to ensure our country's access to cheap oil.

My wood comes from hills that I can see. It's delivered by a guy named Paul, who cuts it with his chainsaw, then loads it in his one-ton pickup. We talk about the weather, the conditions in the woods, and the burning characteristics of different species. We finish by bantering about whether he's delivered a large cord, a medium, or a small. I can't imagine having the same conversation with the man who delivers oil or propane.

Of the different ways that I conduct my life to Power Down, burning wood has the greatest impact. Unfortunately (and this is where I begin to get confused), there is another side to the wood-burning issue. Even the process of taking a step back to supposedly simpler times does not necessarily solve problems or provide the right answers. The many issues involved in something as simple as burning a stick of wood can touch on so many diverse and complex issues that coming to a clear, uncomplicated answer can prove next to impossible.

The woodburner, perhaps more than anyone, appreciates what powerfully concentrated stuff oil is. I know the impact of just one summer of sunlight on my woodpile. Think of the immense quantity of natural energy, then, that would be needed to make such a powerful fuel as oil. It literally is, at least to my simplistic mind, Black Magic.

The National Resources Defense Council (NRDC) frequently assists Real Goods in the development of statistical data. It was the NRDC, for instance, who provided the basic data that lets Real Goods set its goal of preventing a billion pounds of CO_2 from being added to the atmosphere by the year 2000.

It was the NRDC, who, in conjunction with other environmental activists back in the mid-1980s, effectively put the brakes on the growth of woodburning by suing the EPA. This led to government regulations, and the total redesign of woodburning products, a process that was completed (at least for woodstoves) in 1990.

So now I burn my wood in a "clean" stove. Theoretically, I should be able to go back to feeling good about wood, but life is never that simple. Along comes evidence that the emissions from airtight stoves contain insidious carcinogens, and further indications that a home's internal environment can be more adversely affected by burning wood than by passive cigarette smoke. Moreover, my environmentally active friends say that the only good smoke is no smoke, and that even my supposedly high-tech stove is smogging up their skies.

On the positive side of the ledger comes the information that woodburning, in combination with responsible reforestation, actually helps the environment by reversing the greenhouse effect. The oxidation of biomass, whether on the forest floor or in your stove, releases the same amount of carbon into the atmosphere. It makes more sense for wood to be heating my home than contributing to the black skies over Yellowstone.

This is obviously a simplistic analysis, but wood should be a simple subject. Unfortunately, I can offer no solutions — simple or otherwise — within the compass of this discussion, because I haven't fully decided where *I* stand. I'm trying to keep an open mind, and do the right thing. Powering-down is not always as easy in practice as in theory.

In the meantime, I have to keep warm, so I'm choosing to burn wood. In my never-ending quest to power down, I keep coming back to the fact that humans have been burning oil for more than 50 years and wood for five million. Most of our problems, from the planetary perspective, have come during the Age of Oil.

— *Stephen Morris*

Real Goods Hearth Center
Amherst, Wisconsin

In the original Real Goods retail store, there were no solar modules. There were however, hearth products of every style and shape. There were also the accessories and the authoritative advice that enabled a homeowner to address every aspect of their wood-heating challenge. Even in a climate as moderate as northern California's more energy is used in keeping warm than any other household activity.

As Real Goods evolved into a mail order provider, the emphasis on hearth products diminished in favor of developing our expertise in solar electric systems. Hearth products were heavy and difficult to ship. Their installation was often dependent on meeting the needs of local building inspectors, making this a product category more suitable for the full-service local retailer.

The problem for the local retailer, however, is that they do not usually have the room or financial power to stock and display the range of products that the innovative homeowner needs in order to select just the right hearth option. Ironically, this has set the stage for Real Goods to re-surface as the premier provider of hearth products and information.

When Real Goods merged with the former Snow-belt Energy Center in Amherst, Wisconsin, we gained an experienced, knowledgeable staff of home heating experts, with a store full of woodstoves, gas fireplaces, hearth accessories, and solar thermal products. We're pleased to offer a full array of hearth products, and the knowledge of our staff.

Just a few of the brands offered by Real Goods Hearth Center are:

- Vermont Castings
- Ameritech Chimney systems
- Dutchwest
- Heartland
- Energy King
- Stanley
- RSF Energy
- Tulikivi
- Lopi
- Envirotech
- FPX

Call Real Goods Hearth Center at 715-824-5020, or visit them at 286 Wilson St., Amherst, WI, 54406.

WOODBURNING STOVES

The Heartland Cookstove

At the turn of the century, before the advent of central heating, the kitchen was the energy center of the house. The pinnacle of design was achieved with the wood cookstove, a powerhouse of form and function that was the family's source of warmth, food, and even hot water. The Oval Cookstove was originally designed in 1906 and has been a focal point of energy independent homes ever since.

The Oval of today (and its smaller sister the Sweetheart, both produced by the Heartland Company of Canada) features the same stunning appearance and multi-functional versatility its ancestor, but its interior has been steadily improved over the years, to make wood-fired cooking easier than it was in great-grandmother's day.

The Oval can produce up to 55,000 Btu/hr., enough to heat a maximum of 1800 square feet. Its airtight firebox can hold a generous 35-pound load of wood, enough to burn through the night. Its proficiency as a heater is matched by its cooking versatility. The oven can handle a 25-pound turkey, while the six-square-foot cooktop, with six separate temperature zones, gives you flexibility to handle all the needs of an elaborate holiday dinner. There's even a warming cabinet on top for desserts. Water can be heated in a traditional copper-lined reservoir with spigot or with an optional stainless steel water jacket inside the firebox to be plumbed into your hot-water storage. Both the Oval and the Sweetheart have durable, firebrick-lined fireboxes, capable of full-time operation with wood or coal. They are solidly constructed of cast iron, with porcelain finish and easy-to-clean nickel fixtures. Thousands of these classic stoves have been sold worldwide.

Stoves are FOB Ontario, Canada. Call for freight quote; shipping charges will be added to your invoice. Freight is quite reasonable since the manufacturer has a 60% discount with the freight company.

	Sweetheart	Oval
Heating Capacity (square feet)	800-1500	12,001,800
Heat Output (Btu/hr.)	12,000-35,000	18,000-55,000
Firebox Capacity (Hardwood)	22 lbs.	35 lbs.
Wood Length	16"	16"
Stove-Top Dimensions	22" x 32"	24-1/4" x 34"
Cooking Elements or Lids	6	6
Flue Size	6"	6"
Weight (without reservoir)	385 lbs.	500 lbs.

Standard Features: Adjustable summer cooking firebox
Porcelain-enamel finish
Nickel trim
Ash pan on sliding track
Warming cabinet
Certification: Warnock Hersey Laboratories
Optional Extras: Coal grate
Water Reservoir (5 gal./Sweetheart, 7 gal./Oval)
Water jacket

Item #	Description (Shipping Weight)	
Oval		
•61-110	Oval Stove & Reservoir, Almond (550)	$3600
•61-111	Oval Stove & Reservoir, White (550)	$3600
•61-112	Oval Stove, Almond (500)	$3295
•61-113	Oval Stove, White (500)	$3295
Oval Accessories		
•61-118	Oval Stove Heatshield (48)	$175⁰⁰
•61-119	Oval Stove Stainless Water Jacket (15)	$199⁰⁰
•61-120	Oval Stove Coal Grate Package (56)	$169⁰⁰
Sweetheart		
•61-114	Sweetheart Stove & Reservoir, Almond (415)	$2895
•61-115	Sweetheart Stove & Reservoir, White (415)	$2895
•61-116	Sweetheart Stove, Almond (385)	$2595
•61-117	Sweetheart Stove, White (385)	$2595
Sweetheart Accessories		
•61-121	Sweetheart Heatshield (48)	$175⁰⁰
•61-122	Sweetheart Stainless Water Jacket (18)	$199⁰⁰
•61-123	Sweetheart Coal Grate (22)	$169⁰⁰

Freight additional, call for quote.

Heartland also offers these high-quality stoves in gas and electric versions. Call for more information.

The Harrowsmith Country Life Guide To Wood Heat

Dirk Thomas. Wood continues to be a popular fuel in North America, and this book provides a sensible perspective to the practical, safety, economic, and environmental questions that attend woodburning. Topics include the economics of burning wood, environmental impact, buying, storing, and seasoning firewood, how to select and maintain a chain saw, cutting and splitting wood, skidding wood out and bucking it up, woodlot management, and cleaning the chimney. Whether you are a weekend woodburner or rely on wood as a major fuel source, his book will be a valuable resource for years to come. 176 pages, paperback, 1992.

82-167   Guide to Wood Heat $15⁰⁰

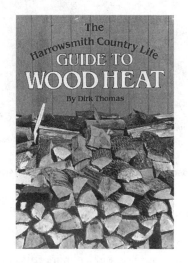

Hudson Bay Camping Axe

Snow & Nealley of Bangor, Maine, have been renowned for their fine axes since 1864. The heads are hand-forged and polished for a lifetime of use. All wood used for handles is chosen, dried, and turned with exceptional care. The Hudson Bay Camping Axe is a highly versatile tool. At 2-1/4 pounds, with a 24-inch handle and slim profile, it's easily taken on a trip or used around your property. Though lightweight enough for chopping kindling, it has the heft and balance to allow felling small trees with minimal effort. One-year manufacturer's guarantee.

63-123   Camping Axe (shown on next page) $29⁰⁰

Mini Maul

Snow & Nealley's superior 3-pound mini maul. The 18-inch handle is long enough for good mechanical advantage when splitting small logs for the stove, yet short and handy enough to allow chopping them into kindling-sized pieces. The flat end of the maul can also drive wedges, pound stakes, or serve as a heavy, sturdy hammer. One-year manufacturer's guarantee.

63-122   Mini Maul $25⁰⁰

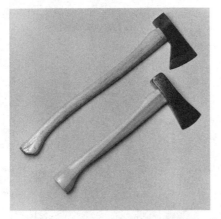

Hudson Bay Camping Axe and Mini Maul

Log Carrier

This professional-strength carrier makes the process of moving logs into your home much more efficient and keeps the dirt and bark off your floor. The Jumbo Log Carrier is made from a double layer of biodegradable jute, the same rugged fabric used for sandbags by the U.S. military. This versatile log hauler is big, tough, and reusable. It also works as a handy, general-purpose carryall. It will also quickly get you where you want to be: sitting by a warm, cozy fire on a cold winter's night. 24" x 12" x 12".

63-465   Log Carrier $19⁰⁰

Dragon Humidifier

Fill this or brass medieval sculpture with water and set it on top of your woodstove. As the stove heats up, the dragon breathes steam out its nostrils creating not only a dazzling sensation but humidifying the air as well.

63-417   Brass Dragon $179⁰⁰

Kettle Humidifier

This traditional cast-iron humidifier looks great on any woodstove and provides a moistening steam to mitigate the dry wood heat. It holds 3 quarts of water and has a brass handle.

63-436   Kettle Humidifier $29⁰⁰

Rugged And Economical Firewood Storage Rack

Properly stored wood is easier to handle and faster starting. This convenient storage rack keeps your firewood off the ground, dry, neat, and tidy. It allows proper air circulation to prevent dampness, rot, and insect infestations. Portable unit is constructed of top quality, heavy gauge, 1-1/2" tubular steel. Durable, baked-on, glossy black finish. It disassembles easily for storage during off-season. Assemble with just a wrench. Made in USA. Eliminates the possibility of woodpile avalanches — especially important to parents of small children!

54-524 Large Firewood Rack (90" x 48" x 14 1/2") $59⁰⁰
54-525 Small Firewood Rack (42" x 48" x 14 1/2") $49⁰⁰

The Snorkel Hot-Tub Heater

The Snorkel Stove is a woodburning hot-tub heater that brings the soothing, therapeutic benefits of the hot-tub experience into the price range of the average individual. Besides being inexpensive, simple to install, and easy to use, it is extremely efficient and heats water quite rapidly. The stove may be used with or without conventional pumps, filters, and chemicals. The average tub with a 450-gallon capacity heats up at the rate of 30° F or more per hour. Once the tub reaches the 100° range, a small fire will maintain a steaming, luxurious hot bath.

Stoves are made of heavy-duty, marine-grade plate aluminum. This material is very light, corrosion resistant, and strong. Aluminum is also a great conductor of heat, three times faster than steel. The Snorkel hot-tub stove will heat up a hot tub 50% faster than an 85,000 Btu/hr. gas heater and three times as fast as a 12-kilowatt electric heater!

We also offer two cedar hot tubs from Snorkel. They are made of the long-lasting Western Red Cedar.

•45-401   Snorkel Stove $595⁰⁰
•45-207   Cedar Hot Tub (6' x 3') $1,295
•45-208   Cedar Hot Tub (7' x 3') $1,595

Shipped freight collect from Washington state.

CHAPTER 7

WATER

WATER. It covers three-quarters of our planet. It makes up 75% of our bodies, too. The food we eat is mostly water. The oceans are wider than any continent and deeper than the height of the tallest mountain. Streams trickle incessantly all over the world, little capillaries carrying their precious liquid from one place to another. Water evaporates, rises into the heavens, only to come down and join us again and play its part in the symphony of life.

Our survival is completely dependent on a good supply of fresh, clean water; we use and interact with water constantly. At home, we use it for cleaning our bodies and our possessions, for cooking and eating. At work, we use it to make products, wash things, conduct experiments, and much more. For recreation we go swimming, boating, and fishing. Water has saved countless lives in many ways, from sterilizing surgical instruments to putting out fires. Our interaction with good old H_2O covers the whole gamut of our experiences here on earth, and makes life possible.

The great irony about water is that, even in its incredible abundance, it is still the most precious and essential thing to us. We find ourselves surrounded by water almost everywhere we go, but somehow there is never enough.

This is due mainly to our mismanagement and abuse of this vital resource. Examples of water misuse fill our society. Each time we flush a toilet, we pour an average of five gallons of good drinking water into the sewage system. A 15-minute shower can use up to 75 gallons of water that has been treated and processed for drinking. I'll never forget the time I was riding my bike past a Mervyn's department store during a huge downpour, and they had their outdoor sprinkler system on. I went inside and told them. They flustered for a while, and finally told me that it was on a timer and they didn't know how to turn it off!

Not only do we make mistakes in the amount of water that we use, but we also have a habit of contaminating the precious water supplies we have, pouring sewage and waste into rivers, lakes, and oceans.

It is clear that, if we are to make our planet sustainable, we must examine the way we use water and propose and implement changes in our water use that can ensure a wet world, abundant with clean-flowing sources of water, for generations to come.

The first step on the path to water sustainability is conservation. Our homes and workplaces are filled with opportunities to use less water and still get the job done. By using drip irrigation and water timers, we can grow the same vegetables with a fraction of the water we used to use. Low-flow showerheads and faucet aerators allow us to get our bodies and dishes just as clean as before, while using much less water.

A promising new development in water conservation is the use of graywater (water recycled for other uses from the bath, shower, or bathroom sinks), which has just been approved for the entire state of California.

Also, since water is the most vital nutrient to our health and survival, it is essential that we take a look at the purity of the water we drink. The quality of city water supplies is questionable at best, and at worst downright toxic. It's up to each of us to make sure that the water we drink is clean and pure.

Water use is very closely tied to energy use. The more water we use for showering, the more energy we need to pump and heat the greater volume of water. If we obtain our water from a well, every gallon we use requires a certain amount of energy for pumping; if from a municipal water supply, who knows how much energy is used for pumping, and what chemicals for treatment? In this light, conserving water is one of the best ways to conserve energy!

In this edition of the *Sourcebook,* we have decided to honor water and give it a complete chapter of its own. Instead of breaking up all the uses and interactions of water and scattering them all over the book, we have grouped them together. Whether you are looking for information on the purification, heating, storage, pumping, or conservation of water, you will find it here in this chapter.

— *Gary Beckwith*

WATER BOOKS

Planning For An Individual Water System

American Association of Vocational Instructional Materials (AAVIM). The definitive book on water systems and one of our all-time favorites. It was a surprise and complete delight when one of our customers pointed it out to us. Incredible graphics and charts. Presents a thorough discourse on all forms of water pumps, and discusses water purity, hardness, chemicals, water pressure, pipe sizing, windmills, freeze protection, and fire protection. If you have a new piece of land or are thinking of developing a water source, you need this book! 160 pages, paperback, 1982.

80-201	Planning for an Individual Water System	$19⁰⁰

Build Your Own Water Tank

Donnie Schatzberg. An informative booklet, that has recently been updated and greatly expanded with more text, drawings, and illustrations. It gives you all the details you need to build your own ferro-cement (iron-reinforced cement) water storage tank. No special tools or skills are required. The information given in this book is accurate and easy to follow, with no loose ends. The author has considerable experience building these tanks, and has gotten all the "bugs" out. 58 pages, paperback, 1995.

80-204	Build Your Own Water Tank	$12⁰⁰

Earth Ponds

Tim Matson. A thorough and attractive treatment of a small subject. The first edition of this book appeared a decade ago, and was instrumental in the sculpting of thousands of ponds around the world. This new edition is much larger and more professionally illustrated. But most important, it has grown through the author's experience with a wealth of anecdotes and research back to the seventeenth century, and attentive love for his subject. This book is utterly indispensable for all pond dreamers, builders, and maintainers everywhere. 150 pages, paperback, 1991.

80-156	Earth Ponds (Book)	$18⁰⁰
80-179	Earth Ponds (Video)	$30⁰⁰

Waterhole: A Guide To Digging Your Own Well

Bob Mellin. This small but thorough book deals with all aspects of one of the simplest of tasks — digging a hole in the ground: site selection, where not to dig, how to dig, how to keep the hole uncontaminated, digging and pumping equipment. This book treats only one type of well, the small-bore auger-dug well, with humor and balance. Offered in the tradition of great Real Goods do-it-yourself books. 72 pages, paperback, 1991.

80-614	Waterhole	$9⁰⁰

Rainwater Collection Video

There is an untapped, inexpensive water source on your own property: rainwater. *Rainwater Collection: Texas Style* tells the story of three Texan hill country families who collect rainwater for all their household needs, including drinking. The half-hour video details how low-quality well water drove a physician and his wife to devise a successful rainwater collection method that inspired others in the area to follow suit. John Dromgoole, from the PBS series *The New Garden,* hosts the video. With it comes a 50-page booklet that elaborates on the systems illustrated in the video and gives information on equipment suppliers and prices. 30 minutes.

80-132   Rainwater Video $30⁰⁰

The Home Water Supply

Stu Campbell. Explains in depth and completeness how to find, filter, store, and conserve one of our most precious commodities. A bit of history; finding the source; the mysteries of ponds, wells, and pumps; filtration and purification; and a good short discussion of the arcana of plumbing. After reading this book, you may still need professional help with pipes and pumps, but you will know what you're talking about. 236 pages, paperback, 1983.

80-205   The Home Water Supply $19⁰⁰

Country Plumbing: Living With A Septic System

Gerry Hartigan. This book is a practical guide that will prepare you for (and maybe help prevent) one of life's great crises — what to do when the drains don't work. Gerry Hartigan clearly and with a persistent sense of humor strips away the shrouds of mystery, superstition, and misinformation that surround septic tanks and the attached home sewage-disposal systems. The text covers the different types of subsurface disposal systems, new installations, and commercial tanks, and describes the problems that can make these systems stop functioning. An entire chapter is devoted to "free advice" on how to prevent problems. 80 pages, paperback, 1984.

82-153   Country Plumbing $10⁰⁰

Cottage Water Systems

An out-of-the-city guide to water sources, pumps, plumbing, water purification, and waste disposal, this lavishly illustrated book covers just about everything concerning water in the country. Each of the 12 chapters tackles a specific subject such as sources, pumps, plumbing how-to, water quality, treatment devices, septic systems, outhouses, alternative toilets, graywater, freeze protection, and a good bibliography for more info. This is the best illustrated, easiest to read, most complete guide to waterworks we've seen yet. 1993, 150 pages, softcover.

80-098   Cottage Water Systems $25⁰⁰

Build Your Own Ram Pump

Utilizing the simple physical laws of inertia, the hydraulic ram can pump water to a higher point using just the energy of falling water.

The operation sequence is detailed with easy-to-understand drawings. Drive pipe calculations, use of a supply cistern, multiple supply pipes, and much more are explained in a clear, concise manner.

The second half of the booklet is devoted to detailed plans and drawings for building your own 1" or 2" ram pump. Constructed out of commonly available cast iron and brass plumbing fittings, the finished ram pump will provide years of low-maintenance water pumping for a total cost of $50 to $75. No tapping, drilling, welding, special tools, or materials are needed. This pump design requires a minimum flow of three to four gallons per minute, and three to five feet of fall. It is capable of lifting as much as 200' with sufficient volume and fall into the pump.

The final section of the booklet contains a set-up and operation manual for the ram pump. Paperback, 1990, 25 pages.

80-501   *All About Hydraulic Ram Pumps* $8⁰⁰

WATER STORAGE

Kolaps-a-Tank

These handy and durable nylon tanks fold into a small package or expand into a very large storage tank. They are approved for drinking water and withstand temperatures to 140° F, and fit into the beds of several truck sizes. They will hold up under very rugged conditions, are self-supported, and can be tied down with D-rings. Our most popular size is the 525-gallon model which fits into a full sized long-bed (5 x 8 ft.) pickup truck.

47-401	73 gal. Kolaps-a-Tank, 40" x 50" x 12"	$335⁰⁰
47-402	275 gal. Kolaps-a-Tank, 80" x 73" x 16"	$445⁰⁰
47-403	525 gal. Kolaps-a-Tank, 65" x 98" x 18"	$525⁰⁰
47-404	800 gal. Kolaps-a-Tank, 6' x 10' x 2'	$795⁰⁰
47-405	1140 gal. Kolaps-a-Tank, 7' x 12' x 2'	$895⁰⁰
47-406	1340 gal. Kolaps-a-Tank, 7' x 14' x 2'	$940⁰⁰

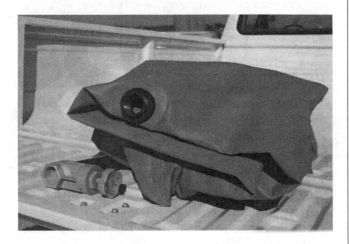

Pond Liners

A pond is a major investment, requiring lots of earth work, soil blending, and compaction. It's a gamble if it will hold water or not, or for how long. With an Everliner, earth work may be kept to a minimum, and your pond will not have a leak. You can establish a pond on any kind of soil without fear of having the water percolate away. This is the highest quality liner on the market, made of very tough 20 mil UV-stabilized polyethylene. Factory seams are as strong as the material itself. It resists punctures, tears, root penetration, and rodents. When covered, it will last indefinitely. It weighs only 80 lbs. per 1,000 square feet, so it's easily handled and installed. Nontoxic, chemically inert, and non-leaching, it is FDA/USDA approved for potable water supplies.

Pond liners are made to order so please have dimensions ready when ordering. *100 sq. ft. minimum order.* Free freight.

47-203	Everliner 100-200 sq. ft.	.53/sq. ft.
47-205	Everliner 201-500 sq. ft.	.46/sq. ft.
47-206	Everliner 501-2000 sq. ft.	.43/sq. ft.
47-207	Everliner 2001-10,000 sq. ft.	.40/sq. ft.
47-208	Everliner 10,001-20,000 sq. ft.	.36/sq. ft.
47-209	Everliner 20,001-30,000 sq. ft.	.33/sq. ft.
47-210	Everliner 30,001-40,000 sq. ft.	.30/sq. ft.
47-204	Everliner Installation Kit	$12⁰⁰
47-212	Everliner Fab Tape 2" x 30'	$20⁰⁰

Renewable Energy Powered Water Pumping
or How to Get Water Up the Hill When You're Far Away From Utility Power

Water is the single most important ingredient in any homestead. Without a dependable water source, we can't call anyplace home for very long. Real Goods offers a wide variety of water pumps that are specifically made for solar-, battery-, wind-, or water-powered pumping. These pumps are designed for long hours of dependable duty in out of the way places where the utility lines don't reach. For instance, ranchers are finding that small solar-powered submersible pumps are far cheaper and more dependable for moderate lift applications than the old wind pumpers we're used to seeing dotting plains. Many State and National Parks use our renewable-energy powered pumps for their backcountry campgrounds. The great majority of pumps we sell are solar- or battery-powered electric models, so we'll cover them first.

Solar-powered pumps tend to be far more efficient than their conventional AC-powered cousins. The initial cost of solar-generated electricity is high. By wringing every watt of energy for all it's worth we keep the start-up costs reasonable. Because of this high initial cost, most solar pumping equipment is scaled toward more modest residential needs, rather than larger commercial or industrial needs.

The Three Components of Every Water System

Every rural water system has three easily identified basic components.

A Source. Which can be a well, a spring, a pond, creek, or surface water.

A Storage Area. Which is sometimes the same as the source, and sometimes is an elevated or pressurized storage tank.

A Delivery System. Which used to be as simple as the bucket at the end of the rope. But given a choice, most of us would prefer to have our water arrive under pressure from a faucet. Hauling water, because it's so heavy, and we use such surprising amounts of it, gets old fast.

The first two components, source and storage, you need to produce locally, however, we can offer a few pointers based on experience and some of the better, and worse, stories we've heard.

Sources:

Wells

The single most common domestic water source is the well, which can be hand-dug, driven, or drilled. Wells are less prone to pick up surface contamination from animals, etc. Hand-dug wells are usually about 3 feet or larger in diameter and rarely more than 40 feet deep. The most common sizes for driven or drilled wells is 4" or 6". Beware of the "do-it-yourself" well drilling rigs which can rarely insert pipe larger than 2". This restricts your pump choices to only the least efficient jet-type pumps. A 4" casing is the smallest into which a submersible pump can be inserted.

Springs

Many folks with country property are lucky enough to have surface springs which can be developed. Springs need to be fenced to keep the local wildlife out, and are often "developed" with either a backhoe and 2' or 3' concrete pipe sections sunk into the ground, or by one of the newer, lightweight horizontal drilling rigs. This helps to ensure that the water supply won't be contaminated by surface runoff or animals. Before developing a spring you should consider that it is probably supporting a large and diverse ecosystem. If there are no other springs nearby, try to insure that some runoff will still continue after development.

Ponds, Streams and Other Surface Water

Ponds are one of the least expensive and most pleasing methods of supplying water. They are often used to hold winter and spring runoff for summertime use. Some water systems are as simple as tossing the pump intake into the lake or stream. Surface water is usually used for livestock or agricultural needs. Don't drink or cook with it unless it's treated first (see the *Sourcebook* section on Water Purification for more help on this topic).

Storage:

The purpose of water storage may be any or all of the following: to get through long dry periods, to provide pressure through elevated storage or pressurized air, to keep drinking water clean and uncontaminated, or to prevent freezing.

Surface Storage

Those with ponds, lakes, streams, or springs may not need any additional storage. However, many systems will need to pump water to a freeze-proof or elevated storage site in order to develop pressure.

Liners

Custom made poly-vinyl liners are available for ponds or leaking tanks. Real Goods carries a brand of very heavy-duty, custom sized liners. These ingenious liners are designed for installation during pond construction and are then buried with six inches of dirt, practically guaranteeing a leak-free pond even when working with sandy or other problem soils. When buried the life expectancy of these liners is 50 years +. These make reliable pond construction possible in locations that normally could not accommodate such inexpensive water storage methods.

Tanks

Covered tanks of one sort or another are the most common and long lasting storage solution. The cover must be screened and tight enough to keep critters like mice and squirrels from drowning in your drinking water (always an exciting discovery). The most common materials are polypropelene, fiberglass, and concrete.

Plastic Tanks

Both fiberglass and polypropelene tanks are commonly available. Both types will suffer slightly from UV degradation in sunlight. Simply painting the tank will stop the UV degradation and will probably make the tank more aesthetically pleasing. With fiberglass make sure the tank is internally coated with an FDA approved material for drinking water. Plastic tanks can usually be found at your local farm supply store. (Shipping costs make out-of-area sales impractical.) Some plastic tanks can be partially or completely buried for freeze protection. Ask before you buy if this is a consideration for you.

Concrete Tanks

One of the best storage solutions, but expensive initially. Concrete tanks are built on site, and can be concrete block, ferro-cement, or monolithic block pours. Monolithic pours require hiring a contractor with specialized forming equipment, though these are usually the most trouble free in the long run. Any concrete tank will need to be coated internally with a special sealer to be watertight. Concrete tanks can be buried for freeze protection, and to keep the water cool in hot climates.

A Mercifully Brief Glossary of Pump Jargon

Flow: *The measure of a pump's capacity to move liquid volume. Given in gallons per hour (gph), gallons per minute (gpm), or for you worldly types who have escaped the shackles of archaic measurement, liters per minute (lpm).*

Foot Valve: *A check valve (one-way valve) with a strainer. Installed at the end of the pump intake line, it prevents loss of prime, and large debris from entering the pump.*

Friction Loss: *The resistance of water to move through a pipe. As flow rates rise and pipe diameter decreases, friction loss can result in significant flow and head loss. See charts at section end.*

Head: *The pressure or effective height a pump is capable of raising water, or the height it actually is raising the water in a particular installation.*

Lift: *Same as* Head. *A measure of pressure, or how high the pump is raising the water. Contrary to the way this term sounds, pumps do not* suck *water, they push it.*

Prime: *A charge of water that fills the pump and intake line, allowing pumping action to start. Centrifugal pumps will not self prime. Positive displacement pumps will usually self-prime if they have a free discharge (no pressure on the output).*

Pressure Tanks

These are used in pumped pressurized systems to store pressurized water so that the pump doesn't have to start for every glass of water. They work by squeezing a captive volume of air, since water doesn't compress. The newer, better types use a diaphragm (sort of a big heavy-duty balloon) so that the water can't absorb the air charge. This was a problem with the older plain pressure tanks. Pressure tanks are rated by their total volume. Draw down volume, the amount of water that can actually be loaded into and withdrawn from the tank under ideal conditions, is typically about 40% of total volume.

Delivery Systems

Delivery systems are what we're going to cover in this section. A few lucky folks will be able to collect and store their water high enough above the level of intended use, so that the delivery system will simply be a pipe, and water's weight will supply the pressure free of charge. Most of us, though, are going to need a pump, or maybe even two, either to get the water up from underground, or to provide pressure. We'll cover electrically-driven solar and battery-powered pumping first, then water-powered and wind-powered pumps.

The standard utility-powered water delivery method is to have the submersible water pump in the well deliver into a pressure tank system. A pressure tank extends the time between pump run cycles by saving up some pressurized water for delivery later. This system usually solves any freezing problems by placing the pump deep inside the well, and the pressure tank indoors. The down side is that the pump must produce enough volume to keep up with any potential demand, or the pressure tank will be depleted and pressure will drop dramatically. This requires a 1/3 hp pump minimally, and usually 1/2 hp or larger. Well drillers will usually sell a larger than necessary pump because it increases their profit and guarantees that no matter how many sprinkler systems you add in the future, the water will be there. This sort of system is fine when you have large amounts of utility power available to meet heavy surge loads, but is very costly to power with a renewable energy system because of the large equipment requirements. We try to work smarter, smaller, and use less expensive resources to get the job done.

Solar-Powered Pumping
Where Efficiency is Everything

PV modules can be quite expensive, and water is surprisingly heavy. These two facts dominate the solar pumping industry. At 8.3 pounds per gallon it can require a lot of energy to move water uphill. Anything we can do to wring a little more work out of every last watt of energy is going to make the system less expensive to set up. Because of these harsh economic realities the solar pumping industry tends to use the most efficient pumps possible. For many applications that means a *positive displacement* pump. In this class of pump there is no possibility of the water slipping from high pressure areas to lower pressure areas inside the pump. Positive displacement pumps also ensure that even when running very slowly, such as under partial light conditions, water will still be pumped. There are some disadvantages to using positive displacement pumps. They tend to be noisier, as the water is expelled in rapid, discrete segments. They usually pump smaller volumes of water, they must start under full load, periodic maintenance is required, and some types won't tolerate running dry. These are reasons we don't find this class of pumps used extensively in the AC-powered pumping industry.

Most common AC-powered pumps are *centrifugal* types. This class of pump is preferred because of easy starting, low noise, smooth output, and minimal maintenance requirements. Centrifugal pumps are good for moving large volumes of water at relatively low pressures. As pressure rises however, the water inside the centrifugal pump "slips" increasingly, until they finally reach a pressure at which no water is actually leaving the pump. This is 0% efficiency. In the solar industry, we use centrifugal pumps for pool pumping, and some circulation duties in hot water systems. But in all applications where pressure exceeds 20 psi, you'll find us recommending the slightly noisier, occasional maintenance-requiring, but vastly more efficient positive displacement type pumps. For instance, an AC submersible pump running at 7%-10% efficiency is considered "good." Our solar sub runs at 30%-35%.

Running *PV-Direct* for Even More Efficiency

We often design solar pumping systems to run *PV-direct*. That is, the pump is connected directly to the photovoltaic (PV) modules with no batteries involved in the system. This is because the electrical to chemical conversion of a battery doesn't operate at 100% efficiency, making batteries less than perfect containers for electrical energy. When we can avoid batteries and deliver our energy directly to the pump, 20% to 25% more water will get pumped everyday. This kind of system is ideal when the water is being pumped into a large storage tank, or is being used immediately for irrigation. It also saves us the maintenance and periodic replacement that batteries require, plus the initial cost of those batteries, charge controllers, and the fusing/safety equipment that batteries demand. PV-direct pumping systems, running all day long, also help us get around the lower gallon per minute output of most positive displacement pumps.

However (every silver lining has its cloud), there is one piece of modern technology we like to use on PV-direct systems that isn't often found on battery systems. A *Linear Current Booster* or LCB for short is a solid-state marvel that will help get a PV-direct pump running earlier in the morning, keep it running later in the evening, and sometimes make running at all a possibility on hazy or cloudy days. An LCB will down convert excess PV voltage into amperage when the modules aren't making quite enough amps for the pump. The pump will run slower than if it had full power, but with positive displacement pumps, if it runs at all, we get water delivered. LCBs will boost water delivery in most PV-direct systems by 20% or more, and we usually recommend a properly sized one with every system.

Direct Current (DC) Motors for Variable Power

Most of our pumps use DC electric motors. DC is the kind of electricity produced by PV modules, and stored by all battery types. DC motors have the advantage of accepting variable voltage input without distress. Common AC motors will overheat if supplied with low voltage. DC motors simply run slow when the voltage drops. This makes them the ideal match to work with PV modules. Day and night, clouds and shadows, these all effect the PV output, and a DC motor simply "goes with the flow!"

Okay, Great, But Which Solar-Powered Pump Do I Want?

That depends on what you're doing with it, and what your climate is. We'll start with the most common and easiest choices, and work our way through to the less common.

Pumping from a Well

Got a well that's cased with 4" or larger pipe, and a static water level that is no more than 230' below the surface? Fine. You probably want the ShurFlo Solar Submersible Pump. We've worked with several solar-powered submersible pump manufacturers over the years and have

Suction Lift: The difference between the source water level and the pump. Theoretical limit is 33 feet; practical limit is 10-15 feet, often less depending on pump type and elevation. Suction lift capability of a pump decreases 1 foot for every 1,000 feet above sea level.

Submersible Pump: A pump with a sealed motor assembly designed to be installed below the water surface. Most commonly used when the water level is more than 15 feet below the surface, or when the pump must be protected from freezing.

Surface Pump: Designed for pumping from surface water supplies such as springs, ponds, tanks or shallow wells. The pump is mounted in a dry, weather-proof location within 10-15 feet of the water surface. Surface pumps cannot be submerged (and be expected to survive).

ShurFlo Sub Pump

come to really appreciate the dependability of this model, and the excellent customer service of this manufacturer. The Solar Sub works with either one or two PV modules (two usually recommended unless minimal delivery is okay). How big the modules are depends on how high you need to lift. See the performance chart and minimum recommended PV wattages at specific lifts accompanying the Solar Sub in the product section. This is a diaphragm-type pump, and unlike AC-powered submersible pumps, it can tolerate running dry. The manufacturer says just don't let it run dry for more than a month or two! This feature makes this pump ideal for many low output wells. 230' is the maximum lift limit for this pump. It's okay to install the pump deeper than 230', so long as the water level itself isn't any deeper than 230'. Remember that water seeks its own level, and we don't have to start exerting energy until we're lifting it above the static level. Complete pumping system including PV modules, mounting structure, LCB, and pump range from $1,300 to $1,900 depending on lift required. Options like float switches that will automatically turn the pump on and off to keep a distant storage tank full are easy and inexpensive to add when using an LCB with remote control as we recommend.

Because this is a low output pump, averaging 1.25 to 1.5 gpm, it won't directly keep up with average household fixtures. For household use we usually recommend, in order of cost and desirability:

1. Pumping into a storage tank elevated at least 50 vertical feet higher than the house, if terrain and climate allow.

2. Pumping into a house-level storage tank if climate allows, and using a booster pump to supply household pressure.

3. Pumping into a storage tank built into the basement for hard-freeze climates, and using a booster pump to supply household pressure.

4. Using battery power from your household renewable energy system to run the ShurFlo submersible pump, and really big pressure tanks (80 gallons minimum).

5. Using a conventional AC-powered submersible pump, large pressure tank(s), and your household renewable energy system with large inverter to run it. (Note that there is a great loss of efficiency running this way, but it's the standard way to get the job done in freezing climates, and your plumber won't have any problems understanding the system. Besides, if you've already **got** the renewable energy system. . .)

Many folks, for a variety of reasons, already have an AC-powered sub pump in their well when they come to us, but are real tired of having to run the generator to get water. For wells with 6" and larger casings it's usually possible to install *both* the existing AC pump and ShurFlo DC pump. If your AC pump is 4" diameter, it's possible to install the DC pump underneath it. The cabling, safety rope, and 1/2" poly delivery pipe from the ShurFlo will slip around the side of the AC pump sitting above it. Just slide both pumps down the hole together. It's often comforting to have emergency back-up for those times, like when it's been cloudy for three weeks straight, or the fire is coming up the hill, and you want a lot of water fast!

Pumping from a Spring, Pond, or Other Surface Source

Your choices are a bit more varied here, depending on how high you need to lift, and how many gallons per minute you want. Surface mounted pumps are not freeze tolerant. If you are in a freezing climate make sure that your installation can be (and **is**!) completely drained before freezing weather sets in. If you need to pump through the hard freeze season, we recommend the ShurFlo Solar Sub described above.

Pumps don't like to pull water up from a source. Or, put simply, *Dr. Doug's # 1 Pumping Rule: **Pumps don't suck, pumps push.*** In order to operate reliably, your surface-mounted pump must be installed as close to the source as practical, and in no case should the pump be more than ten vertical feet above the water level. With some positive displacement pumps a 15-foot suction lift is possible, but not strongly recommended. If you can get the pump closer to the source, and still keep it dry and safe, do it! You'll be rewarded with more dependable service, longer pump life, and less power consumption.

For modest lifts up to 50 or 60 feet and volumes of 1.5 to 3.0 gpm we have found *Diaphragm*-type pumps to be dependable and moderately priced. Specific models may change from time to time, but either the ShurFlo Low-Flow, or the Solar Star 1000 Diaphragm are good examples. These all can tolerate running dry for moderate amounts of time. Diaphragm pumps can be rebuilt quite easily in the field, and both companies offer appropriate parts directly to the consumer. Diaphragm life expectancy is usually one to five years depending on how hard, and how much, the pump is working. Diaphragm pumps will tolerate sand, algae, and debris without damage, but these may stick in the internal check valves reducing or stopping output, and necessitating disassembly to clean out. Who needs the hassle? Filter your intake!

For higher lifts, or more volume, we usually go to a pump type called *Rotary-Vane*. Examples of these pumps are the Solar Star 1075, 1140, and 1450 models. Rotary-vane pumps are capable of lifts up to 400', or volumes up to eight gpm, depending on model. Of all the positive displacement pumps, they are the quietest and smoothest. But they will not tolerate running dry, or abrasives of any kind in the water. It's very important to filter the input of these pumps with a five-micron or finer filter in <u>all</u> applications. Rotary-vane pumps are very long-lived, but will eventually require a pump head overhaul, which must be done at the factory.

Household Water Pressurization

We've got two pumps that are commonly used for this application. The Chevy and the Mercedes models if you will. Any household pressure system requires a pressure tank. A 20-gallon tank is the minimum size we recommend for a small cabin; full-size houses usually have 40-gallon or larger tank(s). Pressure tanks are big, bulky, and expensive to ship. Get one at your local hardware/building center store.

The Chevy model. ShurFlo's Medium-Flow pump is our best-selling pressure pump. It comes with a built-in 20-40 psi pressure switch. With a 2.5 to 3.0 gpm flow rate it will keep up with most household fixtures, garden hoses excluded. The diaphragm pump is reliable, easy to overhaul, but somewhat noisy. We recommend 24" flexible connector in a loop on both sides of this pump, and a pressure tank plumbed in as close as possible to absorb most of the buzz.

The Mercedes model. Solar Star's 1450 rotary vane pump is our smoothest, quietest, largest volume pressure pump. Delivers seven to eight gpm, and is very long lived, but quite expensive initially. (But hey, the good stuff in life usually is.) This pump will easily keep up with garden hoses, sprinklers, and any other normal household stuff.

Solar Hot Water Circulation

Most of the older solar hot water systems installed during the tax-credit heydays of the early 1980s used AC pumps with complex controller/brains and multiple temperature sensors at the collector, tank, plumbing, ambient air, etc. This kind of complexity allows too many opportunities for Murphy's Law. The smarter systems simply used a small PV module wired directly to a DC pump. When the sun shines a little bit, making a small amount of heat, the pump runs slowly. When the sun shines bright and hot, making lots of heat, the pump runs fast. Very simple system control is achieved with an absolute minimum of "stuff." We carry

ShurFlo 115 V AC Pump

Solar Star 1450 Pump

WATER 291

El-Sid Hot Water
Circulation Pump

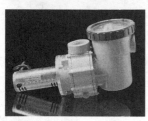

PV Pool Pump

several hot water circulation pumps, the best of which is the new El-Sid pump. This is the first solid-state, brushless DC circulation pump that was designed from scratch for PV-direct applications. It only requires a 5-watt module to drive it, and life expectancy is three to four times longer than any other DC circulation pump. Volume and lift are sharply limited however. The El-Sid is standard equipment in some of the solar hot water systems we sell. See hot water circulation in the products section for all the pump specs.

Swimming Pool Circulation

Yes, it's possible to live off-the-grid and still enjoy luxuries like swimming pools. In fact pool systems dovetail nicely with household systems in many climates. Houses generally require a minimum of PV energy during the summer because of the long daylight hours, yet there is a maximum of energy available. By switching a number of PV modules to pool pumping in the summer, then back to battery charging in the winter we get better utilization of resources. We offer DC pool pumps in three horsepower sizes. DC pumps run somewhat more efficiently than AC pumps, so a slightly smaller DC pump can do the same amount of work as a larger AC pump. Our 1/3 hp pool pump can replace a conventional 1/2 hp. Our 1/2 hp can replace a 3/4 hp, and our 1 hp can replace a 1-1/2 hp. See the product section for specs and prices. Please consult with out technical staff regarding pool filters, PV array sizing, and switching equipment.

Water-Powered Pumps

A few lucky folks have an excess supply of falling water available. This falling water energy can be turned to pumping water for you. Both the *High Lifter* and *Ram* type water pumps use the energy of falling water to force a portion of that water up the hill to a storage tank.

Ram Pumps

Ram pumps have been around for many decades providing reliable water pumping at almost no cost. They are commonly used in the Eastern U.S. where modest falls and large flow rates are the norm, but they will happily work almost anyplace their minimum flow rate can be satisfied. Rams will work with a minimum of 1.5 feet to a maximum of about 20 feet of fall feeding the pump. Minimum flow rates depend on the pump size, see the product section for specs. A flow is started down the drive pipe and then is shut off suddenly. The resulting pressure surge from the momentum of moving water slamming to a stop sends a little squirt of water up the hill. How much of a squirt depends on the pump size, the amount of fall, and the amount of lift. Output charts are with the pumps in the product section. Each ram needs to be carefully tuned for its particular site. If they run short of water they will stop pumping and simply dump incoming water, so don't buy too big. Rams make some noise. A lot less than a gasoline-powered pump for sure, but the constant chunk-chunk-chunk-chunk-chunk is a consideration for some sites. Ram pumps deliver less than 5% of the water that passes through them, and the discharge must be into an unpressurized storage tank or pond. But they work for free and have life expectancies measured in decades.

The High Lifter

This pump is unique. It works by simple mechanical advantage. A large piston at low water pressure pushes a smaller piston at higher water pressure. High Lifters recover a much greater percentage of the available water than ram pumps do, but they require greater fall into the pump generally. This makes them better suited for more mountainous territory. They are available in two ratios, 4.5 to 1 and 9 to 1. So fall to lift ratios and waste water to pumped water ratios are also either 4.5 to 1 or 9 to 1. Note, however, that as the lift ratio gets closer to theoretical maximum the pump is going to slow down and deliver less gallons per day. High Lifters are self-starting. If they run out of water they will simply stop and wait, or slow down to match what water is available. A very handy trait for unattended, or difficult to attend sites. Output charts for this pump are included in the product section.

Wind-Powered Water Pumps

If you've been reading all this way because you want to buy a nostalgic old-time jack-pump windmill, we're going to disappoint you. They <u>are</u> still made, but we don't sell them, and don't recommend them very strongly except in unusual cases. They are <u>very</u> expensive (try $6,000 and up), quite a big deal to set up and install into the well, and require routine yearly service at the top of the tower. This is technology that has largely seen its day, except for very deep remote location pumping. We do have a wind-powered pump that has seen good success.

Bowjon Pumps

The Bowjon pumps work by compressed air. There are two models available depending on lift and volume needs. The simple pole-mounted turbine direct-drives an air compressor. The air is piped down the well and run through a carefully engineered air injector. As it rises back up the supply tube it carries water in-between the bubbles. The lift/submergence ratio of this pump is fairly critical. Approximately 30% of the total lift is the recommended distance for the air injector to be submerged below the static level. Too little submergence and the air will separate from the water, too much and the air will not lift the water, though there is considerable latitude between these performance extremes. This pump isn't bothered by running dry. Output depends on wind speed, naturally, but the larger Rancher model is capable of up to nine gpm at lower lifts, or can lift a maximum of 275'. The air compressor requires an oil level check twice a year. See the product section for more details and output specs.

Wind Generators Running Submersible Pumps

The larger wind generator manufacturers, Bergey and Whisper, both offer options which allow the three-phase wind turbine output to power a three-phase submersible pump directly. These aren't residential-scale systems and are mostly used for large agricultural projects, or village pumping systems for underdeveloped countries. They are relatively expensive initially. Contact our technical staff for more information on these options.

Freeze Protection

In most areas of the country freezing is a major consideration when installing plumbing and storage systems. For outside pipe runs the general rule is to bury the plumbing below frost level. For large storage tanks burial may not be feasible, unless you go with concrete. In moderate climatic zones simply burying the bottom of the tank a foot or two along with the input/output piping is sufficient. In some locations, due to climate or lack of soil depth, outdoor storage tanks simply aren't feasible. In these situations we can go for a smaller storage tank built into a corner of the basement with a separate pressure boosting pump, or we can pump directly into a large pressure tank system.

Other Considerations

Hopefully by this point you've zeroed in on a pump or pumps that seem to be applicable to your situation. If not, our technical staff will be happy to discuss your needs, and recommend an appropriate pumping system (which might <u>not</u> be renewable-energy powered sometimes). At this point another crop of questions usually appears; such as. . .

*1. Pumps prefer to <u>push</u>
water, not <u>pull</u> it. In fact
most pumps are limited to 10
feet or 15 feet of lift on the
suction side. Mother Nature
has a theoretical suction lift of
33.9 feet at sea level, but only
if the pump could produce a
perfect vacuum. Suction lift
drops 1 feet with every 1,000
feet rise in elevation. To put
it simply,* Pumps Don't
Suck, They Push.*
2. Water is heavy, 8.33 lbs.
per gallon. It can require
tremendous amounts of energy
to lift and move.
3. DC electric motors are
generally much more efficient
than AC motors. If you have
a choice, use a DC motor to
pump your water. Not only
can they be directly powered
by solar modules, but your
precious wattage will go
further.
4. Positive displacement
pumps are far more efficient
than centrifugal pumps.
Most of our pumps <u>are</u>
positive displacement types,
AC powered submersibles, jet
pumps, and booster pumps
are centrifugal types. (See the
discussion below for more info.)
5. As much as possible we try
to avoid batteries in pumping
systems. When energy is run
into and out of a battery,
25% is lost. It's more efficient
to take energy directly from
your PV modules and feed it
right into the pump. At the
end of the day, you'll end up
with 25% more water in the
tank.
6. 1 pound per square inch
(psi) of water pressure equals
2.31 feet of lift, a handy
equation.*

How far can I put the modules from the pump?

Often the best water source will be deep in a heavily wooded ravine. It's important that your PV modules have clear, shadow-free access to the sun for as many hours as practical. Even a fist-sized shadow will effectively turn off most PV modules. The hours of ten to two are usually the minimum your modules want to see, and if you can capture full sun from nine to three that's more power to you. If the pump is small, running off one or two modules, then distances up to 200' can be handled economically. Longer distances are always possible, but consult with our tech staff or check out the wire sizing formula in the *Sourcebook* appendix first, as longer distances require large (expensive!) wire. Many pumps routinely come as 24-volt units now, as the higher voltage makes long distance transmission twice as easy. Some of our pumps can be special-ordered as 120-volt units when distances over 500' are involved.

What size wire do I need?

This depends on distance, and the amount of power we're trying to move. We <u>always</u> recommend #10 gauge copper submersible pump wire going down the well to the ShurFlo submersible pump, simply because #10 is the largest commonly available submersible pump wire. For topside wiring consult the wire sizing formula in the appendix, or give our tech staff a call.

What size pipe do I need?

Most of the pumps we offer have modest flow rates of four gpm or less. This makes it okay to use smaller pipe sizes without increasing friction loss. If the pipe run is being used for pumping delivery only, then 3/4" or one inch pipe sizes are usually sufficient. Please note we said for pumping delivery only. There's no reason you can't use the same pipe to take the water up the hill to the storage tank, and also bring it back down to the house or garden. Pipes don't care which way the water's flowing through them. But if you do this, you'll want a larger pipe to avoid friction loss and pressure drop when the higher flow rate of the household or garden fixtures comes into play. We usually recommend at least 1-1/4" pipe for household and garden use, and two inch for fire hose lines.

What size PV modules do I need?

A number of the pumps listed in the product section have performance and wattage tables listed with them. For instance; the Solar Star 1075 rotary vane pump at 40 feet of lift running PV-direct will deliver 2.2 gpm and requires 60 watts. Hey, what's to figure here? All PV modules are rated by how many watts they put out, right? So I just need a 60-watt module. An MSX-60 looks perfect. Wrong, friend. PV modules are rated under *ideal laboratory conditions*, not real life. If you want your pump to work on hot or humid days (heat reduces PV output, water vapor cuts available sunlight), then you must add 30% to the pump wattage when figuring PV wattage. So in our example we actually need 78 watts. Looking at available PV modules we don't find exactly 78-watt modules, which means we buy a 75-watt module if this is a relatively temperate climate, or an 85-watt module if this is a hot (over 80° F) climate. And, of course, an LCB is practically standard equipment with any PV-direct system.

Can I automate the system?

Absolutely! Life's little drudgeries should be automated at every opportunity. The LCBs that we so strongly recommend with all PV-direct systems help us do this. The three models that cover most applications are all supplied with the remote control option. This option allows you to install a float switch at the holding tank, using a pair of tiny #18 gauge wires which can be run as far as 5,000 feet back to the pump/controller/PV modules area. Float switches can be used to either fill or empty a holding tank, and will automatically turn the system off when the job is done. The three common LCBs, plus float switches and other pumping accessories are presented in the product section. LCBs can also be used on battery-powered systems when remote sensing and control would be handy.

Where Do We Go From Here?

If you haven't found all the answers to your remote water pumping needs yet please give our experienced technical staff a call. They'll be happy to work with you selecting the most appropriate pump, power source, and accessories for your needs. We run into situations occasionally where renewable energy sources and pumps simply may not be the best choice, and we'll let you know if that's the case. For 95% of the remote pumping scenarios there is a simple, cost-effective, long-lived renewable energy powered solution, and we can help you develop it.

— Doug Pratt

Dr. Doug's Handy Water Formulas & Tips

1 gallon of water = 8.3 pounds = 231 cubic inches

1 pound of water = 27.7 cubic inches

1 cubic foot of water = 7.5 gallons = 62.5 pounds (sea water = approx. 64.3 lbs)

2.31 vertical feet = 1 pound per square inch (psi), or inversely,

1 vertical foot of water = 0.434 psi

APPROXIMATE DAILY WATER NEEDS FOR HOME AND FARM

Usage	Gallons per Day
Home	
Typical home use, including kitchen, laundry, flush toilets, showers, etc.	100 per person
Typical home use as above, without flush toilet	50–60 per person
Swimming-pool maintenance, per 100 sq. ft. surface area	30
Lawn and Garden	
Lawn sprinkler, per 1000 sq. ft. per sprinkling	600 (approx. 1 inch)
Garden sprinkler, per 1000 sq. ft. per sprinkling	600 (approx. 1 inch)
Farm (maximum needs)	
Dairy cows	20 per head
Dry cows or heifers	15 per head
Calves	7 per head
Beef, yearlings, full feed 90° F ambient	20 per head
Beef, brood cows	12 per head
Sheep or goats	2 per head
Horses or mules	12 per head
Swine, finishing	4 per head
Brood sows, nursing	6 per head
Laying hens (at 90° F)	9 per 100 birds
Broilers (at 90° F)	6 per 100 birds
Turkeys (over 100° F)	25 per 100 birds
Ducks	22 per 100 birds
Dairy sanitation — milk room and milking parlor	500 per day
Flushing floors	10 per 100 sq. ft.
Sanitary hog wallow	100 per day

Information courtesy of *Planning for an Individual Water System* (item #80-201) and *Pocket Ref* (item #80-506)

Friction Loss Charts for Water Pumping

Friction Loss- PVC Class 160 PSI Plastic Pipe
Pressure loss from friction in psi per 100 feet of pipe.

Bold Numbers indicate 5 Feet per Second Velocity

Flow GPM	NOMINAL PIPE DIAMETER IN INCHES										
	1	1.25	1.5	2	2.5	3	4	5	6	8	10
1	0.02	0.01									
2	0.06	0.02	0.01								
3	0.14	0.04	0.02								
4	0.23	0.07	0.04	0.01							
5	0.35	0.11	0.05	0.02							
6	0.49	0.15	0.08	0.03	0.01						
7	0.66	0.20	0.10	0.03	0.01						
8	0.84	0.25	0.13	0.04	0.02						
9	1.05	0.31	0.16	0.05	0.02						
10	1.27	0.38	0.20	0.07	0.03	0.01					
11	1.52	0.45	0.23	0.08	0.03	0.01					
12	1.78	0.53	0.28	0.09	0.04	0.01					
14	2.37	0.71	0.37	0.12	0.05	0.02					
16	**3.04**	0.91	0.47	0.16	0.06	0.02					
18	3.78	1.13	0.58	0.20	0.08	0.03					
20	4.59	1.37	0.71	0.24	0.09	0.04	0.01				
22	5.48	1.64	0.85	0.29	0.11	0.04	0.01				
24	6.44	1.92	1.00	0.34	0.13	0.05	0.02				
26	7.47	2.23	1.15	0.39	0.15	0.06	0.02				
28	8.57	**2.56**	1.32	0.45	0.18	0.07	0.02				
30	9.74	2.91	1.50	0.51	0.20	0.08	0.02				
35		3.87	**2.00**	0.68	0.27	0.10	0.03				
40		4.95	2.56	0.86	0.34	0.13	0.04	0.01			
45		6.16	3.19	1.08	0.42	0.16	0.05	0.02			
50		7.49	3.88	1.31	0.52	0.20	0.06	0.02			
55		8.93	4.62	**1.56**	0.62	0.24	0.07	0.02			
60		10.49	5.43	1.83	0.72	0.28	0.08	0.03	0.01		
65			6.30	2.12	0.84	0.32	0.09	0.03	0.01		
70			7.23	2.44	0.96	0.37	0.11	0.04	0.02		
75			8.21	2.77	1.09	0.42	0.12	0.04	0.02		
80			9.25	3.12	1.23	0.47	0.14	0.05	0.02		
85			10.35	3.49	**1.38**	0.53	0.16	0.06	0.02		
90				3.88	1.53	0.59	0.17	0.06	0.03		
95				4.29	1.69	0.65	0.19	0.07	0.03		
100				4.72	1.86	**0.72**	0.21	0.08	0.03	0.01	
150				10.00	3.94	1.52	0.45	0.16	0.07	0.02	
200					6.72	2.59	**0.76**	0.27	0.12	0.03	0.01
250					10.16	3.91	1.15	0.41	0.18	0.05	0.02
300						5.49	1.61	**0.58**	0.25	0.07	0.02
350						7.30	2.15	0.77	0.33	0.09	0.03
400						9.35	2.75	0.98	0.42	0.12	0.04
450							3.42	1.22	**0.52**	0.14	0.05
500							4.15	1.48	0.63	0.18	0.06
550							4.96	1.77	0.76	0.21	0.07
600							5.82	2.08	0.89	0.25	0.08
650							6.75	2.41	1.03	0.29	0.10
700							7.75	2.77	1.18	0.33	0.11
750							8.80	3.14	1.34	**0.37**	0.13
800								3.54	1.51	0.42	0.14
850								3.96	1.69	0.47	0.16
900								4.41	1.88	0.52	0.18
950								4.87	2.08	0.58	0.20
1000								5.36	2.29	0.63	0.22
1500									4.84	1.34	0.46
2000										2.29	0.78
2500										3.46	1.18
3000											1.66

Friction Loss- Polyethylene (PE) SDR-Pressure Rated Pipe
Pressure loss from friction in psi per 100 feet of pipe.

Numbers in Bold Indicate 5 Feet/Second Velocity

Flow GPM	NOMINAL PIPE DIAMETER IN INCHES							
	0.5	0.75	1	1.25	1.5	2	2.5	3
1	0.49	0.12	0.04	0.01				
2	1.76	0.45	0.14	0.04	0.02			
3	3.73	0.95	0.29	0.08	0.04	0.01		
4	**6.35**	1.62	0.50	0.13	0.06	0.02		
5	9.60	2.44	0.76	0.20	0.09	0.03		
6	13.46	3.43	1.06	0.28	0.13	0.04	0.02	
7	17.91	4.56	1.41	0.37	0.18	0.05	0.02	
8	22.93	**5.84**	1.80	0.47	0.22	0.07	0.03	
9		7.26	2.24	0.59	0.28	0.08	0.03	
10		8.82	2.73	0.72	0.34	0.10	0.04	0.01
12		12.37	**3.82**	1.01	0.48	0.14	0.06	0.02
14		16.46	5.08	1.34	0.63	0.19	0.08	0.03
16			6.51	1.71	0.81	0.24	0.10	0.04
18			8.10	2.13	1.01	0.30	0.13	0.04
20			9.84	2.59	1.22	0.36	0.15	0.05
22			11.74	**3.09**	1.46	0.43	0.18	0.06
24			13.79	3.63	1.72	0.51	0.21	0.07
26			16.00	4.21	1.99	0.59	0.25	0.09
28				4.83	2.28	0.68	0.29	0.10
30				5.49	**2.59**	0.77	0.32	0.11
35				7.31	3.45	1.02	0.43	0.15
40				9.36	4.42	1.31	0.55	0.19
45				11.64	5.50	1.63	0.69	0.24
50				14.14	6.68	**1.98**	0.83	0.29
55					7.97	2.36	0.85	0.35
60					9.36	2.78	1.17	0.41
65					10.36	3.22	1.36	0.47
70					12.46	3.69	**1.56**	0.54
75					14.16	4.20	1.77	0.61
80						4.73	1.99	0.69
85						5.29	2.23	0.77
90						5.88	2.48	0.86
95						6.50	2.74	0.95
100						7.15	3.01	**1.05**
150						15.15	6.38	2.22
200							10.87	3.78
300								8.01

CENTRIFUGAL PUMPS

12V Submersible Pump

This 12-volt submersible pump measures only 1-1/2 inches in diameter and 6-1/2 inches long, and will pump a maximum of 4 gpm. It will pump a maximum vertical head of 42 feet or 18 psi. At 25-feet of head the LVM-105 will deliver about 5/8 gallon per minute. A strainer fitted over the input prevents large particles from entering the pump. The inlets and outlets are 1/2-inch barbed fittings, it comes with 13 feet of cable with battery clips on the end, and it weighs only 1.2 pounds. Designed for intermittent use only. Three-month warranty.

41-671   12V Submersible Pump $79⁰⁰

Bronze 12V Centrifugal Pump

This bronze-bodied and bronze-impeller pump is designed for locations with a large volume of water to move at a minimum lift. The pump is not self-priming and must have a flooded suction. Pump head can be rotated in 90° increments. Pump must be mounted in a dry location where the motor is protected from dampness. Intake and outlet ports are 3/4-inch FIPT suitable for either brass or plastic pipe fittings. One-year warranty.

Performance gpm at total feet of head:

Head	Gpm	Amps
0 ft.	18	4.0
5 ft.	13.5	4.6
10 ft.	7.5	5.6
15 ft.	0.5	6.5
23 ft.	0	8.5

41-301   Bronze 12V Centrifugal Pump $185⁰⁰
Filtering is an absolute must! The pump is coupled to the motor.

March Circulating Pump

Used with hot tub, spa, solar thermal, and woodstove hot water systems, this 1/100 hp pump runs at 1,950 rpm, uses very little power, keeps needed water circulation happening, and has a magnetic drive that eliminates the old-fashioned shaft seal. It is easy to service, requiring only a screwdriver, and the entire motor assembly can be replaced without draining the system. Will withstand temperatures to 200⁰ F. Weight 7.1 lbs. Six-month warranty. Specify 12 volt or 24 volt.

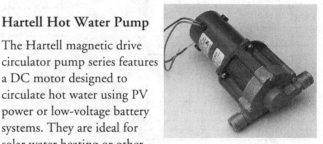

> Max Amps: 1.5
> Max gpm: 1.5
> Max head: 1-1/2

41-505   March Circulating Pump, 1/100 hp $299⁰⁰

Hartell Hot Water Pump

The Hartell magnetic drive circulator pump series features a DC motor designed to circulate hot water using PV power or low-voltage battery systems. They are ideal for solar water heating or other closed-loop, low-flow pumping applications. This model was designed to operate directly off of a solar panel of 15 to 30 watts — the brighter the sun, the faster the water is pumped! It will withstand temperatures to 200⁰ F. Brass pump casing has inline 1/2 inch MIPT inlet and outlets. Six-month warranty.

Amps Min. 1.5/Max. 3.0
Max gpm 6.5
Max head 11 feet

41-526   Hartell Circulator Pump — specify 12 or 24-volt $189⁰⁰

Solar Fountain Pumps

These submersible 12-volt pumps are designed as continuous-duty marine bilge pumps. They won't lift very high, but they move an incredible volume of. The pump motor twist-locks onto the pump base to allow easy cleaning and removal of debris. They can withstand dry running for a limited time. Three-year warranty.

Attwood V625 Pump

Model V625 is a good match with a 30-watt module. (Solarex SX30, #11-511, recommended.) 3/4" hose outlet.

> *Amps: 1.3 @ 3 ft. head*
> *Max gpm: 7.5 @ 3 ft. head*
> *Max head: 5 ft.*

41-157   V625 Solar Fountain Pump $25⁰⁰

Attwood V1250 Pump

Model V1250 is a good match with a 50-watt module. (Siemens M-75 or Kyocera K-51 recommended.)

> *Amps: 2.9 @ 3 ft. head*
> *Max gpm: 15.5 @ 3 ft. head*
> *Max head: 8 ft.*

41-158   V1250 Solar Fountain Pump $35⁰⁰

Rena Fountain Pumps

Simply the finest, most energy efficient, and one of the least expensive, fountain pump designs we've ever seen, these submersible pumps have only one moving part. Has an easily cleanable foam filter, can be quickly disassembled or serviced without any tools, and has a generous 10' power cord. Runs on conventional 120VAC. Spray heads are included, but can easily attach to flexible tube also. Highly recommended for use with our Solar Power Station on pp. 109 for a solar-powered fountain. UL-approved. One-year mfg. warranty.

10-watt Fountain Pump

This larger unit produces a maximum flow of 180 gph, or a maximum spray height of 4.5 feet. The included "flower" spray head comes with four different snap-on spray patterns. The Solar Power Station will run this pump about three hours per sunny day.120VAC.

41-165   10-w Fountain Pump $39⁰⁰

4-watt Fountain Pump

A smaller pump that produces 80 gph, or a maximum spray height of 2 feet. Comes with a variety of extensions and sizes of single nozzle jets. The Solar Power Station can run this pump about six to seven hours per sunny day. 120VAC.

41-166   4-w Fountain Pump $29⁰⁰

El-Sid Hot Water Circulation Pump

The first completely solid state circulation pump. The "motor" has no moving parts! No brushes or bearings to wear out or replace. The rust-proof bronze pump section is magnetically driven, so there are no seals. Life expectancy is three or four times longer than any other DC circulation pump. Requires only 4 to 5 watts of PV to drive. This is only a circulation pump for closed systems, it will not do any significant lift, it just pushes the water around the loop. For simple system control it's hard to beat a PV module wired directly to a circulation pump. The more sun, the faster it circulates. Our 5-watt module, the Solarex SA-5 (item # 11-521, $89) is a perfect match with this pump. Both these components are featured in the Real Goods Basic Hot Water System (item # 45-424, $895). Temperatures up to 240⁰ F are okay. One year mfg. warranty.

41-134   El-Sid Circulation Pump $195⁰⁰

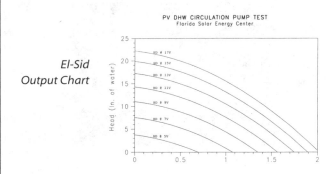

El-Sid Output Chart

Photovoltaic Pool Pump

These solar-powered pool pumps offer reliable and cost-effective alternatives to the standard AC-driven pool pump. A specially designed DC motor is matched with a self-priming centrifugal solar water pump. The pump is constructed of glass reinforced thermoplastic, which ensures lifetime service with no corrosion. The strainer can be removed in seconds for cleaning without loss of prime. All parts in contact with water are stainless steel, carbon/ceramic, or non-corrosive plastic. The pump will circulate from 30 to 50 gpm through a cartridge-type swimming-pool filter. The motor uses permanent magnets and drives a stainless-steel shaft with dual ball bearings. Brushes are externally accessed for easy service.

Building a solar-powered pool filtration system requires, in addition to the pump and motor, a PV array (either six or nine high-power modules) with support, a cartridge type filter (such as Jacuzzi CFT100), plus wiring and plumbing. For pool-water filtering, an annual electrical cost of hundreds of dollars can be replaced with a one-time investment at installation. Tax credits may be applicable for the solar electric power system. The cost of the system will depend on the size of the pool, the kind of pump/filter options currently on the system, and the quality of your solar exposure. Prices range from $2,500 up. Call one of our technicians for an estimate. We'll need to know your pool size (in gallons), what options you have on the pump/filter system (pool sweeps, solar heating, etc.), and what type of filter you're presently running. Low restriction cartridge-type filters are a necessity with these pumps. A diatomaceous earth filter will require more PV modules and possibly a larger pump.

> *Amps: 7A @ 50V max.*
> *Max gpm: 70*
> *Max head: 28 ft.*

41-149   1/3 hp PV Pool Pump $745⁰⁰

*PV Pool
Pump*

1/2 HP PV Pool Pump

For larger pools, or systems with pool sweeps, solar heating, or other applications requiring more volume or greater pressure. Please call for help in sizing.

> *Amps: 5.5A @ 90V max.*
> *Max gpm: 60*
> *Max head: 35 ft.*

41-155   1/2 hp PV Pool Pump $1,195

1 HP PV Pool Pump

For very large pools, or systems with rooftop solar heating, pool sweeps, too many 90° bends, or other complications.

> *Amps: 9.0A @ 90V max.*
> *Max gpm: 90*
> *Max head: 45 ft.*

41-156   1 hp PV Pool Pump $1,655

ROTARY VANE PUMPS

Photocomm DC Powered Rotary Vane Pumps

Photocomm manufactures efficient DC powered rotary vane water pumps that have a brass body with stainless steel internals and composite vanes. Water supply to pump must be clean, filtered to a minimum 150 mesh screen. If any sand or grit is present, 5 micron filtering is an absolute must! The pump is coupled to the motor with a V-band clamping system which can be rotated to any position and allows easy installation. One-year warranty on all models.

Solar Star 1075

This pump, the smallest in the rotary-vane series, is our highest lift water pump, and can be run either panel direct (the most efficient way), or from batteries.

Output Chart for 1075 Pump

Head	PV direct GPM	Watts	GPM	Battery Watts
20 ft.	2.25	50	1.5	40
40 ft.	2.2	60	1.45	45
60 ft.	2.15	72	1.4	50
80 ft.	2.1	83	1.3	57
100 ft.	2.0	96	1.3	65
120 ft.	2.0	104	1.25	74
140 ft.	1.95	115	1.2	84
160 ft.	1.9	126	1.2	92
180 ft.	1.9	135	1.15	97
200 ft.	1.85	148	1.15	106
240 ft.	1.8	172	1.1	126
280 ft.	1.7	192	0.97	145
320 ft.	1.6	219	0.91	165
360 ft.	1.5	242	0.85	185
400 ft.	1.4	257	0.78	209

41-457   Solar Star 1075, 12V $499⁰⁰
41-458   Solar Star 1075, 24V $499⁰⁰

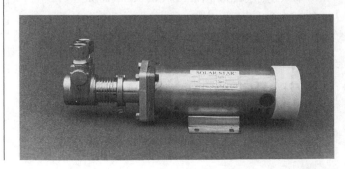

Solar Star 1140

The 1140 pump is larger than the 1075. It will produce more volume, but isn't capable of extreme lifts. Can be run either panel direct (the most efficient way), or from batteries. It, of course, uses more power than the 1075 model.

Output Chart for 1140 Pump

Head	PV direct GPM	Watts	GPM	Battery Watts
20 ft.	3.95	80	2.65	45
40 ft.	3.8	96	2.6	57
60 ft.	3.7	114	2.55	68
80 ft.	3.6	128	2.45	80
100 ft.	3.5	141	2.4	92
120 ft.	3.45	159	2.35	104
140 ft.	3.4	171	2.3	115
160 ft.	3.3	185	2.25	128
180 ft.	3.2	200	2.2	141
200 ft.	3.1	211	2.15	155
240 ft.	3.0	247	2.0	182
280 ft.	2.8	268	1.9	212

41-459	Solar Star 1140, 12V	$499⁰⁰
41-460	Solar Star 1140, 24V	$499⁰⁰

Solar Star 1450

This model is designed for household pressure boosting running off batteries. It is not designed to run PV-direct. At 7 to 8 gpm it has no problem keeping up with multiple fixtures simultaneously. Because it's positive displacement, power use is less than 1/4 what a conventional centrifugal booster pump will cost to run. Rotary vane pumps are the quietest and smoothest of positive displacement pumps for pressure boosting.

Output Chart for 1450 Pump

	1450 at 12 Volts		*1450 @ 24 Volts*	
PSI	GPM	Amps	GPM	Amps
30	7.4	20	7.8	11
40	7.0	24	7.5	13
50	6.7	28	7.3	15

41-159	Solar Star 1450, 12V	$549⁰⁰
41-153	Solar Star 1450, 24V	$549⁰⁰

Solar Star 1450

Rotary Vane Easy Install Kit

This kit gives you everything you'll need to make a quick, safe, protected installation. Fits pump models 1075, 1140, and 1450. The kit includes:

• 2-1/2 ft. of 3/4" nylon braided vinyl tubing for inlet
• 2-1/2 ft. of 1/2" nylon braided vinyl tubing for outlet
• One 1/2" MIPT x 1/2" barbed fitting
• One 1/2" MIPT x 3/4" barbed fitting
• One 3/4" MIPT x 1/2" barbed fitting
• One 3/4" MIPT x 3/4" barbed fitting
• Four stainless steel hose clamps
• One 3/4" combination foot valve/strainer or in-line check valve (can be used as either)
• One 10" filter housing with 5 micron cartridge. Uses readily available replacement cartridges
• One spare replacement 5 micron filter cartridge
• One fuse holder and fuses

41-148	Easy Install Kit #505	$85⁰⁰

Pressure Switch

You must use a pressure switch for pressure tank systems. It will turn the pump off when the tank pressure reaches 40 psi, and then on again when the pressure drops to 20 psi. It also turns the pump off when pressure drops below 10 psi to prevent damage from running the pump dry. The settings are fully adjustable in the range of 5 to 65 psi.

41-140	Pressure Switch	$24⁰⁰

Inline Sediment Filter

A simple 5-micron filter for your rotary vane pump intake, highly recommended, and costs less than a pump head overhaul. Uses common, inexpensive 10" filter cartridges. For cold water only. 3/4" female pipe thread inlet and outlet fittings. Includes a 5-micron filter cartridge, spares sold below.

41-137	Inline Sediment Filter	$74⁰⁰
41-132	Rust & Dirt Cartridge 5-micron	$7⁰⁰

DIAPHRAGM PUMPS

SHURflo Pumps

SHURflo produces a high quality line of positive displacement, diaphragm pumps for RV and remote household pressurization, and for lifting surface water to holding tanks. These pumps will self prime with 10 feet of suction and free discharge. The three chamber design runs quieter than other diaphragm pumps and will tolerate silty or dirty water or running dry with no damage. However, if your water is known to have sand or debris it's best to use an inexpensive inline cartridge filter on the pump intake (order #41-137 filter housing and #41-138 filter cartridge.) Sand sometimes will lodge in the check valves necessitating disassembly for cleaning. No damage, but who needs the hassle? Put a filter on it!

Motors are slow speed DC permanent magnet types (except AC model) for long life and the most efficient performance. All pumps are rated at 100% duty cycle and may be run continuously. There is no shaft seal to fail, the ball bearing pump head is separate from the motor. Pumps are easily field repaired and rebuild kits are available for all models with a toll-free phone call to the factory for Parts and Repair Kits. All pumps have a built-in pressure switch to prevent over-pressurization. Some are adjustable through a limited range. Low flow model has 3/8" MIPT inlets and outlets. Medium and high flow models are 1/2" MIPT. All pumps carry a full one year warranty.

SHURflo Low Flow Pump

For modest water requirements or PV direct applications. Available in 12 volt DC only. Delivers a maximum of 60 psi or 135 feet of head. This is an excellent surface water delivery pump used with the PV Direct Pump Control (25-134RC) and PV modules as necessary for your lift. (Add 30% to pump wattage to figure necessary PV wattage.) 3/8" MIPT inlet and outlet.

SHURflo Low Flow 12V

PSI	GPM	Watts
0	1.75	37
10	1.66	41
20	1.57	50
30	1.48	59
40	1.38	67
50	1.30	76
60	1.23	86

41-450	SHURflo Low Flow Pump	$109⁰⁰

SHURflo Medium Flow Pump

The Medium Flow is available in 12 or 24 volt DC, performance specs are similar for both (the 24V pump does slightly less volume). Maximum pressure is 40 psi or 90 feet of lift for both models. This model is a good choice for low-cost household pressurization, it will keep up with any single fixture. Has a built-in 20 to 40 psi pressure switch, perfect for most households. Buzzy output can be reduced by looping 24" flexible connectors on both inlet and outlet. 1/2" MIPT inlet and outlet.

SHURflo Medium Flow 12V			SHURflo Medium Flow 24V		
PSI	GPM	Amps	PSI	GPM	Amps
0	3.30	5.1	0	3.00	2.5
10	3.02	5.6	10	2.87	3.2
20	2.82	7.0	20	2.64	3.8
30	2.65	8.4	30	2.41	4.3
40	2.48	9.6	40	2.20	4.7

41-451	SHURflo Medium Flow, 12 Volt	$149⁰⁰
41-452	SHURflo Medium Flow, 24 Volt	$165⁰⁰

SHURflo High Flow Pump

This is SHURflo's highest output model. Good for household pressurization systems, if you need higher than normal 40 psi. Has a built-in 30 to 45 psi pressure switch. Buzzy output can be reduced by looping 24" flexible connectors on both inlet and outlet. Available in 12 volt only. Maximum pressure is 45 psi or 104 feet of lift. 1/2" MIPT inlet and outlet.

SHURflo High Flow 12V

PSI	GPM	Amps
0	3.40	6.0
10	3.00	6.9
20	2.75	7.5
30	2.50	8.6
40	2.25	9.5

41-453	SHURflo High Flow Pump	$189⁰⁰

SHURflo Medium Flow Pumps

SHURflo AC Pump, 115 Volts

For those who need to send power more than 150-200 feet away, this pump can save you big $$ on wire costs! The motor is thermally protected and under heavy continuous use will shut off automatically to prevent overheating. This should only happen at maximum psi after approximately 90 minutes of running. Pressure switch is adjustable from 25 to 45 psi. Maximum pressure is 45 psi. Weight 5.8 lbs. 1/2" MIPT inlet and outlet.

SHURflo 115 volt AC Pump

PSI	GPM	Watts
0	3.05	59
10	2.85	65
20	2.60	82
30	2.40	96
40	2.24	108

41-454   SHURflo 115 volt AC Pump $179⁰⁰

SHURflo Solar Submersible Pump

This efficient, submersible, diaphragm pump will deliver up to 230 feet of vertical lift and is ideal for deep well or freeze-prone applications. Remember, you're only lifting from the water *surface*, not from the pump. In a PV-direct application it will yield from 300 to 1,000 gallons per day depending on lift requirements and power supplied. Solar submersible pumps are best used for slow steady water production into a holding tank, but may be used for direct pressurization applications as well. (See suggestions on this subject in the Editorial section a few pages back.) Flow rates range from .5 to 1.8 gpm depending on lift. The pump can be powered from either 12 or 24 volt sources. Output is best with 24 volt panel-direct. The best feature of this pump, other than proven reliability, is easy field serviceability. No special tools are needed. Replacement parts and technical advice are provided to the consumer by the manufacturer, whose toll-free phone number is listed right on the pump.

The pump is 3-3/4 inches in diameter (will fit standard 4" well casings), weighs 6 pounds, uses 1/2" poly delivery pipe, making easy one man installation possible. One year warranty.

In addition to the pump you will also need:

• PV modules & mounting appropriate for your lift (see chart)

• A Controller for panel-direct systems (PV Direct Controller #25-138RC for 24v, #25-134RC for 12v)

• 1/2" poly drop pipe or equivalent

• #10 gauge submersible pump cable (common 3-wire cable okay if flat style)

• Poly safety rope

• Sanitary well seal

Drop pipe, submersible cable, rope, and seal are all commonly available at any plumbing supply store. Watertight splice kit for **flat** cable is included.

SHURFLO SUBMERSIBLE PUMP PERFORMANCE CHART AT 24 VOLTS

(12 volt performance will be approximately 40% of figures listed below)

Total Vertical Lift	Gallons per Hour	Minimum PV Watts	Amps
20	117	58	1.5
40	114	65	1.7
60	109	78	2.1
80	106	89	2.4
100	103	99	2.6
120	101	104	2.8
140	99	115	3.1
160	98	123	3.3
180	93	135	3.6
200	91	141	3.8
230	82	155	4.1

41-455   SHURflo Solar Submersible $695⁰⁰

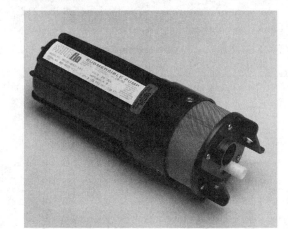

Submersible Pump Options
PV-Direct Pump Controller

A compact weather-tight package with built-in 7-amp linear current booster and remote control capability. The current booster will increase pump performance under marginal light conditions. The pump will keep operating, though more slowly, under conditions when it would normally stall. The remote control terminals can be connected to a float switch, pressure switch, or other remote control. The control wire can be inexpensive 22 gauge phone cable up to 5,000 ft. in length. Note: the remote control terminals are reverse-logic. An open circuit turns the pump ON, a closed circuit turns the pump OFF. The fixed 24-v controller is the right one for most 2-module systems. The tuneable controller can be adjusted to either 12v or 24v.

25-134RC	Fixed 12-v PV Direct Pump Controller	$89⁰⁰
25-138RC	Fixed 24-v PV Direct Pump Controller	$89⁰⁰
25-134TRC	Tuneable PV Direct Pump Controller	$129⁰⁰

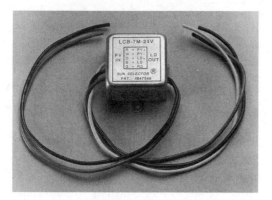

Float Switches

These fully encapsulated mechanical (NO mercury!) float switches make life a little easier by providing automatic control of fluid level in the tank. You regulate the range of water levels by merely lengthening or shortening the power cable. Easily installed, the switches are rated at 15 amps at 12 volts, or 13 amps at 120 volts. The "U" model will fill a tank (closed circuit when float is down, open circuit when up) and the "D" model will drain it (closed circuit when up, open circuit when down). When used with a PV-Direct Pump Controller the "U" and "D" functions are reversed. "D" will fill, "U" will empty.

41-637	Float Switch "U"	$35⁰⁰
41-638	Float Switch "D"	$35⁰⁰

12-Volt to 24-Volt Controller

Want to run your submersible pump from your household 12-volt battery bank, but worried that it will pump too slow? This controller is your salvation. It will draw 12 volts from your batteries and convert it to 24 volts for the pump. Good for up to 10 amps input. Has the same remote control functions as the PV-Direct Controller listed above.

41-633	Controller/Converter, 12V to 24V	$215⁰⁰

Flat 2-Wire Sub Pump Cable

This flat two-conductor 10-gauge submersible cable is sometimes difficult to find locally. Works with the SHURflo Solar Sub splice kit. Cut to length; specify how many feet you need.

26-540	10-2 Sub Pump Cable	$0.64/ft.

WIND POWERED PUMPS

Bowjon Wind-Powered Pumps

Bowjon utilizes a unique "air injection" system that is ideal in areas with high wind velocities. A small air compressor is direct-driven by the turbine. You can place your Bowjon where the wind blows best, as far as 1/4 mile away from your well. The compressed air output is directed to an injector at the bottom of the well. Rising air in the delivery tube carries water bubbles up to a collection tank. It uses five high-torque, heavy-duty aluminum blades. It is quickly installed on a single post tower. One person can assemble it without special rigs, tools, or equipment. Its minimal maintenance only requires checking the compressor oil every six months. The pump has no moving parts, no leathers to replace, and no cylinders or plunger assembly, allowing the Bowjon to run dry without harm.

The Bowjon is available with six- or eight-foot diameter blades, and single or twin in-line compressors. The Bowjon can pump water at 7 gallons per minute. More than one windmill can service the same well. The hub is a massive 17 lbs. of tempered aluminum that will withstand 3,200 psi. Bowjon's five blades are made of 6061-T6 tempered aluminum, each consisting of three layers for the maximum strength that's required in extremely high winds.

Propeller: Tempered triple layer reinforced blades 6'8" or 8'8" across to minimize flex in high winds. Blades have a unique varied pitch to allow low-wind, high-torque start-up. They have high rpm ability.

Compressor: Industrial strength twin or single cylinder depending on model selected for required water demand. Compressor is capable of 5 cubic feet of air volume per minute.

Start-Up: 5 to 8 mph.

Pump: Air injection. No moving parts, no cylinders, valves, rods, or leathers to wear out. Will run dry, accepting silt, sludge, and sand without harm. Length is 5 feet. Minimum well casing diameter is 2 inches.

Air Line: 3/8-inch polyethylene tubing (200 psi rated).

Water Line: 0 to 50 ft. lift: 1-inch tubing or PVC Schedule 40 plastic pipe.

                 50 to 100 ft. lift: 3/4-inch tubing or PVC Schedule 40 plastic pipe.

                 100 ft. lift and over: 1/2-inch tubing or PVC Schedule 40 plastic pipe.

Submersion: At least 30% of the vertical lift distance from static water level in the well to the highest point of delivery or storage. Where minimum recommended submersion is not possible, use of a collector tank and a regulator in the air line will prevent excessive air pressure and volume. Submersions less than 30-feet restrict vertical lift; but where water is only to be lifted a few feet, submersions of 5 or 10-feet with air tank and regulator are practical.

LIFT/SUBMERGENCE RATIOS:

Vertical lift	Submergence below water	Optimum total well depth
50 ft	35 ft	90 ft
80 ft	56 ft	140 ft
120 ft	85 ft	210 ft
200 ft	100 ft (50% of lift)	350 ft

If the submergence is too low for the amount of lift, the air will separate from the water. If the submergence is too high, the air will not lift the water. For these reasons, the submergence of the pump must be calculated carefully.

Homesteader Model Bowjon Pump

The Homesteader model comes with 6'8" diameter blades with a single cylinder compressor. It has a 3/8-inch air line 250 feet long and AL2 air injection pump. One-year warranty.

41-701   Bowjon Pump, Homesteader $1,295
Shipped freight collect from Southern California

Rancher Model Bowjon Pump

The Rancher model comes with 8'8" diameter blades with a twin cylinder compressor. It has a 3/8-inch air line 250 feet long and an AL3 air injection pump. One-year warranty.

41-711   Bowjon Pump, Rancher $1,595
Shipped freight collect from Southern California

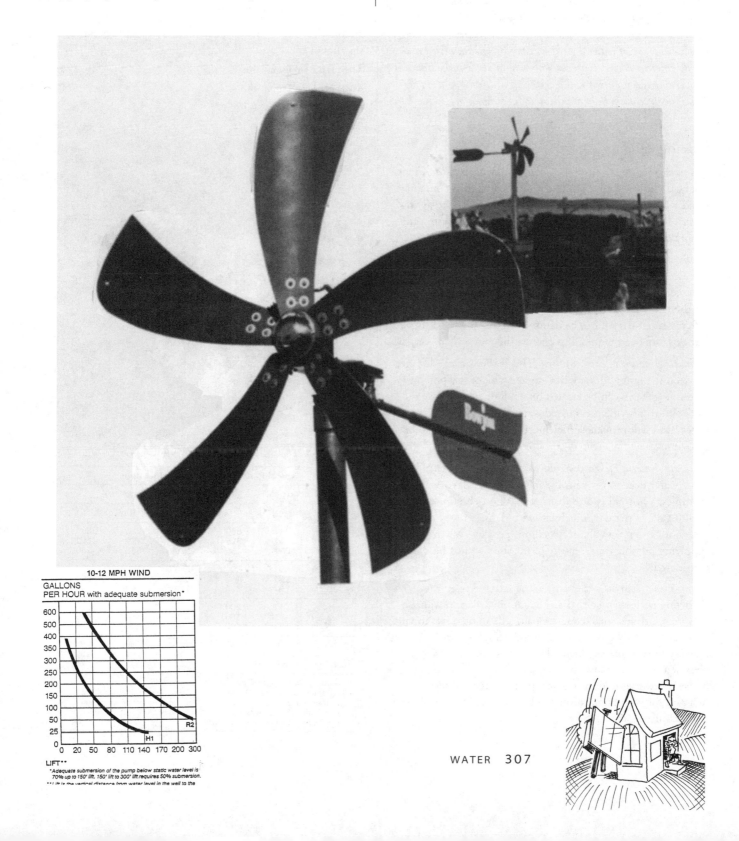

10-12 MPH WIND

GALLONS
PER HOUR with adequate submersion*

LIFT**

*Adequate submersion of the pump below static water level is 70% up to 150' lift. 150' lift to 300' lift requires 50% submersion.

**Lift is the vertical distance from water level in the well to the

WATER-POWERED PUMPS

Ram Pumps

The ram pump works on a hydraulic principle using the liquid itself as a power source. The only moving parts are two valves. In operation, the water flows from a source down a "drive" pipe to the ram. Once each cycle, a valve closes causing the water in the drive pipe to suddenly stop. This causes a water-hammer effect and high pressure in a one-way valve leading to the "delivery" pipe of the ram, thus forcing a small amount of water up the pipe and into a holding tank. In essence, the ram uses the energy of a large amount of water falling a short distance to lift a small amount of water a greater distance. The ram itself is a highly efficient device; however, only 2% to 10% of the liquid is recoverable. Ram pumps will work on as little as 2 gpm supply flow. The maximum head or vertical lift of a ram is about 500 ft.

Selecting A Ram

Estimate Amount of Water Available to Operate the Ram. This can be determined by the rate the source will fill a container. Make sure you've got more than enough water to satisfy the pump. If a ram runs short of water it will stop pumping and simply dump all incoming water.

Estimate Amount of Fall Available. The fall is the vertical distance between the surface of the water source and the selected ram site. Be sure the ram site has suitable drainage for the tailing water. Rams splash big-time when operating! Often a small stream can be dammed to provide the 1-1/2 feet or more head required to operate the ram.

Estimate Amount of Lift Required. This is the vertical distance between the ram and the water storage tank or use point. The storage tank can be located on a hill or stand above the use point to provide pressurized water. Forty or fifty feet water head will provide sufficient pressure for household or garden use.

Estimate Amount of Water Required at the Storage Tank. This is the water needed for your use in gallons per day. As examples, a normal two to three person household uses 100 to 300 gallons per day, much less with conservation. A 20 by 100 foot garden uses about 50 gallons per day. When supplying potable water, purity of the source must be considered.

Using these estimates, the ram can be selected from the following performance charts. The ram installation will also require pouring a small concrete pad, a drive pipe five to ten times as long as the vertical fall, an inlet strainer, and a delivery pipe to the storage tank or use point. These can be obtained from your local hardware or plumbing supply house. Further questions regarding suitability and selection of a ram for your application will be promptly answered by our technical staff.

Aqua Environment Rams

We've sold these fine rams by Aqua Environment for over 15 years now with virtually no problems. Careful attention to design has resulted in extremely reliable rams with the best efficiencies and lift-to-fall ratio available. Working component construction is of all bronze with O-ring seal valves. Air chamber is PVC pipe. The outlet gauge and valve permit easy start up.

Each unit comes with complete installation and operating instructions.

41-811	Ram Pump, 3/4"	$249.00
41-812	Ram Pump, 1"	$249.00
41-813	Ram Pump, 1-1/4"	$299.00
41-814	Ram Pump, 1-1/2"	$299.00

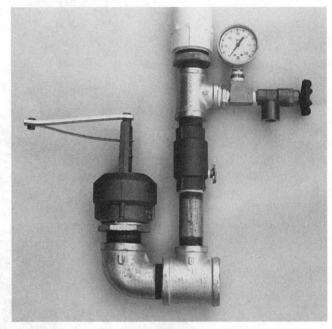

Typical Performance and Specifications

Vertical Fall (feet)	Vertical Lift (feet)	Pump Rate (Gallons/Day)			
		3/4" Ram	1" Ram	1¼" Ram	1½" Ram
20	50	650	1350	2250	3200
20	100	325	670	1120	1600
20	200	150	320	530	750
10	50	300	650	1100	1600
10	100	150	320	530	750
10	150	100	220	340	460
5	30	200	430	690	960
5	50	100	220	340	460
5	100	40	90	150	210
1.5	30	40	80	130	190
1.5	50	20	40	70	100
1.5	100	6	12	18	25

Water Required to Operate Ram

3/4" Ram - 2 gallons/minute	Maximum Fall - 25 feet
1" Ram - 4 gallons/minute	Minimum Fall - 1.5 feet
1¼" Ram - 6 gallons/minute	Maximum Lift - 250 feet
1½" Ram - 8 gallons/minute	

High Lifter Pressure Intensifier Pump

The High Lifter Water Pump offers unique advantages for the rural user. Developed expressly for mountainous terrain and low summertime water flows, this water-powered pump delivers a much greater percentage of the input water than ram pumps can. This pump is available in either 4.5:1 or 9:1 ratios of lift to fall. For example, assume a flow of 2 gallons per minute and a fall of 40-feet from a water source (spring, pond, creek, etc.) to the High Lifter pump with a 200-foot rise (head) from the pump up to a holding tank. The lift-fall ratio between the 40-foot fall and the 200-foot lift would be 5:1. The High Lifter pump (in 9:1 ratio) would deliver approximately 300 gallons of water per day from this working ratio. The 4.5:1 ratio pump wouldn't work at all in this application. The High Lifter is self-starting and self-regulating. If inlet water flow slows or stops the pump will slow or stop, but will self-start when flow starts again.

The High Lifter pump has many advantages over a ram (the only other water-powered pump). Instead of using a "water hammer" effect to lift water as a ram does, the High Lifter is a positive displacement pump that uses pistons to create a kind of hydraulic lever that converts a larger volume of low-pressure water into a smaller volume of high-pressure water. This means that the pump can operate over a broad range of flows and pressures with great mechanical efficiency. This efficiency means more recovered water. While water recovery with a ram is normally about 5% or less, the High Lifter recovers 1 part in 4.5 or 1 part in 9 depending on ratio.

In addition, unlike the ram pump, no "start up tuning" or special drive lines are necessary. This pump is quiet, and will happily run unattended.

The High Lifter pressure intensifier pump is economical compared to gas and electric pumps, because no fuel is used and no extensive water source development is necessary.

A kit to change the working ratio of either pump after purchase is available, as are maintenance kits. Maintenance consists of simply replacing a handful of O-rings. The cleaner your input water, the longer the O-rings last. Choose your model High Lifter pump from the specifications and High Lifter performance curves. One year parts and labor warranty from mfg.

41-801	High Lifter Pump, 4.5:1 Ratio	$895⁰⁰
41-802	High Lifter Pump, 9:1 Ratio	$895⁰⁰
41-803	High Lifter Ratio Conversion Kit	$97⁰⁰
41-804	High Lifter Rebuild Kit, 4.5:1 ratio	$49⁰⁰
41-805	High Lifter Rebuild Kit, 9:1 ratio	$49⁰⁰

TYPICAL APPLICATION

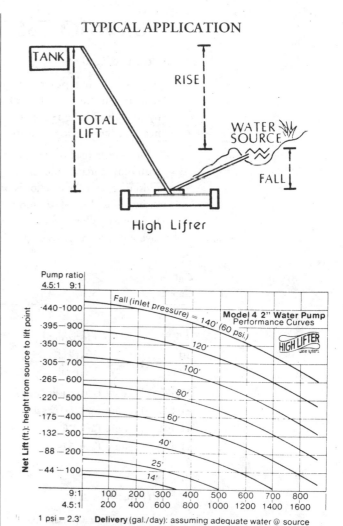

High Lifter

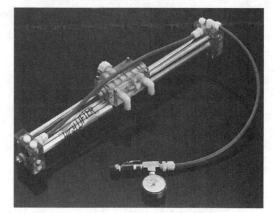

Model 4 2" Water Pump Performance Curves

1 psi = 2.3'   **Delivery** (gal./day): assuming adequate water @ source

PUMP ACCESSORIES

Pressure Tanks

Any pressure pump system requires a pressure tank. It prolongs the life of the pump by reducing the number of on/off demands. Some very small, single faucet water systems can get by with the 2-gallon pressure tank. For anything larger, or any system using a tankless water heater, twenty gallon capacity is the minimum recommended.

Our precharged pressure tanks have a captive bladder air charge which cannot be absorbed as with old standard style tanks. The interior of the tanks are epoxy-coated for corrosion resistance, and the exterior is baked enamel. They are shipped precharged at 20 psi, but are easily adjustable. Maximum working pressure is 100 psi, maximum water temperature is 120° F. Twenty gallon is the largest size that can be shipped by UPS. For larger systems you may save on shipping costs by purchasing locally.

Gallon Capacity	Style	Maximum Drawdown	Connection	Diameter	Height
2	Vertical	1.1 gal	3/4" MIPT	9"	11.25"
8.5	Horizontal	4.3 gal	3/4" MIPT	12.7"	14"
20	Vertical	8 gal	1" FIPT	16"	29.75"
32	Vertical	15 gal	1" FIPT	21"	28-3/4"

41-401	Pressure Tank, 2 gal.	$59⁰⁰
41-406	Pressure Tank, 8-1/2 gal.	$165⁰⁰
41-404	Pressure Tank, 20 gal.	$219⁰⁰
41-407	Pressure Tank, 32 gal.	$399⁰⁰

Pressure Switch

You must use a pressure switch for pressure tank systems. It will turn the pump off when the tank pressure reaches 40 psi, and then on again when the pressure drops to 20 psi. It also turns the pump off when pressure drops below 10 psi to prevent damage from running the pump dry. The settings are fully adjustable in the range of 5 to 65 psi.

41-140	Pressure Switch	$24⁰⁰

Pressure Tank

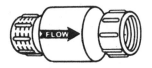

Check Valves

A check valve must be installed after the pump in pressurized systems. Check valves seal the water system between the pump and the faucets in the home. It prevents reverse flow of water back into the pump and maintains pressure switch settings. All of our check valves have accessory holes for a pressure gauge and pressure switch.

41-410	Check Valve, Bronze 3/4"	$18⁰⁰
41-411	Check Valve, Bronze 1"	$19⁰⁰
41-412	Check Valve, Bronze 1-1/4"	$23⁰⁰
41-413	Check Valve, Bronze 1-1/2"	$28⁰⁰
41-414	Check Valve, Bronze 2"	$49⁰⁰

Foot Valves

Priming a pump is necessary to ensure that it isn't run dry. A foot valve prevents the need to do this every time that you start the pump. It is a one-way valve with an attached screen that lets water into the inlet line when the pump is running, but won't let it drain back and lose the prime when the pump is off.

41-420	Foot Valve, Bronze 3/4"	$20⁰⁰
41-421	Foot Valve, Bronze 1"	$21⁰⁰
41-422	Foot Valve, Bronze 1-1/4"	$24⁰⁰
41-423	Foot Valve, Bronze 1-1/2"	$39⁰⁰
41-424	Foot Valve, Bronze 2"	$54⁰⁰

Pressure Gauges

Pressure gauges make system monitoring and troubleshooting much easier. We stock gauges in a variety of pressure ranges. All gauges are 2" diameter with a 1/4 inch MIPT fitting.

41-442	Pressure Gauge 0 to 30 psi	$12⁰⁰
41-443	Pressure Gauge 0 to 60 psi	$12⁰⁰
41-444	Pressure Gauge 0 to 100 psi	$12⁰⁰
41-445	Pressure Gauge 0 to 300 psi	$17⁰⁰

Line Strainers

Most high-head and positive displacement pumps are particularly vulnerable to dirty water. These 80 mesh strainers are all the protection a diaphragm pump will need. <u>Don't</u> use these on rotary vane pumps, they need finer 5-micron protection. These line strainers have clear bowls so you can see if they need to be cleaned. Inlet and outlet are both FIPT.

42-508	Line Strainer, 80 Mesh 3/4"	$44⁰⁰
42-509	Line Strainer, 80 Mesh 1"	$79⁰⁰

Float Switches

These fully encapsulated mechanical (NO mercury!) float switches make life a little easier by providing automatic control of fluid level in the tank. You regulate the range of water levels by merely lengthening or shortening the power cable. Easily installed, the switches are rated at 15 amps/12-volt, or 13 amps at 120-volt. The "U" model will fill a tank (closed circuit when float is down, open circuit when up) and the "D" model will drain it (closed circuit when up, open circuit when down). When used with a PV-Direct Pump Controller, the "U" and "D" functions are reversed. "D" will fill, "U" will empty.

41-637	Float Switch "U"	$35⁰⁰
41-638	Float Switch "D"	$35⁰⁰

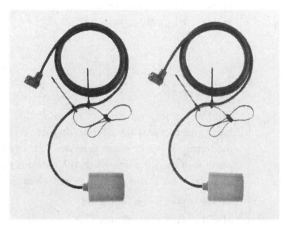

PV-Direct Pump Controller

A compact weather-tight package with built-in 7-amp linear current booster and remote control capability. The current booster will increase pump performance under marginal light conditions. The pump will keep operating, though more slowly, under

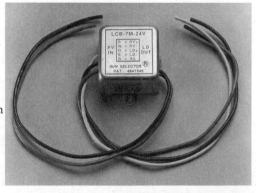

conditions when it would normally stall. The remote control terminals can be connected to a float switch, pressure switch, or other remote control. The control wire can be inexpensive 22 gauge phone cable up to 5,000 ft. in length. Note: the remote control terminals are reverse-logic. An open circuit turns the pump ON, a closed circuit turns the pump OFF. The fixed 24-v controller is the right one for most 2-module systems. The tuneable controller can be adjusted to either 12v or 24v.

25-134RC Fixed 12-v PV-Direct Pump Controller	$89⁰⁰
25-138RC Fixed 24-v PV-Direct Pump Controller	$89⁰⁰
25-134TRC Tuneable PV-Direct Pump Controller	$129⁰⁰

WATER PURIFICATION

Water is vital to our bodies. Like the surface of the Earth, our bodies are mostly water. The average adult contains 40 to 50 quarts of water, which must be renewed every 10 to 15 days. You must drink at least six glasses of water daily to enable your body to function properly. Water is the basis of all life.

Your Body Depends upon What You Drink

The EPA has recently released information stating that, no matter where we live in the U.S., there is likely to be some toxic substance in our groundwater. Even some of the chemicals that are added to our drinking water to protect us (such as chlorine) have been linked to certain cancers and can form toxic compounds (such as trihalomethanes, or THMs).

The adage, "If you want something done, you have to do it yourself," certainly applies to drinking water. The most obvious solution to pollution is a point-of-use water purification device. The tap is the end of the road for water consumed by our families. There are no pipes or conduits to leach undesirable elements into our drinking water beyond this point.

To make the best choice for a water purification system, we must first examine the various problems we can encounter with water quality.

Biological Impurities: Bacteria, Viruses, and Parasites: Modern municipal water supplies are relatively free from harmful organisms because of frequent monitoring and routine disinfection with chlorine or chloramines. This does not mean, however, that municipal water is free of all bacteria.

People with private wells or small rural water systems have reason to be concerned about the possibility of microorganism contamination from septic tanks, animal wastes, or other sources.

Approximately 4000 cases of waterborne diseases are reported every year in the U.S. Many of the minor illnesses and gastrointestinal disorders that go unreported can be traced to organisms found in water supplies.

Inorganic Impurities: Dirt and Sediment, or Turbidity: Most water contains suspended particles of fine sand, clay, silt, and precipitated salts. The cloudiness or muddiness of water is called turbidity. Not only is this turbidity unsightly, but it can also be a source of food and lodging for bacteria, and can interfere with effective disinfection.

Total Dissolved Solids (TDS): Total dissolved solids consist of rock and other compounds from the earth. The entire list could fill this page. The presence and amount of total dissolved solids (TDS) in water represents a point of controversy among water purveyors.

Here are some facts about the consequences of higher levels of TDS in water:
1. High TDS results in undesirable taste, which can be salty, bitter, or metallic.

2. Some individual mineral salts may pose health hazards. The most problematic are nitrates; sodium, barium, and copper sulfates; and fluoride.
3. High TDS interferes with the taste of foods and beverages.
4. High TDS makes ice cubes cloudy, softer, and faster melting.

Toxic Metals or Heavy Metals: Among the greatest threats to health are the presence of toxic metals in drinking water: arsenic, cadmium, lead, mercury, and silver. Maximum limits for each are established by the EPA's Primary Drinking Water Regulations.

Asbestos: Asbestos exists as microscopic suspended mineral fibers in water. Its primary source is asbestos-cement pipe, which was commonly used after World War II for city water supplies. It has been estimated that some 200,000 miles of this pipe are currently in use transporting drinking water. Because pipes wear as water courses through them, asbestos shows up with increasing frequency in drinking water. It has been linked to gastrointestinal cancer.

Organic Impurities, Tastes, and Odors: If water has a disagreeable taste or odor, the likely cause is one or more organic substances, ranging from decaying vegetation to algae, hydrocarbons to phenols.

Pesticides and Herbicides: The increasing use of pesticides and herbicides in agriculture has a profound effect on water quality. Rain and irrigation carry these deadly chemicals into groundwater, as well as into surface waters. Since more than 100 million people in the U.S. depend upon some groundwater source for their drinking water, this increasing contamination is a major concern.

Toxic Organic Chemicals: The most pressing and widespread water contamination problem results from the organic chemicals created by industry. The American Chemical Society listed 4,039,907 distinct chemical compounds as of 1977, and they began their list only in 1965. The list can grow by as many as 6000 chemicals per week!

Every year approximately 115,000 manufacturers produce more than 1.3 billion pounds of different chemicals in the U.S., an industry that is worth $113 billion. This is a very difficult juggernaut to try to stop or change. There is almost certainly some kind of toxic substance in our groundwater, no matter where we live in the U.S.

The EPA says there are 77 billion pounds of hazardous waste generated each year in the U.S., 90% of which is disposed of improperly. This equals 19,192 pounds of hazardous waste for every square mile of land and water surface in the U.S. (including Alaska and Hawaii).

There are 181,000 man-made lagoons at industrial and municipal sites, 75% of which are unlined. Many are within a mile of wells or water supplies. Information on the location of these sites, their condition, and the contaminants they contain ranges from sketchy to nonexistent. Will this be the horror story of the millennium?

Chlorine: Trihalomethanes (THMs) are formed when chlorine, used to disinfect water supplies, interacts with natural organic materials (for instance, by-products of decaying vegetation, algae, etc.). This interaction creates toxic organic chemicals such as chloroform, and bromodichloromethane. Chlorinated water has been linked to cancer, high blood pressure, and anemia. Anemia is caused by the deleterious effect of chlorine on red blood cells.

METHODS FOR SOLVING OUR WATER PROBLEMS

What are the alternatives for the seeker of pure water? There are a mind-boggling number of water systems on the market. The following is a brief analysis of each option's strengths and weaknesses.

Bottled Water

An increasing number of people find their solution for safe drinking water by paying anywhere from 80 cents to $2.00 per gallon to drink water prepared and bottled by someone else. The price reflects the costs of bottling, storage, trucking, fuel expenses, wages, insurance, and advertising, making bottled water extremely cost-ineffective, particularly since some water sources have proved inferior to what comes out of the tap.

Point-of-Use Water Treatment

The most efficient and cost-effective solution to water purity is to treat *only* the water you plan to consume. A point-of-use water system eliminates the middleman costs associated with bottled water, and can provide purified water for pennies per gallon. Devices for point-of-use water treatment are available in a variety of sizes, designs, and capabilities.

Mechanical Filtration: Mechanical filtration acts much like a fine strainer. Particles of suspended dirt, sand, rust, and scale (in other words, turbidity) are trapped and retained, greatly improving the clarity and appeal of water.

When enough of this particulate matter has accumulated, the filter is discarded. This type of filter is called a pre-filter.

Activated Carbon Adsorption: Carbon adsorption is the most widely sold method for home water treatment because of its ability to improve water by removing disagreeable tastes and odors, including objectionable chlorine.

Activated carbon is processed from a variety of carbon-based materials, such as coal, petroleum, nut shells, and fruit pits steamed to high temperatures in the absence of oxygen (the activation process). This process leaves millions of microscopic pores and crevices. One pound of activated carbon provides anywhere from 60 to 150 *acres* of surface area. The pores trap microscopic particles and large organic molecules, while the activated surface areas cling to, or *adsorb*, the smaller organic molecules.

While activated carbon theoretically has the ability to remove numerous

organic chemicals like pesticides, THMs, trichloroethylene (TCE), and PCBs, the actual effectiveness is highly dependent on several factors:

1. The type of carbon and the amount used.
2. The design of the filter and the rate of water flow (contact time).
3. How long the filter has been in use.
4. The types of impurities the filter has previously removed.
5. Water conditions (for example, turbidity, temperature, etc.)

A disadvantage of carbon filters is that they can provide a base for the growth of bacteria. When the carbon is fresh, practically all organic impurities and even some bacteria are removed. Accumulated impurities, though, can become food for bacteria, enabling them to multiply within the filter. The high concentration is considered by some to be a health hazard.

After periods of non-use (such as overnight), a quantity of water should be flushed through a carbon filter to minimize the accumulation of bacteria.

Chemical Recontamination of Carbon Filters: Another weakness of carbon filters is the chemical recontamination which can occur when the carbon surface becomes saturated with the impurities it has adsorbed — a point that is impossible to predict. If use of the carbon continues, the trapped organics can release from the surface and recontaminate the water with more impurities than those contained in the untreated tap water.

To maximize the effectiveness of carbon, it should be kept scrupulously clean of sediment and heavy organic impurities such as the by-products of decayed vegetable matter and microorganisms. These impurities will prematurely consume the carbon's capacity and prevent it from doing what it does best — adsorbing lightweight, toxic organic impurities like THMs and TCE, and undesirable gases, such as chlorine.

Solid Block Carbon: This is obtained when very fine pulverized carbon is compressed and fused with a binding medium (such as a polyethylene plastic) into a solid block. The intricate maze developed within the block insures contact with organic impurities and, therefore, effective removal. The problem of channeling (open paths developing because of the buildup of impurities, and rapid water movement under pressure) in a loose bed of granulated carbon granules is eliminated by solid block filters.

Block filters can be fabricated to have such a fine porous structure that they are capable of mechanically filtering out coliform and associated disease bacteria.

Among the disadvantages of compressed carbon filters are their reduced capacity due to the inert binding agent, and their tendency to plug up with particulate matter, requiring more frequent replacement. They are also substantially more expensive than conventional carbon filters.

Limitations of Carbon Filters: A properly designed carbon filter is capable of removing many toxic organic contaminants, it but falls short of providing protection from the wide spectrum of impurities which have been referred to previously.

1. Carbon filters are not capable of removing any of the excess total dissolved solids (TDS).

2. Only a few solid block or carbon matrix systems have been certified for the removal of lead, asbestos, VOCs (volatile organic compounds), cysts, fecal coliform, and other disease bacteria. Large *suspended* materials will be removed by some filters. However, small *dissolved* materials can't be removed by carbon filtration.

3. Carbon filters have no effect on harmful nitrates, or on high sodium and fluoride levels.

4. For any carbon filter to be effective (even for organic removal), water must pass through the carbon (whether it be granular or compressed) slowly enough so that complete contact is made. This all-important factor is referred to in the industry as *contact time*. At useful flow rates of 0.5 to 1.0 gallon per minute, the flow rate is determined by the amount of carbon and flow restricters used in filtration.

One must carefully read the data sheets provided by responsible filter manufacturers to verify claims. Many companies are certified with the National Sanitation Foundation (NSF), whose circular logo appears on their data sheets.

Minerals in Drinking Water: Manufacturers and sellers of activated carbon filtration systems often claim, "We need minerals in water — they are essential for good health." This statement has *never been proven* by scientific studies. Therefore, the value of minerals in drinking water remains to be established. Filter proponents make their point about minerals to obscure the fact that their product will not remove dissolved solids. The same process that allows dissolved minerals to pass through a filter into your tap water, also allows total dissolved solids, hardness, and some heavy metals.

Carbon Filters in Summary: Activated carbon filters are an important piece of the purification process, although only a piece. Activated carbon removes chemicals and gases, but it will not affect total dissolved solids, hardness, or heavy metals.

Ultrafiltration/Reverse Osmosis: In the *reverse osmosis* (RO) system, water having a lesser concentration of substances is derived from water having a higher concentration of substances. Tap water with dissolved solids is forced by the water pressure inside our pipes against a membrane. The water penetrates the RO membrane, which is carefully formulated to leave up to 99% of the dissolved solids behind.

The RO membrane is the ultimate mechanical filter, with a pore size measuring 2×10^{-8} (two 100-millionths) inch in diameter — too small to be seen even by an optical microscope!

Through the remarkable phenomenon of RO, even particles smaller than water molecules can be removed! The molecules diffuse through the membrane in a purified state, and collect on the opposite side.

Ultrafiltration/RO membranes remove and reject a broad spectrum of impurities from water using *minimal added energy* — just water pressure. RO gives the best water available for the lowest price.

Reverse osmosis effectively reduces the following:
1. Particulate matter, turbidity, sediment
2. Colloidal matter
3. Total dissolved solids (up to 99%)
4. Toxic metals
5. Radium 226/228
6. Microorganisms (potable water only)
7. Asbestos
8. Pesticides and herbicides (when used with activated carbon)

Reverse Osmosis and Activated Carbon Adsorption: Ultrafiltration/RO alone will not remove all of the lighter (low molecular weight) volatile organic compounds, such as THMs, TCE, vinyl chloride, carbon tetrachloride, and others. These compounds are too small to be removed by the straining action of the RO membrane, and their chemical structure is such that they are not repelled by the membrane surface. Since these are some of the most toxic of the contaminants found in tap water, it is very important that a well-designed carbon filter be used in conjunction with the RO membrane.

In some applications, activated carbon is used before the membrane. In *all* quality RO systems, however, there is activated carbon after the membrane. This arrangement means that the post-membrane activated carbon filter doesn't have to contend with bacteria and all of the other materials which cause fouling and impair performance.

Not all RO systems are created equal, either in performance or price. The engineering and experience behind the RO design is critical to overall performance and dependability.

Note: The typical time required to purify one gallon of RO water is three to four hours, since RO uses water to purify water. This time period is known as the *rate of recovery*. RO units will use anywhere from three to nine gallons of brine (wastewater) to make one gallon of purified water.

Brine is necessary to remove excess accumulated materials from the RO membrane; if these materials are not flushed from the system, they will impair efficiency. One can direct brine water outside with a drip line.

The cost of water energy for a fine RO system will be about $1.33 per month if one pays for water at the rate of $1.00 per 100 cubic feet.

HOW TO GUARANTEE THE QUALITY OF YOUR DRINKING WATER

Your primary decision is how comprehensive a water treatment system should you purchase and install. A system which combines technologies will produce better water than a system incorporating just one.

Choose technologies that will meet your needs for a long time. If you are making a commitment to pure water, begin to enjoy the benefits today.

Water Filtration Systems

Are you a candidate for a water filter or purifier, and, if so, how do you decide which system is appropriate for you? These are questions we hope to help you answer.

The majority of people who purchase a water purifier are previous users of bottled water. If you are still using tap water, chances are you haven't "bought in" to the need to spend money on better taste and quality in water. Once a person changes water habits, the two most popular options are to purchase bottled water, or to install a point-of-use drinking water system.

The word "filter" usually refers to a carbon drinking water system, which reduces certain impurities as verified by independent testing. A "purifier" refers to a slower process, such as reverse osmosis, which also greatly reduces dissolved solids, hardness, and organics. A purifier takes several hours to process each gallon of water. Make sure that the claims of any purifier you consider have been verified by independent testing.

Be aware that certain states require third-party documentation of the performance of any water system claiming to reduce contaminants that are known to have adverse health effects. (These contaminants are known collectively as *health-related contaminants.*)

Here is a partial list of items which are considered to be health-related contaminants: total coliform bacteria and turbidity, arsenic, barium, cadmium, chromium, fluoride, lead, mercury, nitrate, selenium, silver, benzene, carbon tetrachloride, 1,2-dichloroethane, 1,1-dichloroethylene, p-dichlorobenzene, endrin, lindane, methoxychlor, total trihalomethanes, toxaphene, trichloroethylene, 1,1,1-trichloroethane, vinyl chloride, 2,4,5-TP (silvex), 2,4-D, and radium 226/228.

FILTER TYPES AND WHAT THEY REMOVE

Sediment Filters: The most common size is 5 microns. This means that any particle 5 microns or larger is trapped and removed from the water. This is often used as a pre-filter in order to keep a carbon or reverse osmosis filter from becoming clogged prematurely.

Activated Carbon: Depending on how it is processed, carbon can remove lead, asbestos, cysts, fecal coliform, volatile organic compounds, and chlorine. The most common use today is for removing lead and chlorine from city or municipal water supplies.

Ceramic: Ceramic filters are used to remove bacteria from the water supply. They typically filter out particles between 0.9 and 0.2 microns in size with 0.2 being the smallest.

Reverse Osmosis (RO): RO will filter total dissolved solids (TDS), particulate matter, turbidity, sediment, radium 226/228, fluoride, asbestos, heavy metals and microorganisms (potable water only).

There are three kinds of bottled water: distilled, purified, and spring. *Distillation* evaporates the water, then recondenses it, thereby theoretically leaving all impurities behind. *Purified water* is usually prepared by reverse osmosis, de-ionization, or a combination of both processes. *Spring water* is either acquired from a mountain spring or artesian well, or is no more than processed tap water. Spring water will generally have higher total dissolved solids than purified water. (Distilled water and purified water are better for battery use and steam irons because of their lower content of total dissolved solids.)

Without trying to overcomplicate the decision-making process, here is a simple exercise to help you determine if you will be happier with a filter or a purifier.

Step 1: Look for total dissolved solids, hardness, and pH written on the label of your favorite bottled water, and write down the results. If not listed on the label, call up the bottler of your water (they often have a toll-free 800 number or can be found through directory assistance) and ask for the information. Specify that you want data both for the normal levels of these substances and for *how high they range.*

Next (you're not quite through): call your own municipal supplier of tap water. Speak with a water chemist at the water company. Remember to ask for the normal amounts as well as how high they range — this is very important. Obtain the same information for these six items:

TAP WATER RESULTS

Total dissolved solids (TDS):
Hardness:
pH:
Residual chlorine:
Iron:
Manganese:

Step 2: Compare the first three bottled water results with the first three tap water results. How close are the numbers of your tap water to the bottled water? If your bottled water is higher in TDS and hardness than your tap water, a good filter should satisfy your needs.

Step 3: If your bottled water is lower in TDS and hardness, then you should consider a good purifier. The odds are against you being satisfied with a filter alone, because your tap water is "heavier" in TDS and hardness than your preferred bottled water.

If you are not drinking bottled water at present, then a good filter should provide enough improvement over tap water to make you thrilled with your new purchase.

If You Are Using Your Own Well Water: Wells can have problems with staining, sediment, hydrogen sulfide (rotten-egg smell), iron, manganese, tannin, etc. These will often be accompanied by telltale signs. These are all pre-treatment problems, and must be corrected before you purchase any point-of-use water treatment device. Knowing this beforehand can save you headaches and money. A water test is advisable if you have pre-treatment problems. If you are unconcerned about coliform bacteria or lead there is a low-priced test available. If, however, you are concerned about bacteria, lead, and other heavy metals, pesticides, and volatile organic compounds, (see the Resource List in the Appendix), then the comprehensive 93-item test conducted by National Testing Laboratories (NTL) is recommended.

If you spend the time to obtain the figures for your tap water as outlined in this article, you will be able to purchase a water treatment system that you'll love and respect.

— Real Goods Staff

WATER PURIFICATION

National Testing Laboratories Extensive Water Test

We work with National Testing Laboratories (NTL) to provide a thorough analysis of your water. Now that the EPA has determined that the groundwater in more than 30 states is seriously polluted, we feel it's essential to thoroughly test your water. NTL has two laboratory facilities, one in Michigan and one in Florida, which perform a full range of drinking water analysis on inorganic, organic, and bacteriological attributes. Their laboratories are certified in 13 states to perform analyses of drinking water, using only U.S. EPA-approved methods, and a strict quality assurance program. We've compared National Testing Laboratories with other testing services and find them to be comprehensive, accurate, and reasonably priced. Their program is simple: you order the test kit from us, we mail it directly to you, you fill the sample bottles and ship the kit to the NTL lab. NTL will analyze the samples according to the test series purchased and will send the full report and explanatory letter back to you within ten working days. Your test will show if your water contains any of the listed pollutants in amounts higher or lower than EPA limits. A cover letter interprets your test and clearly explains what action (if any) is recommended to ensure that your water is safe to drink.

NTL Water Check With Pesticide Option

This is the best analysis of your water available in this price range. Your water will be analyzed for 73 items, plus 20 pesticides. If you have any questions about your water's integrity, this is the test to give you peace of mind. The kit comes with five water sample bottles, a blue gel refrigerant pack (to keep bacterial samples cool for accurate test results), and easy-to-follow sampling instructions. You'll receive back a two-page report showing 93 contaminant levels, together with explanations of which contaminants, if any, are above allowed values. You'll also receive a follow-up letter with a personalized explanation of your test results, plus knowledgeable, unbiased advice on what action you should take if your drinking water contains contaminants above EPA-allowed levels. NTL, located outside of Detroit, must receive the package within 30 hours of sampling in order to provide accurate results. If you live in Alaska or Hawaii, you will need express shipping. Check with your shipper. The test will check for the accompanying parameters and pesticides.

42-003   NTL Check & Pesticide Option $149⁹⁵

METALS
Aluminum
Arsenic
Barium
Cadmium
Chromium
Copper
Iron
Lead
Manganese
Mercury
Nickel
Selenium
Silver
Sodium
Zinc

INORGANICS & PHYSICAL FACTORS
Total alkalinity (as $CaCO_3$)
Chloride
Fluoride
Nitrate (as N)
Nitrite
Sulfate
Hardness (as $CaCO_3$)
pH
Total dissolved solids
Turbidity (NTU)

VOLATILE ORGANICS (VOCs)
Bromoform
Bromodichloromethane
Chloroform
Dibromochloromethane
Total trihalomethanes
Benzene
Vinyl chloride
Carbon tetrachloride
1,2-Dichloroethane
Trichloroethylene (TCE)
1,4-Dichlorobenzene
1,1-Dichloroethylene
1,1,1-Trichloroethane
Acrolein
Acrylonitrile
Bromobenzene
Bromomethane
Chlorobenzene
Chloroethane
Chloromethane
0-Chlorotoluene
P-Chlorotoluene
Dibromochloropropane (DBCP)
Dibromomethane

1,2-Dichlorobenzene
1,3-Dichlorobenzene
trans-1,2-Dichloroethylene
cis-1,2-Dichloroethylene
Dichloromethane
1,1-Dichloroethane
1,1-Dichloropropene
1,2-Dichloropropane
trans-1,3-Dichloropropane
cis-1,3-Dichloropropane
2,2-Dichloropropane
Ethylenedibromide (EDB)
Ethylbenzene
Styrene
1,1,2-Trichloroethane
1,1,1,2-Tetrachloroethane
1,1,2,2-Tetrachloroethane
Tetrachloroethylene (PCE)
1,2,3-Trichloropropane
Toluene
Xylene
Chloroethylvinyl ether
Dichlorodifluoromethane
cis-1,3-Dichloropropene
Trichlorofluoromethane
Trichlorobenzene(s)

MICROBIAL
Coliform bacteria

PESTICIDES & HERBICIDES
Alachlor
Aldrin
Atrazine
Chlordane
Dichloran
Dieldrin
Endrin
Heptachlor
Heptachlor Epoxide
Hexachlorobenzene
Hexachlorapentadiene
Lindane
Methoxychlor
PCBs
Pentachloronitrobenzene
Simazine
Toxaphene
Silvex 2,4,5-TP
2,4-D

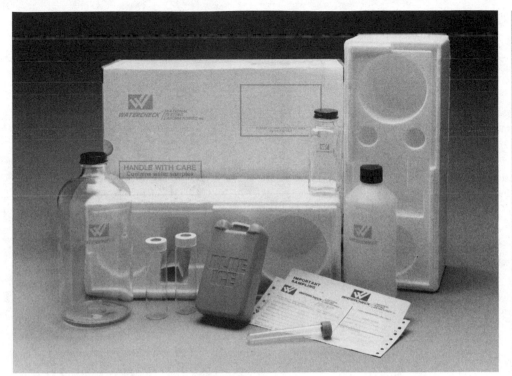

NTL Water Check Unit

Laboratory Water Test
For Private Water Supplies (Private Wells, Springs, Catchment, etc.)

We've found through lots of experience that if you're not on municipal or pretreated water, we can't conscientiously sell you a water purification system without first knowing these specific particulars about the water you have: pH, total dissolved solids (TDS), hardness, iron, manganese, copper, and tannin. This test normally costs $40, but we're providing it to our customers for only $17. This analysis tells us enough about your water so that we can make an informed recommendation to you regarding what system will work best with the water you have.

Upon receipt of your $17 we will send you a questionnaire and a small plastic bottle. Fill out the questionnaire, return the water sample, and you'll have your results back in a few weeks. It's the only way we can both be sure you're getting the right filter for your system! This test does not show contamination problems such as lead, coliform bacteria, or chemicals. Refer to our water check with pesticide option for this.

Remember, if you're on a municipal or pretreated water system we don't need to test your water to recommend a filtration system for you.

42-000   Water Test (non-city water) $19⁹⁵

Build Your Own Solar Distiller

A simple, low-energy way to get purified water from any source, even vegetable matter like grass clippings! The Solar Spring is a set of easy-to-follow plans that will produce a 2-foot by 4-foot solar still. Output quantity averages one to four liters per day (varies widely in response to available sun and temperature). For the highest purity, glass is used on all surfaces that contact water. Well-written with plenty of drawings. Paperback, 10 pages, cottage-industry published.

80-336 The Solar Spring (plans only) $9⁹⁵

Katadyn Pocket Filter

The Katadyn Pocket Filter is standard issue with the International Red Cross and the armed forces of many nations, and is essential equipment for survival kits. Manufactured in Switzerland for over half a century, these filters are of the highest quality imaginable, reminiscent of Swiss watches. The Katadyn system uses an extremely fine 0.2-micron ceramic filter that thoroughly blocks pathological organisms from entering your drinking water. A self-contained and very easy-to-use filter about the size of a 2-cell flashlight (10" x 2"), it will produce a quart of ultra-pure drinking water in 90 seconds with the simple built-in hand pump. It weighs only 23 oz. It comes with its own travel case, and a special brush to clean the ceramic filter. The replaceable ceramic filter can be cleaned 400 times, lasting for many years with average use. This filter is indispensable for campers, backpackers, fishermen, mountaineers, river runners, globetrotters, missionaries, geologists, and workers in disaster areas.

42-608	Katadyn Pocket Filter	$289.00
42-609	Replacement Filter	$139.95

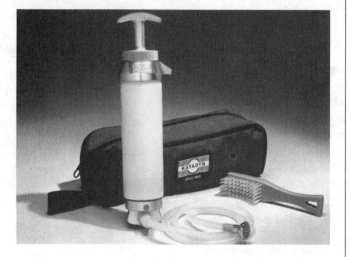

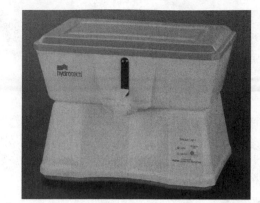

Hydrotech Countertop R.O. Drinking Water System

This state-of-the-art Reverse Osmosis system features the patented SMARTAP water quality monitor that provides instant performance verification at the touch of a button. A split-second power pulse compares the feed water vs. the product water to accurately monitor the membrane performance. Power is supplied by a 9 volt battery which is included with the system.

No installation is required, it simply snaps onto the faucet. The system operates with feed pressures from 35 to 100 psi, in temperatures from 40^0 to 100^0 F, and with pH values from 3 to 11. Maximum TDS is 2000 ppm.

The reservoir has a two gallon capacity with a water level viewing window. Output varies with feed pressure, temperature, and total dissolved solids, but averages from 2 to a maximum of 25 gallons per day. Output continues so long as feed water is on. When reservoir is full it overflows into brinewater drain (no countertop floods).

The advanced four-stage treatment process allows operation with either chlorinated or unchlorinated water supplies. There are both pre-sediment and pre-carbon filters as well as a post-carbon polishing filter. All filters are simple to replace and the cabinet is easy to clean.

As with all R.O. units there is a continual trickle of water that flushes off the TFC membrane. This brinewater is intended to drain into the sink, but can be rerouted and used for watering of gardens or landscaping. The increased total dissolved solids in "brinewater" is minuscule. The diversion rate on this R.O. unit is user-adjustable depending on the source purity, but should be in the range of 3 to 5 gallons of brinewater per 1 gallon of purified water.

Unit dimensions are 13"H x 10"D x 16.5"W, weight is 10 lbs. Warranty is two years on the system, 12 month prorated on the TFC membrane.

42-218	Hydrotech Countertop R.O. Unit	$399.00
42-219	Replacement TFC Membrane for 42-218	$109.00
42-220	Sediment prefilter for 42-218	$ 14.95
42-221	Carbon prefilter for 42-218	$ 14.95
42-222	Carbon postfilter for 42-218	$ 14.95

Hydrotech Undercounter R.O. Drinking Water System

This state-of-the-art undercounter Reverse Osmosis system features the patented SMARTAP water quality monitor that provides instant performance verification at the touch of a button. A split-second power pulse compares the feed water vs. the product water to accurately monitor the membrane performance. Power is supplied by a 9-volt battery which is included with the system.

The patented one-piece injection-molded manifold eliminates several external hoses and fittings where leaks can occur. There are both pre-sediment and pre-carbon filters as well as a post-carbon polishing filter. All filters and the membrane are simple to replace. The drinking water reservoir will hold two to three gallons depending on feed pressure. Stops automatically when the reservoir is full. This R.O. system provides approximately 9 gallons of purified water per day depending on demand.

The advanced four-stage treatment process allows operation with either chlorinated or unchlorinated water supplies. The system operates with feed pressures from 35 to 100 psi, in temperatures from 40^0 to 100^0 F, and with pH values from 3 to 11. Maximum TDS is 2000 ppm.

As with all R.O. units there is a continual trickle of water that flushes off the TFC membrane. This brinewater is routed to the sink drain, but can easily be rerouted and used for watering of gardens or landscaping. The increase in total dissolved solids in "brinewater" is minuscule. The diversion rate on this R.O. unit is automatic, depending on the source purity, but runs in the range of 3 to 5 gallons of brinewater per 1 gallon of purified water.

Has a chrome countertop dispenser, all other components mount under the sink. Two year warranty on the system, 12 month prorated warranty on the TFC membrane.

42-223	Hydrotech Undercounter R.O. System	$449^{00}
42-224	9GPD TFC Membrane for 42-223	$129^{00}
42-225	Sediment Prefilter for 42-223	$ 12^{95}
42-227	Carbon Pre & Postfilter for 42-223	$ 14^{95}

Solar-Powered Water Purifier For Pools, Spas, Fountains, And Ponds

Solar Ionization: It's the natural way to clean, clear, and healthy water. No more red eyes, discolored hair, or bleached suits. The Floatron is safe and nontoxic. It controls algae and bacteria without the need for risky (and toxic) chemicals. While floating on the surface, sunlight is converted into a safe, low-power electrical current, energizing a mineral electrode below waterline. This releases atomic amounts of specific mineral ions into the water which are lethal to microorganisms. Typically, Floatron reduces chemical expenses by an average of 80%. This can mean less chlorine damage to the ozone layer, too. Floatron is designed to last the life of your pool and carries a two-year manufacturer's warranty. The sacrificial electrode will last one to three seasons, depending on conditions, and is replaceable direct from the manufacturer. It takes only a minute to change.

The Floatron measures 12" in diameter and is effective for water volumes up to approximately 40,000 gallons. Use multiple Floatrons for larger capacity applications. Call or write for information packet.

Not for use with ozone systems.

42-801	Floatron	$299^{00}

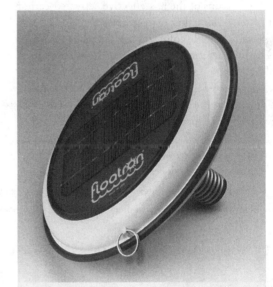

Shower Filter

The Rainshower is a great shower filter that removes chlorine and other contaminants from your water. It is molded out of ABS plastic and is only 5 inches long, including our finest low-flow showerhead. The new filter will remove 90% or more of residual chlorine from shower water.

Chlorine and residual chlorine are very hazardous to hair, skin, eyes, and lungs. We actually can take in more chlorine from one 15-minute shower (*but don't ever waste that much water!*) than from drinking eight glasses of the same water in one day. Chlorine plays havoc with our skin and hair, chemically bonding with the protein in our bodies. It makes hair brittle and dry and can make sensitive skin dry, flaky, and itchy.

"Your chlorine filter for the shower is wonderful. My hair has changed from dry and brittle on the ends to healthy and soft. I've spent hundreds of dollars on assorted hair care products that promised to 'nourish and moisturize.' What they didn't mention was that the water involved . . . was perhaps the most important part. So thank you for keeping me healthy and doing for me what the billion-dollar hair care industry could not."

— Gwendolyn Field, San Francisco, CA

Rainshower's use of electrochemical oxidation technology makes it effective for thousands of gallons of water, not the hundreds of gallons of the less-effective activated-carbon shower systems. The filter converts free chlorine into a harmless water-soluble chloride which washes out of the filter. The product meets FDA standards for copper and zinc in drinking water.

The Rainshower is easy to install and comes with our lowest-flow showerhead. Savings in water and energy costs can pay for this product in less than six months for the average family. Teflon tape is included for a leak-free installation. Each unit comes with a back-flushing adapter. A one-year warranty is provided.

| 42-701 | **Rainshower Shower Filter** | $69^{95} |
| 42-703 | **Rainshower Replacement Filter** | $59^{95} |

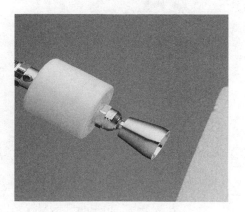

Ozone Purifiers

Clearwater Tech makes several different ozone generators for different applications. The ozone is manufactured in the generator by intaking air, which is composed of 20% oxygen (O_2) and bombarding it with a specific light frequency. This frequency causes the oxygen molecules to disassociate and reassemble as ozone (O_3). Ozone is the most powerful oxidizing agent available. When ozone is drawn into the spa or pool water, it will kill the bacteria, virus, or mold spores that come in contact with it. Ozone has a life expectancy of approximately 20 minutes. Several short cycles through the day are recommended.

S-1200 Ozone Purifier

PR-1300 Ozone Purifier

S-1200 Ozone Purifier

The S-1200 features a polished stainless-steel reaction chamber, thermally protected self-starting ballast, weather-tight (outdoor approved housing), and a 17-inch specially designed high-output ultraviolet lamp. The S-1200 is wired to the pump circuit to be on when your pump is on. The unit has convenient mounting brackets, comes with all necessary fittings, and is easy to install. If the spa is run daily, as recommended by most spa manufacturers, four separate 1-hour cycles in a 24-hour period will generate a sufficient amount of ozone to keep the spa free of biological contamination. The 12-volt S-1200 has the same features, and works the same. It will treat the same capacity of water as the AC model. The S-1200 will purify up to 1,000 gallons for spas and 2,000 gallons for pools.

42-811	S-1200 Ozone Purifier, 110V	$269⁰⁰
42-812	S-1200 Ozone Purifier, 12V	$349⁰⁰

PR-1300 Ozone Purifier

This system comes with its own 24-hour timer and compressor (on the AC system only; the 12-volt unit doesn't have the timer). This means that the PR-1300 can run independently of the circulation pump in your spa or pool. It comes with all necessary tubing, check valve, and fittings for installation. There is also a diffuser stone which can be attached to the ozone delivery line and submerged into any vessel of water.

You can treat yourself to a lavish chlorine-free bath by using the system in your bathroom, or anyone else's bathroom since the system is totally portable. You can treat your friend's spa before you use it with this portable water treatment system. Treating your water with ozone instead of chlorine when you are dealing with bacteria, iron, or other problems in your storage tank is the solution for those who have a spring or catchment system. The PR-1300 features a GFCI (Ground Fault Circuit Interruption) circuit breaker on 110-volt systems only. This gives state-of-the-art electrical protection. It has a weather-tight cabinet, coated with a baked enamel finish for years of corrosion-free service. The UV lamp is encased in a polished stainless-steel reaction chamber, and can be replaced by the homeowner in minutes. The compressor rating at 12 volts is 2.5 psi. Average lamp life is 9000 hours. Power consumption is approximately 80-90 watts. The units are rated up to 1,000 gallons for spas and 2,000 gallons for pools. The size is 20" x 9" x 4". Maximum pressure is 20 psi.

42-821	PR-1300 Ozone Purifier, 110V	$429⁰⁰
42-822	PR-1300 Ozone Purifier, 12V	$489⁰⁰

CS-1400 Ozone Purifier

The CS-1400 is a UV ozone generator designed to be used on swimming pools up to 15,000 gallons or spas up to 2,500 gallons. You can double the output and capacity by adding another CS-1400. It has a polished stainless-steel reaction chamber, a thermally protected self-starting ballast, weather-tight enclosure (outdoor approved), and a 29-inch specially designed high-output ultraviolet lamp. With no moving parts, the CS-1400 requires virtually no maintenance and will provide years of uninterrupted service. Wire the CS-1400 to the pump circuit, so each will work together to keep your pool or spa perfectly clear. It will also work with a compressor (not included). The CS-1400 comes with a 1-1/2-inch venturi injector suitable for use with a single-speed pump.

Each UV lamp is rated at 9,000 hours. Running the S-1200, the PR-1300, or the CS-1400 for four one-hour intervals per day equates to five-years of operation, with an energy-consumption worth of approximately 85 cents per month.

42-831	CS-1400 Ozone Purifier, 110V	$529⁰⁰
42-832	CS-1400 Ozone Purifier, 12V	$579⁰⁰

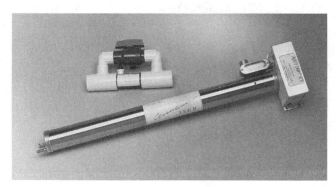

CS-14 Ozone Purifier

Sterling Spring® Euro-Style Water Filter

More and more frequently, news of water contamination leaves consumers concerned about the safety of their municipal water systems. To help ensure clean, safe drinking water for your family, this countertop water filter effectively removes chlorine, lead and waterborne parasites such as cryptosporidium, giardia, and turbidity from your tap water for a full year before cartridge needs to be replaced. Stylish and very easy to install, no tools or plumber required. Awarded the prestigious Gold Seal from the Water Quality Association. Measures 6 .5"W x 5"D x 13.25"H.

42-611   Sterling Spring® Euro-Style Water Filter $129⁹⁵

Sterling Spring® Euro-Style Water Filter

Sterling Spring® Ice & Water Filter

Drinking lots of water is one of the hallmarks of a healthy lifestyle. Now, your family can enjoy safer, great tasting cold water and ice cubes, directly from your refrigerator/freezer. Designed to last for a full year, this compact ice and water filter reduces lead and chlorine, as well as unwanted tastes and odors. Quick and easy install, it fits right into the water line behind your refrigerator. Includes connectors for both plastic and copper tubing. Measures 2.5" diameter x 12"L.

42-612   Sterling Spring® Ice & Water Filter $49⁹⁵

Sterling Spring® Under-Counter Water Filter

Even though your tap water may look clean and clear, it can actually contain unhealthy amounts of lead, chlorine and sediment—as well as unwanted odors and tastes. This easy-to-install filter effectively reduces all of these, so your family enjoys safe, fresh-tasting water for drinking, cooking, preparing baby formula or making beverages like coffee, tea or juice. Ideal for compact spaces, the main unit sits conveniently under the counter, with only an unobtrusive faucet above. Each hard working filter cartridge lasts up to one full year. Measures 4.75" diameter x 12.5"H.

42-610   Sterling Spring® Under-Counter Water Filter $129⁹⁵

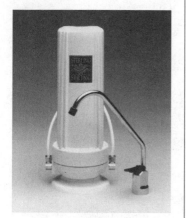

Sterling Spring® Under-Counter Water Filter

Sterling Spring® Tower Water Filter

Rugged, dependable and built to last for years, this convenient water filter is a solid investment in your family's health. An economical, earthwise alternative to bottled water, it removes odors, filters chlorine, lead and sediment from your tap water, leaving behind only a clean, delicious taste. Designed to install in minutes, it sits right on your countertop—no need for bulky faucet attachments or pitchers. Each replaceable filter cartridge lasts up to one full year. Measures 4.75" diameter x 12.5"H.

42-606   Sterling Spring® Tower Water Filter
$99⁹⁵

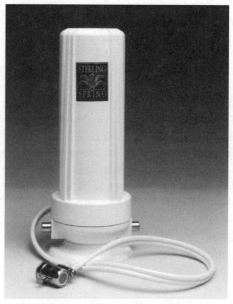

Sterling Spring® Tower Water Filter

Sterling Spring® Lead & Chlorine Replacement Cartridge

This easy to replace filter cartridge lasts one year, providing safer, great tasting drinking water for your family for less than 3 cents a gallon. Filters lead, chlorine, sediment and makes water taste and smell great. Fits all Sterling Spring® Counter Top & Under Counter Water Filters.

42-615 $29⁹⁵

Sterling Spring® Crypto Block® Replacement Cartridge

Sterling Spring® has created an advanced filter cartridge to address a water problem—Cryptosporidium and Giardia—that is becoming a health issue in many cities. These waterborne parasites could be lurking in the water supply. Healthy people exposed to them may feel like they have a bad case of the flu. But for the elderly, the very young, cancer patients, and others with weakened immune systems, Crypto can be deadly.

The Sterling Spring® Crypto Block Replacement Cartridge has the same convenient lead, chlorine, taste, smell and sediment protection as the Sterling Spring Lead & Chlorine cartridge, plus it filters potentially dangerous waterborne parasites—like giardia and cryptosporidium—for one full year. Quick and easy to replace, it fits all Sterling Spring® Counter Top & Under Counter Water Filters.

42-616 $59⁹⁵

Sterling Spring® Ice & Water Filter

STERILIZE YOUR WATER

If you live away from the hustle and bustle of city life, you probably are responsible for your own water supply. Getting that water pure is now a lot easier with the "WaterFixer". Using a three step filtration process, the WaterFixer first uses a 5 micron prefilter to remove suspended particles. A second .5 micron activated carbon prefilter then removes chlorine and other chemicals which effect taste and odor. As well, the .5 micron filtration removes microorganisms such as giardia cysts and cryptosporidium. The WaterFixer then passes the water through a stainless steel chamber and exposes the micron-strained water to an ultraviolet lamp specially tuned to sterilize bacteria and viruses.

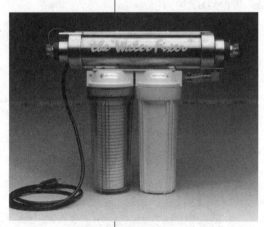

The WaterFixer is a great technology for rural home, boat, and RV owners who want safe drinking water without the addition of chemicals. 120 VAC/30 watts.

42-621 WaterFixer $445⁰⁰

Seagull IV X-1F

Seagull IV X-1D

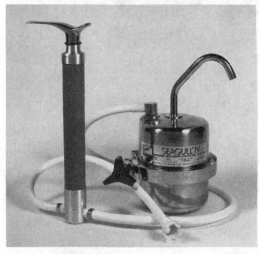

Seagull IV X-1P

Seagull Carbon Filters

To expand your options for the purest, safest water, we are introducing the Seagull IV systems. Seagull IV (X-1) products are among the very few qualified for continuous use aboard international airlines. They are manufactured in the USA from stainless steel and other high-grade components. Water is filtered through a unique microfine structured matrix, with a high flow rate of 1 gpm at standard pressures (30-40 psi). The replaceable cartridge should last anywhere from 9 to 15 months with ordinary use.

Here are some of Seagull IV's unique features:

1. Ultrafine filtration down to 0.1 micron — small enough to remove all visible particles, as well as giardia, cysts, harmful bacteria, and larger parasites.*

2. Molecular sieving and "broad spectrum" adsorption mechanisms remove chlorine and many organic chemicals such as pesticides, herbicides, solvents, lead, and foul tastes and bad odors.*

3. Electrokinetic attraction removes colloids and other particles even smaller than those removed by microfine filtration.*

*See performance data sheet for specific contaminants. Certain states may prohibit health claims as a matter of local or state law. Such claims not in compliance are hereby withdrawn. Please check with appropriate officials as necessary.

Seagull IV X-1F

Under-sink unit, with revolutionary ceramic disc faucet for countertop and accessories for easy installation.

42-650	Seagull IV X-1F	$489⁰⁰
42-657	Seagull Replacement Cartridge	$89⁹⁵

Seagull IV X-1D

Countertop unit, faucet-attaching, diverter unit for apartments, cabins, etc.

42-651	Seagull IV X-1D	$459⁰⁰
42-657	Seagull Replacement Cartridge	$89⁹⁵

Seagull IV X-1P

Same as X-1D with the addition of a high-capacity handpump for use on non-pressurized systems.

42-652	Seagull IV X-1P	$499⁰⁰
42-657	Seagull Replacement Cartridge	$89⁹⁵

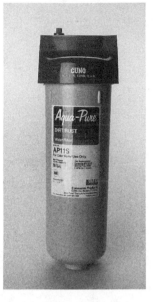

Inline Sediment Filter

Sometimes your water isn't dirty enough to mess with fancy and expensive filtration systems and all you need is a simple filter. Our inline sediment filters accept standard 10-inch filters with 1-inch center holes. They are designed for cold-water lines only and meet National Sanitation Foundation (NSF) standards. Easily installed on any new or existing cold-water line (don't forget the shutoff valve), they feature a sump head of fatigue-resistant Celcon plastic. This head is equipped with a manually-operated pressure release button to relieve internal pressure and simplify cartridge replacement. The housing is rated for 125 psi maximum and 100° F. It comes with a 3/4-inch FIPT inlet and outlet and measures 14 inches high by 4-9/16 inches in diameter. It accepts a 10-inch cartridge and comes with a 5-micron high-density fiber cartridge.

41-137	Inline Sediment Filter	$74^{95}
42-632	Rust & Dirt Cartridge	$6^{95}
41-436	Taste & Odor Cartridge (2)	$39^{95}

First Need

The First Need is the best-selling portable water purification system in America. Weighs less than 16 ounces. Made by the folks who make the incredible Seagull line, and has the same removal effectiveness as the Seagull IV systems. Great for camping, backpacking, and travel.

| 42-653 | First Need | $74^{95} |

WATER HEATING

Most of us take hot water at the turn of a tap for granted. It makes civilized life possible, and we get terribly annoyed when it is absent. Yet most people probably do not realize how much this convenience costs them. The average household spends an astonishing 20% to 40% of its energy budget for water heating. If your hot water heater uses electricity, you are toward the top of that range, and if you heat water with natural gas you are nearer the bottom. There are, however, much cheaper and better ways to heat your home water supply. In this section, we will start with the simple solutions and work our way up through the more complex ones.

INSTANTANEOUS OR TANKLESS WATER HEATERS

This strategy is so obvious it's a no-brainer. Why keep 30 to 80 gallons of water at 120° F all the time? Do you leave your car idling 24 hours a day just in case you want to take a ride? Of course not. You start it when you need it. Our water heaters should work the same way, starting up only when we need hot water. Otherwise we heat the water up, it sits there and radiates the heat away, then we heat it up again, it sits there . . . and soon, ad infinitum. In many cases this waste heat must even be removed from the home with expensive air conditioning! Everybody else in the world figured out the virtues of instantaneous water heating a long time ago. Due to a long honeymoon with cheap power, the United States is one of the few countries that still uses tank-type water heaters. Tank-type heaters use a *minimum* of 20% more energy than instantaneous systems due to heat loss. If yours is a small one- or two-person household, your inefficiencies are even greater, because the hot water spends more time sitting around, waiting to be used.

How Instantaneous Water Heaters Work

Instantaneous or tankless water heaters only go to work when someone turns on the hot water. They sense the water flow through the heater and turn on, heating the passing water. Tankless heaters do not store any hot water for later use, but heat it only as needed. One great advantage of tankless heaters is that you run out of hot water only when either the gas or water is gone. On the other hand, tankless heaters are limited to a fixed output in terms of gallons per minute. Faster water flow will result in lower temperature output. So the smaller tankless heaters are limited to running just one fixture at a time. Larger units, like the Aquastar 125, can run multiple fixtures simultaneously.

The Aquastar and Myson units are thermostatically controlled. At lower flow rates the gas flame is automatically reduced to maintain a stable output temperature. The temperature is front-panel adjustable from roughly 90° F to 140° F. (We consider 120° F to be an optimal setting.) If a second demand comes on line, the heater will respond by increasing the gas flow up to the heater's Btu limits. The Aquastar units all have built-in flow restrictors, so that it is impossible to run very much more water through the unit than it can heat. If too many taps are opened simultaneously, the water pressure will fall and the water will simply run lukewarm (rather than ice-cold).

We prefer the Aquastar to the Paloma models we used to carry for several reasons. Palomas lack a thermostatic control and flow restrictors, both conveniences that make tankless heaters perform virtually the same as conventional tank-types from the user's point of view. Also, the yen/dollar exchange rate has pushed the cost of Paloma heaters through the stratosphere. In short, we find that the Aquastars are technically superior, have better customer-support facilities, and cost less than Palomas. How can you compete with that?

Life Expectancy of a Tankless Water Heater

The other great advantage of tankless heaters is their life expectancy. Instantaneous heaters are designed so that all parts are repairable or replaceable. Both Aquastar and Myson warranty their heat exchanger for a full ten years! Neither model has any corrosive parts that touch water. The manufacturers have toll-free 800 service numbers backed up with full-time technicians and fully stocked parts warehouses. There is no reason why these heaters should not last the rest of your home's life. Compare this to tank-type heaters, which have a 10- to 15-year life expectancy at best. The tank eventually rusts through, and the entire heater must then be replaced.

Solar System Back-Up

Aquastar and Myson tankless heaters can be used with solar- or woodstove-preheated water. As we mentioned above, the standard Aquastar and Myson units sense the outgoing water temperature and adjust the flame accordingly to maintain a steady output temperature. Full zero to 100,000 Btu modulation is standard for the Myson. However, the standard Aquastar units can only modulate the flame down to about 20,000 Btus. The optional "S" units can modulate the flame all the way down to zero. So, if the water is already preheated to whatever output temperature you set on the Aquastar, then the heater will only come on briefly until preheated water reaches it. If your preheated water is at 90° F, and you've got a 120° F output set, then the Aquastar will come on just enough to give 120° F output. We think that's pretty slick.

Where Should I Install a Tankless Heater?

Just like a tank-type heater, the shorter the hot-water pipe, the less the energy loss. However, if you are willing to wait for the hot water to reach the fixture pipe, length is insignificant. A tankless heater does not need to be installed right at the fixture. If you are replacing an existing tank-type heater, it is probably most convenient to install the tankless unit in the same space. Your water and gas piping are already in the vicinity and will require only minor plumbing changes. If this is new construction just pick the most central location. As with all gas appliances the heater requires venting to the outside. In a retrofit installation the existing flue pipe may need to be upsized for the tankless heater. DO NOT reduce the flue size of the tankless unit to fit a smaller flue already in place. Also, do not install a tankless

heater outside or in an unheated space, unless it never freezes in your climate. The pilot light can be blown out easily if the unit is exposed to the wind. Myson heaters must be installed on a outside wall. See the box on Water Heater Installation Tips for more information.

Can I Get a Tankless Electric Water Heater?

Real Goods doesn't sell them, but there are some very small point-of-use, tankless electric water heaters available. These units are primarily for washing hands or other small tasks. They will not support a shower or other household fixtures. A household-sized tankless electric heater would draw more power than the average house is wired to supply.

— *Doug Pratt*

WATER HEATER INSTALLATION TIPS

Gas piping: Most of these heaters require a larger gas supply line than the average tank-type heater. Just adapting a 1/2 inch supply line to 3/4 inch at the heater won't work. If the heater says "3/4 inch supply line," that means all the way back to the regulator. Most of these heaters also have a small regulator on the gas inlet. This is **in addition to** your standard regulator at the tank or gas entrance. Gas inlet is bottom center for all tankless heaters.

Pressure/temperature relief valve: Unlike conventional tank-type heaters, there is no P/T valve port built into the heater. This important safety valve must be plumbed into your hot water outlet during installation. Simply tee into the hot water outlet.

Venting: Tankless heaters **must** be vented to the outside. All models come with draft diverter installed. All Aquastar tankless heaters use conventional double-wall Type B vent pipe. This is the same type of vent as conventional water and space heaters. Vent pipe is **not** included with the heater. This vent pipe is easily available at plumbing, building supply, or hardware stores. The vent piping used with most of these heaters is larger than tank types use. Do not adapt **down** to existing vent size! Replace the vent and cut larger clearance holes as necessary if doing a retrofit. Venting may be run horizontally as much as 10 feet. Maintain 1 inch rise per foot of run. Ten feet of vertical vent is recommended at some point in the system to promote good flow.

The Myson Direct-Vent tankless heater is an exception. It vents directly through the wall behind the heater, and comes complete with its own telescoping vent kit Most installations won't require anything additional, unless you're venting through a flammable wall. Then get the Wall Protection Kit.

Installation location: Don't install your tankless heater outside unless you are in a location where it **never** freezes. Freeze damage is the most common repair on these heaters. Don't tempt fate by trying to save a few bucks on vent pipe. If the heater is installed in a vacation cabin, put tee fittings and drain valves on both the hot and cold plumbing to ensure full drainage. The heat exchanger that works so well to get heat into the water works just as well to remove heat.

WARNING: It is illegal and dangerous to install an instantaneous water heater in an RV or trailer unless the heater is certified for RV service.

Customers frequently ask us about clearances for installing our instantaneous water heaters. We have developed this chart to help you plan your tankless heater installation before you buy.

Water Heater Clearance Chart						
Heater	BTU's/hr	Top	Bottom	Side	Front	Wall to flue centerline
Aquastar	38,700	12"	12"	1"	6"	3-1/2"
Aquastar	77,500	12"	12"	1"	6"	5-1/8"
Myson	100,000	2"	6"	2"	6"	direct vent
Aquastar 125	125,000	12"	12"	6"	36"	5-1/8"
Aquastar 170	165,000	12"	12"	6"	open only	7-3/4"

ASK DR. DOUG

Dear Dr. Doug:
Is it possible to use one of the tankless heaters on my hot tub?

Answer: It has often been done successfully. You need to bear in mind that these heaters are designed for installation into pressurized water systems. For safety reasons, they won't turn on until a certain minimum flow rate is achieved. The flow rate is sensed by a pressure differential between the cold inlet and the hot outlet. Tankless heaters flow at rates of five gallons per minute and less, which is a much lower flow rate than hot tubs. A hot-tub system is very low-pressure, but high-volume. What usually works is to tee the *heater inlet* into the *pump outlet,* and the *heater outlet* into the *pump inlet.* This diverts a portion of the pump output through the heater, and the pressure differential across the pump is usually sufficient to keep the heater happy. Aquastar has a very inexpensive "recirc kit" for its heaters that lowers the pressure differential required. This kit is mandatory! In addition, you'll need an aquastat to regulate the temperature by turning the pump on and off (your tub may already have one if there was a heating system already installed). You'll also need a willingness to experiment and a good dose of ingenuity.

INSTANTANEOUS HEATERS

Aquastar Tankless Heaters

The French-made Aquastar tankless water heater is thermostatically controlled, making it perform very like a tank-type heater, only without the wasteful stand-by radiant heat losses. The Aquastar features a safety thermocouple at the burner and pilot, an overheat fuse, a thermostat control for temperature output (will adjust from approximately 90° to 140° F), and built-in gas shut-off valves.

The Aquastar instantaneous water heater can be used with preheated water systems. The "S" series is designed for use with solar or woodstove preheated water and is "zero modulated." What this means is that if the incoming water is hot enough, the thermostat will dial the Aquastar's burner down to zero. But if your preheated water is coming in a 80° and you've got the Aquastar's thermostat set at 110°, the burner will come on just enough to add 30°.

All Aquastars have a ten-year warranty on the heat exchanger and a two-year warranty on all other parts. Be sure to use correct item number to indicate propane (LP) or natural gas (NG).

Unlike tank-type heaters, all parts on Aquastars are repairable or replaceable. All parts that touch water are non-corrosive brass, copper, or stainless steel. There is a toll-free 800 # for parts and service advice with real technicians on the other end who are backed up with a fully stocked parts warehouse. These folks really do a good job of customer service. There's no reason why your Aquastar can't last as long as your house.

Aquastar 38

The Aquastar 38 (38,700 Btus/hr.) is the perfect water heater for a small cabin with a low-flow shower or anywhere the demand for hot water is low. Will run a shower so long as incoming water temperature is 45° F. or above. The model 38 is equipped with a piezo igniter. Not for use in RVs and trailers! Not AGA listed. Propane only.

> *Btu Input: 38,700 Btu/hr.*
> *Recovery Time: 39 gph*
> *60° F Temp. Rise: 1.1 gpm*
> *90° F Temp. Rise: 0.8 gpm*
> *Minimum Flow: 1/2 gpm*
> *Vent Size: 3" Type B gas vent*
> *Water Connections: 3/8" Copper Sweat*
> *Gas Connections: 1/2" MIPT*
> *Min. Water Pressure: 15 psi*
> *Dimensions: 18-1/2"H x 8-1/2"W x 7"D*
> *Shipping Weight: 18 lbs.*

45-101   **Aquastar 38 LP** $359⁰⁰

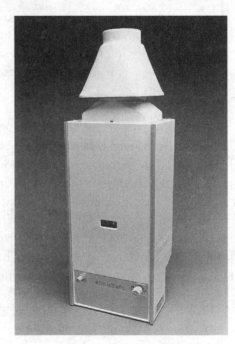

TANKLESS WATER HEATER SELECTION CHART		
Heater Mfg.	Model	Recommended Use
Aquastar	38	Single fixture, small cabin w/ low flow shower
Aquastar	80	Any single fixture, small bathtub, 1- or 2- person household
Aquastar	80 "S"	Same as above, but with possibility of preheated water
Aquastar	125	Conventional home, single shower or multi smaller fixtures simultaneously
Aquastar	125 "S"	Same as above, but with possibility of preheated water
Myson	Direct-Vent	Conventional home, single shower or multi smaller fixtures simultaneously
Aquastar	170	Industrial/Commercial applications, or filling spa tubs

Aquastar 80

The Aquastar 80 is the convenient choice where large volumes are not needed. Will handle any single household fixture. Great for small or single-person homes, cottages, service stations, and workshops. AGA listed. Standard models will only modulate down to approximately 20,000 Btus, "S" models will modulate down to zero for preheated water. Get the "S" model if you have, or plan to have, a solar or woodstove preheating system. Will not affect normal operation (but costs lots more to add as a retrofit).

> Btu Input: 77,500
> Recovery Time: 78 gph
> 60° F Temp. Rise: 1.8 gpm
> 90° F Temp. Rise: 1.3 gpm
> Minimum Flow: 0.75 gpm
> Vent Size: 4" Type B gas vent
> Water Connections: 1/2" Copper Sweat
> Gas Connections: 3/4" MIPT
> Min. Water Pressure: 15 psi
> Dimensions: 27-1/2"H x 12"W x 9-1/2"D
> Shipping Weight: 32 lbs.

45-102-LP Aquastar 80LP		$549[00]
45-102-NG Aquastar 80NG		$549[00]
45-106-LPS Aquastar 80LPS		$599[00]
45-106-NGS Aquastar 80NGS		$599[00]

Aquastar 125

The Aquastar 125 is the perfect home residential heater. It's capable of delivering a constant supply of hot water at the temperature you select. The 125 can support a single shower, or smaller multiple fixtures simultaneously. This model is the best choice for the average home. Appropriate for small volume commercial installations also. Standard models will only modulate down to approximately 20,000 Btu, "S" models will modulate down to zero for preheated water. Get the "S" model if you have, or plan to have, a solar or woodstove preheating system. Will not affect normal operation (but costs lots more to add as a retrofit).

> Btu Input: 125,000
> Recovery Time: 126 gph
> 60° F Temp. Rise: 3.25 gpm
> 90° F Temp. Rise: 2.11 gpm
> Minimum Flow: 0.75 gpm
> Vent Size: 5" Type B gas vent
> Water Connections: 1/2" Copper Sweat
> Gas Connections: 3/4" MIPT
> Min. Water Pressure: 15 psi
> Dimensions: 27-1/2"H x 17"W x 9-1/2"D
> Shipping Weight: 43 lbs.

45-105-LP Aquastar 125LP		$639[00]
45-105-NG Aquastar 125NG		$639[00]
45-107-LPS Aquastar 125LPS		$699[00]
45-107-NGS Aquastar 125NGS		$699[00]

Aquastar 125

Aquastar 170

The Aquastar 170 is the ultimate water heater for commercial or industrial locations which demand high temperatures and/or high volumes. Not usually recommended for residential applications because of the 1 gpm flow rate to turn it on, but is used occasionally to fill large spa tubs quickly. This model comes with standard equipment with "zero modulation," like the optional "S" units in the smaller sizes.

> *Btu Input: 165,000*
> *Recovery Time: 176 gph*
> *60° F Temp. Rise: 4.4 gpm*
> *90° F Temp. Rise: 2.95 gpm*
> *Minimum Flow: 1.1 gpm*
> *Vent Size: 6" Type B gas vent*
> *Water Connections: 3/4" MIPT*
> *Gas Connections: 3/4" MIPT*
> *Min. Water Pressure: 15 psi*
> *Dimensions: 35-1/2"H x 22-1/2"W x 13-1/2"D*
> *Shipping Weight: 75 lbs.*

•45-104-P Aquastar 170LP $949⁰⁰
•45-104-N Aquastar 170NG $949⁰⁰

Model 170 only shipped freight collect from Vermont or California.

Aquastar Recirculation Kit

Aquastars are sometimes used in recirculation applications like hot-tub heating or slab-floor heating. This is a far cry from what they were designed to do. But they are inexpensive, and they do work, particularly if you're running on renewable energy and really want to get rid of that electric heater in the hot tub that needs three days of generator running to bring the tub up to temperature. Just plan on having some creative ingenuity on hand. To help you along Aquastar makes a recirculation kit that allows the heater to turn on at a much lower pressure differential. (It's the difference between the incoming cold and outgoing hot pressures that triggers these units to light up.)

45-103   Aquastar Recirculation Kit $19⁹⁵

Un-Clog-It Descaling Kit

The Un-Clog-It works great on tankless water heaters, as well as spa/hot-tub heaters, humidifiers, ice machines, water coolers, air conditioners, and cooling jackets. It will dissolve harmful mineral buildup from water lines. Minerals occurring in hard water will adhere to the sides of copper pipe, gradually choking the water path and eventually interfering with the normal operation of your water heating and cooling appliances. This kit dissolves the mineral buildup by circulating a safe hot acid solution. The kit contains a submersible pump, 1/2-inch inside diameter clear plastic hoses, heavy-duty brass swivel fittings, and the descaling agent consisting of 2 pounds of sulfamic acid, which will yield four 1-gallon treatments. The kit is available in three options: 1/2-inch, 3/4-inch, or Aquastar fittings. Pump is 120-volt AC only.

•45-491   Un-Clog-It Kit, 1/2" $169⁰⁰
•45-492   Un-Clog-It Kit, 3/4" $169⁰⁰
•45-493   Un-Clog-It Kit, Aquastar $225⁰⁰

(Please note: There is a 25% restocking charge on returns of this item.)

Myson Direct Vent Water Heater

The Myson direct-vent tankless instant water heater is simpler (and probably cheaper) to install, and safer to operate. Mounting on any exterior wall, the Myson heater is completely sealed from inside air. There's no possibility of vent gases being blown or sucked back into your house, and you aren't using your valuable preheated interior air for combustion. For manufactured homes, in many states this may be the only legal tankless water heater available. Because it doesn't steal any indoor air it's the smart choice for any home, manufactured or not. Sized for conventional residential use, with 100,000 Btu input, the Myson can support a single shower, or multiple smaller fixtures such as sinks or washers. Thermostatically controlled, the Myson will smoothly modulate all the way down to zero Btu input to maintain a stable output temperature, just like the optional "S" Aquastar model. If you are preheating water with a solar hot water system, or plan to in the future, the Myson will work with you. A "matchless" piezo pilot ignitor is standard equipment. Myson has a toll-free technical service number for assistance in installation, service, or parts.

The Myson comes complete with a telescoping vent kit for standard 6" to 10" wall thickness. It requires a 12" square vent hole. No additional vent piping is required for installation! An optional Vent Extension Kit is available for walls up to 24" thick. The optional Wall Protection Kit is required if your exterior wall is flammable. The Freeze Protection Kit is a thermostatically controlled 200-watt electric heater that can prevent the heat exchanger from freezing in extreme hard-freeze climates. It requires 110V AC power. Available in Natural Gas, or LPG. Specify when ordering! Mfg. warranty — 1 year on all parts, 10 years on heat exchanger.

Btu Input: 100,000
Recovery Time: 101 gph
60°F Temp. Rise: 2.6 gpm
90°F Temp. Rise: 1.7 gpm
Minimum Flow: 0.75 gpm
Vent Size: vent fittings included
Water Connections: 3/4" NPT
Gas Connection: 3/4" NPT
Min. Water Pressure: 30 psi
Dimensions: 29-3/8"H x 16-5/8"W x 10-7/8"D
Shipping Weight: 60 lbs (2-pkg)

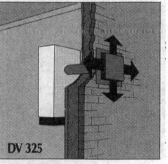

DIRECT VENT

•45-090	Myson Direct Vent Water Heater	$695⁰⁰
•45-092	Myson Wall Protector Kit	$59⁹⁵
•45-091	Myson Vent Extension	$149⁹⁵
•45-093	Myson Freeze Protection Kit	$89⁹⁵

Wisconsin, Oregon, or California? Who gets into the most hot water?
We want you to have the solar thermal system best suited to your needs. We would be glad to discuss domestic hot water systems and small pool systems with you, as we feel we can provide great service for these types of systems. Space heating, custom active storage, and other large solar thermal applications require engineering and consultation which we cannot provide, although we might be able to recommend other sources of information. Real Goods Snowbelt, in Amherst, Wisconsin, is our main distributor for solar thermal products, particularly for hard-freeze climates. Call them at 715-824-3982 for details on the Heliodyne and ProgressivTube domestic hot water systems as well as the Sun Unit solar pool heaters. Our Eugene, Oregon store is a distributor for Sun Family solar systems. Oregon has a very progressive state tax incentive for solar thermal systems. Call Eugene at 541-334-6960. Our technicians in Ukiah, California have extensive experience in solar thermal applications, particularly for milder climates. You can reach them at 707-468-9292 ext. 2210.

SOLAR WATER HEATERS

Heating your water with a tankless heater is a good idea, heating with the sun is an even better one. If tankless heaters save you some energy, a solar hot water heater can save you <u>lots</u> of energy. The domestic hot water solar heaters we sell will meet 30% to 80% of the annual hot water needs for the average household.

Before we go any further, a few words about quality and simplicity. Remember the solar tax credits that Jimmy Carter gave us in the late 1970s and early 1980s? The 50% solar tax credit available during this period gave rise to some pretty despicable, high-pressure door to door solar salesmen, and an abundance of poorly designed and sloppily installed solar hot-water systems, giving a lingering black eye to the whole solar industry for many folks. You, or someone you know may have got stuck with one of these overpriced, breakdown-prone lemons.

We would like to disassociate ourselves and the solar water heaters offered here from this schlocky period of history. All the heaters presented here were designed after the solar tax credit days, and benefited from lessons learned during this period.

The first lesson was simplicity. The systems we sell don't use controllers, sensors, air vents, or draindown valves , all the stuff that experience has shown was most prone to breakdown. These heaters use simple passive designs. The second lesson was quality. All of these heaters are warranted for 5 to 10 years by solid companies that are not going to disappear the day after the solar tax credits expire. Let's look at the weak and strong points of each system.

Batch Heaters

Some solar hot water collectors are called a *batch heater* because it holds a batch of water inside the collector until someone turns on the hot water. Batch heaters have the advantage of simplicity: set a tank of water out in the sun, then wait for it to get hot. A disadvantage is that a tank full of water is very heavy. The batch collector is plumbed between incoming cold water and the cold inlet port of the existing or back-up water heater. Any time hot water is used the back-up heater gets preheated water from the solar collector rather than stone-cold water. Real Goods offers two batch-type systems.

ProgressivTube Collectors

This simple but highly effective 4' x 8' collector contains horizontal 4" copper pipes connected in series. The entire assembly holds 41 gallons. Cold water is introduced at the bottom, as it warms, it will slowly stratify from one selectively coated 4" tube to the next, until only the hottest water will be drawn off the top outlet. The outgoing hot water is never diluted by the incoming cold water. Rated at approximately 30,000 Btu/day, the ProgressivTube collector will supply all the hot water required by the average family. It's all housed in a top-quality rust-proof aluminum box with bronze acrylic finish, dual glazing with tempered low-iron solar glass on the outside, and 96% transmittance Teflon film on the inside. Non-degrading, high-temperature phenolic foam board insulation is used on the sides (R-12.5), between tubes (R-12.5), and the bottom (R-16.7).

The collector and mounting system have been officially tested in 180 mph winds, and 250 units installed in St. Croix were unofficially tested by the 200+ mph winds of Hurricane Hugo in 1989. Only six units suffered minor damage, and that was due to flying debris. Use caution if roof-mounting a ProgressivTube collector. Filled, they weigh close to 600 lbs. Rafters may need to be doubled up. Ground mounting is often a better alternative.

Because of the thermal mass and highly insulated box the collector is unlikely to freeze (the warranty even covers freeze damage in the Sunbelt), but the supply piping is vulnerable. In hard-freeze climates it must be drained for the winter.

Advantages of this unit are a high-quality, lifetime package, simple mounting and connection, and a very good Btu per dollar ratio. Disadvantages are the 650 lb. weight of the filled unit, and need to drain in hard freeze climates.

Sun Family

The Sun Family collector is a unique design having long 5" copper tanks inside 6" glass tubes, with the space between drawn to high vacuum like an oversized thermos bottle. It is supplied in four-tube modules. Each module is rated at approximately 20,000 Btu/day. Up to three modules can be connected in series for residential applications, but it's unusual to need this much heating power. We've found the smallest four-tube PK10 unit to be just fine for up to three-person households. To use the eight-tube PK20 you'd better have four or more people in the household, otherwise you'll be dumping excess hot water. You can add a second or third module at any time in the future.

One disadvantage of batch heaters is that your tank of water cools off at night or during cloudy whether. The Sun Family is a major step ahead of other batch heaters due to the evacuated tube design. With no air molecules, heat cannot be transferred out of the tank except by radiation, which occurs slowly. In a laboratory test a single 50^0 F. water-filled tube placed in a -13^0 F. freezer took 6-1/2 days to freeze! Vacuum makes highly effective insulation, and this heater is just an oversized high quality Thermos bottle!

The Sun Family uses a manifold inside a manifold at the top of the tubes to route the incoming cold water to dip tubes that reach almost to the bottoms of the copper tanks, and to collect the hottest water at the top of the tanks for delivery to the house. The Sun Family must be installed at a 15^0 to 90^0 slope so the hotter water will stratify to the top.

Because of the round design this collector is very good at capturing diffuse radiant energy. During cloudy weather it will capture 80% of the radiant energy available. The round design also strengthens the glass and allows the Sun Family to survive hail up to 1.25".

The primary problem with this heater is that it doesn't know when to quit. It will easily hit 210^0 F. if no water is drawn, and the Pressure/Temperature relief valve will pop off. While not dangerous, it does waste water. The Sun Family benefits from using a small circulation pump and high temperature sensor that will turn on as needed and circulate the hottest water to the storage/back-up tank.

The smaller PK 10 unit weighs 250 lbs., the PK 20 weighs in at 504 lbs. Roof mounting may not be a good idea without first beefing up of the support system. Check with a knowledgeable contractor if you have any doubts. Often wall or ground mounting is a better solution.

The strong points of this heater are a low cost per BTU delivered, good energy production during marginal sunlight conditions, and modular construction. The weak points are vulnerability of the supply and return piping to freezing, a tendency to overheat, and the weight of the filled units.

Active Systems

Active Systems use pumps to move the heated fluid from the collector to a storage tank. We offer two active systems that both feature very simplified control systems. Obviously you don't want the system pump to run unless the collector is hotter than the storage tank. Most active systems accomplish this by using multiple sensors and a little controller "brain". Experience has shown that temperature sensors, and the wires and connections leading to them, are the most

Who gets into the most hot water? Everyone knows that all Californian's lay around in their hot tubs all day while sipping Chardonnay and talking on their cell-phones. No contest! What did you think that splashing noise was in the background when you called for tech help?

trouble-prone part of solar systems, followed closely by controllers. Our systems eliminate controllers, wiring, and sensors by using a small PV module connected directly to the circulation pump. The pump will only run when the sun shines, and circulation speed is in direct proportion to sunlight intensity. Simple elegance!

The Real Goods Solar Hot Water System

The smallest domestic solar hot water system we offer, the Real Goods Solar Hot Water System delivers about 10,000 Btu/day, and is small enough to be inexpensively UPS deliverable. To keep costs reasonable this heater is designed as a three-season system that runs the household water through the collector, and must be drained during hard-freeze weather. 10,000 Btu/day is approximately 1/3 of the average 3.4 person household hot water demand, making this system ideal for smaller, more conservative households. A second panel option is available that will double the daily BTUs collected with no additional plumbing.

The Real Goods Solar Hot Water System is delivered as a do-it-yourself kit. The basic kit contains a single 2' x 5' collector, PV module, DC circulation pump, plus all the fittings, valves, flashings, mounts, and complete well-illustrated instructions to connect to your existing water tank. No sweat soldering is required. 1/2" soft copper pipe and insulation are the only parts not included. And that's only because individual installations vary so widely.

The system includes a Dole freeze protection valve that prevents system freezing during overnight cold snaps. The valve will open at approximately 36⁰ F. running cold water through the collectors, and dumping it out through the Dole valve. This water running off your roof in the morning is your warning that it's time to shut the system down for the winter, which takes less than five minutes with the included shut-off and drain valves.

Strong points of this entry-level solar system are; the low cost, a simple, foolproof control system, and a complete kit package. Weak points are; limited output, the possibility of mineral accumulation in the collector, and the need for seasonal draining.

Heliodyne PV-Powered Systems

Heliodyne is the favorite of any professional who's ever installed one. Though they cost more initially, all the tech staff agrees that Heliodyne offers the best value and most trouble-free service in the long run. Only the best, highest-quality components are used. The closed-loop plumbing system <u>can't</u> freeze in any climate, making the Heliodyne our first (and really only) choice for many parts of the country. The pumps, expansion tank, temperature gauge, fill/drain valves, and heat-exchanger are pre-assembled into the Helio-Pak that cuts plumbing time by 80%. Heliodyne systems are complete, <u>including</u> an appropriately sized storage tank. The two Heliodyne systems we're offering are PV-powered, eliminating the controller and sensors. Systems vary by collector area. The smaller system has a single 4' x 10' collector, rated at approximately 40,000 Btu/day. The larger system has two 4' x 8' collectors rated at approximately 65,000 Btu/day.

Strong points of this system are: top-quality throughout, absolutely freeze-proof design, higher output than other systems, and no possibility of mineral build-up inside the collector. Weak points are: high initial cost.

Official Ratings for Solar Hot Water Systems (plus a little controversy thrown in for spice)

In our product discussions above we've tossed out various Btu/day ratings. Where do these figures come from, are they reliable? There are two primary agencies that rate solar heating panels. The *Florida Solar Energy Center* (FSEC) tends to test actual panels in operation, while the *Solar Rating & Certification Corporation* (SRCC) uses standardized computer simulation tests to rate solar water heaters. Ratings from either organization are used as the yardstick to compare one brand of solar water heater against another, and these are the figures we've quoted. (The technical staff at Real Goods thinks this a splendid idea, and wishes someone would get organized and do the same for photovoltaic modules!)

The Sun Family PK10 is SRCC rated at 18,900 to 23,600 BTUs per day. The variation in the ratings is caused by higher or lower incoming water temperature. If the incoming water is colder, the panel will perform better, as less BTUs will be lost to radiation.

There is currently some industry controversy about the Sun Family ratings. There is reasonable doubt about the SRCC computer models not accurately reflecting true performance of this unique collector design. Colorado State University conducted a series of highly monitored, rigorously scientific tests on a variety of solar hot water collectors. They noted 33% to 41% higher actual collection and delivery BTUs for the Sun Family than the SRCC simulation ratings. That is an enormous variation from the "official" yardstick. This controversy hasn't been resolved yet, but every indication is that the SRCC data is flawed, and Sun Family heaters actually produce substantially more hot water than their "official" ratings would suggest.

— Doug Pratt

SOLAR HEATERS

Sun Family Solar Water Heater

The Sun Family solar heater keeps water hot through the night, the same way a thermos bottle keeps your coffee at the right temperature, with a vacuum layer that drastically reduces heat loss. Made of super-strong, high-tech glass, that can withstand the blow of a 1-1/4" hailstone, the heater utilizes a double tube design that provides a full 360° of heat collecting surface. The metal trim at top and bottom is stainless steel.

Regardless of the sun's angle, the Sun Family unit soaks up the maximum possible amount of heat, morning, noon, and evening, in every season of the year. Even sunlight and heat reflected from the roof is captured and absorbed. Combine a Sun Family with a supplemental instantaneous-demand water heater, and you'll always have all the hot water you need at tremendous savings. Since the PK-20 both heats and stores a generous 42 gallons of water, you need no separate tank. And because it's directly connected to your water supply, you'll always have a high level of pressure, no matter at what height it's installed. It can be mounted at ground level, against a wall, or on the roof. (Watch the weight! A full PK-20 weighs 500 pounds.) Modular design allows interconnecting as many units as required. This makes the system highly successful for large apartment houses, as well as small, individual installations. The Sun Family heater is the most exciting development in water heating to date, and the most cost-effective system on the market. At Real Goods we give it our highest recommendation.

We have found that in most areas of the country the PK-20 just has too much horsepower unless you've got a family of four or more. The PK-10 works great for households of one to three, the PK-20 for households of three or more. Use your judgment; how sunny is the weather, can you arrange to use hot water in the middle of the afternoon, are you raising quintuplets, etc. Call our tech staff for help if we've totally confused you now. In fact you can call even if you're not confused.

We offer two Installation Kit options that contain a variety of plumbing parts you will need to properly install one of the Sun Family heaters. *All installations will require the Pre-Assembled Valve Pak.* This is a compact assembly which includes diversion, isolation, and drain valves; a tempering valve to make sure that 180° water from the heater doesn't go out into the household plumbing; a temperature gauge for the returning preheated water; a T&P relief valve that installs at the collector to keep things safe; an expansion tank to allow for expansion and contraction up at the collector; and provision for the pump option.

The Optional Pump System is for those systems that have a tank-type heater for back-up. This system has a small circulation pump and a hi-temp snap switch that will turn on at 160° collector temperature to circulate very hot water from the collector back down to your standard water-heater tank. This allows more energy collection, and less chance of the collector reaching 210° when the T&P valve will open to dump excess hot water. Those with tankless heaters for back-up won't need the pump system; for everybody else, it's highly recommended. The pump requires 120V AC. Call us for a custom kit if you have a renewable energy system and need to run the circulation pump on low voltage DC.

The PK-20 Valve Pak Kit has the flex connector and the stainless-steel trim pieces to connect the two heater modules.

•45-409	PK-10 Solar Water Heater	$1,195
•45-421	PK-10 Pre-Assembled Valve Pak	$295⁰⁰
•05-216	PK-10 Crating Charge	$30⁰⁰
•45-407	PK-20 Solar Water Heater	$2,390
•45-423	PK-20 Pre-Assembled Valve Pak	$299⁰⁰
•05-211	PK-20 Crating Charge	$60⁰⁰
•45-422	Optional Pump System	$175⁰⁰

Shipped freight collect from Eugene, OR.

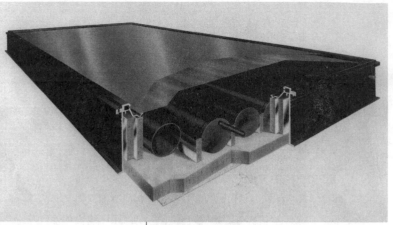

Progressiv Tube
Hot Water Heater

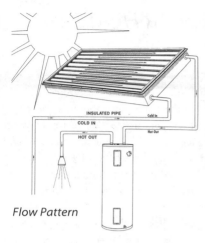

Flow Pattern

ProgressivTube Solar Hot Water Heater

This passive 40-gallon batch-type heater is one of the simplest, yet most elegant and durable water heaters we've seen. Plumbed inline ahead of your conventional water heater's cold inlet it will preheat all incoming water, reducing your conventional heater's work load to near zero. Rated at approximately 30,000 Btu/day, the 4' x 8' collector is housed in a top-quality bronze-finished aluminum box with stainless and aluminum hardware. There are no rusting components. The dual glazing has tempered low-iron solar glass on the outside, and non-yellowing or degrading 96% transmittance Teflon film on the inside. High temperature, non-degrading phenolic foam board insulation is used on the sides (R-12.5), between internal tubes (R-12.5), and the bottom (R-16.7).

The collector is composed of large 4" diameter selective-coated copper pipes aligned horizontally. They are connected in series, with cold water introduced at the bottom, and hot water taken off at the top. As the water warms it stratifies, and gradually works its way from tube to tube toward the top. The outgoing hot water is never diluted by incoming cold water.

The collector and mounting system have been officially tested and approved for 180 mph winds, although 250 units installed in St. Croix were unofficially tested by the 200+ mph winds of Hurricane Hugo in 1989. Only six units suffered minor damage, and that was due to flying debris. Use caution if roof-mounting a ProgressivTube collector. Filled, they weigh close to 600 lbs. Rafters may need to be doubled up. Ground mounting is often a better alternative.

Because of the thermal mass and highly insulated box the collector is unlikely to freeze (the warranty even covers freeze damage in the Sunbelt), but the supply piping is vulnerable. In hard-freeze climates the ProgressivTube must be drained for the winter.

ProgressivTube has the best warranty we've ever seen, ten years. But during the first five years, there is no charge to the homeowner for any warranty claim. The warranty covers labor and transportation, in addition to parts. The second five years covers parts only. Both fixed (roof angle) and tilting mounts with 36" of cut-to-fit back leg tubes are offered. The Valve Kit includes the necessary parts to choose solar pre-heating, or solar by-pass, drain valves, and tempering valve.

- 45-094   ProgressivTube Collector $1310
- 45-095   ProgressivTube Fixed Mounting $55⁰⁰
- 45-096   ProgressivTube Tilt Mounting $66⁰⁰
- 45-097   ProgressivTube Valve Kit $209⁰⁰

Shipped freight collect from Sarasota, FL. Collector pricing includes $50 crating charge.

Heliodyne PV-Powered Systems

If you value quality, or live in a hard freeze climate, Heliodyne is the solar hot water system for you. In business since 1976, Heliodyne survived the slim post-tax credit years because they make the best components, and pre-package them in a *Helio-Pak* that saves more than half the typical plumbing time. Any professional who's worked with Heliodyne loves them. All Heliodyne systems feature freeze-proof closed loop design, in which the fluids which flow through the collectors are a separate closed loop. Food-grade propylene glycol is used as anti-freeze. A counterflow heat exchanger mounted on the *Helio-Pak* transfers the heat to the household water.

Heliodyne's *Gobi* collectors feature durable blackchrome selective coating over nickel plating. The all-copper absorber has full wrap-around bonding to the riser tubes. Glazing is low-iron, tempered, solar glass, and frames are bronze anodized aluminum extrusions that won't bend or sag.

Appropriately sized solar storage tanks, collector(s), flush mounting hardware, *Helio-Pak*, PV module, air vent, thermometer, tempering valve, and installation manual are included with the Heliodyne system.

- 45-098 Heliodyne 1 4x10 System w/65 gal. tank $3280
- 45-099 Heliodyne 2 4x8 System w/80 gal. tank $3950

Shipped freight collect from Richmond, CA. System pricing includes $60 crating charge.

One of these two systems is the correct size for 90% of all applications. Heliodyne offers both smaller and larger systems. Call our Snowbelt staff for information on other sizes @ 715-824-3982.

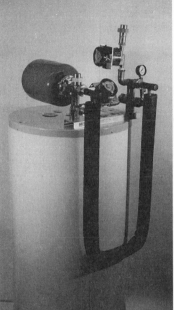

Pre-Assembled Helio-Pak

Heliodyne Sizing Recommendations	
2 to 3 people	one-4x10 system
3 to 4 people	two-4x8 system

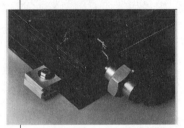

Heliodyne Gobi collector detail

Solar Collection Fluid

A propylene glycol that is food-grade safe and non-flammable. It dilutes 25% to 50% with water depending on climate. Each 4x8 collector needs almost one gallon. Fluid should be changed every five years.

- 45-100 Dyn-O-Flo, 4 single gallon bottles $86⁰⁰

Tilting Rack Options

Flush mounts are included. If you need to tilt your collectors at something other than roof angle, here's the answer. Includes 42" of back leg tubing to allow cutting at needed collector angle.

- 45-109 Tilt Rack for 1 Collector $89⁰⁰
- 45-110 Tilt Rack for 2 Collectors $140⁰⁰
- 45-111 Tilt Rack for 3 Collectors $199⁰⁰

Real Goods Solar Hot Water System

Out of sixty-six entries in a 1991 Florida competition to design the most cost-effective solar hot water system, this was the winning system. The system is not designed to operate during freezing weather, but will protect itself during overnight cold snaps. In many climates with heavy winter cloud cover a three-season solar hot water system is the best possible solution. The included isolation and drain valves make it less than a 5-minute job to shut down and drain the system. The combination of the patented bottom-feed plumbing configuration, with the fool-proof PV-direct pump control eliminates any need for electronic controls. The pump only runs when the sun shines; as the amount of direct sunlight increases, more hot water is produced, and the pump automatically runs faster. All parts except the soft copper pipe and insulation are included, no soldering is required. All parts are high-quality bronze or copper. Your existing water heater is used for storage.

The basic system will supply 30% to 40% of hot water needs for the average American household (3.4 people). For larger families the optional second collector will double this figure with no additional plumbing or valves needed

Many states or local utilities offer significant rebates on solar hot water systems. Rebates are sometimes more than the total system cost. Check with your local energy and utility offices. The complete basic system includes a single 24" x 60" collector, installation and plumbing hardware package, circulator pump, and PV module. It ships by normal UPS. Collectors warranted for five years, all other components warranted for one year.

•45-424   Real Goods Basic Solar Hot Water Systems $895⁰⁰
•45-425   Optional 2nd Collector $395⁰⁰

Sun Unit Pool Heater

Solar Pool Heater

This solar pool heater will extend the length of your swimming season. You'll be able to swim earlier in the spring and keep swimming later in the fall. The Sun Unit™ is easy to install and is a reliable and cost-effective solution to pool heating. You can cut your energy bills by using your existing heater strictly for back-up. Each SunUnit measures a flexible 4 x 22 feet. If your pool has more than 15,000 gallons, you will need two SunUnits. Sun Unit Plus is 30% larger, 4' x 30', for larger pools. The installation is simple — just unroll your SunUnit in a sunny location and connect it inline (on the return hose from the filter and then to the supply line back to the pool). In the winter, you can disconnect the SunUnit, roll it back up, and store it until next spring. Or it can be left out all year. Made of synthetic, rubberized material, it has a three-year warranty.

•45-419   SunUnit $399⁰⁰
•45-420   Dual SunUnits Adaptor Kit $35⁰⁰
•45-426   Sun Unit Plus $599⁰⁰

Solar Roof Flashing

It's often tough to find these high-quality rubber gasketed flashings in the smaller solar plumbing sizes. Each flashing will fit one single 1/2-inch, 3/4-inch, or 1-inch copper pipe. Standard size flashing pan for conventional asphalt shingle roofs. Not for wood shakes. Accommodates 0/12 to 12/12 pitch. A Sun Family installation requires two.

45-414   Solar Flashing $7⁰⁰

Solar Hot-Tub Heater

After numerous requests from our customers for a solar hot-tub heater, we are pleased to introduce this package, which comes complete with pump, controller, mounting hardware, and first-quality collector panel. All you add is the plumbing to suit your site. This single 4' x 10' collector will need a couple days or more to bring the average tub up to temperature after a water change, but can easily maintain a cozy 105° F at almost zero operating cost once it's up. The frames are heavy-duty bronze anodized aluminum extrusion with reinforced corners. The glazing is tempered double-strength 5/32" water white solar glass with weathertight compression fitted gasketing. Has aluminum back sheet. Collector panel is all copper with 360° wrap-around fin-to-tube mechanical bond. Has low emissivity blackchrome selective surface over nickel for highest efficiency. Fittings are large 1" solid brass unions with Viton O-rings for fast assembly and effortless sealing. Insulation is fiberglass over aluminum-backed isocyanurate.

The Hot Tub Kit is designed to mount the collector on a nearby roof and connect with insulated copper piping to and from the hot tub in such a way that all water can drain from the collector and pipes when not operating. Automatic operation is assured with a differential thermostat and sensors. (One at the top of the collector, and two at the hot tub outlet.) The upper temperature limit is factory set at 105° F. The 1/22 hp bronze Grundfos pump is not self priming, and needs to be installed lower than hot-tub water level. The stock pump will lift 1 story; if your collector panel will be higher, order the 2-story kit which has a higher head pump (all else the same).

Standard 1-story Kit includes: 1 Heliodyne 4' x 10' collector with unions, flush mounting hardware, blind unions, differential controller, sensors, air vent, solar 1/22 hp bronze Grundfos pump with unions. Two-story Kit substitutes a bronze 1/12 hp Grundfos pump with flanges. Crating is required for all collectors.

•45-209   Solar Hot Tub 1-Story Kit $1,245
•45-210   Solar Hot Tub 2-Story Kit $1,345
•05-226   Solar Hot Tub Crating Charge $60⁰⁰

Shipped freight collect from San Francisco Bay area.

Five-Gallon Solar Shower

This incredible low-tech invention uses solar energy to heat water for all your washing needs. The large 5-gallon capacity provides ample hot water for several hot showers. On a 70° F day the Solar Shower will heat 60° water to 108° in only three hours for a tingling-hot shower. This unit is built of four-ply construction for greatest durability and efficiency.

90-416   Solar Shower $11⁹⁵

A Wood-Or Oil-Fired Water Heater

This simple 15-gallon Mexican water heater is the low cost answer for many summer camps, hunting cabins, simple hot tubs, or remote homesteads. Provides hot water quickly by burning wood, corn cobs, pine cones, cow chips, or trash. The dual-fuel model, with its 1.75 liter adjustable dribble tank and removable burner pot, can also use kerosene, heating oil, wood alcohol, or diesel (not gasoline!!). Heavy-gauge steel design has a weatherproof, heat treated surface and standard 3" vent stack. Inlet & outlet fittings are 1/2" standard. Tank is pressure tested to 100 psi. Pressure/Temperature relief valve is included for safety. Has drain to prevent freezing, and ash drawer under burner rack.

12" diameter, 49" tall. Outdoor use recommended. Color may vary from picture. Shipped freight collect. Shipping Wt. 45lbs.

•45–427   Wood-Fired Water Heater $279⁰⁰
•45–428   Dual-Fuel Water Heater $299⁰⁰

Breadbox Solar Hot Water Plans

We've gotten lots of requests for simple plans like this over the years. This passive solar water heating system is simple to build using readily available materials; uses no pumps, controllers, or sensors; and works wonderfully as a preheater or even as a stand-alone system. Uses a pair of common 40-gallon water heater tanks, either new or recycled. The system is virtually freeze-proof with a little care during construction. The detailed, easy-to-follow plans include many drawings; a materials list; construction sequence; and options for different tank configurations, reflectors, absorber surfaces, and glazing techniques. Developed by a licensed solar plumber based on 20 years of community education workshops. 1993, 8.5" x 11", 8 pages (folds out).

80-497   Solar Hot Water Plans $12⁰⁰

WATER CONSERVATION

As we said at the beginning of this chapter, water is essential to our survival. The reasons for conserving water are many. Some pertain to the well-being of our planet, society, and ecosystems, while others just make simple, honest, economic sense to us as individuals. The low cost and improved quality and performance of new water-conservation devices today make it easy for anyone to reap all the benefits of saving water.

On a societal level, though, we are depleting many of our water supplies, and damaging ecosystems in the process. One of the worst examples is Mono Lake on the east side of the Sierra Mountain range near Yosemite. Mono Lake is a major rest area for migratory birds and the nesting area for 80% to 90% of the California gull population. Hundreds of species from all over the Western Hemisphere rely on this vital ecosystem. For the past half century, the Los Angeles Department of Water and Power has diverted water that would normally flow into the lake to Southern California's water supply. Since this began, the lake has shrunk to less than half its original size, leaving receding shorelines of toxic alkali dust. In addition, the water level has dropped 47 vertical feet, exposing large calcium carbonate pillars called *tufa*. Towering above the water line, the tufa are ominous indicators of the recession of the water level over the years. Land bridges have formed between the shore and islands which were once safe havens for nesting birds and their young. Now coyotes and other predators easily walk to the islands and feed on the helpless inhabitants.

Should we just sit back and blame the L.A. Department of Water and Power, or is there something else we can do? In past droughts, the L.A. community has used simple water-conservation methods to save more water than has been historically diverted from the lake. If each of us in our own community would conserve water with devices like the ones found in this chapter, we could together prevent environmental disasters like Mono Lake from happening.

On the individual level, the benefits of water conservation are many. Low-cost water-conservation devices make simple economic sense for anyone paying for municipal water. Yet, even if you're not paying for your water, you certainly are paying to *heat* your water. And this means that every gallon of hot water you use costs you money. Using a low-flow showerhead can cut your hot-water use, and your water heating bill, by 50% or more, while still giving you a comfortable shower. For those of you who are still reluctant to get your feet wet (so to speak) with these showerheads, we encourage you to give our Lowest-Flow Showerhead a try. The flow out of the nozzle is much better than most hardware store models — we chose this model over all the others because it adds just enough pressure to make it easier to rinse off, without making it uncomfortable at all. It took me over a year of working here at Real Goods to give it a try, and as soon as I hooked it up, I wondered why I had waited so long! If you are one of the many people who wish they had more pressure from their shower, this is a simple and inexpensive way to get it without buying a pump and pressure tank.

Spring '96 Update:
Mono Lake has been saved.
In a historic negotiated court
case the water level of Mono
Lake has been fixed
approximately 20 feet higher
than present levels. The L.A.
Department of Water and
Power agreed to maintain
this higher lake level.

Mono Lake

A family of four, with each family member taking a five-minute shower every day, will use over 700 gallons of water every week — equal to a three-year supply of drinking water for one person. With a low-flow showerhead, though, the same family would save at least 14,000 gallons of water per year. If only 10,000 families used these low-flow showerheads, we would save 140 million gallons every year. And if it were 100,000 families, the savings would be *1.4 billion gallons!*

Similar water savings can be realized at the toilets and sinks of nearly all American households. Basically every point-of-use for water in the home is a target for water conservation. In the following section, you'll find faucet aerators for the kitchen and bathroom, showerheads, devices for the toilet, and even new toilets that consume less than a gallon per flush. We've scoured the market to find the utmost in quality, value, reliability, and water savings.

— *Gary Beckwith*

WATER CONSERVATION

Toilet Lid Sink

This ingenious water saver supplies clean water for hand washing, then uses it to fill the tank for the next flush. These units have been used in Japan for decades. The unit, constructed of durable plastic, has the appearance of porcelain, fits rectangular tanks (up to 8-inch–wide) and is attached with moisture-resistant Velcro. When the toilet is flushed, incoming fresh water is rerouted through the chrome fixture into the basin, then filtered into the tank and bowl. Water automatically shuts off when it reaches normal fill level in the tank. This sink is a boon to people with limited arm or hand movement, since they need not struggle with faucet handles. And it's so easy that children are more likely to wash their hands. Excellent for small spaces, too; no separate washbasin is needed. Installation is very simple, requiring no tools and only a few minutes of time.

46-121	Toilet Lid Sink, 17" to 20"	$39⁹⁵
46-122	Toilet Lid Sink, 20-1/2" to 22"	$39⁹⁵

Water Conservation Kit

An average family of four will save 30,000 gallons of water per year by simply installing one low-flow showerhead, two faucet aerators, and a set of toilet dams. We've packaged all these water-savers together with a toilet leak detection kit, an instruction card, and a 26-page booklet on saving water. This is a $45 retail value but we're offering this kit at a price that will make saving water irresistible.

46-109   Water Conservation Kit $19⁹⁵

SHOWERHEADS

Water-Saving Showerheads

"If each member of a family of four takes a five-minute shower each day, they will use more than 700 gallons of water every week — the equivalent of a three-year supply of drinking water for one person."— 50 Simple Things You Can Do to Save the Earth

Showers typically account for 32% of home water use. A standard shower head uses about 3 to 5 gallons of water per minute, so even a 5-minute shower can consume 25 gallons. According to the U.S. Department of Energy, heating water is the second largest residential energy user. With a low-flow showerhead, energy use and costs for heating water for showers may drop as much as 50%. Add one of our instantaneous water heaters for even greater savings. Our low-flow showerheads can easily cut shower water usage by 50%. A recent study showed that changing to a low-flow showerhead saved $0.27 of water per day and $0.51 of electricity per day for a family of four. So, besides being good for the Earth, a low-flow shower head will pay for itself in about two months!

Lowest Flow Showerhead

This is by far our best-selling and finest-designed showerhead. It can save up to $250 a year for a family of four by cutting hot and cold water use by up to 70%. At 40 psi it delivers 1.8 gpm, with a maximum at any pressure of 2.4 gpm. Manufactured in the USA of solid brass, and chrome-plated, it exceeds California Energy Commission standards. Built-in on/off button for soaping up. Standard 1/2-inch threads make a wrench the only tool necessary for installation. Fully guaranteed for 10 years. Specified by the City of Los Angeles.

46-104   Lowest Flow Showerhead $11⁹⁵

The Ultimate Shower

We've located the perfect water-saving showerhead for comfort. Certified as a Low-Flow fixture, the Ultimate Shower uses only 2.74 gpm, even with its 3-1/2-inch showerhead that has 127 holes. The all-directional showerhead does not sacrifice spray quality as most available water-saving fixtures do. The wide-covering consistent rainstorm effect that emanates from the head produces a sensation never experienced with conventional showers. The Ultimate Shower allows over 20 inches of height flexibility with its universally adjustable, all-directional arm. Constructed of solid brass, and chrome-plated, the shower includes a wing spanner wrench and a water flow restrictor. Shower arms install to industry standard 1/2-inch pipe thread.

46-102   Ultimate Shower $59⁹⁵

Low Flow, Dual Head Shower

The Shower Sensation's twin-headed shower arm offers great flexibility and control—and it saves water, too. The two-foot extendible arm adjusts to any angle or height, enabling persons of all ages and sizes to shower in comfort. Each low-flow showerhead receives full and equal pressure, and each has an on/off/partial flow valve. Great for targeting sore and aching muscles, for enjoying an all-encompassing spray, or for sharing with a friend. The handicapped or elderly can shower safely sitting down and control or direct the spray without assistance. Solid brass fixture. With chrome-plated finish. Easy to install on any existing shower fixture with standard 1/2" threads.

46-118   Chrome Twin Head Shower $109⁹⁵

LOW-FLOW FAUCET AERATORS

Low-Flow Faucet Aerators

According to Home Energy *magazine, we would save over 250 million gallons of water every day if every American home installed faucet aerators. Installing aerators on kitchen and bathroom sink faucets will cut water use by as much as 280 gallons per month for a typical family of four.*

Bathroom, Kitchen, and Deluxe Faucet Aerators

Bathroom Faucet Aerator

This is one of the aerators included in our Water Conservation Kit. When installed in the bathroom, this aerator will save a typical family of four up to 100 gallons every day. Very simple to install and constructed of solid brass, chrome-plated. Sold in sets of two. Uses 1.5 gpm at 30 psi. Fits both male and female faucets.

46-108   Bathroom Aerator (set of 2) $3⁹⁵

Kitchen Faucet Aerator

This is one of the aerators included in our Water Conservation Kit. With the aerator installed at the kitchen sink, a typical family of four will save up to 50 gallons every day. Very simple to install and constructed of solid brass, chrome-plated. Sold in sets of two. Uses 2.5 gpm at 30 psi. Fits both male and female faucets.

46-110   Kitchen Aerator (set of 2) $3⁹⁵

Deluxe Faucet Aerator

This low-flow faucet aerator with fingertip on-off lever will cut hot and cold water use by up to 60%. It is dual threaded both internally and externally to fit almost all male and female faucets. It limits faucet flow to 2.75 gpm at any water pressure. It's made of solid brass, chrome- plated. It installs simply in seconds and can save thousands of gallons of wasted water every year. The on-off fingertip control lever allows the user to temporarily restrict the flow of water to a trickle without readjusting at the hot and cold controls. It's ideal for shaving, brushing, and washing dishes.

46-103   Deluxe Faucet Aerator $7⁹⁵

Spradius Kitchen Aerator

Here is a two-position faucet aerator that saves lots of water. It's rated at 2.5 gpm, and it swivels 360 degrees to direct spray or stream to every part of your sink. When pushed up, it functions as an aerator, keeping water from splashing out of the sink by injecting air into the water to soften it. When pulled down for spray, it saves water with a great spray pattern that's handy for cleaning or rinsing. It installs quickly and without tools. The double swivel design allows you to rinse the entire sink all the way up to the edges, and eliminates the need for an expensive sprayer hose.

46-131   Spradius Kitchen Aerator $10⁹⁵

COMPOSTING TOILETS

A composting toilet is a treatment system for toilet wastes that does not use a conventional septic system. Composting toilets were developed in Scandinavia, where residents sometimes have a great view of the fjord, but no soils suitable for conventional septic systems. A composting toilet is basically a large, warm, externally-vented container with a diverse community of aerobic microbes that live inside and break down the waste materials. At the end of the process, a dry, fluffy, odorless compost is left. Flowers and orchards love it. The composting process does not smell. Since the aerobic decomposition takes place in the presence of oxygen, this is a good-smelling process, like a well-turned garden compost pile. This is the opposite of the process that takes place in smelly outhouses, which work by anaerobic decomposition. If a composting toilet smells bad it means something is wrong.

How Do They Work?

Ninety percent of what goes into a composting toilet is water, which is dealt with by evaporation. To speed up the evaporation process, we need warmth and air flow through the unit. The solids are dealt with by a diverse microbiological community — in other words, by being composted. To keep this biological community happy and working hard we need warmth and plenty of oxygen, the same things that are required for evaporation. Composting toilets work best at temperatures of 70° F or higher; at temperatures below 50° F the biological process slows to a crawl. It is okay to let a composting toilet freeze, and normal activity will resume when the temperature rises again. A cup or so of peat moss added daily helps soak up excess moisture, makes lots of little wicks to aid in evaporation, and creates air passages that prevent anaerobic pockets from forming.

The Sun-Mar composting toilets all use a rotating drum design that allows the compost pile to be turned and mixed easily (and remotely!). This ensures that all parts of the pile get enough oxygen, and that no anaerobic pockets form.

Sun-Mar produces both self-contained composting toilets and centralized units, in which the compost chamber is located outside the bathroom and connected to an ultra-low flush toilet with standard three-inch waste pipe.

All electric models use a thermostatically controlled heating element to assist with evaporation, and a small fan to ensure fresh air flow and negative air pressure inside the compost chamber. We strongly recommend the electric models for customers who have utility power available. For those with intermittent AC power there are AC/DC models available, which take advantage of AC power when it's available, but can operate adequately without it as well. Finally, there are nonelectric models that don't need any power (though use is more limited).

Will My Health Department Love It as Much as I Do?

Possibly. It depends mainly on the enlightenment quotient of your local health official. Some health departments will welcome composting toilets with open arms, while others will deny their very existence. The Excel model has the NSF (National Sanitation Foundation) seal of approval, which makes acceptance easier for local officials, but doesn't mandate local acceptance. It is really up to your local sanitarian. He can play God within his own district and there is nothing that you, Real Goods, or Sun-Mar can do about it. Sun-Mar engineers have been successful at convincing local building and health inspectors to approve their toilets, but it is not a given. Be courteous! Also, be aware that in all cases composting toilets *only* deal with toilet wastes, colorfully called *blackwater*. You will still be required to treat your graywater wastes — all the shower, tub, sink, and washer water. We have often found that, in lakeside summer-cabin situations, health officials greatly prefer to see composting toilets rather than pit privies or poorly working septic systems that leach into the lake.

Installation Tips

Compost piles like to be warm because it makes them work faster. Similarly, composting toilets work best in warm, moderately dry environments. If your installation is in a summer-use cabin, then a composting toilet is ideal. A western Washington outdoor installation in wintertime, however, would *not* be a good idea. If you plan to use the composter year-round, then install it in a heated space on an insulated floor. The electric heater inside the compost chamber will *not* keep it warm in a outdoor winter environment, but it *will* run your electric bill up about $25 a month trying. Talk to one of Real Goods' technicians if you have any doubts about proper model selection or installation.

If you have utility power, by all means use one of the electrically assisted units; they have far fewer problems overall. If you have intermittent AC power from a generator or other source, use one of the AC/DC, or hybrid, units. Anything that adds warmth or helps push ventilation will help these units do their work.

The compost chamber draws in fresh air and exhausts it to the outside. If the composter is inside a house with a woodstove and/or gas water heater that are also competing for inside air, you can have lower air pressure inside the house than outside. This can pull cold air down the composter vent and into the house. Composters shouldn't smell bad, and something is wrong if they do, but you do not want to flavor your house with their gaseous by-products, either. Do the smart thing: give your woodstove outside air for combustion.

Insulate the composter's vent stack when it passes through any unheated spaces. The warm, humid air passing up the stack will condense on cold vent walls, running back down into the compost chamber again. The self-contained Sun-Mar units are supplied complete with vent stack, which includes six feet of insulated pipe. Add more insulation if needed for your particular installation.

With central units such as the Centrex and WCM models, where the composting chamber is separate from the ultra-low–flush toilet, the slope of any horizontal waste pipe run is critical. The standard three-inch ABS pipe needs to have precisely 1/4 inch of drop per foot of horizontal run. More drop per foot allows the liquids to run off quickly, leaving the solids high and dry. Vertical runs are no problem, so if you want a toilet on the second floor, go ahead. Sun-Mar recommends horizontal runs of 18 feet or less if possible, but customers have successfully used runs of well over 20 feet. Play it safe and give yourself clean-out plugs at any elbow.

— Doug Pratt

COMPOSTING TOILETS

Sun-Mar Composting Toilets

We have now supplied well over 2,500 Sun-Mar toilets (formerly Bowlis) to our customers, and we are pleased to report that these toilets have exceeded our expectations and the manufacturer's specifications. We're convinced that this is the best small composting toilet system on the market. The old problem of resistance from doubting building inspectors has virtually been eliminated with the NSF approval (on the XL).

The heart of the Sun-Mar system lies in the revolutionary "Bio-Drum" composting process. The toilet's inventors are the same people who were involved in the original Swedish composting systems 27 years ago. The inventor was the recipient of the Gold Medal for the best invention at the International Environmental Exhibition in Geneva, Switzerland.

Sun-Mar composting toilets work like a compost pile. Human waste, peat moss, and kitchen scraps are introduced into the toilet; heat and oxygen transform this mixture into good fertilizing soil. The Bio-Drum ensures effective aeration and sterilization, killing anaerobic microbes and mixing the compost well. Turning the drum periodically maintains the aerobic composting process. Oxygen is provided by the ventilation system. The material entering the toilet is approximately 90% water, which is evaporated into water vapor and carried outside through the venting system. The remaining waste is transformed into an inoffensive earthlike substance: compost!

Freezing temperatures will not damage the toilet or the compost; however, in temperatures below 50° F the composting action decreases. Toilet paper is composted along with the rest of the material. Composted material is removed one to four times per year, depending on use. Residential use may require removal slightly more often. This residual compost is the best fertilizer you can get for use on lawns, orchards, or ornamental beds.

How can the compost be dumped out for use when fresh waste is present? The device's bottom drawer serves the purpose: when the drum is approximately half to two-thirds full, some of the compost is cranked into the bottom drawer where final composting action occurs prior to use.

We have only rarely had a Sun-Mar customer complain about odor. The air flow provides a negative pressure which ensure no back draft. The air is admitted through ventilation holes in the front. The rotation and aeration by the Bio-Drum or shaft-mixers along with the addition of organic material ensures a fast, odorless, aerobic breakdown of the compost.

All Sun-Mar toilets need a minimum of maintenance. All that needs to be added is a cup of peat moss per person per day, plus, if available, some other organic material such as vegetable cuttings, greens, and old bread. If the toilet is used continuously, the compost needs to be aerated and mixed once every third day. This is simply done by giving the handle a few turns.

All Sun-Mar toilets have a full two-year parts warranty, and come with the vent stack and everything necessary for a do-it-yourself installation. All units are certified by Canadian authorities, and the XL is fully tested and certified by NSF (National Sanitation Foundation).

Composting Toilet Comparision Chart

Model	Summer Use (over 60⁰ F) Maximum # of People		Winter Use (over 60⁰ F) Maximum # of People		Seat Height	Compost Chamber Inside	Compost Chamber Outside	Needs AC Power	Overflow Drain Required	DC Exhaust Fan Recommended
	Full-time	Weekend	Full-time	Weekend						
Sun-Mar XL	3-4	5-6	3-4	5-6	28"	●		●		
Sun-Mar Centrex	5-6	6-7	4-5	6-7	15"A		●	●		
Sun-Mar Centrex AC-DC	3-4	4-5	1-2	3-4	15"A		●	O	●	I
Sun-Mar Centrex NE	1-2	2-3	NR	NR	15"A		●		●	I
AE Hybrid	2-3	3-4	1-2	2-3	29"	●		O	●	●
Sun-Mar NE	1-2	3-4	NR	NR	28"	●			●	●
Biolet NE	2-3	3-UP	Use extra bins		21"	●			●	●

A- Dimension for Sealand Toilet, Aquamagic is 16"   NR- Not Recommended   O- Optional   I- Included

Sun-Mar Centrex

The Centrex is the model for those who want the composting chamber outside of the bathroom. This is the re-engineered, slightly smaller, more serviceable version of the WCM model. The Centrex has all controls, access, and venting on the front panel for ease of service. Inlet access can be from any direction. Front venting makes cottage installation much easier, particularly when using the NE (nonelectric) model, which requires a straight vent with no bends.

The Centrex uses an RV toilet with an ultra-low flush of approximately 1 pint. Toilets available are either the Aqua Magic in standard plastic, or the Sealand in fine china. Both toilets use standard 3-inch ABS waste pipe for hook-up to the composting unit. Horizontal pipe run should be no more than 18 feet with precisely 1/4-inch drop per foot. Vertical run can be any distance.

Available in three models:

Centrex NE for installations without any electricity. Has a single 4-inch vent stack. Good for one to two people full-time, three to five people intermittent.

Centrex AC/DC for installations with limited intermittent AC power available. Has twin vents, 2-inch and 4-inch with a 1.4-watt 12-volt fan installed in the 4-inch stack. Good for two to three people full-time, or four to six people intermittent.

Centrex for installations with utility AC power. Has a single 2 inch vent stack. Good for two to four people full-time, or four to six people intermittent.

> *Dimensions: 33"W x 26-1/2"H x 25"L (N.E. 25"H)*
>
> *Weight: 57 lbs., shipping weight 95 lbs.*
>
> *Vent & Drain Systems: (Pipe & fitting supplied with unit.) 2" vent on Centrex & Centrex AC/DC. 4" vent on Centrex NE & Centrex AC/DC. 3/4" drain fitting on all models.*
>
> *Electrical Power Requirements: AC models 110 volts, 2.5 amps maximum. 30-watt fan, 250-watt heater with thermostat. Approximate average demand, 125 watts. AC/DC model in DC mode — 1.4 watts, 12 volts. (24-volt units available by special order)*

•44-106	Sun-Mar Centrex N.E.	$999
•44-107	Sun-Mar Centrex AC/DC	$1,279
•44-105	Sun-Mar Centrex	$1,199
•44-203	Aqua Magic Toilet	$149[00]
•44-405	Sealand China Toilet	$199[00]

Toilets shipped freight collect when ordered with a Sun-Mar unit, shipped UPS when ordered separately.

Sun-Mar Centrex

Aqua Magic Toilet

Sealand China Toilet

Sun-Mar Compact

This is a scaled-down version of the XL model. For those who don't need the larger capacity of the XL and have access to 120-volt power, this is the recommended unit. The working components are the same as the XL: a bio-drum for mixing and aeration, a thermostatically controlled base heater, and a small fan for positive air movement. What makes the Compact model different is a smaller variable diameter bio-drum which allows a smaller overall size, an attractive rounded design, and most importantly, no more footrest! Also, the handle for rotating the bio-drum is now hinged, and folds into the body when not being used. Recommended for one person in full-time residential use, or two to four people in intermittent cottage use.

Dimensions: 22-1/2"W x 31"H x 32"L

Installation length to remove drawer: 51"

Seat height: 21-1/2"

Weight: Shipping weight 70 lbs., product weight 45 lbs.

Vent & Drain: 2" vent pipe & fitting (supplied with unit), 1/2" emergency overload drain (optional hookup).

Electrical Power Requirements: 115-volt, 2.5-amp. 30-watt fan, 250-watt heater with replaceable thermostat. Approximate average demand: 125 watts.

•44-104 Sun-Mar Compact	$1,149

Shipped freight collect from Buffalo, NY.

Sealand China Toilet

We offer the Sealand low-flush vitreous china toilet with the Centrex, which uses less than one pint of water to flush. This compact low-flush toilet has a very effective flushing mechanism with a 360-degree rim flush. All Sealand toilets are covered by a two-year warranty and are built in the USA. Recommended for Centrex or RV use only, these toilets cannot be connected to a sewer system. They require as little as 5 psi water pressure (roof-top gravity feed), and 2 gallons per minute flow rate.

•44-405 Sealand China Toilet	$199⁰⁰

Shipped freight collect when ordered with Sun-Mar toilets; shipped UPS when ordered alone.

Real Goods AE Hybrid

The Alternative Energy Hybrid was co-developed by Real Goods and Sun-Mar specifically for Real Goods' off-the-grid customers who derive part of their power from generators. The AE is identical to an XL, except that it is fitted with an NE drain, and with an additional 4-inch NE vent installed next to the XL's 1-1/2-inch vent stack. The AE provides the increased capacity of an XL unit when a generator is running, but operates as a non-electric unit when 120-volt electricity is not available. For residential use, a built-in 12-volt fan is recommended when the AE is running in a non-electric mode. System does not include 12-volt fan. Dimensions: 23-1/2"W x 29-1/2"H x 33"L.

•44-103 Sun-Mar AEHybrid	$1,279
•44-803 Sun-Mar 12-Volt Fan	$46⁰⁰
•44-804 Sun-Mar 24-Volt Fan	$46⁰⁰

Sun-Mar Non-Electric (NE)

The Non-Electric is our most popular composting toilet. It's perfect for many of our customers living off-the-grid and not wanting to be dependent on their inverters. The tremendous aeration and mixing action of the Bio-Drum, coupled with the help of a 4-inch vent pipe and the heat from the compost, creates a "chimney" effect which draws air through the system in a manner similar to that of a woodstove. In residential use a 12-volt fan is recommended to improve capacity, aeration, and evaporation. Dimensions: 22-1/2"W x 28"H x 33"L.

•44-101   Sun-Mar N.E. Toilet $999⁰⁰

Sun-Mar XL

At sites with 120-volt electricity, the XL is a high-capacity unit ideal for year-round or seasonal use, using a maximum of 280 watts for the thermostatically controlled heater and fan. Comes with 1-1/2-inch vent pipe. Dimensions: 22-1/2"W x 29-1/2"H x 33"L.

•44-102   Sun-Mar XL Toilet $1,199

Sun-Mar Video

Want to learn more about composting toilets? This is a 27-minute video on how a Sun-Mar composting toilet works, proper model selection, installation, maintenance, and interviews with customers.

44-108   Sun-Mar Video $15⁰⁰

12-Volt Toilet Exhaust Fan

At last, a powerful 12-volt fan designed to fit easily over a standard 4-inch ABS or PVC vent pipe! When used as a standard bathroom ventilation fan, it can be wired into your 12-volt lighting system via a switch. It can also be used on a continuous daytime basis by connecting it to a 10-watt solar module, such as the Solarex SX-10 (#11-513) or the Amorphous 14-watt (#11-514). When used for ventilating composting toilets, such as the Sun-Mar NE, it should be connected to a battery to provide round-the-clock ventilation. It is also a good system for greenhouse ventilation.

The bathroom fan uses only 9 watts (3/4 amp at 12 volts), and is rated at 90 cfm. The quiet, brushless fan motor is encapsulated to protect it from corrosion. The exhaust fan is highly recommended for residential use and in locations where downdraft can occur, such as areas surrounded by mountains or high trees. It will greatly improve airflow and capacity.

44-802   12-Volt Toilet Exhaust Fan $69⁰⁰

Sun-Mar RV/Marine Composting Toilet

A brand new model from Sun-Mar, the Ecolet is the first composting toilet designed especially for RV and marine use. Imagine the freedom to travel where you want, when you want, without having to worry about dump stations or pumpouts. At 19.5" wide by 23" deep, the Ecolet is only slightly larger than a regular sized toilet seat. It will fit almost anywhere! Can handle 3 to 5 people in weekend use, or 1 to 2 people in residential use. Has a 12-volt stack fan (3.4 watts) to assist air flow and ensure no smells inside, plus thermostatically controlled 120 watt 12v and 120v heaters when excess power is available to speed up evaporation and composting. Constructed of high quality fiberglass and marine grade stainless steel. Not damaged by freezing, although composting activity ceases below 50° F. Stands 29" high and uses a 3" vent stack which is supplied with a deck outlet fitting. Mfg. warranty: 3 yr. parts, 25 yr. body. Ships freight collect from Buffalo, NY. Shipping Wt. 70 lbs. Unit Wt. 45 lbs.

• 44–207   Sun–Mar Ecolet Toilet $1,049

Clivus Multrum Composting Toilets

Clivus is the company that originally brought composting toilet technology to North America in the early 1970s. Until recently Clivus has focused their expertise on large, heavy-use automated composters designed for public parks and recreation areas. These large composters can accommodate multiple toilets and are designed for many thousands of uses per year, sometimes even thousands of uses per day.

Clivus composting toilets are National Sanitation Foundation (NSF) approved, and are currently in successful use in many National and State Parks, military installations, ski resorts, U.S. Forest Service installations, and thousands of residential homes.

Clivus produces a full line of composting toilets to accommodate from 2 to 10,000 people per day. They can provide anything from a basic composter tank, through specialized foam-flush toilets, complete buildings, and even foundations. They can also provide complete graywater systems for sinks and showers.

Because larger public systems usually need to be engineered for the specific site, Real Goods will put you directly in contact with the Clivus staff if you have a large public project in mind. Please write or call for further information.

If you are looking for something smaller for your remote homestead or full-time cabin, Clivus is offering their experience and proven designs in two models for the home-scale market.

The Multrum 1 and Multrum 2 composting toilets are designed for two to six full-time residents. These units are more expensive initially, but less trouble to live with, than the Sun Mar composters.

The advantages are:

Almost zero maintenance. Other than emptying 2 to 3 gallons of finished compost per user, per year, and the removal of a high-quality, odorless liquid fertilizer once or twice yearly, there is no routine maintenance.

A completely finished compost product. Because compost spends more time in the Clivus, and because moisture levels are maintained for best composting, the finished product is completely composted and pathogen-free. Smaller composters partially compost, and then simply dry whatever is left.

Constructed of 100% recycled and recyclable polyethylene. The solid waste composting chamber and the liquid storage cradle are both made of sturdy, high density, 3/8" thick, 100% recycled material that is corrosion-and leak-proof.

Minimal reliance on mechanical or electrical parts. Has no heating elements, thermostats, or moving parts to break down. Only uses a small 35-watt fan.

Handles high peak loads. Because of the larger compost chamber the Clivus can handle parties and occasional overloads.

Low power use. Uses only a 35-watt maximum ventilation fan with variable speed control. Can be supported on renewable energy systems.

Can be used with waterless, 1-pint flush, or 3-oz. foam flush toilets. The variety of input toilets available make a variety of installations easy. Will accept multiple toilets.

Five-year warranty. The composting chamber and liquid storage cradle are warranted against defects in material and workmanship for a full five years. Electrical parts are warranted for one year.

The Clivus Multrum composters are sold as a complete package with your choice of waterless, Sealand low flush, or Nepon foam toilets.

Each Clivus Multrum package includes:

- compost chamber & liquid fertilizer storage tank
- ventilation system
- compost moistening system
- all necessary hardware for installation
- starter bed
- toilet of choice (waterless, Sealand low flush, or Nepon foam)
- Multrum Bacteria, Composting Worms, & Cleaner
- Installation Manual

Shipped freight collect from Massachusetts. Allow 4 to 6 weeks for delivery.

The Multrum 1

The Multrum 2

CM Waterless Toilet

Multrum 1

2 to 4 persons residential use
6 to 8 persons intermittent use

This is the smallest Clivus composter. The tank is designed for basement or crawl space installation. Either waterless, foam, or 1-pint flush toilet packages can be ordered. It has a built-in moistening system with a programmable timer. The fan-forced ventilation system maintains a negative pressure inside the compost chamber for odor-free operation. The vent fan has variable speed control and uses a maximum of 35 watts. A solar powered fan is optional. 18 feet of 4" flexible vent ducting and 2-1/2 feet of 4" PVC pipe with bird screen cap and rubber storm collar are provided. Additional ducting is easily available if needed.

> *Tank Dimensions: 60"L x 29"W x 46-1/4"H*
> *Tank Weight: 175 lbs.*
> *Tank Volume: 32 cu.ft.*
> *Liquid Cradle Dimensions: 65"L x 32-1/2"W x 15-1/2"H.*
> *Cradle Weight: 85 lbs.*
> *Cradle Volume: 7 cu.ft. (44 gal.)*

•44-407   Waterless Multrum 1 package $2,495

Includes 36" stainless chute section, order additional chute sections as needed

•44-413   Clivus Stainless Toilet Chute 36" section $50⁰⁰
•44-414   Clivus Stainless Toilet Chute 24" section $40⁰⁰

Multrum 2

3 to 5 persons residential use
6 to 8 persons intermittent use

The Multrum 2 is a slightly larger version of the Multrum 1 allowing a higher level of continuous use. It uses the same 35 watt ventilation fan, moistening system, and also comes with 18 feet of 4" flexible vent ducting and 2-1/2 feet of 4" PVC pipe with bird screen cap and rubber storm collar. Tank will fit through most 30" or larger doorways.

> *Tank Dimensions: 60"L x 29"W x 64-1/2"H*
> *Tank Weight: 95 lbs.*
> *Volume: 46 cu.ft.*
> *Liquid Cradle Dimensions: 65"L x 32-1/2"W x 15-1/2"H.*
> *Cradle Weight: 85 lbs.*
> *Cradle Volume: 7 cu.ft. (44 gal.)*

•44-410   Waterless Multrum 2 package $2,995

Includes 36" stainless chute section, order additional chute sections as needed

•44-413   Clivus Stainless Toilet Chute 36" section $50⁰⁰
•44-414   Clivus Stainless Toilet Chute 24" section $40⁰⁰

Other Options: Clivus Multrum Bacteria

A 1 lb. bag of compost enhancer. Included with the above packages to get your Clivus Multrum compost pile off to a balanced start. Keep your composter happy with a teaspoonful occasionally.

•44-415   Multrum Bacteria $24⁰⁰

The Humanure Handbook

A Guide to Composting Human Manure (emphasizing minimum technology and maximum hygienic safety)

For those who wish to close the nutrient cycle, and don't mind getting a little more personal about it than our manufactured composting toilets require, here's the book for you. Provides basic and detailed information about the ways and means of recycling human excrement, without chemicals, technology, or environmental pollution. Includes detailed analysis of the dangers involved and how to overcome them. The author has been safely composting his family waste for the past 15 years, and with humor and intelligence has passed his education on to us. Published 1994, softcover, 198 pages, indexed, with glossary, appendices, 63 tables & figures, 9 sidebars, and a few bad jokes.

80-498   The Humanure Handbook $19⁰⁰

The Toilet Papers

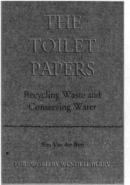

A '70s classic is back in print. One of the favorites of the back to the land movement, *The Toilet Papers* provides an informative, inspiring, and irreverent look at how people have dealt with human wastes over the centuries, and what safe designs are available today that reduce water consumption and avert the necessity for expensive treatment systems. Van der Ryn provides homeowners plans for several types of dry toilets, compost privies and graywater systems. He also covers the history and philosophy of turning organic wastes into a rich humus, linking us to the fertility of the soil, and ensuring our ultimate well-being.

80-064   The Toilet Papers $11⁰⁰

BioLet Composting Toilets

Made in Sweden, BioLet biological toilets are some of the best-selling composting toilets in the world. We have added BioLet to our product line because they offer excellent performance and a more compact design for homes that have utility power available.

Both of the BioLet models share some common features.

• The seat, at 21", is close to conventional height. No stairsteps neded.
• The compost chamber is hidden from view until you sit. Weight on the seat opens a pair of clamshell doors below the seat, when standing the doors close again.
• The composting process works from the top toward the bottom, just like gravity. The freshest material goes in the top, the oldest, most digested material goes out the bottom to the finishing tray.
• The finishing tray at the bottom is sealed. If there's excess liquid build-up it won't leak. A sight tube at the bottom allows you to see if liquid is building up.
• If liquid accumulates, you can adjust the thermostat to increase the warmth of the air flow. Incoming air is heated before being routed through the composter and out the vent stack.
• Warm air flow encourages faster evaporation, and ensures that finished material isn't baked in the finishing tray by simple bottom heaters.
• The design has all the mechanical and electrical parts mounted on a one-piece cassette in the top of the composter. If any fan, heater, thermostat, or mixer motor fails it can be easily serviced on site.
• The sealed casing is high-impact ABS plastic. Mixing arms and other important metal components are stainless steel.

BioLet toilets include an installation kit consisting of: 14 feet of 2" white vent pipe, 39" of insulation, 39" of black outer pipe jacket to cover insulation, one black reducing coupling (the top), and one black flexible roof flashing. There is a full two-year warranty, and a toll-free 800# service line for all BioLets.

BioLet XL Model

BioLet's top-of-the-line model is National Sanitation Foundation (NSF) approved for residential use. Has 40% larger composting capacity than the Manual Model. The thermostatically-controlled heating element is 305 watts. The mixer motor runs automatically for 30 seconds every time the lid is lifted and replaced. The quiet 25-watt fan runs continuously, the heating elements run as required by thermostat setting and room temperature. The BioLet XL measures 26.5"H x 25.6"W x 33"D. It requires 26" x 54" of floorspace for the finishing tray to slide out the front. For homes with utility power. Weighs approx. 50 lbs. Two-year mfg. warr.

•44-109 BioLet XL Model $1,559
Shipped freight collect from Massachusetts or Texas. Shipping weight approx. 75 lbs.

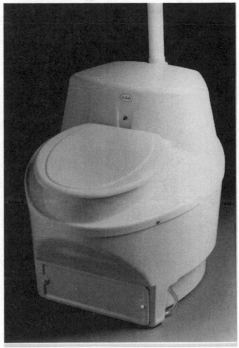

XL Model

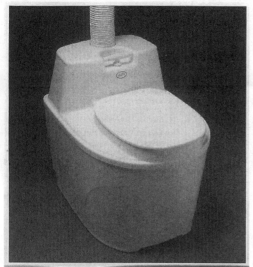

BioLet Non-Electric

The BioLet Non-Electric is our simplest, and lowest-cost composting toilet. This model uses no electricity, water, or chemicals, and has no moving parts, making it very inexpensive to operate. Inside the BioLet Non-Electric are two composting containers. When the container under the seat is filled, the composter seat/top lifts off, and the first container is slid to the back of the toilet for further composting. The second container is now moved to the front. When the second container is filled, it is time to dispose of the finished compost in the back container. Under favorable (warm) conditions this composter will handle three adults full-time. Under colder conditions, or when more people are using it, capacity can be inexpensively increased by purchasing extra compost bins.

There is a small price to pay for the lower cost and increased capacity of the BioLet Non-Electric. There is a two-minute weekly maintenance. Humus mix must be added and manually mixed with the included mixing rod every week. The weekly mixing of the solid waste, paper, and humus-starter will keep the compost porous and moist, and increase the oxygen supply, which speeds the compost process. A drain to a septic system, leaching pit, or other excess liquid collection system is also required. The Owner's Manual gives several examples.

BioLet Non-Electric

Included with the installation kit are 6.5 ft. of flexible 4" vent hose, 8 ft. of rigid 4" vent with 40" of insulation, 40" of 6" outer pipe, rain cap, insect netting, roof flashing, connectors, clamps, and 40" of 1" drainage hose. Size is 22"W x 29"D x 27"H. Seat height is 20". Weight is 50 lbs. Mfg. warranty is two years. *Shipped freight collect from Texas.*

•44-113 BioLet Non-Electric	$949⁰⁰
•44-114 Spare Composting Bin	$74⁰⁰

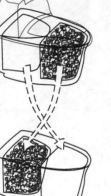

BioLet Manual Model

This model, BioLet's best seller, has a smaller composting chamber and a thermostatically-controlled 250-watt heating element. Mixing is done manually with the handle on top; instead of flushing you simply twirl the handle a few times. The mechanism is geared down ten to one, so little effort is needed. The quiet 25-watt fan runs continuously, the heater element runs as required by thermostat setting and room temperature. The BioLet Manual measures 27"H x 22"W x 29"D. It requires 22" x 40" of floorspace for the finishing tray to slide out the front. For homes with utility power. Weighs approx. 50 lbs. Two-year mfg. warr.

•44-110 BioLet Manual Model $1,169
Shipped freight collect from Massachusetts or Texas. Shipping weight approx. 75 lbs.

Manual Model

A GRAYT WAY TO WATER

For decades it's been illegal in most places to use *graywater* (very loosely defined as all household wastewater except spent toilet water, which is called *blackwater*) for anything other than emptying into a septic tank with a trailing leach field. Yet homesteaders, God-fearing Christians, back-to-the-landers, farmers, and even writers and schoolteachers have long been violating this law on a daily basis. These otherwise upstanding citizens are usually driven to this life of "crime" by failing leach fields and overflowing septic tanks. Their heinous deed? Using graywater on their landscapes.

While such diversion will lessen the load on a septic system, it's usually a far cry from proper irrigation of plants. When properly channeled and used on the landscape, graywater can be both beneficial and resource-efficient. Unfortunately, many people just throw a hose from the washing machine out a window to dispose of discharge water; this isn't irrigation, it's just surface disposal. There's nothing environmentally kosher, groovy, or cool about hurling a bunch of dirty water on top of some poor unsuspecting plants. The graywater may be kept out of the septic tank, but its unregulated dumping will certainly mess up the soil structure, flood and collapse its pore space, brown root hairs, and make clays more sticky and unworkable. In addition, "water dumping" from washers, sinks, and showers exposes people and animals to possible bacterial, viral, and parasitic infections, provides a draw for mosquitoes and other insects, and represents a very poor way to make use of this valuable resource.

Better than Well Water, Better than City Water

Thanks to the recent, much-ballyhooed seven-year California drought (actually nothing compared to the 20-, 40-, and 60-year droughts documented since the 1400s) we know a lot more about graywater and its prerequisite plumbing. The most important lesson we learned also brought good news for gardeners: properly managed, graywater makes plants grow better than either city or well water does. With a slight change in cleansers or detergents used in the house, all the stuff in graywater, which is often dismissed as filth, dirt, or pathogens, can become wonderful plant food.

The common experience of many of us graywater "criminals" has been that, with no more attention to watering than with any normal garden-irrigating schedule, our plants appear to grow better than they do with so-called pure water. Whether this effect is simply the result of our paying more attention to the plants, or whether it is attributable to the nutrient "filth" in the graywater itself has not yet been studied, but logic dictates that when a reasonable detergent (especially one *with* phosphates) washes garden loam and compost out of your jeans, the resulting graywater is bound to "taste" much better to root hairs than chlorinated city water or even well water. This plant-promoting effect means that, drought or no drought, *in almost any climate*, graywater is as much a garden-enhancer as it is a water-

conservation measure. Graywater systems are now becoming legal in more and more places, and they are increasingly being built as *permanent* additions to homes.

Designing and Installing a Graywater System

Graywater systems can be perfectly safe and convenient. The best sources of graywater are the washing machine, tub, shower, and bathroom sinks. Everything *but* the kitchen sink, that is (since even in a vegetarian household with a sink disposal unit, there's too much oil and too many food particles coming from the kitchen sink for practical filtering). Wastewater from the toilets and the automatic dishwasher are also verboten and should never be used in a graywater system.

After seven years of field-testing tens of thousands of graywater systems all over the West, we've learned some basic do's and don'ts. Here are the most important guidelines for the healthiest use of graywater, for both you and your plants.

1. **Water in, water out.** *Never* store graywater, not even for a day. The drums you see in properly designed graywater systems are called "surge" or "buffer" tanks. Their purpose is to accumulate graywater as several or more sources are generating it at the same time, and to hold it *temporarily* until a hose or pump can distribute it to the plants. Storing graywater on warm or hot summer days can quickly produce a septic bacterial soup, which is likely to stink and to breed disease pathogens.

2. **The surge tank must have an overflow port into the main sewage line,** in case the system can't distribute the water fast enough during peak loading. If possible, this overflow line should rely upon gravity flow and connect to the blackwater (sewage) line with a minimum 1/4-inch drop for every ten feet of run. Add a one-way swing check-valve between the sewer line and the surge tank to allow graywater to drain into the sewer, and to prevent a stopped-up sewer, God forbid, from filling the surge tank. (It's important that the valve be installed with the embossed arrow pointing *downstream* toward the sewer or septic tank.)

3. **Never apply graywater with a sprinkler or onto a lawn.** Graywater should never "daylight" pool, puddle, or lie on the surface of the landscape. Following this advice prevents insects from breeding in puddled graywater, animals from drinking it, and kids from being attracted to splash in it.

4. **Always plumb your graywater system with a set of two ball-valves or a single three-way valve.** This allows for convenient switching back to the sewer or septic tank when necessary; for example, whenever it rains too hard for the graywater to seep into the soil, or during the winter in cold climates, or if someone becomes ill with a dangerous communicable disease. The switching can be done with a remote-controlled electric solenoid three-way valve. One flip of the switch and the water is diverted to either the graywater system or the sewer.

5. **Distribute the graywater to different sections of the landscape every couple of days,** depending upon your yard's soil structure. Never apply graywater exclusively and continuously to the same spot, or the soil will be destroyed. A set of simple manual valves in the landscape will allow you to

rotate the graywater to a different section every couple of days, depending upon how much of it your household generates. That way plants only get graywatered every week or two.

6. **Avoid using graywater on acid-loving plants,** because the water will be slightly alkaline, due to soaps and detergents. Most shade-loving plants prefer an acid soil; examples include: blueberries (*Vaccinium* spp.), heath (*Erica* spp.), heather (*Calluna* spp.), spruce trees (*Picea* spp.), pin oak (*Quercus palustris),* crepe-myrtle (*Lagerstroemia indica),* lily-of-the-valley (*Convallaria majalis),* holly (*Ilex* spp.), mountain ash (*Sorbus* spp.) and hemlock (*Tsuga* spp.).

7. **Be cautious about which soaps, detergents, and cleansers you use in the laundry and bathroom.** Use only those products which are considered safe and nontoxic for people. (A good reference book is *Nontoxic & Natural,* by Debra Lynn Dadd.) Avoid products with boron, such as Borateam and Boraxo. Limit your use of bleach as much as possible, or, even better, cut out bleaches altogether. If you have hard water, add 1/8- to 1/4-cup hexametaphosphate per five gallons of laundry water in place of bleach, and then use one-half the amount of your normal detergent. For kitchen and bathroom cleaning, use simple scouring powders such as Bon Ami™ Cleaning Powder.

8. **In especially arid locations, ones that receive less than 15 inches of rain per season, sodium may begin to accumulate in the soil** no matter how careful you are in selecting cleansers and detergents. While this has not been studied enough, early indications are that homeowners in such areas may need to flush the salts deep into the soil (below the root zone) with a deep freshwater irrigation once every three to five years. Where the rainfall is greater than 15 inches, the seasonal accumulation of salts may be adequately flushed by rainfall.

If you install a system with these guidelines in mind, your plants will have a grayt time; in fact, every time you shower, you'll trickle them silly.

— *Robert Kourik*

DRIP IRRIGATION FOR A LUSH GARDEN

The term drip *irrigation* has slipped into our modern gardening lexicon, just as drip irrigation hardware has crept into more and more gardens. Sadly, all too often this phrase has become synonymous with the dreaded "D" word — drought. While it's no surprise that drip irrigation helps gardens flourish during periods of severe drought, drip irrigation is much more than just a last-ditch strategy for a parched landscape — it's a way to help *all* gardens prosper. Gardens with well-designed drip systems display plentiful foliage growth, a tangible increase in bloom, higher vegetable and tree crop yields, and a marked reduction in diseases such as mildew, crown rot, and rust — and all of this with a 30% to 70% reduction in the amount of water used. Plus, the gardener has more leisure time (or time for other gardening tasks) due to the elimination of the time-consuming task of hand-watering.

Drip irrigation technology originated in Israel during the early 1960s, and gets its name from the action of *emitters,* small devices that regulate the flow of water to tiny droplets, which slowly moisten the ground without flooding. Drip emitters release water very slowly and form a wet spot beneath the soil's surface. The shape of the moist area is affected differently by each type of garden soil and ranges from a long, carrot-like shape in sandy soil to a squat and beet-shaped profile in heavy clay soils.

Why Drip Irrigation Works

To understand why drip irrigation promotes luxurious growth, we need to take a fresh look at how roots absorb moisture and nutrients and how plants respond to wet and dry periods in an irrigation cycle. Most gardening books and radio garden show "experts" recommend deep watering on an infrequent basis — every couple of weeks or even once a month. This "ancient wisdom" obviously hasn't killed every tree or shrub, because millions of people continue to follow this advice. But recent research shows this approach to watering does *not* in fact promote optimum growth.

While deep irrigations *are* useful for the survival of trees during times of drought, this watering regime is far from ideal in terms of encouraging both the quality and quantity of plant growth. Many studies have shown that the healthiest trees, with the biggest canopies and greatest productivity, are those which receive more frequent, regular, and shallow irrigations. Infrequent irrigations, on the other hand, tend to produce significantly less growth and reduced yields of fruits, nuts, and vegetables.

Why Moist Surface Soils Promote Growth

The upper layers of the soil are the most aerobic, with the highest population of air-loving bacteria and soil flora. These valuable decomposers are nature's fertilizing machines and must have plenty of oxygen to fuel their activity. It is mostly these bacteria and flora that are responsible for the liberation of mineralized (unavailable) nutrients into a soluble form that the plant can absorb. The soil life stimulates and aids in the production and renewal of humus, which holds onto and releases many of the nutrients plants utilize. Thus, the upper, aerobic horizon of the soil is where the greatest amount of nutrients is liberated. Most studies conclude that, for the sake of yields, the upper one to two feet of the soil accounts for over 50% of all the nutrients a plant absorbs.

Plants primarily absorb these nutrients in a soil-water solution. If the soil is too dry, then nutrient uptake is inhibited, because the soil life can't thrive in a soil that's too dry. Allowing the upper soil to dry out between infrequent irrigations means that nutrient uptake also "dries up." Then, when plenty of water is supplied all at once, the soil becomes saturated to the point that roots and air-loving soil life may be stressed or killed from *too* much water. Nutrient uptake is also inhibited by the lack of air, which would normally sustain the humus-producing bacteria and support active root hairs. It takes some time for the air-loving soil bacteria to repopulate either too-wet or too-dry soil, so there is a biological lag of hours, days, or weeks before the roots of plants can get their best meals. Infrequent and deep irrigations tend to produce two points in the watering cycle where the soil life is damaged enough to reduce or prevent growth.

Frequent Watering Is Best for Growth
I like to think of frequent irrigations as "topping off the tank." Starting with an ideal soil moisture, not too wet and not too dry, the goal is to replace, as frequently as once each day, exactly the amount of moisture that has been lost due to evapora-

PROS AND CONS OF DRIP SYSTEMS

All garden techniques and tools have both good points and drawbacks, and drip irrigation is no exception. Here are the pluses and minuses of drip irrigation:

The Benefits of Drip Irrigation
Uses water efficiently. Sprinklers waste water as a result of the wind, evaporation, runoff, or deep leaching.
Provides precise water control. Every part of a drip irrigation system is constructed with an exact flow rate so you can control the amount down to the ounce.
Increases yields. Drip irrigation easily maintains an ideal soil moisture level, promoting more abundant foliage, greater bloom, and higher yields than any other method of irrigation.
Provides better control of saline water. Saline water applied to a plant's foliage can cause leaf burn. Also, frequent use of drip irrigation helps to keep the salts in solution in the soil so they don't affect plant roots adversely.
Improves fertilization. A fertilizer injector (or proportioner) can easily apply dissolved or liquid fertilizers without leaching the fertilizer beyond desired root zones.
Encourages fewer weeds in dry summer climates. The emitters make only a small moist spot while the larger dry areas between emitters remain too dry for weed seeds to sprout.
Saves time and labor. Drip irrigation systems eliminate tedious and inefficient hand-watering, especially with automated systems.
Reduces disease problems. Plants are less likely to develop sprinkler-stimulated diseases, such as powdery mildew, leaf spot, anthracnose, crown rot, shothole fungus, fireblight, and scab.
Provides better water distribution on slopes. Drip emitters apply the water slowly enough to allow all the moisture to soak in, regardless of slope.
Promotes better soil structure. Drip-applied water gradually soaks into the ground and helps maintain a healthy aerobic soil which retains its loamy structure.
Conserves energy. The low pressure of a drip irrigation system means lower pumping costs with both municipal and private water supplies.
Uses low-flow rates. Drip emitters can water larger areas than sprinkler systems with the same amount of water.

tion from the soil and transpiration from the plant's leaves (called the *ET rate*, for "evapotranspiration"), plus an amount that represents enough extra water for higher yields or more gorgeous ornamental foliage. From Richard Harris' preeminent book on trees, *Arboriculture*, we read: "In contrast to other systems, drip irrigation must be frequent; waterings should occur daily or every two days during the main growing season . . . the amount of water applied should equal water lost through evapotranspiration." This doesn't mean you'll be wasting tons of extra

Is more economical than permanent sprinkler systems. Drip irrigation systems usually cost less than fixed underground sprinkler systems.

The Limitations of Drip Irrigation

Eliminates soothing hand-watering. For some gardeners, the act of hand-watering is more valuable than any form of therapy or meditation, and drip irrigation is counterproductive.

Initial costs are high. A simple oscillating sprinkler will always be cheaper than any drip system.

Can clog. Many early models of emitters were more prone to clogging and gave the industry a bad reputation. With the correct modern emitter, clogging is no longer a serious problem.

May restrict root development. With only one or two emitters per plant, root growth can be greatly restricted. However, with the proper placement of emitters, root growth will be uniform, expansive, and healthy.

Rodents can perforate the tubing. Gophers may chew on the drip hose for a drink. Occasionally, even mice and wood rats will chew through the hose.

Isn't compatible with green manures and cover crops. The growth of a green manure crop gets all tangled up with the drip tubing, thus prohibiting the usual tilling-under of the plants.

Weeding can be difficult. Care must be taken not to damage an exposed drip system while weeding. (Mulching to suppress weeds can mitigate this problem.)

Requires greater maintenance. Drip irrigation requires more routine maintenance than a hose or sprinkler to sustain its high level of efficiency, but the maintenance is relatively simple.

Doesn't cleanse the foliage. Plants with leaves which require an occasional sprinkling should be grown with low-flow sprinklers.

Doesn't create humidity. Plants which are humidity-loving prefer misters or sprinklers.

You can't see the system working. With a well-mulched drip system, the emitters quietly go about their work hidden from view. For some gardeners this is the beauty of the system. For others, not being able to watch the watering can be unsettling.

water. Often it means just applying the same amount of water on a monthly basis, but in a very different pattern of application.

An ET-based irrigation schedule actually means the same amount of water is applied during any given period. If the daily water requirement (ET) is 0.125 gallons per square foot, then a weekly watering schedule would apply seven times this amount (or 0.88 g/sq. ft.), and a monthly watering would apply 30 times this amount (or 3.75 g/sq. ft.) — in each case, the rate is equal to the same daily amount of water.

Sometimes, daily irrigation can actually use less water per month than other methods. For example, I once planted a drought-resistant garden for a neighbor, with plants such as lavender, santolina, rockroses, and rosemary. After the risk of transplant shock was over, the irrigation line was turned on each day for only eight minutes. While the system comes on daily, each emitter in the line only passes 0.5 gallon of water per hour (gph). This means that each emitter was distributing the paltry amount of 7.5 *tablespoons* of water, per emitter, per day. Since the entire line has 400 emitters, capable of passing 200 gph, the line only uses 25 gallons per day — for over 600 square feet of landscaping. By contrast, a garden near my neighbor's, and in a similar soil, is arbitrarily irrigated twice a month for two hours with 0.5 gph emitters. This amounts to as much as two gallons per emitter for the two-week period, or over 18 tablespoons of water per day — more than twice the water applied in daily doses to my neighbor's garden.

Watering rates and schedules do have to be adapted to the soil type. Even clay soils can be watered daily if the length of irrigation is short enough — often less than a minute or two. In fact, daily irrigation of clayey soil with minute amounts of water may be beneficial, because the pore spaces are less likely to flood. This preserves, in the words of Sunset's *Western Garden Book*, "a healthy air-to-water ratio for plant roots." The goal is to time the irrigation to maintain a healthy soil moisture — *not* damp, *not* wet.

All Plants Are Not Created Equal

Many western plants, especially those from desert and chaparral communities, are quite sensitive to soils that are watered in any way during the normally dry summer season. A primary cause of death with such drought-resistant plants is crown or root rot caused by *Phytothphora* and other kinds of fungi that are present in moist, warm soils. With a bit of planning, though, drip irrigation is compatible with many of these moisture-sensitive plants. First and foremost, the emitters should be located 18 to 24 inches away from the stem or trunk of the plant, so the crown of the root system will lie between the moist wet spots near the soil's surface and remains permanently dry. Second, either less water is applied or the interval between irrigations is extended to maintain a drier soil. Finally, some especially sensitive plants, such as flannelbush *(Fremontodendron californicum)* may be best fall-planted in well-drained soil and left wholly unirrigated thereafter.

The biggest controversy with the daily irrigation concept has come from xeriscaping and native plant advocates. There are plenty of plants that can be killed by improper irrigation. Yet the range of options for native, drought-resistant plants is greater than many people assume. Consider the new cultivar garden at Rancho

Santa Ana Botanic Garden in Pasadena, California. This garden contains hundreds of plants, representing dozens of genuses of California native plants. The garden purchased thousands of dollars of in-line emitter, drip-irrigation tubing. The tubing was laid out on a clayey soil in a conservative pattern, which allocated only two to three emitters per plant. While they didn't irrigate daily, for the first year the system came on twice weekly for two minutes at a time — far more than the common conception of "occasional summer watering." According to Bart O'Brien, staff horticulturalist, "the growth has been tremendous." The success is due to the proper combination of emitter placement, irrigation intervals, and length of irrigation. In the second year, they stepped the irrigation back to once every five days, with an irrigation cycle of: on for 1 minute, off for 30 minutes, on for 1 minute, off for 30 minutes, and on again for 1 minute. This fancy scheduling has allowed them to deal with a fairly clayey soil.

Regardless of the location, plan your landscape around zones of compatible irrigation needs, and use the best irrigation method and timing for each grouping of plants.

Parts Are Simply Parts

The plumbing required for a drip irrigation system is well within the means and skills of most gardeners. Whatever mistakes you make in plumbing a drip system are easily rectified and virtually harmless, providing you correctly install a backflow preventer to protect the purity of your home's drinking water. Drip irrigation hardware is as simple to work with as Tinker Toys. After you've installed a drip system, your garden will thank you with an abundance of bloom, productivity, and foliage.

— *Robert Kourik*

Agwa Systems Automated Greywater

California Code Approved!

It took six years of drought, but it's finally legal! Turn grey into green! Using greywater from tubs, sinks, showers, and washing machines is an intelligent method of water conservation, and is now fully approved by the California Plumbing Code. Greywater has tiny amounts of fertilizers (phosphates from the laundry and protein from dead skin cells) which help promote healthy plant growth. Up to 70% of all water used inside a home can be used again outside. The average Los Angeles family produces 160 gallons of greywater per day, and would save $350 per year with the Agwa System (based on LA DWP water & sewer rates 4/95).

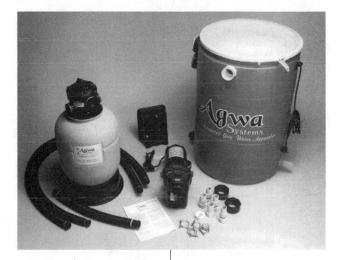

The Agwa System M-100 consists of three major components, a 55-gallon collection tank, 1/2 hp pool pump, and a sand filter. When the water level in the tank reaches a certain level a float switch turns on the pump which forces the greywater through the sand filter, and out to the irrigation system. All major components are off-the-shelf swimming pool technology, so service is minimal, easy, and initial costs are moderate. The M-100 comes complete with color-coded connection hardware, and assembly takes less than one hour. Plugs into a standard 120V outlet. Maintenance is a simple five-minute procedure every month or two depending on usage, and won't get you dirty. Simply fill the collection tank with clean water from a hose, switch one valve to backwash, and plug it in.

A Sub-Surface Irrigation Package properly sized for your house is required with the M-100. This Package includes the minimum amount of Rootguard® irrigation line required by the State of California. This irrigation line has pre-installed emitters every 15". Each emitter has a tiny amount of mild herbicide injected, which prevents root fouling. The Package also contains compression adapters, 1 1" Y-filter (150 mesh), 2 pressure regulators, 2 flush valves, & 2 air relief valves.

•45-120	Agwa Systems M-100	$995⁰⁰

Sub-Surface Irrigation Packages

•45-121	2-Bedroom Package (750 ft.)	$395⁰⁰
•45-122	3-Bedroom Package (937 ft.)	$445⁰⁰
•45-123	4-Bedroom Package (1125 ft.)	$495⁰⁰
•45-124	5-Bedroom Package (1312 ft.)	$565⁰⁰

•Drop shipped from manufacturer.

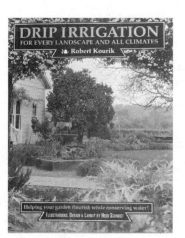

Drip Irrigation For Every Landscape And All Climates

Robert Kourik. The author, a friend and neighbor of ours in nearby Sonoma county, has spent years developing the most trouble-free and maintainable strategies for this efficient way to water. Thorough, well-illustrated, witty, and engaging. We liked his gradual, empirical approach, and were proud to earn a Dripologist diploma at the end. 128 pages, paperback, 1992.

80-202	Drip Irrigation	$12⁰⁰

Create An Oasis With Greywater
A Complete Guide To Managing Greywater

Simply the best book about greywater we've ever seen. This brand-new edition incorporates the considerable experience gained from California's recent drought. The concise, readable format is long enough to cover everything you need to consider, and short enough to stay interesting. Plenty of charts and drawings cover health considerations, sources, designs, what works and what doesn't, bio-compatible cleaners, maintenance, preserving soil quality, the list goes on... There's also an extensive bibliography. Made this reviewer want to run right out and set up a greywater system on the washing machine. 1994, 47 pages, paperback.

80-340   Create An Oasis With Greywater $7⁰⁰

Building Professional's Greywater Guide

A companion to Create an Oasis with Greywater

This one's written for professionals or just homeowners who want all the straight poop (sorry). Will help you successfully include greywater systems in new construction or remodeling. Includes reasons to install or not install a greywater system, flowcharts for choosing an appropriate system, dealing with inspectors, legal requirements checklist, design and maintenance tips, and the complete text of the new California greywater code with extensive (plain English!) annotations. This California code is very similar to the greywater appendix in the Uniform Plumbing Code, which applies to much of the US. 1995, 45 pages, paperback.

80–097   Building Professional's Greywater Guide $13⁰⁰

Arkal Filters

The Arkal filter is probably the best-designed filter for drip irrigation systems. It also works famously for all filtration systems. Instead of using the traditional cartridge or fine mesh screen for filtering, the Arkal uses an assembly of thin rings. A spring holds the rings tightly together when the filter cover is on and lets them separate when the cover is removed. This makes cleaning the filter very quick and easy, as well as avoiding the expense of replacing cartridges. Another advantage of this system is that it can maintain a higher gpm and psi than with a cartridge or mesh system. The filters have built-in valves that can be used both to turn your water flow off and to regulate pressure up to 120 psi. They have a 140 screen mesh.

	3/4" Filter	1" Filter
Max. flow rate:	18gpm	27gpm
Filtering vol.:	6 cu. in.	27 cu. in.
Filtering area:	25 sq. in.	47 sq.in.

43-201   Arkal Filter & Shutoff, 3/4" $39⁰⁰

43-202   Arkal Filter, 3/4" $22⁰⁰

43-203   Arkal Filter, 1" $79⁰⁰

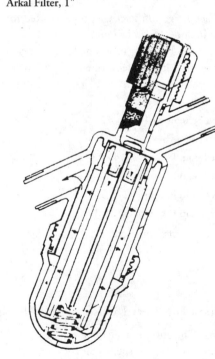

CHAPTER 8

ENERGY CONSERVATION

WHEN WE TALK ABOUT energy conservation, the first question we must ask is: Why do we want to conserve energy?

There are actually many reasons to conserve energy. Some pertain to us as individuals, while others are grounded on societal and global concerns.

If we first take a look at our global environment, we must consider what happens when we turn on an appliance in our house. When any appliance such as a lightbulb, toaster, or television begins to draw electricity, it is all too easy for us to be unaware of what is making that appliance do its job. Yet, if we could project ourselves "upstream" through the wires that lead behind that outlet in the wall, we would travel through the wall cavity to the "breaker box" or distribution center of the house, then on to the watt-hour meter, which measures how much electricity our house uses. And we would have scarcely begun our journey.

After we leave the house, we make our way into a transformer, and we are met by other cables from the other houses in our neighborhood. Then we travel for miles, often hundreds or even thousands of miles, through a network of transformers, transmission lines, and related equipment through which electric power is delivered, transformed, transferred, and traded. It is the source of this long journey which is of utmost importance. Ultimately, at the many sources feeding these networks, we find a few different types of power plants. These are huge "manufacturers" of electricity, most of which convert some form of energy into electricity to feed the networks. Finally, we have found the other end of the wire. At one end we have our little lightbulb in the house; on the other end, several hundred miles away, we find a huge power plant.

Our discussion of the global (and national) reasons for conserving energy is rooted in an understanding of what is going on at those power plants and what effects these processes have. Over 50% of the power in our country comes from coal-fired power plants. In order for these plants to produce their power, the coal first has to be mined from the earth. Even this step isn't pretty — coal miners suffer from lung diseases, and the earth is left with mines that often leach toxic runoff into vital water supplies. Further pollution occurs when the coal is transported to the power plants. When it finally arrives at the plant, the coal is burned in a huge furnace, and the heat from the fire is converted into electricity. Only about 35% of the energy that is contained in the coal reaches our house. Coal-fired plants, even ones of the most modern design and construction, are dangerous and dirty to work in, and difficult both to start up and shut down. This last defect, termed "lack of dispatchability," often puts the power company in the bind of having to dump cheap power and artificially stimulate demand in order to keep their baseload plants operating.

When any substance that contains carbon, such as coal, is burned in combination with oxygen, the main by-product is carbon dioxide. Other by-products of this process include sulfur and nitrous oxides. All of these compounds have negative effects on our environment, the most important of which may be that of carbon dioxide, which is a greenhouse gas. As the concentration of CO_2 increases, it increases the atmosphere's tendency to allow the sun's energy to come in, then keep it trapped inside. Most scientists agree that an increased concentration of CO_2 in our atmosphere leads to a greenhouse effect on the planet. This is the same process that makes greenhouses and solar ovens work: Solar radiation comes in and its heat is trapped inside. Where scientists disagree is on how the greenhouse effect will change life and our environment, and when it will happen. At Real Goods, we say, "Why wait and see, when there are better alternatives which can reverse this process, and we know they work? We are making a solar oven out of our planet!"

For this discussion of energy conservation, it is important for us to understand that every watt of energy an appliance uses can be linked to a certain quantity of carbon dioxide and sulfur and nitrous oxides, which are being emitted into our atmosphere through the use of coal-fired power plants. Even as the debate over global warming vs. global cooling rages, and the role of carbon dioxide appears less obvious, it has become abundantly clear that the sulfur and nitrous oxides, the SO_2es and NO_2es, are worrisome chemicals with wide-reaching negative effects. Sulfur oxides, for instance, are a main contributor to acid rain. The less electricity we use, the fewer of them spew into our air.

For example, according to Amory Lovins of the Rocky Mountain Institute, a single 18-watt compact fluorescent lightbulb, which produces the same amount of light as a 75-watt incandescent and lasts 10 to 13 times as long, will eliminate the emission of about one ton of carbon dioxide and eight kilograms of sulfur dioxide, along with nitrous oxides and other heavy metals.

Not only do these gases produce a greenhouse effect, but they also pollute the air we breathe. This affects our health, and the entire health-care system. Clearly, some of the hidden costs of health care come from a polluted environment, and energy conservation should be one of the pieces in the puzzle we call health-care reform.

Other sources of fuel for power plants include nuclear, oil, natural gas, geothermal, and hydroelectric. Oil and natural gas both produce carbon dioxide, just as coal does. Nuclear power generates tremendous amounts of toxic waste that have to be stored for tens of thousands of years. If we go back to the compact fluorescent lightbulb example above, the amount of energy that the lamp saves could prevent half a curie (a unit of radioactivity) of strontium 90 and cesium 137 from being generated, plus about 25 milligrams of plutonium, equal to the explosive power contained in 385 kilograms of TNT!

Even if we ignore the ever present possibility of a nuclear accident, we must think about all of this toxic waste we are generating, and its amazingly long life. No one wants to build one of these nuclear waste sites in his or her backyard, so we are shamefully turning to Indian reservations and poor communities as a last resort. Ironically, one of the biggest problems engineers face in designing nuclear waste sites is how to warn peoples of future civilizations to stay away from these toxic warehouses. No culture or language has ever survived as long as the toxic compounds will, and finding a way to keep people away ten thousand years from now is a real challenge. Again, we feel that, since there are alternatives that work, why waste our time trying to answer such questions, and why threaten future civilizations unnecessarily?

I'll never forget the time when Mrs. Gottermeier, my eighth-grade social studies teacher, taught us the concept of the "dollar vote." She explained that, whenever we buy something, we are, in a sense, voting for it. Spending the dollar is like casting a ballot, expressing our approval for all the processes involved in the making of that product. In the case of energy, the money we spend supports the production of greenhouse gases, acid rain, and toxic nuclear waste. But, unfortunately, the alternatives to using utility power require an initial investment that most of us cannot afford. Still, we can cast our monetary "ballots" for appliances that use less of this energy, thus voting for a reduction in the use of conventional power plants.

To sum up the societal and global reasons for conserving energy, we can see that every watt of energy that we conserve prevents a certain amount of destructive by-products from being generated and emitted into our atmosphere. Therefore, energy conservation is very beneficial to our community — globally, nationally, and locally.

On the individual level, the reasons for conserving energy are basically economic. For the majority of us who live on the grid, it makes financial sense to conserve energy because we have to pay for it. The less we use, the less we spend. But investing in energy conservation is worthwhile economically only if the investment saves more money over its lifetime than its up-front cost.

For those of us who are off the grid, energy conservation makes sense because we have to buy our own solar modules, wind generators, or hydroelectric turbines, and the more energy we want to use, the more we have to spend on these devices. For off-the-gridders, the individual economic argument for conserving energy is clear and simple: The more we conserve, the less we spend. And the savings are often quite dramatic.

There are two ways to conserve energy. The first is simply to use our appliances less, being conscientious about turning them off when not in use. The other way is to use more efficient appliances. A more efficient appliance consumes less energy than a conventional appliance performing the same job. For instance, an 18-watt compact fluorescent lightbulb uses only 18 watts, but illuminates as well as a 75-watt incandescent bulb.

Neither of the ways to conserve energy requires a change in lifestyle. Switching to a more energy-efficient kind of lightbulb still gives you the light you need, and turning off appliances that are not being used is just a simple, commonsense habit we all need to adopt.

The nice thing about energy conservation is that your reason for conserving does not matter. You might be more concerned about global matters and the environment, or you might just be trying to save money. Either way, you get both benefits. You help the environment *and* you save money.

In this chapter you will find energy conservation products that use less energy than conventional appliances. As stated above, an energy-efficient appliance makes economic sense only if it saves more money over its lifetime than it costs to purchase, use, and maintain. With every energy-conservation product we offer, we make sure that it not only saves energy (and therefore benefits our environment), but that it also makes good economic sense. Whoever said that economics and the environment don't go hand in hand obviously didn't have a *Sourcebook!*

—*Gary Beckwith*

AC LIGHTS

A revolution has taken place in commercial and residential lighting. Until recently, we have been bound by the ingenious but crude discoveries that Thomas Edison made over 100 years ago.

Mr. Edison was never concerned with how much energy his incandescent bulb consumed. Today, however, economical, environmental, and resource issues force us to look at energy consumption much more critically. *Incandescent* literally means "to give off light as a result of being heated." With standard lightbulbs, typically 90% of the energy consumed is given off as heat; the light is only a small by-product.

New technologies are available today that improve on Edison's invention in terms of both reliability and energy savings. Although the initial price for these new, energy-efficient products is higher than that of standard lightbulbs, their cost-effectiveness is far superior in the long run. (See the discussion of whole-life cost analysis on page 14.) When people understand just how much money and energy can be saved with compact fluorescents, they often replace regular bulbs that haven't even burned out yet.

The types of efficient lighting on the market include new-design incandescents, quartz-halogens, standard fluorescents, and compact fluorescents.

"Energy-Saving" Incandescent Lights

Inside an incandescent light is a filament that is heated, giving off light in the process. The filament is delicate and eventually burns out. Some incandescents incorporate heavy-duty filaments or introduce special gases into the bulb to increase life. While this does not increase efficiency, it increases longevity by up to four times. Most manufacturers are now offering "energy-saving" incandescent bulbs. These are nothing more than lower-wattage bulbs that are marketed as having "essentially the same light output as . . ." In fact, General Electric just lost a multimillion-dollar lawsuit relating to this very issue. A 52-watt incandescent bulb does *not* put out the same quantity of light that a 60-watt bulb produces. An "energy-saving" incandescent bulb will use less power, true, but it will also give you less light. This specious advertising claim is just a megacorporation excuse to avoid retooling by selling you a wimpier lightbulb at a higher price. Don't be hood-winked; you can do much better.

Quartz-Halogen Lights

Tungsten-halogen (or quartz) lamps are really just "turbocharged" incandescents. They are typically only 10% to 15% more efficient than standard incandescents; a step in the right direction, but nothing to write home about. Compared to standard incandescents, halogen fixtures produce a brighter, whiter light and are more energy-efficient because they operate their tungsten filaments at higher temperatures than standard incandescents. In addition, unlike the standard incandescent lightbulb which loses approximately 25% of its light output before it burns out, a halogen light's output depreciates very little over its life, typically less than 10%.

To make these gains, lamp manufacturers enclose the tungsten filament inside a relatively small, quartz-glass envelope filled with halogen gas. During normal operation, the particles that evaporate from the filament combine with the halogen gas and are eventually redeposited back on the filament, minimizing bulb blackening and extending bulb life. Where halogen lamps are used on dimmers, they need to be occasionally operated at full output to allow this regenerative process to take place.

Tungsten-halogen lamps produce slightly more light per watt than standard incandescents and last longer, having useful service lives ranging from 2000 to 2500 hours, depending on the model. Tungsten-halogen lamps are very sensitive to operating voltage.

There are several styles of halogens available for both 12-volt and 120-volt applications. The most popular low-voltage models for residential applications include the miniature multifaceted reflector lamp (product #33–108) and the halogen version of the conventional A-type incandescent lamp, which incorporates a small halogen light capsule within a protective outer glass globe. The reflector lamp offers a light source with very precise light-beam control, allowing you to distribute light exactly where you need it without wasteful spillover, while the latter option is intended for general ambient-lighting applications. A new family of 120-volt AC, sealed-beam reflector halogen lamps are now on the market to replace standard incandescent reflectors.

The higher operating temperatures used in halogen lamps produce a whiter light, which eliminates the yellow-reddish tinge associated with standard incandescents. This makes them an excellent light source for applications where good color rendition is important or fine-detail work is performed. Because tungsten-halogen lamps are relatively expensive compared to standard incandescents, they are best suited for applications where the optical precision possible with the compact reflector models can be effectively utilized.

Never touch the quartz-glass envelope of a halogen lamp with your bare hands. The natural oils in your skin will react with the quartz glass and cause it to fail prematurely. Because of this phenomenon, and for safety reasons, many manufacturers incorporate the halogen lamp capsule (generally about the size of a large flashlight bulb) within a larger outer globe.

Fluorescent Lights

Fluorescent lights are still trying to overcome a bad reputation. For many people the term "fluorescent" connotes a long tube emitting a blue-white light with an annoying flicker. However, these past limitations have been overcome with recent technological improvements.

The fluorescent tube, no matter how it is shaped, contains a special gas at low pressure. When an arc is struck between the lamp's electrodes, electrons collide with atoms in the gas to produce ultraviolet radiation. This, in turn, excites the white phosphors (the white coating on the inside of the tube), which emit light at visible wavelengths. The quality of the light fluorescents produce depends largely on the blend of chemical ingredients used in making the phosphors; there are dozens of different phosphor blends available. The most common and least expensive are "cool white" and "warm white." These, however, provide a light with relatively poor color-rendering capabilities, making colors appear washed out, lacking luster and richness.

The tube may be long and straight, as with standard fluorescents, or there may be a series of smaller tubes, in a configuration that can be screwed into a common light fixture. These latter tubes are called *compact fluorescent* (or CF) lights.

The growth of the compact-fluorescent market in the past five years has been almost overwhelming. There is now a good variety of lamp wattages, sizes, shapes, and aesthetic packages. This is a rapidly changing and growing marketplace, and the more frequently published Real Goods catalogs will undoubtably feature lamp types not described here in this *Sourcebook.*

Color Quality: The color of the light emitted from a fluorescent bulb is determined by the phosphors that coat the inside surface. The terms "color temperature" and "color rendering" are the technical terms used to describe light. *Color temperature,* or the color of the light that is emitted, is measured in degrees Kelvin (° K) on the absolute temperature scale, ranging from 9000° K (which appears blue) down to 1500° K (which appears orange-red). The *color rendering index* (CRI) of a lamp, on a scale of 0 to 100, rates the ability of a light to render an object's true color when compared to sunlight. 100 is perfect, 0 is a cave. The table on page 380 will demystify the quality of light debate.

Phosphor blends are available that not only render colors better, but also produce light more efficiently. Most notable of these are the fluorescent lamps using tri-stimulus phosphors, which have CRIs in the eighties. These incorporate relatively expensive phosphors with peak luminance in the blue, green, and red portions of the visible spectrum (those which people are most sensitive to), and

LIGHT, CRI, AND COLOR TEMP		
Type of Light	CRI	Deg. K
Incandescent	90-95	2700
Cool white fluorescents	62	4100
Warm white fluorescents	51	3000
Compact fluorescents	82	2700

produce about 15% more visible light than standard phosphors. Wherever people spend much time around fluorescent lighting, specify lamps with higher (80+ CRI) color rendering ratings.

Ballast Comparisons: All fluorescent lights require a *ballast* to operate, in addition to the bulb. The ballast regulates the voltage and current delivered to a fluorescent lamp, and is essential for proper lamp operation. The electrical input requirements vary for each type of compact fluorescent lamp, and so each type/wattage requires a ballast specifically designed to drive it. There are two types of ballasts that operate on AC: *core-coil* and *electronic.* The *core-coil ballast,* the standard since fluorescent lighting was first developed, uses electromagnetic technology. The *electronic ballast,* only recently developed, uses solid-state technology. All DC ballasts are electronic devices.

Compared to the core-coil ballast, electronic ballasts weigh less, operate lamps at a higher frequency (20,000+ cycles per second vs. 60 cycles), are silent, generate less heat, and are more energy-efficient. However, electronic ballasts cost more, particularly DC units, because they are manufactured only in very small numbers. With few exceptions, electronic ballasts for AC lamps are presently only available as a part of integral units. This is because the ballasts can be designed to last about as long (approximately 10,000 hours) as the lights they drive. Magnetic core-coil ballasts, on the other hand, can last up to 50,000 hours, and often incorporate replaceable bulbs.

All of the compact fluorescent lamp assemblies and prewired ballasts reviewed in the *Sourcebook* are equipped with standard medium, screw-in bases (like normal household incandescent lamps.) The ballast portion, however, is wider than an incandescent light bulb *just above the screw-in base.* Therefore, fixtures having constricted necks or deeply recessed sockets may require a socket "extender" (to extend the lamp beyond the constrictions). These are readily available either in our catalogs or at most hardware stores.

Savings: The main justification for buying fluorescent lights is to save money. The new compact fluorescents provide opportunities for tremendous savings without any inconveniences. Simple payback calculations prove their cost-effectiveness (see the table at the bottom of page 381). **Fluorescent lights typically last ten times longer and use one-fourth the energy of standard incandescent lights.**

LIFETIMES OF LIGHTS (IN HOURS)	
Standard incandescents	1,000
Long life incandescents	3,000
Quartz-halogen	2,250
Compact fluorescents	10,000

EFFICACY OF LIGHTS (LUMENS PER WATT - APPROXIMATE)	
Incandescents	16
Compact fluorescents	60

The fixture applications in these charts are generalities. There are always exceptions to various applications. Use the full-sized cutouts in the Appendix for exact fit. Usually, if it fits and will stay dry — use it!

The next question, of course, is, "How much do compact fluorescents save?" The more a light is used, the more that can be saved by replacing it with a compact fluorescent. The calculations in the table below assume that the light is on for an average of six hours per day and power costs $0.10 per kWh (approximate national average). Note that the Saving Over Life is calculated *after* the compact fluorescent bulb has paid for itself; it returns another $30 to your pocket.

Health Effects: Migraine headaches, loss of concentration, and general irritation have all been blamed on fluorescent lights. These problems are caused not by the lights themselves, but by the way in which they are run. Common core-coil ballasts run the lamps at the same 60 cycles per second that is delivered by our electrical grid. This causes the lamps to flicker noticeably 120 times per second, every time the alternating current electricity switches direction. Approximately a third of the

LIGHT SAVINGS*	Phillips SLS 20W	Incandescent 75-watt bulb
Cost of bulb	$29	$0.50
Product life	4.5 years	167 days
Watts used annually	20	75
Energy cost	$4.38	$16.42
Bulbs replaced in 4.5 years	0	10
Total cost	$48.71	$78.89
Savings over life	$30.18	

*Computed at 6.0 hours lamp use per day; electricity rates of $0.10/kWh.

human population is sensitive to this flicker on a subliminal level. Research conducted by Canadian education authorities has suggested that this effect has nothing to do with the light itself, but is caused by electronic static produced by the ballasts, which can be sensed by humans. (The static appeared to cause learning problems in blind students at the same rate as with seeing students; carefully shielding the offensive ballasts reduced the problem dramatically.)

The cure is to use electronic ballasts, which operate at around 30,000 cycles per second. This rapid cycling totally eliminates perceptible flicker and avoids the ensuing health complaints.

AC Fluorescents and Remote Energy Systems: When planning a remote energy system, one must make numerous decisions as to whether AC or DC power is more appropriate for specific appliances. Fluorescent lights are available in both AC and DC. Most people using an inverter choose AC lights because of the wider selection, a significant quality advantage, and lower price. Some older inverters may have problems running some magnetic-ballasted lights. If you have an older Heart inverter, buy one light to try it first.

Most inverters operate compact fluorescent lamps satisfactorily. However, because all but a few specialized inverters produce an alternating current having a modified sine wave (versus a pure sinusoidal waveform), they will not drive compact fluorescents which use electromagnetic, core-coil ballasts as efficiently or "cleanly" as possible, and some lamps may emit an annoying buzz. Electronic-ballasted compact fluorescents, on the other hand, are tolerant of the modified sine-wave input, and will provide better performance, silently.

Compact Fluorescent Applications: Due to the need for a ballast, a compact fluorescent lightbulb is shaped differently from an "Edison" incandescent. This is the biggest obstacle in retrofitting light fixtures. Compact fluorescents are longer, heavier, and sometimes wider. The ballast, the widest part, is located at the base, right above the screw-in adapter. We recommend using a ruler to measure the available space, and referring to the sizing chart before purchasing; or use the full-sized cutout drawings in the Appendix.

Compact fluorescents have many household applications — table lamps, recessed cans, desk lamps, bathroom vanities, and more. As manufacturers become attuned to this relatively new market, light fixtures suited for compact fluorescents are becoming more available. We now offer a limited selection of fixtures; watch future catalogs for the latest offerings in this rapidly expanding field.

With **table lamps**, the metal "harp" that supports the lampshade is sometimes not long or wide enough to accommodate these bulbs. We offer an inexpensive replacement harp that can solve this problem. Beware also that heavier CFs may change a light but stable lamp into a top-heavy one. Table lamps are an excellent application for compact fluorescents.

Recessed cans are limited by diameter, and sometimes cannot accept the wide ballast at the base. The base depth is adjustable on most recessed cans. There are a number of special bulbs available just for recessed cans now.

Desk lamps can be difficult to retrofit, but complete desk lights that incorporate compact fluorescents are readily available.

Hanging fixtures are one of the easiest applications for compact fluorescent lights. One of the best applications is directly over a kitchen or dining-room table. Usually the shade is so wide that it couldn't get in the way. It may even be possible to use a Y-shaped two-socket adapter (available at hardware stores) and screw in two lights if more light is desired.

Track lighting is one of the most common forms of lighting today. When selecting your tracking system, choose a fixture that will not interfere with the ballasts located right above the screw-in base. It is best to use one of the reflector lamps, such as the Euertron 2000L, or the SLS/R series on page 391.

Dimming: The current generation of compact fluorescent lights should *never* be dimmed. Using these lights on a dimmer switch may even pose a hazard. If you have a fixture with a dimmer switch that you wish to retrofit, your best option is to replace the switch with a conventional wall switch. This is a simple, inexpensive procedure. Remember to turn off the power first!

Dimmable compact fluorescent lights are a reality in Europe, but have hit a snag with the FCC in the U.S. They use a small radio wave transmitter to excite the phosphors, which hasn't received offical approval. Refer to the Real Goods color catalogs for the most up-to-date information and product selection.

Starting Time: The start-up time for compact fluorescent lamps varies. It is normal for most core-coil compact fluorescents to flicker for up to several seconds when first turned on while they attempt to strike an arc. Most electronically-ballasted units start their lamps instantly. All fluorescent lamps start at a lower light output; depending on the ambient temperature, it may take anywhere from several seconds to several minutes for the lamps to come up to full brightness.

The very brief start-up time, which is only apparent in some of the fluorescent lights, is a very small price to pay for the energy savings and the subsequent good feeling about doing less harm to our fragile environment.

Cold Weather/Outdoor Applications: Fluorescent lighting systems are sensitive to temperature. Manufacturers rate the ability of lamps and ballasts to start and operate at various temperatures. Light output and system efficiency both fall off significantly when lamps are operated above or below the temperature range at which they were designed to operate. The temperature at which any given fluorescent system will effectively work varies greatly, depending on the specific lamp and ballast combination. The optimum operating temperature for most fluorescents is

between 60° and 70° F, though most standard double-ended types will operate satisfactorily between 50° and 120° F. Special lamps and ballasts must be specified for applications outside of these temperature ranges, such as low-ambient ballasts for cold applications.

Temperatures below freezing inhibit compact fluorescent lamps in starting and in attaining full brightness (normally reached between 50° to 70° F). In general, the lower the ambient temperature, the greater the difficulty in starting and attaining full brightness.

Manufacturers are usually conservative with their temperature ratings. We have found that most lamps will start at temperatures 10° to 20° F lower than those stated by the manufacturer.

To help improve the operation of compact fluorescents used in cold temperatures, we recommend that they be installed in enclosed fixtures. This insulates them, improving both their ability to start and to attain full brightness. It is also very important that the fixture be well grounded; a grounded metal fixture or reflector near the lamp, ideally within 1/4 inch of the lamp, helps it to start (that is, to strike an arc of electrons between the electrodes). If this is not possible, or if the lamp flickers for more than several seconds before coming on, a thin strip of copper can be loosely inserted between the tubes of the lamp and attached to a grounded point.

Most compact fluorescent lamps are not designed for use in wet applications (for example, in showers or in open outdoor fixtures). In such environments, the lamp should be installed in a fixture rated for wet use.

Service Life: All lamp-life estimates are based on a three-hour duty cycle, meaning that the lamps are tested by turning them on and off once every three hours until half of the test batch of lamps burn out. *Turning lamps on and off more frequently decreases lamp life while keeping them on longer increases lamp life.* Generally, it is more cost-effective to turn fluorescent lights off whenever you leave a room for more than ten minutes, since the greatest cost associated with operating a light is for the electricity it uses versus the cost to replace it. Please note, however, that the rated lamp life represents the *average life* of a lamp; as a result, some will have longer lives and some shorter. A typical compact fluorescent lamp will last as long as (or longer than) ten standard incandescent AC lightbulbs or five standard incandescent AC floodlights, saving you the cost of numerous bulbs, in addition to much electricity due to the greater efficiency.

Three-Way Sockets: Compact fluorescents can be screwed into any three-way light socket, but, just like a standard bulb in a three-way socket, they will only operate on full light output.

Full-Spectrum Fluorescents: Many customers have inquired as to why we don't sell full-spectrum fluorescents. First, we are not convinced that they perform as promised. The ultraviolet (UV) end of the lighting spectrum is the first part of the lamp's spectrum to disappear, lasting only a fraction of the overall lamp life. Therefore, the purported benefits of full-spectrum lighting are extremely short lived and thus overrated, and do not justify the high price of these lamps. Second, shipping small quantities of something as fragile as fluorescent tubes through the mail is problematic. We sell fluorescent tubes only when included with a fixture. We recommend that you purchase *all* your lamp tubes locally. There is no difference between a 120-volt tube and a 12-volt tube; only the ballast is different.

Differences between Core-Coil and Electronic Ballasts: Core-coil ballasts generally last about 50,000, hours and many have replaceable bulbs, which last about 10,000 hours each. Core-coil ballasts flicker when starting and take a few seconds to get going. They also run the lamp at 60 cycles per second, and some people are affected in a negative way by this flicker.

Electronic ballasts last about 10,000 hours, the same as the bulb, and most do not have replaceable bulbs. Electronic ballasts start almost instantly with no flickering. They run the lamp at about 30,000 cycles per second. For the many people who suffer from the "60-cycle blues," this is the energy-efficient lamp you have dreamed of.

— Doug Pratt

A GREAT WAY TO SAVE MONEY AND
HELP THE ENVIRONMENT

Cut your lighting costs by 75%. Compact fluorescents use only 1/4 of the energy of a standard incandescent bulb. And since compact fluorescents last 10 to 13 times longer than standard incandescents, you won't have to change them as often and you will produce less waste. If every household in the U.S. used just one compact fluorescent, we'd save enough energy to eliminate one Chernobyl-sized nuclear power plant.

Things to Consider . . .
- Choose a bulb that emits the approximate equivalent lumens to the incandescent wattage you are used to. For example, choose a 1650 lumen, 27-watt compact fluorescent to replace a 1750 lumen, 100-watt incandescent.
- Check the measurements of the compact fluorescent bulb to be sure it fits your fixture. Be sure to turn the power off first!
- For quickest payback, use compact fluorescents in your most-used light fixtures. In lesser-used ones, consider using an efficient halogen option.

Lamp Life
Life of a fluorescent bulb is 10,000 plus hours.
Average life of an incandescent bulb is from 750 to 1,000 hours.

A RULE OF THUMB

Wattage	Wattage
25W	7W
40W	11W
60W	15W
75W	20W
100W	25W
Incandescent Bulb	Fluorescent Bulb

Actual equivalency depends on lumen output, and varies with bulb type and from manufacturer to manufacturer.

Appropriate Uses
Shaded Lamps: Table and floor lamps
Enclosed Fixture: Ceiling and wall fixtures. Exterior yard post lights, wall brackets and hanging fixtures. (Check minimum starting temperatures).
Open Indoor: Hanging pendant and ceiling lamps and wall fixtures
Recessed Can: Recessed ceiling-down lights or track lights

INSTALL A 30-WATT CIRCULAR LAMP FOR
MAXIMUM BRIGHTNESS AND SAVINGS

	30W Circular Fluorescent	135W Incandescent
Bulbs totalling 10,000 hrs. of light	1	10
Bulb Cost	$35.00	10 x $0.75 = $7.50
Energy Cost for 10,000 hrs. of light*	$30.00	$135.00
Total Cost	$65.00	$142.50

Save up to $77.50 on a bulb. Compact fluorescents can yield up to 54% return on your investment over the life of the bulb.

* Calculated at $0.10/kWh for electric power.

What is a ballast?

Ballasts regulate the power delivered to fluorescent lights. There are two types of ballasts: magnetic and electronic. Compact fluorescent lights with magnetic ballasts are usually less expensive, but they may start slowly and flicker for a moment when first turned on. Magnetic ballasts operate at 60 cycles a second, which creates a subliminal "strobe" effect that some people find annoying. In contrast, electronic ballasts start instantly, have no "strobe" effect, and weigh less.

Can I use a dimmer?

The current generation of compact fluorescents will not work on dimmers. Our halogens are a good choice for use with dimmers.

Can I put a compact fluorescent in a three-way light socket?

Yes, but it will produce only one light level.

COMPACT FLUORESCENT LIGHTING

These Adapters Fit In Your Table Lamps

Until now, adapting table and floor lamps to compact fluorescents almost always required changing the harps, which were designed to accommodate incandescent bulbs. But because these CF's ballasts are mounted to the side of the socket, it will fit within all lamp harps. The 13-watt bulb with an output of 900 lumens replaces a 60-watt incandescent. 6" L x 2" W x 2-3/4" D. The 22-watt bulb with an output of 1300 lumens replaces a 75-watt incandescent. 7-1/2" L x 2" W x 3-3/16" D. Great for outdoor applications, too. The 13-watt lamp starts down to 0° F, the 22-watt is good all the way down to –26° F! Not to be used with dimmers.

- *Magnetic ballast*
- *Ballast life: 45,000 hours*
- *Bulb life: 10,000 hours*
- *Operating temperature: 0° to 140° F*

36-607	13-watt Adapter with Bulb	$19^{95}
31-122	Replacement Quad-13 Bulb	$11^{95}
36-609	22-watt Adapter with Bulb	$22^{95}
31-123	Replacement Quad-22 Bulb	$14^{95}

Unbeatable Combo: More Light, Longer Life, Lowest Cost

The ProLight 13-watt lamp makes conversion to compact fluorescent lighting more cost effective than ever. After about 10,000 hours of operation, you replace just the lamp. 2.4" W x 9.6" (PL13) or 7.5" L (Quad-13). Bulbs sold separately. May not be used with dimmers.

- *Magnetic ballast*
- *Ballast life: 40,000 hours*
- *Bulb life: 10,000 hours*
- *Lumens: 900*
- *Operating temperature: 0° to 140° F*

36-608	Screw Base Adapter	$9^{95}
31-104	PL-13 Bulb	$4^{95}
31-122	Quad-13 Bulb	$11^{95}

Panasonic G-16 and T-16

Panasonic T-16

Compact fluorescents are continuing to get smaller. Panasonic's break-through bulb fits easily in a table lamp without changing the harp. The same pleasing shape as the previous generation T-15, but smaller and electronically ballasted. The T-16 is 2.7" wide by 5.4" long. May not be used with dimmers.

- *60-watt equivalent*
- *800 lumens*
- *Electronic ballast*
- *Lamp life: 10,000 hours*
- *Operating temperature: 14° to 122° F*

36-503   Panasonic T-16 $29⁹⁵

Panasonic G-16

Panasonic has shrunk its light capsule with the new G-16. Only 5.1" long, it will fit in many traditional fixtures. Its width is 3.7", so it still requires a harp adapter for most table lamps. May not be used with dimmers.

- *60-watt equivalent*
- *800 lumens*
- *Electronic ballast*
- *Lamp life: 10,000 hours*
- *Operating temperature: 14° to 122° F*

36-504   Panasonic G-16 $29⁹⁵

SLS 11-Watt
New! This Lamp Is Great For Sconces

Philips Lighting keeps improving your choices. The new SLS series has added an 11-watt lamp that is smaller than any compact fluorescent previously. This tiny lamp, which is the equivalent of a standard 40-watt bulb, will fit virtually anyplace an incandescent bulb will. Measuring just 1-1/2" square at the base, by 5-13/16" long, this lamp will fit directly into many desk lamps and other fixtures with narrow necks.

- *40-watt equivalent*
- *450 lumens*
- *Electronic ballast*
- *Operating temperature: 0° to 140° F*
- *Lamp life: 10,000 hours*

36-210 SLS 11-watt $24⁹⁵

Fluorever 13
A Super-Efficient Compact Lamp

The Fluorever is one of the smallest compact fluorescents available. Its super-efficient ballast makes it a good choice for people who are generating their own electricity. The two-piece 13-watt lamp (940 lumens) is equivalent to 60 watts and is 7" L x 2-1/16" W.

- *Electronic ballast*
- *Operating temperature: –15° to 140° F*
- *Bulb life: 10,000 hours*
- *Ballast life: 40,000 hours*

36-209	Fluorever 13-watt	$29⁹⁵
31-122	Replacement Quad-13 Bulb	$11⁹⁵

LOA Quads With Replaceable Bulbs

These commercial-grade two-piece electronic compact fluorescents are built to last by Lights of America. They feature a long-lived electronic ballast and a replaceable bulb. Both lamps are 3" in diameter. The 18-watt bulb (1200 lumens) is 7-1/2" long, and is equivalent to a 75-watt incandescent. The 27-watt bulb (1650 lumens) is 8-1/4" long, and is equivalent to a 100-watt incandescent. When the bulb burns out, you can replace it without having to discard the ballast. May not be used with dimmers.

- *Electronic ballast*
- *Ballast life: 65,000 hours*
- *Lamp life: 10,000 hours*
- *Operating temperature: 0° to 140° F*

36-107	LOA Quad 18-Watt	$32⁹⁵
36-108	LOA Quad 27-Watt	$34⁹⁵
36-109	LOA 18-Watt Replacement Bulb	$14⁹⁵
36-110	LOA 27-Watt Replacement Bulb	$14⁹⁵

Philips Triple Tube

Philips Lighting has taken compact fluorescent technology into the 21st century with its innovative, new triple-tube design. The electronically ballasted triple design makes these the smallest energy-saving lights available. They will fit in table lamps with harps as small as 8 inches for the SLS 15 and SLS 20, and 9 inches for the SLS 25. The 15-watt SLS 15 replaces a 60-watt incandescent (900 lumens) and is almost the same size as an incandescent (5" long); the 20-watt SLS 20 replaces a 75-watt incandescent (1200 lumens) and is only 5-1/2" long; and the 25-watt SLS 25 replaces a 90-watt incandescent (1550 lumens) and is 6" long. All bulbs are 2-1/4" wide at the base. May not be used with dimmers.

- *Electronic ballast*
- *Lamp life: 10,000 hours*
- *Operating temperature: 0° to 140°*

36-115	Philips SLS 15-Watt	$24⁹⁵
36-116	Philips SLS 20-Watt	$24⁹⁵
36-117	Philips SLS 25-Watt	$24⁹⁵

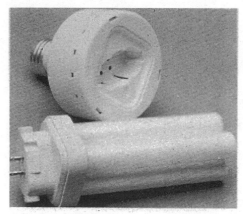

LOA Quad Lamp

A Great Choice for Recessed Cans

This intelligently designed floodlamp uses only 20% of the power and lasts 10 times as long as the 75-watt floodlamp it replaces. That means lower electrical and maintenance costs. The bulb is replaceable at a fraction of the complete lamp cost. It has a metallic reflector to insure all the light goes where you want it. The compact size fits most recessed cans. 6.5"L x 5"W. The 13-watt lamp produces 860 lumens, and is roughly equivalent to a 75-watt floodlamp.

- *Magnetic ballast*
- *Ballast life: 45,000 hours*
- *Lamp life: 10,000 hours*
- *Operating temperature: 00,1400 F*

31-109	Enertron 2000-L Flood	$29⁹⁵
31-122	Quad 13 Repl. Bulb	$11⁹⁵

Energy Saving Floodlights

These two new floodlights from Philips are ideal for use in recessed cans, track lighting and weather-protected outdoor fixtures. The reflector snaps onto a standard SLS lamp. The 15-watt flood produces 600 lumens, is 6" long, and replaces a 50-watt R-30 incandescent reflector. The 20-watt version produces 875 lumens, is 6.6" long, and replaces a 75-watt R-40.

- *Electronic ballast*
- *Lamp life: 12,000 hrs.*
- *Operating temperature: -10⁰ to 140⁰*

36-190	SLS/R-30 15-watt Flood	$34⁹⁵
36-191	SLS/R-40 20-watt Flood	$34⁹⁵

ProLight Spots And Floods

These well-designed compact fluorescent spots and floods from ProLight represent a new generation of technology. A standard PL-13 DC bulb plugs into a reflector housing with a commercial-grade, long-life ballast. The reflector design features a "cluster focus," which produces nearly 50% more candlepower than other floodlights. Choose between the flood-light diffuser lens or the spot-light directional lens. Compact overall size allows these lamps to fit in almost any downlight application. They measure 6-1/2 inches long by 4-3/4 inches wide, and produce 860 lumens (equivalent to a 75-watt incandescent). The bulb is replaceable. Cannot be used with dimmers.

- *Magnetic ballast*
- *Ballast life: 100,000 hours*
- *Lamp life: 10,000 hours*
- *Operating temperature: 32° to 140° F*

36-306	ProLight QCR-38 Spot	$49⁹⁵
36-307	ProLight QCR-38 Flood	$49⁹⁵
31-122	Quad-13 Replacement Bulb	$11⁹⁵

Brightest Compact Fluorescent on the Market

The new 39-watt D-lamp from GE is the highest output compact fluorescent lamp available. It gives nearly the light output of a 150-watt incandescent bulb at 2780 lumens. The 22-watt is equivalent to more than a 75-watt incandescent bulb at 1300 lumens. The unique D-shaped design gives an even overall soft-white light distribution, which is ideal for reading or other close work. 39-watt is 8" long, 4.3" high; 22-watt is 5.5" long, 4" high.

- *Electronic Ballast*
- *Lamp Life: 10,000 hours*
- *Operating temperature: -10⁰ to 104⁰ F*

31-119   39-watt D Bulb $34⁹⁵
31-120   22-watt D Bulb $26⁹⁵

39-Watt D Bulb

Circular Fluorescent

Designed to last 12,000 hours, this circular fluorescent made by Lights of America is designed for locations where maximum illumination is required. The Tri Phosphor tube provides a 10% brighter and warmer light that is far closer to sunlight than standard cool white circular fluorescents. Circular lamps are ideal for retrofitting your table lamp with a bright bulb for reading. Or use it as an attractive ceiling fixture. These two-piece units have commercial grade ballasts with a replaceable lamp. The 22-watt lamp (1550 lumens) is equivalent to a 90-watt incandescent and is 8" wide. The 30-watt (2400 lumens) is equivalent to 135 watts and is 9" wide. Both are 3-1/4" high. May not be used with dimmers.

- *Electronic ballast*
- *Ballast life: 65,000 hours*
- *Lamp life: 12,000 hours*
- *Operating temperature: 0° to 140° F*

36-147   22-watt Circular $24⁹⁵
36-145   30-watt Circular $34⁹⁵

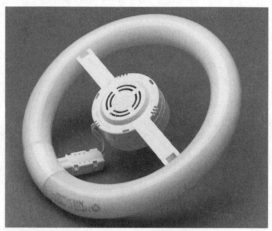

Circular Fluorescent

Light Comes On When The Power Goes Out

Our Emergency Light won't let you down when your electricity goes off. Keep the Home-Guard light plugged into a wall socket and, the next time the power goes out, it will come on automatically. The super-bright replaceable 5-watt fluorescent bulb illuminates up to a 10' x 10' room. You can unplug it whenever you need a portable light. Lasts approximately 1-1/2 hours on a charge. With the fold-out stand on the back, you can put it wherever you need it.

25-409   **Home-Guard Light** $39⁹⁵
25-414   Replacement Bulb $11⁹⁵

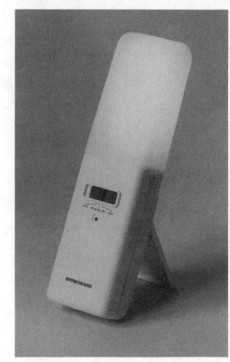

Home Guard Light

An Energy Tale, or Real Goods Walks Its Talk

In May of '95 Real Goods headquarters moved into a new-to-us 10,000 sq. ft. office building. Built in the early '60s, the building had conventional four-foot fixtures, each holding four 40-watt cool white tubes. The dreary, glaring "office standard." Light levels were much brighter than recommended when employees are using computers a lot. Plus, a number of our employees were having headache, energy level, and "attitude" problems after moving, probably caused by the 60-cycle flicker from the old magnetic ballasts. Reducing overall light levels for less eye strain was needed, as was giving employees a healthier working environment. And if we could save some money too, it wouldn't hurt.

*Our solution was to retrofit the existing fixtures with the Elba electronic ballasts sold below for no-flicker lighting, install specular aluminum reflectors, and convert from four cool white 40-watt T-12 lamps per fixture to two warm-colored 32-watt T-8 lamps. Power use per fixture was reduced by over 60%, but desktop light levels, because of the reflectors and more efficient lamps and ballasts, were reduced by only 30% or less. The entire retrofit cost about $6,000 (not counting the $2,300 rebate from our local utility), yet our electrical savings alone are close to $5,000 per year. Plus our employees enjoy flicker-free, warm-colored, non-glaring light and greatly reduced EMF levels. **The entire project pays for itself in less than a year! Improved working conditions come as a freebie!***

Elba Electronic Ballast

This high quality UL-listed electronic ballast does not produce any discernable flicker when running fluorescent lights, nor will it hum when powered by an inverter. Months of testing have

proved the Elba superior to all competitors. This exceptional ballast has won many design awards, including a gold medal in 1982 from the New York International Inventors Expo.

We offer the Elba in several versions. All are designed to run standard 4-foot fluorescent lamps. The first model is for standard T-12, 40-watt lamps. If your lamp or fixture was installed anytime prior to 1985 this is almost surely what you have. But if you're changing the ballast anyway, we strongly recommend that you upgrade to the new T-8, 32-watt tubes at the same time. (In fluorescent tube designations the T-number refers to the diameter of the tube.) The T-8 tube is the same length, same pin configuration, same light output (although color rendition is better), and they use 20% less power.

We offer three models of Elba ballasts for T-8 lamps, in 2, 3, and 4 lamp per fixture configurations. All Elba ballasts run the lamps in parallel. If one lamp burns out the remainder stay lit. The Elba also is UL-listed for single lamp operation.

We don't offer T-8 bulbs because of the high breakage rates in shipping small quantities of fluorescent tubes. They are available locally at any electrical supply or good hardware store.

Approved by California Energy Commission, New York State Energy Office, and UL-listed. Sound rating A. Replaces any standard 120-volt AC ballast. Magnetic radiation is a super-low 1 milligauss at 8". Install the Elba in any standard 4-foot 120-volt AC fluorescent fixture. Three year warranty, 15-year expected life.

36-204	Elba Ballast for 2 F40T12 tubes	$39⁹⁵
36-205	Elba Ballast for 2 F32T8 tubes	$39⁹⁵
36-206	Elba Ballast for 3 F32T8 tubes	$45⁹⁵
36-207	Elba Ballast for 4 F32T8 tubes	$49⁹⁵

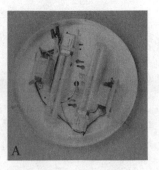

A Pair Of Bright, Low-Profile Drum Fixtures For Your Ceilings

Many standard ceiling fixtures are too small to accept screw-in compact fluorescent bulbs. These stylish fluorescent drum fixtures are hard-wire replacements for your inefficient existing fixtures. We're supplying these fixtures with warm-colored lamps, the same color as 60-watt incandescent bulbs. Available in two sizes, the 26-watt unit (11" x 3") is the equivalent of two 60-watt incandescent bulbs for average sized rooms. The 54-watt unit (14" x 4") is approximately equivalent to 225 watts of incandescent lamps for larger rooms, or areas like the kitchen or bathroom where more light is needed. The drums are high-impact white acrylic; the pans are powder-coated stamped steel. Magnetic ballasts, 10,000-hour lamp life, both are UL-listed.

A. 36-338 26-Watt Drum Fixture $45^{95}
B. 36-339 54-Watt Drum Fixture $65^{95}
31-104 Replacement 13-watt Bulb $4^{95}

High-Tech Night Light Uses 99% Less Energy

This safe, energy-efficient, 2-3/4" square night light casts a soft, green glow almost forever. It provides soothing illumination in dark bedrooms, bathrooms, and hallways. Space age lighting material is completely cool to the touch, so there's no danger of burned fingers. Uses less than 1% of the electricity of standard night lights. No bulbs to replace. Slim, 1/4" thickness prevents curious children from prying it out of electrical outlet. Unobtrusive high-tech design fits into any decor. Packs easily for trips away from home. Lifetime guarantee. If it ever burns out, just return to manufacturer for replacement. Costs 2¢ per year to run!

25-417 Limelite Night Light $8^{95}
2 for $16^{95}

An Energy-Saving Night Light And Wall Switch In One

Moving around the house at night is safe and easy with this high-tech lighting duo. An energy-efficient night light and compact wall switch in one, it costs less than 2¢ a year to run. Since it provides light without taking up an outlet, it's especially useful for bathrooms, guest rooms, and hallways. For long-lasting illumination, it uses a space-age material that stays completely cool to the touch — no bulbs are required. The night light has its own control, so you can use the wall switch with or without the light. Fits a standard rocker switch plate. UL-listed. Measures 2-3/4" W x 4-1/2" L x 1-1/2" D.

25-421 Limelite Switch $29^{95}

No More Fumbling In The Dark

Cordless Sensor Light gives you a glowing welcome as you step into your darkened entry hall with an armload of groceries. It only goes on when it senses your motion and its 2-1/2-watt krypton bulb will light your way to your garage, workshop, or stairway. Conserves energy in places where people often forget to turn out the light: bathrooms, corridors, garage, etc. In auto mode, it comes on for 15 seconds when the infrared detector senses motion up to 10 feet away. Also has on and off settings. Place your Sensor Light on a desk, table, counter, floor, or wall-mount it. Requires 4 "C" batteries (not included). Imported.

25-413   Cordless Sensor Light $39⁹⁵

50-107   Rechargeable "C" Nicad Battery $4.75 ea.

Sensor Socket Light-Sensitive Timer Helps Cut Your Electric Bill

An innovative way to reduce your electric bill, this weather-resistant timer/sensor has a built-in photo eye that automatically turns a light on at dusk, then turns it off six hours latter. It is the only one that works with compact fluorescents. By using light judiciously, it prolongs the life of your bulbs, keeps your home secure, and offers maximum energy savings. The compact size fits most smaller enclosed indoor or outdoor light fixtures. Screws into a standard-sized socket — no wiring or tools required. 1-1/2" x 1-1/2" x 2-3/4". UL-listed.

36-402   Sensor Socket $19⁹⁵

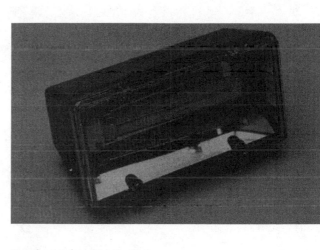

Bright Light, Big Savings For Security, Safety, Or Sports...

Entertain outdoors or keep intruders away. Providing a wide beam of light equivalent to a 75-watt incandescent floodlight, this 13-watt unit lives up to its name. Suitable for patio, yard, driveway, or parking lot.

Dimensions: 4" D x 4-1/2" H x 8-1/2" W
- *Replaceable PL-13 bulb*
- *Ballast life: 100,000 hours*
- *10,000-hour bulb life*
- *Suitable for use outdoors to 32°*
- *UL-approved for wet areas*
- *Adjustable 1/2" threaded swivel mount*

36-127 13-Watt Floodlight $39⁹⁵

31-104 Replacement PL-13 Bulb $ 4⁹⁵

Easy-To-Install Screw-In Fixture

Now you can retrofit a standard light bulb fixture with a simple twist of your wrist. Just screw this 20-watt all-in-one fixture into the socket and you are done. The warm-white bulb is covered by a soft-white acrylic diffuser for a pleasant overall room light Ideal for utility rooms, closet and small bedrooms or bath. Lamp is easily replaceable, but with average usage will last about eight years. Width is 8.2" and height is 4.2".
- *90-watt incandescent equivalent*
- *1400 lumens output*
- *Electronic ballast*
- *Lamp life: 12,000 hours*
- *Ballast life: 75,000 hours*
- *Operating temperature: 0° to 150° F*

32-104 Ready Lite Screw-In Fixture $35⁹⁵

Lamp Harp Retrofits

Some compact fluorescent lamps are larger than standard incandescent bulbs and will not fit ordinary floor or table lamps unless you change the harp. The harp is the rigid piece that holds the shade over the bulb. We offer 10" and 12" harps to satisfy most lamp conversion needs. They are very easy to install, requiring no tools.

36-403	10" Lampshade Harp	$2^{95}
36-404	12" Lampshade Harp	$2^{95}

Lamp Harp Retrofit

Socket Extender

Sometimes your compact fluorescent lamps may have a neck that is too wide for your fixture. This socket extender increases the length of the base. The extender is also useful if a bulb sits too deeply inside a track lighting or can fixture, loosing its lighting effectiveness. Extends lamp 1-3/16".

36-405	Socket Extender	$2^{95}

Socket Extender

Shed Some Light On Your Workspace

What makes this desk lamp so superior? A weighted precision gear system that really works. Move the 25" arm to optimum lighting position—it stays put indefinitely. Lamp head is adjustable, too. Simply twist the thumb screws to change the lighting angle. Slick and streamlined, this high-quality, black matte finish desk lamp powers an energy-efficient, bright-white 50-watt halogen bulb. Imported. UL-listed. Overall length 29" L x 11" H.

35-337	Adjustable Halogen Desk Lamp	$49^{95}
31-105	Repl. 50-watt Bulb	$5^{95}

Adjustable Halogen Desk Lamp

HALOGENS

Halogen Bulbs

These quartz halogen bulbs use about 10% to 20% less energy than an ordinary incandescent and last more than twice as long. They will fit in fixtures where compact fluorescents are too big. They can also be used with dimmer switches. Our sturdy halogens are about the same size as a standard incandescent and they screw right into ordinary sockets. Their crisp, white light is excellent for reading lamps and hobby work. We offer three choices: 42-watt (50-watt equivalent), 52-watt (60-watt equivalent) and 72-watt (100-watt equivalent). All bulbs are 4-1/4 inches long by 2-1/4 inches wide.

33-114	42-Watt Halogen (set of 2)	$9⁹⁵
33-115	52-Watt Halogen (set of 2)	$9⁹⁵
33-116	72-Watt Halogen (set of 2)	$9⁹⁵

Halogen Par 38 AC Lights

These efficient halogens from Osram Sylvania supersede standard incandescent PAR 38 spot and flood lights. They produce a whiter, brighter light while consuming 40% less power. And they last two to three times as long; average life is 2,000 hours. Refit your store with them, or use them around your home, and watch your power bills drop. The 45-watt lamp replaces a standard 75-watt PAR 38, and the 90-watt lamp replaces a 150-watt PAR 38. They're diode-free, nonflickering, and dimmable. The efficient halogen burner produces a white tungsten light, 3000° K, without lamp blackening.

36-123	Halogen PAR 38 Spot, 45 Watt	$9⁹⁵
36-124	Halogen PAR 38 Spot, 90 Watt	$9⁹⁵
36-125	Halogen PAR 38 Flood, 45 Watt	$9⁹⁵
36-126	Halogen PAR 38 Flood, 90 Watt	$9⁹⁵

Mini-Sized Halogen, Maxi-Light

Enjoy superior quality lighting and reduce your electricity bill. Halogens provide bright indoor illumination for eco-conscious consumers. This small PAR 20 50-watt halogen indoor flood is three times brighter than an ordinary 50-watt reflector — without using any additional energy. Whiter and brighter light makes halogen lighting a favorite of interior designer,s too. Use it to spotlight your favorite painting, photos, and wall decor with museum quality light. Fits all standard and track sockets. Suitable for outdoor use in enclosed fixtures. 3-1/4" long, 2-1/2" deep. Average life is 2,000 hours.

33-117	PAR20 Halogen Flood	$9⁹⁵

DC LIGHTS

When designing an independent power system, it is essential to use the most efficient appliances possible. Fluorescent lighting is four to five times more efficient than incandescent lighting. This means you can produce the same quantity of light with only 20% to 25% of the power use. When the power source is expensive, like PV modules, or noisy *and* expensive, like a generator, this becomes terribly important. Fluorescent lighting is quite simply the best and most efficient way to light your house.

Usually we recommend against using DC lights in remote home applications except in very small systems where there is no inverter. The greater selection, lower prices, and higher quality of AC products has made it hard to justify DC lights. However, if you are seriously concerned about EMF (electromagnetic field) radiation, you might want to consider a DC lighting system. DC systems produce low to nonexistent EMF levels. But wiring and fixture costs are significantly higher than with AC lighting, and lamp choices are limited. You might also want to consider putting in a few DC lights for load minimization and emergency back-up. A small incandescent DC night light uses less than its AC counterpart plus inverter overhead, and a single AC device may not draw enough power to be found by the inverter's load-seeking mode. Inverter failures are rare (under 1%), but they do happen occasionally.

Since the introduction of efficient, high-quality, long-lasting inverters, the low-voltage DC appliance and lighting industries have almost disappeared except for RV equipment, which tends to be of lower quality. This is because most RV equipment is designed for intermittent use, where short life or lower quality isn't as noticeable.

One of the greatest reasons not to use DC lights has been the poor quality of the ballasts made for DC compact fluorescent bulbs. In our search to overcome this problem, we have uncovered a couple of new sources for premium-quality DC ballasts. The new Osram 12-volt ballast is absolutely the best available, but it only works with one lamp type. We've also found the Iota ballasts to be far more reliable than the Sunalux units we used to recommend.

Tube-type fluorescents, such as the Thin-Lite series, are great for kitchen counters and larger areas. These are RV units, with inferior life expectancy compared to conventional AC units. Replacement ballasts for these units are available, and the fluorescent tubes are standard hardware-store stock. We also offer incandescent and quartz-halogen DC lights, but suggest caution in using these for alternative energy systems, where every watt is precious.

— *Doug Pratt*

12-VOLT COMPACT FLUORESCENTS

12-Volt PL Lamp Assemblies

These modular, preassembled, compact fluorescent lamp assemblies screw into standard light-bulb sockets. They incorporate a 12-volt solid-state ballast, a standard edison base, a compact fluorescent lamp socket, and a long-lived 10,000 hour life PL lamp, which can be replaced when it burns out. Remember, these lamps are for 12-volt use only! For all PBS series lights, the center tip pin must be positive. The ballast on the side of these assemblies sometimes makes them difficult to screw into some fixtures. For those applications see the RK kits below.

31-661	PBS-PL5 12V Bulb, 25W equiv. 5-1/4" x 2"	$59⁰⁰
31-662	PBS-PL7 12V Bulb, 40W equiv. 6-1/2" x 2"	$59⁰⁰
31-663	PBS-PL9 12V Bulb, 50W equiv. 7-3/4" x 2"	$59⁰⁰
31-664	PBS-PL13 12V Bulb, 60W equiv. 8-1/4" x 2"	$65⁰⁰
31-665	PBS-Quad9 12V Bulb, 50W equiv. 5-1/2" x 2"	$65⁰⁰
31-666	PBS-Quad13 12V Bulb, 60W equiv. 5-3/4" x 2"	$69⁰⁰

12-Volt Ballast/PL Lamp Kits

Use this kit to install energy-efficient DC powered compact fluorescent lighting into fixtures that cannot accommodate the bulkier size of the PBS series, or where the bare lamp may be visible and a neater look than that of the PBS series is desired. The RK-PL kits come with the appropriate pieces: Iota hard-wire ballast, compact fluorescent bulb, and PL bulb/edison screw-in base adapter. You put them together on site. The ballast measures 3.2"L x 1.75"W x 1.37"D and may be installed anyplace within about 10' of the bulb. *Please note: Switching of power to the lamp must be done on the positive input lead to the ballast, so that the ballast is not energized when the light is off.* The height of the PL lamp/ edison adapter base is the same as for the above comparable PBS series units, and the width is 1-1/4 inches.

For 12-volt DC use only.

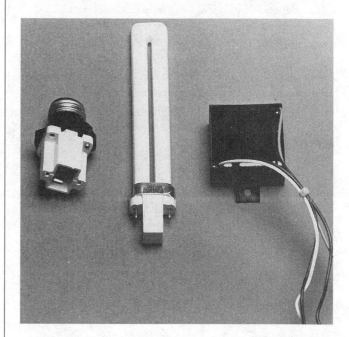

31-671	RK-PL 12V Bulb	$43⁰⁰
31-672	RK-PL7 12V Bulb	$43⁰⁰
31-673	RK-PL9 12V Bulb	$43⁰⁰
31-674	RK-PL13 12V Bulb	$43⁰⁰
31-675	RK-Quad9 12V Bulb	$49⁰⁰
31-676	RK-Quad13 12V Bulb	$49⁰⁰

Iota 12-Volt Ballasts

These high-quality ballasts are a major step up in reliability. Minimum recommended starting temperature is 50° F. According to the manufacturer, the recommended operating voltage for 12-volt models is 10.5-volt to 14-volt, although we have seen them work at higher voltages. Approximate lifetime is 10,000 hours, and all units come with a one-year warranty.

31-205	12V DC Ballast/PL5	$34⁰⁰
31-206	12V DC Ballast/PL7, 9, Quad 9	$34⁰⁰
31-207	12V DC Ballast/PL13, Quad 13	$34⁰⁰
31-208	24V DC Ballast/PL5	$55⁰⁰
31-209	24V DC Ballast/PL7, 9, Quad 9	$55⁰⁰
31-210	24V DC Ballast/PL13, Quad 13	$55⁰⁰
31-211	12V DC Ballast/F40	$69⁰⁰

Iota 12-Volt Ballast

12-Volt Ballast/Bulbs

This is a state-of-the-art DC ballast from Osram Sylvania. Though somewhat limited in application, they are the best DC ballasts we have found. The main limitation is that they only can operate 4-pin, 9-watt compact fluorescent bulbs. These electronic ballasts are extremely efficient, drawing only 1 watt, will start instantly down to 0° F, and eliminate the need for copper grounding strips. Our preliminary testing has amazed us! Since they must be used with the four-pin bulbs, there is no way to adapt them to a screw-in base to insert them into a common edison plug. For convenience, we are selling these as complete kits, with a ballast, 4-pin PL hardwire socket and a bulb.

31-650	Osram Ballast/Bulb	$49⁰⁰

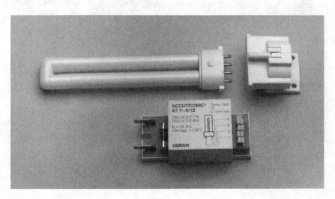

Osram Ballast and Bulb

PL Twin-Tube Fluorescent Bulbs

PL bulbs are available in 5-, 7-, 9-, and 13-watt versions. These quad tube bulbs answer the common complaint that bulbs are too long. The Quad-9 consists of two 5-watt PL bulbs and the Quad-13 consists of two 7-watt PL bulbs. You can also order the edison base adapter separately that accepts a pin-based PL bulb on one end and screws into a standard edison base fixture. Also available are hardwire PL sockets for those who choose not to employ the edison base. *All PL bulbs must use a ballast!*

31-101	PL-5 Bulb, Pin Base	$6⁹⁵
31-102	PL-7 Bulb, Pin Base	$6⁹⁵
31-103	PL-9 Bulb, Pin Base	$6⁹⁵
31-104	PL-13 Bulb, Pin Base	$4⁹⁵
31-121	Quad 9 Bulb, Pin Base	$10⁹⁵
31-122	Quad 13 Bulb, Pin Base	$11⁹⁵
31-402	Edison to PL adapter	$5⁰⁰
31-406	Hardwire Socket (PL13 & Quad13)	$3⁰⁰

PL Twin-Tube Fluorescent Bulb & Edison to PL Adaptor

INDOOR FLUORESCENTS

While not always as efficient as the new compact fluorescent lamps, these standard indoor fluorescent lamps are nonetheless far more efficient than incandescents and halogens, and are reasonably priced and well-built.

Thin-Lites

Thin-Lites are made by REC Specialties, Inc. Their 12-volt DC fluorescents are built to last, and are all UL-listed. Easy to install, they have one-piece metal construction, non-yellowing acrylic lenses, and computer-grade rocker switches. A baked white enamel finish, along with attractive woodgrain trim completes the long-lasting fixtures. Two-year warranty. All use easy-to-find, standard fluorescent tubes, powered by REC's highly efficient inverter ballast. For those who didn't know, bulbs used in DC fixtures are exactly the same as those used in AC fixtures and can be purchased at any local hardware store — only the ballast is different.

30-Watt 12-Volt DC Light

Uses two F15T8/CW fluorescent tubes (included).

- *18" x 5-1/2" x 1-3/8"*
- *1.9 amps*
- *1760 lumens*

32-116　　Thin-Lite #116, 30-Watt　　　　　　　$45⁰⁰

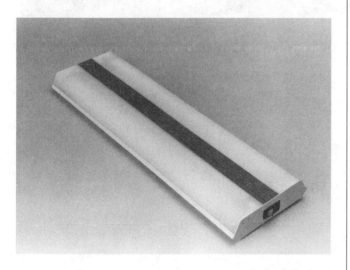

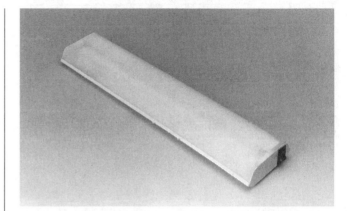

15-Watt 12-Volt DC Light

Uses one F15T8/CW fluorescent tube (included).

- *18" x 4" x 1-3/8"*
- *1.26 amps*
- *800 lumens*

32-115　　Thin-Lite #115, 15-Watt　　　　　　　$37⁰⁰

22-Watt 12-Volt DC Circline Light

Uses one FC8T9/CW fluorescent tube (included).

- *9-1/2" diameter x 1-1/2"*
- *1.9 amps*
- *1100 lumens*

32-109　　Thin-Lite #109C, 22-Watt　　　　　　$44⁰⁰

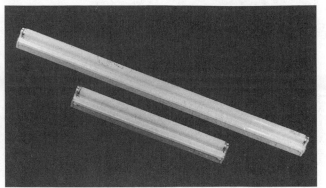

Commercial Thin-Lite

20-Watt Commercial 12-Volt DC Light

The Thin-Lite 20-watt surface mount fluorescent is designed for practical lighting in commercial, industrial, and remote-site applications. It features anodized aluminum housings in a 2-foot length. It is shipped without the fluorescent tubes. This fixture uses one F20T12/CW fluorescent tube (not included), which is a standard AC 20-watt fluorescent tube readily available worldwide. It is designed for remote switching.

- *24" x 3-3/8" x 1-5/8"*
- *1.6 amps*
- *1250 lumens*

•32-151   Thin-Lite #151, 20-Watt $45⁰⁰

Hi-Tech Styles

Thin-Lite Hi-Tech styles were developed for both efficient and attractive lighting where maximum light is required. Anodized aluminum housings and clear acrylic diffuser lenses provide high light output on three sides. They are designed for commercial and industrial vehicles, and for use in remote area housing, schools, and medical facilities in conjunction with alternative sources of energy. Fixtures have almond-colored end caps.

8-Watt 12-Volt DC Light

Uses one F8T5/CW fluorescent tube (included).

- *12-3/8" x 2-1/4" x 2-7/16"*
- *0.9 amps*
- *400 lumens*

•32-191   Thin-Lite #191, 8-Watt $42⁰⁰

15-Watt 12-Volt DC Light

Uses one F15T8/CW fluorescent tube (included).

- *18-1/8" x 2-1/4" x 2-7/16"*
- *1.3 amps*
- *870 lumens*

•32-193   Thin-Lite #193, 15-Watt $39⁰⁰

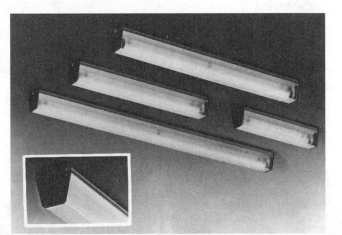

Hi-Tech Thin-Lite

Surface-Mount Lights

These are our most popular Thin-Lites. They are economically priced, practical lights that feature pre-painted aluminum housings and acrylic diffuser lenses. They are available as large as 4 feet long with two standard 40-watt AC fluorescent tubes to meet maximum lighting requirements. Where practicality is the principal consideration, the ST 130 series suits the requirement perfectly.

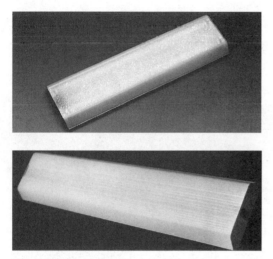

Thin-Lite #134, top. Thin-Lite #138, bottom

30-Watt 12-Volt DC Light

Uses two F15T8/CW fluorescent tubes (included).

- *18" x 5-3/8" x 1-3/4"*
- *2.1 amps*
- *1760 lumens*

•32-134   Thin-Lite #134, 30-Watt $49⁰⁰

40-Watt 12-Volt DC Light

Uses two F20T12/CW fluorescent tubes (included).

- *24" x 5-3/8" x 1-3/4"*
- *2.5 amps*
- *2500 lumens*

•32-138   Thin-Lite #138, 40-Watt $65⁰⁰

40-Watt 12-Volt DC Light

Uses one F40T12/CW fluorescent tube (included).

- *48" x 5-3/8" x 1-3/4"*
- *2.9 amps*
- *3150 lumens*

•32-139   Thin-Lite #139, 40-Watt $80⁰⁰

Twin Light & Magnetic Twin Light

Twin Lights

A permanent fixture for boat or cabin, this 12-volt DC unit has an extra energy-saving feature: a selection switch that allows use of only one or both of the two 12-inch, 8-watt fluorescent tubes. It's equipped with screws set for mounting to a fixture, and connecting wire for easy installation. Lamps included.

- *14.3" x 1.6" x 2"*
- *1.8 amps*
- *800 lumens*

32-107   Twin Light, 8-Watt $15⁹⁵

Magnetic Twin Light

The Magnetic Twin Light has two 8-watt, 12-inch fluorescent tubes, but includes magnetic discs at the rear for attaching to any metal surface. It also comes with a folding metal hanger for easy hook-up. The high-intensity light source comes with a 15-foot 12-volt DC power cord with cigarette lighter plug, making it useful for camping, repairs and in-car illumination. Lamps included.

- *14.3" x 2" x 2"*
- *1.8 amps*
- *800 lumens*

32-108   Magnetic Twin Light, 8-Watt $19⁹⁵

THIN-LITE BALLASTS

Thin-Lite Ballasts

Thin-Lite ballasts convert a wide range of standard 120-volt AC fluorescent fixtures to 12-volt operation. One ballast is required for each fluorescent tube adaptation in a fixture unless otherwise noted. As a general rule, to find the amp draw of a particular light or inverter ballast, divide the watts by volts (W/V = A). As an example, the 12-volt DC circline table lamp adapter Model 107 is rated at 22 watts; 22 watts divided by 12 volts = 1.8 amps.

Ballast dimensions: 5-1/2" L x 1-1/4" W x 1" H

Note: Thin-Lite ballasts are available by special order only; allow two to four weeks for delivery.

•32-203	22-Watt, Single Lamp	$29⁰⁰
•32-209	32-Watt, Single Lamp	$29⁰⁰
•32-210	8-Watt, Single Lamp	$29⁰⁰
•32-211	8-Watt, Dual Lamp	$29⁰⁰
•32-212	14-Watt, Single Lamp	$29⁰⁰
•32-213	14-Watt, Dual Lamp	$29⁰⁰
•32-214	15-Watt, Single Lamp	$29⁰⁰
•32-216	15-Watt, Dual Lamp	$29⁰⁰
•32-225	13-Watt, Single Lamp	$29⁰⁰
•32-226	13-Watt, Dual Lamp	$29⁰⁰
•32-247	30-Watt, Single Lamp	$29⁰⁰
•32-251	20-Watt, Single Lamp	$29⁰⁰
•32-252	20-Watt, Dual Lamp	$29⁰⁰
•32-253	40-Watt, Single Lamp	$29⁰⁰

Note: Lamps and sockets not included.

INDOOR LIGHTING

Tail-Light Bulb Adapters

This simple adapter has a standard medium edison base and accepts a standard automotive-type bulb. It's a very easy way to convert lamps to 12-volt with this 1/2-inch long adapter. Bulb pictured but not included. Get tail-light or turn-signal bulbs at your local auto supply store.

33-404	Tail-Light Bulb Adapter	$7⁵⁰

Variable Speed Switches

Solid-state variable speed control switches are extremely useful for fans, dimming DC incandescent lights and halogens (don't attempt to use on ballasted fluorescent lamps!), using less than 1/2 watt of power to control them. They are available in a 4-amp and 8-amp models. The 4-amp Forward/Reverse model is perfect for DC ceiling fans. This control works on 12- or 24-volt systems.

24-101	Variable Speed Switch, 4A	$27⁰⁰
24-102	Variable Speed Switch, 4A, Forward/Reverse	$31⁰⁰
24-103	Variable Speed Switch, 8A	$39⁰⁰

12-Volt Christmas Tree Lights

It just doesn't seem like Christmas without lights on the tree and these 12-volt lights will dazzle any old fir bush. These lights are actually great all year round for decorating porches, decks, and accenting homes and showrooms. The light strand is 20 feet long and consists of 35 mini, colored, non-blinking lights (draw = 1.2 amps) with a 12-volt cigarette lighter socket on the end. Note: you can't connect them in series like you can with many 120-volt lights. Uses conventional 3-volt mini-lights, same as any 120-volt mini-light string, just wired for 12-volt input.

37-301	12-Volt Christmas Tree Lights	$11⁹⁵

HALOGEN LIGHTING

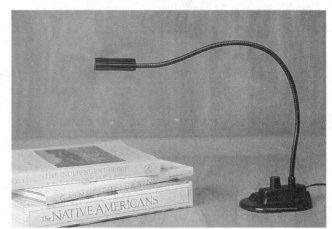

Littlite with Weighted Base

Littlite High-Intensity Lamp

The Littlite is a great gooseneck lamp for your desk, stereo, headboard, or worktable. It's available in a 12- inch or an 18-inch length. The "A" series comes with base and dimmer, 6-foot cord, gooseneck, hood, and halogen bulb. Two pieces of snap mount are included for permanent mounting. Options include a weighted base for a movable light source, plastic snap mount with adhesive pads, and an adjustable mounting clip that adjusts from 1/16- to 3/4-inch wide. Also available is a replacement halogen bulb. The power transformer is available for 120-volt users to convert the Littlite from 12-volt to standard house current. Available with a cigarette lighter plug (but won't work with weighted base). 12-volt DC only.

33-306	12" Littlite	$45⁹⁵
33-307	18" Littlite	$45⁹⁵
33-308	12" Littlite, 12V plug	$49⁹⁵
33-309	18" Littlite, 12V plug	$49⁹⁵
33-401	Littlite Weighted Base	$11⁹⁵
33-402	Littlite Snap Mount	$3⁹⁵
33-403	Littlite Mounting Clip	$12⁹⁵
33-109	Replacement Bulb	$9⁹⁵
33-501	120V Transformer	$14⁹⁵

Quartz Halogen Lamps Inside Frosted Globes

A small cottage industry in the Northeast manufactures these incredibly ingenious 12-volt light bulbs. On the outside they appear identical to an incandescent, but they have the increased efficiency and longevity of quartz halogens. Multiply the watts on halogens by 1.5 and you'll get an idea of their equivalent light output compared to an incandescent lamp. 12-volt DC only!

33-102	Quartz Lamp, 10W/0.8amp	$14⁹⁵
33-103	Quartz Lamp, 20W/1.7amp	$14⁹⁵
33-104	Quartz Lamp, 35W/2.9amp	$14⁹⁵
33-105	Quartz Lamp, 50W/4.2amp	$14⁹⁵

Halogen Flood Lamp

The MR-16 is a very bright flood lamp with a faceted reflector and a quartz halogen bulb in the center, and comes covered with a glass lens. It is ideal for situations where a bright light is desired but not overly directed to a pinpoint. The bulb is a 20-watt halogen which draws 1.7 amps but is far brighter than you'd expect. Fits into an edison base. 12-volt DC only!

33-108	20-Watt Halogen Flood, 12V	$25⁹⁵

SOLAR OUTDOOR LIGHTING

Siemens Pathmarker

The Pathmarker is used to outline or define driveways, paths, walkways, patios, or steps. It comes with an 8" stake for easy mounting. A large textured acrylic lens emits a soft red glow by way of a specially designed high-brightness LED that will never burn out. The LED is driven by premium quality AA nicad batteries through environmentally sealed circuitry. The batteries are charged by a Siemens thin film silicon industrial solar panel which is fully protected by a polycarbonate lens cover. The Pathmarker's body and stake are injection molded from premium-quality black ABS treated with UV inhibitors to prevent the fading and brittleness associated with untreated plastics in long-term sunlight exposure. It operates up to 12 hours every night because of the low amp-draw of the red light. The soft red glow promotes a feeling of safety. One-year warranty and three-year battery life. (Battery is replaceable.)

34-321	Pathmarker Solar Light (set of 2)	$39⁹⁵

Pathway Light

The Pathway Light is an economical pagoda-style light that mounts atop a two-piece 12" stake. A 1/2-watt flourescent tube provides a soft glow within a unique prismatic lens. The bulb is driven by high temperature nicad batteries. The batteries are charged by a solar cell. The Pathway Light accents driveways, walkways, patios, and gardens, with up to six hours nightly run time. Full one-year warranty. Three-year battery life. Batteries and bulbs both replaceable.

34-323	Pathway Light	$59⁹⁵

Siemens Prime Light

The Siemens Prime Light, with its 1-watt compact fluorescent lamp, provides a clear, bright light, ideal for a patio or pool area. This light works well mounted on a pole about 8-feet above the ground, or lower along walkways. The round 1.8-watt photovoltaic cell can produce up to 30% more power than square cells, making the Prime Light one of our brightest garden lights. The high/low switch offers you a choice of duration and brightness; nine hours on low and five hours on high. A full charge takes six hours. The rechargeable, replaceable nicad battery has a three-year life. Instant-on to -20° C. 8" W x 8" H, it comes without a pole so that you can mount it any height you like. Two-year warranty.

34-320	Prime Light	$79⁹⁵
34-332	Replacement Bulb	$11⁹⁵

Prime Light

Light Your Garden With The Power Of The Sun

The Classic Prime Light produces a bright light with its 1-watt compact fluorescent lamp. The classical styling makes it an ideal choice to illuminate garden areas, walkways, driveways, and patios. It has a 1.8-watt photovoltaic cell. High/low switch offers you a choice of duration and brightness — 9 hours on low and 5 hours on high. A full charge takes 6 hours. The rechargeable, replaceable nicad battery has a three-year life. Instant-on to 0°F. 7" W x 7" H. It comes with two 8" stakes that screw together for easy mounting. 2-year warranty.

34-325	Classic Prime Light	$75⁹⁵
34-332	Replacement Bulb	$11⁹⁵

Classic Prime Light

Siemens Sensor Light

The Sensor Light eliminates the need to wastefully burn an outside light while you're away. The solar-charged detection circuit automatically turns on the light to welcome you when triggered by heat or motion at approximately 35 feet, then turns off after you leave the area. This can also make it effective in deterring prowlers. The bright 20-watt quartz-halogen bulb will last over 100 hours and lights instantly at even the coldest temperatures. There are no timers and only one switch to set (off, charge only, on). An adjustable sensitivity control reduces false triggers by your cat, a raccoon, tiny UFOs, etc. The unit mounts easily to wall, fascia, soffit, or roof eaves. No wiring or electrician required. Has a 14-foot cord on solar module for best sun exposure. Extra battery capacity allows up to two weeks of operation without sun. Comes with mounting bracket hardware and bulb included. Replaceable battery and bulb. Full one-year warranty.

34-310	Sensor Light	$125

Siemens Sensor Lightt

FLASHLIGHTS

Palm-Sized Flashlight For Your Car's Cigarette Lighter

Dashlite is Australia's hottest-selling car accessory, and we can see why. This palm-sized lamp plugs into any 12-volt lighter socket and a little LED shows it's charging. Flip the switch for a surprisingly bright light. It will operate for 1-1/2 hours away from the 12-volt charging source. Great for map reading in your car and for a compact pocket flashlight.

| 37-310 | Dashlite Flashlight | $14⁹⁵ |

Solar Flashlight

This solar flashlight is very compact and lightweight. Place in a sunny spot and its two AA nicad batteries will charge from the built-in solar panel. Panel output is approximately 90 mA. Charging time is 17-1/2 hours in full sun. Running time on a full charge is about 2-1/2 hours. The flashlight also doubles as a solar AA nicad charger, charging two AA batteries at a time. Batteries are not included.

| 90-473 | Solar Flashlight | $19⁹⁵ |
| 50-106 | AA Nicads (each) | $2²⁵ |

Dynalite Flashlight

We have had numerous requests for a manual flashlight. With continuous squeezing action the Dynalite produces a small steady light, perfect for anyone who does not want to rely on batteries. The light may not be as bright as a conventional flashlight's, but it will not let you down in an emergency. This flashlight is compact, shockproof, and has a shatterproof lens.

| 37-308 | Dynalite Flashlight | $9⁹⁵ |
| 37-312 | Replacement Bulbs (pack of two) | $4⁹⁵ |

Portable Solar Lantern

If you like to camp and read or play cards after the sun sets, this is the solar light for you. All-purpose solar light is perfect for using in your tent to brighten all your camping trips. It can safely illuminate a large area with its 6-watt compact fluorescent bulb. Side solar panel charges a built-in lead acid battery. When fully charged energy-efficient lantern provides 3 hours of running time. Compact and lightweight, strap it on your backpack or lay it on your car shelf for recharging. Stand it up to bring light where you need it. Great lantern to keep on hand for emergencies, too. 12-1/4" H x 5-1/2" W x 2-1/4" D. Weighs 3 lbs.

| 37-302 Solar Lantern | $79⁹⁵ |

Rugged Solar Lantern
Charges Anywhere With Sunlight

This rugged lantern features a pair of powerful 6-watt fluorescent lamps with high/low switching. Runs up to eight hours on low, or four hours on high. This is the longest run time, with the highest wattage lamps we've seen yet on a solar lantern. The foldable 4-watt solar panel plugs in with a six-foot cord for recharging. Full recharge time is 16 hours. Water resistant, 1 yr. mfg. warranty.

| 37-329 Siemens Solar Lantern | $89⁹⁵ |

Dashlite Flashlight

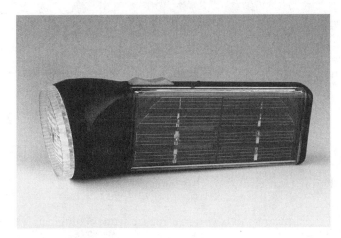

Solar Flashlight

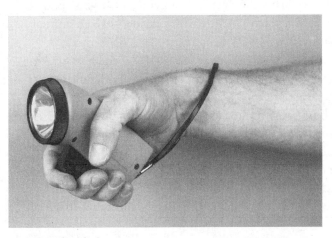

Dynalite Flashlight

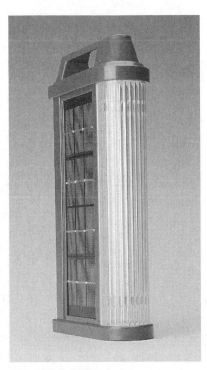

Portable Solar Lantern

Siemens Solar Lantern

INVESTING IN CONSERVATION

The utility companies have discovered that energy conservation pays back large dividends. In 1990 alone, California's Pacific Gas & Electric (PG&E) earned $15 million by saving 280 million kilowatt-hours of electricity. As an added bonus this energy savings also eliminated the production of 200,000 pounds of CO_2 (the most prevalent greenhouse gas) and kept it from entering the atmosphere. Between 1972 and 1986, we Americans saved more energy with conservation than the entire domestic oil industry produced! According to Amory Lovins, "The best technologies on the market today can save about three-quarters of all electricity now used in the United States, while providing unchanged or improved services. The cost of that quadrupled efficiency is about 0.06 cents per kilowatt-hour, far cheaper than just running a coal or nuclear plant (let alone building it). Thus the global warming, acid rain, and other side-effects of that power plant can be abated, not at an extra cost but at a profit. It's cheaper to save fuel than to burn it."

This idea of making a profit by reducing energy consumption and helping the environment at the same time sounds too good to be true. But it *is* true, and not just for the utility companies, but for us as consumers as well. Anyone who can afford to spend $10 can buy into the efficiency market. In today's money or stock market a 15% annual return on investment (ARI) is practically unheard of. In fact, any wise investor would jump at the chance to earn 15% ARI. But energy-efficient products routinely return paybacks in excess of 15%. The fact that you can receive no higher guaranteed return on your money from any other investment is a compelling reason to purchase energy-efficient products. Socially responsible investors can now reap benefits, not only for themselves, but also for our economy and environment.

Here are a few examples of the amazing ARIs you can realize from ecological, zero-risk investments:

Radiant barrier products use thin aluminum films to reflect radiant heat. Installed on the underside of roof rafters, a radiant barrier reflects 97% of all radiant heat out of your home in the summer. This will reduce the typical air-conditioning bill by 10% to 15%. Depending on roof shading, solar gain, and whole house insulation, at 14 cents a square foot radiant barriers have an annual return on investment of between 20% and 50%.

Low-emissivity (low-E) window films reflect at least 55% of the heat striking the window (outside, summer — inside, winter), and over 90% of the ultraviolet radiation. Low-E films also reduce glare by 60%. At a cost of about 75 cents a square foot, the payback for the average home (1,600 to 2,000 square feet.) is 12 to 18 months. Even at a two-year payback, the annual return on your investment is 50%. Low-E window films are a fraction of the cost of "factory high-tech" Low-E windows, which have a payback period as long as 38 years. Life of the film is about 15 years, and you also get shatter-resistant glass out of the deal. Another benefit of this product is that it decreases the fading of rugs, upholstery, and other materials inside the house.

One of my personal favorites is the **Sun Oven**. At home we do most of our oven cooking with the sun. Not only does food taste better, but you can't burn anything in a Sun Oven. And, by using the Sun Oven in the summertime, you can avoid heating up your kitchen the way a conventional oven does. Just don't forget that it works great in the winter, too!

If you are feeling frustrated with low returns on your investments, make your money work for you. Energy-efficient products make excellent economic and environmental sense.

— Jeff Oldham

ENERGY CONSERVATION

Reflectix Insulation

Reflectix is a wonderful insulating material with dozens of uses. It is lightweight, clean, and requires no gloves, respirators, or protective clothing for installation. It is a 5/16-inch–thick reflective insulation that comes in rolls and is made up of seven layers. Two outer layers of aluminum foil reflect most of the heat that hits them. Each layer of foil is bonded to a layer of tough polyethylene for strength. Two inner layers of bubble pack resist heat flow, and a center layer of polyethylene gives Reflectix additional strength.

The standard thickness has an R-value comparable to standard insulation (see chart). Multiple layers of Reflectix will not increase the R-value significantly. Reflectix inhibits or eliminates moisture condensation and provides no nesting qualities for birds, rodents, or insects. Other benefits include reductions in heating and cooling costs that accelerate the payback time of the cost of installation over ordinary insulations. It is Class A Class 1 Fire Rated and nontoxic.

Reflectix BP (bubble pack) is used in retrofit installations. Proper installation requires a 3/4-inch air space on both sides of any Reflectix products. The best application for Reflectix products is preventing unwanted heat gain in the summertime. Reflectix is also a great add-on to already insulated walls, providing a radiant barrier, vapor barrier, and R-value.

R-Value Table For Reflectix

R-Value ratings need to be clarified for different applications and more specifically for the direction of heat flow. "Up" refers to heat escaping through the roof in the winter or heat infiltration up through the floor in the summer. "Down" refers to preventing solar heat gain through the roof in the summer or heat loss through the floor in the winter. "Horizontal" refers to heat transfer through the walls.

Up	Down	Horizontal
8.3	14.3	9.8

As well as its most common usage as a building insulator, Reflectix has a myriad of other uses: pipe wrap, water heater wrap, duct wrap, window coverings, garage doors, as a camping blanket or beach blanket, cooler liner, windshield cover, stadium heating pad, camper shell insulation, behind refrigerator coils, and a camera-bag liner.

Reflectix BP (bubble pack) comes in 16", 24", and 48" widths and in lengths of 50 feet & 125 feet. (We cannot ship rolls larger than 125 feet by UPS.)

- •56-503-BP Reflectix 16" x 50' (66.66 sq. ft) $34⁹⁵
- •56-502-BP Reflectix 16" x 125' (166.66 sq. ft) $85⁹⁵
- •56-511-BP Reflectix 24" x 50' (100 sq. ft) $49⁹⁵
- •56-512-BP Reflectix 24" x 125' (250 sq. ft) $125⁰⁰
- •56-521-BP Reflectix 48" x 50' (200 sq. ft) $109⁰⁰
- •56-522-BP Reflectix 48" x 125' (500 sq. ft) $259⁰⁰

We can only ship Reflectix at standard rates within the continental U.S. For air shipments to AK, HI or foreign countries, call for freight quote before ordering.

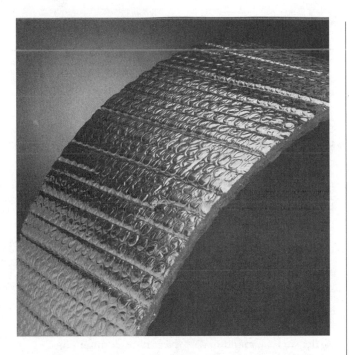

Bubble Pak Staple-Tab Reflectix

This Reflectix product is the same bubble-pack insulation as the standard Reflectix insulation with the addition of staple tabs which are made to be installed between framing members as opposed to on the surface. This product makes installation far easier as it eliminates the need to add furring strips that are needed to form an air space with standard Reflectix. It comes with easy-to-install staple tabs. Same pricing as BP insulation above, but be sure to specify Staple-Tab by ending the product number with ST instead of BP.

•56-503-ST	Reflectix Staple-Tab 16" x 50' (66 sq. ft)	$34⁹⁵
•56-502-ST	Reflectix Staple-Tab 16" x 125' (166 sq. ft)	$85⁹⁵
•56-511-ST	Reflectix Staple-Tab 24" x 50' (100 sq. ft)	$49⁹⁵
•56-512-ST	Reflectix Staple-Tab 24" x 125' (250 sq. ft)	$125⁰⁰
•56-521-ST	Reflectix Staple-Tab 48" x 50' (200 sq. ft)	$109⁰⁰
•56-522-ST	Reflectix Staple-Tab 48" x 125' (500 sq. ft)	$259⁰⁰

We can only ship Reflectix at standard rates within the continental U.S. For air shipments to AK, HI or foreign countries, call for freight quote before ordering.

Tape For Reflectix

Reflectix makes a 2-inch aluminum tape that is excellent for bonding two courses of Reflectix together. It works far better than masking tape or duct tape. It also has reflectability and is highly recommended for any Reflectix installation. Two sizes of tape rolls are available: 30 feet and 150 feet.

56-531	2" Reflectix Tape, 30' Roll	$2⁹⁵
56-532	2" Reflectix Tape, 150' Roll	$8⁹⁵

Low-E Insulating Film

Installing this Low-E window film brings year-round comfort and energy savings. It reflects at least 55% of solar heat in summer; retains 55% of interior warmth in winter. By blocking a minimum 90% of ultraviolet light, it can reduce costly fading and deterioration of furniture and fabrics. There is evidence that the visible spectrum causes as much or more fading than UV. Glare is cut by up to 60%, thereby enhancing the view. Yet high light transmission makes the film ideal for kitchens, sunrooms, and atriums. You can use it on most single pane (except for 1/4"-glass) and dual-pane glass windows with great results. A margin of safety is added by making the window shatter-resistant. For the average 1,600-2,000 square foot home, it only takes 18-24 months to pay back your initial investment in the film. It is a fraction of the cost of factory applied Low-E films. And the film has a lifetime of about 15 years. Installs easily, like wallpaper, using a squeegee (available in the optional Application Kit) and other common household tools. The scratch-resistant surface allows use of normal household window cleaners.

56-495	Low-E Insulator, 36" x 15'	$35⁹⁵
56-496	Low-E Insulator, 48" x 15'	$44⁹⁵
56-497	Low-E Insulator, 24" x 15'	$29⁹⁵
56-498	Low-E Insulator Application Kit	$3⁹⁵

Superior Perforated Radiant Barrier
Made especially for under floors

Radiant barrier, the high performance, low cost, easy-to-install insulation product, has revolutionized the insulation industry, but there's been a problem. Air and water vapor can't pass through it, and if installed improperly, like under existing floors, moisture could accumulate and cause damage. Here's the solution.

Superior Radiant Barrier now has a perforated, reinforced version for use under floors, or other applications where a vapor barrier would create condensation problems. There is no loss of R-value; under a floor, or other application where heat is being reflected upward, this product still has an R-14 rating. That's performance you can't beat at the price.

Installation ease? Which would you rather deal with while lying on your back; scratchy, itchy fiberglass that's falling into your eyes, or a smooth roll of 33.5" wide aluminum-coated building paper that you simply staple to the bottom of your joists? This new perforated radiant barrier spans two conventional 16" on-center joists, with enough extra to make stapling easy. It is also reinforced with mineral fiber scrim to make it almost impossible to tear. Meets California standards for Bleeding, Pliability, and Heat Resistance. Class A fire rated, UL-Classified. Ships by conventional UPS in 500 sq. ft. rolls.

56-504   Perforated Superior Radiant Barrier 33.5" x 180' $109.00
Add $10 additional freight for this overweight item.

Superior Radiant Barrier

About the only complaint we ever get about radiant barriers is that they can be a "bear" to install in the wind. Superior Radiant Barrier, as the name implies, is the best radiant barrier we have found to date. Near pure aluminum is bonded to both sides of 100# Kraft paper, this heavyweight stock is quite manageable in windy conditions. It also has the best price we have ever seen for radiant barriers. Now, you can't afford not to use it! It saves you money on heating and cooling costs for your home or business. Install on the underside of roof rafters, floor joists, or the bottom of the top cord of roof trusses. It even makes a good building wrap. Radiant barriers reflect 97% of all radiant heat out of your home in the summer and back into your home in the winter. It improves the efficiency of your insulation, greatly reduces the operating time of air conditioning and heating equipment, and helps maintain a constant temperature throughout your home. Environmentally safe, nontoxic, and hypoallergenic. Meets California standards for Bleeding, Pliability, and Heat Resistance. Class A fire rated, UL-Classified. A truly cost-effective energy saver and comfort enhancer. Both sizes listed below come as a 500 square foot rolls, length varies according to width. Ships by conventional UPS.

•56-493   Superior Radiant Barrier 26" x 250' $89.95
•56-494   Superior Radiant Barrier 52" x 125' $89.95
Add $10 additional freight for these overweight items.

RECHARGEABLE BATTERIES AND CHARGERS

Every form of energy has an inherent cost that can be measured in terms of dollars (or cents) per kilowatt-hour. This cost is a simple way to compare the different types of energy and their effects on our pocketbooks. For example, utility power costs on average ten cents per kilowatt-hour to consumers. Photovoltaics, or solar power (installed), costs about 25 cents per kilowatt-hour, when amortized over a 20-year period. This is the primary reason that most people still use utility power today. Of course, this figure does not take environmental effects into account, or other factors like independence.

If we look at the choices we have in using common AA, C, and D cell batteries, and compare their cost, we could again look at cost per kilowatt-hour. We find that disposable batteries are by far the most expensive way to use energy. Depending on the type, capacity, and cost of the battery, disposables carry a price tag for energy use ranging from $400 to $1,000 per kilowatt-hour. In contrast, the cost for using rechargeable batteries is less than $1.00 per kilowatt-hour! Each charge of an efficient AA nicad battery costs only a fraction of a penny — you would have to charge the battery about 12 times before you would have spent a whole penny on the energy you've used! Compare that to the 75 cents to $1.00 for each throwaway alkaline battery, and you'll see how quickly these little powerhouses can pay for themselves. (This neglects the inefficiency built into the charging device, which may be ten times the actual energy pumped into the battery. Moral: Learn how long it takes to charge the batteries, then unplug the charger when not in use. Or, better and cheaper yet, use a solar-powered recharger!)

Economics is not the only reason to use rechargeable batteries. Alkaline batteries are toxic waste, and there are very few places in the country that will take them. Most people solve this problem by disposing of batteries casually: millions of these hazardous batteries get thrown in with common trash every day. This problem has become so severe in Europe, where traces of battery pollution are found in even the highest alpine lakes, that these batteries are outlawed, or so we have been told.

Nicad batteries are another story. Not only do you get to use them up to 1,000 times before you even have to *think* about disposal, but when that time does arrive, you can send them to us at Real Goods, and we will have them recycled for you. What's more, if you return our special rechargeable batteries, we'll send you brand-new batteries to replace them at no charge! And we'll keep doing this for as long as you live. (You must keep your receipt to receive replacements, so please read the warranty.)

A new entrant in the battery sweepstakes, nickel metal hydride batteries are another type of rechargeable that we recommend. They are nontoxic and don't have to be recycled.

Types of Rechargeable Batteries

Using rechargeable batteries is easy, but it takes a little getting used to. The following is information you will need to choose the right batteries, and to make sure you get the best performance from them.

Nicad Cells: The full name of this cell type is nickel-cadmium. Nicads were first produced around 1900 and came into common use in the 1950s. Nicads are made of nickel, cadmium, and potassium hydroxide in a water solution. Cadmium is a toxic material and should be recycled as such. There are two basic ways in which nicad cells are made: Sintered-plate cells, which are common in AAA, AA, C, and D cells; and pocket-plate cells, which are available in the same sizes and also as wet-cell storage batteries. All nicads, regardless of construction or size, produce about 1.25 volts per cell. Voltage is stable during charging and discharging, but this stability can present problems when you want to know how much energy is left in the battery. In lead-acid battery types, checking the voltage will give you an idea of the cell's state of charge. Not so with nicads, which display the same voltage from 10% to 90% of capacity.

Sintered-Plate Nicads: This type of construction is the most common in nicad rechargeable batteries because it costs the least to manufacture. Powdered cadmium is pressed into nickel support plates. These cells are sealed and operate in any position (remember, the electrolyte is a liquid), and can withstand over 100 psi internally. They do have a one-shot safety vent in case of massive overcharge. This vent prevents the cell from exploding, but, once the vent is breached, the electrolyte dries out and the cell is ruined. The capacity of nicad cells varies according to how much active material is used. Many manufacturers just take an AA size cell and put it in a C or D size can, which results in very low capacity. Of course, we never recommend batteries made this way! Weight is a good indicator of how much material the cell really contains. In the laboratory, sintered-plate cells are capable of 500 discharge/recharge cycles. In the real world, where cells may be abused by "fast" high-charge rate chargers, life expectancy may be as low as 100 cycles. The sintered-plate cell has a calendar lifetime of about ten years under "float charge" conditions (meaning that they are kept almost completely charged at all times).

Pocket-Plate Nicads: In this construction type, the nickel support plates are covered with perforated cavities. These weblike pockets contain the cadmium, constraining it from thermal expansion and contraction. This reduces loss of active materials, making the cell live longer. Pocket-plate cells may be manufactured as sealed cells (AAA, AA, C, D, or cordless tool cells) or as wet, vented cells (large storage battery systems). Sealed cells have a one-shot vent like the sintered-plate cells. Sealed cells have a life of 1,000 cycles, vented cells up to 2,000 cycles.

Since both sintered-plate and the better-quality pocket-plate cells are sold as sealed batteries, how do you tell them apart? If the cell can be charged in three hours or less, then it's a pocket-plate. Sintered-plate cells are incapable of fast charging. The Golden Power cells we recommend are *all* pocket-plates. Many cordless tool and video-camera batteries are pocket-plate for better performance, quick charging, and longer life expectancy. Calendar life under float conditions is about 15 years.

Nickel-Metal Hydride Cells: Nickel-Metal Hydride (NMH) cells are the new darlings of the environmental movement. Compared to nicads they offer better life expectancy, more capacity per cell, and no toxic chemicals. The cadmium is replaced with metal hydrides, which are environmentally benign. In the lab, NMH batteries have done as many as 400,000 cycles; out in the real world we can expect considerably less — typical life expectancies are 300 to 500 cycles. Nickel-metal hydride batteries have almost double the energy density of nicads (they hold almost twice the power per given size). Like the nicad chemistry, NMH cells produce 1.25 volts per cell. The major disadvantages of this cell chemistry are high initial costs and high self-discharge rates. NMH batteries discharge fairly rapidly when left sitting. Our experience has been that these cells will fully discharge in four to five weeks — so they're not a good choice for the flashlight in your glove box. So far NMH batteries are available only in AA size; as they catch on we may see increased availability.

Rechargeable Alkaline Batteries: This technology became available from Ray-O-Vac in 1994, and will probably be available from other manufacturers in the near future. You can bet that they will all claim to have the very best technology available. Let us try to sort out the truth from all the flashy advertising and PR fluff. Rechargeable alkalines have some great advantages over other rechargeable chemistries, and they also have some serious problems.

Advantages

Greater Capacity. Initially, rechargeable alkalines have two to three times the capacity of a nicad battery.

Longer Shelf Life. Rechargeable alkalines self-discharge very slowly. Their shelf life can be as long as five years.

Arrives Fully Charged. Because of their long shelf life, rechargeable alkalines are shipped fully charged, and don't need charging before the first use. You pop them right into the toy or device and *go*.

Higher Voltage. Alkaline chemistry produces the 1.5 volts per cell output that many battery-driven appliances and toys prefer. Nicads often provide limited or wimpy performance in comparison.

No Cadmium. The Ray-O-Vac product can be disposed of safely in conventional landfills.

Disadvantages

Diminished Capacity with Each Recharge. Rechargeable alkalines lose some of their capacity each cycle. How much capacity is lost depends on how deeply you cycle the battery.

Full Cycling Reduces Life Expectancy. The deeper the cycle, the more capacity loss. You can lose up to 10% capacity per cycle, if you run these batteries completely down. So, really using the greater capacity of rechargeable alkalines has a hidden cost.

Highest Cost per Cycle. Because of their high initial cost and sharply limited life expectancy, rechargeable alkalines are significantly more expensive per cycle than nicads (although they are still a big step up from throwaway alkalines).

No Solar Charger Available. So far the only chargers available for rechargeable alkalines are plug-in AC-powered types. You *must* use the special charger to recharge these batteries.

In conclusion, rechargeable alkaline batteries are an emerging technology. At this point, though, nicad batteries are still the more cost-effective, environmentally correct small-battery solution. This may change as better rechargeable alkaline technology comes to market, but for now we recommend using nicads wherever you can, and limiting rechargeable alkaline use to those toys and appliances that demand the higher voltage or longer shelf life of the alkalines.

In the meantime, our Eco charger will recharge any alkaline battery for those appliances which demand 1.5 volts per cell. Performance will be similar, if not identical, to the high-cost Ray-o-vac system.

Differences between Rechargeables and Throwaway Batteries

When you take the plunge into the rechargeable battery world, you should be aware of a few differences between them and "throw away" batteries. Nicad and NMH batteries have steady operating voltages of 1.2 volts per cell, whereas standard alkaline batteries have an initial voltage of 1.5 volts, which gradually declines as charge is used. While the difference is inconsequential in most appliances, it can affect performance in some devices.

Rechargeables typically store one-half the energy of the best alkaline. Comparisons can be difficult, as standard batteries rarely list capacities on their labels, but simply claim to "keep going, and going, and . . ." Battery capacity is rated in amp-hours (Ah) or, in the case of small batteries, milliamp-hours (mAh), there being 1,000 milliamps in a whole amp. Nicad and NMH manufacturers typically label how much energy is storable and at what rate the energy should be recharged. Recent advances made in rechargeable batteries have increased their capacity, and these are the batteries that Real Goods recommends. In purchasing rechargeables, be sure to compare capacity as well as cost. Consider purchasing several sets, one to operate while the others are charging. Having fully charged batteries ready to go will avoid the frustration that may be caused by the lower capacity of rechargeable batteries.

CHOOSING THE RIGHT BATTERY CHARGER

Now that you have seen the choices you have in rechargeable batteries, you will also need to decide which charger to use. The three types of chargers are: AC charger (plugs into a wall socket), solar charger (operates in the sun), and the 12-volt charger (plugs into your car's cigarette lighter or your home's renewable energy system). It is important to know about these different types of chargers.

Solar Chargers

The great benefit of solar battery chargers is that you don't need an outlet to plug them in — you just need the sun! They can be used just about anywhere. However, these chargers work more slowly than plug-in types, and it often takes several days to charge up your batteries. For this reason, we strongly recommend that you buy *an extra set* of batteries with your solar charger, so you can charge one set while using the other. Another advantage of the solar chargers is that you can leave the batteries in there forever without worrying about overcharging them. Given time and patience (prerequisites for a prudent, sustainable life in any case), solar chargers are definitely the best way to go.

DC Chargers

Choose the 12-volt charger to charge your batteries from the cigarette lighter of your car, or from your home's 12-volt renewable energy system. The 12-volt charger we recommend is actually the fastest charger we have found. It's also great because it puts the right amount of power into different sizes of batteries. The AC charger puts the same amount of power into the battery, regardless of what size it is (AA, C, or D). Even though the 12-volt charger is a fast charger, it will not drain your car battery too much, so you can leave it in there overnight or even for a couple days. This charger *can* overcharge batteries, however, so pay close attention to how long it takes to charge them up, and don't leave them in there too much longer. Leaving them in the 12-volt charger for one-and-a-half times the recommended duration will not harm the batteries. See the chart below for charging times.

AC Chargers

The AC (plug in the wall) chargers are by far the most convenient for the conventional utility-powered house. Most of us are surrounded by AC outlets all day long and all we have to do is plug in the charger and let it go to work. Observe the recommended charge times listed below.

Determining Charge Time

It is important to understand that nicad batteries *can* be overcharged. It is vital not to leave them in the charger too long.

There are two ways to determine the charge time of your batteries.

First, you can use a formula. You will need to know the capacity of the battery being charged, and the output of the charger. This is the more involved way of finding out, but it is educational. Here is the information you will need:

We haven't listed the Eco charger here, as this ultra-smart charger does a better job of figuring charge time than any human. Shuts off automatically when done too.

Charger	Number of Batteries in Charger	Output per Battery
4AA Solar (50-201)	1 battery	90mA
	2 batteries	45mA/battery
	3 batteries	30mA/battery
	4 batteries	22mA/battery
SNC Solar (50-202)	1 battery	150mA
	2 batteries	75mA/battery
	3 batteries	50mA/battery
	4 batteries	37mA/battery
Solar Super Charger Kit	1 battery	300mA
(50-212)	2 batteries	150mA/battery
AC Big Charger (50-223)	140	mAh/battery
DC 12V Charger (50-214)	AAAs	50mA/battery
	AAs	120mA/battery
	Cs or Ds	240mA/battery

Now that you have all the information, all you need is the simple formula. To find out how long it takes to charge up your battery(ies), use this simple equation:

$$1.25 \times \{(\text{Battery Capacity}) \div (\text{Charger Output})\} = \text{Charge Time}$$

So much for the hard way of figuring out how long it takes to charge up your batteries. Having covered the hard way, we have a shortcut for you. We did all the calculations for all the batteries and chargers we recommend, added the best of our invaluable experience, and came up with the following chart.

1. Charge time for solar chargers is listed in *days*. This is based on a good sunny day, and will only be accurate if the charger is placed *directly* in the sun with the panel positioned perpendicular to the sun. Cloudy days count as a quarter of a day. Indirect light counts for nothing at all.

2. The times given for all the chargers assumes that the battery(ies) are almost completely discharged. This is the best time to recharge your battery(ies). At the first sign of weakness, remove the battery from the device and charge for the appropriate amount of time.

	Golden Power days	NMH days
4AA Solar Charger*50-201		
1 AA	2.0	3.5
2 AA	4.0	6.8
3 AA	6.0	10.0
4 AA	9.0	14.0
SNC* 50-202		
1 AA	1.3	2.0
2 AA	2.6	4.0
3 AA	4.0	6.1
4 AA	5.3	8.2
1 C	3.3	6.5
2 Cs	6.6	13.0
1 D	7.5	
2 Ds	15.0	
Solar Super Charger Kit* (50-212)		
2 AA	1.3	not recommended
2 Cs	3.3	6.5
2 Ds	7.5	
AC Big Charger (50-223)		
AA	6.25 hrs	
C	16 hrs	31 hrs
D	36 hrs	
12V DC Charger (50-214)		
AAA	6 hrs	
AA	7.3 hrs	11.5 hrs
C	9.4 hrs	18.3 hrs
D	21 hrs	18.3 hrs

*Note: charge time for solar chargers is listed in days. This is based on a good sunny day, and will only be accurate if the charger is placed directly in the sun and when the panel is positioned toward the sun. Cloudy days count as a quarter of a day. Indirect light doesn't count at all.

The times given for all the chargers assume that the battery(ies) are almost completely discharged. This is the best time to recharge your battery(ies). At the first sign of weakness, remove the battery from the device and charge for appropriate amount of time.

COMMON RECHARGEABLE BATTERY QUESTIONS AND ANSWERS

What about the new Ray-O-Vac rechargeable alkalines? Should I use them or nicads?

Rechargeable alkalines are a new product. At this point the higher cycle cost and lower life expectancy (10 to 20 cycles on average) of the Ray-O-Vac product make it the preferred choice only for those toys and appliances that demand a full 1.5-volt output or long shelf life. For most applications the nicad is still the most cost-effective and environmentally correct choice.

Is it true that my batteries can develop a "memory"?

We get a lot of questions about the memory effect of rechargeable batteries. If you only use a little of the battery's energy *every time,* and then charge it up, the battery will eventually develop a "memory," and it won't be able to store as much

energy as it could before. This is only true with nicads, not with NMHs. Again, this is only a problem if you use only a little of the battery's energy almost all the time. Doing this every once in a while is fine.

How do I tell when a nicad or NMH battery is charged?

Because these batteries always show the same voltage from 10% to 90% of capacity, this is tricky. A voltmeter tells little or nothing about the battery's state of charge. The only half-sure way is to calculate the recommended charging time by dividing the battery capacity by the charger capacity and adding 25% for charging inefficiency, then charging the batteries accordingly. For instance: a battery of 1100 mA capacity in a charger that puts out 100 mA will need 11 hours, plus another three, for a total of 14 hours to completely recharge. This assumes, of course, that the battery was discharged completely to begin with. With plug-in chargers it doesn't matter if there is one or eight batteries — the current flow will be 100 mA (in our example) for each battery. With the solar chargers, the PV panel output is *divided* by the number of batteries in the charger. A cruder method to use is the touch method. When the battery gets warm, it's finished charging. (Note: This method is only applicable with the plug-in chargers.)

My new nicad (or NMH) batteries don't seem to take a charge in the solar-powered charger.

All batteries are chemically "stiff" when new. Rechargables, which are shipped uncharged from the factory, are difficult to charge for the first few cycles. The solar chargers may not have enough "grunt" (power) to overcome the battery's internal resistance — hence, no charge. The solution is to use a plug-in charger for the first few cycles to break in the battery. A better solution is to install *only one* battery at a time in the solar charger.

We cannot emphasize this question too heavily! We get lots of complaints about "dead" solar chargers and nicads, and the problem is almost always in overcoming the battery's shiny new stiffness. Follow the suggestions above, and, if your battery will still not charge up, we will cheerfully help. Once the batteries have been broken in, there is no problem. These rechargeables will give you many, many years of use for no additional cost with solar charging.

My tape player [or other device] doesn't seem to work when I put nicads in. What's wrong?

Some battery-powered equipment will not function on the lower-voltage nicad batteries, which produce 1.25 volts vs. the 1.5 volts produced by *fresh* alkalines. Sony Walkmans are the biggest offenders because the manufacturer has built in a low-voltage cutoff. What's worse, if the cutoff won't let you use nicads, it's also forcing you to buy new alkalines when they're only 50% depleted! This kind of appliance is bad news all around!

Sometimes with devices that require a large number of batteries (six or more), the 1/4-volt difference between each nicad and alkaline gets magnified to a serious problem. Even a fully charged set of nicads will be seen as a depleted battery pack by the appliance.

SMALL BATTERY COMPARISON CHART

Battery	Capacity
AAA nicad	240 mAh
AA nicad	700 mAh
C nicad	1800 mAh
D nicad	4000 mAh
AA NMH	1000 mAh
C NMH	3500 mAh

Note: capacities listed are for the Golden Power nicads that Real Goods sells.
If you have some other brand, check the battery label for capacity. It's in really tiny print.

Battery Capacities	Voltage	cell milli-Amp hours			
		AAA	AA	C	D
General purpose	1.50	300	900	2000	4000
Alkaline	1.50	500	1500	5000	10,000
Panasonic (consumer)	1.50	n/a	500	1200	1200
Golden Power (Real Goods)	1.25	240	700	1800	4000
Millennium	1.25	180	700	1800	1800
Power-Sonic	1.25	n/a	500	1800	4000
SAFT (no name industrial)	1.25	180	500	1800	4000
Nickel-Metal Hydride	1.25	n/a	1100	3500	n/a

All figures given in milliamp-hours of capacity
(1000 milliamp-hours = 1 amp-hour)
General-purpose and alkaline battery capacities are average figures and are not based on any single brand name.

Do I have to run the nicad completely dead every time?

No. For best life expectancy it's best to recharge the nicad when you first sense the voltage is dropping. You'll be down to 10% or 15% capacity at this point. Further draining is not needed or recommended.

— *Gary Beckwith*

RECHARGEABLE BATTERIES

Rechargeables — Batteries That Never Die!

Americans use two billion disposable batteries every year. Rechargeable nicad batteries cut down your battery costs from 9¢ to 1/10 of a cent per hour. The ideal way to buy nicad batteries is to buy two sets. That way you have one set in your charger at all times and one set in the appliance. When the appliance gets weak, just swap batteries with the fresh ones. If you change batteries once a week, they will last for 17 years. With our Infinity Guarantee, your nicads will last forever.

Nicad Rechargeable Batteries

These are the batteries you've been waiting for. Strong enough to make "rechargeable living" more practical and convenient than ever. Designed to keep performing so long that they'll almost become family heirlooms, yet priced to allow you to fulfill all your power requirements. Our high-powered AAs have a capacity of 700 mAh (compared to 500 mAh for most others); the C is 1800 mAh (compared to 1200 mAh for most others); and the D is a full 4000 mAh, more than three times greater than most competitors. We also offer a AAA with 240 mAh. These higher-capacity nicads will last much longer on a charge than other nicads. Other nicad batteries often don't even list their capacity. Our nicads are also closer in power to alkaline disposable batteries. All our batteries are covered by our Real Goods Infinity Guarantee, to last your entire lifetime. We recycle them for free. 1.25 volts.

50-109	AAA Nicads, 240 mAh	$2²⁵
50-106	AA Nicads, 700 mAh	$2²⁵
50-107	C Nicads, 1800 mAh	$4⁷⁵
50-108	D Nicads, 4000 mAh	$6⁷⁵

Rechargeable AA Batteries

Our Nickel Metal Hydride batteries are often referred to as "green" rechargeable batteries, because they don't contain toxic materials such as cadmium, lead, mercury, or lithium. At 1000 mAh for the AA, they store nearly twice the capacity of standard rechargeables. They need to be charged at a slower rate than nicad batteries, so use any of our solar chargers or our 12-volt charger. 1.25 volts.

50-103	Nickel Metal Hydride AA Battery	$5⁰⁰

AA Nicad Charger

This 4 AA Solar Nicad Charger is our most popular solar battery charger. We've sold over 30,000 of them. It couldn't be more simple to operate. Just leave the unit in the sun for two to five days and your dead AA nicads will become fully recharged. The most common uses are for portable tape players and camera flash attachments, flashlights, and toys. With a set of AA cells in the tape player and a set of batteries in the charger, you'll never be without sounds. Not designed to get wet. 90 mA output.

50-201   4 AA Charger $13⁹⁵

Improved Solar Super Charger

Our new high-powered Solar Super Charger will charge any two nicad batteries at a time. The holder fits AAA, AA, C, or D sizes. The improved single-crystal solar module measures 3-1/2" x 10", is waterproof, nearly unbreakable, and has a 16" wire lead. Output is 4.5 volts at 250 mA. Will easily charge a pair of our high-capacity AA nicads in 6 hours or less.

50-212 Improved Solar Super Charger $25⁹⁵

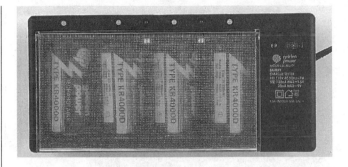

120-Volt Nicad Charger

Our universal charger recharges any size nicad from a standard 120-volt outlet in a day to a day and a half. This 4-cell model features LED lights (to indicate charging is in progress) and has reverse polarity protection, in case you accidentally put your nicads in backwards. Delivers 130 mA to any battery size. It will charge any combination of four AAA, AA, C or D nicad batteries simultaneously or two 9-volt batteries. This charger is ideal for our high-capacity nicads.

50-223   4-Cell AC Charger $15⁹⁵

Solar Button Battery Charger

Our Solar Button Battery Charger is designed especially for charging mercury button-type batteries that power hearing aids, cameras, alarms, calculators, watches, hand-held electronic games, and many other appliances. Charging time is two to six hours depending on light intensity, size, and condition of battery. This charger will keep a great deal of mercury out of the landfills. Includes suction cup for sunny window mounting and a complete instruction booklet. Measures 2-1/4" L x 1-3/8" W x 3/8" D.

50-205   Solar Button Battery Charger $19⁹⁵

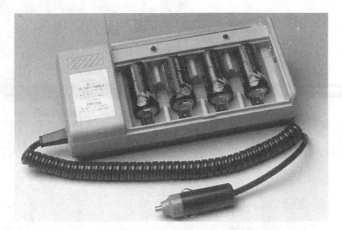

12-Volt Nicad Charger

This 12-volt charger will recharge AAA, AA, C, or D nicad batteries in 10 to 20 hours from your 12-volt power source. It will charge up to four batteries simultaneously. Batteries must be charged in pairs. Has a clever contact system on the positive end that will deliver different current flow depending on battery size. AAAs will receive 50 mA, AAs will receive 130 mA, Cs and Ds will receive 240 mA. It comes with a 12-volt cigarette lighter plug to go into any 12-volt socket. Make sure the cigarette lighter works with the key off if using in your car. One or two charge cycles will not discharge the car's battery.

50-214   12-Volt Nicad Charger $24⁹⁵

A Compact Eco Charger

Featuring the same high quality testing, monitoring, and charging electronics, with the same informative display panel in a smaller package with a smaller price, the Compact model will charge up to four AA or AAA cells at a time. Will recharge any nicad or standard alkaline AA or AAA size battery. UL-listed. 5" L x 3-3/4" W x 2-1/4" H.

50-198   Compact Eco Charger $39⁹⁵

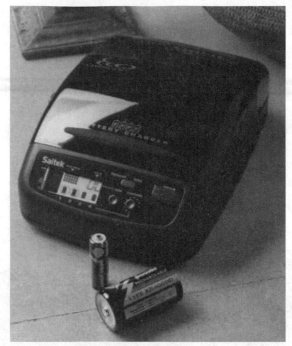

New! State-Of-The-Art Charger
Works With All Batteries

The intelligent Eco-Charger is a sophisticated charger that will test and safely recharge any type of AAA, AA, C, or D battery, including nicads, alkalines, or standard zinc-carbons. Ordinary alkaline batteries can be safely recharged 8 to 20 times, depending on depth of discharge. The built-in microprocessor analyzes each cell's capacity and charge level by simulating actual working conditions. Defective or worn-out cells will be rejected. Each battery is tested and managed individually, so up to four cells of different sizes or discharge levels may be loaded together. The microprocessor continues to monitor and test each cell during charging to assure the fullest possible charge without over or under charging. Charge times vary widely with individual cell needs, but generally, nicads take 3 to 60 hours, and alkalines take 3 to 96 hours. The LCD display shows your current state of charge, time remaining to full charge, and graphically displays charge progress. Charging stops, and a finished symbol displays when each battery is done. This is simply the finest, most intelligent battery charger we have ever seen. UL-listed. 10-1/4" L, 6-1/2" W, 3-1/2" H. One year mfg. warranty.

50-199   Eco Charger $59⁹⁵

12-VOLT APPLIANCES

Since the advent of modern, super-efficient, long-living inverters, the 12-volt appliance market has shrunk to the point of near invisibility. Many of the remaining 12-volt appliances are designed for the recreational vehicle market, which usually means low quality products with limited durability. This isn't the sort of equipment that we want Real Goods to be known for, so we've weeded out this type of schlock.

We realize that some folks need nothing more than a very small, simple 12-volt power system without an inverter. For these installations we maintain a selection of dependable 12-volt products. Some appliances, like the bedwarmer, simply run more efficiently without an inverter.

If you use a good 12-volt appliance that you think we might want to offer, drop us a line, we'll be happy to look it over.

12-Volt Hairdryer/Defroster

Our 12-volt hairdryer is most commonly used as a 12-volt defroster. It beats using your credit card for scraping icy windshields, and is found in many of our customers' glove boxes. It has a 54-inch cord, measures 4-1/2" x 4-1/2", and draws 12 amps from your 12-volt battery.

63-323   12-Volt Hairdryer/Defroster $16⁹⁵

12-Volt Bedwarmers

When it gets really cold, there is no other bed-warming appliance that can put so much heat in your bed as fast as our bedwarmers. These units are far more efficient than electric blankets because they put all the heat in the bed rather than on top of the bed where it dissipates rapidly. The radiant heat keeps the bedding dry, and warms the mattress and spaces beside you. It gives you a soothing feeling of relaxation, like sitting in warm bathwater. Three widths to choose from, all five feet long. The rated amp draws are only valid while the heater is cycling on, otherwise they draw nothing! 12-volt only.

63-328   12V Bedwarmer, 24" Bunk, 5.0 amps $55⁹⁵
63-329   12V Bedwarmer, 48" Double, 6.7 amps $65⁹⁵
63-330   12V Bedwarmer, 60" Queen, 6.7 amps $89⁹⁵

12-Volt Waring Blender

This is by far the strongest, most durable, and most attractive 12-volt blender ever made. It was developed by Waring's commercial division. Not an RV toy. It draws up to 11.5 amps to blend a very hefty load. The base is chrome-plated metal. The 45-ounce shatter-resistant plastic carafe has a snug-fitting vinyl lid with a removable center insert (so you can add ingredients while blending) which doubles as a 2-ounce measure. Stainless steel blades are removable for easy cleaning. A 15-foot cord with attached cigarette lighter plug are included.

63-314   12-Volt Waring Blender $139⁹⁵

Many other 12-volt products are listed in other sections of the *Sourcebook*. This section lists only those products and appliances that did not easily fall into some other category. Find other 12-volt products in the following chapters:

Chapter 4 — Harvesting Energy
Photovoltaic modules (commonly called "solar panels")
Wind Generators
Hydroelectric Generators

Chapter 5 — Managing Energy Systems
Batteries
Controls, Metering, and Safety Equipment
Inverters
Wire, Adapters, and Outlets

Chapter 6 — Heating and Cooling
Refrigerators
Evaporative Coolers
Fans

Chapter 7 — Water
Water Pumps
Pump Controls
Water Purification

EDUCATIONAL TOYS

Most of us want to share our enthusiasm for natural forces and physical knowledge with our children. Unfortunately, one of the things our children learn from a marketplace filled with disposable toys is that things are seldom what they seem. "Educational toys" are too often shoddily made of inferior materials. There are two possible strategies we can follow: buy real tools and explain, or buy toys and compensate. The first course served me well in the raising of my manually dexterous and scientifically inclined daughters, and served my parents well, too, I guess, in that it produced a tool and tech lover. By choosing tools, devices, and instruments that can really be used (this *Sourcebook* and the local hardware and Radio Shack stores have many such tools) and working *with your children* to ensure that they appreciate the value of quality as well as the principles demonstrated by these real-world objects, you pass along a legacy based in your own understanding of lessons learned over a lifetime; no education can be more precious.

If your children yearn for lessons that you consider to be outside of your field of expertise, I suggest that you start with books that are within your shared comprehension as a family; the quality and good spirit of books constantly astonishes me. If, along the path to discovery, there is a certain amount of stumbling and frustration, this should be understood as very much part of the educational process; we seldom get it completely right the first time. As an example, for the child interested in electrons I would recommend an array of rechargeable batteries, a small photovoltaic module, a plastic battery box, a modest array of wire-working tools, a small multimeter, and a book on elementary electricity. Even if this interest proves short-lived, the tools will be useful around the homestead.

If you must rely on ready-made kits for presenting scientific and educational precepts, I encourage you to apply the same values to your children's learning tools that you do to your own: if a part or tool proves unsatisfactory, replace it with one of better quality. Do not let explorations fail because of frustration with poorly designed and fitted materials. Do not expect these kits to be "knowledge in a box;" help your children get to the success that can be found in abundance with packaged science kits. If their interest proves to be enduring, reward them by replacing the toy-grade equipment with the kind of gear used by real practitioners.

If your children explore beyond the limits of your own comfort with technology or materials, seek the advice of experts. Often, local high-school science teachers will be graciously willing to help talented and inquisitive young ones expand their grasp of a subject. It takes a village to raise a child, and most villages are abundantly endowed with adults who can remember their own childhood and who are eager to help communicate their skills and enthusiasm.

As a jumping-off point, the educational toys and kits collected in this section offer the best platform we can find for beginning the systematic exploration of the real world.

— Michael Potts

EDUCATIONAL TOYS

Solar Wooden Model Kits

Our most popular toy. Our wooden model kits have been incorporated into many other popular product catalogs. These easy-to-assemble (a few hours) models are a perfect demonstration to children (and adults) of the wonder of solar power. The parts snap together easily; glue is included. They make great science projects. We've sold over 4,000 of these over the last two years. Ages nine and up.

90-402	Airplane Kit	$19⁹⁵
90-403	Helicopter Kit	$19⁹⁵
90-404	Windmill Kit	$19⁹⁵

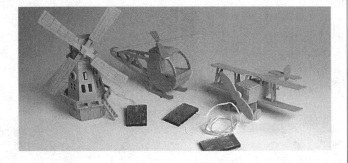

Educational Solar-Energy Kit

This kit helps prepare your child for the coming world society based on renewable energy. Young people, eight years old and up will learn how solar energy works in a practical way. Comes with eight mini-photovoltaic modules that can be wired in endless combinations of series and parallel circuits. They'll get a sense of accomplishment from assembling a solar circuit, and seeing how electricity functions, plus the excitement of actually powering a radio, calculator, battery charger, or the fan that comes with the kit. Our favorite solar experiments kit.

90-470	Educational Solar-Energy Kit	$19⁹⁵

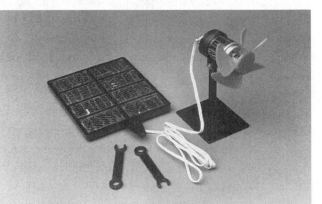

The Best Kits Catalog: Top-Of-The-Line, Time-Tested Kits For Constructing Practically Anything

Frank Coffee. An extraordinary range of products is available in kit form that can save you money and can provide the satisfaction of building something yourself, and this catalog has something for everyone — hooked rugs, a rocking horse, furniture, guitars, canoes and kayaks, "classic" cars, even an airplane! The author offers tips and observations, and takes the reader step-by-step through the construction of a variety of kit projects, including a grandfather clock, a Chippendale end table, enchanting birdhouses, a Shaker wall clock, and a Victorian dollhouse. 264 pages, paperback, 1993.

82-155   The Best Kits Catalog $15⁰⁰

Race With The Power Of The Sun

This fun and educational toy runs on sun power. Solar racer runs fast — watch it race up to 10 feet in just 6 seconds. Assembles quickly and easily, too. Our model was assembled by an 11- year-old in just 45 minutes, and he loved every minute! Powered by its own solar cell, this complete kit also includes a ready-to-cut clear plastic body and chassis, metal motor, 3 wheels, and all necessary hardware. Color it with your own design and paints. No batteries or glue required — just scissors and a screwdriver. Made in USA; 90-day limited warranty. A great way to learn about tomorrow's solar electric cars, today. Ages ten and up.

90-318 Solar Racer Model $24⁹⁵

KEROSENE LAMPS

The Technology

The 1800s in Europe saw a period of revolution in indoor living after dark. The state of the art peaked with the invention of the Kosmos burner system and its derivative, the Matador flame-spreader burner, featuring symmetrical central drafting and an area maximizing circular burning surface. Each of our lamps is equipped with these burners. With only the occasional replacement of wick or chimney, these lamps will provide decades of daily illumination. All use kerosene or lamp oil.

These very high-quality kerosene lamps are imported from France. Every lamp is time tested. Spare parts and service are readily available.

Aladdin Lamp

Aladdin Lamp

Today's beautiful Aladdin lamps differ little in look from early oil lamps. However, the patented mantle and burner assembly are the most efficient ever designed, creating an adjustable pure white light equal to a 60-watt incandescent bulb. The lamp is perfect for use in off-the-grid homes; for illumination during power failures; for use in cabins, camps, RVs; or just for dining on the patio. Burning 12 hours on a quart of lamp oil, it is safe, smokeless, and odorless. No pumping or startup is necessary. The 6-inch–diameter base is hand-blown glass, and the entire lamp stands 20" high.

35-329	Kerosene Lamp, Aladdin	$49⁹⁵
35-329A	Spare Mantle for Aladdin	$4⁹⁵
35-329B	Spare Chimney, Aladdin	$11⁹⁵
35-329C	Spare Wick for Aladdin	$7⁹⁵
35-329D	Insect Screen for Aladdin	$2⁹⁵

Vintners Lamp

The Vintners Lamp is a small kerosene lamp perfect to light your way when no other lighting is possible. One 6-ounce filling will last 20 hours. This is a functional and attractive addition to any vacation or solar home or for Off-The-Grid Day!

•35-326	Kerosene Lamp, Vintners	$39⁹⁵

Lamp Patronne

The Patronne has a Kosmos burner and a beautiful 6-inch–diameter etched glass ball that fits over the glass chimney. Overall height is 19 inches.

•35-320	Kerosene Lamp, Patronne	$75⁹⁵
•35-324	Replacement Etched Ball, Patronne	$29⁹⁵
•35-321	Spare Chimney, Patronne	$6⁹⁵
•35-323	Spare Burner, Patronne	$16⁹⁵

Vintners Lamp

Patronne Lamp

PROPANE LIGHTING

Humphrey Propane Lights

Propane lighting is very bright. It is a good alternative if you don't have an electrical system. One propane lamp emits the equivalent of 50 watts of incandescent light while burning only one quart of propane for 12 hours use. The Humphrey 9T contains a burner nose, a tie-on mantle, and a #4 Pyrex globe. The color is "pebble gray." The mantles seem to last around three months, and replacements are cheap. Note: Propane mantles emit low-level ionizing radiation due to their thorium content.

35-101	Humphrey Propane Lamp (9T)	$54.95
35-102	Propane Mantle (each)	$1.95

Lamp Concierge

The Concierge has a Kosmos burner and a beautiful 6-inch–diameter etched glass ball. The pleasing profile reveals a design with a very low center of gravity, plus a carrying handle that is slotted for wall mounting. Overall height is 15 inches.

•35-340	Kerosene Lamp, Concierge	$79.95
•35-344	Replacement Etched Ball, Concierge	$29.95
•35-341	Spare Chimney	$6.95
•35-343	Spare Burner	$10.95

Lamp Spare Parts Packages

To assure uninterrupted service from your oil lamp, we recommend that you include a spare parts package with your purchase. Each package has two wicks and two chimneys.

•35-327	Lamp Parts, Vintners	$9.95
•35-328	Lamp Parts, Patronnne	$14.95
•35-346	Lamp Parts, Concierge	$14.95

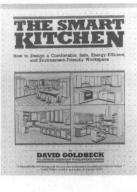

The Smart Kitchen

David Goldbeck. This book caused a stir in architecture circles when it hit the stands two years ago, but until now, it's been hard to find. Excellent chapters on in-kitchen recycling, lighting, and refrigeration will help you save money, but it is the wealth of thought and research, such as a two-page bibliography on refrigerators alone, that has won this book such praise from *Fine Homebuilding* and Amory Lovins's Rocky Mountain Institute. If you're building or remodeling a kitchen, don't start without this book. 134 pages, paperback, revised 1994.

80-144　　The Smart Kitchen　　　　　　　　　　　　　$17⁰⁰

Simple Food For The Good Life
A Collection Of Random Cooking Practices And Pithy Quotations

Helen Nearing. A wonderful collection of quips, quotes, and one-of-a-kind recipes meant to amuse and intrigue all those who find themselves in the kitchen, by the author of *The Good Life* books. Recipes such as Horse Chow, Scott's Emulsion, Crusty Carrot Croakers, Raw Beet Borscht, Creamy Blueberry Soup and Super Salad for a Crowd are sure to improve the disposition as well as the appetite of any guest. More than a collection of recipes, this book encompasses a day-to-day lifestyle that has evolved over the years through careful thought and diligent application. "The funniest, crankiest, most ambivalent cookbook you'll ever read." — *Food & Wine*. 309 pages, paperback, 1980.

82-229　　Simple Food for The Good Life　　　　　　　$10⁰⁰

Diet For A New America

John Robbins. An extraordinary expose of the "Great American Food Machine." Neither preachy nor a list of do's and don'ts, the book will help you select and enjoy foods that will make you healthier and happier. Extremely well-researched, this book is a profound critique of how the standard American habit of eating high on the food chain results in tremendous environmental destruction. 423 pages, paperback, 1987.

"Every so often a book comes along which has the capacity to awaken the conscience of a nation. *Silent Spring* was one such book; I believe John Robbins' volume is destined to be another." — Cleveland Amory.

82-231　　Diet for a New America　　　　　　　　　　$14⁰⁰

The Sun Oven

The Sun Oven is a great solar cooker. We have been using it for several years now, and love it. The Sun Oven is very portable (one piece) and weighs only 21 pounds. It is ruggedly built with a strong, insulated fiberglass case and tempered glass door. The reflector folds up and secures for easy portability on picnics, etc. It is completely adjustable, with a unique leveling device that keeps food level at all times. The interior oven dimensions are 14" W x 14" D x 9" H. It ranges in temperature from 360° to 400° F. This is a very easy oven to use and it will cook anything! After preheating, the Sun Oven will cook one cup of rice in 35 to 45 minutes.

63-421　　Sun Oven　　　　　　　　　　　　　　　　$199⁰⁰

Villager Institutional Sun Oven

Here is an institutional sun oven that's big enough for a whole village, school, hospital, bakery, or refugee camp. In many areas of the world deforestation caused by fuel gathering has reached crisis proportions. In Haiti the situation is so bad that seedlings are snatched up almost immediately for fuel. The Villager may be the greatest invention yet for people living in developing nations. It can bake up to 50 loaves of bread per hour, purify hundreds of gallons of water per day, or sterilize hospital instruments, while relieving the fuel-gatherers for more constructive work. Temperatures inside the oven can reach 400° F in about ten minutes. It has an outside dial thermometer to monitor oven temperature, and a rugged, simple-to-operate "easy-trak" tracking system. May be ordered in either a stationary or mobile (with trailer) mode, and with or without a propane gas standby option. The Villager weighs 1,100 pounds in stationary mode, and 1,350 pounds in the mobile mode. Shipping is freight collect by common carrier. Crating extra.

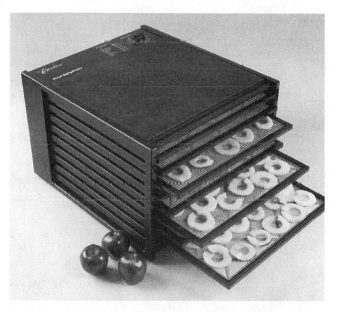

- •63-356 Villager Institutional Sun Oven — Stationary $8,665
- •63-357 Villager Institutional Sun Oven — Mobile $9,265
- •05-228 Sun Oven Crating Charge $475⁰⁰

FOB Milwaukee, Wisconsin. Crated on a pallet; pallet jack or forklift needed for unloading.

Food Dehydrators

Preserve your excess produce the easy way with this food dehydrator. Enjoy healthful snacks, such as apple rings, dried tomatoes, squash, carrots, and herbs for rich flavor in soups and sauces. Think of the money you'll save, not buying snacks and other foodstuffs! Within the dehydrator's attractive black case, a fan forces air horizontally across the plastic trays, removing moisture rapidly, so that flavor, color, and nutrients are locked in. And there's no need to rotate trays. The 120-volt AC unit has a draw of 252 watts for the 4-tray, 430 watts for the 5-tray, and 650 watts for the 9-tray, but wattage is not continuous because the unit's thermostat regulates temperature up to 145° F by turning the power off and on. Preparation guide and recipes are included. One-year limited warranty.

- •63-369 Food Dehydrator, 4-Tray $129⁰⁰
- •63-368 Food Dehydrator, 5-Tray $189⁰⁰
- •63-367 Food Dehydrator, 9-Tray $229⁰⁰

See the Solar Dehydrator plans on page 504 also.

Champion Juicer

Champion Juicers

The Champion Juicer is the finest, most reliable, and most versatile juicer on the market. It works great on carrots, all vegetables, apples, and also makes nut butters. It is 120-volt AC only and runs great off of Trace inverters (812 and larger). Draw is 5.7 amps from a 120-volt source. Available in Almond or White. All units come with a one-year warranty on the motor and a five-year warranty on parts. Specify color.

•63-401   Champion Juicer $279⁰⁰
•63-402   Champion Grain Mill Attachment $89⁹⁵

Hand Wheatgrass Juicer

Hand Wheatgrass Juicer

Among the many claims for the benefits of wheatgrass juice is that it acts as a blood purifier and general detoxifier, due to the highly available, live chlorophyll it contains. It's touted as a high-energy, high-protein, health-promoting drink, rich in vitamins A, B-complex, C, E, and K. But most juicers are unable to extract the juice from wheatgrass. This manually operated, 15-inch–high model is specially designed for the job, and can handle leafy vegetables, sprouts, and herbs as well. Made of fine cast iron, hot tin dipped for a durable finish, it's precision machined for superior juice extraction and smooth operation. The strainer is stainless steel. There is a clamp attachment for portability, and a screw-slotted base for semi-permanent mounting.

•63-366   Hand Wheatgrass Juicer $99⁹⁵

Reconditioned Kirby Vacuum

Reconditioned Kirby Vacuum

One of our long-time customers has a business reconditioning Kirby Vacuum cleaners. Kirbys, like lots of other American appliances, were made better in the fifties and sixties than today. Of even greater significance to Real Goods customers is that these older models are lots more efficient, using only 3 or 4 amps compared to today's electricity hogs that use up to 7 amps! That means they'll run on a 600- or 700-watt inverter handily and be gentler on your batteries if you have a 12-volt system. For those of you not familiar with Kirby, it is quite simply the best vacuum ever made, and it uses no replaceable bags! Two replacement belts and a spare guide light are included as well as an instruction booklet and a six-month warranty. (Reconditioned.)

The Kirby Tool Kit includes a crevice tool, duster brush, utility air nozzle, portable handle, shoulder strap, suction coupler, curved extension tube, 5-1/2–foot flexible hose, and two straight extension tubes equalling 36 inches.

•63-320   Reconditioned Kirby Vacuum $219⁰⁰
•63-318   Kirby Tool Kit $49⁹⁵

Bissell Carpet Sweeper

Do you equate vacuuming with obnoxious noise, plus the never-ending hassles of over-flowing dust bags, clogged tubes, and broken fan belts? Who needs it! This updated professional carpet sweeper from Bissell cleans all types of floors and carpets. It's quiet. It requires no electricity. It saves you money. Heavy-duty construction, extra-wide cleaning path, and large twin dirt pans make this sweeper a righteous cleaning pro. Six-wheel system automatically adjusts to changes in floor surfaces. Six brushes scoop up crumbs, suck up pet hair, and all the rest. Gets into corners and along baseboards, too. Because this sturdy sweeper is lightweight and easy to maneuver, you expend minimal energy. Saves on expensive vacuum replacement bags. Cleans like a quiet storm. Lifetime warranty.

63-497   Bissell Carpet Sweeper $44⁹⁵

Bissell Carpet Sweeper

Uncle Bill's Tweezers

The "Sliver Grippers" made by Uncle Bill's Tweezer Company are quite simply the finest tweezers you'll ever use. Made of spring-tempered stainless steel, the precision points are accurately ground and hand-dressed. With these tweezers it's easy to find and grip even the tiniest splinter or stinger. No pocket, purse, first-aid kit, or tool box should be without a pair! All tweezers come with a lifetime money back guarantee and a convenient holder that fits on your keychain! Our local Lyme Disease Control Center is now recommending Uncle Bill's Tweezers for removing ticks.

63-428   Uncle Bill's Tweezers (set of 3) $11⁹⁵

Uncle Bill's Tweezers

James Washers

The James hand-washing machine is made of high-grade stainless steel with a galvanized lid. It uses a pendulum agitator that sweeps in an arc around the bottom of the tub and prevents clothes from lodging in the corner or floating on the surface. This ensures that hot suds are thoroughly mixed with the clothes. The James is sturdily built. The corners are electrically spot-welded. All moving parts slide on nylon surfaces, reducing wear. The faucet at the bottom permits easy drainage. Capacity is about 17 gallons. Wringer attachment pictured with the washer is available at an additional charge.

63-411   James Washer $315⁰⁰

Hand Wringer

The hand wringer will remove 90% of the water, while automatic washers remove only 45%. It has a rustproof, all-steel frame and a very strong handle. Hard maple bearings never need oil. Pressure is balanced over the entire length of the roller by a single adjustable screw. We've sold these wringers without a problem for over 13 years.

63-412   Hand Wringer $129⁰⁰

James Washer with Hand Wringer

Wooden Clothes Drying Rack

Wooden Clothes Drying Rack

Electric and gas clothes dryers are big-time energy hogs. Even worse, the rough spin ages clothes prematurely. Our solar dryer uses sun and wind to gently dry clothes, leaving them with a naturally fresh scent that can't be duplicated by chemical additives. The large accordion laundry rack, 48" x 36" x 27", gives you over 40 feet of drying space. It's made of poplar and birch; the hardwood and extra-thick dowels provide superior strength and durability. We've found that, even with two children in cloth diapers, this rack can do the job.

•63-435 Wooden Clothes Drying Rack $84⁹⁵

Save Space And Conserve Energy With A Fold-Out Clothes Dryer

Mount this no-energy dryer anywhere you need it. In winter, use for drying damp gloves and scarves. Helps you get organized. Hang coats from handy wooden pegs. Store hats and sunglasses on the all-purpose top shelf. Wooden rack is easy to install on any wall surface. Position unit on porch for solar drying, near back door for easy access, or in laundry room for indoor drying. The perfect drying rack for flowers, herbs, and pasta, too! Space-saving dryer protrudes only 6-1/4" when folded into wall. Measures 18" x 30". Extends to 30" when in use. Constructed of sturdy Eastern white pine with durable birch hardwood dowels and four Shaker-style pegs. Comes fully assembled. Includes mounting hardware.

63-500 Wall Shelf Drying Rack $69⁹⁵

Mighty Mule Solar Gate Opener

This is a heavy-duty, durable automatic gate opener. The Mighty Mule will open gates up to 16 feet wide or up to 250 pounds. It is very easy to install and doesn't require an electrician. It comes with an adjustable timer for automatically closing the gate, and will automatically latch the gate shut. A 12-volt, 6.5 amp-hour battery is included that can be easily charged with the optional 5-watt solar module. As shipped, the Mighty Mule is powered by 120-volt AC with a step-down transformer. If you don't have AC power available at the gate site, then you need the solar-panel option. All units come with a wireless remote transmitter so that you can open the gate without getting out of your vehicle. Additional transmitters are often purchased so that you can keep one in each car. The controller will accept up to three inputs, radio transmitter, digital keypad, simple pushbutton, or any combination up to three. It takes 18 seconds for the gate to open and 18 seconds to close.

Mighty Mule's electronics are state-of-the-art. The gears are all metal, not nylon or plastic. The highly efficient motor provides enough power to operate ornamental iron and commercial chain link gates up to 250 pounds. Even in high cycle applications the Mighty Mule will not overheat. Solar panel mount not included.

Does not include mounting hardware.

•63-126	Mighty Mule	$735⁰⁰	•63-144 Mighty Mule Extra Transmitter	$39⁹⁵
11-521	Mighty Mule 5-Watt Solar Panel Option	$89⁹⁵	•63-146 Mighty Mule Digital Keypad	$79⁹⁵
•63-142	Mighty Mule Horizontal Gate Latch	$169⁰⁰	*Send SASE for free brochure.*	

Parmak Solar Fence Charger

The 6-volt Parmak will operate for 21 days in total darkness and will charge up to 25 miles of fence. It includes a solar panel and a 6-volt sealed, leakproof, low internal resistance gel battery. It's made of 100% solid-state construction with no moving parts. Fully weatherproof, it has a full two-year warranty.

63-128	Parmak Solar Fence Charger	$195⁰⁰
•63-138	Replacement Battery for Parmak, 6V	$39⁹⁵

Pocket Reference

Thomas J. Glover. This amazing book, measuring just 3.2 by 5.4 inches, is like a set of encyclopedias in your shirt pocket! Here is a very small sampling of the hundreds of tables, maps, and charts within: battery charging, lumber sizes and grades, floor joist span limits, insulation R values, periodic table, computer ASCII codes, IBM PC error codes, printer control codes, electric wire size vs. load, resistor color codes, U.S. holidays, Morse code, telephone area codes, time zones, sun and planet data, earthquake scales, nail sizes, geometry formulae, currency exchange rates, saw-blade sizes, water friction losses, and a detailed index! Have we said enough? Indispensable! 480 pages, paperback, 1992.

80-506	Pocket Ref	$10⁰⁰

Building Stone Walls

John Vivian. This concise and elegant little book tells all you need to know to build a monument to yourself. The simple approach, detailed pictures and emphasis on safety and forethought set a fitting tone for the monumental task. Wall making is a contemplative as well as a physical endeavor, and the author captures the spirit and passes it along admirably — a classic little book. 105 pages, paperback, 1993.

80-180	Building Stone Walls	$9⁰⁰

Fences, Gates, And Bridges: A Practical Manual

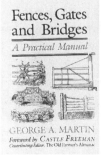

George A. Martin. Fences should be strong, inexpensive, and they should last. Those were the desired virtues when this book was first published more than 100 years ago, and we shouldn't expect any less today. Discussed in this comprehensive and unique text are hedges, rail and other primitive fences, and fences made of boards, stone, barbed wire, and even sod. Special applications are covered in detail: Streams and gullies, making and setting posts, gates and fastenings, and wickets and stiles. Liberally illustrated with 295 drawings. 189 pages, paperback, 1992.

82-152	Fences, Gates & Bridges	$10⁰⁰

Better Beer And How To Brew It

M. R. Reese. This book will get you started, with clear instructions and step-by-step photographs of the homebrewing process. Very much created for the basic brewer whose justification for the hobby is economic, this book demystifies and debunks any misconceptions about beermaking. If you want good, inexpensive brews, this book will help you create them. Includes 19 recipes. 122 pages, paperback, 1978.

82-220   Better Beer & How to Brew It $10⁰⁰

Brewing The World's Great Beer: A Step-By-Step Guide

Dave Miller. If simply brewing good beers isn't enough, author Dave Miller tells in this book how to create some of the most exotic beers in the world. Step-by-step instructions, amply illustrated, leads you through 85 easy-to-follow recipes that cover every beer style from California Common Beer to Munich Dunkel. This volume is most appropriate to those who are prepared to sharpen their palates in anticipation and appreciation of the the world's classic beer styles. 150 pages, paperback, 1992.

82-221   Brewing the World's Great Beers $13⁰⁰

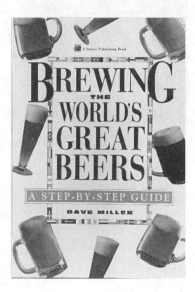

Shiitake Mushroom Kit

Grow Shiitake mushrooms indoors all year. Kit contains a sterilized, enriched sawdust growing medium that is pre-inoculated with mushroom spawn. The Shiitake mushroom patch is prolific and economical. Harvest these fungi, prized for their health and culinary value, at 2-week intervals for up to 16 weeks. When the kits stop producing, they can be used to inoculate hardwood logs. You can be certain your homegrown mushrooms are pesticide-free and fresh. And they represent a beneficial alternative to the common white market mushroom, *Agaricus bisporus*, which contains a natural carcinogen.

•63-395   Shiitake Mushroom Patch $27⁹⁵

ENTERTAINMENT, ELECTRONICS, AND SOLAR APPLIANCES

Solar Safari Hat

Not since Dr. Livingston explored the Congo has technological innovation touched the pith helmet. Now, a perennial favorite has been made much better. The Solar Safari Hat, with its open-weave construction, adjustable headband, and a fan powered by the sun (or 2 AA battery back-up), is worn by gardeners, fishermen, boaters, and African explorers. Tan only. Batteries not included.

90-411	Solar Safari Hat	$39⁹⁵
50-106	AA Nicads, 700 mAh	$2²⁵

Solar Cool Caps

These caps work as miniature evaporative coolers. The cool breeze blowing on your moist forehead quickly acts to cool down your entire body. A switch allows you to select solar or battery, and a battery compartment is included for two AA cells (not included). Definitely more than a gadget or conversation piece — they really work! We equipped an entire roofing company with them for the hot Ukiah summers. Specify your first and second choice of colors: Red 90-410-R, Blue 90-410-B, or White 90-410-W.

90-410	Solar Cool Cap	$29⁹⁵

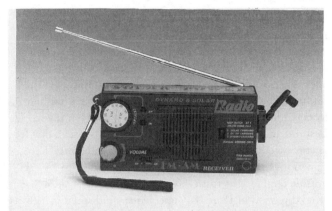

AM/FM Radio Recharges By Hand Or In The Sun

Turn the Dynamo AM/FM receiver's crank for just one minute to hear ten minutes of music, news, or sports. The radio contains a miniature dynamo and a small solar panel that charges the built-in nicads. So no household current is needed, no batteries that die at the beach. The back-up battery compartment holds two AA nicads for days when you feel really lazy. Flashlight on end can be used with a white lens cover as a steady beam, or with a red lens cover as an emergency blinker.

90-319	Dynamo Radio	$34⁹⁵
50-106	AA Nicad ea.	$2²⁵

Weather Wizard III

The Weather Wizard III is a professional-quality station that monitors indoor and outdoor weather at the touch of a button. Everyone lives and works in microclimates that may vary in a short distance from "frost pocket" to "banana belt," and where conditions may differ greatly from those reported for the general area. So Weather Wizard is a useful tool for farming, for the dedicated home gardener, and for any business affected by weather. Additionally, it's fascinating for anyone with a scientific bent or a closeness to nature. Functions include inside temperature from 32° F to 140° F; outside temperature from -50° F to 140° F; highs and lows; wind direction in 1° or 10° increments; wind speed to 126 mph; high wind speed; wind chill to -134° F; low wind chill; highs and lows with time and date; temperature, wind speed, wind chill and time alarms; 12/24-hour clock; date. Measures 5-1/4" x 5-7/8" x 3"; can be wall mounted. AC operation with battery backup or 12-volt operation with optional lighter cord (#63-386, listed below). Comes with anemometer with 40 feet of cable; external temperature sensor with 25 feet of cable; junction box with 8 feet of cable; AC power adapter.

63-381   Weather Wizard III $233⁰⁰

Weather Monitor II

This top-of-the-line model combines all the operational abilities of Weather Wizard III and Perception II, without any increase in size: it, too, is just 5-1/4" x 5-7/8" x 3". Glance at the display, and see wind direction and wind speed on the compass rose. Check the barometric trend arrow to see if pressure is rising or falling. The unit provides inside temperature readout from 32° F to 140° F; outside temperature from -50° F to 140° F; high and low temperature; wind direction in 1° or 10° increments; wind speed to 126 mph; high wind speed; wind chill to -134° F; low wind chill; barometric pressure (with memory recall); inside humidity; high and low humidity; timed and dated highs and lows; alarms for temperature, wind speed, wind chill, humidity, and time; barometric trend alarm; 12/24-hour clock; date. Comes with anemometer with 40 feet of cable; external temperature sensor with 25 feet of cable, junction box with 8 feet of cable; AC power adapter. Optional 12-volt DC lighter cord (listed below).

63-382   Weather Monitor II $395⁰⁰

Weather Wizard III

Extension Cable

Gives more latitude in the placement of the anemometer, external temperature sensor, Rain Collector, or Weatherlink, by adding 40 feet to the length of the standard cables. Depending on the electromagnetic and radio frequency interference in the area, cables may be linked together for runs from 80-160 feet.

63-387   4-Conductor 40-Foot Cable $19⁹⁵
63-388   6-Conductor 40-Foot Cable $24⁹⁵

Weather Monitor II

Weatherlink

For in-depth weather studies, Weatherlink teams the Perception II, Weather Wizard III, or Weather Monitor with an IBM-compatible computer, to create graphs, calculate average weather conditions, generate summaries, analyze trends, and more. For example, over time one might be able to trace weather effects from global warming, fluctuations in atmospheric and oceanic currents, or sunspot activity. Weatherlink stores data until it is transferred into the PC. Data may be exported to *Lotus 1-2-3* or *dBase III* compatible spreadsheet or database software.

Software Features: instant weather bulletin displays the weather on one screen. Graph any function on a daily, weekly, monthly, or yearly basis. Graph two days, weeks, months, or years on the same screen. Display two different functions on the same graph. Track information from two or more weather stations (one Weatherlink required for each station).

Installation: Fit Weatherlink inside the mounting base of the weather station and plug it in. Then run the cable to a serial port on the PC. To monitor weather conditions in remote locations, use the Modem Adapter with a Hayes or compatible modem.

Specifications: For IBM PC, XT, AT, PS/2 or compatible personal computers with 512K conventional memory. Requires Hercules monochrome, CGA, EGA, VGA or compatible video graphics adapter and monitor, MS-DOS or PC-DOS 2.1 or higher, and one serial port. Supports RS-232 serial ports 1, 2, 3, or 4 and most dot matrix and laser jet printers. Comes with 9-pin and 25-pin RS-232 serial port adapters, 8 feet of cable, and 5-1/4" 360K floppy disk.

63-383	Weatherlink Module	$175⁰⁰
63-384	Modem Adapter	$8⁹⁵

Rain Collector

Since rainfall is one of the major components of climate, Weather Wizard III and Weather Monitor II achieve their maximum potential when joined with Rain Collector. It allows reading both daily and accumulated rainfall totals. The self-emptying receptacle measures precipitation in 0.1" (3mm) increments with exceptional accuracy. The ruggedly built unit has an easy plug-in connection; comes with 40 feet of cable and mounting hardware.

63-385	Rain Collector	$75⁹⁵

DC Lighter Cord

A product for climate monitors, allowing greater flexibility in weather research. For travelers by car, truck, RV, or boat who like to keep track of weather conditions. And for those who live and/or work off the grid. This cord with cigarette lighter plug replaces the standard AC adapter, allowing all the weather stations to operate on 12-volt DC power.

63-386	Weather Station 12-Volt Cord	$9⁹⁵

TOOLS FOR RECYCLING

"Nobody made a greater mistake than he who did nothing because he could only do a little."

— *Edmund Burke*

We have all fallen heir to a throwaway society, one nurtured by wastefulness and old habits. The results are all around us — overflowing landfills, trash on our beaches, air and groundwater pollution. Studies show that:

- In the course of a lifetime, the average American will throw away 600 times his or her weight in garbage. This means that a 150-pound person will leave 90,000 pounds of trash to the next generation.
- Every year we trash an incredible amount of goods that could be replaced with reusables — approximately 16 billion plastic diapers (five million tons), and two billion disposable razors.

According to Jon Naar in his book *Design for a Livable Planet*, up to 88% of our trash could be recycled. Countries like Japan, Germany, and the Netherlands are way ahead of the U.S. when it comes to energy consciousness. They currently recover more than twice the solid waste that we do. The benefits of recycling are great. When we look at what is gained when we recycle, the small effort it takes seems inconsequential. Recycling saves natural resources, because we use the waste stream as a supply for raw materials instead of depleting the beautiful and limited supply the earth gives us. Recycling helps reduce pollution by keeping hazardous substances from being burned or discarded into landfills where they can leach into the ground. Recycling also saves energy, which in turn helps prevent global warming. Did you know that:

- Recycling aluminum requires only one-tenth as much energy as making the same aluminum from virgin bauxite ore? Throwing away an aluminum can wastes as much energy as if you filled the can half full of gasoline, then poured it out on the ground.
- Recycled glass uses only two-thirds the energy needed to manufacture glass from scratch? This means that, for every soda bottle recycled, enough energy is saved to run a TV set for an hour and a half.
- Manufacturing one ton of recycled paper consumes 64% less energy, reduces air pollution by 34%, uses 58% less water, and saves 17 trees?

There is no question that recycling benefits the environment. But there are other aspects to consider as well. Examining our real needs, reducing the amount of unnecessary items in our life, reusing, and *pre*cycling (conscious buying), go hand-in-hand with recycling. When you realize that packaging makes up one-third of our trash, this becomes especially meaningful!

Every little bit makes a difference. The following steps are simple things you can do to reduce waste:

- Reduce consumption of products you don't truly need.
- Purchase products in recyclable containers. Choose packaging you can recycle locally.
- Avoid overpackaged products; you are simply buying space at the landfill.
- Reuse the bag you received yesterday; better yet, take your own reusable cloth bags when you shop.
- Buy food in bulk. This avoids unnecessary packaging.
- Purchase reusable products, such as cloth napkins, diapers, and rechargeable batteries.
- Make a compost pile to recycle your kitchen scraps and yard waste. Composting is nature's perfect recycling method. Food and yard waste traditionally make up 25% to 30% of our waste stream. Plus, the compost you make is great fertilizer, and it's free!
- Set up a worm bin (a practice known as *vermicomposting*) for composting indoors or in limited space.
- Use recycled paper products at home and in your office.
- Photocopy paper on both sides. Use both sides of each piece of paper.
- Purchase goods made from post-consumer recycled materials. This means the product was manufactured from something previously bought and used by someone else. If the label just says "recycled contents," generally it has been made from scrap left over from the normal manufacturing process — business as usual in a well-run factory.
- And, last but not least, recycle!

Another reason for recycling is that it makes simple economic sense. Because recycled materials can be used for making new products, they are valuable, and worth money. You will find that you can get cash for most of the things that you recycle if you bring them to the right place. More and more people are starting recycling businesses nowadays, providing a place for people to bring their materials, and turning trash into cash.

Anything can be recycled, and would be, if the demand for what it could be made into were great enough. Let your grocers and store owners know that you are basing your buying decisions on the environmental impact of products. Be a wise consumer and realize that you *can* make a difference!

Here are some hints on where and how to recycle your recyclables:

- **Recycling centers** *(found in Yellow Pages)*: Aluminum cans, glass jars and bottles, juice boxes, tin cans, aluminum foil, scrap metal, plastic bottles, newspaper, office paper, magazines, glossy paper, cardboard, and paperboard. If you aren't successful in locating a recycling center nearby, then call the EDF for information on your area (see below).
- **Grocery stores:** Bring back your plastic bags. Many grocery stores also recycle cans and bottles.
- **Auto-part stores and garages:** Used car batteries and used engine oil.
- **Local mail-order companies and shippers:** When you buy something that comes in a box filled with Styrofoam, don't throw it away; most places that

have to ship things will gladly accept your cardboard boxes and packing materials. This way the materials are re*used,* which is even better than recycling.

For more information, contact:

Environmental Defense Fund
257 Park Ave. South, New York, NY 10010
(212) 505-2100
(800) CALL-EDF (statewide recycling hotline)

Forest and Paper Information Center
(800) 878-8878
Free pamphlets on recycling paper.

California Waste Materials Exchange Program
(CALMAX)
(916) 255-2369

Mail Boxes, Etc.
Check your local phone book.
They will recycle clean Styrofoam peanuts.

Mercury Refining Co.
(518) 459-0820
Dry-cell battery recycling.

INMETCO
(412) 758-5515
Nicad battery recycling.

Mercury Technologies International
(800) 628-3675
Fluorescent glass tube or bulb recycling.

FulCircle Ballast Recyclers
(800) 775-1516
Magnetic ballast recycling.

— *Karen Hensley*

RECYCLING TOOLS & RECYCLED PRODUCTS

Toilet Paper

Made from 100% recycled paper, and using 95% post-consumer waste, this two-ply toilet paper is unbleached and dioxin-free. It has 500 sheets to the roll compared to 300 for standard virgin toilet paper. Because of excessive weight, we must charge extra for Alaska and Hawaii.

51-102	Toilet Paper Full Case (96 rolls)	$65^{95}
51-101	Toilet Paper Half Case (45 rolls)	$35^{95}
51-107	Toilet Paper Sampler (12 rolls)	$8^{95}

We only ship toilet paper in the continental US west of the Rockies.

Unbleached Paper Towels

These towels are a nice healthy brown, no useless bleach or dyes. Made from 100% recycled, 95% post-consumer waste, these 2-ply paper towels are extremely strong and absorbent. You can have the convenience of a paper towel with less impact on the planet. 90 sheets per roll; each one 11" x 9".

51-113	Paper Towels Full Case (30 rolls)	$39^{95}

Add $10 additional freight for AK & HI

51-114	Paper Towels Half Case (15 rolls)	$21^{95}

Add $6 additional freight for AK & HI

51-115	Paper Towels Sampler (6 rolls)	$8^{95}

Unbleached Napkins

Paper napkins made of bleach-free, recycled paper with 100% post-consumer content. Each bag contains 500 folded napkins 6" x 6-1/2" in size. We ship you two bags.

51-118	Napkins (1,000 count)	$7^{95}

Facial Tissue

Made from 100% recycled paper, these two-ply facial tissues are extremely soft and gentle. These tissues come 100 sheets to the box.

51-106	Facial Tissue Full Case (30 boxes)	$27^{95}
51-105	Facial Tissue Half Case (15 boxes)	$14^{95}
51-109	Facial Tissue Sampler (6 boxes)	$7^{95}

Five-In-One Shopping Bag

Our five-in-one bag is an efficient, eco-protective shopping system. Within the roomy bag are pockets and four supplemental bags with easy-carry handles. Bottles, canned goods, and magazines also fit nicely into the pockets. The canvas "mother sack" has a flat bottom. Even if you're presently recycling paper and plastic market bags, their conversion takes energy that could be saved by using these bags. An elegant addition to all your shopping trips.

51-905	Five-In-One Bag	$29^{95}

Thermalwhiz Cooler

No more struggling with bulky fiberglass coolers on your picnics. The Thermalwhiz is an ingenious collapsible nylon fabric cooler. Super-insulated by Therma-flect™, featuring a reflective waterproof covering, it will keep food either hot or cold for long periods of time. You can use it with or without ice. With its removable plastic inner liner you can divide the cooler into two compartments. It also makes for easy cleaning. For small loads it folds into a compact triangle. After your picnic is done, it folds flat for use as a cushion or for storage. Velcro tabs on the outside make it easy to attach to a car seat, trunk, or boat. It measures 10" x 12" x 11" and can easily hold dinner for four.

51-223   Thermalwhiz Cooler $29⁹⁵

Lifetime Coffee Filter

Once again, the sustainable way of living proves to be the most economical in the long run. This permanent filter is guaranteed to last a lifetime. By not buying any more disposable paper coffee filters, ever, you'll help preserve trees and decrease the solid-waste problem, while saving a considerable amount of money over the years. The stainless-steel filter can handle any grind of coffee, with outstanding results. You may choose cone or basket model; either will work for 2-, 4-, or 6-cup capacity.

51-202   Coffee Filter, Cone $21⁹⁵
51-203   Coffee Life Filter, Basket $17⁹⁵

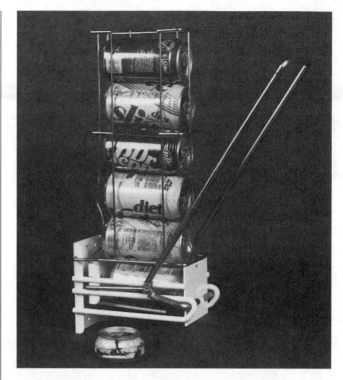

Multi-Crush Can Crusher

Try as you may to limit your purchases of canned drinks, the empties seem to pile up as if they were mating and reproducing. Multi-Crush reduces the size of the problem. It makes recycling aluminum cans faster and easier than any other can compactor. The unit holds up to six cans at once, automatically feeds into a crushing chamber, and ejects flattened cans without interruption. Just mount it on the wall and place your can receptacle beneath. Multi-Crush features all-steel construction and a lifetime guarantee.

51-210   Multi-Crush Can Crusher $24⁹⁵

CHAPTER 9

SAFER LIVING

WHEN WE THINK OF HOME, the impression that comes to mind is a place that is safe from the hazards of the outside world — a place where we are loved, where we can feel secure, and where our surroundings nurture our spirit and renew our health. But most modern homes are not as safe as we imagine them to be, for many common household products can pose health hazards from toxic substances and low-level exposure to radiation.

The human body is an amazingly resilient organism that can adapt to and withstand exposure to many things. In our modern world, however, there are many new, manmade substances not otherwise found in nature, and our bodies have not developed means to identify or assimilate them. These chemicals are used in nearly every industry and every type of consumer product.

There are two types of toxic exposure: acute and chronic.

Acute toxicity is poisoning as a result of a one-time exposure. It is the concern for most consumer products that have warning labels, and the reason we have poison control centers. Every year five to ten million household poisonings are reported as the result of accidental exposure to toxic products in the home. Some are fatal, and most of the victims are children.

Chronic toxicity is illness as the result of many repeated exposures to small amounts of a chemical, over a long period of time; this extended, cumulative effect makes it difficult to identify some toxins, and to recognize we are being affected by them. It is easy to see the effects when someone spills the drain cleaner on his or her hand and the skin burns. The effects of chronic toxicity, though, may not show up for years. For example, numerous common household products can cause cancer, which is not an immediate effect because carcinogenic substances take 20 years or more to act. Other household chemicals are mutagenic: they can alter

449

genetic material and lead to health problems. Still others are known to be teratogenic, and the high incidence of birth defects continues to remind us that not all household substances have been tested for this danger.

Toxic chemicals are made biologically by nature and in the laboratory by humans. Just because something is "natural" doesn't mean it's not toxic. Some of the most toxic substances in the world are found in plant and animal species, having evolved as protection against predators. We don't generally encounter natural toxins in our everyday world, but we are exposed to many man-made substances in everyday products that have potentially toxic effects.

In 1987 the United States produced almost 400 billion pounds of synthetic organic chemicals — over four pounds of man-made substances for every person in this country each day. Worldwide, about 70,000 synthetic chemicals are in use, with nearly 1,000 new ones added every year. Some of these chemicals are considered safe for human use, but the vast majority have not been tested.

Next to nothing is currently known about the human toxic effects of almost 80% of more than 48,000 chemicals listed by the EPA. Fewer than 1,000 have been tested for immediate acute effects, and only about 500 have been tested for their ability to cause long-term chronic health problems such as cancer, birth defects, and genetic changes. A National Research Council study found that complete health-hazard evaluations were available for only 10% of pesticides and 18% of drugs used in this country.

Almost no tests have been undertaken to evaluate the possible synergistic effects that occur when chemicals are combined in food, water, or air, or when chemicals react with other chemicals in the human body. The few studies that have been done indicate that such effects increase risks dramatically. Because scientists do not understand the ultimate effects of these chemicals, the government cannot begin to effectively regulate their use.

The average American home is literally filled with products made from these inadequately tested petrochemical substances; we use more chemicals in our homes today than were found in a typical chemistry lab at the turn of the century. When professionals use chemicals in industrial settings, they are subject to strict health and safety codes; yet we use some of these same chemicals at home without guidance or restriction.

As further research is done, scientists are finding that many household products we assumed to be safe are actually toxic to some degree or another. A multitude of common symptoms can be related to exposure to household toxins, such as headaches and depression. Insomnia, for example, is listed in toxicology books as a common symptom from exposure to formaldehyde — the same formaldehyde used in your bed sheets to keep them wrinkle-free.

But just because a substance has potentially toxic effects doesn't mean that it is toxic for everyone. Whether or not a particular substance actually creates a toxic effect in *you* depends on:

- the quantity of the substance you are exposed to
- the strength of the substance (a small amount of one substance might be much more harmful than a large amount of another)

- the method of exposure (ingestion, inhalation, or skin absorption)
- how frequently you are exposed — many substances are safe to inhale, but not to eat or rub on your skin; others are dangerous regardless of how you are exposed
- your own individual tolerance for a substance

After years of studying all the complexities of the toxicity of consumer products, the best advice I can give is: rather than trying to figure out if something that is inherently toxic is safe for you, just use as many nontoxic alternatives as you can. The more we stay away from toxins, the better off we will be, in terms of both our own health and that of the environment.

Virtually everything we use in our homes may have some toxic component to it, but don't despair: for almost every product there is either a nontoxic alternative available, or a way to make what you have safer. It *is* possible to have a completely safe home. But don't try to change everything at once; start with those items that take a long time to produce chronic illness.

The most toxic products that most people have in their homes are cleaning products. They are so toxic, in fact, that they are regulated by the Hazardous Substances Act and require warning labels as to their dangers. But there are also many nontoxic and natural cleaning products on the market now, and you can always make your own from simple, safe substances such as baking soda, vinegar, lemon juice, and plain soap. To clean your windows, just mix vinegar half-and-half with water in a squirt bottle, squirt it on the glass, and rub with a cotton cloth or old newspaper. Baking soda works great for scouring sinks — just sprinkle it on right from the box and rub with a sponge.

Next most toxic are pesticides, which also can be replaced with far safer methods of pest control. First, figure out how pests are getting into your home and do something to keep them out. You might need to screen your windows to block flying insects or fill holes and cracks in your building structure that mice can crawl through. Then, take steps to remove things from your home that attract pests: take away their food supply by sweeping up crumbs, repair leaky faucets to dry up their water supply, and clear away clutter they can hide in. And you can always use traps and repellents if necessary.

Tap water is full of toxic substances. The EPA has identified more than 700 pollutants that occur regularly in drinking water. At least 22, including the ubiquitous chlorine and fluoride, are known to cause cancer (though not all have been tested). Every home should have adequate water filtration to remove contaminants from water used for both drinking and bathing.

Many personal-care products contain toxic substances that are absorbed through the skin or inhaled. While the most common complaint is skin rash (which can range in intensity from moderately irritating to painful and disfiguring), there are even more worrisome problems. Some common ingredients in lipstick are carcinogenic, just a swallow of perfume can kill a child, and talcum powder may be contaminated with carcinogenic asbestos. Check out the cosmetic products at your natural food store . . . they're nontoxic and pleasant to use.

Though our food supply is well regulated for acute hazards (few people are

immediately sickened from food poisoning or toxic residues), long-term chronic illness can result from the food additives, pesticides, antibiotics, hormones, and other unnatural substances used in growing and processing food. Buy (or grow) organically grown food, and prepare your meals from fresh ingredients.

Even our clothing and bedding can be toxic. Formaldehyde-based no-iron fabric finishes are common, and synthetic fibers can cause skin rashes. All kinds of textile products are now available made from natural fibers (clothing, bed and bath linens, mattresses), and some are even organically grown.

There are a few particular hazards that every home should be checked for: lead, formaldehyde, combustion by-products from burning gas, and radon. These can cause harm to almost anyone, and if they are present in the home, immediate action should be taken to fix the problem. Home test kits and monitors can help you determine if these substances are in your home.

With a little common sense, awareness of toxic hazards, home test kits, and nontoxic products, you can have a healthy, happy, and safe home.

— *Debra Dadd-Redalia*

Debra Dadd-Redalia is the author of The Nontoxic Home & Office *(#82-172) and the consumer guidebook* Nontoxic, Natural, & Earthwise *(#80-817).*

SAFER LIVING

Interior Concerns Resource Guide

Victoria Schomer. An annually updated resource guide of environmentally sensitive and low-toxic alternative products for designers and builders — the best we've found. Well-organized and easy-to-use, it presents a surprising number of sane alternatives to notoriously gross materials. 222 pages, 3-ring binder.

80-608   Interior Concerns $39⁰⁰

Nuclear-Free Smoke Alarms

Here is a long-lasting, non-radioactive photoelectric smoke alarm in either 120-volt AC or 9-volt battery powered. These environmentally friendly smoke alarms sense only visible smoke particles. Invisible smoke particles such as those created by cooking or car exhaust will not trigger the alarm. They react to smoldering fires, the most common of household fires, 24 to 68 minutes faster than the radioactive models. An LED indicates power on, the 85 decibel alarm exceeds UL requirements, and the circuitry/alarm test button makes sure everything is working properly. Up to 12 of the AC powered alarms may be interconnected using just three wires. When one alarm sounds, they all do. The 9-volt model will beep at 30-second intervals for up to a month when the battery needs replacement. The cover can't be installed if the battery is out. Some local fire departments require a 120-volt alarm with a 9-volt backup. Check on your community's regulations before ordering. Battery included.

57-122   120V Smoke Alarm $34⁹⁵
57-123   9V Smoke Alarm $34⁹⁵

Car Ionizer

The interior of a car can be more polluted than the air outside, due to gas and exhaust fumes, emanations from plastics, dust, and cigarettes. Plug the N-3 Car Ionizer into the cigarette lighter socket, and it will produce negative ions that bond with pollutants and odors, removing them from the air. The most compact ionizer for automobiles, the N-3 is both highly effective and extremely durable; with normal use and care, it should last indefinitely, requiring no maintenance. It generates a very low radio frequency that can sometimes affect the lower AM bands.

53-706   Car Ionizer $34⁹⁵

A Roomful Of Clean Air — Every 3 Minutes

Pure air is one of your body's simplest and most vital needs. The Healthmate makes sure the air you breathe is the cleanest possible. This environmentally-sensitive air purifying system uses a unique HEPA (High Efficiency Particulate Arresting) filter. Guaranteed for five years of normal use, this innovative filter has a 360⁰ intake with filtering efficiency at 99.97% for all airborne particles, including pollen, dust mites, dander, and smoke. Cleaning the air throughout a room every three minutes, the powerful motor is whisper quiet and twice as energy efficient as other comparably sized models. Casters provide easy mobility. Cleans indoor areas from 150 to 1,500 sq. ft. and has three speeds of air flow. Measures 23 1/2" H x 14 1/2" W x 14 1/2" D and weighs 45 lbs. UL listed.

•53-668 Healthmate Air Cleaner $399⁰⁰

Shipped from manufacturer

NewAire Heat Exchangers   *Shown with cover off*   *HE-2500 Model*

NewAire Air To Air Heat Exchangers — Fresh Air Without Heat (Or Cool) Loss

As building technology gets better at giving us tight, energy efficient buildings with the use of vapor barriers, building wraps, and weather stripping, our indoor air quality has deteriorated. With less fresh air leaking into a building, problems with moisture, radon, chemicals used in construction and furnishings, and carbon monoxide from gas appliances and wood stoves can accumulate. Testing by independent University programs in the U.S. and Canada have found that over 70% of homes built in the last 10 years have insufficient air exchange rates. These aren't "energy efficient" homes, just good average tract homes. Air to air heat exchangers provide the necessary solution. These simple devices use a pair of small blowers, one pushing stale, moistureladen air out, and the other pulling fresh air in. Both air flows are passed through a heat exchanger, so a very minimum of heat (or cool) is lost. We've selected the NewAire line because they feature a much higher than average 78% heat recovery at reasonable cost. NewAire's model # cleverly indicates the maximum recommended square footage for each model.

HE–1800

A compact heat recovery ventilator for recessed installation in insulated ceilings. White eggcrate grill. 70-cfm, 55-watts, 6" duct for outside air, 40 lb., 18"x 18"x 13" deep.

•63-160 NewAire HE–1800 $665⁰⁰

HE–2500

The most popular model. Sized to ventilate most homes, up to 2,500 sq. ft. Fully ducted for basement or attic mounting. 110-cfm, 120-watts, 6" inlet & outlet ducts, 50 lb., 30"x 20"x 12" deep.

•63-161 NewAire HE–2500 $845⁰⁰

HE–5000

A fully ducted whole house ventilator for homes up to 5,000 sq. ft. 210-cfm, 240-watts, 8" inlet & outlet ducts, 70 lb., 30"x 20"x 21" deep.

•63-162 NewAire HE–5000 $1,295

Control Options for NewAire Heat Exchangers

A variety of controls are offered. The right one for you depends on what your indoor air problem is. All control options fit standard single-gang electrical boxes.

A. Dehumidistat Control

Automatic humidity sensing control. Adjustable from 20% to 80% humidity. Must be seasonally adjusted.

•63-166 NewAire Dehumidistat $80⁰⁰

A.

B. 24-Hour Timer

For use where regular ventilation is required regardless of humidity levels. Good for radon or chemical control. Programmable in 20-minute increments.

•63-164 NewAire 24-Hr. Timer $55⁰⁰

B.

C. Delay-Timer Control

A dual purpose control, looks like, and is, a light switch, but also turns on ventilator. When light is turned off ventilator continues to run for preset time. Adjustable from 5 to 60 minutes.

•63-165 NewAire Delay-Timer $105⁰⁰

C.

D. Spring Wound Timer

The simplest control. Most often used in a bathroom or spa room for humidity control. A quiet, attractive, 60-minute timer. White trim.

•63-163 NewAire S-W Timer $33⁹⁵

D.

Outside Wall Caps

Most installations will require a pair of wall caps, fresh in, stale out.

6" Round Wall Cap

Vinyl caps suitable for installation with HE–1800 or HE–2500. Louvered and screened to keep out rain and critters. 8" square exterior.

•63-167 NewAire 6" wall cap $11⁹⁵

8" Round Wall Cap

Suitable for installation with the HE–5000. Galvannealed and primed for painting. Hooded and screened. 10" exterior.

•63-168 NewAire 8" wall cap $29⁹⁵

ELECTROMAGNETIC FIELDS

"Our present AC distribution network is putting the entire planet at risk."
— *Dr. Robert O. Becker*

We as a nation have been living in an ever-thickening web of electromagnetic fields (EMFs) ever since the Rural Electrification Act of the 1930s. A correlation is hard to prove, but our nation's cancer rate has also increased steadily since the 1930s.

It seems that, for almost every study showing that EMFs do biological harm to humans, there is a threatened industry that will commission a (damage-control) study to show opposing results. The goal of the counterstudy is to muddy the waters and keep the results of the original study inconclusive for as long as possible. We saw this go on for almost 20 years between medical doctors and the tobacco industry. In the 1970s and 1980s similar scenarios have been played out between dedicated medical doctors like Robert O. Becker and representatives of the electronics and power industries.

If you look at the *motives* of those protecting the power and electronics industries and the *results* of the studies produced by EMF health scientists, it becomes very clear that we should avoid long-term exposure to EMFs. While the protectionists quibble over how many milligauss (a unit of magnetic induction) the body can safely be exposed to and which frequencies are the most harmful, the rest of us are out here being silently exposed to the stuff on a daily basis.

EMFs probably won't make you sick overnight. However, long-term exposure can cause health problems, especially for the very young and the very old; the very young have rapidly dividing cells, and the very old have less functional immune systems.

As with most other types of radiation, a person is unaware of the EMF field when he or she is in it. Doctor Becker has written two books describing how electromagnetic fields can influence the body and interfere with human biological functions such as cell reproduction and immune response. In his book, *The Body Electric,* on page 327 Dr. Becker states:

> *ELF electromagnetic fields vibrating at about 30 to 100 hertz[1], even if they're weaker than the earth's magnetic field, interfere with the cues that keep our biological cycles properly timed; chronic stress and impaired disease resistance result . . . the available evidence strongly suggests that regulation of cellular growth processes is impaired by electropollution, increasing cancer rates and producing serious reproductive problems.*

Dr. Becker's later book, *Cross Currents,* published in 1990, talks about such things as the link between 60-Hz (household) electromagnetic fields and cancer. On page

---

[1]One hertz (abbreviated Hz) is equal to one cycle per second. Common household AC electricity vibrates at 60 cycles per second. This can also be written as 60 Hz, 60 hertz, or 60 cps. It all means the same thing.

206 he states, "At this time, the scientific evidence is absolutely conclusive: 60 Hz magnetic fields cause human cancer cells to permanently increase their rate of growth by as much as 1600 percent and to develop more malignant characteristics."

The question then becomes how to detect the source and eliminate EMFs from the home and workplace. The EMFs at the high and low ends of the frequency spectrum appear to do the most biological damage. Almost everyone knows that microwaves and radar are unhealthy, but many are still not aware that EMFs generated by common household lights and appliances are unhealthy also. Electric blankets can wrap you in EMFs all night long. So, at the very time when your body is supposed to be repairing itself, your immune system is having its communications jammed by the low-frequency EMFs from the electric blanket. The same goes for bedside clocks with those little red or blue numbers that glow all night. Many of those plug-in clocks will radiate a magnetic field for up to six feet in all directions. So don't put the clock near a wall that has someone sleeping on the other side, because EMF radiation goes through walls as if they were not there. The best solution is to buy a battery-powered clock with a dial that doesn't glow all the time.

The easiest way to find EMF hot spots is with an EMF meter. Using an EMF meter is also a great way to get your kids to sit farther back from the TV set. Just give them the meter and let them find the safe distance. It will make more sense to them when they can see what is happening. It can be fun too, seeing that there is this invisible stuff in the air that shows up on the meter when the TV set is turned on and then disappears when it is turned off (in some cases you may have to unplug the TV or use a switched power strip to turn it all the way off).

Hair driers and electric shavers can produce strong EMFs, though exposure to these appliances is usually brief. It is most important to look at areas of your life where there is long-term exposure. The bedroom and workplace, at about eight hours each, account for about two-thirds of where most people spend their lives.

Remember to check computers and fluorescent lights at work. Have someone turn the lights on and off while you are in your work area with the EMF meter. Some of the older-style ballasts radiate as far as 18 feet but the low-radiation Elba fluorescent ballast radiates less than 2 feet. It also uses less power, does not produce that annoying fluorescent flicker, and, if you use the new T-8 tubes, it will produce twice as much light as the old-style ballast with the standard tubes. EMFs below one milligauss (1 mg) are "generally recognized as safe;" however, the safest is a zero meter reading.

What it all boils down to is this: If you want to live a long and healthy life you should avoid EMFs. Remember, it took 20 years and study after study before the cigarette industry, dragged kicking and screaming into the lab, finally admitted that cigarettes cause cancer. Does it really make sense to wait around in hopes that some of these counterstudies funded by companies with a lot to lose will ultimately prove to be correct? Better to err on the side of caution.

— *Ross Burkhardt*

RADIATION SAFETY

Trifield Meter

Low-level electromagnetic radiation may be a hazard to human health; many studies have raised that suspicion. And the dangers of microwaves are well known. The versatile TriField meter will let you discover the calm backwaters of the magnetic swamp in which we live. With it you can test microwave oven door
gaskets, appliances, cellular phones, computers, and other electronic equipment. You'll be able to construct a magnetic safety "map," so as to move couches, chairs, beds, and especially, cribs out of "hot spots" (children seem to be more susceptible to radiation damage). The meter has three easy-to-read scales, red-lined to show field levels that may be harmful. It reads magnetic, electric, and microwave pollution quickly and accurately without the need to correct for the influence of your own body. Since the human body absorbs radiation from all directions, a meter has to read on all three possible axes to get an accurate measurement. Other meters require tipping and turning; the Trifield reads the X, Y, and Z axes simultaneously. It also compensates for different magnetic field frequencies, to accurately indicate the electrical currents induced within the human body. It will measure fields through the non-ionizing range from 30 Hz to 3 billion Hz. 9-volt battery included.

57-108   TriField Meter $159⁰⁰

Returns are accepted only on truly defective meters, otherwise we charge a 25% restocking charge.

Cross Currents

Dr. Robert Becker. This book discusses the hazards of electromagnetic radiation from ordinary household appliances, utility power lines, radar, and microwave transmitters. This book is fascinating reading and will give you background information and motivation to clean up the magnetic pollution in your local environment! 336 pages, paperback, 1990.

80-612   Cross Currents $14⁰⁰

Monitor 4 Pocket Geiger Counter

The Monitor 4 is the best low-cost nuclear radiation monitor we could find for giving instantaneous readings. Other nuclear monitors cost thousands of dollars, are bulky to carry, and use hard-to-find 22.5-volt batteries. The Monitor 4 is small enough to fit in a shirt pocket and uses a standard 9-volt battery. It's a must-have item if you live near a nuclear plant. (Plant operators don't usually tell you about spills and accidents until it's too late!) The Monitor 4 reads ranges from 0.5 to 50 mR/hour. It has a switch-selectable audio/visual count indicator.

57-105   Geiger Counter $319⁰⁰

Returns are accepted only on truly defective meters, otherwise we charge a 25% restocking charge.

Radalert Digital Radiation Monitor

The Radalert is a high-quality digital Geiger counter sensitive to alpha, beta, gamma, and X-radiation. The product of choice for environmentalists; it measures subtle low-level radiation changes, a must if you live near a nuclear power plant. Digital counting technique and accumulated counts mode allow greater sensitivity in measuring low dose radiation compared to other Geiger counters. Utility is increased with an accumulated counts mode for comparing indoor and outdoor averages and an adjustable alert to signal radiation changes. Operating range is 0 to 19,999 counts per minute (0 to 20 mR/hour). An output jack lets you connect the radiation monitor to a computer for hard copy needs. Runs on 9-volt battery for months of use. Belt loop carrying case included. Battery included.
25% restocking fee on all returns.

57-106   Radalert $369⁰⁰

Radon Gas Detector

Radon is a natural radioactive gas formed by the decay of uranium in the soil. It is tasteless, odorless, and invisible. It works its way to the surface through cracks and porous soil. In the outside atmosphere, it poses little danger. But if it seeps through gaps in a home's foundation or insulation and accumulates, it can be harmful. The EPA estimates that long-term exposure is responsible for up to 20,000 lung cancer deaths annually. Radon Alert monitors gas levels in the home continuously when plugged into a 120-volt outlet. After the first 48 hours, it will display an accurate radon gas evaluation in picocuries. The reading is updated every two hours thereafter. Above 4 picocuries, the red LED hazard light comes on. Gas levels can vary due to weather, ventilation, and other factors. If monitoring shows a continuing hazardous level of radon, action must be taken to block the gas from entering one's home.

57-118   Radon Alert $149⁰⁰

Limited supply due to high demand. Phone for availability before ordering please.

NoRad Shield

If you are concerned about the electric field radiation or high-frequency magnetic radiation emitted from your computer terminal, the NoRad Shield is for you. 99.99% of electric fields in the range of 30 Hz to 1 GHz is blocked, along with 50% of magnetic fields above 10 MHz. It also eliminates static electricity, glare, and reflections. The NoRad shield is the highest resolution screen available, at 260 lines/inch in both directions. It works on color or monochrome monitors with 13" to 14" diagonal screens, and it's easy to install.

57-113   NoRad Shield 13" to 14" $139⁰⁰
57-133   15" NoRad Shield $139⁰⁰

NONTOXIC PRODUCTS

The typical American home has dozens of aerosol cans containing a multitude of carcinogenic ingredients, enough to scare even the most reckless of toxic-waste handlers. Fortunately, though, there are alternatives to the "cancer-in-a-bottle" that you find on the supermarket shelf. In the following pages you will find products that are safe for humans and for the environment, and that do their jobs at least as well, if not better, than their toxic counterparts.

There are cleaning products, for instance, that can handle some of the toughest jobs. For laundry, the nastiest stains come right out with Ossengal spot remover and Oasis biocompatible detergent, leaving behind water that is laced only with fertilizer, and that can be used safely on plants. For lightly soiled everyday clothing, try the new Laundry Discs. California has finally legalized the use of graywater for subterranean irrigation. Watch for other states to follow. General household cleaning can be wiped up with Oasis All-Purpose Cleaner. Citra-Glow cleaner does an amazing job on "heavy" housework. For the really nasty stuff, use Citra-Solv; this great product will even clean paint brushes and inks. In the pages that follow, you will find toilet cleaners, drain cleaners, cedar carpet deodorizers — even a car wash that uses no water!

Pesky critters can be dealt with without pesticides. We have yellow jacket and mouse traps, and mole chasers. You will find cedar blocks and hangers to keep moths away from your clothes, natural mosquito repellent, solar mosquito guards, and citronella candles to keep those little biters away from your body. Nature's best flying-insect abatement system, bats, will consume bugs with an appetite that puts Garfield to shame. Bats can be attracted to your home with our Bat House made from locally reclaimed wood and mill ends.

We have a radon detector that constantly monitors radon levels, and a less expensive tester for short-term sampling. We find it incredible that parents often welcome their new baby into the world with a freshly painted, newly carpeted room, furnished with new furniture spewing out formaldehyde and volatile organic compounds (VOCs); all of which attack the baby's highly sensitive respiratory and nervous systems. We recommend using Water Seal, which seals up and prevents outgassing in particleboard, plywood, and other composite materials. For carpeting, there is Carpet Guard to seal in the outgassing chemicals. When you've done all that you can to clean up your indoor environment, the Friedrich Air Filter will help scrub the rest away.

How about a solar-powered pool ionizer that can eliminate at least 80% of the chlorine normally needed in a pool, or an ozone generator for pool or spa? You'll also find a smoke detector that is not radioactive. In this section we've included a wide range of products to safeguard your family's health. And, by using these products, you will enjoy the satisfaction of knowing you are leaving a lighter footprint on our fragile planet.

— *Jeff Oldham*

NONTOXIC HOUSEHOLD PRODUCTS

Don't Mask Odors, Get Rid Of Them

Rid your home of unpleasant odors without toxic chemicals. Smells Begone is a nontoxic, odorless, and non-allergenic deodorant system that actually eliminates malodors. Instead of simply masking unpleasant odors with noxious scented sprays, this environmentally responsible product gets rid of them for good. Industrial strength formulation tackles heavy-duty odors like tobacco smoke or skunk spray, yet is safe for use in kitchen or nursery. Natural and nontoxic way to freshen any living environment. 12 oz. pump.

54-102 Smells Begone $11⁹⁵

Laundry Disks

This new laundry product from Japan eliminates detergents. We found it hard to believe at first, but tested them at home and they do work. Drop these three 2-1/2" disks in your washing machine with your clothes. Each disk is filled with electrically charged, ceramic chips. During the washing cycle when they're vigorously agitated, they break the water molecules up into ions, allowing the water to deeply penetrate fabrics and remove dirt and odors. Laundry disks are kindest to the environment since you can wash your clothes without pouring harsh detergent chemicals down the drain. The disks are good for over 500 washes, which is more than two years for most families. Recommended for everyday washing in warm water. (You may need to pre-wash or add a couple tablespoons of detergent for tough stains.)

54-169   Laundry Disks (3 disks) $49⁹⁵

Pure Soap

Our Pure Soap is gentle, natural soap that your whole family will enjoy using. Gentle enough for babies, it is made of the highest grade vegetable oils, pure cocoa butter oil, and a few drops of almond oil for a natural fragrance. It cleans and refreshes your skin without the drying alkali residue that many soaps leave. Each bar is firm, so that it will last a long time on your sink, yet it gives a rich, cleansing lather. You can even use it as a shaving cream.

90-567   Pure Soap (12 bars, 3 oz. ea.) $9⁹⁵
90-568   Pure Soap (24 bars, 3 oz. ea.) $16⁹⁵

Yellow Jacket Inn

Since so many of our customers live in rural situations, we thought it appropriate to offer some relief from the yellowjacket menace. These reusable traps capture incredibly large numbers of the stinging pests, enough to make outdoor eating a pleasure again. They use natural bait such as tuna (only the dolphin-safe variety!) or chicken (or chicken-scented tofu). They're ideal for taking camping or fishing as they weigh only 6 ounces and can be reused over and over. These traps are made by Seabright Laboratories, a company dedicated to nontoxic devices for dealing with pests.

54-201   Yellow Jacket Inn   1/$5⁹⁵ ea.
                                              2/$4⁹⁵ ea.

Ossengal Stick Spot Remover

Ossengal Stick is wonderful stuff. It gets rid of spots and stains better than anything we've ever tried and it's organic, odorless, free of industrial chemicals, and easy to use. It's made from the purified gall of oxen by the Dutch, and when you use it you can understand why they're famous for getting things clean. We asked a chemist why Ossengal works so well. He sent a page of words like "short chain organic acids, phospholipids, enzymes . . ." The gist of it seems to be that an ox needs tough-acting stuff to break down what it eats, and it acts the same way on spots and stains on fabric.

Ossengal is a pure white stick of soaplike substance, packaged in a small pocket-size pushup tube. To use it you just moisten the spot, rub it well with the stick, brush or knead it in, then rinse. Most stains just disappear.

54-101   Ossengal Stick Spot Remover   1/$6⁹⁵ ea.   2/$5⁹⁵ ea.   3/$4⁹⁵ ea.   12/$3⁹⁵ ea.

Citra-Solv Natural Citrus Solvent

Citra-Solv is one of our favorite products. It will handle nearly all your cleaning needs. It dissolves grease, oil, tar, ink, gum, blood, fresh paint, and stains, to name just a few. It replaces carcinogenic solvents, such as lacquer and paint thinner, toxic drain cleaners, and caustic oven and grill cleaners. You won't believe how well it cleans your oven! Use Citra-Solv in place of soap scum removers, which use bleaching agents. Citra-Solv is composed of natural citrus extracts derived from the peels and pulp of oranges. It is highly concentrated and can be used on almost any fiber or surface in the home or workplace, except plastic. Citra-Solv is 100% biodegradable, and the packaging for Citra-Solv is 100% recyclable.

54-140 Citra-Solv (pint) $9⁹⁵   54-141 Citra-Solv (quart) $15⁹⁵   54-135 Citra-Solv (gallon) $45⁹⁵

Citri-Glow Cleaning Concentrate

Throw out those poisonous household products: new Mia Rose Citri-Glow is a total home cleaner that really works. It's been formulated with the health of your family and the preservation of our natural environment in mind. Use it in your kitchen and bathroom with perfect safety: it's nontoxic, noncaustic, color safe, bleach-free, with no animal testing or products used. It's amazingly effective and economical: 1 quart makes up to 5 gallons. We don't usually go for hyperbole, but Citri-Glow is truly miraculous.

54-142   Citri-Glow (quart) $9⁹⁵
54-148   Citri-Glow (gallon) $19⁹⁵

Oasis Biocompatible All-Purpose Cleaner

A true superconcentrate, Oasis is excellent for hand dishwashing, hand soap, and general cleaning. Perfect for camping, it can also be used for hand laundry and diluted for body soap or shampoo. Like Oasis Laundry Detergent, it biodegrades entirely into plant nutrients with no plant or soil toxins.

54-155   Oasis All-Purpose Cleaner (quart) $7⁹⁵
54-156   Oasis All-Purpose Cleaner (gallon) $21⁹⁵

Oasis Biocompatible Laundry Detergent

Wash water containing Oasis Laundry Detergent can be used in the garden without harming plants. It can even produce better growth than plain water, because it biodegrades into a mix of essential plant nutrients. By running a hose from your washing machine to your garden, you can capture up to 1000 gallons a month headed right down the drain. Your clothes will come out clean (we tested it) and your fruit trees will grow stronger with Oasis! Concentrated, one gallon does 64 loads (1/8 to 1/4 cup per load). Contains no phosphates.

54-151   Laundry Detergent (quart) $11⁹⁵
54-152   Laundry Detergent (gallon) $25⁹⁵

Solar Mosquito Guard

Our Solar Mosquito Guard puts out a high-frequency audible wave that drives away most species of mosquitoes in a 12' radius. There's an on/off switch, so you don't have to activate it until they arrive. The battery will recharge in three hours of sunlight. We've received raves from many customers, even in mosquito-dense areas like Alaska and Hawaii.

90-419   Solar Mosquito Guard $8⁹⁵

3/$7⁹⁵ ea.

6/$6⁹⁵ ea.

Humane Mouse Trap

Often, the first interaction that a child has with "wild" animals is a mouse caught and deformed in a standard mouse trap. The Smart Mouse Trap teaches a child the idea of working in harmony with animals and the ecosystem. This is a humane and effective trap, a forgiving answer to unwanted visitors. Why kill? For a little more effort, you can take the mouse to the nearest brush or wooded area and release it to live out its tiny life. It is a joy to feel the shared compassion with a child watching a trapped mouse escape to freedom. This trap is simple to set, the bait is a soda cracker, it will catch mice that standard traps can't, and you can use it again and again. The trap is make of Kodar plastic with two stainless-steel springs. It measures 2" by 3" by 7".

54-202   Smart Mouse Trap $9⁹⁵

Bats Are The Ultimate House Guests

Mosquitoes don't have a chance around bats. Because they devour up to 600 night-flying insects in an hour and rarely bother humans, it makes sense to welcome bats to take up nearby residence. Made of Western Red Cedar sawmill trim, this handsome slatted bat conservatory is 30% larger than our original pine bat house, providing shelter for approximately 40 bats. Measures 24-1/2" x 16" x 5-1/4"; attaches easily to your home, barn, or nearby tree. Put out the welcome mat for our native North American bats, and these fascinating creatures will reward your hospitality year after year. No more nasty mosquito repellents!

54-185   Bat Conservatory $44⁹⁵

America's Neighborhood Bats

Merlin Tuttle. This book is a wonderful companion to your bat house. It provides a wealth of helpful information on bat behavior, biology, and habitat. It includes range maps, a source list, and easy-to-understand text. Spectacular color photos! 128 pages, paperback, 1988.

80-830   Bat Book $10⁰⁰

The Bat House Builder's Handbook

Also from Merlin Tuttle. Provides plans for building your own bat houses, in the most "user-friendly" fashion possible. These are proven plans that the bats simply love. Has detailed plans and drawings with all dimensions and materials listed. 36 pages, paperback, 1993.

80-856   Bat House Builder's Handbook $7⁰⁰

Least Toxic Home Pest Control

Dan Stein. Finally, an exterminator who believes in controlling critters by outsmarting them. Intelligently written and beautifully illustrated, this book explains life cycles and vulnerabilities of pests, and shows ways to exploit those weaknesses without overkilling, or polluting our homes. A masterpiece of environmental sensitivity. 87 pages, paperback, 1991.

80-826   Least Toxic Home Pest Control $9⁰⁰

Nontoxic, Natural, & Earthwise

Debra Lynn Dadd. What's healthful in the home can be healthful to the planet as well. This book lists more than 2000 products that are not only nontoxic but also otherwise commendable: recyclable or containing recycled material, organically grown, kind to animals, nonpolluting, packaged responsibly, socially conscious, energy efficient, biodegradable, or made from renewable resources. These products include air and water filters, cleaning products, energy-saving appliances, baby clothes, pest controls, and gardening supplies. 360 pages, paperback, 1990.

80-817   Nontoxic, Natural, & Earthwise $13⁰⁰

The Nontoxic Home & Office

Debra Lynn Dadd. We spend 90 percent of our adult lives in the home or office. In updating her 1986 classic, *The Nontoxic Home,* Debra Lynn Dadd has added information about detoxifying the office environment. The book provides an A to Z compendium of healthful alternatives to the hazardous products that surround us — food, bedding, housecleaning and personal care products, to items and equipment used in the office; and it also tells how to recognize pollution in your home or office. 212 pages, paperback, 1992.

82-172   The Nontoxic Home & Office $11⁰⁰

DISASTER PREPAREDNESS

Stay Alive No Matter What

It's smart to plan for your next meal. No need to go hungry, even in the worst of times. The Ark III survival kit keeps you supplied with food and water anywhere disaster strikes. Light and compact, each kit includes a three-day supply of food rations for one person, six packets of emergency water, and one mylar emergency blanket. Stock up Stash one in your car. Buy one for your mother. Give them to your children to store in their school lockers. Bring it along on your next camping trip. This serious survival kit withstands extreme heat and cold, and is US Coast Guard approved.

03-803 Ark III Survival Kit $19⁹⁵

Our Real Goods Natural Disaster Kits Offer The Basic Tools For Survival

The Super Survival Kit includes the Ark III with enough food and water for one person for three days, a mylar emergency blanket plus a five-gallon Solar Shower, a Nuwick 44-hour Candle, and a handy hand-powered Dynalite flashlight.

Our Natural Disaster Defense Kit includes all the items in the Super Survival Kit plus the Dynamo radio. Stay in touch by listening to important news updates.

The Deluxe Natural Disaster Defense Kit includes the Ceramic Filter Pump, in addition to all the above. We hope you never need to use this compact survival kit. But the peace of mind alone makes it worth every penny.

63-469 Super Survival Kit $39⁹⁵
63-485 Natural Disaster Defense Kit $65⁹⁵
63-486 Deluxe Natural Disaster Defense Kit $89⁹⁵

Deluxe Natural Disaster Defense Kit

Natural Disaster Defense Kit

Portable Ceramic Filter Pump Provides Safe Drinking Water

If you are unsure of the purity of your drinking water following an earthquake or flood, use this advanced ceramic filter pump for complete peace of mind. Draws water from contaminated sources and purifies it in seconds. Peerless ceramic filter pump removes bacteria, giardia, cysts, and other harmful pathogens down to 0.9 absolute micron level. Filter traps dangrous microbiological organisms, preventing their passage through the thick ceramic wall. Simple pump action quickly siphons one ounce of water per stroke. Filter up to 500 gallons, depending on water conditions. Requires no chemicals. Lightweight, compact, and easy-to-clean. Perfect for camping too. Includes handy clip for canteen or water carrier.

46-233   Ceramic Filter Pump $29⁹⁵

Our Emergency Candle Will Cook Your Dinner

This candle will fit in your backpack, your car trunk, or on a shelf for emergencies. It's enclosed in a metal can with a lid so you can take it anywhere. Use it for lighting, heating, and cooking for up to 120 hours. The secret is its six long-burning movable wicks. Light just one for illumination, two or three for heating food. The specially formulated paraffin is nontoxic, FDA approved (food grade).

63-452 120-Hour Candle $9⁹⁵

CHAPTER 10

HOME AND MARKET GARDENING

IN A RECENT SURVEY of our customers, we found out that one of their top interests was gardening. It came as no great surprise to us that gardening interest ran high among people who live independently, so we have added this new section to our *Sourcebook.*

It is also just as natural that people who love gardening tend to support small, local farms, or are small farmers themselves. I would bet that a large percentage of the shoppers at community farmers' markets also grow some of their own food . . . just not enough to provide all their own needs. And, they *don't* support the continuation of our current unsustainable food system in which the food on America's plates has travelled an average of 2,500 miles to get there.

What can Real Goods offer the serious home gardener that is not already offered elsewhere in other catalogs or in local garden stores? How about Real Tools, along with some inspiration? Are you as turned off as I am by "garden tinsel," as Gene Logsdon (see page 488 for his article on small farming) once called it? You might also call it "garden stuff". . . endless varieties of gardening knick-knacks and outdoor paddiwacks that have nothing to do with *real* gardening and everything to do with *play* gardening.

I mean, what do you really need to garden? A few trusty garden tools and a few feet of open ground. You're set. Dig it!

We went looking for the best garden tools we could find for Real Goods gardeners and cottage farmers. You might be able to find similar tools elsewhere, if you look hard enough and can wade through the clutter of garden tinsel surrounding it. But you won't find better, sturdier tools anywhere. As with all our products, we stand behind them 100%.

For some good, up-to-date information and inspiration, we asked a couple of "dirt-under-the-nails, callouses-on-the-hands," real good practioners to fill us in on the latest garden and small farm information. Robert Kourik and Gene Logsdon don't sit around their computer screens trying to drum up another story. They are out and about, practicing their crafts, and are able and willing to share their hard-earned wisdom. We welcome them to the Real Goods family.

— Dave Smith

LETTUCE CONSIDER THE STATE OF THE GARDEN

As I sat writing a rough proposal for a book on edible landscaping in 1982, I gazed across a well-clipped Long Island estate lawn to a recently abandoned tennis court. After just a few seasons of neglect, every fissure in the asphalt had been filled with grasses and a wealth of herbaceous weeds. All the human effort we put into culti-vated gardens is dwarfed by even the casual momentum of nature. Left untended, even the most environmentally kosher garden soon returns to nature's bosom. This is because within this steady, powerful momentum are the seeds of nature's renewal — which are more powerful than any human garden.

Bearing this in mind, what follows is a personal discussion of various styles of "natural" gardening and where I think each style is headed.

Permaculture Piety?

In 1978, I read *Permaculture One,* by Bill Mollison and David Holmgren. The term *permaculture* was coined as a compound word, linking the concepts of "per-manent" and "agriculture." A good permaculture is supposed to be a food-produc-ing ecosystem (garden) that is humanly designed, requires little work to sustain, mimics the diversity and complexity of a forest (or other natural system), is heavily based upon perennial food plants, and is self-perpetuating and permanent. With Bill Mollison's first U.S. lecture in 1980, sponsored by the Farallones Institute (where I was then directing the Edible Landscape Program), interest in permaculture took off like lamb's-quarters on a heap of moist horse manure.

In the late 1970s, I was very excited about permaculture — especially its attempt to develop integrated, sustainable food gardens. Gradually, though, my enthusiasm waned. Like most of the people I've watched cycle through the permaculture "experience" over the past 16 years, I found the details to be either lacking or counterproductive.

One of the big draws of permaculture, especially to well-educated nongardeners, is the lure of less- or no-work gardening, bountiful yields, and the soft fuzzy glow of knowing that the garden will continue to live on without you. Yet these same "advantages" often prove to be the biggest letdown for many people.

Another disappointment comes when the young "permie" (permaculturist) realizes not too much can be grown in a forest. In reality, forests, whether in the tropical or temperate zones, are not the place where most of the foods we like to eat come from. Forests are a natural result of the evolution of grass and shrub lands. The vegetables and fruits we crave — and most of the flowers, too — come from meadow and forest border environments. In most American states it is

> *To forget how to dig the earth and to tend the soil is to forget ourselves.*
>
> *— Mahatma Gandhi*

necessary to take away some of the forest area in order to create an artificial and ecologically degraded environment for the sake of our favorite foods. Luckily, early white settlers did most of the damage, so today we can pretend that it didn't really happen and that the resulting landscape isn't unnatural. In actuality, though, most gardeners must still hold back the ecologic momentum of nature in order to raise food. For as soon as one stops weeding, pruning, or mowing, the reclamation process begins. All human food gardens require continuous stewardship. Sometimes the labor is minimal, such as the yearly burning of grasses and pine seedlings beneath oak trees in the Yosemite valley by native peoples prior to the European invaders. Other times, the work is sweaty, filthy, character-building toil.

As I write today, the third wave of interest in permaculture has arrived. Even *Landscape Architecture* magazine, which has seldom been known to promote any environmentally responsive landscape design, is reviewing the topic. Mostly, I'm glad that permaculture is around to intrigue a new audience. Permaculture will continue to be a worthwhile intellectual hook, one that captivates and lures mainly cerebral types into the fuzzy logic of the garden. Permaculture is like a beneficial fungus in your brain, which attaches to your brain cells but eventually roots into the duff and soil. Once a person is gardening and getting really dirty, the dictates of the permacultural religion fall away like layers of a molting caterpillar.

A GARDEN DEFINED

Let me tell you what a garden is.

First, you work really hard on it for years, fighting cold and insects and storms, drought and cats who use the new lettuce bed as a litter box. You worry a lot and sometimes watch helplessly as small plants die. You plant things and then pull them out because they didn't look the way you expected. You go out when it's raining to knock down basins and you pull weeds when you're too tired to even take off your shoes.

Then one day, you're standing there eating a warm, slightly dusty apple off the tree and you find yourself crying because you've suddenly realized that this lively place is part of you, like your face or the way you laugh. You're crying and you don't even want to stop, because you've been transported unexpectedly to a special part of your spirit where you've never been before, and it's like a wonderful dream, like waking up in a warm, safe place where you knew you always wanted to be and knowing you can stay forever if you want to.

That's what a garden is.

— *Owen E. Dell*

Reprinted courtesy of the author and Green Prints *magazine. Used by permission.*

Can You Double-Dig It?

Perhaps the most widely accepted "new" idea in gardening in the past 20 years has been double-digging. As an interpretation of Alan Chadwick's work at the University of California Santa Cruz campus in the 1970s and 1980s, John Jeavons was successful in popularizing raised-bed gardening (see *How To Grow More Vegetables . . .* on page 502). Actually, Jeavons meant to promote a very astute program of gardening called the biodynamic French Intensive (BFI) method. The BFI approach is an integrated set of techniques, tools, plants, and environmentally sensitive management. Alas, the popularization of the method has led to some very diluted and confused concepts.

The BFI method uses a variety of techniques to deeply texturize the soil so that more plants can be grown closer together, but it does not really intend to cultivate in the sense of "turning over" the soil. The true BFI method of cultivation very precisely loosens soil without inverting any soil layers. This allows the beneficial bacteria to thrive in the zone of soil for which they spent thousands of years evolving and adapting. The oversimplistic interpretations of this method have led many people to assume that the more they churn and the deeper they dig, the better their gardens will be. This is not necessarily the case. In fact, excessive cultivation is often detrimental. Any digging, in the sense of churning, kills microbes and slows the absorption of nutrients.

In the process of texturizing the soil, the soil's surface is raised because of the incorporation of some air and organic matter. The raised-bed look to the cultivated area is thus simply a *by-product* of the soil-improving techniques — not a goal in and of itself. Conversely, heaping a bunch of dirt to make the shape of a raised bed may not produce any desirable results. Sometimes, it's true, the mere heaping or boxing of a good topsoil will provide enough extra drainage to promote a healthier and more productive crop. But, while a heaped-up raised bed may look like a BFI garden, it's only a distant ecological cousin.

Often, raised beds do have a net gain in productivity over old-fashioned rototilled dirt gardens. So I'm happy to see even the bastardized versions of raised and double-dug beds proliferate around the country. I often use boxed, raised vegetable beds in edible landscape designs for their tidiness, enhanced drainage, convenience, and permanence. But, I do not pretend that I've duplicated the delicate and not-so-subtle ecological dynamics of a BFI garden.

The simplified forms of raised beds and double-digging will continue to spread, especially as busy schedules dictate more efficient gardens. Hopefully, more gardeners will explore the roots of these superficial methods and discover the challenging nuances of the BFI approach. Their gardens will only improve.

Edible Landscaping

Many people still identify me as one of the "founders" of the "edible landscaping" movement. The generic form of edible landscaping is meant to reintroduce food plants back into the yard around the home in an aesthetically pleasing fashion. I have never suggested that *every* plant in the yard has to be digestible to have a true

edible landscape. The idea is to blend some utilitarian plants into a visually pleasing design. Edible landscaping doesn't really dictate organic versus chemical gardening. However, my book on the topic, *Designing and Maintaining Your Edible Landscape Naturally* (see page 499), provides a thorough, nondogmatic, and scientific basis for edible landscapes, while offering a smorgasbord of organic techniques for their care.

To some, the idea of good-looking landscapes with a high functionality or productivity was pretentious — maybe even elitist. The rush of enthusiasm for edible landscapes, mostly during the 1980s, was fueled by the realization that long, straight rows of ratty corn don't have to be the order of the day. Previously, the landscape architecture department of a state university had never considered talking to the agriculture or pomology departments. Now, these various ivory-tower "divisions" more routinely mingle.

Lately, people have been asking me, "Whatever happened to edible landscaping; is it dead?" While interest has dwindled in California compared to the early 1980s, the seed of the idea has multiplied and disseminated like dandelions — (which are themselves edible). While some believe that eternal popularity is the true mark of success, I take an altogether different view. Like the successional growth of grasses and lupines cloaking a recent landslide and yielding to the choking shade of woody shrubs some decades later, all gardening styles and methods should go through natural deaths, rebirths, and renewals. Edible landscaping is no exception.

Today edible landscapes have begun to fully integrate into the web of suburban life — as they were meant to. My favorite example of the success of edible landscaping comes from one of those women's magazines found clustered around the supermarket cashier like leering buzzards. It is a cartoon. In the background, three men are on all fours "grazing" on the lawn. One guy is picking from a hedge. Another is on a ladder in a fruit tree. One woman is leaning over the fence to tell her neighbor, "It's one of those new edible landscapes . . . saves me hours in the kitchen."

Mainstream America is making fun of edible landscaping — we have arrived! So many of my friends from the sustainable organic alternative-energy nonprofit groups of the 1980s actually tried *not* to have their work introduced into "those silly suburbs." I always pleaded the opposite. Newsstand satire of edible landscaping is the mark of market penetration. This means the foot is finally in the door.

I'm not worried if the phrase "edible landscape" should disappear completely from the lexicon like a head of lettuce dissolving into a worm bin. While the spread of edible landscaping now seems slow and plodding, it merely represents a period of completely healthy and natural assimilation by our culture.

In fact, the gradual spread of edible landscaping is much preferred over the exotic fads some have fostered in an effort to prolong media exposure. The culinary gyrations found with some edible flower recipes and snooty cuisine are more worthy of laughter than serious eating. While I do appreciate an elegant "gastronomical experience," I don't think of it as anything more than a wonderful, hedonistic extravagance — nothing that will shift the culture's center of gravity. Real people eat real food. Good barbecue will probably always triumph in numbers (pounds or cholesterol) over carved mushroom caps and fried squash blossoms stuffed with weird cheeses.

Ruth Stout's "No-Work Gardening" Legacy

Ruth Stout died 14 years ago. But her incredibly buoyant, gregarious spirit and large-handed, sturdy body — like a Julia Child of the garden — is easily remembered by all who met her or saw the few videos about her. She was, and still is, *the* lady of the no-till, toss-the-straw-and-sit-back garden. Few people realize that she had been gardening in the conventional spade-over-the-dirt manner for decades before she was inspired to retire her shovel. Even fewer realize that in her younger days she was an ardent prohibitionist, but later in life enjoyed her glass of wine and loved gardening in the nude! Yet the results of her method were still quite fantastic. Ruth was really on to something. And her legacy lives on in the occasional article about no-till vegetable gardening.

Most gardeners are attached to digging from the purely arbitrary assumption that you're not "working with the earth" unless you shovel some of it around, yet the landslide is the only natural model for cultivation. And landslides are rather rare phenomena in most regions of the country (though, unhappily, not in California). Therefore, cultivation via tillage is actually a rather rude imposition on the natural position of soil.

More practically, the yields from a no-till garden can be as good as — or, in rare situations, better than — a single-dug garden (though I doubt if no-till yields could compete with the productivity from an intensive double-dug bed). Properly tended, the no-till garden will require considerably less labor and none of the strain of double-digging. With a bit of practice, some smarts, and a little more space, no-till gardens can be grown without hauling in any manures or fertilizers from outside the yard. A decent harvest. Less work. No costly fertilizers. You'd think this would take off like free beer at a baseball game or complimentary tofu at a vegan convention. I certainly expected more interest in and practice of no-till gardening techniques by now. But I suspect that the culture is still too attached to digging as a self-serving measure of gardening accomplishment, and mulched gardens look untidy to most people. It is mildly amusing — and, on some days, slightly depressing — that all the environmentalists so intent on "saving the planet" are so quick to grab a shovel. What they are really doing is degrading the soil's structure, sending some of the planet downstream into a lake or the ocean, and reveling in what is actually quite an abnormal act.

My fondest wish is that the coming years will bring a renewal of this valuable gardening method. The soil, at least, will breathe a sigh of relief.

The Organic Movement

The organic gardening movement has begun to receive the respect it has deserved for so many years. What's more, the organic method is finally coming out of the dark closet of wishful, undocumented thinking. Good science, empirical observation, replicated field trials, and a level-headed knowledge of what we don't know are some of the elements balancing out the starry-eyed visionaries of the 1970s.

Some things happen faster than others. Ladybugs are now, after just a decade, culturally associated with organic gardening. Thousands of people buy ladybugs thinking they will actually control all the garden's pests, or at least the aphids. Alas,

purchased ladybugs are a very poor, if not completely ineffectual, way to control any pest. Even *The New Yorker* ran an extensive article (October 7, 1991) on the science behind the futility of buying ladybugs for any other purpose than fattening the wallets of the seller. Yet, only a year later, a prominent "leader" of the California Green Industry (a misleading title for an association of landscape contractors and conventional nurseries) tells a conference that she is so happy to sell ladybugs to every client, and, that they "make such great symbols of the environment." Symbols of deceptive advertising, perhaps. The only ladybugs worth buying are those guaranteed to have been defatted prior to sale. Still, these little ladies are programmed to wander off. A much more effective and behaved critter for consuming aphids is the green lacewing.

A common theme in the sustainable organic world is the premise that "diversity equals stability." I wish the continual references to a generic "diversity" in the popular press would end. While diversity within an ecosystem does allow for a complex interaction between all the elements (plants, animals, and people), it is not a panacea. Randomly chosen diversity or complexity doesn't necessarily provide any special benefits, and using tropical diversity as a model is simply foolish. While the tropics often have lots of vertically integrated plants, the temperate American landscape, with its hardwood forests, meadows, and prairies, is less vertically complex and is more adapted to the growth of annual plants than the tropics.

This is not to say that our gardens, and particularly our farms, should not be diverse. Most gardens have plenty of room for more well-chosen diversity, which may enrich the biological atmosphere and the environmental dynamics for a more self-modulating garden. Such gardens are perhaps more natural, but still a far cry from actual native ecosystems.

Our garden's diversity must be composed of the proper plants. For example, kudzu is theoretically a multipurpose plant with many edible parts and useful fibers. Ergo, planting kudzu would be another logical step in diversifying your garden. But if you do, you'd better leave town before your neighbors come after you, since the kudzu vines will entangle everything that's moving slower than 25 miles per hour! To use a less outrageous example, you might think twice about nurturing lamb's-quarters if you're growing lots of heirloom tomato plants, because this tasty edible "weed" also harbors verticillium wilt. Too much diversity, or the wrong kind, only promotes that universal dynamic called chaos.

Nonetheless, the organic movement will only continue to grow both literally and figuratively. So my main organic precepts have become: gardeners are not as important to nature as we think; nature eventually takes back everything; moderation furthers; and we should endeavor to stay humble and in our place, relative to the grander scheme of things. As George Carlin so eloquently observed in a recent HBO special: "Save the *planet*!? Are these f —ing people kidding me? There's nothing wrong with the planet that it can't fix. We still haven't learned how to care for one another — and we're gonna *save* the f—ing planet? So, take care of yourself. And take care of somebody else."

Certainly, one of the best ways to help folks is to garden together. It might not save the planet, but it sure is good for the soul.

— *Robert Kourik*

GETTIN' DOWN

Let's face it. The main problem with gardens is that you can't walk into the K-Mart, pick one out, and take it on home — all set up and ready to go. It seems that nature is very patient and works more slowly than you are used to working yourself. Have you ever watched tomatoes grow? 'Course not. Too slow. You want results, and you want them now!

DOWN AND DIRTY

You figure . . .
have a cappuccino
turn the soil over
throw in the fertilizer
drop in the seeds
    where the hell's the tomatoes?

the tomatoes are down the street
waiting for you at the Safeway
    if you want 'em fast . . .

those tomatoes
have been on a speed trip:
    grown fast
    harvested too early
    packaged quickly
    shipped expeditiously
    stocked in a hurry . . .
they've been
shot up with:
    herbicides
    pesticides
    fungicides . . .
        dicloran
        permethrin
        acephate
        lindane . . .

they've been
polluted and
chemotherapied . . .
they've been
hyped up and
strung out on
    factory farm dope . . .
but they ain't all
    they're cracked up to be . .

no, this is a whole different trip

slow down
slowin down

gardening is about you and dirt
    dirty knees and dirty fingernails
      growing great dirt to get
great plants

look around you . . .
    do you see anything that hasn't
    come out of the ground,
    out of the dirt?
they've dug it up
    melted it
        forged it
            into vehicles and
computers . . .
    it's like magic . . .
      everything comes out of the dirt
like you

gardening is about you and dirt
    gettin slow
    gettin down
      gettin dirty . . .

— *Dave Smith*

COMPOST PILES

The key to good organic gardening is good composting. There are several good reasons. The demand for most organic fertilizers is increasing, while the supply available to each person in the world is decreasing. Soon, few fertilizers may be available at reasonable prices. Yet compost can be produced cheaply in a sustained way by living soils. Around 96% of the total amount of nutrients needed for plant growth processes are obtained as plants use the sun's energy to work on elements already in the air and water. Soil and compost provide the rest.

Compost is created from various forms of decomposing plant life. This organic matter is a tiny fraction of the total material that makes up the soil, yet it is absolutely essential for soil life and fertility. Microscopic life-forms digest and decompose organic matter, and then die, resulting in humus, which helps both to retain and release nutrients to the plant roots. Plants move along the "cafeteria line," checking out what's on the menu, and pulling off whatever combination of food nutrients they choose from the organic smorgasbord in the soil. Good organic gardening practices rely on this natural, continual process to produce abundant and healthy plants.

You should be aware, however, that plants, kitchen scraps, and other green matter used for composting but grown in deficient soil can result in deficient compost. For example, if the soil your kitchen food is grown in lacks certain trace minerals, your kitchen scraps will not be able to add these missing nutrients to your compost. So it is important that the soil you grow food in is balanced and fertile, and that food you have purchased that later becomes compost is also nutritionally balanced.

Soil with good humus virtually pulsates in an orgy of eating, drinking, hot sex, birth, and death. There are protozoa, the smallest forms of animal life, and eel-like nematodes. There are yeasts, fungi, algae, bacteria, and actinomycetes. There can be more individual microbes in a handful of compost than the entire human population of the world.

Procreation is so furious that a single microbe reaching maturity and dividing within less than half an hour, can, in the course of a single day, grow into 300 million more individuals, and in another day amount to more than the total number of human beings who have ever lived. Microbes are the kitchen cooks of the soil, making recipes, mixing ingredients, and feeding the plants. They excrete a kind of soil glue, called *humic acids* or *colloids,* as they work, which helps to build the soil structure. Microbes also store up soil food in their tiny bodies, releasing it to the plants slowly as they die and decompose. There is something awesome in the harmony and cooperation that the life forces bring to bear in creating living soil.

Compost helps build strong soil in nine different ways

- It improves structure by loosening tight clay and clods, and by binding together loose, sandy soil.
- It retains moisture by holding six times its own weight in water.
- It allows roots to breathe by keeping the soil loose.

- It stores needed nitrogen.
- It fertilizes with phosphorus, potassium, magnesium, sulfur, and trace elements needed for plant and animal health.
- It neutralizes soil toxins.
- It releases nutrients for plants.
- It provides food for microbial life to fight soil pests.
- It recycles materials.

There are many ways to make compost. Most methods try to make compost as quickly as possible. To do that you must skillfully bring the internal temperature of the compost pile to a certain degree of heat, so you know that the feverish activity described above is really working to break down the materials fast. You also have to turn the pile over periodically so that everything gets all mixed together and decomposes completely, and you have to monitor the pile to make sure that it stays at a certain temperature and doesn't get too hot, or lose heat too quickly.

Such bother!

What's all this hustle and bustle and attentiveness and sweat and strain? I thought gardening was supposed to be a slow, gentle, relaxed, cultural experience. Too much fuss . . . too much attention to detail! I have enough of that at work!

When compared to the anal-retentive, micromanaged "quick compost" technique, "cold" compost piles are absurdly simple. They are much easier to maintain, and provide more organic matter than "hot" piles, but they do take longer to produce. It's so easy to build and maintain compost the Real Goods way, it's a no-brainer. That's why we call them Gomer Piles (with sincere apologies to Jim Nabors).

BASIC GOMER PILE RECIPE

We learned this method from Steve Rioch, who runs a small soil testing service (see page 492) and teaches biointensive gardening at Ohio University.

We put together compost in thirds (by weight):

- one-third **dry vegetation**, such as dry leaves, dry grass, stems, and straw
- one-third **green vegetation**, such as kitchen wastes (tea leaves, coffee grounds, citrus rinds — but no meat, fish, dairy, or sizable amounts of oily material), newly cut grass, and hedge trimmings
- one-third **soil**

Gomer Piles use no manures or other animal products . . . they are vegetarian!

Building a Gomer Pile in Six Easy Steps
The minimum size of your pile should be 3 cubic feet (3' x 3' x 3') in order for the composting process to begin (colder climates may require 4 cubic feet of material to get things cooking). *Maximum* size should be 5 feet high and 10 feet wide at the base. Any larger pile creates compaction problems because of its weight. The best times to build a compost pile are spring and autumn, when biological activity is at its highest.

1. Loosen the soil 1 foot deep.
2. Spread a layer of dry vegetation, 1 to 2 inches deep.
3. Add a layer of green vegetation, 1 to 2 inches deep.
4. Now add a soil layer, 1/3 inch deep.
5. Continue building up in this sequence and these proportions, or as materials become available.
6. Water lightly after adding each layer.

That's it. Your Gomer Pile will yield finished, ready-to-use compost in six to nine months.

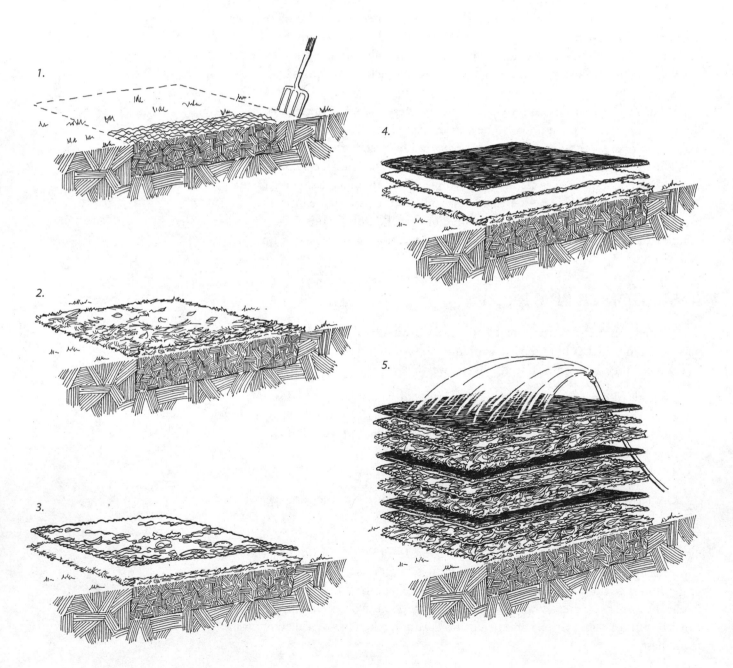

Hot Tips for Cold Piles

Here are a few helpful pointers to keep your Gomer Piles problem-free:

- You may need to collect kitchen scraps in a container for several days before you have enough to add. When you do add them to the pile, immediately throw the soil layer on top of the new layer to avoid flies and odor.
- Water your pile when necessary as you water the garden. The particles in the pile should glisten in the sunlight, glimmer in the moonlight.
- Cover your pile during the rainy season so it does not get saturated. You do not need a container for your compost pile, nor do you need to add manures, or packaged preparations that are supposed to accelerate decomposition. Just sprinkle a pound of compost from your last cured pile throughout the new one you are building.
- Compost is ready to use when it is dark, rich, smells good, and crumbles in your hand.
- Spread a half-inch layer of compost each year over your whole garden. Purchased organic compost may be used for the first year if you do not have a ready supply of your own.
- Don't fuss over your compost pile and don't worry about it . . . everything will turn out fine.

— Dave Smith

REAL GOOD MOUNDS

This is a variation on the theme of "cold-compost" Gomer Piles, but a bit more complicated. I call them Real Good Mounds.

Have you got lots of yard waste and a compost bin too small to digest it all? Afraid that all the woody waste will eat up all your garden's nitrogen? Worry no more. You can heap piles of garden "waste" *and* grow beautiful ornamentals and vegetables, mounding your way to gardening success. The mounding approach is like building a very large, slow-release or low-heat compost pile. Then, you plant directly into a safe, neutral soil "cap" of topsoil, aged compost, or potting soil mix located on top of the pile. There are no hard-and-fast rules about this novel, money- and labor-saving method — experiment to see which organic matter materials work best.

The mounding method has many names and is used in many parts of the world. In my book, *Designing and Maintaining Your Edible Landscape Naturally* (see page 499), I refer to mounding as "free berms, made from large helpings of brush and prunings." Mounds are also much like great heaps of "sheet compost." They are similar to the toss-and-tumble approach to soil development practiced by Ruth Stout. In Germany, the use of mounds for vegetable gardening has a long history. Bill Mollison of permaculture fame refers to the technique as "an instant garden" — a reference to how quickly everything can be thrown together, once you've collected all the raw ingredients. The method also resembles a common technique used throughout the tropics. Heaps of slash (forest prunings) are partially buried or heaped on slopes, covered with a small amount of soil and planted to food crops. This allows the shifting agriculturist to utilize the foliage of the forest as nutrients while developing no-cost terraces.

Here are a few guidelines: Stockpile wood chips (not sawdust, but large, chunky chips and shredded leaves from a tree chipper) from local tree-trimming services. The best wood chips are hardwoods — especially oak, which really seems to attract worms. Avoid any shredded material that might easily resprout, such as willow, acacia, alder, bay laurel, or various vines. Avoid highly allelopathic foliage — in other words, those plants whose foliage stunts or kills the growth of other plants that try to grow nearby. Examples of allelopathic plants include black walnut, sagebrush, and mesquite, which is allelopathic to tomato plants.

Because the chips are so high in carbon, layer or mix the chips with some manure. Start the mound with the chunkiest, largest-sized pieces to keep the bottom of the pile from settling too much. For all of the subsequent layers, use a blend of smaller chips, half inch to six inches long, and manure. Experiment with a ratio of one part manure to three or four parts chips. You can also use other types of high-nitrogen materials to help decompose the woody chips: fresh grass clippings; green manure crops, such as buckwheat, vetch, bell beans, and clover; wet kitchen garbage/scraps; and sludge. The more nitrogen you add, the faster the mound will decompose and the greater the nitrogen supply for the growing plants. Pile this mixture of high-carbon and high-nitrogen materials *at least* one-third higher than you want your final soil level. Depending upon the types of materials used, the mound may settle as much as 50%.

Make the soil cap at least four inches thick; the thicker, the better. Use a mixture of 50% well-rotted poultry bedding or compost and 50% native soil. This will ensure good drainage, a neutral soil temperature, balanced nutrition, and good initial growth. Once the mound starts rotting, the root hairs of the plants will follow right behind the decomposition, taking advantage of the newly available nutrients. The roots won't grow into areas that are too hot due to thermophilic decomposition (a fancy name for hot composting). The plants do get some benefit from the subtle amounts of heat that the mound may generate, much like the enhanced growth seen in bottom-heated greenhouse flats or inside a hot bed frame.

Until the plants get their roots into the untouched, native soil, a mound will often require more irrigation than conventional soil-based beds. Yet there are exceptions. Fall-planted Mediterranean plants on a large, 24-inch-high mound may need only two irrigations the entire second summer.

Eventually the whole mound settles to a lower level, the chips fully decompose, the shrubs root into the native soil and the hummock becomes a wonderful, sensuous feature in the landscape. Mounded plantings seem to be a heap of contradiction at first, but they'll soon become one of your preferred techniques for quick soil development, a cheap way to create "instant" soil drainage, and a simple method for creating an aesthetically pleasing, meandering landscape.

— *Robert Kourik*

IT'S SO EASY

It's so easy
driving to the market
wandering the aisles
not thinking about what you're buying
or who you're buying it from
        from where it was flown
        where or how it was grown
it was flying the friendly skies
and hitting the freeway
        weeks and thousands of miles ago

plant it
spray it
cut it
cool it
preserve it
        salt it
        sugar it
        package it
        can it
                bottle it
                label it
                store it
                ship it
                stock it
                bag it
                      zap it
                      serve it

out of sight
out of mind
        but not out of body

if it wasn't for shelf-life
it wouldn't have no life at all

— *Dave Smith*

ORGANIC PRODUCE IS A BARGAIN

The price of organically grown food reflects the true cost of raising food — what it costs for a farmer simply to make a living. This current U.S. farm crisis, with 35% to 40% of all family farms in foreclosure, is a symptom of our unrealistic agricultural economy. The payments farmers receive for their crops are not enough to cover the cost of farming. These payments have not changed much in the last 15 years, while costs have increased as much as four or five times. Oil, gasoline, agricultural chemicals, food, energy, and the real-estate prices as well as interest rates have all skyrocketed, while farmers still get the same range of prices for their products! Often, the grower sells the product at a loss.

When we buy lettuce, the grower may receive $2.50 to $3.00 for 24 heads. The labor to pick and pack a box of 24 heads is, say, $1.50, and the box itself costs $1.25. What the farmer gets paid may scarcely cover the picking and packing costs, while costs for planting, cultivation, fertilization, weed, pest, and disease control, machinery upkeep, fuel, seed, and irrigation are not covered at all. No wonder the farmer's reward is foreclosure.

The prices for organic food are fair in terms of giving the farmer a chance to make a living. Organic food practices, in other words, are not only sustainable agriculturally but also economically. The price we pay for organic food allows the farmer to put more back into the soil, rather than using the fastest and cheapest "fertilizers" (which do the opposite of making the land fertile). The organic matter used by organic growers brings fertility back to the soil, which had been neglected by conventional farmers. While the rise of pesticides and synthetic fertilizers has increased *tenfold* in the last 40 years, crop losses to insects have *doubled.* Organic methods, on the other hand, build up the soil, creating stronger, more disease-resistant plants.

That seems fair to me! What good does it do any of us if farming techniques turn our once-vital farmlands to empty desert? Given the price of soil abuse and the price of commercial produce anywhere else in the world, organic food in America is an *incredible bargain.*

When we request and buy organic produce, we are lending our support to the replenishing of the lands that yield our food. Why would anyone want those lands to be depleted? Organic produce is not only safe and pesticide-free, but also an investment in our future.

— Marilyn Diamond
from The American Vegetarian Cookbook
(Warner Books, 1990). Reprinted by permission.

ADRIEN AND GENE DUNCAN

Adrien Duncan is a 70-year-old sculptor, painter, and teacher who lives with her retired husband Gene on one-fifth of an acre in the San Francisco Bay area. Their living area covers most of the land they're on, and the rest of it is devoted to raising their own food . . . most of it in four raised beds that they double-dig every year.

When I was growing up, I was influenced by my father's sisters. They would sit around the table and talk about what they wanted in life, and the main thing they wanted was to have a nursery, or at least a greenhouse — someplace to grow things. They were marvelously gifted women who could raise all kinds of plants in the harsh environment of New York State.

Gene and I wanted to purchase a half-acre so we could raise enough food for ourselves. I complained to my Japanese gardener that I wanted to grow food, but that we did not have enough land. He told me that we had more than enough land . . . in fact, if we really got into it, we had enough land to feed the whole street. He said that on our one-fifth acre we had plenty of room, and it proved to be true. We took some classes from John Jeavons [see his book on page 502], and never saw anything so successful in our lives.

We've lived here about 35 years. I was interested in having fresh fruit year-round, so I have cultivated fruits that will follow one another. Starting with boysenberries and strawberries, we progress through cherries, raspberries, apricots, Bartlett pears, d'Anjou pears, four kinds of apples (grafted on one root stock), figs, kiwi, oranges, and grapefruit. With careful storage the apples usually last from July through December and the grapefruit are good from December through May. We give our surplus fruit and vegetables (and there actually are some, even after canning, preserving, and freezing) to kitchens for the homeless.

We have double-dug the beds every year with compost since they were established 12 years ago. The ground was like concrete when we started, with gravel fill and dense clay underneath, but the double-digging and composting have turned our yard into a beautiful, bountiful, food-producing area. Over the years, the beds have become more and more productive. Compared to the conventional planting in rows, we get three times the yield in half the space.

Because of the excellent climate here, some beds produce three crops per year, including "winter" crops such as brassicas, peas, and lettuces. Everything is rotated and grown in succession to be ready for our table. During planting or harvesting time, we may spend about two hours a day in the garden, but the rest of the time weeding, watering, and doing other chores only takes about 20 or 30 minutes a day.

We like to experiment. For example, one year we planted a bed of Bolivian corn. It grew 16 feet high, requiring a ten-foot ladder to reach the mature ears. They tasted delicious.

We grow and use most of our own seeds, rather than buying seeds and seedlings each year. The only foods we buy are bread, meat, dairy products, and wine.

We just don't buy much food anymore. We are constantly amazed at the volume of food that our own little yard can produce.

Almost everything we have on our table comes out of our backyard. We'll have two vegetables with our potatoes, and maybe a little bit of purchased meat. The herbs used in the kitchen are ours. Dessert will often be our fruit, either fresh or preserved. There isn't a whole lot left that needs to be purchased: maybe the napkins, a bottle of wine . . . and we haven't grown coffee beans yet.

Our neighbors think we're nuts. We tell them we have a garden and raise most of our own food, and they say, "Oh, sure, you raise your own food." Then we show them the wall-to-wall produce stored in our garage, and they become believers.

— *Interviewed by Dave Smith*

ANNA RANSOME

Anna runs her two-acre Foggy Bottom Farm in Northern California.

I got into farming because I love food and every aspect of it. I was a chef for 11 years. Instead of going to school and learning how to farm, I took a course in farming at the local junior college and then figured, instead of taking classes for two years, I would learn by actually doing it for two years. I really hadn't gardened much previously because I was so busy cooking to make a living. My main desire was to farm so I could supply good food to restaurants.

The first things I learned when we moved here four years ago were how extreme the temperature changes were, getting very cold at night and very hot during the day, and how poor the soil was. The soil was very sandy, like being on a beach, and had practically no organic matter in it. So we had to work on the soil and find out what grew well here. The temperature extremes really stressed out plants, but the plants that made it had very strong cell structures. We do best with basil and leeks.

I love doing this. I'm happier doing this than I've ever been doing anything in my life. The only thing I miss about working in a restaurant is that I love working cooperatively with a large group of people. But I'm really happy doing this. I love working in the soil and I love growing things. It's a real inspiration to see things grow and to go through the cycles of nature and feel you're in touch with the seasonal aspects of life and your own life, too. The changes in your own life don't feel so intimidating. You don't look forwards or backwards so much . . . you're more part of daily life as it is and feel in place. Part of that is working hard. For me, it's really beneficial to work and produce something that you feel is contributing. Food is an important need, so it feels like a real enterprise that you're dealing with. Cooking felt that way, too.

There seems to be more respect coming from the urban restaurants than the country ones around here. We've had chefs who are very knowledgeable come out here from the big cities touring the farms. Restaurants really appreciate quality and good farming — what it takes to produce good food and how difficult it is to predict what you're going to have in the next two or three weeks. Chefs appreciate having relationships with farmers and have become much more in tune with the seasons and don't expect to have fresh tomatoes off the local farms 12 months a year. You appreciate things more when you become part of the seasonal cycles. When you've waited for months and then taste that first tomato, or that first raspberry of the season, it's worth the wait. You are really experiencing something. Consumers have grown so accustomed to having any food they want year-round that they are no longer in touch with what grows in what season. It's kind of scary that people are so detached from how food is grown.

What's been happening in California these past few years is creating a tradition of food. It took centuries to establish that in Europe. It's a really big move. I'm not interested in the mass-produced market. I'm more interested in that European mindset, where you have a relationship with a community that knows you and knows your food and expects certain things from you. When I go into a Safeway and see the vegetables displayed perfectly and looking so glossy, I'm not tempted. I know it's cardboard. At the farmer's market, you see all the imperfect little potatoes people have grown, and you just know they're going to taste good.

Everybody should be growing some of their own food, even if it's just a few plants. Only by doing that can a person begin to understand how food is really produced, and appreciate how good food can taste. We've been fooled for too long by fast foods and the quick fix.

One thing that's wonderful about farming that really appeals to me is that the farm is like a little system . . . everything gets used. The most bizarre things we have out in our shed will be used eventually around the farm. We're always looking for things at garage sales we can use to make our job easier. All the elements on the farm are moving around all the time, getting recycled or composted, or being used for something other than their original purpose. You feel like you're more responsible for what you're doing and for the impact you're having on the planet when you're directly working with nature.

— interviewed by Dave Smith

MICHAEL MALTAS

Fetzer Winery in Mendocino County, California, runs the Valley Oaks Cooking School with famed executive chef, John Ash. It holds workshops on cooking for professionals and the general public using food and herbs from its garden, and wine from its vineyards. In 1985, Michael Maltas, a former "Midwestern Gardener of the Year" named by Organic Gardening *magazine, was given the "crummiest" piece of land out of the 3,000 acres Fetzer now owns, for a four and one-half–acre garden. His biointensive garden has grown so spectacularly out of the compacted clay soil swamp it was started on, with over a thousand varieties of fruits, vegetables, herbs, and flowers, that the winery is now converting all of their vineyards toward organic, leading the way among large wineries in a general movement to organic viticulture. Michael's raised beds are fed by compost made from the grape pomace discarded by the winery. Michael, who has now left Fetzer for other challenges, has also had a major impact on agriculture in general in Mendocino County and elsewhere.*

I was born and raised in Zimbabwe in central Africa, which was then known as Rhodesia. I have no gardening or farming background, although my mother had an extensive flower garden that must have influenced me, because I used to watch our black gardeners a lot and had great respect for them. I had a lot of skin problems when I was growing up . . . had these lumps on my skin . . . and one day someone told me that I was eating too much fruit. Although I didn't understand it at the time, what she meant was that I was eating too much fruit that had a lot of pesticides on it.

My formal education led to a Bachelor of Science degree from the University of Cape Town. It was only after I came to America and got involved in communes that I started growing food, making compost, and realizing that I really liked doing that. I would be eating organic food at the communes and my skin problems would go away. But if I went into the city for a couple of weeks and ate food out of the grocery store, my skin problems came back. When the connection finally dawned on me, I decided I wanted to raise organic food seriously.

When my visa ran out, I went back to England and apprenticed myself on a 600-acre mixed organic farm. I drove the tractor, looked after the cows, ran the market garden, and worked in the bakery. Then I went to Emerson College for two terms and took some biodynamic farm training. They gave me a good background in machinery, activities in tillage, and taught me to ask questions about what was really going on in the soil. That gave me a much better sense of ecology as a whole related to the farm as a foundation.

When I came back to America, I renovated, then ran, a Waldorf school garden in Sacramento for three years, got mugged by people stealing melons, moved to Missouri, and ran a small nursery/market garden and biodynamic dairy farm that made cheese and shipped it all over the U.S. We had this beautiful demonstration biodynamic mini-farm in Missouri, but no one there seemed to care.

So in 1986 I decided to come out to California and eventually to Fetzer. After a very tough first year, having to learn about working with a corporation (where they want "everything yesterday," versus the "let's take some time and build it" organic philosophy), we eventually worked into a mutual respect. The garden is now in its eighth year.

The concept here initially started very loosely. Mendocino County has a strong tradition of organic farms . . . Alan Chadwick, John Jeavons, the farmers over in Anderson Valley. We wanted an organic situation, but Fetzer didn't know if they wanted to grow food or flowers or herbs or what. They only knew that they wanted it to look nice and be clean. What we finally thrashed out was a stand for selling wine as a food, and selling food with wine, because Fetzer was trying to say that you don't just buy wine, put it in a brown bag and drink it on a street corner. Drink it with a meal. So we would do all their variety trials and they would say, "Wow, we never knew there were 25 varieties of melons."

And I would tell them that we just had a small sample of the 250 varieties that were available. The whole public relations and education aspect started to grow, because we were doing things that people hadn't seen before. The food and wine center started to take shape as more people started to get interested in organics and different varieties, and as the California cuisine thing started happening. Then a pavilion was built as a demonstration kitchen to bring trade people and representatives to Fetzer for public relations.

Currently, we are matching up wine with herbs, and experimenting with fresh herbs in cooking. John Ash is our Executive Chef, and he develops recipes from the varieties out of the garden and serves gourmet meals. We have trained people from the Sheraton Hotels who want to upgrade their wine cuisine program. They come here to learn about wine and food . . . how to use fresh produce and how to order in season. Then they make a meal and are judged on it. We show them how to grow food from the garden, or how to make window boxes for the top of their hotels, or patio gardens to grow garnishes in.

We are also a resource for farmers in this area in terms of eco-agriculture and organics. It's a wine-subsidized garden, but the education outreach is much broader. We're part of the Seed Savers Exchange, helping to preserve heirloom seed varieties, and we're an information resource for a lot of people, both in this area and nationally.

The marriage I've always been interested in is between the best of farming and the best of gardening. Farmers often look at gardeners and don't see the applicability . . . gardeners look at farmers and don't see anything they can learn. Both use techniques and think about things that are useful to each other. But very seldom do they mix. So, even though we do use machines here, especially at the beginning, we are aware of the problems they create and use them as sensitively and sparingly as possible. There are places where the machines are useful, like deeply digging the beds to start things off. We rip the soil as deep as we can and follow it with lime, because subsoils are usually acid. To do this by hand on this four and one-half–acre site, we would have to hire 15 people and double-dig for months on end. But after the initial plow, we don't use much machinery, because we've found that it's actually faster without it. A small team of people, with good-quality tools, can do the job better and faster than firing up the tractor and having to deal with the mechanics, fuel, and fumes. Rototilling does not leave the soil in very good shape like hand-digging does. Trained people with hand tools can work very fast, if you've set up a system that is well-organized and efficient.

The soil here was extremely poor when we started. The garden is doing so well now, it scares me to think of what can be done on a site with really good soil.

Our beds are 42 inches wide, which is the best width for me to reach into and not step on the bed. Because this was a swamp here initially, there were a lot of wet weeds. The key to what we did at the beginning, and the one thing that gardeners and farmers don't often do, is to clean the beds. We prepare a bed, making it completely ready to plant, and then water it and leave it. When the weeds come up, we go through and rake them under and sow our seeds or transplants. That reduces our weed population down to 15% or 20%. The amount of energy you save by having a clean soil condition versus doing all your weeding later is quite substantial. In just a few days, we've eliminated 90% of our work. We don't have to do that anymore because the soil is now essentially weed-free. Over time, the weeds eventually disappear. Raised beds give you good drainage, so weeds don't do as well, and on the beds you can really see them and get them all out. The few weeds we get now come in through our compost. If we rush our compost and use it after four or five months instead of waiting a year, then we'll get some weeds. But other than that, the weeds are gone.

We're trying to grow the best-tasting produce possible for this particular bioregion. All gardens are different, and we're experimenting to find what works best in this particular one. The extension agent comes here and tells me about how many farmers around here seem to like to struggle, compared to what we're doing here. There are other farms now beginning to adopt these techniques.

When I started this project, the ranch manager on this farm, who was a traditional, well-trained, chemical-agriculture farmer, came up to me and said, "You know, this is going to get eaten up within a year. I've been here for a while, and I know." So I said, "Well, let's just wait and see." After a couple of years, with him wandering around in the garden in the evening when he didn't think I was here, he came up and said, "I didn't believe this would work and it has." He quit his job here on the ranch and went and started his own organic farm.

— Interviewed by Dave Smith

GROWING YOUR OWN FOOD CAN BE VERY PROFITABLE

A commercial farmer is lucky to clear $800 an acre on high-value fruits and vegetables, or $100 an acre from livestock, or $25 an acre from grain, but a backyard food producer can make $2,500 an acre, and do it in spare time to boot. And if you think about it deeply enough, there is a real possibility of making a whole lot more than that from a garden, like maybe creating a new career or a more meaningful life for yourself.

One reason for the greater per-acre profitability of small gardens is that backyarders can figure the value of the food they produce at what they would otherwise pay for it at supermarket or restaurant prices, not the less-than-wholesale price that the commercial farmer receives. Secondly, the backyarder does not have to pay out-of-pocket cash for labor to produce the food. Third, the backyard land is also not an out-of-pocket cost, since it would be there regardless, whether it is used for badminton or berries.

Trying to determine how much a typical American family spends on food these days is extremely difficult because of the many variables involved. The average family grocery food bill seems to float around $100 a week. But some grocery food bills may actually be declining, as more people eat out. If you grow, say, three-fourths of your own food, you save nearly three-fourths of your grocery bill, and that can be a significant chunk of money. But if you are so un-American as to eat your own food rather than dine at McDonald's frequently — let alone an expensive restaurant — the value of your home food jumps right out of the frying pan. What you get for ten bucks at a fast-food restaurant, you could produce on a mini-farm for a dollar. The dab of meat and vegetables that you pay $50 for at a fine French restaurant, you could raise yourself for 50 cents.

You (hopefully) know already what your grocery/restaurant bill amounts to annually. I'm going to make a calculated guess that a family of two adults and two

children, a teenager and a grade schooler, spends $4,000 a year for three-fourths of all their purchased food. I use the three-fourths calculation as the amount you could reasonably produce yourself. I have found no way to grow Tanqueray gin martinis, for example.

You can raise three-fourths of a family's supply of fruits and vegetables on one-fifth of an acre. Actually a good French Intensive gardener could do a lot better than that. Assuming that you spend $500 a year on these two items, that puts an acre's worth of garden at about $2,500, which is how I arrive at the figure in the opening paragraph.

If you have another two and four-fifths acres on your property, bringing your total to three acres, you can turn them into a mini-farm and raise most of the beef, pork, chicken, eggs, and milk for your family if the soil is fertile and you are skilled in the farming arts. Three acres would then produce the $4,000 that you are now spending for food that you could raise yourself. And I'm sure that for most families the savings would be even greater. Furthermore, I know from experience that an operation like this would produce a surplus of eggs, milk, tomatoes, apples, squash, and other foods that you could sell to neighbors.

There are, of course, expenses in kitchen processing and storage of these foods for winter use. A stove and refrigerator you will have anyway, but a freezer and canner will also be necessary, plus the gas or electricity to run them. Amortized over 20 to 30 years, these costs would amount to several hundred dollars a year. Also, if you get into livestock, you will have to figure the costs of sheds and fencing, a small riding tractor or walk-behind tractor and mower, a tiller, and other minor tools, all of which can be amortized over a lifetime. But hold those figures in your head while you consider the many other savings that do not appear on agricultural spreadsheets.

What would you be doing with those hours you spend gardening and farming? Golfing? About $15 an hour, not counting the cost of clubs, right? Boating? Even a cheap boat will pay the grocery bill for a year. Attending a professional football game? A ticket to watch the Cleveland Browns lose again costs $98, and don't forget the mileage to and from the stadium, the beer and hot dogs, and the ticket for speeding, which will cost considerably more than the game ticket. The point is that if you consider your "hobby" of garden-farming as recreation, it can save you one nice little compost heap of money. If your hobby otherwise would be rooting yourself to a bar stool, figure two nice heaps.

And speaking of compost, if you get into food production you can save the money you are now paying to have your leaves and grass clippings hauled away. These yard wastes become your principle source of mulch and fertilizer.

Nor is it facetious to suggest that gardening, as exercise and as tension-reliever, promotes health. Some doctors encourage, if not prescribe gardening for that reason. Health-care costs being as horrendous as they are, gardening might just be the best investment you make in a lifetime. What are ten extra years of quality life worth to you?

Moreover, the extra-fresh, high-quality, pesticide- and hormone-free food you can raise yourself is also a wise investment in health, and therefore money. If you have read *Modern Meat* by Orville Schell, you may conclude as I have, that raising my own beef, pork, and chicken is the single most effective dietary rule of health maintenance. In fact, if you don't know how your meat is raised, vegetarianism might be an even better investment.

What is it worth to have something productive for your children to do — the equivalent of the traditional farm chores so many of us remember doing? Yes, we weren't overjoyed with having to care for the farm animals before and after school, but it taught us responsibility and it taught us *caring*. The animals had to be fed and watered every day just like us. And many people discover that they actually *like* to do this kind of work. One of the sad facts of modern life is that, while children are given so many "experiences" today, one they rarely get is husbandry, and millions of people go through life vaguely discontent because they were born to be nurturing farmers and don't even know it.

But chores won't teach kids anything if the parents don't work right along side them and convince them that this work is important and meaningful to the family's and the community's security. It might be wise to let kids sell the surplus and keep some of the money. They might learn something more valuable than that $100,000 college education you think they need.

And that suggests the really big payoff in raising some or most of your own food. Your backyard mini-farm can be a college education in itself, preparing you for a successful and lucrative business of market gardening or specialty farm products, or, in fact, any kind of food-production enterprise. Prevailing sentiment in our society holds that successful gardening and farming require a weak mind and a strong back. This notion is terribly wrong, particularly when it is extended to *commercial* food production. To grow for market, even just in your own neighborhood, requires a mingling of art and science that will challenge a genius. In fact, most of the people society considers geniuses couldn't do it.

Trying to predict the future is almost always self-serving baloney, but because of the very fact that the future usually surprises us, it might be well to consider agriculture today. All the learned think-tanks and universities and other armchair experts believe that farms will continue to get larger, indefinitely, which of course is ludicrous because that would mean only one huge farm eventually, and the old Soviet Union has proved conclusively that won't work. In fact, I believe just the opposite will occur. There will always be a few very large farms, but the dynamic action of the future will be in an increase in small farms. That's what I hear now, out in the fields. Bob Birkenfeld, an extremely thoughtful Texas farmer of 2,500 acres and 2,500 stocker cattle just told me he believes he could make more net profit with a lot less stress on his home of 180 acres with family labor, and he is seriously considering doing just that. Ward Sinclair, who left a very successful career in journalism to become a full-time market gardener in south-central Pennsylvania, tells me there is not a doubt in his mind at all that small-scale farming is already on the comeback trail, and he proves it every day.

There is yet another way that gardening and small-scale farming can become a lot more lucrative than the computer inputs reveal. By learning how to raise your own food, you provide yourself with invaluable information that you can then teach to others. My gardening and farming has enabled me to make the greater part of my living by writing about new food production ideas. And no matter what happens to commercial farms, this opportunity will only get better. Most of us are as ignorant of the manual arts as we are of nuclear physics. Fifty years from now, if not sooner, there will be a great demand for people who can show the rest of society, including the nuclear physicists, how to keep from starving to death.

— Gene Logsdon

ATTENTION: SERIOUS GARDENERS AND COTTAGE FARMERS!

Building fertile and balanced soil is crucial to growing healthy food, and only regular soil testing can tell you how your soil building program is progressing. Our friend, Steve Rioch, of Albany, Ohio, runs Timberleaf Farm, a private agriculture research and educational facility. It's the best little soil test company around. Steve has consistently grown in excess of $10,000 worth of high-quality produce per year on less than one-fourth acre of ground in a four-month growing season without using chemical fertilizers and poisonous sprays. His program emphasizes the analysis of two general areas: the testing of soil elements and the close examination of cultivation practices.

Most soil testing is generally concerned only with the chemical elements available to plants in order to grow a crop. This accounts for only about 4% of what plants need for optimum growth. The other 96% of the plants' needs come from sunshine, soil organic matter, soil oxygen, and soil moisture. All of these, except sunshine, are manageable by cultivation practices. Only by combining the techniques of balancing soil elements with good cultivation practices can you reach maximum fertility, regardless of your soil type. The proof of this approach has been verified by experienced gardeners and farmers scattered across the country, who have celebrated the success of Steve's program with very high yields of quality produce.

Steve also teaches an intensive gardening course every summer at Ohio University. There is a one-week workshop and a one-week accredited course, and it is offered for both basic and advanced gardeners and small farmers . . . or those who want to become one. I attended the course recently and came away with an in-depth knowledge of soils and their chemistry, as well as having gained many hours of hands-on experience in composting, double-digging, growing seedlings, planting, and other skills needed by intensive gardeners and farmers. I highly recommend this course.

If you are interested in either the soil testing service, or the Ohio University Intensive Gardening Course, write to Steve Rioch at 5569 State St., Albany OH 45710. A 10% discount is available for *Sourcebook* readers on the multiple soil-testing program.

— *Dave Smith*

Earthworm Composter

The lowly earthworm eats our vegetable garbage and turns it into a rich, crumbly, life-giving compost. Quicker than traditional composting methods, vermicompost is enriched by worm castings. Earthworms consume their body weight in food daily and reproduce rapidly. The "worm tea" that accumulates at the bottom of the bin is a nutrient-loaded fertilizer. Our worm composter can be used both indoors or out in a shady spot that doesn't freeze, without objectionable odors. The composter is portable, with a lockable raccoon- and rodent-resistant lid and a drain-spout for drawing off worm tea. 27" x 17" x 12". Order a worm coupon, and we'll ship it with your composter. Mail it to our worm supplier and in about a week, and you'll receive 1,000-1,200 (1 pound) healthy, wiggly worms for your composter (or next fishing trip!)

| 54-307 | Worm Composter | $69⁹⁵ |
| 54-308 | Worm Card (Good for 1 lb. worms) | $24⁹⁵ |

Worms Eat My Garbage

Mary Appelhof. A clever, witty introduction to the fine art of vermiculture. Explains sizes and kinds of containers, bedding materials, maintenance, and more. 102 pages, paperback, 1982.

| 80-853 | Worms Eat My Garbage | $9⁰⁰ |

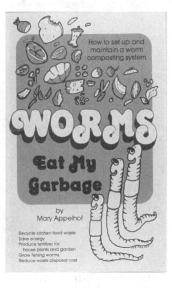

A Spicy Solution To Insect Invasions

Fiery jalapenos and habaneros may be your idea of gourmet treats, but when you use potent Hot Pepper Wax in your garden, you'll give bugs the hot foot. Mix the nontoxic, organic liquid concentrate with water to create a biodegradable spray that wards off pests for up to 30 days. Hot Pepper Wax may be used both in and out of doors, and won't impart flavor to treated fruits or vegetables. Made from capsaicin extract of hot peppers, refined food grade paraffin wax, kelp, eucalyptus oil, herbs, and citrus extracts. 16 oz. bottle.

54-117 Hot Pepper Wax $11⁹⁵
Not available for sale in CA

Sprout A Plant From The Sunday Comics

This unique wooden tool allows you to recycle old newspapers into perfect starter pots for young transplants and seedlings. Ideal for both kids and adults, it's fun and easy-to-use. Just cut newspapers into strips, roll, and press — no glue is required. In minutes you've created an earth-friendly pot that looks great on your windowsill or patio. And you've kept one more plastic pot from cluttering our already overburdened landfills. Providing a conversation-provoking new use for yesterday's sports section, the PotMaker is an inspired gift for your favorite gardening aficionado or creative conservationist. 2-1/8" diameter.

63-225 PotMaker $11⁹⁵

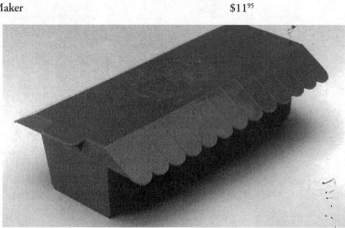

Fatal Attraction For Slugs And Snails

Beer is the safe alternative for killing destructive slugs and snails. Instead of using a toxic chemical bait, simply fill this ingenious slug trap with stale beer. Slimy garden pests can't resist bellying up for their very last sip of suds. Sturdy 6" plastic basin has a hinged roof for easy cleaning. Snap it closed to keep weather out and slugs in. Why endanger children, pets, and wildlife with toxins in the never-ending battle against slugs and snails? Pesticide-free, this effective solution is an organic gardener's dream come true. Set of four.

54-186 Slug & Snail Traps (set of four) $11⁹⁵

An Age-Old Way To Protect Your Garden

Since ancient times, garlic has been used as a natural, effective insect repellent for growing things. Safe, economical, and easy-to-use, a single pint of biodegradable Garlic Barrier liquid concentrate makes enough spray to cover an eighth of an acre. The garlic scent disappears a minute or so after spraying, and no garlic taste is imparted to sprayed fruits or vegetables. Use the Garlic Barrier to keep plants, trees, and vine crops pest-free, or spray the greenery around your property to repel mosquitoes for up to three months. Nontoxic. Comes in a recyclable bottle.

54-119 Garlic Barrier (pt.) $10⁹⁵
54-118 Garlic Barrier (qt.) $16⁹⁵

State-Of-The Art Soil Test Kit

To grow the healthiest plants and vegetables, your soil should have a neutral pH and contain generous amounts of essential nutrients. This comprehensive soil testing kit helps you create an ideal growing environment by using simple procedures to determine pH (30 tests) and concentrations of nitrogen, phosphorus, and potassium (15 tests each). The Garden Guide Manual and Soil Handbook help you interpret test results, and offer lime and fertilizer recommendations. The kit also includes diagrammed instructions and laminated color charts, as well as everything you need for testing: reagents, pouring spouts, graduated test tubes, and holder.

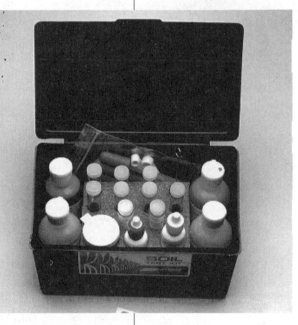

57-132 Soil Testing Kit $39⁹⁵

Plant Your Tomatoes Early

Superior to hot-caps and bell jars which concentrate heat to extremes, the Wall-O-Water® provides a moderating influence. This season-extending tool is so popular today that it hardly requires an introduction. But for those who have yet to discover the advantages to be had from this modern-day "cloche," imagine ripening tomatoes six weeks early or maturing peppers in a marginal climate. Plastic film is greenish, unlike picture, for better heat gathering and to prevent hot lens effect on plants.

63-295 Wall-O-Water (Set of 6) $16⁹⁵

Sturdy Folding Utility Cart Saves Space

This utility cart is lightweight and folds in a matter of seconds without any sacrifice of strength. Haul your firewood, garden soil, leaves, garbage cans... you'll find dozens of uses. Load this high-quality cart up to its 330-lb. rated capacity and it will still be stable and easy to maneuver. Made from marine-grade aluminum and zinc-plated steel for rust-resistant long life. It breaks down easily and uses less than 2 square feet of floor space. At only 36 lbs., you can hang it on your wall for storage or load it in your car for work at other locations. The 20-inch pneumatic knobby tires are mounted on reinforced nylon wheels for excellent balance, even on rough terrain. The bed measures 38" L x 23" W x 11-1/2" H for almost 5-1/2 cu. ft. of capacity. Opened, the cart measures 55" L x 31" W x 23" H. Folded, with handles retracted, it measures 41-1/2" L x 10-1/2" W x 23" H. It arrives fully assembled and ready for use.

•63-513 Folding Utility Cart $199⁰⁰

Use the Sun to Grow & Cut Your Lawn, Quietly!

Gasoline-powered mowers are one of the worst polluters. An hour of mowing equals the pollution load of several hundred miles driving your car. Not to mention the noise, vibration, disturbed neighbors, toxic fuels and oils, messy maintenance, and short life expectancy. Here's a better idea. Our solar mulching mower recharges from sunlight and is powered by an electric motor comparable to a 5-horsepower gasoline engine. It has a 21" cutting radius, single lever height adjustment, safety bar that stops the blade in less than one second, and no cord to tangle with. The durable, impact-resistant construction is waterproof and has a sealed battery with a life expectancy of seven years with virtually no maintenance (you might have to sharpen the blades occasionally). The LCD display on the back indicates battery state of charge and solar power input. A full charge takes about 10 hours of sunlight or 3 hours with the optional AC charger, and yields 1 to 1.5 hours of mowing time. Weight 80 lbs. Three years mfg. warranty. Made in the USA.

•63-650 Solar Mower $899⁰⁰
•63-651 AC Charger $99⁰⁰
Shipped From Manufacturer

The World's First Solar Robotic Lawnmower

"During spring time, I place my mower on the grass. It is light, totally silent. It climbs slopes, goes around obstacles, adapts itself to the grass conditions and keeps my lawn fit during the summer season, all by itself." Sounds like an Isaac Asimov fantasy, doesn't it?

This mower IS reality, a solar-powered robotic lawn mower, developed by Belgium's finest engineers and marketed in the United States by WeedEater. The Solar Robotic Mower utilizes an electronic brain programmed by fuzzy logic, which permits it to recognize different grass conditions and to adjust its cutting speed. When the grass is too wet, it will stop and restart when conditions are drier. It recognizes its boundaries either by touch, or with a perimeter wire placed either on or under the ground and connected to a remote solar generator (included). When it comes within 4" of the wire, it turns away. It can maintain an area up to 13,000 square feet. Cutting height is variable from 1.2" to 2.4". The body is constructed of solid polycarbonate on an aluminum frame. Power is generated from a 45-watt panel. Average consumption is 20 watts. The long-life DC brushless motor operates at high efficiency. The Robotic Mower can travel at a maximum speed of 20 inches per second and can climb up to a 20-degree slope. It works best on an open, sunny lawn. Weighing only 12-1/2 lbs., the mower is easy to carry and store. It's 42" long (28" folded), 25" wide, and 8" tall. Two-year mfg. warranty.

•63-529 Solar Robotic Mower $1,995

Create A Cobblestone Walk In An Afternoon

Create the charming effect of a cobblestone path — for a fraction of the cost. This ingenious 100% recycled plastic form resembles multi-shaped stones. Makes the task easy, fast, and economical. All you do is level the soil, position the form, fill it with pre-mixed concrete, and smooth with a trowel. Allow concrete to dry for just one minute, then remove the form and begin the next interlocking section. Fill each 2' x 2' form with one 80 lb. bag of pre-mix. Unlimited design potential — makes curves, corners, and circles. Use pigments, wood chips, and sand to dramatically enhance surface color and texture. No previous experience required. One Walk Maker form is all you need to create an attractive and economical walkway or patio.

54-510 Walk Maker $35⁹⁵

Brick Masonry Made Easy

They look like bricks, but they're actually pre-mixed concrete. And they're amazingly easy and economical to make yourself. Just fill the recycled plastic mold with pre-mixed concrete, smooth the surface, add color (if you wish), and remove the mold minutes later. One 80 lb. bag of concrete fills a 2' x 2' mold. Kit also contains a trim and border mold and 1 lb. of brick red surface color.

54-511 Walk Maker Brick Kit $35⁹⁵
54-512 Brick Red Colorant (1 lb. bag) $5⁹⁵

Mix Concrete Fast In Innovative Recycled Mixer

Now you can mix perfect concrete in 30 seconds flat — with minimal effort. Odd Job's patented design speeds and simplifies the process. The canister's patented internal baffle design creates a corkscrew effect which tumbles and mixes ingredients, end to end. Simply pour in water measured on the calibrated lid, add pre-mix, and secure the lid. Now roll the canister on its side, back and forth. The result is perfect concrete every time, ready in less than a minute. Made of durable recycled plastic, it's easy to clean and versatile, too. Useful for mixing fertilizer, animal feeds, soil mixes, compost, and paint.

54-513 Odd Job Mixer $39⁹⁵

Manual Lawn Mower

We've found a great manual lawn mower made by the oldest lawn mower manufacturer in the U.S. This mower is safe, lightweight, and very easy to push. It's perfect for small lawns and hard-to-cut landscaping. The reel mower provides a better cut than power mowers, keeping lawns healthy and green, and it doesn't create harmful fumes or noise pollution. The short grass clippings from the mower can be left on the lawn as natural fertilizer, or you can purchase the optional grass catcher and add the grass to your compost pile. Cutting width is 16 inches. It has 10-inch adjustable wheels, five blades, and a ball bearing reel.

63-505   Lawn Mower $119⁰⁰

Add $10 for additional shipping

63-506   Grass Catcher $19⁰⁰

Rechargeable Electric Mower

It's cordless. It's gasless. It's bagless. And it's a mulching mower. There's no troublesome cord to trip over or cut, and it's quiet. No gas fumes to breathe. Since it chops the grass clippings ultra-fine, there's no raking, no bagging. With continual mulching providing nutrients, your lawn becomes thicker, greener, and healthier. Mows up to 1/2 acre on a single charge. Plug it into AC house current, and the powerful 24V battery recharges in 16 hours. Runs about one hour on a single charge. To make it even more environmentally friendly, choose the optional solar charging system described below. This is one mower that starts fast and easily every time. The blade cuts an 18" swath. Wheel height is controlled by a single lever. Handle folds for storage; and the 75-pound unit can stand on end, requiring only a 17" x 20" space.

63-514   Cordless Mulching Mower, 24V $399⁰⁰

Solar Charging Option for Ryobi Mower

Our solar charger is a small, powerful, unbreakable solar panel. The 28-watt, 24-volt module will recharge the Ryobi in about three days with full sun. No danger of overcharging, it's self-regulating, and includes all the necessary interconnects and instructions. Imported.

11-561   Ryobi Solar Charger $219⁰⁰

Hydrosource Polymer

Ever lost a beloved plant to your own neglect? Never again. The Rolls-Royce of water-absorbing polymers is now available for home use. Originally designed for landscaping and professional agriculture applications, Hydrosource can absorb and store enormous amounts of fluid (up to 400 times its weight!), so that you can significantly decrease water usage and extend the time between waterings. Plant roots grow right through the polymer "reservoirs" and tap the nourishment, resulting in better health and increased yields. Many customers have reported watering their houseplants only once a month after using Hydrosource. For gardens and lawns use 2 to 20 pounds per 1,000 square feet. For potted plants plan 1 to 2 ounces per cubic foot of soil. Lasts for eight years or more. Completely safe, decomposes into carbon dioxide, water, ammonium, and minute amounts of lactic acid.

46-142   Hydrosource Polymer (1#) $9⁹⁵   46-143   Hydrosource Polymer (5#) $39⁹⁵

46-144   Hydrosource Polymer (50#) $275⁰⁰

GARDEN AND COTTAGE FARMING BOOKS

We offer a mix of up-to-date garden books and some classics. Some gardeners want a "recipe" approach to gardening, i.e., "Just tell me what seeds to buy and how to plant them and when I can eat them." But Nature is much too complex to offer gardening-by-the-numbers, and its wonders become even more miraculous as you learn some of the science underlying soil and plants. You will become a better gardener the longer you can work in the same place and begin understanding the patterns of your own weather and soils, and can begin identifying the many insects and other life that make up a thriving garden ecology. And you will excel if you have handy references that explain how and why soil, plants, and insects behave the way they do. So we have included some books that are not usually found in the corner bookstore gardening section and which go into greater depth than is usually found.

Designing and Maintaining Your Edible Landscape Naturally

Robert Kourik. This is much more than an edible landscape book. *Mother Earth News* wrote that it ". . . could be the most comprehensive guide in existence to growing vegetables, fruits, flowers, and herbs for both ornamental and culinary purposes." Poet Gary Snyder says, "This is simply the best book I've seen that tells how your orchard, yard, and garden actually work, and how you can learn to play along with them." And Helen Nearing writes, "This is an important book . . . to read and own. I am going to inaugurate such a system of natural planting."

"Hay bale gardening is inspired laziness, and makes the closest thing I know to an instant garden. Here, we really let nature take over. Simply put, hay bales become both the compost pile and the growing medium for vegetables. The bulk of the bale is reduced to plant food by a slow decomposition that feeds the hungry searching roots of the crop. In the end, you have loamy compost, the legacy of the bale, and a tasty crop."

Sourcebook contributor Robert Kourik has turned his lifelong learning and experience into a classic. Encyclopedic and written with style and humor, if there was one gardening book we would recommend above all others, this is it. Extensively illustrated with color photographs, charts, drawings. 370 pages, paperback, 1986.

80-324　Designing and Maintaining Your Edible
　　　　Landscape Naturally　　　　　　　　$25⁰⁰

Gene Logsdon

Our friend, Gene Logsdon, wrote a bunch of great books for Rodale years ago about homesteading and practical country skills. Then he kind of disappeared from the book trade for awhile as the roaring 80s made homesteading passe and the market for this information dried up. Meanwhile, he has been happily working and living on his small farm in Ohio, and has been writing for his old friend, Jerry Goldstein, at In Business *magazine. He has also continued to write some great essays and opinionated columns about country life. Unless you were involved in organic farming or read obscure, low circulation magazines, you haven't heard much from Gene since the early 80s. Now he's back. As we write, a lovely book of his essays has just been published and a book of his farm and country wisdom is hot off the press from Chelsea Green. Gene has also become a columnist for our own* Real Goods News.

If you tried to compare him with other writers, you might say that Logsdon is a combination of Scott Nearing, Wendell Berry, and Edward Abbey. He resembles Nearing in his independent living and in having the strength and skills to live that way. He resembles Berry in his scathing attacks on modern agriculture and his poetic memory for farm life as it once was, and still is on his own small farm. He resembles Abbey for pure orneriness and personal outrage at the way things are turning out. But as a writer, as in his daily life, he is his own man. From his small farm in Ohio, he yearns not for the world that used to be, but for the world that could be now if our values were right, our laws just, and our community life truly savored.

At Nature's Pace

Gene Logsdon. These sometimes-lovely, sometimes-provocative essays evoke admiration for those who still cling to the best of what once was our nation's heritage . . . country life and the small, cottage farm. There are gentle memories, informed opinions and a not-so-hidden anger at the stupidity and short-sightedness of industrial agriculture.

"Another lesson the Amish teach, flowing directly from their philosophy of small scale, be it in farming or manufacturing, is how they integrate business into society. Most shops, even most factories, are embraced by the owners' farms. Or, at least, the owners' homes sit right next to their factories. All that separates Atlee Kaufman's house and factory is a luscious vegetable garden. Wanye Wengerd's children play in the shadow of his factory. 'And when that is the case,' he says, 'you make sure there is no pollution.' Instead of industrial parks, suburban enclaves, and huge stretches of lonely farmland, the Amish community blends all three activities into a harmonious social pattern. There is no distancing of work from family life that breeds the idea that what one does at work is not bound by the same moral code as what one does at home."

208 pages, hardback, 1994.

80-421   At Nature's Pace $23⁰⁰

The Contrary Farmer: A Real Goods Independent Living Book

Gene Logsdon. A fascinating look at what he has learned over many years as a cottage farmer, and why and how an educated man chose to return to small farming as a way of life. The general reader will find wisdom and humor on every page . . . and those who farm or want to farm will find inspiration, hands-on advice, and a bull-headed optimism. Where big farming, since the turn of the century, has gobbled up almost every small operation, Gene finds the seedlings of hope for the inevitable renaissance of small farming that is everywhere visible but seldom recognized.

"The value of a husbandry-driven infrastructure of small cottage farms across the whole nation is also incalculable. If healthy, such a rural culture could mean who knows how many people retreating gladly and willingly to the countryside, relieving the population pressures that are turning cities into heat sinks of human frustration. Spreading out the population to share the life of shepherd and cowboy would hopefully generate a renewed emphasis on traditional rural virtues and give families a reason to work and play and love together again.

There is another aspect to such romantic visions of a pastoral economy that is not at all as impractical as the foregoing sounds. As national columnist Richard Reeves has been thoughtfully suggesting, we may be looking at a future where there is not enough full-time, *salaried* work to go around. It is scary enough when big companies lay off workers right and left, and others flee to developing nations in search of cheap labor. But when Procter and Gamble says it must lay off workers not because the company is losing money but so that it can continue to make money, we may be hearing the first announcement of the end of the industrial society — the end of a society based on an ever-increasing scale of wages and an ever-enlarging rate of production. Whether one looks at that possibility in the practical short term — what will I do if I can't find a steady job? — or in the theoretical long term — what if the number of good-paying jobs, especially white-collar jobs, continues to decrease? — the cottage farm and workshop begin to look more and more appealing as a safe refuge. Within hardly a 2-mile radius of our farm, in a county with a population of only 26,000 people, in a landscape of widely dispersed houses and farms, there are 17 home businesses, not counting farming. Kathryn Stafford, associate professor of family resource management at Ohio State University, who participated in a recent nine-state survey of home businesses, informs me that such numbers are not unusual today. Most of the businesses in our neighborhood generate supportive income, not main income, but most of them are also hedges against possible interruption of the family's main source of paid wages. I don't describe those of us maintaining our livelihoods at home as having a sentimental yearning of the past, but a very practical vision of a post-modern pastoral economy where real goods count for a lot more than money."

236 pages, hardback, 1994.

80-241   The Contrary Farmer $22⁰⁰

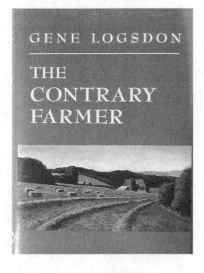

Your Organic Garden

Jeff Cox. This book from Rodale is a good introduction to organic gardening. Like the PBS series of the same name, it is full of practical tips on growing healthy plants, labor-saving techniques, and step-by-step directions you can really use right in your own backyard. It features: complete information on composting and organic soil care; no-fail seed-starting and plant propagation techniques; plant-by-plant growing guides for vegetables and fruits; troubleshooting tips and organic controls for all major pests and diseases; information on growing healthy perennials, annuals, bulbs, roses, and more; easy-to-use low-maintenance techniques for every part of your yard; and guidelines for choosing the right plants for your yard and keeping them healthy.

"Evidence from scientific studies and gardeners' experimentation indicates several possible benefits from companion planting: masking or hiding a crop from pests; producing odors that confuse and deter pests; serving as trap crops that draw pest insects away from other plants; acting as 'nurse plants' that provide breeding grounds for beneficial insects; providing food to sustain beneficial insects as they search for pests; creating a habitat for beneficial insects."

344 pages, paperback, 1994.

80-422   Your Organic Garden $16⁰⁰

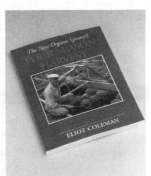

Four-Season Harvest

Eliot Coleman. Seekers of the Independent Life, rejoice! Eliot Coleman, whose *The New Organic Grower* showed a generation of home and market gardeners that you do not have to make your garden a toxic waste dump to produce blemish-free tomatoes, has now produced a work that challenges the premise that home food production is limited by the seasons. In Four-Season Harvest Coleman introduces a system that insures the availability of fresh, delicious, organic produce all year long, no matter where you live. Best of all, Coleman's system does not rely on high-tech, fancy greenhouses or expensive equipment. By matching crops to the seasons, by strengthening the soil, and by extending (here's the secret) *not* the growing season, but the harvesting season, he shows that the unlikely is not only possible but feasible for any gardener willing to spend a little time each day on food production. Coleman's low-tech, inexpensive solutions to seasonal limitations will inspire all gardeners to understand that it's not what you grow that counts, it's what you harvest. This book is a must for anyone who dislikes chemically-grown, artificially-ripened vegetables that have been shipped halfway around the globe. 224 pages, paperback, 1992.

80-843   Four-Season Harvest $18⁰⁰

The New Organic Grower

Eliot Coleman. This is an excellent companion book to Coleman's new work, *Four-Season Harvest*. With 50,000 copies already in print, Coleman has established himself as one of the leading voices pointing a way toward solving the current ecological crisis in agriculture. Crop rotation, green manures, and garden pests are covered in depth, as are seeding, transplanting, and cultivation. Whether you garden for home consumption or for the market, this is required reading for those who want to plant and reap without pillaging and raping. 310 pages, 128 illustrations, paperback, 1993.

82-109   The New Organic Grower $20⁰⁰

How to Grow More Vegetables
(Than You Ever Thought Possible On Less Land Than You Can Imagine)

John Jeavons. Out of an intense desire to teach people of developing nations how to feed themselves, and seeing population growth shrinking the land available to do so, John Jeavons and Ecology Action have been researching and teaching a system of gardening that grows the largest amount of food on the smallest amount of land while building the soil. It's called biointensive gardening and is in use in many countries around the world where land is scarce and hunger plentiful. Its application in the U.S., with abundant food and land, is sniffed at in some gardening circles, but many have found it to be a most efficient way to produce food. In our time-constrained culture, learning biointensive skills can reap great benefits. Others have found it the only way to garden on their small plots of land, and have been able to produce most of their own food in backyard suburban plots (see Duncan story on p. 469). *How To Grow More Vegetables* has been published in five languages, is used in 107 countries, and is the "Biointensive Bible."

"After the soil has been initially prepared [by double-digging] you will find the biointensive method requires less work than the gardening technique you presently use. The Irish call this the "lazy bed" method of food raising. In addition, you will receive good-tasting vegetables and an average of four times as many vegetables to eat! Or, if you wish to raise only the same amount of food as last year, 1/4 the area will have to be dug, weeded, and watered . . . For all-around ease, D-handled flat spades and D-handled spading forks of good temper are usually used for bed preparation. (Poor tools will wear out rapidly while the garden area is being prepared.) D-handles allow the gardener to stand straight with the tool directly in front. A long-handled tool must frequently be held to the side of the gardener. This position does not allow for simple, direct posture and leverage."

185 pages, paperback, 1991.

80-616   How to Grow More Vegetables $15⁰⁰

Solar Gardening: A Real Goods Independent Living Book

By Leandre Poisson and Gretchen V. Poisson, *Solar Gardening* shows how to increase the effects of the sun during the coldest months of the year and how to protect tender plants from the intensity of the scorching sun during the hottest months through the use of solar "mini-greenhouses." The book includes instructions for building a variety of solar appliances plus descriptions of more than 90 different crops, with charts showing when to plant and harvest each. The result is a year-round harvest even from a small garden. In *Solar Gardening* the Poissons show you how to:

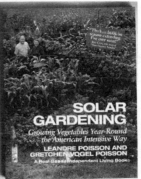

• Dramatically increase the annual square-foot yield of your garden.

• Extend the growing and harvest seasons for nearly every kind of vegetable.

• Select crops that will thrive in the coldest and hottest months of the year, without artificial heating or cooling systems.

• Build solar appliances for your own garden, including detailed instructions on how to build Solar Cones, Pods, and Pod Extenders that provide an ideal growing environment and protect plants from both extreme heat and cold.

Armed with nothing but this book and a few simple tools, even novice gardeners can quickly learn to extend their growing season and increase their yields without increasing the size of their garden plot. 296 pages, paperback, 1994.

80-247 Solar Gardening $25⁰⁰

The Original "Have-More" Plan

Ed & Carolyn Robinson. Much as *The Mother Earth News* and *Whole Earth Catalog* helped spark the back-to-the-land movement of the seventies, this book, subtitled "A Little Land — A Lot of Living" was responsible for a lot of the migration back to the country in the forties. It was published to aid millions of Americans in search of greater personal independence. In addition to its obvious historical and nostalgia value, it contains many practical tips that are still very useful today . . . a compendium of homesteading wisdom.

   "We faced the fact that we knew absolutely nothing about raising any part of what our family needed to live. In fact, our utter and absolute dependency on my job was appalling. If I should lose my job — even temporarily — we would have no money to pay our rent — the landlord would put us out . . . no money to buy groceries or pay the butcher and we wouldn't eat. If there were another depression — and I were to lose my job like millions in the last depression — then there wouldn't be a thing to do but stand in line and beg the government for 'surplus commodities' . . . rent money . . . relief clothing until things got better again — which might be years! . . . A friend once said to me, 'Ed, why do you bother with other people? Why don't you settle down and just enjoy your own job and your Have-More homestead? Why try to spread it all over the country?' I may sound silly trying to tell you why. But I feel, somehow, that in the years to come the U.S. is going to need all the help it can get toward happiness and peace and security. We aren't always going to have a boom going on. I've got a boy and I want to see him grow up in a good country, and if ten- or twenty-million American families can get set as well as we Robinsons are I don't think anything can hurt this nation.
   "70 pages, paperback, 1943.

82-204   The Have-More Plan $8⁰⁰

Who loves a garden still his Eden keeps, Perennial Pleasures plants, and wholesome harvests reaps.

—Amos Bronson Alcott

Common Sense Pest Control

Olkowski, Daar, & Olkowski. This weighty tome provides a thorough review of the literature of least-toxic pest control in 715 large-format, small-print pages: everything you ever wanted to know about controlling pests, and much, much more. The authors advocate a close-the-barn-door-before-the-horse-escapes approach rather than quick-fix chemical overkill, the much beloved panacea of the petrochemical poison industry. There is an excellent short chapter on diversity, trap-crops, and companion planting in the garden. 715 pages, hardback, 1993.

80-837   Common Sense Pest Control $40⁰⁰

A Gardener's Journal
A Ten-Year Chronicle of Your Garden

This handsome, hardcover volume honors the joys of gardening. A Gardener's Journal is designed as a ten-year perpetual diary which you may begin at any time. Record your gardening triumphs, the struggle of occasional setbacks, sketches of your garden to help maintain good crop rotation, key dates such as planting and harvesting, and personal jottings about day-to-day observations. One page is given to each day of the year; each page is then divided into ten sections, making it easy to track and compare weather conditions and other relative influences. Specific enough to be useful, but general enough to be personalized, this journal is the perfect interactive gardening tool for both the most practical and the reflective gardener. Includes intriguing notes on the history of plant lore and appealing pen and ink illustrations. 544 pages. This journal will be a treasured gift for any gardener.

80-341 A Gardener's Journal $35⁰⁰

Seed To Seed: Best Book Ever On Seed
Saving For Home Gardeners

Comprehensive and easy to understand, *Seed To Seed* is an outstanding reference work that enables you to save your own vegetable seeds for over 250 plants from all continents and climates. Excellent photographs teach step-by-step hand pollination. Clear instructions cover all aspects of seed production. Published by Seed Savers Exchange, the world renowned network of gardeners who relentlessly pursue and preserve our eroding genetic heritage. 8-1/2" x 11", 222 pages, paperback, 1991.

80-255 Seed To Seed $20⁰⁰

Solar Dehydrator Plans

Drying is the oldest, simplest, and most natural method of preserving food. Solar food dehydration does not require added energy to presere your foods. This set of plans makes it easy to build your own professional-quality dryer. Totally enclosed, your food is protected from the direct rays of the sun and from insects. The air that dries the food is heated on the curved solar collector panel and moves through the food by natural convection. The dehydrator has 20 square feet of tray area and on an average sunny summer day maintains a temperature of 115° F. It takes two or three days for most food to dry.

80-229   Solar Dehydrator Plans $18⁰⁰

CHAPTER 11

MOBILITY AND ELECTRIC VEHICLES

IN PLANNING THIS NEW SOURCEBOOK, we decided to drop the Mobility chapter. We are fascinated by the topic, but it never seems to attract much attention from readers in our publications or in the press. It is a guilty secret for most of us, who drive more than we should, and so we try to ignore it. Denial, as they say, is not just a river in Egypt.

Then along came three fiesty submissions from three of our favorite writers on energy and technology, J. Baldwin, tech editor for the Whole Earth folks, energy guru Amory Lovins, of the Rocky Mountain Institute, who promises us that by following his advice we "will be able to butter our bread on all six sides," and Michael Hackleman, electric vehicle enthusiast extraordinaire.

It is long past time for us to begin to reform our energy habits. During this long century of ever-expanding mobility, we have developed a national case of *spielkes* that will be hard to cure. Many of us who live in carefully designed energy-conservative homes and recycle paper as if it were gold also think nothing of driving 80 miles to and from work five days a week and half as far to shop on the weekend, thereby blowing away in a single week the savings gained at home in a year. As J., Amory, and Michael tell us, it doesn't need to go on that way forever.

— *John Schaeffer*

ZERO EMISSIONS, ZERO REALITY

"Whose side are you on, anyway?" A California state bureaucrat was chiding me for speaking against the statute requiring that, by 1998, any company annually selling more than 5000 cars in the state must make at least 2% of them "zero-emission" vehicles. She insisted that it was a start, and better than nothing. I agree that incentives and standards are a good idea, but I regard this particular legislation as simple-minded, punitive, and unlikely to bring about the desired result. It may even make things worse. I'll bet the law gets diluted, delayed, or demolished — and not necessarily by the bad guys — by the time you read this. Here's why.

The very phrase "zero emissions" is hustle and hype of the sort found so tiresomely in advertising. This is a question of physics: A zero-emissions vehicle is not physically possible. It might more accurately be termed an "elsewhere emissions" vehicle, and that's only considering the emissions from the propulsion system on board. Where does that propulsion energy come from? Perhaps from an environment-trashing coal mine or a sinister nuclear plant. A hydroelectric dam causes ongoing environmental degradation. Making and installing photovoltaics makes a mess somewhere. Even hydrogen isn't perfect. It isn't ready, either.

All vehicles and their support systems inevitably generate pollution from their manufacture, distribution, fuel source, use, maintenance, and eventual junking or recycling. For example, the wearing-down of tires injects thousands of tons of undesirable particles into the atmosphere, soil, water, and us. The so-called zero emissions cars may represent an environmental improvement, but they won't actually produce zero emissions even locally, especially if there are millions of them.

It is important to be honest about this; otherwise, people will assume there is nothing to worry about once there are no more tailpipe emissions. And cars are only one part of a vast system. If all vehicles were "zero emission," the air might be cleaner, but most of the problems associated with automobiles would remain. There would be little change in accidents, congestion, land use, or depletion of resources (other than oil). Indeed, those particular aspects of our automotive culture might actually become worse if improved acceptability caused more cars to be sold and more miles to be driven. And new technologies always bring unforeseen problems. "Unforeseen" is the polite way of saying that, in hindsight, we didn't know what we were doing.

But the term "zero emissions" is not just misleading. Like all cliches and slogans (especially political ones, which this is), it short-circuits careful thought and free discourse. Wide use of this catchy phrase has already resulted in a kind of zealotry that tolerates only one correct solution — in this case, electric cars — to what is actually a tangle of interrelated problems. "Hybrid" vehicles (those with both fueled and electric propulsion) that promise lower emissions overall than do pure electrics are forbidden. Note that word "pure"! Such arguments sound all too often like confrontations with religious fundamentalists whose beliefs take precedence over physics.

Simple solutions to complex problems always leave out something important because they are too narrowly focused. When written into law, they discourage exploration of new ideas, especially at the beginning of a new effort, when seething inventors and entrepreneurs are most productive and most needed by society. The zero emissions requirement is in effect a code much like our antiquated specification building codes. It causes the same sort of frustrations and hassles that have delayed the wide acceptance of solar and energy-efficient homes for decades. Older readers may remember when avid California environmentalists of good intent, but inadequate understanding, successfully campaigned against big, energy-wasting "picture windows." The subsequent laws came close to making passive solar buildings illegal, and made life difficult for the more imaginative solar architects.

There is also the matter of coercion; that is, forcing manufacturers to comply with a deadline chosen by politicians. It is true that automobile manufacturers have vigorously opposed progress when the larger public good was at stake. Fripperous gadgets have come more easily than essentials such as airbags, seatbelts, antilock brakes, and energy efficiency. The industry has earned our suspicion, and it is easily seen as a hive of villains. But an emotionally satisfying, retributive vindictive is not going to help the environment. There is no way around the fact that we all have helped construct a society that needs cars, and that we obviously desire vehicles with pretty much the same capabilities as the ones we now own. (How did you get to where you are now sitting?) The best-selling single model of vehicle in the world is the full-sized Ford pickup! Neither crisis nor legislation can instantly change such a ponderous market reality, however illogical and profligate it may be.

It usually takes three tries to get anything (not just cars) right. After simulations, a competently designed and engineered first prototype will be built. Testing will reveal its obvious weaknesses. Prototype Two is the first model with the obvious flaws corrected. The third prototype starts fresh with what has been learned. In the case of cars, a hand-built fleet of Prototype Threes is road-tested and perfected under both laboratory and real-life conditions. This procedure usually takes about four years, the time needed to set up the factory. The process is much more time- and money-consuming if the design involved is completely new and thus cannot take advantage of existing parts and more than a century of internal-combustion experience. With electric cars, there are new unknowns. For example, what happens if you dunk the batteries while fording a stream? What happens in a wreck? Electrics have no existing infrastructure with a knowledgeable sales force, refueling stations, or repair shops with experienced mechanics and parts bins. Some of that must be standardized for all makers. (Imagine if you could only fuel today's Fords from a Ford-brand gasoline pump!) All that infrastructure must be in place the same day the cars go on sale. Somebody has to pay for it.

Billions of dollars (that's with a "B") are involved in mere restyling, let alone a radical new design employing unfamiliar technology. The development of the utterly ordinary (except for its plastic body) Saturn car required more billions of dollars and years of development time than our most advanced jet fighter. Whatever the excuse for that, with so much money at risk, car makers must be sure that the public really wants zero-emission cars enough to buy them. Corporate bean-counters remember the Edsel, a car that extensive consumer polls showed to be very desirable. People said they wanted it, but they didn't buy what they saw in the showrooms. A failure of that magnitude would seriously damage a large automaker. It would mean the demise of smaller manufacturers, the ones best able to serve a relatively specialized market (even assuming they could get development capital). If you were the CEO of a car company, would you risk the company and the jobs of thousands of workers on a car that cost you a fortune to develop for a market that might not exist? Worse, at this early stage, a rival's technological breakthrough might make your new car obsolete before you even sold one. A scary prospect to say the least. On the other hand, a less restrictive mandate than the zero-emissions law might well result in a machine with a broader appeal and a much larger potential market. More sales would mean a lower price, and put more eco-cars on the road.

As is true for anything else, cars don't exist separately. They are part of a vast system that includes everything from bridge building to the bottle caps for the ink for the label on the box containing the meter maid's chalk. Threatening the jobs of millions of people will not accelerate progress, and might instead delay a smooth transition to the supercars called for by Amory Lovins.

History has repeatedly shown that change is difficult to mandate unless backed by an informed and agreeable majority. A minimum-emissions vehicle would be better. This would represent a direction rather than an unachievable, phantom goal. The required minimum could be edged down gradually as research keeps pace with our environmental and social goals. Technology doesn't happen instantly; it evolves as it learns, taking along the people and stuff associated with it. As Buckminster Fuller said, "Evolution makes many starts." If laws are necessary to spur environmental responsibility, they should encourage those many starts in every possible way.

— *J. Baldwin*

J. Baldwin has a degree in automobile design, teaches environmental design, and has been technical editor of the Whole Earth Review *and the* Whole Earth Catalog *for 25 years. He and his horticulturist/jeweler wife, Liz, currently live in a rented non-solar converted chicken coop while they brood on a more appropriate nest.*

HYPER CARS: AMORY LOVINS' VISION OF THE FUTURE OF VEHICULAR TRANSPORTATION

In early March of 1994, John Schaeffer and I sat at Amory Lovins' feet while he presented his talk on Hypercars. American car behavior has long been a puzzle for me, a sort of emblem of our infatuation with speed and consumption. All right, I confess: I love my fast little car, which is fairly efficient by present-day standards. John has recently persuaded Real Goods to let him drive the "company car," an electric reproduction of a classic Porsche Spyder, which goes fast enough (but not far enough) to  get in trouble, but, as J. Baldwin says, electric vehicles are "elsewhere-emission" vehicles. The problem is in the solitary commute, a lemming-like motorheaded meditation peculiar to workers with cars.

Well, Amory has a solution for part of the problem: cars need not be so ridiculously consumptive. He is worried, and rightly so in my view, that by making the pleasure of car travel less guilty, our problems with *traffic* will worsen. In this view, the electric vehicle (EV) is like a topical medicine: it may make the itch go away, but it does not cure the basic ill. Nevertheless, amidst his usual off-handed brilliance and casual wittiness, Amory spoke about his current study, which we have excerpted here from an extensively refereed scholarly preprint, a copy of which may be had from Amory and Hunter Lovins' Rocky Mountain Institute, 1739 Snowmass Creek Road, Snowmass, CO 81654-9199.

— *Michael Potts*

Ultralight four-passenger cars with modern hybrid-electric drives could achieve less than 1.6 liters per 100 kilometers (more than 150 miles per U.S. gallon) with demonstrated technologies such as switched-reluctance motors, conventional buffer batteries, and compact gasoline engines. Consumption as low as 240 to 300 mpg is probably achievable with advanced technologies expected to be demonstrated shortly, such as monolithic solid-oxide fuel cells, carbon-fiber flywheels, and small adiabatic diesels. Far from sacrificing other attributes for efficiency, though, ultralight hybrids could be *more* safe, peppy, clean, durable, reliable, quiet, comfortable, and beautiful than existing cars, yet be priced about the same or less. The key improvements required — chiefly aerodynamic drag and mass 56% to 57% below those of present U.S. production cars — have been demonstrated, and, further, two- or three-time reductions in drag-mass product appear feasible. New-shape materials, chiefly polymer composites, could achieve these goals while cutting production costs through materials savings, with a hundredfold fewer parts, tenfold less assembly labor and space, and halved tooling costs. Epoxy dies, lay-in-the-mold color, and other innovations permit extremely short product cycles, just-in-time local manufacturing with direct delivery (hence the same retail price, even if production costs were considerably higher), and on-site maintenance. This would fundamentally change how cars are made and sold. It could be the biggest change in industrial structure since the microchip.

Such "supercars" face serious cultural obstacles in the car industry, and institutional barriers in the marketplace. Supercars' immense societal value merits policy intervention to help speed and smooth this challenging transition, making them appear less a hardship than a lucrative business opportunity. Supercars could also buy time to implement (though not obviate the need for) fundamental transportation and land-use reforms.

The Fallacy of Incrementalism

Troubled car industries now weaken many national economies, while inefficient light vehicles and their ever-increasing use are major causes of oil dependence, air pollution, noise, climatic threats, and other important social costs. These problems demand transportation and land-use innovations, combined with cleaner, more efficient vehicles. Yet the conventional wisdom framing the U.S. car-efficiency debate is that the doubling of new-car efficiency between 1973 and 1986 virtually depleted the "low-hanging fruit" — opportunities for fuel economy consistent with affordability, safety, and performance.

We shall argue that, on the contrary, the next doubling will be easier than the first was, because it will come from very different sources: not from incremental refinement of today's cars, but from replacing them altogether with a different and functionally superior concept of car design, manufacture, and sales. We shall attempt to describe an auto-industry transformation that seems technologically plausible and commercially attractive in the 1990s and beyond, initially for niche and later for general markets, suggesting also analogs in other kinds of vehicles. The implications of this transformation are not all welcome, but the issue seems less whether it will happen than who will do it first and best, and whether it will be done thoughtfully.

New U.S.-made cars halved their fuel intensity during the period from 1973 to 1986, from an 13.4 mpg average to a European-like 27.6 mpg; approximately 4% of the savings came from making the cars smaller inside, and about 96% from making them lighter and better. Although that gradual decoupling of mass from size reached a temporary plateau using conventional materials, many other refinements are far from saturated. Further incremental improvements therefore yield a supply curve extended 24% from the U.S. Department of Energy's estimates by adding two further measures, idle-off and aggressive transmission management. The curve shows cumulative gains in new-car fuel economy, and their empirical marginal costs, from fully deploying a limited list of 17 well-quantified technologies already used in mass-produced platforms, without changing the size, ride, or acceleration of average U.S. 1987 cars. Most of the measures are conventional; for example, front-wheel drive, four valves per cylinder, overhead cams, and five-speed overdrive transmissions.

Ledbetter and Ross (1990) found that this approach can cut 1987-base fuel intensity in the year 2000 by about 35%, to 34 mpg actual (45 mpg rated). That would just counterbalance projected U.S. growth in vehicle-miles travelled by the year 2010.

Conventional cars, like other technologies, have entered their era of greatest refinement just as they may have become obsolete. Imagine that a seventh of the GNP in, say, the United States were devoted to manufacturing typewriters. The Big Three typewriter manufacturers have gradually moved from manual to electric to typeball models. Now they are making the delicate little refinements somewhere between a Selectric 16 and a Selectric 17. Their typewriters are excellent and even profitable. People buy over 10 million of them every year. The only trouble is that the competition is working on subnotebook computers.

That, we suggest, is where the global auto industry is today — painstakingly refining designs that may soon be swept away, perhaps with terrifying speed, by the integration of very different technologies already in or entering the market, notably in advanced materials, software, motors, microelectronics, power electronics, electric storage devices, and computer-aided design and manufacturing. This paper attempts to sketch the outlines of that potential transformation.

The Ultralight Strategy

The incremental approach to improvements saves so little fuel because it focuses disproportionately on fine points of engine and transmission design, while comparatively neglecting the basic strategy of making the car very light and aerodynamically very "slippery." This strategy rests on the basic physics of cars: in urban driving on a level road, drivewheel energy — typically only about 15% to 20% of

Ely Schless sits astride the NoPed, a fast, long-range electric prototype built on a bicycle chassis.

fuel input energy — is devoted about one-third to heating the brakes when they stop, one-third to heating the air they push aside, and one-third to heating the tires and the road. On the highway, air resistance, proportional to the square of speed, accounts for about 60% to 70% of tractive energy needs. The keys to automotive fuel economy, therefore, are braking and downhill-coasting recovery, aerodynamic drag, tire rolling resistance, and mass. Benefits from improving any one of these are limited, but benefits from improving all of them together are striking, and they often reinforce each other.

Beyond the Iron Age: Net-Shape Materials

A typical steel part's cost is only about 15% for steel; the other 85% is for shaping and finishing that raw material. Steel is so ubiquitous, and the success of highly evolved steel-car manufacture — one of the most remarkable engineering and managerial feats in human history — makes its very high design, tooling, fabrication, and finishing costs so familiar that we overlook how they outweigh its cheapness. An electrocoating plant costs some 250 million dollars; a paint shop, 500 million dollars; and complete tooling for one car model, upwards of one billion dollars. Making a steel car requires thousands of engineers to spend a year designing and a year building a football-fieldful of million-dollar steel dies that are used as long as possible (ideally decades), then thrown away. That inflexible, costly tooling in turn means huge production runs, high risks of stranded investments, and long amortization times, time-to-market, and product cycles — all of which crimp flexibility and innovation. Thus, today's most "modern" cars are really the cutting edge of old technology. Yet new, nonmetallic materials are not just a substitute for steel, as they have been used so far; they can transform the very nature of cars, manufacturing, and marketing. And in the process, they also support the ultralight strategy.

The most pessimistic of experts we have consulted estimate mass-production cost of ultralight, carbon-fiber cars at one to two times that of steel cars today. The most detailed assessments, however, suggest breakeven at carbon-fiber costs widely expected to prevail by 2000 if not before. Moreover, carbon is only about half the total mass of carbon-fiber composites, and there are many other kinds of far cheaper fibers that can make excellent ultralight cars .

The resulting revolution of car design, production, and operation — as profound as the electronics-driven transformation was in the 1970s — has just begun. The challenge to metal might come surprisingly quickly. U.S. passenger cars' bodies switched from 85% wood in 1920 to over 70% steel only six years later, making possible the modern assembly line. Mainly in the 1960s, composites rapidly displaced wood and metal in boatbuilding, as they are now doing in aerospace niche markets. Today, the switch to molded synthetic materials could support a "major breakthrough in the technological development of the automobile industry" (Amendola, 1990), making at long last an agile, short-cycle competitor.

Hybrid-Electric Drives

Net-shape ultralight car platforms, then, will probably cost about the same as steel platforms or less. But adding a further step can make them still cheaper and radically simpler: hybrid-electric powertrains. A really successful hybrid car cannot be made out of steel, for the same reason that a successful airplane cannot be made out of cast iron. But net-shape ultralight materials and hybrid drives are strongly synergistic, because a hybrid's design and performance depend critically on mass, drag, and rolling resistance, and because mass savings compound more quickly with hybrids than with conventional powertrains.

Pure-electric, externally recharged cars work poorly when scaled up to carry four to five passengers rather than one, because the battery mass, like any other vehicle mass, compounds: too much energy and power are needed to haul the heavy batteries, requiring even heavier batteries to store that extra energy, and so on. In all, each unit of added battery mass increases total vehicle mass by a factor conventionally assumed to be about 1.5 in heavy cars and often about 5 in ultralights. Electric hybrids, however, scale well to both large and small sizes *of ultralights.* With ultralight construction, the car's size has little to do with its body's mass: going from two to four passengers adds less than 100 kilograms, not counting suspension and powertrain. And with low drag and regenerative braking, the energy needed to propel a larger vehicle's greater mass is largely recovered, although heavier equipment is needed to accelerate the greater mass.

In the simplest (series-hybrid) concept, the wheels are always driven electrically, but the electricity is made onboard as needed by a low-power Otto, diesel, or gas turbine engine, or by a fuel cell. This has four key advantages over direct mechanical drive:

- The engine is sized to the average load, not the peak load, because a small buffer store between the engine-driven generator and the traction motor(s) stores energy for hill-climbing and acceleration.
- The engine drives a generator, not the wheels, so it runs only at its optimal condition. Just this collapsing of the engine performance map to a point *doubles* an Otto engine's practical efficiency, and permits simultaneous optimization for emissions, too.
- The engine never idles; when not running at its "sweet spot," it turns off.
- *Regenerative braking* (recovering most of the braking energy into the buffer store) improves conventional platforms' fuel economy by an additional 25%; in the USEPA urban cycle, 23% of the time is spent braking.

Together, these features permit the fuel tank/engine/generator to be inherently smaller, lighter, cheaper, and longer-lived than the 300 or 400 kilograms they typically displace in a pure-electric car; and those mass savings then compound. Several-fold lighter, though costlier, batteries of more exotic kinds are becoming available, but the hybrid's chemical fuel will still win out, because it has on the order of 100 times the energy density of lead-acid batteries. It thus permits longer range with lower total mass, cost, and refueling inconvenience.

Electric-hybrid vehicle designers differ on the ideal traction motor, and impressive progress has enabled both asynchronous and DC motors to achieve goals set forth in the following section. Yet, especially in the U.S., most experts, while familiar with those achievements, have overlooked the potential advantages of modern switched-reluctance drives. A recent review (Lovins and Howe, 1992)

GM Ultralite

suggests that, for fundamental reasons, properly designed switched-reluctance drives can outperform all other types — including electronically commutated permanent-magnet motors — in size, mass, efficiency, versatility, reliability, ruggedness, fault-tolerance, and cost.

These remarkably strong, light servomotors can, but need not, be integrated into each wheel-hub, eliminating all gears and saving net weight. Depending on failure-mode analysis and the ability of the buffer store to accept high inrush currents, it may be possible to eliminate mechanical brakes. At least in principle, differential wheel-speed, integrated with electronic suspension to lean into turns, may also permit an ultralight car with hard, narrow tires to steer without angling the front wheels. The only disadvantage of switched reluctance drives is that they are an order of magnitude harder to design than traditional types: excellent design demands a level of system (especially software) integration and numerical simulation that only a few dozen people have mastered thus far.

Integrated Design of Ultralight Hybrids

Redesigning an ultralight-*and*-hybrid car from scratch, using aerospace systems concepts, could yield an elegantly frugal and unusually attractive vehicle. A four-passenger, family-car version would start with low mass (<700 kg now, <500 kg soon, perhaps about 400 kg ultimately), and could achieve high crashworthiness with special materials and design. Like an aircraft, it would be designed for high payload/curb-weight ratio, perhaps above the Peugeot 205XL's 0.56. It would use switched-reluctance actuators, of which Ford cruise controls now use on the order of 4000 units a day; and would control them by fiber optics ("fly-by-light/power-by-wire"). It would combine a drag coefficient of <0.2 now and about 0.1 later with a "smart" active suspension and advanced tires. Its hybrid drive would initially use a small internal-combustion engine, on the order of 10 to 15 kilowatts — probably an advanced stratified-charge engine, high pressure-injection diesel, Elsbett engine, or small gas turbine — directly driving a switched-reluctance generator. Buffer storage would be provided initially by a few kilowatt-hours of improved conventional batteries, such as nickel/metal-hydride, lithium, or sodium-sulphur, driving two to four switched-reluctance motors (possibly hub-integrated). This design — at least if a series hybrid — eliminates the transmission, universal joints, differential, perhaps axles, and possibly brakes.

Meanwhile, accessory loads would be rigorously reduced, starting with the air conditioner, which, in a typical U.S. car, is now sized to cool an Atlanta house. A single high-intensity–discharge light source, such as those recently introduced by Philips, Hella, and GE, could provide all exterior and cabin lights via fiber optics and light pipes. Electric loads and mass would be minimized everywhere. All the powertrain friction reductions available, down to the last bearing and advanced lubricant, would be systematically exploited. Standard parametric analysis suggests that this sevenfold reduction in drag-mass product from the 1990 U.S. new-car mean would correspond to an approximately five-fold gain in fuel economy. Thus a car spacious enough for four adults with luggage could achieve 150 miles per gallon. Of course, capturing even part of this goal would be richly rewarding; but capturing all of it seems well within reach of the technologies already individually proven and only awaiting proper integration.

The next generation of technologies that should emerge from the laboratory during the mid-1990s shows strong promise of an even more surprising technological edge-of-envelope early in the next decade. Three look particularly important: advanced kinds of fuel cells (which convert hydrogen directly into electricity and water), "electromechanical batteries" using composite superflywheels, and possibly ultracapacitors. With plausible further progress, an early 21st-century hybrid car might, for example, have under the hood a grapefruit-sized fuel cell wrapped in a <40-liter envelope and user-selected to the proper modular size, which could even be temporarily modified for special applications; a melon-sized package of power electronics, also modular (plug in an extra "slice" for higher performance), an orange-sized computer; perhaps an optional breadbox-sized space-conditioning package; and virtually nothing else. So why have a hood at all? There could be two trunks for extra storage and crush space. . . .

An early priority should be assessing the transferability of these concepts to vans and light trucks. This is urgent in the U.S., whose light trucks are not only 20% less efficient than cars, but are also driven farther for much longer. Some analogues are also evident for heavy trucks — traditionally, inefficient vehicles with efficient diesel engines.

Safety, Performance, and Aesthetics

A common generic objection to fuel-efficient cars is their alleged crash risk. But this confuses fuel economy, mass, size, and design.

Fuel economy and light weight need not compromise safety. There is no correlation, far less a causal relationship, between present cars' crash-test performance and their mass, nor between their fuel economy and their on-the-road death rate. That is chiefly because occupant protection systems are lightweight, and because a vehicle's design and materials are vastly more important than its mass. It may also be partly because light cars can avoid more accidents by stopping sooner and handling more nimbly than heavier cars.

Theoretically, collisions between two cars identical except in mass tend to damage the lighter car more. (Practically, this is often incorrect because other, unequal factors such as design dominate. The National Highway Traffic Safety Administration sought to show the danger of light cars in recent light/heavy crash tests; the light cars reportedly came off better until stronger heavy cars and flimsier light cars were substituted.) This idealized theory leads some to propose that you should drive a heavier car — thus reducing the risk from such collisions to yourself while raising others' risk correspondingly. But the right answer is to make *all* cars safe whatever their weight, without putting all the adjustment burden on light cars. Better control of destructive driver behavior, such as drunkenness, is often crucial: behavior may be up to a thousand more times risk-determining than the car itself, and only about 5% of crashes do not involve driver factors.

Composites and other ultrastrong, net-shape materials — many stronger than the familiarly durable but lower-grade carbon-fiber fishing rods, skis, etc. — would dominate in a supercar. They would bounce without damage in minor fender-bender collisions: most deformations of carbon-fiber composite panels simply pop out again with little or no damage. Under severe loads, composite structures fail very differently than metal. Even under compressive loading, though — often considered composites' weak point, "[they] show high and in many cases better energy absorption performance than comparable metal structures." (Kindervater, 1991).

An ultralight car using ultrastrong materials, modern airbag restraint systems, and crash-energy–managing design can weigh less than half as much as today's platforms — as the GM's Ultralite does — yet still be far safer than any car now sold. That is why race drivers are rarely killed nowadays when composite cars hit walls at speed in excess of 350 km/hr: as tens of millions of Americans saw on their 1992 TV news, the composite car flies to bits, failing at "trigger" sections specifically designed to initiate such breakaway and absorbing extensive crash energy through controlled failure modes, but the "survival capsule" remains intact and the driver generally limps away with perhaps a broken foot.

The main potential disadvantages of ultralight hybrids are that:

- with their low drag and low or absent engine noise, pedestrians may not hear them coming unless a noisemaker is added that somehow warns them without being objectionable; and
- obstacles such as small trees, crash barriers, and lampposts, against which a heavy car can dissipate energy by breaking or deforming them, may instead stop a light car or make it bounce off, increasing deceleration and perhaps bounceback acceleration forces on passengers.

But beyond their general crashworthiness, "supercars" also offer important safety features:

- The two- or (with series hybrids) four-wheel switched-reluctance drives offer full-time anti-lock braking and anti-skid traction, but with far greater balance, response speed, and effectiveness than today's methods.
- Supercars' light weight means faster starts and stops; their stiff shell means quicker and more precise handling.
- Carbon-fiber designs can be so stiff and bouncy that an ultralight car, if broadsided by a heavy truck, could go flying — like kicking an empty coffee can. The very unfavorable momentum transfer would go not into mashing the ultralight car but into launching it. Yet occupants restrained by belts, bags, and headrests and protected from intrusion into their protection space might well survive unless accelerated by more than the often survivable approximately 40 to 60 g range — in which case they'd be dead anyway in any car today, light or heavy, steel or composite.
- In the rare accidents that are so severe as to crush the composite shell (usually in hammer-and-anvil fashion), the occupants would be far less likely to be injured by the composite fragments than by intruding torn metal edges in a steel car.
- Victims' extrication would be much faster (a crucial element of critical medical care — most victims not dead on the spot can be saved if brought to the hospital within an hour): the doors are likelier to function, the composite shell can provide easier access, cutting it with a rotary wheel is quick and makes no sparks to ignite fuel vapors, and breakaway, energy-absorbing main components would no longer impede access to the passenger compartment.
- Hydroplaning risk should not rise and may in fact fall, because the car weighs less but has narrower tires.
- Small powertrain volume and raked hood design are consistent with improved visibility.
- With careful design, composites' (especially foam-core composites') excellent attenuation of noise and vibration could yield an extremely quiet ride — important because road noise is no longer masked by engine noise. This, plus the virtual absence of wind noise, should make driving less fatiguing, potentially boosting driver alertness.
- The whole car is so simple, reliable, corrosion-resistant, fault-tolerant, and failsafe-designable that dangerous mechanical failures are far less likely.

For all these reasons, the design approach described here could yield substantially improved safety. Supercars could also offer ample comfort, unprecedented durability and ease of repair, exceptional quietness, and beautiful finish and styling while retaining significant stylistic flexibility, impeccable fit and weatherproofness, high performance (lightweight means faster acceleration), unmatched reliability, and probably low cost.

One caveat is in order, however. Especially in the litigious United States, innovation is deterred by the threat that makers of new and hence initially "unproven" technologies may have to pay damage claims, even for accidents in which they are blameless. Absent tort reform, removing this important barrier to market entry may require some kind of government indemnity or coinsurance to makers of supercars that meet a national safety standard, at least until the actuarial experience has field-validated the supercars' theoretical ability to match or exceed the safety of today's cars.

[Editor's note: For space reasons, sections on reforming the automotive industry and public policy have not been reproduced here.]

Acknowledgments

We are not "car guys," but have greatly benefitted from detailed discussions with dozens of car designers and other technologists from Australia, Britain, Canada, Germany, Italy, Japan, Sweden, Switzerland, Russia, and the United States. One of us (ABL) has especially learnt from a major automaker, whose senior technical staff, starting in mid-1991, graciously took his car education in hand. This paper is consistent with these informants' teachings but uses no proprietary information. We are especially grateful for the inspiration of Dr. Paul MacCready's Sunraycer and Mr. Jerry Palmer's Ultralite; for Dr. Steve Rohde's kind provision of the spreadsheet for the parametrics; and for the opportunity, first accorded by the July 1991 Irvine hearings of the U.S. National Research Council panel, to draw these views, two decades in formation, into coherent form. More than 30 peer reviewers (including RMI's Don Chen and Daniel Yoon) provided invaluable counsel and corrected many errors; any remaining are our responsibility.

— excerpted by permission from a refereed and documented technical paper, "Supercars: The Coming Light-Vehicle Revolution" by Amory B. Lovins, John W. Barnett, and L. Hunter Lovins.

ELECTRIC VEHICLES: THE SILENT REVOLUTION

Almost two decades ago, I was busy building towers and installing wind-electric machines, solar modules, biogas containers, and waterwheels. I had land, a research center, a family, animals, orchards, and gardens. Still, I was not self-reliant. No matter how little I drove them, two machines parked out in the driveway, a car and a truck, needed their fill of gasoline. So I began the search for an alternative.

I first investigated alternative fuels. Methane (biogas). CNG (compressed natural gas). Alcohol. Hydrogen. At the time, I found them wanting. They were all good possibilities, but the quantities required were awesome; more, at any rate, than my farm could manufacture on-site. I quickly learned why this was so. Gasoline is rocket fuel! A gallon of it stores 125,000 Btus of energy. The alternative fuels contained roughly one-half as much energy (by volume) as gasoline.

Something was definitely wrong here. Using engineering books, I had calculated the amount of energy needed to accelerate my car to 60 mph and hold it at that speed for 20 minutes (to equal 20 mpg). Assuming 100% efficiency, my figures showed that less energy would be used — by more than a whole decimal point. I had just discovered a sad truth about internal combustion engine (ICE) technology in cars. The engine itself was the bottleneck, wasting 90% of the energy of *any* fuel it consumed before it did work.

Meanwhile, my research uncovered steam technology. A fuel was still consumed to do work, but it was burned *externally,* transferring its energy to water and producing steam to power pistons. This design allowed fuel (of any type) to combust more completely, reducing the type and amount of emissions. Steam engines, I learned, had undergone a tremendous leap in performance due largely to the efforts of Bill Lear (of LearJet fame) in his attempt to commercialize a steam car. (Like other innovators, he found competing with the Big Three to be a challenge.) At this point, I was on the verge of converting my vehicle to steam power.

Yet it was a footnote buried somewhere in these technical papers that connected me with electric propulsion. At the turn of the century, this footnote stated, steam cars and electric cars, not gasoline cars, had dominated the roads. An electric car? I daydreamed about plugging a car into one of my wind-electric machines. To this end, I paid $200 for an industrial EV in a Goodwill salvage yard, and modified it for farm use. It plugged directly into my 32-volt Wincharger. I was driving on wind-watts! And I was powering a 32-volt drill, saw, and welder at remote sites. In April 1977, *Popular Science* published my brief article on our 'Lectric Ox, paying me $200 for it. The circle was complete. I felt happily self-reliant. I had an alternative to the ICE car and gasoline.

The Basics of Electric Propulsion

Virtually unknown 17 years ago, today "electric vehicle" is a household phrase in the U.S.

Have you ever seen one up close? Let me describe it to you. An electric vehicle is one that uses an electric motor instead of an engine, and batteries instead of a fuel tank and gasoline. The electric motor is the size of a five-gallon water bottle and bolts right to the stock transmission. The batteries are similar in size and shape to the one used to start your car's engine. There are just many more of them. The accelerator pedal is connected to a "potbox," which operates the electronic controller. Pressing the accelerator smoothly delivers power to the motor in proportion to the amount of pedal you give it.

The electric car is only one example of an electric vehicle. An electric-assist bicycle is also an EV. An electric motorcycle is an EV. A trolley and a San Francisco BART subway train are EVs. A solar-powered car is an EV. A Formula electric racer is an EV. Scratchbuilts and conversions are EVs. The only thing they have in common is electric propulsion.

The world is more than ready for EVs. In fact, it uses electric motors everywhere. Electric propulsion is not a new technology. The modern locomotive is diesel-electric; its diesel engine-generator supplies the electricity to power the electric motor drivetrain. Electric motors also power elevators, industrial assembly lines, ventilation and air-conditioning units, refrigerators, blow dryers, washers and dryers, computers and printers, CD players and autotape decks, and pumps. Ironically, an electric motor is also needed to start an automobile's engine. Wherever silent, efficient, reliable service is needed, you will find an electric motor at work.

How Much Better Is an Electric Car?

Electric propulsion would easily win out over engines for automobiles on any level playing field. What advantages do EVs offer?

- EVs produce zero emissions at the point of use.
- An electric motor is 400% to 600% more efficient than an internal combustion engine.
- An EV, per mile, uses one-half the fossil-fuel resources an ICE consumes.
- An EV produces only 5% to 10% of the emissions of an ICE per mile traveled. *All* of the EV's emissions occur at a (oil- or coal-fueled) power plant, which runs 400% to 500% more efficiently than an ICE *and* scrubs its own exhaust.
- EVs can use electricity from *anywhere* including sustainable energy resources (wind and sun).
- EVs are simple, silent, and affordable to operate.

Zero-Emission or Emission-Elsewhere?

It is a *major* step for an automobile to no longer emit exhaust gases. In fact, it is nearly inconceivable. That's why consumers will love zero-emission vehicles. They may not be driving a solar-powered car, but they will be helping the sun to once more shine through clear skies.

In the literal sense, of course, the EV is the "emission-elsewhere" car. The electricity to power the car has to be generated *somewhere*. And, though some energy is available from sustainable energy sources like wind, solar, and water power, the bulk comes from coal- and oil-fueled power plants. Isn't this just transplanting the problem somewhere else?

The answer is both yes and no. Yes, the pollution is transferred to another region. However, this matter has been scrutinized extensively by the U.S. Department of Energy and several California agencies charged with air-quality management. The bulleted list above reflects their findings.

EVs are very efficient. They have to be. A pound of battery has 1/100 of the energy of a pound of gasoline. On average, a 30 mpg ICE car uses only 5% to 10% of the energy of its fuel, whereas the EV converts 70% to 80% of the battery's pack into propulsion, for the same mile.

I join others in their concern about EVs bringing back nuclear power. Two factors mitigate this fear. One, the electric utilities bore the brunt of the financial disasters (problems, waste disposal, Three Mile Island, etc.) of nuclear power plants and are unlikely to repeat the mistake. Two, it has been estimated that 30 million EVs could be added (at night) to the grid nationwide without causing the need for a single new power plant to be built. That represents a lot of surplus energy that is currently available off-peak.

In addition, energy from sustainable resources, such as wind and solar energy, is gaining ground. California alone has 1.5 gigawatts (one gigawatt equals 1000 megawatts) of generating capacity from wind machines scattered throughout the state. These produce enough power annually to power all of San Francisco all year long. Southern California Edison's (LUZ-built) solar-powered plants supply electricity to 500,000 people. These are 24-hour power plants, using CNG (compressed natural gas) to complement solar energy during stormy weather and at night.

The Politics of Oil

Electric propulsion seems to offer *significant* advantages over ICE technology through a broad spectrum of criteria. Can such a simple technology address so many problems *and* be environmentally friendly, too? Why, you may ask, aren't EVs in widespread use?

The answer is: Politics and the balance of power. Oil-based technology is entrenched worldwide. The oil business developed rapidly as a result of two world wars. Oil is, however, a finite resource. The planet's supply will run out. Yet it is being gobbled up as if there was no tomorrow. In the minds of the people who work in oil, there is only today.

The reality is there's money to be made in oil. Gasoline is made from oil. It is priced low enough to ensure that sustainable alternatives cannot compete with it. Think about it. A gallon of bottled water costs more than a gallon of gasoline. In turn, cheap gasoline helps to sell cars. Cars are *big* business. Car sales. Car parts. Service and repair. Collision damage. Insurance. Highway construction. Parking lots. Parking meters. Traffic tickets. Health services. The lack of mass transit forces everyone to own a car. Or two. And the "more cars" policy, in turn, sells more gasoline.

What are some of the side effects of our oil-based energy policy? Pollution. Wars (to control the supply). Health problems. Humans killed and maimed in accidents (in the U.S. alone, 50,000 deaths per year, with another 1 million maimed, crippled, or injured). None of these are included in the gas pump price.

A Greenpeace poster with a picture of the Exxon *Valdez*'s captain summed up the current situation in one brief statement: "It wasn't *his* driving that caused the Alaskan oil spill. It was *yours.*"

Oil and cars, together and individually, have proven hazardous to the planet. It's time to usher in the alternatives.

Test-Drive an EV

Fortunately, EVs aren't something we *have* to drive; they're something we should *want* to adopt! And not just for the planet, its species, or even our own species. People may start driving an EV to make a statement, but they will continue to drive one regularly because they *like* it.

See for yourself what the rave is all about! Test-drive an EV. There are dozens of EV clubs nationwide. The members of these clubs are common folk, who are usually happy to demonstrate their pride and joy.

Imagine that I've just offered to let you drive my EV. Your brain may be pumping questions through your mouth, but driving an EV for the first time will hit you on a gut level. It *feels* right. Turn the key. No whirrrr of the starter motor, no plume of exhaust, no roar of the engine, no vibration. Just a reassuring little thunk. That's the main contactor powering up the control system. Then, silence. This car is waiting for you to do something, just like a computer. Take your time. No power is being consumed.

At this point, I'll be reassuring you that driving an EV is identical to driving an ICE car — with earplugs.

Of course there *are* other differences. For instance, you don't have to "warm up the motor" in an EV. Unlike IC engines, electric motors work best when they are cold. So, are you ready to go?

After all these years, it still delights me that this silent vehicle actually moves when I press down on the accelerator. Spectators are always awestruck over this feature. You'll feel it, too.

Quickly, we're up to speed. If the EV has gauges, the next surprise awaits the moment you lift your foot off the pedal. Suddenly, no power is flowing, but you're still zipping along. "Gliding" is a better word for it, though. Everytime you lift your foot off of the accelerator, it is as though the vehicle's transmission has slipped into neutral. On a level surface, you'll be surprised at how gradual the slowing is, and at how far you can "coast." Once you've perfected this art by anticipating traffic flow, the overall range of the EV increases dramatically.

More delights lie ahead. Bring the vehicle to a stop, at a signal light, stop sign, or parking place. Silence. At first, you'll worry that you've stalled the motor. But you haven't. It's just not running. No wasted fuel, no emissions, no power consumed, no vibration. Silence.

Do you prefer an automatic transmission? Cars with automatic transmissions

A junior-high school class builds an Electrathon racer as part of its environmental studies program.

don't make good conversions (lousy efficiency). The beauty of the typical EV is that it lets you drive your manual transmission as though it *were* an automatic. Just stick the lever in second gear and forget the clutch. That's right. Push the accelerator and away you go. This will likely work all the way up to 45 to 55 mph, perfect for just driving around the town. Again, leaving the clutch alone, bring the EV to a stop. It will do it silky-smooth. It feels strange, but it's the truth. You can't stall an electric motor!

The fun isn't over. See that service station? Pass it by! And keep on doing it (if only to bug the owner). I recently rented a car while doing an energy exposition in another city, and promptly ran it out of gas. I have completely gotten out of the habit of stopping at service stations! When I added the can of gasoline and, later, when I refilled the tank at the station, I realized I had forgotten that refueling a "gas-hole" has a *smell* to it, too.

A frequent question I'm asked is, "How long does the car take to recharge?" If I completely discharge the pack, it will take six hours to put 85% of the energy back in. In reality, I always plug the EV into the socket when I get home. It's an automatic gesture. That way, when I'm ready to drive next, I'm usually starting out with a fully charged pack.

People who are afraid of getting caught out in the boonies with a dead battery pack don't drive EVs. Yet, for the frequent EV user, this fear quickly dissipates. There are three reasons why. First, the EV has a distinctive range. It varies, of course, with speed. But it's *there*. Measurable, repeatable. Second, the slower the speed, the greater the range of an EV. If I need to "full range" my car in order to take care of business, I go into "conserve" mode. This means I'm careful on acceleration, I anticipate traffic more, and I take advantage of coasting. And, third, if I've been my usual absent-minded self and the gauges say I'm low on juice, I go into my ultra-conserve mode. Friends and businesses I frequent do not begrudge me the 15 cents per hour of electricity I might need if I'm a little low. If I do run out of juice, though, I simply pull over and read a book for five minutes. The batteries will then "recover" and allow me a few extra miles. The bottom line: I've *never* been stuck out in an EV.

Transportation has changed for me. Driving an electric car has taught me some humility. I don't mindlessly power myself around anymore. I always think about what I need to do, and how I will do it. Somehow, everything gets done. For the first year I drove an EV, I had a backup vehicle. Each time I wanted to go somewhere, I could choose between my electric Honda and my VW van. Alas, the van sat forlornly at the curbside the whole time! I always found a way to make the electric work!

The Limitations of EVs

People who don't drive EVs like to talk about all of their limitations. It's time for my confession. What do I miss with my EV?

I miss making appointments to have my gas-powered car serviced or repaired. I miss the oil and grunge in the "motor" compartment. Or on my clothes and hands after I work around an engine. I miss the oil spots on the driveway. I miss periodically replacing oil filters, air filters, fuel filters, fan belts, plugs, points, and plug wires. Or checking, adding, or changing the oil. Or adding coolant. Or adjusting the plugs, points, timing, and carburetor. I miss the tangle of pollution-abatement equipment. I miss smog checks. I miss waiting in gas lines. I miss pumping gas.

Automakers (those who don't make EVs) claim that electric vehicles are low-performing and limited in range. They claim that, since the battery pack must be replaced every two to three years, electric vehicles are too expensive to be competitive with gasoline-powered cars. They claim that technological breakthroughs in batteries and motors are required before the EV will meet performance standards that the driving public has come to expect from automobiles.

Give me a break! Every sixth commercial on the major TV networks is about cars. That's a lot of hammering about the freedom, success, sex, power, and prestige a new car will bring us. After 50 years, the message is still seductive. But it's also wearing thin. What most of us want is reliable, affordable, safe transportation in our daily lives. Something that doesn't mess up the planet, foul the air we breathe, and make us want to retire to the mountains to "get away from it all." Let's examine some of the auto industry's anti-EV claims more closely.

How Far Do You Drive?: The daily trip length for over 80% of the population in the urban sprawl of Los Angeles is less than 20 miles. At freeway speeds, the average converted EV will go twice that distance. The better conversions have a strong 60-mile range. Want more than 40 to 60 miles? Would 80 to 100 miles work for you? The onboard battery charger permits worksite recharging from a standard wall socket. Overnight charging, then, gets you to work, and recharging during work hours gets you back home. *If* you have that far to go. Eighty miles a day for $2.50 of electricity is pretty good.

Another thing to consider is that a new "superhighway" is coming: telecommuting. You work at home, or at a nearby "station" that connects you with your place of employment. Or you order goods, or move goods around, without going anywhere yourself. The need for a daily commuter car thus diminishes.

Operating Costs: The annual operating costs of an EV are quite low. Even high rates for electricity still produce a "fuel" bill that is 30% to 50% that of gasoline. Other than watering the batteries four to six times per year, there is no other component (connected with propulsion) that is likely to need adjustment, repair, or replacement. If your EV does not have regenerative braking, brake jobs will occur more frequently (the car is typically heavier) than with an ICE car. Conversely, an EV *with* regenerative braking will require less brake work than its ICE counterpart; plus the batteries will get the power normally dissipated as heat in the brakes.

The battery pack will need replacement every two to three years, depending on your driving habits. Heavy acceleration, daily use, high speeds, and full-ranging the EV will take a toll on battery service life. Battery replacement costs $1,200 to $1,800, depending on the size of the pack. Despite this large expenditure, careful record-keeping by many EV owners clearly shows that the operating cost of an EV is often less than for a gas-powered car. Why? ICE cars accumulate repair and maintenance bills with more frequent visits to the shop over the same number of years.

The deep-cycle lead-acid battery is considered by automakers to be too primitive and low-performing. Most are betting on exotic types now under development. However, this casual rejection ignores that a whole industry exists right now to recycle lead-acid batteries at a fraction of the cost and toxicity of other battery technologies. What's more, lead-acid battery technology has yet to reach its full potential — improvements will come quickly when there is incentive to do so! The EV provides that incentive. The Advanced Battery Consortium has ignored the lead-acid battery. Consequently, a Lead-Acid Battery Consortium has been formed to continue this work.

In this midst of this rivalry, it is important to remember, in the greater view, that a better battery is *not* the priority. Clean air is.

Performance: Electrics have something that engines don't until they're screaming: peak torque at takeoff. With utmost elegance, most EVs can beat almost any ICE car off the line at the traffic light. Whether you indulge yourself or not, it's nice to know that you *can* do it. If an EV is observed accelerating slowly or going slower than the traffic around it, it doesn't mean it can't. Joggers don't sprint through their morning routine. Like the EV driver, they are simply conserving energy for the long haul.

Range: EV owners know that range is not the major issue that automakers insist it is. EV owners *know* about range. Most EV owners drove gas-holes before they discovered EVs. Then, they just thought they knew how far they had driven. Or how far away certain places were. Like everyone else, they described distances in terms of time. "L.A. airport is half an hour away," they might tell an inquiring driver. Rarely did they know the number of miles. Driving an EV, they know exactly how many miles away *everything* is. Or they can make a good guess.

"What if you need to go to Phoenix?" is a predictable response when I tell a group the range of my EV. "I fly!" I answer. Locally, I walk, ride a bicycle, or drive my EV up to 35 miles away. Anything beyond that, I ride the bus or rail, or fly.

It is unlikely that better batteries (at a reasonable cost) will improve the range of an electric car. Fast-charging stations are also a questionable solution. The benefits of such a fast-service utility would likely be outweighed by the decrease in service life of the battery itself.

Building the Unlimited-Range EV

Will service stations go away? Is an EV of unlimited range possible? The answer to both questions is yes.

For two years running (1992 and 1993), the Hackleman-Schless team had a perfect record with its Formula electric racer, grabbing four first-place trophies in the Open class at the annual Phoenix Solar & Electric races in Arizona. This feat may be attributed to four factors: lightweight design, aerodynamic body, reliability, and a fast-swap battery pack. With four people, we exchanged two battery modules (total: 500 pounds) in pit stops in less than 20 seconds. Repeatedly.

After the 1992 wins, we wanted to apply our fast battery-swap scheme to a standard car. A new 1992 Geo Metro was converted to electric propulsion. The seats were removed and a rectangular hole was cut through the bottom of the vehicle. A steel receptacle for a battery "module" was welded in place and the seats custom-fit. Thirty low-profile Trojan batteries (three paralleled 120-volt strings) were arranged and wired into a steel case. A stock 12-volt winch was installed in the trunk and a clever arrangement of pulleys routed steel cables to four removable anchor points on the module. With this setup, we could raise, lower, or hold the battery pack in place.

When the *GeoMetric*'s driver wanted a fresh battery pack, he (or she) drove up behind a fresh module, parked, opened the trunk, and operated the winch, dropping the depleted pack (a matter of three seconds). A quick walk around the car unhooked the anchors (ten seconds). The car was then pushed (or "motored," using the onboard auxiliary battery) to a position directly over the fresh pack (15 seconds). The anchors were re-affixed to the module (ten seconds). The winch was operated, pulling the module up into the receptacle flush with the bottom of the car (four seconds). Using this method, a 1200-pound battery could be exchanged in 42 seconds time.

The reasoning behind this project was that "refueling" an EV should be as convenient as changing the battery pack in a video camera. Consider the implications of a fast exchange of batteries:

The Schless-built GeoMetric demonstrates the simplicity of exchanging large battery modules in large EVs.

- This EV is *not* range-limited. Whether you drive long distances routinely or infrequently, you can go the distance whenever you want, for only a small service fee.
- Any front-wheel–drive, manual-transmission car can use battery-exchange.
- With offboard lifting devices in a carwash arrangement at a service station, the driver need only insert a credit card, and the exchange (selection, credit for remaining charge in old pack, and billing) will occur automatically.
- The batteries are located at the center of gravity (no adverse effect on steering or handling), positioned low (rollover protection), clustered together (easy access, better operation), inaccessible from the passenger compartment (safety, liability), and helpful in overall structural integrity (stiffened sidewalls at bumper height).
- The EV's owner does not *own* the battery pack, but leases it instead. This makes the initial EV purchase less expensive.
- The EV owner does not have to water, test, or clean the battery pack. This avoids exposure to shock, hydrogen gas, dilute sulfuric acid, or other hazards. Corporate lawyers will love how this handles liability issues.
- The EV owner can choose to recharge only at home, exchanging packs every few months only to meet the maintenance mandated in the lease.
- The battery pack is checked and serviced whenever it is exchanged. Problems are spotted early, and bad cells and connectors can be replaced. The whole job is performed by trained personnel.
- The exchanged packs are recharged at night, when utility rates are low and energy is abundant. This avoids adding to the daytime peak utility load.

Infrastructure

In truth, one-half of the infrastructure is already in place to make EVs work. Virtually every home, office, and business uses electricity. The installation of unmetered, 20-amp household circuits is adequate for most EVs. A little bit of "opportunity charging" helps batteries in a big way, yet an hour at 15 amps amounts to little more than 20 cents of power. Restaurants and businesses can easily afford this for their customers. Or, they could charge 50 cents and make a profit! EV owners will gladly pay this and more.

Two years ago, a nationwide contest was sponsored in the U.S. to solicit specific ideas on ways to integrate alternative fuels and EVs into existing cities and communities. The results were remarkable, indicating that there is great enthusiasm and solid potential for integrating these ideas into mainstream cities, communities, and society.

The 1998 2% ZEV Mandate

A milestone is looming four years off. By 1998, 2% of all new cars in California must be ZEV (zero-emission vehicles, such as EVs, flywheel cars, hydrogen cars, etc.). In other words, this means an auto manufacturer must sell two EVs out of every hundred vehicles it sells. There will be a $5,000 penalty for *each* non-ZEV car sold beyond this ratio. And, importantly, this ratio will be based upon actual consumer sales: cramming a big, heavy, boxy van full of batteries won't get the manufacturer off the hook with a "nobody wanted or could afford it" argument. Other states have also adopted this mandate. Even Canada is close to joining the ZEV club.

Several large automakers are balking. The 1998 mandate, they claim, isn't a milestone. It's a wall. Influence-peddling is in high gear to kill, delay, or weaken the mandate. It generally takes four years for a product to go from an auto company drawing board to the showroom floor. If the mandate isn't altered, auto companies must commit to doing the work now.

Much of the heat is unofficially aimed at CARB (California Air Resources Board), the agency charged with enforcing the 1998 ZEV mandate. At first, one of the giants in the auto industry suggested that they were "counting on the flexibility of CARB in this matter." CARB chairperson Jananne Sharpless's response was a suggestion that they "step outside in the real world and take a look around." Will CARB back off? "Any car company that can't deliver ZEVs in 1998," replied Sharpless, "won't be selling cars in California!" This is a *big* slice of pie. The 2% ZEV mandate for 1998 represents a 40,000 EV market in California alone!

General Motors, owner of the prototype Impact, a stunning EV designed specifically to run on electricity, is one of the protestors. While they are building 50 Impacts for customer testing, large-scale Impact production was killed as part of a general cutback. In its unveiling, the Impact was an immediate hit, surprising GM marketing personnel, who "promoted" the vehicle in an uncharacteristically apologetic manner. With sports-car acceleration, good range, and a cruising efficiency that tops a 90 mpg equivalency, this validates thinking "light and aerodynamic." For a comparable size of car, the Impact has chopped both drag and weight by 50 percent without sacrificing crashworthiness or safety.

How does CARB know that an affordable ZEV is possible? Jerry Martin says, "ZEVs are already being produced in California by smaller companies that meet our expectations and are comparably priced." CARB already certifies conversions (cars converted from engines to electric drive) and kits from Electro Automotive, Solectria, Solar Car Corporation, and others. This includes Geo Metros, VW Rabbits, Chevrolet S-10 pickups, and imports like the Kewet (from Green Motor Works).

It is the perception of most EV advocates that automakers have been dragging their feet on the issue of EVs for *many* decades. In the U.S., some of this foot-dragging has occurred at the taxpayer's expense. Prototypes have ranged from embarrassingly awful to good. Despite favorable reviews in magazines and newspapers, though, there has been little follow-up. Automakers have all too quickly dropped taxpayer-sponsored EV work when gas prices dropped.

Comments like "We can't do it," or "We need more time!" are especially revealing when you consider that the 1998 mandate for 2% ZEVs was actually announced in 1990. A joke circulating in EV circles is that a certain U.S. automaker hired a thousand lawyers to *fight* the mandate while a certain Japanese automaker hired a thousand engineers to figure out how to *meet* it.

Further evidence tends to reflect the disdain automakers feel about anything that doesn't have an engine:

• The Vehma-built GVan and Chrysler TEVan, 8000 and 6000 pounds, respectively, and priced at $75,000 to $125,000 (each) to utilities, represent conversions that no self-respecting EV designer would ever build. Every official I have talked to from a utility that owns a GVan considers it a bad joke. It's like putting wings on a brick. You can do it, but don't expect it to fly.

• GM, Chrysler, and Ford recently accepted a commitment of several billion dollars from the Clinton administration, promising a 90 mpg car in ten years' time. The GM Impact, with half the weight and half the aerodynamic drag of an equivalent car, could probably meet that goal today with a fuel-efficient engine installed. (The Impact was largely designed and built by AeroVironment.) Still, there is definite resistance to building an electric version of the same model.

• Are four years really needed to produce a viable ZEV? Stock conversions by Solectria and Solar Car Corporation of a variety of late-model, four-passenger cars and trucks sold by GM, Chrysler, Ford, and Honda have already paved the way. Many more automakers are represented in conversions done by individuals throughout the U.S. The truth is, EV prototypes and production cars have already been built and tested worldwide, anywhere that gasoline prices are high. The U.S. market, with the cheapest gasoline in the world, is merely the last market for these EVs to penetrate. It seems as though automakers have a very poor opinion of the American driving public. Or is it fear? In these other countries, the wisdom of rail and electric bus dominates, minimizing the need for cars. When the American public awakens, will they follow suit?

A company may feel reluctant to incorporate *real* improvements in this year's product, lest it make last year's model obsolete. In today's competitive and rapidly changing market, though, such foot-dragging is more likely to ensure that a current product is made instantly obsolete by a competitor's product!

The Supercars Are Coming

A magnificent paper, written by Amory Lovins and the staff at RMI (Rocky Mountain Institute) and recently presented at conferences and in magazine articles, has shaken the automotive industry worldwide. (The excerpted piece is reprinted starting on page 509.) At last, someone has *quantified* the research findings from projects all over the world into an enlightening vision of what *is* possible relative to transportation. Anyone who has ever designed and built a safe, practical, affordable, lightweight, aerodynamic vehicle will appreciate this work. While it is not specifically written for the layperson, the paper's message is clear: the old way of building cars won't work anymore. Lovins is a master at both cranking out the numbers and providing the analogies that will help executives, engineers, designers, and entrepreneurs see the handwriting on the wall. For anyone who wants to be a part of the transportation future, I strongly advise getting and reading the full text of the paper.) In it, the physics of energy, mass, velocity, acceleration, and staid thinking are clearly revealed! Most of us know that square functions get big fast, but here's the proof that square functions get small even faster! The power of "less is best" is formidable. For example, in a formula, if x is one-fourth, and y is one-third, and z is one-half, what's the value of xyz? One-twenty fourth! This basic math is the secret of the GM Impact's success. Paul MacCready (and team) used the same principle to make the age-old dream of human-powered flight possible with their Gossamer series.

Supercars, like the ones Lovins envisions, combined with electric propulsion, would drive the Btu/mile consumption of our resources so low that they would appear insignificant compared with today's use (or rather, abuse).

The City-el, a European production vehicle, is tested at SMUD's facilities in Sacramento, California.

The Crashworthiness of EVs

Electric propulsion is an elegant technology that is nicely complemented by a lightweight chassis and aerodynamic body. Predictably, consumers are wary of small, light cars because of the issue of crashworthiness. Ultralight aircraft was the first industry to demonstrate superior strength for a fraction of the weight, using lightweight composites as structural materials. Today's race cars, TV viewers may recall, provide further proof of the crashworthiness of lightweight vehicles. Again and again, drivers spin their Indy cars into the wall at 200 mph — and frequently walk away or sustain only minor injuries. Not so widely publicized is the fact that today's small car is no longer the easy loser when it tangles with a big car.

Lovins touts the advantages of carbon fiber and fiberglass construction, but there are relatively nontoxic alternatives, too. In the 1930s, Henry Ford actually built a car body out of *soybeans* that was easy to shape, paint, and repair. Imagine a car body that you could shred and toss onto the compost heap! There is also a growing movement in the U.S. to legalize hemp, another source for the cloth-and-resin material needed for composite bodies. While the Hemp Initiative is hampered by the ancient plant's association with marijuana, we may someday again have an organic, environmentally friendly alternative to oil and plastics, drugs and timber.

SUPERRAIL

The United States is very focused (some would say fixated) on the automobile. Unfortunately, even major improvements — using lightweight and low-cost materials, aerodynamic and crashworthy bodies, and efficient propulsion — will not circumvent other problems associated with the automobile. Cars, not just engines, contribute to the planet's environmental problems.

Fortunately, what we perfect for the EV on *roads,* we also perfect for EVs on *rail.* It is basic physics that steel-on-steel offers only 10% of the rolling resistance of a tire on a road. More than 30% of the propulsive effort of a car (ICE or EV) is lost in rolling resistance, which is evident in the heat and noise of tire friction. Rail, then, can transport people and freight for a fraction of the energy and resource costs normally consumed by cars on streets and highways.

Rail transport and trains seem to have an image problem with the majority of the American driving public. Why? Major reasons include the supposed "inconvenience" of train travel: the lack of privacy, the loss of time, the need for transfers, the absence of service to some areas, the ignorance of schedules, the need for change for tokens, the inherent baggage limitation, and the fear of crime and violence.

In the recent Los Angeles earthquake, however, many highway spans collapsed, creating a commuting nightmare for millions of people. A fairly new, little used MetroRail service saved the day for many commuters, leaping to full capacity overnight. Undoubtedly, many people will continue to use this service long after the roads have been repaired. Apparently, change seems unattractive unless compelled by some emergency or crisis. Then fear succumbs to convenience. People use what works and what's convenient, and rail offers both.

In many European countries, rail is the preferred mode of transportation. Many generations of people have been raised in its presence. In the U.S., though, rail was lost early in this century. State ownership of rail, like today's road systems, would have saved rail transport, but it was enacted too late. Weakened by operating costs, maintenance slipped, repairs went undone, and modernization became impossible. The system was easily overwhelmed by the aggressive promotion of automobiles and the outlaw tactics of sponsoring corporations. That influence extends into the modern day. Cities looking to install locomotives, train cars, trolleys, as well as light-rail and monorail systems must purchase their rolling stock outside the U.S. because (with the exception of a few small companies like Bombardier) it is not manufactured domestically.

The nationwide loss of railroad right of ways (many are now fragmented or privately owned) suggests that the way back to rail will be difficult. However, the same technology used to build the Supercar can be readily applied to SuperRail. This raises some interesting prospects for communities. The loss (or lack) of rail right of ways is not an issue with SuperRail.

Imagine dedicating every seventh street to rail (north-south *and* east-west). An ultralight rail (ULR) car could easily transport 40 passengers on the energy required by two electric cars. With a GVWR (gross vehicle weight rating) of 7500 pounds (only 15% of the weight of light-rail in use today), the ULR car would run on specially built, fast-laying, modular track. Since both the track and the ULR car are relatively lightweight, a standard asphalt road is strong enough to act as a roadbed. Anchor bolts (shot from a gun or screwed in) secure the track to the roadway surface. Power, signal, and control wiring is built in (at the factory) to the low-profile track modules. This design makes it easy to electrify each section of modular rail *only* as the ULR car passes onto it. Fault-detection devices warn of shorts. A relatively small battery pack carried onboard the ULR car supplies propulsive power at pedestrian crossover points, or during a power outage.

Each ULR vehicle works as a stand-alone car. However, when two or more ULRs are coupled together, control is transferred to the lead vehicle. The weight and cost of the locomotive is eliminated in this scheme, since propulsion is supplied uniformly throughout the length of the "train." Traction, then, increases in direct proportion to the load of passengers and/or freight.

The beauty of ULR technology is the speed with which it can be installed. A whole mile of track could be installed overnight. It's easily *undone*, too. If the trial period suggests an alternate layout, the track could be easily removed, the anchor holes patched, and the track rerouted.

A major advantage of transporting people and goods by rail is safety. Department of Transportation reports indicate that 95% of all car accidents are *driver*-related (i.e, drunk driving, speeding, fatigue, etc.). While state transportation departments maintain the roads and highways, they do not (and cannot) inspect the cars that travel on them. A state-owned rail system would inspect and maintain both the rail and ULR cars, and greatly reduce the possibility of driver-induced accidents. This suggests that tens of thousands of auto-related fatalities might be prevented each year by switching to rail in the USA.

A SYSTEM APPROACH TO TRANSPORTATION

Transportation has long been managed by means of crisis and politics, with short-term gains emphasized over long-term solutions. If the public has repeatedly shown its impatience for change, it is likely due to the betrayal they feel when told how much better their lives are going to be once a new highway or an additional lane is built. Apathy is about promises broken.

Transportation cries out for a system, not a collection of niches. For example, the city of Honolulu wonders why the tourist trade has fallen off. Each hotel and tourist bureau uses its own buses to hastily transport guests to and from hotels and through tour routes. Each room has its own air conditioner. The whine of thousands of air conditioning units permeates the nighttime sky, and the stench of diesel wafts through the streets night and day. Island fragrances and vistas of sea, land, and air are crushed in the stampede of human flesh to visit or profit.

Turning the Tide

A tide is turning in our nation's transportation policies. What can we do in the near and long term to move it along?

- Pray for the survival of the "2% ZEV by 1998" mandate.
- Adopt transportation systems appropriate to the application.
- Scrap ICEs planetwide.
- Aggressively limit the use of carbon-based fuels, including the new RFG (reformulated gasoline) fuel.
- Subsidize sustainable energy resource development.
- Fast-track ULR technology.
- Adopt CNG and alcohol fuel (from grains) as interim fuels.
- Initiate education on energy and transportation alternatives in the media, on the screen, and in the classroom.

It is easy to feel dispirited about the challenges that lie ahead. The fear of change is strong. Leaders who argue that the issues are too complex, that the oil and automotive industries are too entrenched, or that trades like freeway construction will be destroyed by using alternatives must be replaced with people of vision, compassion, and fortitude. The voices of young people are important. They inherit the pollution and garbage of decades of abuse and squandered resources. Only they have the stamina to see us through.

Oil-based technology continues to impact the environment. The party is over. It is time for all of us to show the fabled "right stuff," moving transportation and the human species into an alignment with the environment.

Let's do it while we still have choices.

— *Michael Hackleman*

Michael Hackleman is the editor of the Go Power section in Home Power *magazine and an EV designer and consultant.*

MOBILITY BOOKS & VIDEOS

Convert It

Mike Brown. This book, the 1994 edition, is a step-by-step manual for converting a gas car to an electric-powered car. It's a very readable and practical manual for the do-it-yourselfer wanting the fun and educational experience of converting a conventional automobile into an electric vehicle that will be both practical and economical to operate. Included are a generous number of illustrations of an actual conversion, along with instructions for testing and operating the completed vehicle. Mike Brown shares a wealth of his own personal experience in making conversions and includes many practical tips for both safety and ease of construction that will be useful for the novice or the experienced home mechanic alike. Brown does a great job of explaining the cost of the conversion components. This book has been strongly recommended by the Electric Auto Association, Electric Vehicle Progress, and Alternate Energy Transportation. Recently republished and expanded to better than twice the size with a lower price! 126 pages, paperback, 1993.

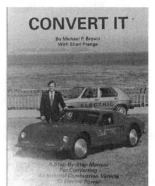

80-404   Convert It $25⁰⁰

Electric Vehicle Components Primer

Electro Automotive Information Services. Yes, electric cars are simple. But should you use a series DC motor, brushless DC, or an AC? What about aircraft generators? In this new video, Mike Brown, author of *Convert It*, sorts out all the choices. It's a good starting point if you are considering building an electric vehicle, or just want to know more about how they work. Brown's business, Electro AutoMotive, has been supplying components for electric vehicles since 1979, and is known for equipping the winners of countless races and rallies. The tape, with footage of dozens of cars and components, runs 92 minutes.

80-125   Electric Vehicle Components Primer $39⁰⁰

Hydrogen Fuel Cells: Key To The Future

This video features a presentation by Dr. Roger Billings of the American Academy of Science, on the hydrogen fuel cell. This detailed video outlines the design and construction of a fuel cell and its uses; in particular, its application to electric vehicles. Dr. Billings discusses range and what kind of infrastructure it would take to incorporate this exciting technology into our world today. Cleaner, safer, and more efficient alternatives to our present power grid distribution system are also offered. There is footage of the Laser Cell I electric vehicle, and a demonstration of a fuel cell. This is a must see for anyone interested in hydrogen. 65 minutes.

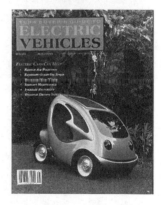

80-303   Hydrogen Fuels Video $35⁰⁰

1995 Electric Vehicle Directory

Philip Terpstra. The seventh edition of this high-quality publication. The **only** current and complete guide to the EV industry, featuring highway-legal electric cars available today. 80 pages, paperback. *Re-published periodically, no '96 Directory will be published. This is still the current edition.*

80-257   1995 Electric Vehicle Directory $14⁰⁰

The Real Goods Company Car, a Beck Industries replica of the classic 1955 Porsche 555 Spyder fitted with a 120-volt electric drive by MendoMotive. Capable of 50 to 70 miles of regular highway driving between recharges. This zippy little convertible can go from zero to 60 in less than 13 seconds, and operates at speeds up to 95 miles per hour. An on-board charger allows for overnight recharging (ten hours to recharge from total discharge) or topping off the batteries wherever a standard electrical outlet can be found. At the April, 1994 Los Angeles Grand Prix, the Real Goods Spyder blew away the competition in acceleration and braking trials. Builder Stephen Heckeroth can build one for you for about $32,500.

Drive The World's Most Exciting Electric Car

Soar silently down the highway in this replica of the 1955 Porsche Spyder. MendoMotive, an innovative design team located on the Mendocino coast, has created a 120-volt electric drive conversion, utilizing a high quality fiberglass replica. This car will accelerate from zero to 60 in under 13 seconds, and has a range of 60 to 100 miles — (our record is 130 miles!) — depending on the terrain. The on-board recharger plugs into any 110-volt outlet, letting you recharge wherever you park.

The Electric Spyder gives all the thrills without the noise and pollution.

Your cost of acquiring an electric Spyder is about $32,500. This vehicle qualifies for federal tax credits of up to 10% and no California sales tax on the electric drive components or installation. (Outside of California, check with your state tax bureau for available tax credits and/or sales tax due.)

MendoMotive also offers other economical conversion options:

1. Passenger vehicles: conversions are available for any lightweight vehicle such as Honda Accord, Civic, or Rabbit; Dasher or Geo Metro. Prices range from $9,000 and up for the conversion.
2. Utility vehicles: including lightweight pick-ups, four-wheel drives, and delivery vans.
3. Electric tractors and rototillers: ideal for organic farming.

Real Goods can design a solar, wind, or hydro system to recharge your vehicle with clean, renewable power.

For complete information on the conversion options available, call 1-800-762-7325 for our free brochure.

Conversions II: Case Studies

This is an excellent video on electric vehicle conversions, without a doubt the best we have ever seen. It is broken into three presentations — the first by Mike Brown (Electro Automotive). Mike brings his many years of EV experience to the screen. Mike covers **everything** from the ground up; this is absolutely the best advice on tape on EV conversions. He knows all the common mistakes everyone makes, and saves you big headaches bringing them to your attention. The second part is a moving presentation by Bruce Burk (St. Johnsbury Academy) on his trials and tribulations working with a high-school class to build an EV that won the 1991 American Tour de Sol open class. Again, there is much to be learned from the experiences of others. The final section is on "Safety Issues for Conversions" by Gary Carr (Vehicle Development Group). The title tells it all, and Gary does a great job of it. Some of the subjects covered are: battery placement, weight distribution, suspension, wheel base design, and crash protection. Combined, these three experts have made the most informative video on EV conversions you can get your hands on. You can't afford to pass up this video if you are considering either building or buying an EV. One hour, 47 minutes.

80-304   Conversions II Video $39.00

ZAP PowerBikes Electric Vehicles that are FUN

The ZAP (for Zero Air Pollution) electric-assist power system transforms bicycle commuting from sweating drudgery to a push-of-the-button breeze. The electric motors, battery, and handlebar-mounted controls work with almost any full-size bicycle to become a cost-effective electric vehicle. Non-pedal speeds with a 180 lb. rider are approximately 10/18 mph, faster if you pedal. Range depends greatly on terrain and rider assistance. A battery charge lasts approximately 45 to 60 minutes, and range is typically 8 to 20 miles. The ZAP auto-engagement feature allows the bike to be pedaled and coast normally, saving the electric power till a boost is needed. Regeneration allows recharging the battery when going downhill, or when used with an exercise stand.

The basic ZAP Power System consists of dual, permanent-magnet, ceramic drive, 600-watt motors, a sealed, maintenance-free, 12-volt/17-amphour/12.5 lb. battery mounted in a Cordura nylon zippered bag at a low center of gravity, all necessary switches, wiring, and mounting hardware, and an automatic shut-off, three-hour, quick charger.

You can install the Power System on your own bike in one to two hours with a few basic hand tools, or you can buy the System already installed on a high-quality PowerBike'. This comfortable, upright seating, chromoly framed mountain bike is equipped with 18-speed Shimano SIS index shifting, alloy rims with semi-slick road tires, and a Top Gear Ultra-Soft saddle for electri-cruising in comfort and style.

Several options are available to increase the utility, safety, or security of your electric bike.

The Dual Battery Upgrade provides a second battery and pack. You'll always have a charged battery ready to go. The quick-change battery pack swaps in about 20 seconds.

The Headlight Package provides a powerful 25-watt halogen driving light. Not a wimpy "bike light" that dimly illuminates what you're about to crash into, but a real headlight.

The Rear Cargo Rack is a standard unit that fits any full-size bike.

The Security System offers peace of mind when parking. Moving the heavy-duty vinyl-coated, key-operated U-lock lock for more than four seconds, tampering with the battery door, or cutting the cord 8-foot cord for looping through the wheels will set off the 107 decibel alarm.

All items manufacturer warranted for one year. Requires some assembly.

63-246	ZAP PowerBike	$989	63-249	Headlight Package	$85
63-247	ZAP Power System	$499	63-250	Security System	$79
63-248	Dual Battery Upgrade	$99	63-251	Rear Cargo Rack	$29

CHAPTER 12

LIVELIHOOD AND LEARNING

BUSINESS AND EDUCATION leaders are now telling us that, in order to compete in the world economy, we each must become lifelong learners. No longer can we depend on a career path that asks us to get our degrees, enter the job market, find a good, stable company, and gradually climb the ladder of job and salary advancement until the day we retire: getting our gold watch, trading it in for a couple of solar panels, and moving off-the-grid to the country on our pensions.

No more. Careers and industries are beginning to change so rapidly that a young person entering the job market for the first time today can expect his or her career to change completely several times before retirement; the only possible way to compete will be to remain constantly curious, constantly learning . . . and infinitely adaptable to whatever the future may bring. As unnerving as that may sound to some people, what will also be needed is a total revolution in how we educate ourselves and how information and learning tools are provided.

At Real Goods, we welcome and encourage any advancement that fosters independence and self-teaching. We are hopeful that the hugely hyped "Information Superhighway" will open vast storehouses of information that will be accessible (inexpensively) to everyone, and that the new communications technologies will enable our pioneering friends and customers in remote locations to enter into worlds that have previously been available only to city and suburban dwellers.

In our surveys and on the telephone, we have found that our customers are not lacking in curiosity. You, our customers, are constantly asking, probing, questioning, suggesting, proposing, and advising. There is a constant flow of information between you, our customers, and us, your company. Just as it is important for an individual to learn, a progressive company must remain curious . . . constantly learning, and constantly changing.

— Dave Smith

GOOD WORK: AT HOME IN THE BRIARPATCH

The Briarpatch, an informal association of friends in business, is one of those great ideas that came out of the social movements of the 1960s. It was started by several friends in 1974 who found themselves running small businesses, albeit in an unusual way. They tended to be honest, open about their financial information, and willing to share information about their businesses with their customers, suppliers, and others in their fields.

The association takes its name from Joel Chandler Harris' Uncle Remus tales of community life in the briarpatch. Brer and Sis Rabbit live and work in the safety of thorny brambles, where Brer Fox can't reach them.

The "Briarpatch Community" as an idea was originally formulated by Dick Raymond, who was the catalyst for several community-based groups around the Portola Institute in Northern California, publisher of the original *Whole Earth Catalog.* It grew out of the image of a dinosaur-like demise of large corporations. In his early visions of the Briarpatch, Dick saw big businesses unable to find enough food for their enormous profit appetites. He visualized nothing short of a business apocalypse, with forces and events in our national economy not unlike those we have seen operating through the 1980s and into the 1990s.

As early as 1972 Briars foresaw the failures we are witnessing today of S&Ls, banks, and insurance companies. They also anticipated corporate bankruptcies and downsizing, as well as the inability of society to provide equal access to jobs, education, and health care. And they saw that the best strategy for surviving this decline was to create small businesses occupying market niches that were too small or not profitable enough to interest mainstream companies.

Briars belongs to a community of businesspeople who believe in cooperating with one another, in being honest and open, in putting the quality of their products and services above pure profit, and in protecting the environment.

Briar businesses come in every size and kind. There are million-dollar restaurants (Greens in San Francisco), magazines *(Whole Earth Review),* and publishing companies (Nolo Press). There is a liberal-arts college (New College of California), a Japanese tea ceremony school (Urasenke), and a school of traditional Japanese arts (California Institute of Traditional Japanese Arts), as well as several alternative grammar and high schools. Members include accountants, acupuncturists, architects, archivists, artists, bodyworkers, bookkeepers, caterers, consultants, chiropractors, dancers, designers, doctors, editors, financial planners, floor finishers, historians, lawyers, organic farmers, manufacturers, market researchers, martial-arts instructors, musicians, painters, plumbers, photographers, printers, retailers, stockbrokers, teachers, therapists, weavers, and writers . . . to name just a few occupations.

When the Briarpatch first began in 1974, the kinds of precepts that members chose to follow were considered suspect, even weird. From the beginnings of the Industrial Revolution, over 250 years ago, businesspeople have assumed that you have to use cutthroat tactics to succeed. The very idea of sharing, and caring for one another, was considered "bad for business."

Life without Principle: It is remarkable that there is little or nothing to be remembered written on the subject of getting a living; how to make getting a living not merely honest and honorable, but altogether inviting and glorious; for if getting a living is not so, then living is not.

—Wendell Berry,
"Life without Principle"

Today, of course, mainstream corporations and consultants pepper their language with words like "excellence," "ethics," and "total quality." But finding people who really *live* these ideals is a challenge. And finding business advice that really is "values-focused" is just about impossible.

But not quite. Members of the Briarpatch have been receiving values-focused advice and delivering the best possible products and services they could for almost 20 years.

The sudden burst of businesses claiming to be socially and environmentally responsible has been enough to make the most trusting among us a bit suspicious. Just changing the look or size of a package and using the words "green" and "environment" in an advertising campaign won't really slow down environmental devastation nor provide local communities with needed social services. But open, honest, community-oriented businesses *can* help create a sustainable, socially responsible economy.

Before "quality circles" and "total quality" there was Briarpatch. Before "excellence and ethics" there was Briarpatch. Before "green business" and "green marketing" there was Briarpatch.

What if you want to start a Briarpatch of your own? In response to the hundreds of questions like this one that we receive each year, Briars Mart Pearson and Lynn Gravestock have developed the following answer:

> "Once upon a time, we were simply a small group of friends. Several of us had small businesses, and large questions. 'How do you stay alive in business if you don't necessarily want to become rich, famous, impersonal, or focused on creating a larger business empire?' 'What do the words say when you advertise without hype? And where then do you advertise?' 'What parts of business let you afford informality, and when does informality invite disaster?'
>
> "You already know that you can go to school forever and you probably won't get told or shown the answers that a Briarpatch business sometimes has a need to know. Out of our own frustration and loneliness here, we turned to each other for help. We taught each other, made suggestions, loaned equipment, traded skills, and generally struggled to create the world we wished had been there all along.
>
> "The key to a successful Briarpatch network is ongoing personal contact. If you own your own business and want to create or tap into a helpful, supportive community, just do as we did — ask the people you already know the questions you need help with. Your pool of friends and acquaintances is an invaluable resource. You will find that in most groups of 20 to 30 people almost all needs will be met. For those that aren't, at least strong leads will be provided.
>
> "One danger that small business owners are particularly susceptible to is that of becoming a store 'shut-in.' We can't stress enough how important it is to keep lines of communication open. If you don't have time to get out and visit much, then make your customers part of your support group. You might be surprised at how much they have to offer, and how willing they are to share it.
>
> "Once you've got your business and a group of supportive friends, you've got a Briarpatch. Over time your community will grow and grow and grow. We

did this for some years, and then we were given a small taste of national and international fame. Letters started pouring in. On the surface they said, 'Send me anything you have. Can I join your group?' Underneath they asked, 'What do you know that will give me a sense of belonging?' 'How can I be a part of a small business where people care for each other?'"

Briarpatch meets the kinds of needs that Mart and Lynn describe by providing emotional support and shared advice on management, taxes, bookkeeping, legal questions, and marketing. As coordinator of the network, I receive a dozen calls a week from members. One call may be about the details of writing a partnership agreement. I might refer the caller to a do-it-yourself book from Nolo Press on partnership agreements. Another call might be a question about a statement of terms and conditions for clients. I'll pull some samples that have been donated by past members and mail them to the caller. Another call might be from a landscape contractor or a jewelry manufacturer, either of whom may want help in establishing a line of credit with their bank. I'll schedule an advisory session, in which the volunteer team of technical consultants evaluates the business' financial condition and makes recommendations about how to present the member's case to the bank.

This kind of technical assistance is available to individual businesses from a team of volunteers, some of whom are experts in their own field and all of whom have common sense. One day each week this team is available to visit businesses to assess their problems and strengths.

"You learn more about the health of a business through visiting it than you can in an hour's conversation with someone," says one volunteer. "If the back room is a mess and there are no records being kept, you know instantly that the business is in trouble."

In addition to the volunteer consultants, advisory boards are formed on an ad hoc basis to provide particular assistance to a business, especially if it is facing problems or challenges or an emergency situation. Whether it is technical consultants or a board of advisors, the owners of the business are required to provide a short summary of the business' financial condition and its problems as they perceive them.

"In our culture," comments an advisor, "if you get into trouble you don't tell anyone. In the Briarpatch we do just the opposite." Briars believe it's easier for a brainstorming group of informed and sympathetic advisors to come up with a creative solution to almost any problem, one that will be better in the long run than a plan thought up by a single isolated business owner.

Of the members, 55% are women and 89% are people of color. Some 62% are between 30 and 50 years of age and 83% have been in business more than five years. One-person businesses make up 58% of the membership, and another 34% have fewer than ten employees. Some 17% of members work with nonprofits, and 52% provide products or services primarily for individuals or professionals. Around 94% practice recycling or other activities to lower their impact on the environment, and 6% are businesses that deal directly with environmental issues. Since 1974 over 1000 businesses have been members. Today there are about 300 member businesses in seven countries. About 150 of them are active in the Bay Area, and the others live and work around the U.S. and in Japan, Sweden, New Zealand, and elsewhere.

The Briarpatch does not advertise its existence, does not have corporate sponsors, and does not do any fundraising. It is entirely supported by voluntary dues payments and a team of volunteer technical advisors. Dues payments to the Briarpatch treasury help provide: 1) monthly meetings where new and old members meet to offer each other emotional and brainstorming support (from 10 to 20 people routinely attend); 2) weekly visits to member businesses by a team of volunteer technical consultants (an average of 50 businesses a year are visited); 3) ongoing emotional support and resource referrals by telephone (the coordinator routinely fields 20 to 50 calls a month); 4) a 50-page member directory with Yellow Pages index and descriptions (issued every couple of years); and 5) a newsletter featuring member news and technical tips (published at least once a year). The newsletter is used by members to promote an event, pose a question, or make a request of other Briars. The network coordinator spends between 300 and 400 hours a year — mostly fielding questions and bringing Briars together so that they can share information, resources, and support.

We hope that you find this information helpful, and that your own network, whether you knew it existed or not, grows to meet whatever your future needs may be. Let a thousand Briarpatch networks blossom.

— Claude F. Whitmyer

Claude Whitmyer is the current Briarpatch Coordinator. He is an internationally known right livelihood and values-focused business consultant and founder of The Center for Good Work.

Claude is also Associate Professor and Director of the Master of Arts in Business program at the California Institute of Integral Studies. He is the coauthor, with Salli Rasberry, of Running A One-Person Business *(see page 547), and the editor of two anthologies:* In the Company of Others: Making Community in the Modern World *(Tarcher/Perigee, 1993), and* Mindfulness and Meaningful Work: Explorations in Right Livelihood *(Parallax, 1994).*

LIVING IN YOUR OFFICE

It is a clement January day in Caspar, California; sunlight alternates with lazy curtains of rain whirling lazily in from the ocean to the southwest. You can almost see the native grasses growing, casting off the last golden shadings of autumn with bumptious spring green. Our squadron of bucks cannot keep ahead of the growth, which will be deep enough to lose a six-year-old in by April unless I intervene. Mowing season again, at least to clear enough field for a good Frisbee toss. Grazing cattle, as my neighbor does, keeps the grass neatly clipped, but the meadow muffins left behind by what my father calls "hooved locusts" add unwelcome obstacles to Frisbeeing. For the last several years, I have mowed with a noisome, cantankerous

---

The real work of planet-saving will be small, humble, and humbling, and (insofar as it involves love) pleasing and rewarding. Its jobs will be too many to count, too many to report, too many to be publicly noticed or rewarded, too small to make anyone rich or famous.

—Wendell Berry

dead-dinosaur-swilling internal combustion pig. Reputedly the dirtiest form of petroleum consumption known, excepting chain saws and Saddam Hussein.

During a good sunny slice around noon I tried out my new solar-charged mower. Bliss! Less than half the noise of the old gas hog, and with an altogether more satisfying electrical hum. This new mower has more than enough power for our tough volunteer turf. Noontime aerobic mowing, even behind the infernal combustion pig, has been a part of my working regimen for years. Problems line themselves up among the wild radish clumps, and I methodically mow them down. This new mower enables meditation at a higher level, and I return to work mentally and spiritually renewed.

The mower's video owner's manual says, "For best results when cutting tall grass, walk more slowly." Fijians walk more slowly when the hill steepens, and kindly notice that Europeans (which means white folks, with no slur intended) walk grimly faster or doggedly maintain our pace.

There is something almost sacramental about the year's first mowing. This year I welcome as an old friend the recognition that my body is a year older but the grass grows just the same. I feel a little tougher, while grass still yields to a sharpened knife. The electric mower's stamina surprises me by exercising me well into my aerobic range and keeping me there for half an hour.

Back inside there's a league playoff game coming up on TV, but it's a workday like any other for me. The tasks of office, woodlot, meadow, kitchen, building site, and nest-maker blend smoothly together, erasing the distinction between life and work, weekday and weekend. I'm always integrating new tools into my overall operating system — this week a PV-recharged mower, while on my computer there's a new disk operating system and attached, a new color printer. We Americans allot more time to digesting innovation, I'll wager, than people at any other time in history. The new mower's indoor home needs an electrical outlet; Rochelle brings the tools and findings I need, and the line extension in the tool shed goes swiftly. The action on the field is briefly interrupted for halftime so I interrupt it completely to power down and connect the new wires to the system.

How can we be sure that we innovate appropriately? For eight years in matters of alternative energy I and thousands of other people have turned to this *Sourcebook* for advice. Over those years we have evolved our home offices and our domestic rhythms, trying to follow as few dead-end paths as possible. Despite our best efforts, though, plagues of Edsels, Osborne computers, and other cutting-edge technologies have sometimes amputated us from chunks of our grubstakes and left us feeling burned and stupid. Once again, we bought technology that looked too good to be true . . . and it was. I am now working on a game plan, though, to help choose winners and shun dead ends. Here's my first rule: Never eat anything bigger than your head. Computers constantly teeter on the brink of being smarter than me. Sometimes, on the verge of being buffaloed by a pile of silicon and rarified metals, I hear my father's advice, "You've just got to be smarter than it is." I have learned to go to bed defeated, trusting that I'll prevail eventually, with a little help from my friends.

Another rule: Ask for help when you need it. Here in Caspar, my neighbors and I work different sectors of the fireline of progress. Those who are working the same virtual pitch are often miles, hundreds of miles, *time zones* away. At times I miss having "face-time" with those fellow firefighters, although my reclusive working style suits me: most of my colleagues are voices in my ear, images on the fax, a shared screen and keyboard across the miles via telephonic lifeline. One moment I am working intensively with someone, then alone again with the click of a button a moment later. Some friends and coworkers would wilt without the office community, or would work-dance into oblivion. E.M. Forster provided in his epigraph to *Howards End* a favorite precept: "Only connect." My technological batting average (TBA) has bloomed as I, and like-minded millions, have discovered the on-line community. Neo-Luddite purists complain that this "pseudo community" carries us away from our localities, but for me it creates a new orb for communication. Before buying a tool I try to cruise the electronic byways to see what pioneers have encountered in their explorations. I count on finding good advice, encouraging stories of difficulty at the beginning crowned by eventual success. The general level of dedication, friendliness, and expertise sometimes leaves me feeling like a dilettante. Knowledge launches itself so far into the regions of intriguing complexity that I must constantly exercise restraint lest I stray too far (and expensively) from my proper tasks. If I seem to be ranging alone, I post a question, and return a few days later to find that some patient soul who solved the problem months ago has taken time from his or her busy life to lend a hand. Or I may find a queue of fellow-sufferers, each of whom is more impatient than the last that the technology is so funky and unbiddable. Sometimes a sheepish admission from the responsible technologue follows: a problem exists, and will be solved. Real Soon Now, as they say. A latter-day version of "the check is in the mail."

What we need most desperately, I think, as I wade through my weekly E-mail, is a good knowledge navigator, a cybernaut capable of fetching precisely what I need without my intensive and prolonged coaching. It is on this front where we need the next breakthrough; we can make lead acid batteries work well enough to get us through the night. I dream of launching my electronic amanuensis onto the nets like a peregrine from the gauntlet, then awaiting her return with just the morsel of data I need. Whenever two or three gather: if my clone meets yours on the nets, do we not comprise a community?

The cost of staying *au courant* in the home-office game is an invisible and frequently understated cost of doing business. I have already mentioned the cost of time lost during innovation, learning to use new tools well enough not to do mischief. Dave Katz overstates the matter only slightly when he says, "Be prepared to spend three hours before you can do anything useful with a new tool." New tools had better cost-justify themselves, and, in this era when planetary correctness is *de rigueur,* the money cost of maintenance upgrades, new models, and other constant pressures to keep dangerously close enough to the cutting edge is joined by a new factor in reckoning effectiveness. I have resolved (another rule) that new technology must reduce my reliance on imported energy. At the end of the road, even without a daily commute, I drive too much and almost every call is long distance. For best results when cutting tall grass, walk more slowly.

As you see, most of the practices of ecologically aware business scale down to home office size, and I must allocate time to man the Eco Desk wearing my reduce-reuse-recycler's hat. (If anyone can suggest a nonsexist way to say that, please tell me. I would love to be able to say, "Danger: in times of rapture this vehicle will be unpersoned," but I cannot).

After evolving a home-office living system for decades, I catch brief but gratifying glimpses of a self-sustaining future. Today I found a two-foot length of romex wire just right for extending the house current to the mower's new back-up socket. Recently I discovered draft-quality paper made from hemp and cereal waste without trees or bleaches. My family and coworkers got tired a long time ago of me chiding them. "See this half-used piece of paper in the wastebasket? I buy you the *expensive* kind of paper, the kind with *two sides* for your writing enjoyment. Please use both."

Recommended Products for the Home Office

To operate a home office is to be a jack of many trades with a strange box of tools. Such a toolbox might well contain the following very diverse objects . . .

The Stapleless Stapler: Staples are an abomination: paper shufflers and the post office hate them, and the energy embodied in them, while minuscule, is still unpardonable. This clever little plastic device pokes a neat paper hinge in the folded corner of a thin sheaf (up to five sheets, or eight if you're using fax or tree-free paper), which holds the pages together nicely.

The Ryobi Mulchinator: A self-powered electric mower, which means no trailing extension cord to trip over or slice. Two 12-volt maintenance-free batteries wired in series produce 24 volts of power, enough for tough wild radish stalks and deep grass, and are recharged by two solar modules or can be charged from house current. Keep the office grass under control and solve problems at the same time.

Tree-Free EcoPaper: Thin and lacking tooth, this utility-grade paper is perfect for office tasks. It's too soon to tell if this paper will stand the test of time, but I feel better about using it for long reports and first drafts. One side is slightly rougher than the other, and takes ink-jet printing as well as more expensive paper formulated for the task. An issue to consider: hemp and cereal crops are annuals which tend to deplete the soil; is it better to use 100% recycled paper?

Tools for Feeling like You Aren't Isolated in Your Home Office
Well-Connected

In the last two years, electronic mail (email) has become my most important medium for collaborative work. Email combines some of the best qualities of several other media: the immediacy of a phone call, the permanence of a letter, the convenience of a fax, and editability: words come to me as 'recycled electrons' which I can easily copy and paste using familiar electronic tools. And I am *NOT* consuming reams of paper! Some hardware is required to use email, which critics claim makes email elitist, but many local libraries and schools offer easy, inexpensive access.

To receive email, you need an address, which means you need to open an account with a local Internet Service Provider. (ISP; you will find the 'Net world littered with acronyms.) Many communities are devising ways to connect inexpensively. For some rural folks, the cost of a daily long-distance call may make email prohibitively expensive. There are many national services, and the scene is changing daily. Compuserve, long the champion, is ebbing, and the other commercial services, America Online (AOL) and Prodigy, are too expensive and in my view consumerist, like pay TV. AOL sports a spiffy user interface, but it is forever messing with your system, randomly downloading fonts and advertising graphics onto your computer at your expense. What I dislike about these dinosaur services is that *they* control me. The Internet is anarchic, chaotic and protean, a locally-based international network, and the best points of access are local. Local computer-savvy friends will happily help you find the best gateway, as there is often a reward for new users. Expert, conscientious neighbors will be happy to get your hardware and software working together and connected to the 'Net reasonably, somewhere on the order of a dollar a day. You will find yourself sending, and receiving, many more than three letters a day. For many people, the Web's two-way computing is the first good reason to buy a computer.

Once you are on the 'Net you have access to much more than email. The World Wide Web (WWW) is the hottest phenomenon since the 9th Sourcebook. Those http:/www.blahblah.-com addresses on TV and magazines. Real Goods has a homepage, http://www.realgoods.com/ and we have listed some of our other energy favorites below. As I write this (February 1996) the number of users of the World Wide Web has been compounding at the frightening rate of 10% per *month* for fifteen months – nothing but cancer proliferates faster. Certainly no medium has ever gone from zero to sixty in so quickly. As you might expect, there are gems, some sparkling, some flawed, to be found amongst the self-serving Web trash.

For anyone who wants to live knowingly on the planet, the Web is a dream tool. Never has such current, valuable information been available, especially those of us who live on the fringes. You still have to work hard to find just the information you need. For me, weather is very important in the winter. Planning my day, I consulted an image downloaded from the Web: water vapor in the North Pacific and Gulf of Alaska, our storm brewer, as seen by the new weather satellite GOES9 less than 30 minutes before. I predicted to within minutes when the crew would be driven off the roof by the oncoming front which would shut down outdoor work until quitting time today.

My favorite weather information comes from the Space Science and Engineering Center's Real Time Data resource at http://www.ssec.wisc.edu/data/index.html which can give you colorful sea surface temperature maps and computer-modeled forecasts as well as satellite imagery.

I frequent two virtual communities, finding within their many pages all the issues and contacts we can handle. Jump into the WELL at http://www.well.net/ or look into the Global Action and Information Network at http://igc.apc.org/gain/#About . All the periods floating around in these paragraphs are *not* part of the addresses.

Solar issues are addressed encyclopedically by The Center for Renewable Energy and Sustainable Technology in Washington, DC at http://solstice.crest.org . For a more Pacific Rim perspective, try http://www.solarnet.org . Once again, a tip of my EcoDesk Hat to Will Sugg and friends at *The Green Disk*. Some information, especially the kind on the bi-monthly *Green Disk*, is best distributed electronically. As Donella Meadows has written, "sustainability workers are hard on trees" but in fact, those of us who are really trying to use more recycled electrons are having a great time!

— Michael Potts

THE SUSTAINABLE WORKPLACE

The average working person in America spends close to half of his or her waking hours at work. The energy we use and the waste we generate at work has the same effect on the environment as it does at home. Our workplace represents an often neglected area filled with opportunities for reducing our impact on the delicate ecosystem in which we find ourselves.

As the American public is striving to create a sustainable society, it is time to focus our attention on our workplace. Often, we hesitate to do or suggest anything at work because, with its complicated infrastructure, the institutionalized workplace seems stubbornly resistant to change. But the truth is that the workplace is the best place in our society to spark necessary changes, because people with different backgrounds, concerns, and lifestyles all come together there to interact. It is a wonderful place to exchange ideas. And, often, the best way to ignite that spark of change is simply by setting an example. Once you start doing one little thing just a tiny bit differently from everyone else, it is only a matter of time until someone asks you what you are doing, and why. You have sparked interest in someone's mind. Pretty soon, a ripple effect sends a wave of consciousness across the whole organization. Sure, everyone might not change all at once, but in order for change to be real, it has to be by choice, not force. Even just getting coworkers to *think* *about* an issue in many cases represents a giant step forward and a real accomplishment. This "passive" approach is an effective way to make changes without having to propose major shifts in procedures, to which administrators are often resistant. We encourage you to start out small — pick something really simple at first.

At Real Goods, we always strive to keep our office space as low-impact as possible. The Eco Desk, the company's environmental conscience, helps keep us on our toes. As the rising tide of American awareness gains momentum, new products are constantly becoming available, and new things we can think of to do. We've added this section to the *Sourcebook* to help others move the American workplace toward a sustainable future.

Of course, paper is one of the most heavily used resources in the workplace. For this reason, it offers the greatest opportunity in reducing our impact on the earth. The best place to start is with the "three Rs" of sustainability:

- **Reduce:** Store information on computers instead of paper.
- **Reuse:** How about the other side of that paper you're about to throw away? Or use the maximum post-consumer content paper you can find. (For an excellent selection of recycled papers, call for a copy of our Earth Care catalog; 1-800-347-0070.)
- **Recycle:** Every type of paper from glossy to office memos can be recycled in most towns, and of course buying recycled paper or paper made from renewable sources like hemp or kenaf will complete the loop.

You'd be amazed at how much you can accomplish just by following those three principles. (Contrary to some bumper stickers seen around lately, *trees are NOT America's renewable resource!*)

ENERGY CONVERSION STATISTICS

It takes 11,000 BTUs to generate 1 kilowatt-hour (kWh)

Carbon dioxide emissions per kWh = 1.5 lb

Coal required to produce 1 kWh = 1 lb

57% of U.S. electricity comes from coal-fired plants

6.25 million BTUs = 1 barrel of oil

Average U.S. cost of 1 kWh = $0.08

1 car = 500 gallons gasoline per year

1 low-flow shower head = 466 kWh/year heat savings

1 low-flow shower head = 14,000 gal/year water savings

1 faucet aerator = 4000 gal/year water savings

1 set toilet dams = 5475 gal/year water savings

135 trees = 1 ton of pulp

Saving energy is another easy way to reduce our environmental impact at work. Try to get a copy of your last electricity bill at work. Find how many kilowatt-hours you used, and multiply that number by 1.5. On a national average, this will tell you approximately how many pounds of carbon dioxide your business put into the air just using electricity. Multiply the number of kilowatt-hours by about 9 and you'll get how many grams of sulfur dioxide were emitted, and, if you multiply kilowatt-hours by about 2.2, you'll see how many grams of nitrous oxides entered our atmosphere. But fret not! — there is something you can do about it. Every single watt of energy you save through conservation measures and energy efficiency reduces these numbers and lowers your company's electricity bill. In this chapter, you will find a variety of ways to reduce your use of electricity at work. You should also look through the rest of this book for ideas, since most of the energy conservation devices we offer will work just as well at work as they do in the home. Actually, since the rule of thumb for these devices is, "The more they are used, the more they will save," the savings can be even greater at work than at home, because appliances are often on for more hours per day there.

Everyday, more and more products made from recycled materials are becoming available. We can envision a completely recycled office space in the near future. In this chapter, we've chosen some of the best and hardest-to-find recycled products that are commonly used in the workplace.

And, of course, no workplace is sustainable unless it is safe and healthy. You cannot get a lot done without healthy people! As the use of computers in the workplace has exploded in the last ten years, this rise is being followed by waves of concern about computer monitors and electromagnetic radiation. Here we offer a meter to determine a safe distance from your monitor (and other appliances in the workplace), and we also have a product that reduces the field strength in front of your monitor (see page 459).

Whether you work at a surf shop or a fine gourmet restaurant, as an assistant to a secretary or as CEO of a multinational conglomerate, you can play a vital role at work, and there are many things each of us can do to help make the American workplace sustainable. We invite you to think of just a few . . . and, if you are on to something special you think we should share, please let us know!

— *Gary Beckwith*

LIVELIHOOD & LEARNING

Running A One-Person Business

Claude Whitmyer, Salli Rasberry, & Michael Phillips. Claude Whitmyer, author of *At Home in the Briarpatch* (see p. 537) is one of the co-authors of this book. Although the word "Briarpatch" is never mentioned in the book, it comes out of the Briarpatch Network in the San Francisco Bay Area, started in 1974 and still ticking. It begins with several profiles of successful one-person businesses and continues with chapters on Bookkeeping, Financial Strategies, Information Management, Time Management, Setting Up Shop, Choosing Office Equipment, Marketing, Legal Matters, Emotional Support Systems, and Staying a One-Person Business. Here is a brief excerpt:

"Managing other people adds a significant complication to the business that makes it much more difficult to run. Even one additional person is a major increase in complexity. It changes you from being your own boss into being a manager, which is an additional skill that you may not have. Once you become a manager, you will have the same state of mind for managing one person or five. Your freedom of movement and financial flexibility will be severely restricted... "Because of all the federal regulations and employee taxes and benefits and all, if you eliminate employees you cut your overhead way down. When I had employees it took a lot of work just to keep them all busy. I had to plan and organize so that everybody could have something to do when they arrived at work. In order to get to a point where I could make money and pay my employees, I had to produce about $50,000 a month. This meant more employees, a bigger facility, bigger debts. I decided to get smaller, and it's worked out wonderfully for me." New Second Edition, 224 pages, paperback, 1994.

82-287   **Running a One-Person Business** $14⁰⁰

Ecopreneuring

Steven J. Bennett. This is the first book written on how to start an environmental business. It is a start-up guide offering business opportunities in ecologically sound packaging, child and baby care products, household cleaning products, and foods and beverages, education and training, and more. It goes through many case histories of successful eco-businesses (including Real Goods). 320 pages, paperback, 1991.

80-507   **Ecopreneuring** $20⁰⁰

"Staple" Papers Securely Without Wasting Staples

This efficient, environmentally correct office stapler is so simple, you'll wonder why they ever invented metal staples. This palm-sized crimper neatly and securely crimps up to 5 sheets of paper together, staple-free. Speeds up paper recycling, saves your nails and kids' fingers from harm, too. And it never needs refilling. An ideal gift for the office staff or friends.

51-904   Stapleless Stapler $7⁹⁵

Small-Time Operator

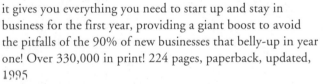

Bernard Kamaroff, C.P.A. This is probably the best book ever written on starting up a small business. In fact, it's the book we at Real Goods used to start our first store in 1978. It's the book that nurtures your first ideals of going into business yet sets your reality firmly in concrete. Written in a personal, nontechnical style, it gives you everything you need to start up and stay in business for the first year, providing a giant boost to avoid the pitfalls of the 90% of new businesses that belly-up in year one! Over 330,000 in print! 224 pages, paperback, updated, 1995

80-502   Small-Time Operator $15⁰⁰

ECOTOOLS

"When tools become toys, then work becomes play."

— *Bernie de Koven*

For some students, computers have become the tool of choice for exploring our world, and especially for understanding how the world works ecologically and environmentally.

We think that the program called C-TEC (The Center for Technology, Environment, and Communication), an alternative learning community at Piner High School in Santa Rosa, California, is a model for how schools of the future should work. It began in Fall 1991 with 90 tenth-grade students and continued in 1993–94 with 180 sophomores, juniors, and seniors participating. C-TEC emphasizes the following:

- The whole community as a classroom, laboratory, and study site.
- Science and technology professionals as mentors, collaborating with teachers to provide extensive real-world experience to students.
- Students learning science by doing science in the real world.
- Integration of curriculum, unifying traditional disciplines around relevant real-world projects.

All students use Word, Excel, and other computer power tools. C-TEC has a network system with more than 70 workstations distributed throughout different classrooms. In any classroom, a student can access a tool, create a new file, load an old file, and generally use a computer in the same way that people use computers in the real world.

C-TEC's three-year program (grades 10 through 12) emphasizes real-world science, math, and technology. Students who choose C-TEC know that they will have to work a little harder and put in extra hours. They also know that they have much more control over how and what they learn than students in a traditional high-school environment.

First-year (sophomore) course offerings include integrated science (biology, earth sciences, physics, and chemistry), English, and math. Major themes of the first-year courses include water quality, energy, and the human body. In the second semester, all sophomores spend six weeks in an integrated science, math, and English energy curriculum. The Real Goods *Sourcebook* is a primary resource for this investigation. Second-year courses include:

- Environmental Biology. Ecology, biology, English, math, and technology applied to environmental issues in the real world around us.
- Engineering and Design. Physical sciences, engineering/design technology, and math applied to challenging problems.
- Anatomy, Physiology, and Sports Medicine.

All courses cover basic theory, but have a heavy emphasis on the exploration and investigation of challenging problems in the community.

Projects are at the center of C-TEC. Every student is a member of a year-long project team. Science and technology professionals, collaborating with teachers, act as mentors in helping students work on real-world projects. Project work can be done at school, at study sites off-campus, or at the mentor's workplace. Students spend 4.5 hours per week of in-school time on projects, plus outside time. Many projects include before-school, after-school, and weekend work at community sites.

The E in C-TEC stands for Environment. Concern for, and study of, the environment is a major strand at C-TEC. More than half of the year-long projects are related to environmental issues. Environmental investigations pervade the curriculum. Students use many "EcoTools" to get things done.

Our EcoTools Backpack

Some people carry stuff in briefcases, suitcases, tote bags, and other encumbrance conveyors. At C-TEC we use backpacks. We have several backpacks in different colors, and each color-coded pack carries tools and toys appropriate for an adventure in learning or some other adventure. We choose our backpacks carefully and stuff them with useful things, playful things, works-in-progress, and, well, some junk. Our green backpack contains EcoTools: books, articles, apparatus, instruments, and a notebook computer with great software.

At home, we use a desktop computer with lots of tools. At school, the same tools are available on the network. Elsewhere, we carry the same tools in an IBM ThinkPad™ notebook computer. A computer in hand is worth two on a desk! Here are some favorite computer EcoTools:

- **Special-purpose tools:** Microsoft Word for word processing and desktop publishing. Word works on both PCs and Macs. We also carry several tutorials that we are evaluating for classroom use or using to increase our knowledge.
- **Multipurpose tools:** Microsoft Excel for number-crunching, charts and graphs, databases, and slide shows. Excel is the greatest computer power tool for math and science that we have found in 32 years of educational computing. Excel works and looks alike on both PCs and Macs.
- **General-purpose tools:** BASIC is a computer programming language, a very general-purpose tool. You can use it to do most anything that a computer can do. True BASIC, by the inventors of BASIC, runs on both PCs and Macs. QBasic is bundled with MS-DOS 5.0 and 6.0 for PCs. If you have a Windows computer, you probably have QBasic. Visual Basic runs on PCs. It will soon become the de facto standard programming language in the real world, embedded in the next generation of spreadsheets, word processors, and databases.

This has been a brief description of what some teachers and students are doing to begin the process of lifelong learning. For more information about C-TEC, send a self-addressed, stamped envelope to James Gonzalez, Chair, C-TEC, 1700 Fulton Road, Santa Rosa, CA 95403.

— *Bob Albrecht and George Firedrake*

REAL GOODS' COMPUTER TOOLS

We would be remiss if we did not bring readers along in the continuing story of Real Goods' love-hate relationship with computers.

First, we believe, just as Bob Albrecht and George Firedrake have implied, that computers are tools, or rather that the programs they run are tools. We look for elegance in our tools, defining *elegance* mathematically: clear and without and wasted motion.

More and more, we find ourselves trapped on the uncertain ground between two religious camps, the IBM-Cloners and the Macheads. The former have been our business computers of choice since our Founder brought one back from Hong Kong. Its cost and performance have been the baseline against which every new Real Goods computer has been compared since. Our business computing network consists now of most of our original IBM clones attached to a true blue RISC 6000 minicomputer that keeps things together on the Unix side. Most of our writing takes place in Word Perfect, where we are again trapped on uncertain ground between three versions. Some of our older computers are very slow running some of the new feature-heavy programs, or just cannot carry the load. We will have to replace most of them as our keyboard-happy people switch to Windows. We've even joked that IBM stands for "I'm becoming Macintosh."

There's an ecumenical movement continuing in the computer world, and it gets easier by the week for our business machines to get along with the predominance of Macintoshes in the graphics and publishing industries. Our copublisher, Chelsea Green, assembled this book electronically using Macs, PageMaker, and MacWrite and working with text files originally written on PCs using WordPerfects 5.1, 6.0, and 6.0 for Windows, amongst which they freely circulated without much mishap. Moving tables from WP to Mac PageMaker proved an ugly chore; ditto our CorelDraw files. Small potholes on a fairly smooth road. By the next time we meet, I hope we'll be even smoother.

We're beginning to see a flutter of a movement to unite small communities like Caspar, Boonville, and foothill towns in Tulare County, all in California, to the Internet *for free* in the manner of a modern public library system. To those of us who lurk at the end of the road on the fringes of society with our fingertips on the keyboard, *on-line* is likely to be where we'll find our most important new tools in the next year, but we are communicationally disadvantaged out here, a long-distance call away from electronic civilization. The more we connect, of course, the more the boundaries between denominations of computers will shimmer and virtually disappear. Real Goods will again contribute to the information freely on-line by placing this *Sourcebook* on the nets — pictures, warts and all. The last edition of the *Sourcebook* was electronically published (without pictures) on EcoNet and the Daily Planet, with an unknown number of echoes heard around the planet. We know that this has made our information freely available in some weird and wonderful places, because we get E-mail inquiries. We're glad to know you are out there, and this seems a perfectly appropriate way for us to get together. Get through to us electronically at *realgood@real goods.com*

— *Michael Potts*

LEARNING

For many of us who work the frontiers of renewable energy and solar living, learning and livelihood coalesce into a constant ferment of new technologies, ideas, tasks, and skills. It is easy to remember a time when adults attained a vocation, and then pursued it their whole lives; not too much earlier, men who had taken up their father's work expected their boys to do likewise. Stirred by cheap energy, global belligerence, and curiosity about the good life on the other side of the mountain, Americans have uprooted themselves and cast themselves so far from their origins that even those who have stayed in one place can hardly recognize their home town.

Now, as the inherited energy tide ebbs, we look to re-establish "ancestral homes" and re-root ourselves where we have landed. In this chapter you will find, first, some ideas about helping our young ones learn a strong regard for place and for knowledge, and then a selection of tools that we have found helpful in our work to bring livelihood and learning together.

—Michael Potts

HOMESCHOOLING

*"A life worth living and work worth doing — that is what
I want for children (and all people), not just, or not even, something
called 'a better education.'"*

— John Holt, 1983

There is probably no more persistent myth in American life, indeed in all of Western culture, than our belief that schooling, or "education," is a good, fundamental, and absolutely necessary part of living. It is a hard myth to shake, mainly because most of us have been thoroughly schooled — or overschooled — to a ridiculous degree (no pun intended). What we don't seem to realize is that, just as the renewable energy movement has shown that it is no longer necessary to depend on multinational energy corporations and public utilities for our power needs, neither is it necessary to rely upon the state with its massive school bureaucracy to meet our learning needs.

Schools are a relatively recent phenomenon in human history. Before their invention children were always with us in the world. In the United States, it was not until the late 1870s that public schools were established. Massachusetts' compulsory attendance law sparked an armed rebellion and literacy rates (close to 100%) went down in its aftermath (and probably they are still falling). In a burgeoning industrial society, schools served to shape an industrial workforce, transforming an agrarian population that essentially worked for itself into urbanized factory beings, people who could stay on task, obey authority, move at the sound of a bell, and otherwise become conditioned, "moral" members of society (especially those pesky immigrants!).

During the twentieth century the educational sector has become professionalized, specialized, bureaucratized, and has grown to massive proportions (with an annual cost at $400 billion and rising). Credentialism makes it necessary to go to school in order to be considered for employment, yet it's very hard to prove that schools actually prepare a person for employment, or, for any other important function in life (except maybe to go to school some more). Pouring knowledge into an empty head — the metaphor most basic to the style of teaching that goes on in schools — does not produce self-motivated learners. Research shows that the most important learning takes place in context, almost always in the context of the real world; the kind of education is usually mediated through some person important to the student (a parent or mentor), and it is functional. Probably the most important thing we all learn — language — takes place at home during the first four years of life. Yet it is never really "taught" to us. Just imagine if schools were put in charge of teaching us how to talk! And so it goes with the rest of what we need to know.

John Holt, a one-time teacher and the author of many important books on how children learn, finally concluded that schools, even the most progressive ones, were simply not a good idea. No one, he decided, could really make other people learn what they didn't want to know in the first place, and what was "learned" in order to pass a test was quickly forgotten (studies have shown that less than 40% of information is retained after two days). What *is* really learned in schools is the "hidden curriculum": how to obey orders, how to compete, how to accept authority unquestioningly, how to fail (the majority of students), and for the few "smart" ones, how to succeed and take their rightful place in an elite profession. One of the things that most irritated Holt about schools was that they always took the credit for the successful students, while blaming poor students for their own failure. Holt ended up advocating homeschooling, or "unschooling" as it is sometimes called, and he is considered by many to be the founder of the homeschool movement. His book, *Teach Your Own,* is one of the basic guides to homeschooling, and through his nonprofit organization, Holt Associates (2269 Massachusetts Ave., Cambridge, MA 02140), he started the first homeschooling newsletter, *Growing without Schooling.*

The homeschooling movement is growing rapidly. It is conservatively estimated that over half a million young people are being homeschooled today. More and more parents are finding that they are perfectly capable of handling their children's education, even though there are as many different ways of doing it as there are homeschooling families. Some parents set up well-defined schedules and a rigorous curriculum; others let the children decide what and when they want to learn and act more as resources, helping them set up apprenticeships and making available the books, computers, libraries, etc. that their children need to investigate the world. Most incorporate music, art, movement, and other skill-learning as well as academics, so that the whole child, not just the left brain, is engaged. But no matter what the specific curriculum or style of "teaching," studies show that homeschoolers score higher on standardized tests and show a higher level of academic achievement — and this happens regardless of parents' educational background and with median family income lower than average. Even more important, homeschooling families find out that they don't have to sacrifice to teach their children at home. On the contrary, they discover that everyone in the family relaxes and starts to enjoy each other again; family life is resurrected. And all this happens in less than half the hours devoted to traditional schooling.

Ivan Illich, one of modern society's most eloquent critics and author of *Deschooling Society* and *Tools,* writes that we've become "schooled to confuse teaching with learning, grade advancement with education, a diploma with competence, and fluency with the ability to say something new . . . to accept service in place of value." This "institutionalization of values" leads us to accept medical care instead of health, prisons instead of public safety, welfare instead of community life, and so forth. Deschooling society means that we invert our institutional bureaucracies, literally turn them upside down, and start a process of "convivial reconstruction": learning webs, skill exchanges, peer matching. It means getting rid of credentialism and creating "tools for conviviality," those Real Goods that work *for* and *with* people and independent enterprise rather than against them. It means meeting nonmaterial needs nonmaterially within the sustainable physical limits of the Earth.

So what and where are the "unschools" of tomorrow, the convivial classrooms in which we can all learn when and what we want? How do we throw off the myths that chain us to wanting "a better education" for our kids, rather than a "life worth living" and "work worth doing"? Perhaps they're as near as our gardens, our solar living rooms, our home offices, our craft places, our computer networks, our community meeting spaces, our dance halls and churches; perhaps they're already here in our hearts and souls to keep our children with us in this world and not institutionalized for the first 17 to 22 years of their lives. Read all of John Holt and find out; read the books presented here and get going. The school system, like the entire industrial system it serves, can only be transformed if it fails to find more customers.

Finally, as Michael Potts in *The Independent Home* says, "Human values, natural values, will at last bring our focus home. There we will find ways to continue re-civilization of our society, to empower and enfranchise, and to broaden our regard for every form of life." Homeschooling is one such way.

— *Margo Baldwin*

THE COURAGE OF THE HOMESCHOOLER

I was going to start out by admitting that my wife and I did not homeschool our children back in the 1960s, but upon a little reflection, I realize that we actually did without knowing it. The only hitch was that we were afraid not to send them to a conventional school, too. Because we made it a point to live as a family, we were almost always home together on evenings and weekends and holidays. We did not send them to preschool or kindergarten, which we deplored as potentially dangerous babysitting services. Neither my wife nor I had social or recreational interests outside the home that did not involve the children, and we hardly ever went anywhere without them. And while we were all together, on weekends, holidays, and from 4 P.M. to bedtime on weekdays — in other words, whenever the kids were not in school — we were passively and actively teaching them. Or more accurately, they were passively and actively learning from us. When they did start in first grade, they already knew the rudiments of reading, writing, and math equal to what their classmates knew. They also were well grounded in storytelling, music appreciation, playing many kinds of games, had experience in gardening beyond their years, and knew more about nature than the average child twice their age. They made up marvelously imaginative play-work, as I called it, mimicking the work they saw my wife and me do. After they started going to school, most of their formal learning time evidently took place at home, too, judging from the piles of homework their teachers inflicted on them. My wife and I often participated, teaching the kids lessons that the teachers evidently didn't have time or didn't know how to teach. We were homeschooling.

God only knows what the kids did in school. Spent a lot of time practicing getting in lines after the bell rang, for one thing. Suffered through lengthy harangues from distraught teachers for another. Picked up bad language and bad attitudes from the other kids. All this, along with the constant danger of picking up pinworms and head lice. Our daughter had a poor teacher for a while, who harmed her sense of self-esteem so much that it took us a year (plus help from a good teacher) to undo it. The only concrete thing that school brought to us was television. We had to buy one because teachers were giving assignments based on television programs!

At that point, if we had been raising our children today, we would most certainly have had the courage to sever the family from the umbilical cord of conventional schooling. For one thing, I had been a teacher myself, and knew that there was no special mystique about the learning process. Teaching in fact would be easy with electronic aids. Much to my surprise, there were some good things on TV, and both my wife and I knew that we could use these programs as effective "learning tools" for the kids because we were already doing the same thing with books. Moreover, today we could buy video monitors that homeschooling suppliers now sell without commercial television receivers built into them. Then we could have run any of the zillion videos now available by means of a VCR and not have to worry about the intrusion of the trash on commercial television. The range of videos now available — from good, classic dramas to instructional material of every imaginable kind — can make any home a classroom, and any even moderately imaginative parent a teacher. All you really have to do is stand aside, enjoy the video, discuss it with the children, perhaps play it several times, and then do some kind of exercise using the material presented.

We didn't quit conventional school because we lacked the courage to do so. Fundamental changes always require courage. Not only does the "Age of Information" mean you have to start doing things differently, which is always difficult, but you also have to risk criticism from friends and neighbors, and sometimes downright persecution from those whose self-interests are tied in with conventional schooling. Very often the worst opposition comes from grandparents, strangely enough.

Before refuting the usual objections to homeschooling, it is important for your courage to see clearly why conventional schooling in institutional classrooms is unnecessary and obsolete and will finally fade away anyhow. School buildings were invented and teachers put into them in the days when there were no other ways to transfer information. We are still using this primitive system. It makes no more sense to build school buildings today than it would be to build canals. Today's typical parent is many times more informed than yesterday's typical teacher or, very often, even today's typical teacher. Many parents lack only courage to be suitable teachers. Some just don't want to be bothered. Often both spouses feel that they must work at outside jobs, a travesty in itself. Many homeschoolers get around this problem by accepting jobs or starting businesses that they can do at home. These kinds of work opportunities are growing fast for the same reason that home-schooling is growing fast. Telephone, radio, television, video and audio cassette, newspapers, books, magazines, fax machine, modem, and computer make working at home just as practical as they make schooling at home practical. Electronic technology pours information down the so-called information superhighway faster than it can be assimilated. We don't *need* all those teachers and formal classrooms anymore. (We don't even need all that information.)

But what discourages me is how abjectly the obsolete classroom educators try to appropriate the electronic teachers and then pretend to teach kids how to use them, when in fact the kids prove over and over again that, with a little basic help, they can learn to operate computers and other tools at home all by themselves, just as millions of adults have learned to do. And mockery of mockeries, the schools are then able, at astronomical expense, to use their appropriation of these machines to force the students to continue to drive or be driven by an obsolete busing system to the obsolete classrooms, when the whole idea of electronic technology is to move information to people, not people to information. We don't *need* to jam students together in huge university complexes anymore, with all the social problems and high expense that these complexes generate. And even if my view when applied to the university is not yet so obvious to everyone, surely, it is preeminently obvious for primary school children.

The first oft-heard objection to homeschooling is that "not every parent will be a good teacher." That is true. Not every professional teacher is a good teacher either, and I have seen many good teachers burn out and just drift along their last few years, yearning for their retirement. There are some parents who are not only terrible teachers but terrible parents, and that is also true of professional teachers. These are not problems of schooling but of human nature. To negate home-schooling because some parents are poor teachers is like saying we shouldn't have automobiles because some people are poor drivers.

The problem of inept or unwilling parents will be solved, I believe, when the pioneering work of homeschools leads to tiny neighborhood schools, where willing parents take turns teaching the children. I have had the good fortune to observe closely the neighborhood Amish schools and have been completely won over by them. The teachers are young adults who themselves have had only the grade-school education that the Amish allow. The ones I know I will match against *any* teacher in the public grade-school system. They are bright, well-informed (the

Amish read incessantly because they don't have radios and TVs) and have grown up taking care of younger siblings, so they are adept at handling a roomful of as many as six different grades. Watching the children laugh and play as they walk to school in the quiet morning through the farm countryside (no buses necessary) is a wonderfully uplifting experience. Some days, the parents come at noon with wagons and food, and the whole school goes for a picnic over noon hour and then joins in extremely zestful ballgames. These people are so far ahead of the rest of us in establishing a proper environment for schooling little children that it makes me weep for the children of mainstream society.

A second much-treasured objection to homeschooling is that the children will be isolated from other children and not learn how to "get along" with others. Quite the opposite is true. Children who spend a good part of their time with adults are *much* more conversant, outgoing, polite, and considerate than those constantly in the company of their peers. Moreover, homeschooled children get plenty of chances to play with other children. To believe otherwise is simply to be looking for an excuse not to homeschool.

A third fear of homeschooling is that the environment is too informal and that the child will not "progress" in learning like they "should." This fear is born out of a lack of confidence in and respect for the human mind, the result of the parents being taught by an erroneous philosophy of learning themselves. Parents need to understand that teaching is not something mysterious, or formal, or requiring special training. Parents teach their children many, many things simply by their example, which is why example is so important. In formal ways, parents teach all kinds of skills from singing songs to reciting nursery rhymes, from brushing teeth to tying shoes. And children learn many, many more things from each other or on their own. My sister taught her slightly younger sisters how to read and write before they went to school. Boys learn fantastic volumes of sports statistics from baseball cards. They could as easily learn all the dates of every event of the Revolutionary War; they don't, because they correctly understand that these dates are not important to them. If such trivia does become important in later life, then they will learn it. A clever teacher would put all the leading people of the Revolution and their statistics on collectible cards.

The idea that teachers need to be trained is a myth of the teaching fraternity to perpetuate itself at ever-higher salaries. Humans not only love to learn but, in fact, *there is no way to prevent them from learning.*

A woman who operates a very small experimental school told me rather confidentially that she had never really held formal "classroom hours" with her own children like she was supposed to do, and she tended to do little of it with her students now. "With my own kids we just sort of went along — living, working, reading, talking together — and, sure enough, when they reached the age of 16 they knew just about everything schooled kids knew and many more things that the public schools seldom can provide. They went to college without any formal schooling at all."

But if young people can prepare themselves for college with informal learning, why not do college learning that way, too? More and more students are asking themselves why they should spend $100,000 for a college education when most of the information they will gain there is available almost free at home. Articles in popular magazines and newspapers are appearing regularly now under the title "Is College Really Necessary?" For many people, the answer is no. In fact, the answer is yes only for those professional careers for which college certification is required, as for doctors, lawyers, etc. As a staff editor for *In Business* magazine, I have learned that a very significant number of successful businesspeople today do not go to college or quit before graduation. Moreover, certification is sometimes merely a way to force people to go to college. After teaching college history for two years, I applied for a job teaching high-school history, but could not be accepted unless I took courses in "teaching skills." Absurd.

Furthermore, it seems to me that schooling everywhere is over-emphasizing the worth of particularized knowledge, especially for little children. I have forgotten at least half of what I learned in school — haven't you? Of what I remember, a third is scientifically or philosophically false or trivial, and another third will soon become obsolete. Learning is an ongoing and never-ending process, and more and more we update our skills and fact banks *by means of information transferral outside traditional schooling.* What we need most of all to help this process along is the gift of curiosity and intellectual aliveness, which conventional schooling so often squelches.

— *Gene Logsdon*

FRINGE SCHOOLING

Good parenting is much more important than good schooling. Almost all of us, especially young ones, love to learn. Left to our own devices in a loving, juicy environment, we learners — and I mean the readers of this book as well as their children — derive wondrous and deep insight and pleasure from discovering new views. Somehow, despite impersonal classrooms and brutal experiences, most of us have prescribed for ourselves the lessons we needed, and have kept our synapses flexible.

Young animals teach themselves, if only they are given good tools. Margo Baldwin's materials and Gene Logsdon's model of the small village school offer perfect examples of how this may happen. In an environment made rich with good books and intentions, parents with a gift for teaching (chiefly the ability to stand back and let learning happen, intervening gracefully and briefly only when a child is on the brink of frustration) are not rare, and seem to emerge when needed. Others who lack this gift or can sustain it only for short periods are freed to support the teaching in other ways, administering, fund-raising, offering occasional lessons. The only danger I have seen in this plan is that a one-sided teacher, for example one who learns auditorally (by ear), or who is in love with language but mathophobic, or who is afraid to take risks, can do a disservice to children who learn through other modes, or are mathematically gifted, or thrive on the edge. This is the danger of the too-small pond; for young students, it is not hard to notice this problem and fix it in time.

The problem of pond-size becomes acute at puberty, when the village school is not big enough to prepare children for the complexities of life. Gene Logsdon has it about right: two thirds of what I learned in school was forgettable or obsolete by the time I needed it. As a teenager, I and many of my friends simply could not learn from our parents. It is unreasonable to expect teenagers to keep their senses of humor intact enough to be able to learn from stodgy parents. What a trial it has been, to experience the "generation gap" from both sides!

Speaking as a veteran high-school teacher (and the allusion to war is deliberate), I believe that the important education taking place from age 13 to 18 has little or nothing to do with classrooms, textbooks, curriculum, and the official stuff of schooling. The important part is in the clique-making and breaking, the social backing and filling, the discovery of sex and love, of mobility, self-determination, anger and denial — the full palette of adult life. This process seems to require a sizeable village of half a thousand or so healthy young animals working each other over. Even if a village school could provide a safe place for this ferment, the responsible adults, if they are also parents to some of the children, could scarcely bear to let it take its necessary course. The sailing is smoother for parents, teenagers, and teachers who understand that secondary school is simply a forum, where studying is an undercurrent on which other, more urgent business rides. With this understanding, the public school system can be a fine place to work on growing up in America. The big lesson for teenagers is *consequence.* Of course these ebullient young animals will take risks, and must learn to accept the results. As parents, we hope that the risks will not be too great, and that the consequences will be manageable. We know that, no matter what, we will always love our children, and so we try to help them shoulder the burden of responsibility for their own lives. If a certain amount of calculus, physics, and home economics are also learned at school, we rejoice. In my opinion, most of the civics, constitution, and other lawyer-mandated nonsense confirms our children's suspicion that government seldom tells the truth.

The key to good schooling for any child is *enrichment.* Public schools are meant to warehouse young ones until they are old enough to work, and hopefully allow them to wick up some useful learning while they wait. Most of us survived this experience, mostly intact, inventing ourselves along the way. Looking backward in memory, I see a very few remarkable teachers, some intense attachments to other people, acres of boredom punctuated by periodic epiphanies of knowledge. Absolutely the best of times, and the worst, too, were the times with my friends and my hopelessly retro parents. I recall quite clearly that the most useful and lasting lessons I taught myself, usually after my parents and friends put me in the way of some juicy situation fraught with learning potential. If the family is working, where the child is schooled becomes less important. I think Gene Logsdon's approach is perfect: send them to school, but make sure learning takes place at home.

For older children, insularism, the sense of living on a small, protected island, is the chief danger of too much homeschooling. On such an island, some children pine for lack of friends to bounce off of, while others thrive as biddable, personable humans. Neither group is prepared for the hurly burly of life; they are eggheads, in the sense that they are fragile. We may deplore the directions and values of modern life, of which public education is a slow mirror, but few of us would care to give up the freedoms modern life gives us. We learned those freedoms at school, not just in the classroom, but in the process of making new friends, exploring new horizons; making judgments, then taking credit for being right or paying consequences for being wrong; extending our personal boundaries. Despite our best intentions to shield our children from the evils and brutalities of school life, if we keep our children home too long, they may miss this enfranchisement.

When we were in school, we thought it very presumptuous of parents and teachers to tell us how to act in this modern world on which they seemed to have such a tenuous hold. Then and now we laugh at the lines, "What was good enough for my grandfather..." and "When I was a kid, I walked two miles through snow-drifts..." In a world of virtual reality, MTV, gangster rap, looming ecological disaster, staggering national debt, and a vocational picture light years ahead of what most schools are prepared to teach, are we not being equally presumptuous in expecting our children to be "conversant, out-going, polite, and considerate"? Can we honestly say that we have a better grasp on what the next 50 years will require than our children do?

For all children, however schooled, let me recommend two sovereign remedies. The drastic course first: consider Montessori's *erdkinderschule*, or its domestic equivalent, where the teenaged child (with help) builds his or her own house, rustles up his or her own grub, and comes face to face with necessity. This requires one of the most difficult of parental arts, tough love; I wish you good luck. The other sovereign: travel. All children benefit from extended horizons, particularly if experienced in the company of open and loving guides. The absolutely best thing I ever did for myself and my children was to take a year off from work and school to travel the Pacific Rim. Talk about risk taking! We all embarked (well, actually we emplaned) worrying that our lives as we knew them were over. After six weeks, we were relieved to discover this was true, as our lives unfolded like the wings of a new butterfly: we found a strong core of interdependence and shared delight which sustains us undimmed to this day five years later. When we returned after ten months, the old lives beckoned, but we all found much more suitable and stimulating things to do. Isn't that what education is about?

—*Michael Potts*

EDUCATIONAL AND TEACHING SUPPLIES

Crosby, Stills, and Nash told us over 20 years ago to "Teach your children well," and their words could hardly be truer than they are today. In virtually all aspects of our society, we shape the future by the example we set for young people and the lessons we teach them.

As easy as it is for us as adults to become disillusioned by the problems in our society, it is even easier for the children of today. It is up to us to portray these problems as opportunities for solutions, not only in our own minds, but most importantly in the hearts and minds of young people we relate with. As parents, grandparents, teachers, and mentors, we must keep the ray of hope and determination that we have, and pass it on to the next generations to come.

Teachers, in particular, have a special opportunity to sow these seeds of hope, and help them grow. Teachers from both private and public schools, from kindergarten to college level, frequently call us up asking for ideas for class projects that teach about environmental and renewable-energy issues. We have added this short section to our *Sourcebook* this year, offering teaching aids, workbooks, and projects for the classroom arena.

We have found that the information is best passed on without a lot of philosophical baggage. Rather than trying to tell young people what is right and wrong, it is more effective to give them the facts and information that they need to make their own choices. In this light, we can simply explain some of the problems that confront our society, show them how renewable energy works first hand, and remind them of the benefits of its use. They will take it from there.

As the standard curriculum in today's schools tends to neglect such issues as environmental concerns and renewable energy, it is up to individual teachers to take the initiative. Here we offer the tools for you to do this. If you are a teacher, or you have an opportunity to teach, look through the following section for ideas. We've searched and found an array of workbooks and materials that will help. Let us know of your success in the classroom.

Of course, sales tax is not required on orders to government agencies such as schools, and Real Goods also accepts purchase orders from such organizations.

— *Gary Beckwith*

The Home School Reader

Edited by Mark & Helen Hegener. Published by the editors of *Home Education* Magazine, this is a unique collection of some of the best writing on homeschooling by writers such as John Holt, Nancy Wallace, Mario Pagnoli, Susannah Sheffer, Jane Williams, and others. This book presents a wide range of homeschooling perspectives and philosophies. Designed to appeal to the long-time homeschooler as well as the newcomer, articles are included on legal considerations, socialization, acountability, compulsory education, selecting curriculum and resource materials, higher education, and more. 162 pages, paperback, 1988.

82-280   The Home School Reader $11⁰⁰

The Home School Sourcebook

Donn Reed. Gathered during 18 years of homeschooling his four offspring, Donn Reed has filled this book with ideas, opinions, books, games, materials, and resources for enriching the homeschooling environment. As it becomes increasingly clear that the difference between a mediocre education and an adequate one is parental involvement, even parents who send their children to public school need to supplement their children's learning experience, and this book is a great collection of basic tools. 226 pages, paperback, 1991.

80-185   The Home School Sourcebook $15⁰⁰

Hands-On Nature:
Information And Activities For Exploring The Environment With Children

Edited by Jenifer Lingelbach. Adults and children will share discoveries of wildflowers, trees, birds, bugs, animal life, and more with this highly successful illustrated guide to using local fields, woods, parks, and backyards for fun activities that sharpen kids' environmental awareness. Published by the Vermont Institute of Natural Science, this book enables novice leaders to teach nature subjects successfully. "*Hands-On Nature* is an exciting new book that brings to educators and lay persons a wealth of experience and activities for teaching children to understand and appreciate the natural world."—Carol Snell, Energy Management Center. Illustrated with drawings, 233 pages, paperback, 1986.

82-263   Hands-On Nature $19⁰⁰

Family Matters:
Why Homeschooling Makes Sense

David Guterson. While David Guterson teaches his neighbors' kids in his high school classroom, he and his wife teach their own at home. With one foot in each world, he examines life at school and the inexhaustable, inspiring opportunites offered by learning outside it. The most important lesson he has to teach is that family matters, and homeschooling is just one way of reaffirming the bond between parents and their children. Addressing the questions that any parent would ask: Aren't you abandoning the schools? Is it legal? What about socialization? Guterson also provides a broader context: the astonishing academic success of homeschooled children, the history of public schools, philosophies of education, the psychology of learning, and how other societies have handled educating their children. Throughout, he evokes the priorities and values that should be at the heart of any discussion of education: family life, individual fulfillment, democracy, community. 264 pages, paperback, 1993.

82-279   Family Matters $11⁰⁰

The principal goal of education is to create people who are capable of doing new things, not simply of repeating what other generations have done.

—Jean Piaget

APPENDIX

5000 sq. foot straw bale showroom under construction at Solar Living Center, Hopland, California.

REAL GOODS MISSION STATEMENT

Through our products, and educational demonstrations,
Real Goods Trading Corp. promotes and inspires an
environmentally healthy and sustainable future.

THE ROBERT RODALE ENVIRONMENTAL
■ ACHIEVEMENT AWARD ■

Real Goods is very proud to be the 1996 winner of this highly prestigious award. The Rodale Award, given by the Direct Marketing Association, honors businesses making positive contributions to the environment. The award presents a $5,000 contribution in Real Goods name, which we distributed as; $3,000 to Greenpeace to kick off their solar project portion of their Climate Change Campaign, $1,000 to Rocky Mountain Institute for their renewable energy projects, and $1,000 to the Trees Foundation to  support their efforts to preserve and protect the Headwaters Forest, the largest stand of privately held virgin redwoods in the world.

THE HOPLAND SOLAR LIVING CENTER

Imagine a destination where business is conducted amidst a fruitful landscape designed to inspire with its variety and sense of place. As you explore the grounds, you are surrounded by the sounds of water flowing through a natural revitalizing cycle. Envision a building of sweeping beauty, created to take every advantage of the sun throughout its seasonal phases. Picture offices and a five-thousand square-foot retail showroom powered purely by the energy of the sun and wind. You're picturing the Solar Living Center in Hopland, California.

Our Solar Living Center began as a shared vision among the workers of Real Goods Trading Corporation and company founder and President, John Schaeffer. It was a vision of an oasis where the company could demonstrate the culture and technology of solar living. With that goal in mind, the SLC was designed to embody the philosophy described in Real Goods' various catalogs and publications. With the opening of the center in April of 1996, the vision is now a reality in Hopland, less than two hours north of San Francisco, in California's wine country.

Passersby on busy Highway 101 are bound to notice the striking appearance of the company showroom. This does not look like business as usual! Both the building design and the construction materials were selected with an eye toward their efficiency of function and educational value. The result, not unexpectedly, is also a place of great beauty.

The architect chosen to design the building was Sim Van der Ryn of the Ecological Design Institute of Sausalito, California. His associate, David Arkin served as project architect and Jeff Oldham of Real Goods, managed the building of the project. Their creation is a tall and gracefully curving single story building which is so adept in its capture of the varying hourly and seasonal angles of the sun that additional heat and light are nearly unnecessary. Wood burning stoves, which we also sell in the showroom, provide backup heating for the coldest winter mornings and solar powered florescent lighting is available, but will rarely be needed. Through a combination of overhangs and manually controlled hemp awnings, excess insolation during the hot weather months has been avoided. Solar powered evaporative coolers are in place as a backup air-conditioning system, and are also used to flush the building with cool night air, storing "coolth" in the six-hundred tons of thermal mass of the building's walls, columns and floor. Grape arbors and a central fountain with "drip ring" for evaporative cooling are positioned along the southern exposure of the building to serve as a first line of defense against the many over-one-hundred-degree days which occur during the summer in this part of California.

Many of the materials used in the construction of the building were donated by companies and providers with similar interests to those of Real Goods. As an example, the walls of the SLC were built with more than 600 ricestraw bales donated by the California Rice Industries Association. Previously, rice straw has been disposed of by open burning, a practice that contributes to the production of carbon dioxide, the so called "greenhouse gas" which is the leading cause of global warming. By using this agricultural by-product as a building material everyone benefits. The farmers receive income for their straw bales, no carbon dioxide is produced, and the builder benefits from a low cost, highly efficient building material that minimizes energy consumption.

At the SLC, visitors can experience the practicality of numerous applications of solar power, including the generation of electricity and solar water pumping. The electrical system for the facility comprises ten kilowatts of photovoltaic power and three kilowatts of wind generated power. Through an intertie with the Pacific Gas and Electric Company, the SLC sells the excess power it generates to the electric company and buys it back only when necessary. Once again, like-minded companies have shared in the costs of developing the Solar Living Center as a demonstration site. Siemens Solar has donated more than 10 kilowatts of the latest state-of-the-art photovoltaic modules to the center and intends to use the SLC as a test site for new

Winding entrance path is lined with photovoltaic arrays and welcomes you to the Solar Living Center

The Agave cooling tower taking shape

Stone pillars mark solstice and equinox sunrise and sunset points

563

View to the south as you enter the Solar Living Center

Fire pit rocks by pond

View to the west across ponds, landscaping, and showroom roof. Duncan's Peak in the distance

Siemens modules in the years to come. Trace Engineering contributed four inverters, which will be on display behind the glass window of the SLC's "engine room" so that visitors may see the inner workings of the electrical system.

Educational opportunities are interspersed throughout the Solar Living Center, beginning with a facility that is open and revealing by design, with the windowed view into the workings of the renewable energy power center and a "truth window" which exposes the inner nature of the strawbale walls. On tours, either self-guided or with SLC tour guides, visitors are informed of the guiding principles of sustainable living and are given a chance to appreciate the beauty that lies in the details of the project. An education center features video programs on the design and building of the Solar Living Center and related topics of interest. This building also provides a meeting place for presentations by guest speakers and a space for workshops and special events. Additionally, it serves as the main classroom for Real Goods' popular Institute for Solar Living.

Learning potential is also inherent in the (soon to be award-winning) landscape, which was designed by Chris and Stephanie Tebbutt of Land and Place. On this project, their design was the first phase of the construction and did much to establish the character of the site. This is a radically different approach than is often seen in commercial building projects, where the landscaping appears to be a cosmetic afterthought. At the SLC, a majority of the plantings produce edible and/or useful crops and they are utilized to maximize energy efficiency while portraying the dramatic aspects of the solar year. An array of sundials and solar calendars help visitors establish a feeling for the relationship between this location and the sun. Guests observe stones and plantings marking the lines of both sunrise and sunset for each equinox and solstice, emanating from the sundial at the exact center of the oasis. This is conceived as a way to keep perspective on seasonal shifts throughout the year.

Unique to these gardens is a series of what are called "Living Structures" which reveal architectural gestures, seasonally. Through annual pruning, plants are coaxed into various dynamic forms, such as a willow dome, a hops tipi or a pyramid of timber bamboo. These living structures grow, quite literally, out of the garden itself. There is also a "cooling tower" where overheated visitors stroll through a gentle mist under the shade of vines and agave plants. Another unusual feature is the "memorial car grove," where the rusting hulks of 50's and 60's "gas hog" cars have been turned into planter boxes for trees!

The Solar Living Center's garden follows the Sun's journey through the seasons, with zones planted to represent the ecosystems of different latitudes. Woodland, Wetland, Grassland and Dry zones are manifested through plantings moving from North to South, with the availability of water as the collaborative element. Trees are planted to indicate the four cardinal directions. The fruit garden, perennial beds, herbs and grasses reflect the abundance and fertility of a home-based garden economy.

In case this all sounds too serious, it should be pointed out that the SLC is a wonderful place to play! Upon entering the showroom, one is greeted by a delightful rainbow spectrum created by the large prism that is mounted in the roof of the building. Outside, the designers have provided interactive games and play areas, focused on, but not only for the young. There is a bicycle connected to a light bulb where visitors feel how much muscle energy is required for illumination and compare it to the ease with which the same amount of energy is harvested simultaneously from the sun with a solar panel. Or kids can try the sand and water area, where a solar powered pump provides a water source which can then be channeled, diverted, dammed and flooded through whatever sandy topography has been constructed. Block the sun reaching the solar panel and you stop the flow of water! It's a combination of learning engineering and social skills! There are also a variety of wonderful artistic statements in design and landscape which will dawn on you while you relax on the grounds, as well as simple pleasures, such as a place to get your feet wet or enjoy a picnic.

The Solar Living Center is located ninety miles north of San Francisco on Highway 101 and is open every day except Thanksgiving, Christmas and New Year's day. There is no admission charge to visit. Custom tours are available for groups with particular interests and a visit is recommended for anyone who has the desire to raise either their consciousness, their spirits or both!

For structured learning opportunities, the Solar Living Center also serves as the campus of the Institute for Solar Living, which offers intensive, hands-on, one-day seminars on a variety of renewable energy and sustainable living topics. Please see the following section to learn more about the Institute.

Institute students set up a tracking PV array to power water pumping

THE INSTITUTE FOR SOLAR LIVING

You are reading the textbook of Real Goods' educational division, the Institute for Solar Living founded in 1992. Over the years, many students have attended classes and the Institute has truly come into its own. Although we reach many more people through publications like this one, our Institutes are very close to our hearts, because they give us a chance to meet people, work beside them, and build the family up close and hands-on.

Institute sessions blend a nice balance of technological independence, rural simplicity, and community building. The secret to the Institute's success, as near as we can figure it, consists of three ingredients perfectly combined: the wonderfully real Real Goods customers who come from far and wide, the allure and promise of living independently with energy and sustenance harvested under our own power, and the practical, logical skills and ideas offered by the Institute's teachers and staff. One graduate expressed it best, "I can't decide if I've just had the best short vacation of my life, or the best learning experience! Could it be both?"

This year the Institute has found its home at the new Real Goods Solar Living Center in Hopland, California. The Solar Living Center, with its mission as the premier model for sustainable development, restorative landscaping, and renewable technologies provides an immensely inspirational setting as the Institute's campus. Each of our varying workshop titles relate directly to technologies employed at the Solar Living Center.

Day-long Workshops

These intensive one day workshops are compressed— a good way to survey the terrain of a sustainable lifestyle quickly while exploring the energy aspects more intensively. At the end of each session, graduates are informed consumers, able to ask the right questions and to understand the answers. They report that the experience inspires them to go farther forward along the path toward independent living. Those who arrive with some technical skill leave qualified to put together a small independent energy system, to begin the design of an energy efficient home, to know what to look for in a piece of property, or to create balance and harmony in their own garden.

Energy is the heart of our series: rational, comfortable, lasting systems for homes and businesses on and off the electric grid. Most of our courses emphasize direct hands-on experience. Our mission is to share excitement, sense of purpose, and knowledge with others interested in living sensibly and lightly on the earth. For the 1996 season we are proud and excited to offer the following workshops.

Planning and Building Your Solar Electric System

Taught by the Real Goods Renewable Energy technical staff, the basics of energy, electricity, load analysis, system sizing, and the components of various systems will be covered. Each student will have the opportunity to wire together a small photovoltaic system. An in-depth tour of the energy systems at the Solar Living Center is also included.

Realizing the Dream- Planning and Buying the Perfect Country Property and Developing Your Homestead

The Real Goods Renewable Energy technical staff and a local Realtor will show you what to look for when purchasing the perfect piece of country property. They will cover development costs, roads, laws, water, waste, how to determine power potential, and site orientation of your home.

Organic Gardening and Drip Irrigation

The directors of Fetzer's Valley Oaks Garden in Hopland, CA will show you how to create balance and beauty in your garden. They will cover soil fertility, composting, companion planting, and beneficial plants and insects. Another local specialist will cover the basics of drip irrigation. A tour of the renowned Valley Oaks garden is included.

Sustainable Building and Eco Design

Real Goods Solar Living Center Project Manager and a local solar architect will discuss the environmental and ergonomic considerations for achieving a healthy, sustainable building. Recycled and non-toxic materials, active and passive solar home design and alternative methods of building will be covered. A tour of the Solar Living Center will focus on site features and building design.

Strawbale Construction

Real Goods veteran Ross Burkhardt of Ukiah will teach basic strawbale construction methods, coating techniques, and supervise the construction of a temporary structure. Ross and hi new company Bale Builders has been active in state legislation issues regarding permitting of strawbale homes. Code issues, moisture control, and different construction methods will be covered in this intensive hands-on class.

Throughout all sessions, give and take between instructors and attendees is a topic in itself. We learn to know each other, and value the efforts at independence we're contemplating or have already begun to bring to our lifestyles. If the instructors don't have an answer, surprisingly, someone in the workshop usually does.

Product demonstrations, cost benefit analyses, and a visit to the Real Goods store provide a practical grasp on the tools and techniques available for taking our energy lives in hand. More resource museum than store, this is the center of testing and evaluation for the Real Goods enterprise, and many visitors are glad to get a chance to see the place and talk to the product specialists.

For a few dedicated folks, time to consult with the Real Goods instructors to devise a strategy for a more independent lifestyle is a logical next step. Appointments can be made for consultations any day of the week during normal business hours.

Feedback from Institute Graduates

We've benefited greatly from the willingness of Institute attendees to tell us what we do well and what could be improved.

It's the helpful hints and learning from mistakes that others have made that brought me to your workshop—in addition to wanting to meet the people I will someday do substantial business with (major purchases). So, less time with generalities and more specifics, hands-on or applied theory with real life examples/pictures/equipment of success and failures. That is why your people are different— real experience, not hypotheticals!

I needed to get beyond the reading stage and become more familiar with the subjects. I feel we got a very well thought-out and balanced presentation on many subjects, and I am very appreciative that you folks are taking the lead and setting a good example in presenting the information so lovingly and enthusiastically. Thank you!

My wife is leading me toward a better lifestyle. I arrived as a skeptic and am leaving with the feeling this is possible for me.

I am so excited to get home and get started that I can't sit still. One night I was sizing my system all night. Last night I dreamt I had five different PV systems laid out around the top of a green meadow. All night I was choosing the best one—a combination—for my needs. Jeff and Ross and Nancy were there talking to other people then answering all my questions!

An honest possibility even for a conservative Republican that voted twice for Reagan.

It is not as difficult as others want you to believe.

Good spiritual company, great food, wonderful setting, excellent information.

You guys are great people, and your knowledge and experience is invaluable to the rest of us—that is why we came to you. Please concentrate on conveying what is unique to you and your experience. The rest we can get from the books you sell.

Institute Staff

Staff for the Institute are technicians from the Real Goods Renewables Division and other local experts invited to share their particular expertise. In early March of 1995, the first Instructor Retreat took place in Mendocino County. Veterans got a chance to share what they had learned over the years, learn about new materials and equipment, and refine their presentations along the lines suggested by graduates. Veteran Institute faculty are already making sure that the 1996 sessions are nothing less than brilliant.

Information and Registration

To register, or for more information on Institute sessions or availability, call Real Goods at 800-762-7325. An Institute brochure and schedule is available upon request.

Cancellations prior to one month of the workshop date are subject to a 25% processing charge. Any cancellations within a month of the workshop are non-refundable unless we are able to re-sell your space.

As soon as you are registered, we will send you written confirmation which will include information on and directions to the site as well as lodging and restaurant options for the area.

Institute students learning the finer points of organic gardening

FOREWORD TO THE ALTERNATIVE ENERGY SOURCEBOOK, SEVENTH EDITION

The following piece by Amory Lovins was originally printed as the foreword to the seventh edition of the Real Goods Alternative Energy Sourcebook. We think the Lovins message is as right now as it was then, and are proud to include this piece in our new book.

From 1979 through 1986, the United States got more than seven times as much new energy from savings as from all net increases in supply. Even more astoundingly, of those increases in supply, more came from sun, wind, water, and wood than from oil, gas, coal and uranium. Even as glossy magazine ads were dismissing renewable energy as unripe to contribute much of anything in this century, renewables came to provide some 11% to 12% of the nation's total primary energy (about twice as much as nuclear power), and the fastest-growing part. The only energy source growing faster was efficiency. Just the increase in renewable energy supplies during those years came to provide each year more energy than all the oil we bought from the Arabs. And efficiency, during 1973–86, came to represent an annual energy source two-fifths bigger than the entire domestic oil industry which had taken a century to build; yet oil had rising costs, falling output, and dwindling reserves, while efficiency had falling costs, rising output, and expanding reserves.

To be sure, during about 1986–88 — the later years of what, in the telling phrase used in their own country by Soviet commentators, may be fairly called "the period of stagnation" — this momentum declined and in some respects stalled. The impressive successes of efficiency and renewables often fell victim to official hostility and collapsing oil prices (which were largely driven by the very success of energy efficiency). The Reagan Administration's rollback of efficiency standards for light vehicles immediately doubled oil imports from the Persian Gulf, effectively wasting exactly as much oil as the government hoped could be extracted each year from beneath the Arctic National Wildlife Refuge. While some electric utilities pressed ahead with good efficiency programs, additional electrical usage spurred by deliberate power-marketing efforts was officially projected, by the year 2000, to wipe out about two-thirds of the resulting baseload savings. The same Administration that touted the virtues of the free market pressed home its strenuous efforts to deny citizens the information they needed to make intelligent choices. And Federal tax credits meant to help offset the generally much larger subsidies — totaling at least $50 billion per year — given to renewable's competitors were generally abolished, while most of the subsidies to depletable and harmful energy technologies were maintained or increased, tilting the unlevel playing field even

568

further. These and other distortions of fair competition gravely harmed many sectors of the renewable energy industries, often drying up distribution channels so that even sound, cost-effective options could no longer reach their customers.

Leadership in some key R&D areas passed from America to Japan and Germany. Cynics began writing premature obituaries of the latest solar flash-in-the-pan. By 1990, the savings achieved since 1979 were no longer seven, but only four and a half times as big as all the net increases in energy supply, and the fraction of that new supply coming from renewables was no longer a bit over half, but only one-third. There were bright spots — Maine, for example, raised the privately generated, almost all renewable, fraction of its electricity from 2% in 1984 to 35% in 1991 — but many other efforts and firms faltered or even went under.

Yet throughout that decade's rise, leveling, and sometimes stumbling of the keys to a safe, sane, and least-cost energy future — high energy productivity and appropriate renewable sources — a band of pioneers in Northern California sustained their vision of a way to give everyone fair access to a diverse tool kit for energy self-sufficiency. Through their dedication, thousands of people have had the privilege of discovering that the energy problem, far from being too complex and technical for ordinary people to understand, is perhaps on the contrary too simple and political for some technical experts to understand. The Real Goods team gave, and gives today, an equal opportunity to solve your piece of the energy problem from the bottom up.

Many did exactly that. They discovered that solar showers feel better, because you're not stealing anything from your kids. They found how to get greater security and high-quality energy services from a judicious blend of efficiency and renewables. Having a high do-to-talk ratio, they worked through trial and error, celebrated their inevitable mistakes, and found in Real Goods an effective way to share their experience with a wide audience. (There is, after all, no point repeating someone else's same old dumb mistakes when you can make interesting new mistakes instead.) Piece by piece, with that uniquely American blend of idealism and intense pragmatism, they quietly built, and continue to build a grassroots energy revolution.

The tools of elegant frugality, the technical options that let you demystify energy and live lightly within your energy income, have long been available to anyone in principle. (There's an old Russian joke about the guy who asks whether he can buy various hard-to-find goods. Tired of being always told, "In principle, yes," he exclaims, "Sure, but where's this [Principle] shop that you keep talking about?") Real Goods turns the principle into practice: a useful selection of the things you were looking for, with essential background information, at fair prices, guaranteed, anywhere you want them delivered.

It's especially gratifying to see this catalog's nice blend of renewable energy supply options with increasingly efficient ways to use the energy. In most cases, the best buy is to get the most efficient end-use device you can, then just enough renewable energy supply to meet that greatly reduced demand. Our own house/indoor farm/research center of nearly 4,000 square feet, for example, by harnessing some of the efficiency options in this catalog, uses so little electricity for the household lights and appliances — about $5 worth per month — that only about 400–500 peak watts of photovoltaics, a couple of thousand bucks' worth, could entirely meet those needs. Indeed, superinsulation and superwindows, by providing 99% of our space-heating passively in a climate that can get as cold as -47° F, raised our construction costs less than they saved us up front by eliminating the furnace and ductwork. All together, saving 99+ percent of our space and water heating load (the latter is virtually all passive and active solar, again made much cheaper by strong efficiency improvements), 90-odd percent of our electricity, and half our water together raised total construction costs only one percent, and paid back in the first ten months. It looks nice, too: we tramp in and out of a blizzard to be greeted by a jungleful of jasmine, bougainvillea, and a big iguana offering advanced lizarding lessons under the banana tree — then we remember, as we wipe the steam off our glasses, that there's no heating system; nothing's being used up. And that's all with nine-year-old technology. Now you can do even better.

Today, in fact, the best technologies on the market can save about three-quarters of all electricity now used in the United States, while providing unchanged or improved services. The cost of that quadrupled efficiency: about 0.6 cents/kWh, far cheaper than just running a coal or nuclear plant (let alone building it). Thus the global warming, acid rain, and other side-effects of that power plant can be abated not at an extra cost but at a profit: it's cheaper to save fuel than to burn it. Similarly, saving four-fifths of U.S. oil by using the best technologies now demonstrated (about half of which are already on the market) costs only about $3 per barrel, less than just drilling to look for more. Such technologies obviously beat any kind of energy supply hands down. But what's the next best buy — the supply choices you need to make in partnership with efficiency in order to live happily ever after? Usually, if you buy the efficiency first, the next best buy will be the appropriate renewable sources. And if you count — as our children surely will — the environmental damage and insecurity that fossil and nuclear fuels cause, then well-designed renewables look even better.

That's not to say that every renewable technology makes sense anywhere. Wind machines and small hydro only work well in good sites. Photovoltaics can look competitive with grid electricity (which, by the way, receives tens of billions of dollars' annual direct subsidies and doesn't bear many of its costs to the earth) only if you're upwards of a certain distance (where we are, about a quarter-mile) from a power line — or closer, right down to zero, if you count such benefits as greater reliability and higher power quality. (The Federal Aviation Administration is switching hundreds of ground avionics stations to photovoltaics for exactly those reasons.) Every renewable application is site-specific and user-specific. No single renewable source can solve every energy problem. That's part of their strength and their charm.

But wherever you are, if you use energy in a way that saves you money, it's quite likely that some kind of renewable energy can further cut your bills, work as well or better, give you a good feeling, and set an example. What more can you ask?

Well, maybe something for the earth — and you get that too. A single compact-fluorescent lamp (see lighting section) replacing, say, 75 watts of incandescent lighting with 14–18 watts (but yielding the same light and lasting about 13 times as long) will, over its life, keep out of the air a ton of carbon dioxide (a major cause of global warming), twenty pounds of sulfur oxides (which cause acid rain), and various other nasty things. And far from costing extra, that lamp will make you tens of dollars richer by saving more than it costs for utility fuel, replacement lamps, and installation labor. Now put that saving into changing over to renewable energy supplies and you're doing even better. Then take the time you used to put into working to pay your electric bill and put it instead into your garden, your compost pile, a walk, a fishing trip. Take the time you used to work to pay your medical bills and build a greenhouse. Invite your neighbors over to help munch your fresh tomatoes in February and tell them how you did it all. Then . . .

You get the idea. Here's a treasure-house of things you can use to improve both your own life and everyone else's. Read, enjoy, use, learn, and tell: implementation is left as an exercise for the reader.

Amory Lovins is currently director of research for Rocky Mountain Institute. He is a consulting experimental physicist educated at Harvard and Oxford. Mr. Lovins has consulted extensively worldwide and has briefed nine heads of state. He has published a dozen books and hundreds of papers (be sure to see his commentary on "Supercars" in this book's Mobility chapter). Newsweek has called Lovins "one of the western world's most influential energy thinkers."

He has our vote for the next U.S. Energy Secretary!

You can reach the Rocky Mountain Institute at 1739 Snowmass Creek Road, Snowmass, CO 81654-9199, phone: 970/927-3851.

REAL GOODS COMPANY POLICIES

The Real Goods Story — Way Back When
(The Ride to Boonville)

As did many of his contemporaries in the 1960s and early 1970s, John Schaeffer, founder of Real Goods, experimented with an alternative lifestyle. After graduating from U.C. Berkeley, where he was exposed to nearly every strand of the lunatic fringe, he moved to a commune outside of Boonville, California, a mountain community as picturesque as it is isolated. There he lived a life in pursuit of self-sufficiency.

Despite the idyllic surroundings, John soon found that certain key elements of life were missing — specifically, money and power. Not that John needed a *lot* of money and power, but he wanted some small amount to strike a balance between what he had grown up with and complete deprivation. Self-sufficiency, in other words, was a more appealing concept than reality.

John discovered power in the sun. A photovoltaic module hooked up to a storage battery, he found, would power lights, a radio, and even a television. Despite his departure from a purist's lifestyle, John became the most popular person on the commune to visit whenever it was time for "Saturday Night Live."

The money came from a job. Unfortunately, the only job John could find was as a computer operator in Ukiah, some 35 twisty miles from Boonville.

Knowing that John would be making the daily trek over the mountain to the big city, it was not unusual for his friends to ask him to buy supplies that they needed on the commune. As a conscientious, thrifty person, John spent many hours plying the hardware stores and home centers of Ukiah, searching for the best deals on fertilizer, bone meal, tools, and other goods related to the communards' close-to-the-earth lifestyle.

One day, while driving his Volvo (what else?) back to Boonville after a particularly vexing shopping trip, a thought occurred to John. "Wouldn't it be great," he mused, "if there was *one* store that sold all the products needed for independent living, and sold them at fair prices?"

The idea of Real Goods was born.

In recent years the focus of Real Goods has been energy. The company now can claim to be the oldest and largest catalog firm devoted to the sale and service of alternative energy products. Yet it would be wrong to think of the company as being either large or old. Real Goods is still devoted to the same principles that guided its founding way back in 1978 — quality products for fair prices, and customer service with courtesy and dignity.

Many of the Real Goods staff, like John Schaeffer, are veterans of a movement that had the right inclinations, but the wrong technology. Today, whether you are a full-time off-the-gridder or have simply added a note of energy sanity to your life, the Real Goods staff is committed to helping you achieve the independence you desire.

The Real Goods Story —
Flash Forward

From its humble beginnings Real Goods has become a real business, with Real Employees serving Real Customers, and even with Real Shareowners. That Real Goods is also a socially committed and environmentally responsible business can be seen from some of the plaudits and honors we have garnered along the way: Corporate Conscience Awards (from the Council on Economic Priorities); inclusion in *Inc.* magazine's 1993 list of America's 500 Fastest-Growing Companies; the Robert Rodale Award for Environmental Education; Northern California Small Business of the Year Winner for 1994; finalist for entrepeneur of the year; and news coverage in *Time, Fortune, The Wall Street Journal,* plus several thick scrapbooks full of press clippings. But has all this attention gone to our heads?

You bet it has.

The reason that Real Goods has been considered so newsworthy is not because we have pioneered new ways of doing business. Our methods do not reflect the latest trends in corporate or business-school thinking. Instead, led on by a certain naiveté and a basic grasp of simple principles, Real Goods has rediscovered some simple ways of conducting business that, by comparison to the "straight" business world, are wildly innovative. This has not, repeat *not,* been the work of commercial gurus or public relations mavens, but rather the result of thinking that business need not be so complicated that the average person cannot understand its workings. Following here are the five key principles that govern our corporate culture here at Real Goods.

PRINCIPLE #1: THIS IS A BUSINESS.

And a business is, first and foremost, a financial institution. You can have the most noble social mission on the planet, but if you cannot maintain financial viability, you cease to exist, and so does your mission. The survival instinct is very strong at Real Goods, and that reality governs many decisions.

Quantity Discounts

We have come to the simple conclusion that we are not well prepared to sell Real Goods products to others for resale. The

Servicing other retailers requires personnel, equipment, and large profit margins that, as a catalog company, we do not have at the present time.

There are several instances in which we are currently able to offer discounts:

1. **To Lifetime Members** (see next section for details)
   Members receive a 5% discount on all purchases. Members currently pay a one-time fee of $50, and then can renew their subscription annually at no additional charge.

2. **To Shareowners**
   Owners of Real Goods stock, purchased during the original two public offerings, are accorded the same status as Members and can take advantage of the same great prices.

3. **To purchasers of multiple units**
   Many products have case-lot prices, in which case our savings in labor and packaging are passed along to the buyer. These prices are available to anyone who satisfies the quantity requirement. Real Goods' phone agents have up-to-the-minute information on case-lot availability and pricing.

Purchasing in case lots or subscribing are the best options for buyers who would like to import Real Goods products to foreign markets. This also applies to schools, businesses, organizations, and government institutions who would like to make quantity purchases.

Regional Centers

In 1993 Real Goods established its first regional center in Amherst, Wisconsin. Amherst is a picturesque little town, a veritable slice of Americana, located in the midst of the lakes region of Wisconsin. It's worth a trip for many reasons, not the least of which is the helpful staff at Real Goods Snow-Belt..

Amherst is also the site of the annual Midwest Renewable Energy Fair, held each June. This exhibition is generally recognized as the best of its type in the U.S.

Real Goods Snow-Belt also serves as Real Goods National Hearth Center. A complete line of hearth products are on display, and you will find the staff knowledgeable on all hearth options. The summer edition of the Real Goods News features over 50 pages of woodstoves and hearth products available by mail order. Request a free copy from the toll-free number.

Real Goods second remote outlet is in the "wilds" of downtown Eugene, Oregon. Besides offering the full range of catalog products and knowledgeable technical advice from the factory-trained staff, this store has the distinction of being Real Goods' official Outlet Store. This is a bargain-hunter's paradise with overstocks, discontinued items, and great prices.

We receive many inquiries about opening up Real Goods stores. If you have reason to believe that a Real Goods regional center would be successful, please make an inquiry (in writing) to the attention of the Public Relations Department, 555 Leslie St., Ukiah, CA 95482. Please, business realists only!

The Demonstration Home

This program made its debut in the Spring of 1994. This is designed for the homeowner or business owner who wants to make a serious personal commitment to independent living, but who has no interest in entering the alternative-energy products business.

Real Goods will extend to Demonstration Home participants an incentive discount to consolidate their purchasing through Real Goods, in exchange for a commitment to install all products to the highest standards of safety and code compliance and a willingness to allow other interested persons to visit the site.

Our intent is to establish a roster of homes and workspaces that will demonstrate the practical capabilities of renewable energy. We try to make it financially feasible for interested individuals by extending a discount, and they in turn help spread the word by agreeing to talk with others.

For details, call our business office (707-468-9292) and speak with the Demo Home coordinator.

PRINCIPLE #2: KNOW YOUR STUFF

Our social and environmental missions are only as strong as the equipment that makes possible the independent lifestyle that we advocate. The technologies work, to be sure, but they often require a degree of interaction that has been largely forgotten during our half-century binge on cheap power, in which convenience has been valued above all else.

At Real Goods we sell the equipment, but along with it comes the knowledge and the service that ensure your equipment will perform to your expectations:

Technical Services

Our technical services are world-class, and we find ourselves doing increasing amounts of design work for customers. Newcomers to PV want to make sure they select and install the right system, while solar sophisticates know that we have the latest data and the best access to solar information.

Technical services are available by phone (800) 919-2400 from 9 A.M. to 5 P.M., Pacific Time, Monday through Friday. We try very hard to answer technical mail within two weeks.

We welcome the opportunity to design and plan entire systems. Here's how our technical services work:

1. To assess your needs, capabilities, limitations, and working budget, we ask you to complete a specially created worksheet (see the Appendix). The informa-

tion required includes a complete list of your energy needs, an inventory of desired appliances, site information, and potential for hydroelectric and wind development.

2. A member of our technical staff will determine your needs, and design an appropriate system with you.

3. At this point, we will begin tracking the time we spend in helping you plan the details of your installation. Your personal tech rep will work with you on an unlimited time basis until your system has been completely designed and refined. He will order parts and talk you through assembly, assuring you that you get precisely what you need. He will also consult and work with your licensed contractor, if need be. first hour is free and part of our service. Beyond this initial consultation, time will be billed in ten-minute intervals and billed at the rate of $60 per hour. If the recommended parts and equipment are purchased from Real Goods, however, this time will be provided *at no charge*.

Local Pros

As proud as we are of our technical services, we cannot provide the on-the-site service that is sometimes required for equipment installation. To assist customers in linking up with local service professionals we offer a no-charge referral program to licensed contractors, called "Local Pro's" who specialize in renewable energy.

Professional contractors are encouraged to contact Real Goods for a Local Pro Application. For a modest fee you will get all the purchasing privileges of a Hard Corps member, special offers, and local referrals.

Real Goods has provided no training to the individuals who are registered as "local pros," and our referral is in no way intended to recommend or document any level of qualifications. Referrals are provided purely as a convenience to our customers. We recommend that you check references before using *any* installer.

To some ways of thinking the Local Pro program defies business logic. Why risk having customers exposed to the sirens of competition when you might be able to sell a product directly? The answer is simple. If you define business as being the achievement of customer satisfaction, then there are times when Local Pro is the best solution.

Company Publications

We at Real Goods are verbose in our communications, operating on the theory that, if we think something is interesting, so will others who share our curiosity and concerns. Here is the basic roster of our company publications:

The Real Goods Color Catalog. Published 4-6 times a year, this bright, colorful piece has energy-saving information and merchandise for everyone.

The Real Goods News, published three times each year is crammed with articles, customer profiles, unclassified ads, and hard-core information for the person who lives Off the Grid, or might someday. The *News* contains our ever-popular, raucous *Readers' Forum* where we turn the soap box over to some of the most interesting people we know- our customers. Each issue focuses on a technical component of our business- renewable energy, hearth products, or information. Paid subscriptions are available, but the smart money knows that the *News* comes free to Lifetime members.

The *Real Stuff newsletter* for *Lifetime Members and Shareowners.* This newsletter is published periodically and features behind-the-scenes information about what is happening at Real Goods, specialized technical information, and special discounts.

The *Solar Living Sourcebook* (formerly the *Alternative Energy Sourcebook*) is sent to new members. The nature of the *Sourcebook* should be self-explanatory, since you're perusing it at this very moment. We update the *Sourcebook* more or less annually, whenever we find it is outdated.

As one of the leading information providers in the realm of solar living, Real Goods has co-published *Wind Power For Home and Business* by Paul Gipe, and *The Independent Home,* by Michael Potts which provided the first-ever national portrait of today's energy pioneers.

Other titles include *The Contrary Farmer* by Gene Logsdon and *Eco Renovation* by Edward Harland. *Solar Gardening* (Lea and Gretchen Poisson)demonstrates how seasons can be extended and yields improved through deployment of innovative devices that help the gardener to better manage sunlight, *Renewable Are Ready* by the Union of Concerned Scientists portrays the heroes of communities where successful renewable energy projects have been successfully undertaken.

Innovative housing techniques are part of the Real Goods lexicon as well. *The Straw Bale House* and *The Rammed Earth House* each show how beautiful, durable homes with a unique sense of place can be created from the most common materials.

What these disparate titles have in common is a strong thread to the greater issue of sustainable living, a subject that involves "everything under the sun." Real Goods, in conjunction with publishing partner Chelsea Green, is seeking to provide the information that makes energy independent a tangible reality.

PRINCIPLE #3: GET INVOLVED!

Our shareowners are among our best customers, or is it the other way around? Our employees, through a stock option plan, are also encouraged to own Real Goods stock. The lines between who owns the company, who works for it, and who benefits from it, are fuzzy at Real Goods — intentionally so.

The Real Goods Lifetime Membership Program

The easiest way to become involved with the Real Goods community is to become a member of our preferred customer club.

We have always tried to find ways to reward, and even glorify, those dedicated customers who read our publications from cover to cover. These folks are the lifeblood of our business. They take energy independence seriously and are anxious to share experiences with others who have made a similar commitment.

Members receive the full roster of Real Goods publications, as well as a 5% discount on all purchases, including sale merchandise. A domestic membership costs $50, but will fluctuate depending on what new items we find to include. Membership privileges continue indefinitely, with annual renewal. In other words, we will keep all the news from Real Goods coming, but you have to tell us that you want it.

Due to the high cost of foreign postage, we can no longer offer foreign membership.

Own the Company: Buy Stock in Real Goods

Real Goods Trading Corporation has made two public offerings of its common stock. Both offerings were oversubscribed.

Proceeds from the offerings are funding business expansion, including retail stores and expansion of the company's catalog business. Our new headquarters at the Solar Living Center will showcase the technologies that make sustainable living a feasible reality.

Real Goods shares began trading April 11, 1994 on the Pacific Stock Exchange. The Stock symbol is RGT or RGT.P to show the exchange, and you can access trading information through computer on-line services or by telephone from securities brokerage firms. In newspaper financial pages, trading information may be under "Pacific" or "Regional Markets." For information about Real Goods shareownership, please write or call our Shareowner Relations Coordinator, at 707/468-9292, ext.2143.

Tell Us What to Sell

Several years ago we instituted the New Product Suggestion Program, figuring that, who could tell us what customers want better than customers themselves? We think of it as 300,000 investigative buyers working to broaden our product mix, seeking new products to enhance the quality of

energy-independent life.

The program works like this: Send us ideas for new products that you think we should carry in our catalog; include the manufacturer's name, address, and phone number, along with any personal experience you have had with the product. If your letter is the first recommendation we receive, and we add your suggested product to our catalog, you will receive a $25 merchandise credit. It's a great way to turn all our creative minds to the benefit of the entire Real Goods family. Send all your new product ideas to our regular address, Attn: Catalog Manager/New Products.

PRINCIPLE #4: TAKE A CHANCE, TAKE A STAND.

Many businesses treat risk as the financial equivalent of the Black Plague. This leads to the mealy-mouthed, corporate jargon that has become synonymous with the term "public relations." At Real Goods we try to take a stance that is consistent with our social and environmental mission. Occasionally, such as when president John Schaeffer published his views on abortion in the *Real Goods News*, we make a mistake. Our customers tell us about it, and if we are honest enough to admit our transgressions, they forgive us.

We are not perfect. (We don't even claim to be more virtuous than average.)

The Real Goods Eco Desk

Our Eco Desk has been among our most successful venues for social interaction. The mission of the Eco Desk is simple: make sure our actions are as noble as our words.

The Real Goods Eco Desk is not made of special material or anything fancy. To tell the truth, it's a regular ol' desk with a phone and a computer on it that is staffed by one of our senior customer-service representatives. The Eco Desk is the clearinghouse for many of the special events, projects, and services that Real Goods becomes involved with.

The Eco-desk conducts our annual eco-audit. Call to request a copy of our latest audit. It is the responsibility of the Eco Desk to make sure that, as a company, we practice what we preach, whether it's the way we use paper in the copier or the type of packaging we use in our shipments.

The Declaration of Energy Independence

In the spirit of our rebellious forebears, Real Goods (in conjunction with our cohorts at the Solar Energy Expo and Rally) wrote a document called "The Declaration of Energy Independence." Thousands of our customers signed. We promised to deliver the signed declarations to the White House in an electric car. As it happened, we even went one better: we got presidential candidate (and ex-Governor of California) Jerry Brown to deliver them for us. The event was filled with sound bites and photo ops — in fact, it was unabashed hype — but it did succeed it bringing the nation's energy situation to the attention of millions.

The Greening of the White House

Because of the company's well-established reputation as renewable energy zealots, Real Goods conceived of and was invited to participate with an elite group of industry experts in a volunteer effort to make the White House a showcase of energy efficiency. If we do the job well enough, we might one day have the White House included on the Tour of Independent Homes.

Real Relief

Unfortunately, most people do not confront their complete reliance on energy until they have been deprived of it. In recent years the crushing reality of our energy vulnerability has become apparent to residents of Florida (Hurricane Andrew), Hawaii (Hurricane Iniki), the Midwest (the floods of 1993), and Los Angeles (earthquakes, riots, mudslides, etc.)

Our homes and businesses need to be prepared for independence, even if it comes as the result of natural disaster. The necessity of energy preparedness is one of the main messages of Real Goods. That having been said, the company has instituted a policy to assist anyone who has been victimized by natural disaster. Called "Real Relief," the program offers a discount and shipping priority to anyone living in a federal disaster area. It goes into effect whenever and wherever a disaster is declared. To place orders, simply contact our customer-service agents through our toll-free number: (800) 762-7325.

PRINCIPLE #5: IT'S OKAY TO HAVE FUN!

Maybe it's all those somber, three-piece suits that create the image of the workplace as a world devoid of humor, but we have found that you can actually have fun doing business.

At our Shareowner's Party in August, we try to convince company owners that we are doing our best by their interests, but within a context that is both casual and convivial.

We never cease to be amazed at the vigor of customer response. When we invited Lifetime members to submit lists of songs that had themes having to do with the sun, several score responded with lists up to 800 songs long. All this for a prize that was billed as "some piece of junk from our warehouse."

Real Goods Institute for Solar Living

See page 565 for a complete description.

Gift Certificates

If you are unable to decide upon a gift, we offer Real Goods Gift Certificates. We will be happy to send a gift certificate in any amount to the person of your choice along with our latest edition of the Real Goods Catalog. Many of our products make great gifts for new households.

00-001   Specify $ Amount

Wedding Registry

If your dream of nuptial bliss includes independent living, then you can register with Real Goods, so that well-wishing friends can honor you with the gift of a much-needed inverter rather than the sterling tea service that will end up in the closet.

Is this a joke? Not really. A number of young married couples have embarked upon their dream of living lightly on the planet, and the Real Goods Wedding Registry has given them a way to get off to a running start. To register, call or write our Eco Desk.

Off-the-Grid Day

The first Off-the-Grid Day was held in 1991, as we encouraged people from coast to coast to flip their breakers for one day to experience life without power. As time passes, we will be using Off-the-Grid Day in different ways to promote awareness of various facets of energy independence. In 1993, Off-the-Grid Day was celebrated in October, when as we organized the nation's first-ever Tour of Independent Homes.

You are free to use the Off-the-Grid concept to promote energy awareness in conjunction with fundraising, political activism, a fun event — whatever you feel is most appropriate. We will provide information and access to materials to help make your event a smashing success. Call our public relations department for information.

Real Goods for Real Kids on Real Planets

We receive many requests to support nonprofit organizations. There are just too many good causes for a small company like ours to support in a meaningful way. Because of this, we have decided that the best financial contribution we can make to a nonprofit is to actively assist in its local fundraising efforts by making available our logistical support The program, called "Real Goods for Real Kids on Real Planets," is designed to harness "kid power" in the most productive way — by helping the community and the planet. In doing so, we can provide your qualifying organization with a fundraising vehicle that sells earth-friendly products in your community. Resulting orders are processed and fulfilled by Real Goods, with your organization receiving a rebate for each order placed.

Full details and an application form are available by writing to Real Goods, Attn: Real Kids. Help your community, help your organization, and help the planet. And let Real Goods help you do it!

The following Suggested Practices Guide is a "plain english" rendition of the 1996 National Electrical Code. In particular, those sections of the '96 Code that deal with photovoltaic systems and components. In the belief that safer, more professional installations are good for everyone in our industry, Real Goods is pleased to be able to present this important information.

Real Goods wishes to thank the Southwest Region Experiment Station, and John C. Wiles in particular for their cooperation in making reproduction of this safety information possible.

DRAFT
PHOTOVOLTAIC POWER SYSTEMS &
THE NATIONAL ELECTRIC CODE

SUGGESTED PRACTICES

The SWRES Testing Facility

December 1995
A "For Comment" Draft
A publication for the Photovoltaic Design Assistance Center

Sandia National Laboratories

A United States Department of Energy Multiprogram National Laboratory

PHOTOVOLTAIC POWER SYSTEMS
AND
THE *NATIONAL ELECTRICAL CODE*
SUGGESTED PRACTICES

December 1995

A Publication of the Photovoltaic Design Assistance Center
Sandia National Laboratories
A U.S. Department of Energy Multiprogram National Laboratory

Prepared by

Southwest Region Experiment Station
Southwest Technology Development Institute
New Mexico State University
P.O. Box 30001/Dept. 3 SOLAR
Las Cruces, New Mexico 88003-0001

Technical Comments To:

John C. Wiles
SWRES/NMSU
P.O. Box 30001/Dept. 3 SOLAR
1505 Payne Street
Las Cruces, New Mexico 88003

Request for copies to Photovoltaic Design Assistance Center
505-844-3698

The original research and drafts of this guide were funded under U. S. Department of Energy Contract DE-AS04-90AL57510.

PURPOSE

This guide provides information on the *National Electrical Code® (NEC)* and how it applies to photovoltaic (PV) systems. It is not intended to supplant or replace the *NEC*. It merely paraphrases the *NEC* and aligns information contained in the *NEC* with PV systems. Any PV system designer, equipment manufacturer, or installer should have a thorough knowledge of the *NEC* and a full understanding of the engineering principles and hazards associated with electrical and photovoltaic power systems. This material is not intended to be a design guide nor an instruction manual for an untrained person. Furthermore, this guide is not intended to cover all aspects of the *NEC* or PV systems—it must be used in conjunction with the full text of the *National Electrical Code*. This guide will be revised and updated as needed. Suggestions should be sent to the address on the front cover.

The *National Electrical Code* including the 1996 *National Electrical Code* is published and updated every three years by the National Fire Protection Association (NFPA), Batterymarch Park, Quincy, Massachusetts 02269. The *National Electrical Code* and the term *NEC* are registered trademarks of the National Fire Protection Association and may not be used without their permission. Copies of the 1996 *National Electrical Code* are available from the NFPA at the above address, most electrical supply distributors, and many bookstores.

In most locations, all electrical wiring including photovoltaic power systems must be accomplished by a licensed electrician and inspected by a designated local authority. Some municipalities have more stringent codes that supplement or replace the *NEC*. The local inspector has the final say on what is acceptable. In some areas, compliance with codes is not required.

DISCLAIMER

Neither the authors, the Southwest Region Experiment Station, the Southwest Technology Development Institute, New Mexico State University, Sandia National Labs, the U.S. Department of Energy, nor the National Fire Protection Association assume any liability resulting from the use of information presented in this manual. This information is believed to be the best available at the time of publication and is believed to be technically accurate. Any application of this information and results obtained from the use of this information are solely the responsibility of the reader.

ACKNOWLEDGMENTS

Numerous persons throughout the photovoltaic industry reviewed the drafts of this manual and provided comments which are incorporated in this version. Particular thanks go to Joel Davidson; Mike McGoey; George Peroni, Hydrocap; Tim Ball, Solar Engineering; Bob Nicholson, Glasstech Solar; Ward Bower, Sandia National Laboratories; Steve Willey, Backwoods Solar; Tom Lundtveit, Underwriters Laboratories; and all those who provided useful information at seminars on the subject. Appendix E is dedicated to John Stevens and Mike Thomas at Sandia National Laboratories. Document editing and layout by Ronald Donaghe, Southwest Technology Development Institute.

TABLE OF CONTENTS

LIST OF FIGURES

THE *NATIONAL ELECTRICAL CODE*

Although numerous portions of the *National Electrical Code* apply to photovoltaic power systems, those listed below are of particular significance.

Article	Contents
90	Introduction
100	Definitions
110	Requirements
200	Grounded Conductors
210	Branch Circuits
240	Overcurrent Protection
250	Grounding
300	Wiring Methods
310	Conductors
331	Electrical Nonmetallic Tubing
336	Nonmetallic Sheathed Cable
338	Service Entrance Cable
339	Underground Feeders
348	Electrical Metallic Tubing
374	Auxiliary Gutters
384	Switchboards and Panel Boards
445	Generators
480	Storage Batteries
690	PV Systems
705	Interconnected Electric Power Production Sources
710	Over 600 Volts, Nominal, General
720	Low-Voltage Systems

PHOTOVOLTAIC POWER SYSTEMS AND THE *NATIONAL ELECTRICAL CODE*

SUGGESTED PRACTICES

OBJECTIVE

- SAFE, RELIABLE, DURABLE PHOTOVOLTAIC POWER SYSTEMS
- KNOWLEDGEABLE MANUFACTURERS, DEALERS, INSTALLERS, CONSUMERS, AND INSPECTORS

METHOD

- WIDE DISSEMINATION OF THESE SUGGESTIONS
- TECHNICAL INTERCHANGE BETWEEN INTERESTED PARTIES

INTRODUCTION

The National Fire Protection Association has acted as sponsor of the *National Electrical Code (NEC)* since 1911. The original Code document was developed in 1897. With few exceptions, electrical power systems installed in the United States in this century have had to comply with the *NEC*. This includes many photovoltaic (PV) power systems. In 1984, Article 690, which addresses safety standards for installation of PV systems, was added to the Code. This article has been revised and expanded in the 1987, 1990, 1993, and 1996 editions.

Many of the PV systems in use and being installed today may not be in compliance with the *NEC* and other local codes. There are several contributing factors to this situation:

- The PV industry has a strong "grass roots," do-it-yourself faction that is not fully aware of the dangers associated with low-voltage, direct-current (dc), PV-power systems.

- Some people in the PV community may believe that PV systems below 50 volts are not covered by the *NEC*.

- Electrical inspectors have not had significant experience with direct-current portions of the Code or PV power systems.

- The electrical equipment industries do not advertise or widely distribute equipment suitable for dc use that meets *NEC* requirements.

- Popular publications are presenting information to the public that implies that PV systems are easily installed, modified, and maintained by untrained personnel.

- Photovoltaic equipment manufacturers have been generally unable to afford the costs associated with testing and listing by approved testing laboratories like Underwriters Laboratories or ETL.

- Photovoltaic installers and dealers in many cases have not had significant experience installing ac residential and/or commercial power systems.

Not all systems are unsafe. Some PV installers in the United States are licensed or use licensed electrical contractors and are familiar with all sections of the *NEC*. These installer/contractors are installing reliable PV systems that meet the *National Electrical Code* and minimize the hazards associated with electrical power systems. However, many PV installations have numerous defects and may not meet the 1996 Code. Some of the more prominent problems are listed below.

- Improper ampacity of conductors
- Improper insulation on conductors
- Unsafe wiring methods
- No overcurrent protection on many conductors
- Inadequate number and placement of disconnects
- Improper application of listed equipment
- No short-circuit current protection on battery systems
- Use of non-approved components when approved components are available
- Improper system grounding
- Lack of equipment grounding
- Use of underrated components
- Unsafe use of batteries
- Use of ac components (fuses and switches) in dc applications

The Code may apply to any PV systems regardless of size or location. A single PV module may not present a hazard, and a small system in a remote location may present few safety hazards because people are seldom in the area. On the other hand, two or three modules connected to a battery can be lethal if not installed and operated properly. A single deep-cycle storage battery (6 volts, 220 amp-hours) can discharge about 8,000 amps into a short-circuit. Systems with voltages of 50 volts or higher present shock hazards. Short circuits on lower voltage systems present fire and equipment hazards. Storage batteries can be dangerous; hydrogen gas and acid residue from lead-acid batteries must be dealt with safely.

The problems are compounded because, unlike ac systems, there are few *UL*-Listed components that can be easily "plugged" together to make a PV system. Connectors and devices do not have mating inputs or outputs, and the knowledge and understanding of "what works with what" is not second nature to the installer. The dc "cookbook" of knowledge does not yet exist.

To meet the objective of safe, reliable, durable photovoltaic power systems, the following suggestions are made:

- Dealer-installers of PV systems become familiar with the *NEC* methods of wiring residential and commercial ac power systems.

- All PV installations be inspected, where required, by the local inspection authority in the same manner as other equivalent electrical systems.

- Photovoltaic equipment manufacturers build equipment to *UL* or other recognized standards and have equipment tested and listed when practical.

- Listed or recognized subcomponents be used in assembled equipment where formal testing and listing is not possible.

- Electrical equipment manufacturers produce, distribute, and advertise, listed, reasonably priced, dc-rated components.

- Electrical inspectors become familiar with dc and PV systems.

- The PV industry educate the public, modify advertising, and encourage all installers to comply with the *NEC*.

- All persons installing PV systems obtain and study the current *National Electrical Code*.

- Existing PV installations be upgraded to comply with the *NEC* or modified to meet minimum safety standards.

RECOMMENDED PRACTICES

Scope and Purpose of the *NEC*

Some local inspection authorities use regional electrical codes, but most jurisdictions use the *National Electrical Code*—sometimes with slight modifications. The *NEC* states that adherence to the recommendations made will reduce the hazards associated with electrical installations. The *NEC* also says these installations may not necessarily be efficient, convenient, or adequate for good service or future expansion of electrical use [90-1]. (Numbers in brackets refer to sections in the *NEC*.)

The *National Electrical Code* addresses nearly all PV power installations, even those with voltages less than 50 volts. It covers stand-alone and grid-connected systems. It covers billboards, other remote applications, floating buildings, and recreational vehicles (RV) [90-2a, 690, 720]. The Code deals with any PV system that produces power and has external wiring or electrical components or contacts accessible to the untrained and unqualified person.

There are some exceptions. The *National Electrical Code* does not cover installations in automobiles, railway cars, boats, or on utility company properties used for power generation [90-2b]. It also does not cover micropower systems used in watches, calculators, or self-contained electronic equipment that have no external electrical wiring or contacts.

Article 690 of the *NEC* specifically deals with PV systems, but many other sections of the *NEC* contain requirements for any electrical system including PV systems [90-2, 720]. When there is a conflict between Article 690 of the *NEC* and any other article, Article 690 takes precedence [690-3].

The *NEC* suggests, and most inspection officials require, that equipment identified, listed, labeled, or tested by an approved testing laboratory be used when available [90-6,100,110-3]. Three of the several national testing organizations are the *Underwriters Laboratories (UL),* Factory Mutual Research (FM), and ETL Testing Laboratories, Inc. *Underwriters Laboratories* and *UL* are registered trademarks of Underwriters Laboratories Inc., 333 Pfingsten Road, Northbrook, IL 60062.

Most building and electrical inspectors expect to see *UL* on electrical products used in electrical systems in the United States. This presents a problem for the PV industry, because low production rates do not yet justify the costs of testing and listing by *UL* or other laboratory. Some manufacturers claim their product specifications exceed those required by the testing organizations, but inspectors readily admit to not having the expertise, time, or funding to validate these unlabeled items.

THIS GUIDE

The recommended installation practices contained in this guide progress from the photovoltaic modules to the electrical outlets. For each component, *NEC* requirements are addressed, and the appropriate Code sections are referenced in brackets. A sentence, phrase, or paragraph followed by a *NEC* reference refers to a requirement established by the *NEC*. The words "**will**," "**shall**," or "**must**" also refer to *NEC* requirements. Suggestions based on field experience with PV systems are worded as such and will use the word "should." The availability of approved components is noted, and alternatives are discussed.

Appendix A lists sources for dc-rated and identified, listed, or approved products, and reference to the products is made as they are discussed.

Appendix B presents diagrams for PV systems of varying sizes showing suggested connection and wiring methods.

Appendix C addresses areas particular to grid-connected systems.

Photovoltaic Modules

Five manufacturers, ASE Americas, Photocomm Solavolt, Siemens, Solarex, and Tideland Signal Corp., offer listed modules at the present time. Other manufacturers are considering having their PV modules listed by an approved national testing laboratory.

Methods of connecting wiring to the modules vary from manufacturer to manufacturer. The *NEC* does not require conduit, but local jurisdictions, particularly in commercial installations, may require conduit. The Code requires strain relief be

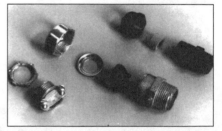

Figure 1. Strain Reliefs.

provided for connecting wires. If the module has a closed weatherproof junction box, strain relief and moisture-tight clamps should be used in any knockouts provided for field wiring. Where the weather-resistant gaskets are a part of the junction box, the manufacturer's instructions must be followed to ensure proper strain relief and weatherproofing [110-3(b), *UL* Standard 1703]. Figure 1 shows various types of strain reliefs. The one on the left is a basic cable clamp for interior use with nonmetallic sheathed cable (Romex). The clamps in the center (T&B) and on the right are watertight and can be used with either single or multiconductor cable—depending on the insert. The plastic unit on the right is made by Heyco (Appendix A).

Module Marking

Certain electrical information **must** appear on each module. If modules are not factory marked (required by the listing agency—*UL*), then they should be marked at the site to facilitate inspection and to allow the inspector to determine the requirements for conductor ampacity and rating of overcurrent devices. The information supplied by the manufacturer **will** include the following items:

• Polarity of output terminals or leads   • Maximum overcurrent device rating for module protection   • Rated open-circuit voltage   • Rated operating voltage   • Rated operating current   • Rated short-circuit current   • Rated maximum power   • Maximum permissible system voltage [690-51]

Figure 2.
Lable on Typical PV Module

Although not required by the *NEC*, the temperature rating of the module terminals must be known to determine the temperature rating of the insulation of the conductors and how the ampacity of those conductors must be derated for temperature [110-14(c)]. Figure 2 shows a typical label that appears on the back of a module.

Module Interconnections

Copper conductors are recommended for almost all photovoltaic system wiring [110-5]. Copper conductors have lower voltage drops and maximum resistance to corrosion. Aluminum or copper-clad aluminum wires can be used in certain applications, but the use of such cables is not recommended—particularly in dwellings. All wire sizes presented in this guide refer to copper conductors.

The *NEC* requires No. 12 American Wire Gage (AWG) or larger conductors to be used with systems under 50 volts [720-4]. Article 690 ampacity calculations yielding a smaller conductor size might override Article 720 considerations, but some inspectors are using the Article 720 requirement for dc circuits, and the Code has little information for conductor sizes smaller than No. 14 AWG, but Section 690-31d provides some guidance.

Single-conductor, Type UF (Underground Feeder — Identified as Sunlight Resistant), Type SE (Service Entrance), or Type USE (Underground Service Entrance) cables are permitted for module interconnect wiring [690-31b]. Stranded wire is suggested to ease servicing of the modules after installation and for durability [690-34]. Unfortunately, single-conductor, stranded, UF sunlight-resistant cable is not readily available, and there is some question about using the PVC insulation found on UF cables in dc circuits in the presence of moisture [310-13 FPN]. Most UF cable has insulation rated at 60°C. This insulation is not suitable for long-term exposure to direct sunlight at temperatures likely to occur on roofs near PV modules. Such wire has shown signs of deterioration after four years of exposure. Temperatures exceeding 60°C in the vicinity of the modules will preclude the use of conductors with 60°C insulation.

The widely available Underground Service Entrance Cable (USE) is suggested as the best cable to use for module interconnects. When made to the *UL* standard, It has a 90°C temperature rating and is sunlight resistant even though not commonly marked as such. Additional markings indicating XLP or XLPE (cross-linked polyethylene) and RHW-2 (90°C insulation when wet) ensure that the highest quality cable is being used [Tables 310-13,16, and 17]. It is acceptable to most electrical inspectors. The RHH and RHW-2 designations frequently found on USE-2 cable allow its use in conduit inside buildings. USE cable, without the other markings, does not have the fire-retardant additives that SE cable has and cannot be used inside buildings.

Where No. 10 AWG meets ampacity considerations, it is a good compromise between ease of installation and minimizing the voltage drop in the array wiring. Where modules are connected in parallel, the ampacity of the conductors will have to be adjusted accordingly. The temperature derated ampacity of conductors at any point must be at least 125 percent of the module (or array of parallel modules) rated short-circuit current at that point [690-8a, b1]. If flexible, two-conductor cable is needed, electrical tray cable (Type TC) is available but must be supported in a specific   manner as outlined in the *NEC* [318 and 340]. TC is sunlight resistant and is generally marked as such. Although frequently used for module interconnections, SO, SOJ, and similar flexible, portable cables and cordage may not be sunlight resistant and are not approved for fixed (non-portable) installations [400-7, 8]. A "WA" marking on these or the SEO hard-service cables indicates that they have some sunlight resistance and are listed for outdoor use. The 1996 *NEC* provides data that will enable these flexible cables to be properly derated for the high temperatures found near modules [Table 690-31c]. Type SEO, SO, and other flexible cables have not been tested for use in conduit.

Tracking Modules

Where there are moving parts of an array, such as a flat-plate tracker or concentrating modules, the *NEC* does allow the use of Article 400, flexible cords and cables [400-7(a)]. When these types of cables are used, they should be selected for extra hard usage with full outdoor ratings [marked "WA" on the cable]. They should not be used in conduit. Temperature derating information is provided by Table 690-31c. A derating factor in the range of 0.33 to 0.58 should be used for flexible cables used as module interconnects.

Another possibility is the use of extra flexible (475 strands) building cable type USE-RHH-RHW. This cable is available from the major wire distributors (Appendix A).

Terminals

Crimped-ring terminals are recommended for use in the module junction box to ensure that all strands of the conductor are connected to the screw terminal. If captive screws are used, then fork-type crimped terminals can be used, but no more than two should be used on any one screw. Crimping and soldering the ring or fork terminal to the wire is recommended—particularly in areas of high humidity—but some consideration should be given to the weakening of the conductor when subjected to the high temperatures of soldering. The conductors used with soldered terminals should be provided with strain relief to avoid stressing the heat-weakened conductor.

Light-duty crimping tools designed for crimping smaller wires used in electronic components usually do not provide sufficient force to make long-lasting crimps on connectors for PV installations even though they may be sized for No. 12-10 AWG. Insulated terminals crimped with these light-duty crimpers frequently develop high-resistance connections in a short time and may even fail as the wire pulls out of the terminal. It is strongly suggested that only heavy-duty industrial-type crimpers be used for PV system wiring. Figure 3 shows four styles of crimpers. On the far left is a stripper/crimper used for electronics work that will crimp only insulated terminals. Second from the left is a stripper/crimper that can make crimps on both insulated and uninsulated terminals. The pen points to the dies used for uninsulated terminals. With some care, this crimper can be used to crimp

Figure3. Terminal Crimpers

uninsulated terminals on PV systems if the terminals are soldered after the crimp. The two crimpers on the right are heavy-duty industrial designs with ratcheting jaws and interchangeable dies that will provide the highest quality connections. They are usually available from electrical supply houses.

Figure 4 shows some examples of insulated and uninsulated terminals. In general, uninsulated terminals are preferred (with insulation applied later if required), but care must be exercised to obtain the heavier, more reliable *UL*-Listed terminals and not unlisted electronic or automotive grades. Again, an electrical supply house rather than an electronic or automotive parts store is the place to find the required items. Although time consuming, the crimping and soldering technique should be considered to ensure the connections last as long as the modules themselves. If the junction box provides box-style pressure terminals, it is not necessary to use the crimped and soldered terminals.

Transition Wiring

Figure 4. Insulated and Uninsulated Terminals.

Because of the relatively high cost of USE and TC cables and wire, they are usually connected to less expensive cable at the first junction box leading to an interior location. All PV system wiring must be made using one of the methods included in the *NEC* [690-31, Chapter 3]. Single-conductor, exposed wiring is not permitted except for module wiring or with special permission [Chapter 3]. The most common methods used for PV systems are individual conductors in rigid metallic and nonmetallic conduit and nonmetallic sheathed cable.

Where individual conductors are used in conduit, they should be conductors with at least 90°C insulation such as RHW-2 or XHHW-2. Conduits installed in exposed locations are considered to be installed in wet locations [100-Locations]. These conduits may have water entrapped in low spots and therefore only conductors with wet ratings should be installed in conduits that are in exposed locations. The conduit can be either thick-wall or thin-wall electrical metallic tubing (EMT) [348], and if nonmetallic conduit is used, electrical (gray) PVC rather than plumbing (white) PVC tubing **must** be used [347].

Two-conductor (with ground) UF cable that is marked sunlight resistant is frequently used between the module interconnect wiring and the PV disconnect device. Black is the preferred color because of higher resistance to ultraviolet light, but the gray color seems durable because of the insulation associated with the jacket on the cable. Splices from the stranded wire to this wire when located outside **must** be protected in rain-proof junction boxes such as NEMA style 3R. Cable clamps must also be used. Figure 5 shows a rain-proof box with a pressure connector terminal strip installed for module wiring connections. The box penetrations (holes for screws) should be sealed with silicon rubber. It has indoor cable clamps, but if the cable clamps were listed for outdoor use, the box could be used outdoors.

Interior exposed cable runs can be made only with sheathed cable types such as NM, NMC, and UF. The cable should not be subjected to physical abuse. If abuse is possible, physical protection must be provided [300-4, 336 B, 339]. Exposed single-conductor cable (commonly used between batteries and inverters) **shall not** be used—except as module interconnect conductors [300-3a].

Figure 5. Rain-proof Junction Box. Shown with Custom Terminal Strip and Interior-Rated Cable Clamps.

WIRING
Module Connectors

Module connectors that are concealed at the time of installation **must** be able to resist the environment, be polarized, and be able to handle the short-circuit current. They **shall** also be of a latching design with the terminals guarded. The equipment-grounding member, if used, **shall** make first and break last [690-32, 33]. The *UL* standard also requires that the connectors for positive and negative conductors **shall not** be interchangeable.

Module Connection Access

All junction boxes and other locations where module wiring connections are made **shall** be accessible. Removable modules and stranded wiring will allow accessibility [690-34]. This means modules should not be permanently fixed (welded) to mounting frames, and solid wire that could break when modules are moved to service the junction boxes should not be used. Open spaces behind the modules would allow access to the junction boxes.

Splices

All splices (other than the connectors mentioned above) must be made in approved junction boxes with an approved splicing method. Conductors must be twisted firmly to make a good electrical and mechanical connection, then brazed, welded, or soldered, and then taped [110-14b]. Although solder has a higher resistivity than copper, a rosin-fluxed, soldered splice will have slightly lower electrical resistance, and potentially higher resistance to corrosion than an unsoldered splice. Mechanical splicing devices such as split-bolt connectors or terminal strips are also acceptable. Crimped splicing connectors may also be used if heavy-duty crimpers are used.

If the highest reliability is needed, then ultrasonic welding should be used for splices. Also, properly used box-type pressure connectors (Figure 7) give high reliability. Fuse blocks, fused disconnects, and circuit breakers are available with these pressure connectors.

Twist-on wire connectors (approved for splicing wires) have not proved adequate when used on low-voltage (12-50 volts) or high-current PV systems because of thermal stress and oxidation of the contacts.

Where several modules are connected in series and parallel, a terminal block or bus bar arrangement **must** be used so that one source circuit can be disconnected without disconnecting the grounded (on grounded systems) conductor of other source circuits [690-4c]. On grounded systems, this indicates that the popular "Daisy Chain" method of connecting modules may not always be acceptable, because removing one module in the chain may disconnect the grounded conductor for all of those modules in other parallel chains or source circuits. This becomes more critical on larger systems where paralleled sets of long series strings of modules are used. Figure 6 shows unacceptable and acceptable methods. Generally, 12- and 24-volt systems can be daisy chained, but higher voltage systems should not be.

Several different types of terminal blocks and strips are shown in Figure 7. The larger blocks are made by Marathon (Appendix A).

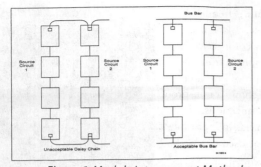

Figure 6. Module Interconnect Methods

Figure 7.
Power Splicing
Blocks and
Terminal Strips.

Conductor Color Codes

The *NEC* established color codes for electrical power systems many years before either the automobile or electronics industries were started. PV systems are being installed in the arena covered by the *NEC* and, therefore, must comply with *NEC* standards that apply to both ac and dc power systems. In a system where one conductor is grounded, the insulation on all grounded conductors must be white or natural gray or be any color except green if marked with white plastic tape or paint at each termination (marking allowed only on conductors larger than 6 AWG). Conductors used for module frame grounding and other exposed metal equipment grounding must be bare (no insulation) or have green or green with yellow-striped insulation or identification [200-6, 7; 210-5].

The *NEC* requirements specify that the grounded conductor be white. In most PV-powered systems that are grounded, the grounded conductor turns out to be the negative conductor. A prominent exception is the telephone system, which uses a positive ground. In a PV system where the array is center tapped, the center tap or neutral must be grounded [690-41], and this becomes the white conductor. There is no *NEC* requirement designating the color of the ungrounded conductor, but the convention in power wiring is that the first two ungrounded conductors are colored black and red. This suggests that in two-wire, negative-grounded PV systems, the positive conductor could be red or any color with a red marking except green or white, and the negative grounded conductor **must** be white. In a three-wire, center-tapped system, the positive conductor could be red, the grounded center tap conductor **must** be white and the negative conductor could be black.

The 1996 *NEC* allows grounded (non-white) array conductors, such as USE or SE that are smaller than No. 6 AWG, to be marked with a white marker [200-6].

Battery Cables

Battery cables, even though they can be No. 2/0 AWG and larger, must be of standard building-wire type conductor [Chapter 3]. Welding and automobile "battery" cables are not allowed. Flexible, highly-stranded, building-wire type cables (USE-RWH) are available for this use. Flexible cables, identified in Section 400 of the *NEC* are allowed from the battery terminals to a nearby junction box and between battery cells [690-74].

GROUND-FAULT PROTECTION AND ARRAY DISABLEMENT

Ground-Faults

Article 690-5 of the *NEC* requires a ground-fault detection, interruption, and array disablement (GFID) device for fire protection if the PV arrays are mounted on roofs of dwellings. Ground-mounted arrays are not required to have this device. Several devices to meet this requirement are under development, but none are commercially available. These particular devices may require that the system grounding conductor be routed through the device. To keep costs to a minimum, the devices under development may replace the PV disconnect switch and may incorporate the following functions:

• Manual PV disconnect switch   • Ground-fault detection   • Ground-fault interruption

• Array disablement   • Array wiring overcurrent protection

588

Ground-fault detection, interruption, and array disablement devices might, depending on the particular design, accomplish the following actions automatically:

- Sense ground-fault currents exceeding a specified value • Interrupt or significantly reduce the fault currents
- Open the circuit between the array and the load • Short the array or subarray

These actions would reduce the array voltages to nearly zero (minimizing human shock hazards and equipment damage) and would serve to direct the fault currents away from the fault path and back into the normal conductors. For fault location and repair, the array shorting device would have to be opened.

Ground-fault devices have been developed for some grid-tied inverters and stand-alone systems, and others are under development. If a version of the *NEC* specifies equipment that is not commercially available, the preceding edition(s) of the Code may be used with the approval of the inspecting authority [90-4]. In this case, the 1987 *NEC* did not require a GFID device.

Array Disablement

Article 690-18 requires that a mechanism be provided to disable portions of the array or the entire array. The term "disable" has several meanings, and the *NEC* is not clear on what is intended. The *NEC* Handbook does elaborate. Disable can be defined several ways:

- Prevent the PV system from producing any output
- Reduce the output voltage to zero
- Reduce the output current to zero

The output could be measured at either the PV source terminals or at the load terminals.

Fire fighters are reluctant to fight a fire in a high-voltage battery room because there is no way to turn off a battery bank unless you can somehow remove the

electrolyte. In a similar manner, the only way a PV system can have zero output at the array terminals is by preventing light from illuminating the modules. The output voltage may be reduced to zero by shorting the PV module or array terminals. When this is done, short-circuit current will flow through the shorting conductor, which in a properly wired system with bypass diodes, does no harm. The output current may be reduced to zero by disconnecting the PV system from any load. The PV disconnect switch would accomplish this action, but open-circuit voltages would still be present on the array wiring and in the disconnect box. On a large system, 100 amps of short-circuit current (with a shorted array) can be as difficult to handle as an open-circuit voltage of 600 volts.

During PV module installations, the individual PV modules can be covered to disable them. For a system in use, the PV disconnect switch is opened during maintenance, and the array is either short circuited or left open circuited depending on the circumstances. In practical terms, for a large array, some provision (switch or bolted connection) should be made to disconnect portions of the array from other sections for servicing. As individual modules or sets of modules are serviced, they may be covered and/or isolated and shorted to reduce the potential for electrical shock. Aside from measuring short-circuit current, there is little that can be serviced on a module or array when it is shorted. The circuit is usually open circuited for repairs.

GROUNDING

The subject of grounding is one of the most confusing issues in electrical installations. Definitions from Article 100 of the *NEC* will clarify the situation.

Grounded: Connected to the earth.

Grounded Conductor: A system conductor that normally carries current and is intentionally grounded. In PV systems, one conductor (normally the negative) of a two-conductor system or the center-tapped wire of a bipolar system is grounded.

Grounding Conductor, 1 & 2: A conductor not normally carrying current used to: (1) connect the exposed metal portions of equipment to the grounding electrode system or the grounded conductor, or (2) connect the grounded conductor to the grounding electrode or grounding electrode system.

Equipment Grounding Conductor: See Grounding Conductor (1), above.

Grounding Electrode Conductor: See Grounding Conductor (2), above.

Grounding—System

For a two-wire PV system over 50 volts (open-circuit PV-output voltage), one dc conductor **shall** be grounded. In a three-wire system, the neutral or center tap of the dc system **shall** be grounded [690-7, 41]. These requirements apply to both stand-alone and grid-tied systems. Such system grounding will enhance personnel safety and minimize the effects of lightning and other induced surges on equipment. Also, grounding of all PV systems will reduce radio frequency noise from dc-operated fluorescent lights and inverters.

Size of Grounding Electrode Conductor

The direct-current system-grounding electrode conductor **shall** not be smaller than No. 8 AWG or the largest conductor supplied by the system [250-93]. If the conductors between the battery and inverter are 4/0 AWG (for example) then the conductor from the negative conductor (assuming that this is the grounded conductor) to the ground rod **must** be 4/0 AWG. The 1996 *NEC* allows exceptions to this large grounding conductor requirement. Many PV systems can use a No. 6 AWG grounding electrode conductor if that is the only connection to the ground rod [250-93].

Point of Connection

The system grounding electrode conductor for the direct-current portion of a PV system **shall** be connected to the PV-output circuits as close to the modules as possible [690-42, 250-22]. When this connection is made as close as possible to the modules, added protection from surges is afforded. Disconnect switches **must** not open grounded conductors [690-13]. In stand-alone PV systems, the charge controller may be considered a part of the PV-output circuit, and the point of connecting the grounding electrode conductor could be before or after the charge controller. But this grounding conductor may be a very large conductor (e.g., 4/0 AWG) while the conductors to and from the charge controller may be No. 10 AWG or smaller. Connecting the 4/0 AWG grounding conductor on the array side of the charge controller, while providing some degree of enhanced surge suppression from lightning induced surges, does not meet the intent of the grounding requirement. Connecting the grounding conductor to the system on the battery side of the charge controller at a point where the system conductors are at the largest size will provide the best system grounding. Figure 8 shows two possible locations for the grounding conductor.

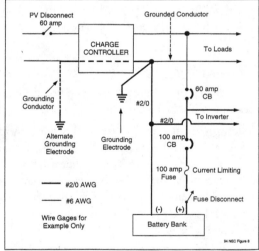

Figure 8. Typical System: Possible Grounding Conductor Location

The *NEC* does not specifically define where the PV-output circuits end. Circuits from the battery toward the load are definitely load circuits. Since the heaviest conductors are from the battery to the inverter, and either end of these conductors is at the same potential, then either end could be considered a point for connecting the grounding conductor. The negative dc input to the inverter is connected to the metal case in some stand-alone inverter designs, but this is not an appropriate place to connect the grounding electrode conductor and other equipment-grounding conductors, since this circuit is a dc-branch circuit and not a PV-output circuit. Connection of the grounding electrode conductor to the negative battery terminal would avoid the "large-wire/small-wire" problem outlined above.

It is imperative that there be no more than one grounding connection to the negative conductor of a PV system. Failure to limit the connections to one (1) will allow currents to flow in uninsulated conductors and will create unintentional ground faults in the grounded conductor [250-21]. Future ground-fault interrupter systems may require that this single grounding connection be made at a specific location.

590

Unusual Grounding Situations

Some inverter designs use the entire chassis as part of the negative circuit. Also, the same situation exists in certain radios—automobile and shortwave. These designs will not pass the current *UL* standards for consumer electrical equipment and will probably require modification in the future since they do not provide electrical isolation between the exterior metal surfaces and the current-carrying conductors. They also create the very real potential for multiple grounding-conductor connections to earth ground.

Since the case of these non-listed inverters is connected to the negative conductor and that case must be grounded as part of the equipment ground described below, the user has no choice whether or not the system is to be grounded. The system will be grounded even if the voltage is less than 50 volts and the point of system ground is the negative input terminal on the inverter.

Some telephone systems ground the positive conductor, and this may cause problems for the PV system. An isolated-ground, dc-to-dc converter may be used to power subsystems that have different grounding polarities from the main system. In the ac realm, an isolation transformer will serve the same purpose.

In larger utility-tied systems and some stand-alone systems, high impedance grounding systems might be used in lieu of or in addition to the required hard ground. The discussion and design of these systems are beyond the scope of this guide. Grounding of grid-tied systems will be discussed in Appendix C.

Charge Controllers—System Grounding

In a grounded system, it is important that the charge controller have no signal processing in the grounded conductor. Relays or transistors in the grounded conductor create a situation where the grounded conductor is not at ground potential at times when the charge controller is operating. This condition violates provisions of the *NEC* that require all conductors identified as grounded conductors always be at the same potential (i.e. grounded). A shunt in the grounded conductor is equivalent to a wire if properly sized, but the user of such a charge controller runs the risk of having the shunt bypassed when inadvertent grounds occur in the system. The best charge controller design has only a straight-through conductor between the input and output terminals for the grounded current-carrying conductor (usually the negative conductor).

Grounding—Equipment

All noncurrent-carrying exposed metal parts of junction boxes, equipment, and appliances in the entire PV and dc load system **shall** be grounded [690-43, 250 E, 720-1 & 10]. All PV systems, regardless of voltage, **must** have an equipment-grounding system for exposed metal surfaces (e.g., module frames and inverter cases) [690-43]. The grounding conductor **shall** be sized as required by Article 690-43 or 250-95. Generally, this will mean an equipment-grounding conductor size based on the size of the overcurrent device protecting each conductor. Table 250-95 in the *NEC* gives the sizes. For example, if the inverter to battery conductors are protected by a 400-amp fuse or circuit breaker, then at least No. 3 AWG conductor **must** be used for the equipment ground for that circuit. If the current-carrying conductors have been oversized to lower voltage drop, then the size of the equipment-grounding conductor **must** also be proportionately adjusted [250-95]. In the PV source circuits, if the array can provide short-circuit currents that are less than twice the rating of a particular overcurrent device for the array circuits, then equipment-grounding conductors **must** be used that are sized the same as the array current-carrying conductors [690-43]. In other situations, Table 250-95 of the *NEC* applies.

Equipment Grounds for Non-Listed Inverters

Many non-listed inverters do not have provisions for the equipment ground connection required by the *NEC*. It is suggested that one of the holes used to mount the inverter have the paint scraped off and the mounting bolt with internal toothed lock washers be used to connect the equipment-grounding conductor. If the inverter has the case connected to the negative terminal, then the negative input terminal can be used for the equipment ground and the system ground. An appropriately sized conductor must be used.

Some listed inverters have provisions for only a small-gage ac output-side equipment-grounding conductor. This equipment-grounding conductor would probably be vaporized if the ground fault were to occur on the dc side of the wiring. It is suggested that the inverter equipment-grounding conductor be sized for the dc input.

Inverter AC Outputs

The inverter output (120 or 240 volts) must be connected to the ac distribution system in a manner that does not create parallel grounding paths. The *NEC* requires that both the green equipment-grounding conductor and the white neutral conductor be grounded. The Code also requires that current not normally flow in the green wires. If the inverter has ac grounding receptacles as outputs, the grounding and neutral conductors are most likely connected to the chassis and, hence, to ground inside the inverter. This configuration allows plug-in devices to be used safely. However, if the outlets on the inverter are plug and cord connected to an ac load center used as a distribution device, then problems can occur.

The ac load center usually has the neutral and equipment-grounding conductors connected to the same bus bar which is connected to the case where they are grounded. Parallel current paths are created with neutral currents flowing in the equipment-grounding conductors. This problem can be avoided by using a load center with an isolated/insulated neutral bus bar which is separated from the equipment-grounding bus bar.

Inverters with hard-wired outputs may or may not have internal connections. Some inverters with ground-fault circuit interrupters (GFCIs) for outputs must be connected in a manner that allows proper functioning of the GFCI. A case-by-case analysis will be required.

Auxiliary Generators

Auxiliary generators used for battery charging pose problems similar to using inverters and load centers. The generators usually have ac outlets which may have the neutral and grounding conductors bonded to the generator frame. When the generator is connected to the system through a load center, to a standby inverter with battery charger, or to an external battery charger, parallel ground paths are likely. These problems must be addressed on a case-by-case basis. A PV system, in any operating mode, **must not** have currents in the equipment-grounding conductors [250-21].

Suggested AC Grounding

Auxiliary ac generators and inverters should be hard-wired to the ac-load center. Neither should have an internal bond between the neutral and grounding conductors. Neither should have any receptacle outlets that can be used when the generator or inverter is operated when disconnected from the load center. The single bond between the neutral and ground will be made in the load center. If receptacle outlets are desired on the generator or the inverter, they should be ground-fault-circuit-interrupting devices (GFCI).

Grounding Electrode

The dc system grounding electrode **shall** be common with, or bonded to, the ac-grounding electrode (if any) [690-44, 250-26c]. The system-grounding conductor and the equipment-grounding conductor **shall** be tied to the same grounding electrode or grounding electrode system. Even if the PV system is ungrounded (optional at less than 50 volts), the equipment-grounding conductor must be connected to a grounding electrode [250-50]. The grounding electrode **shall** be a corrosion resistant rod, a minimum of 5/8 inch in diameter with at least 8 feet driven into the soil at an angle no greater than 45 degrees from the vertical [250-83]. Metal water pipes and other metallic structures as well as concrete encased electrodes may also be used in some circumstances [250-26c, 250-81, 250-83]. Listed connectors must be used to connect the grounding conductor to the ground rod.

A bare-metal well casing makes a good grounding electrode. If it is distant from the PV array or the main disconnect, it should be part of a grounding electrode system. The central pipe to the well should not be used for grounding, because it is sometimes removed for servicing.

For maximum protection against lightning-induced surges, it is suggested that a grounding electrode **system** be used with at least two grounding electrodes bonded together. One electrode would be the main system grounding electrode as described above. The other would be a supplemental grounding electrode located as close to the PV array as practical. The module frames and array frames would be connected directly to this electrode to provide as short a path as possible for lightning-induced surges to reach the earth. This electrode **must** be bonded with a conductor to the main system grounding electrode [250-81]. The size of the bonding or jumper cable must be related to the ampacity of the overcurrent device protecting the PV source circuits. This

592

bonding jumper is an auxiliary to the module frame grounding that is required to be grounded with an equipment-grounding conductor. *NEC* Table 250-95 gives the requirements. Generally, it must be no smaller than No. 8 AWG to comply with bonding jumper requirements. Equipment-grounding conductors are allowed to be smaller than circuit conductors when the circuit conductors become very large. Article 250 of the *NEC* elaborates on these requirements.

Do not connect the negative current-carrying conductor to the grounding electrode, to the equipment-grounding conductor, or to the frame at the modules. There should be one and only one point in the system where the grounding electrode conductor is attached to the system-grounded conductor. See Figure 9 for clarification. The wire sizes shown are for illustration only and will vary depending on system size. Chapter 3 of the *NEC* specifies the ampacity of various types and sizes of conductors.

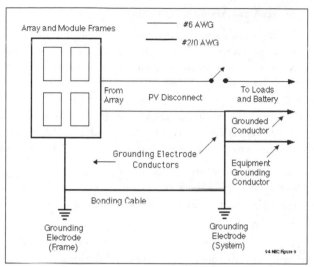

Figure 9. Example Grounding Electrode System

Conductor Ampacity

Photovoltaic modules are limited in their ability to deliver current. The short-circuit current capability of a module is nominally 10 to 15 percent higher than the operating current. Enhanced irradiance (sunlight) due to snow and clouds may increase PV-output current another 15 - 25 percent. These increased currents must be considered in the ampacity calculations. Another problem for PV systems is that the conductors may operate at temperatures as high as 65-75°C when the modules are mounted close to a structure, there are no winds, and the ambient temperatures are high. Temperatures in module junction boxes frequently occur within this range. This **will** require that the ampacity of the conductors be derated or corrected with factors given in *NEC* Table 310-16 or 310-17. For example, a No. 10 AWG USE/RHW-2 single-conductor cable used for module interconnections in conduit has a 90°C insulation and an ampacity of 40 amps in an ambient temperature of 26-30°C. When it is used in ambient temperatures of 61-70°C, the ampacity of this cable is reduced to 23.2 amps.

It should be noted that the ampacity values associated with conductors having 90°C insulation can only be used if the module terminals are rated at 90°C [110-14(c)]. If they are rated at only 75°C, then the ampacity values associated with 75°C insulation **must** be used, even when conductors with 90°C insulation are being used.

The ampacity of conductors in PV source circuits **shall** be at least 125 percent of the rated module or parallel-connected modules short-circuit current [690-8]. The ampacity of the PV-output circuit conductors **shall** be at least 125 percent of the short-circuit output current [690-8]. The ampacity of conductors to and from an inverter or power conditioning system **shall** be 125 percent of the rated operating current for that device [690-8]. In a similar manner, other conductors in the system should have an ampacity of 125 percent of the rated operating current to allow for long duration operation at full power [220-3a]. These *NEC* requirements are to ensure that the connected overcurrent devices or panel boards operate at no more than 80 percent of their ampacity. Operation when snow or cloud enhancement increases the PV output above normal may require additional ampacity. With reflective ground cover such as sand or snow and with reflections from clouds, PV output may reach 125 percent of rated output for short periods of time (minutes).

A 1989 revision to the *UL* Standard 1703 for PV modules requires that module installation instructions include an additional 25 percent of the 25°C ratings for short-circuit current and open-circuit voltage to allow for expected daily peak irradiance and colder temperatures. Conservative design practices require oversizing wire and increasing the ratings of overcurrent devices on PV source and output circuits. However, the rating of the overcurrent device should always be less than, or equal to, the ampacity of the cable. The *NEC* makes only infrequent exceptions to this rule [240-36].

The ampacity of conductors and the sizing of overcurrent devices is an area that demands careful attention by the PV system designer/installer. Temperatures and wiring methods must be addressed for each site. Start with the 125 percent value code requirement, then derate the cable ampacity for temperature. Include an additional 125 percent to comply with the UL requirements. See Appendix E for examples.

Stand-Alone Systems—Inverters

In stand-alone systems, inverters are frequently used to change the direct current (dc) from a battery bank to 120-volt or 240-volt, 60-Hertz (Hz) alternating current (ac). The conductors between the inverter and the battery **must** have properly rated overcurrent protection and disconnect mechanisms [240, 690-8b(3), -8b(4), -15]. These inverters frequently have short duration (tens of seconds) surge capabilities that are four to six times the rated output. For example, a 2,500-watt inverter might be required to surge to 10,000 watts for 5 seconds when a motor load must be started. The *NEC* requires the ampacity of the conductors between the battery and the inverter to be sized by the rated 2,500-watt output of the inverter. For example, in a 24-volt system, a 2,500-watt inverter would draw 105 amps at full load and 420 amps for motor-starting surges. The ampacity of the conductors between the battery **must** be 125 percent of the 105 amps or 131 amps.

To minimize steady-state voltage drops, account for surge-induced voltage drops, and increase to system efficiency, most well-designed systems have conductors several sizes larger than required by the *NEC*. When the current-carrying conductors are oversized, the equipment-grounding conductor **must** also be oversized proportionally [250-95].

When the battery bank is tapped to provide multiple voltages (i.e., 12 and 24 volts from a 24-volt battery bank), the common negative conductor will carry the **sum** of all of the simultaneous load currents. The negative conductor **must** have an **ampacity at least equal to the sum** of all the amp ratings of the overcurrent devices protecting the positive conductors or have an ampacity equal to the sum of the ampacities of the positive conductors [690-8(c)].

The *NEC* does not allow paralleling conductors for added ampacity, except that cables 1/0 AWG or larger may be paralleled under certain conditions [310-4]. DC-rated switchgear, overcurrent devices, and conductors cost significantly more when rated to carry more than 100 amps. It is suggested that large PV arrays be broken down into subarrays, each having a short-circuit output of less than 80 amps. This will allow use of 100-amp-rated equipment (125 percent of 80 amps) on each source circuit.

Overcurrent Protection

The *NEC* requires that every ungrounded conductor be protected by an overcurrent device [240-20]. In a PV system with multiple sources of power (PV modules, batteries, battery chargers, generators, power conditioning systems, etc.), the overcurrent device must protect the conductor from overcurrent from any source connected to that conductor [690-9]. Blocking diodes, charge controllers, and inverters are not considered as overcurrent devices and must be considered as zero-resistance wires when assessing overcurrent sources (690-9 FPN). If the PV system is directly connected to the load without battery storage or other source of power, then no overcurrent protection is required if the conductors are sized at 125 percent of the short-circuit current [690-8].

When circuits are opened in dc systems, arcs are sustained much longer than they are in ac systems. This presents additional burdens on overcurrent-protection devices rated for dc operation. Such devices must carry the rated load current and sense overcurrent situations as well as be able to safely interrupt dc currents. AC overcurrent devices have the same requirements, but the interrupt function is considerably easier.

Ampere Rating

The PV source circuits **shall** have overcurrent devices rated at least 125 percent of the parallel module short-circuit current. The PV-output circuit overcurrent devices **shall** be rated at least 125 percent of the short-circuit PV currents [690-8]. Some installations have experienced the blowing of fuses and loosening of terminals for unknown reasons. Good engineering practice calls for increasing the rating of these overcurrent devices <u>and</u> the ampacity of the conductors they protect to 156 percent of the short-circuit current. This practice agrees with the *UL* requirements mentioned above (1.25 x 1.25 = 1.56). Time-delay fuses or circuit breakers would minimize nuisance tripping or blowing. In all cases, dc-rated devices having the appropriate dc-voltage rating must be used and adequate ventilation must be provided.

All ungrounded conductors from the PV array **shall** be protected with overcurrent devices [Article 240, Diagram 690-1]. Grounded conductors (not shown in Diagram 690-1) should not have overcurrent devices since the independent opening of such a device would unground the system. Since PV module outputs are current limited, these overcurrent devices are actually protecting the array wiring from battery or power conditioning system short circuits.

Because the conductors and overcurrent devices are sized to deal with 125 percent of the short-circuit current for that

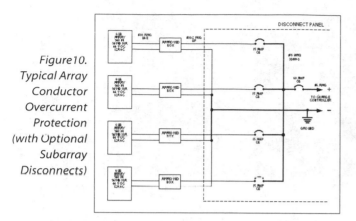

Figure 10. Typical Array Conductor Overcurrent Protection (with Optional Subarray Disconnects)

Figure 11. Listed Branch-Circuit Breakers

particular PV circuit, overcurrents from those modules or PV sources, which are limited to the short-circuit current, cannot trip the overcurrent device in this circuit. The overcurrent devices in these circuits protect the conductors from overcurrents from parallel connected sets of modules (diode failure) or overcurrents from the battery bank (diode or charge controller failure). In standby systems or grid-connected systems, these array overcurrent devices protect the array wiring from overcurrents from parallel strings of modules, the battery, or from the generator or ac utility power when the battery charger or inverter fails.

Often, PV modules or series strings of modules are connected in parallel. As the conductor size used in the array wiring increases to accommodate the higher short-circuit currents of paralleled modules, each conductor size must be protected by an appropriately sized overcurrent device. This device must be placed nearest the source of the largest potential overcurrent for that conductor [240-21]. Figure 10 shows an example of array conductor overcurrent protection for a medium-size array broken into subarrays. The cable sizes and types shown are examples only. The actual sizes will depend on the ampacity needed.

Either fuses or circuit breakers are acceptable for overcurrent devices provided they are rated for their intended uses—i.e., they have dc ratings when used in dc circuits, the ampacity is correct, and they can interrupt the necessary currents when short circuits occur [240 E, F, G]. Figure 11 shows dc-rated, *UL*-Listed circuit breakers being used in a PV power center for overcurrent protection and disconnects. Circuit breakers are manufactured by Heinemann (Appendix A). The *NEC* allows the use of listed (recognized) supplemental overcurrent devices only for PV source circuit protection.

Testing and Approval

The *NEC* requires that listed devices be used for overcurrent protection. A <u>listed</u> device by *UL* or other approved testing laboratory is tested against an appropriate *UL* standard. A <u>recognized</u> device is tested by *UL* or other approved testing laboratory to standards established by the device manufacturer. In most cases, the standards established by the manufacturer are less rigorous than those established by *UL*. Many inspectors will not accept recognized devices, particularly where they are required for overcurrent protection.

Since PV systems may have transients—lightning and motor starting as well as others—inverse time circuit breakers (the standard type) or time-delay fuses should be used in most cases. In circuits where no transients are anticipated, fast-acting fuses can be used. They should be used if relays and other switchgear in dc systems are to be protected. Time delay fuses that can also respond very quickly to short-circuit currents may also be used for system protection.

Branch Circuits

DC branch circuits in stand-alone systems start at the battery and go to the receptacles supplying the dc loads or to the dc loads that are hard wired. In direct-connected systems, the PV output circuits go to the power controller or master power switch and a branch circuit goes from these to the load. In utility-intertie systems, the circuit between the inverter and the ac-load center may be considered a branch circuit.

Fuses used to protect dc or ac branch (load) circuits must be tested and rated for that use. They must also be of different sizes and markings for each amperage and voltage group to prevent unintentional interchange [240F]. DC-rated fuses that meet the requirements of the *NEC* are becoming more prevalent. Figure 12 shows *UL*-Listed, dc-rated, time-delay fuses on the left that are acceptable for branch circuit use, which would include the battery fuse. Acceptable dc-rated, *UL*-Listed fast-acting supplementary fuses are shown on the right and can be used in the PV source circuits. The fuses shown are made by Littelfuse (Appendix A), and the fuse holders are made by Marathon (Appendix A). Other manufacturers, such as Bussman and Gould, are beginning to obtain *UL*-Listed dc ratings on the types of fuses that are needed in PV systems. These particular requirements eliminate the use of glass, ceramic, and plastic automotive fuses as branch-circuit overcurrent devices because they areneither tested nor rated for this application.

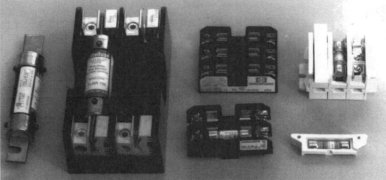

Figure 12. Listed and Recognized Fuses.

Automotive fuses have no dc rating by the fuse industry or the testing laboratories and **should not be used in PV systems**. When rated by the manufacturer, they have only a 32-volt maximum rating, which is less than the open-circuit voltage from a 24-volt PV array. Furthermore, these fuses have no rating for interrupt current, nor are they generally marked with all of the information required for branch-circuit fuses. They are not considered supplemental fuses under the *UL* listing or component recognition programs. Figure 13 shows unacceptable automotive fuses on the left and listed supplemental fuses on the right. Unfortunately, even the supplemental fuses are intended for ac use and frequently have no dc ratings.

Figure 13. Non-acceptable (left) and Acceptable (right) Fuses.

Circuit breakers also have specific requirements when used in branch circuits, but they are generally available with the needed dc ratings [240 G]. Figure 14 shows examples of dc-rated, *UL*-Recognized circuit breakers (supplemental) on the left. They may be used in the PV source circuits for disconnects and overcurrent protection, but they are not listed devices and may not be allowed by the inspector. The larger units are dc-rated, *UL*-Listed branch-circuit rated circuit breakers that can be used in dc-load centers for branch-circuit protection. The breakers shown are produced by Square D and Heinemann. Airpax also produces dc *UL*-Listed circuit breakers, and Potter Brumfield and others produce dc-rated, *UL*-Recognized, supplemental breakers.

To provide maximum protection and performance (lowest voltage drop) on branch circuits (particularly on 12 and 24-volt systems), the ampacity of the conductors might be increased, but the rating of the overcurrent devices protecting that cable should be as low as possible consistent with load currents. A general formula for cable ampacity and overcurrent device rating is 100 percent of the noncontinuous loads and 125 percent of the continuous loads anticipated. Nuisance tripping should determine the lower limit.

Figure 14. UL-Recognized and Listed Circuit Breakers

Amperes Interrupting Rating (AIR)—Short-Circuit Conditions

Overcurrent devices—both fuses and circuit breakers—must be able to safely open circuits with short-circuit currents flowing in them. Since PV arrays are inherently current limited, high short-circuit currents from the PV array are not a problem when the conductors are sized as outlined above. In stand-alone systems with storage batteries, however, the short-circuit problem is very severe. A single 220 amp-hour, 6-volt, deep-discharge, lead-acid battery may produce short-circuit currents as high as 8,000 amps for a fraction of a second and as much as 6,000 amps for a few seconds in a direct terminal-to-terminal short circuit. Such high currents can generate excessive thermal and magnetic forces that can cause an underrated device to burn or blow apart. Two paralleled batteries would generate twice as much current, and larger capacity batteries would be able to deliver proportionately more current under a short-circuit condition. In dc systems, particularly stand-alone systems with batteries, the interrupt capability of every overcurrent device is important. This interrupt capability is specified as Amperes Interrupting Rating (AIR) and sometimes AIC.

Most dc-rated, *UL*-Listed, branch circuit breakers that can be used in PV systems have an AIR of 5,000 amps. However, Heinemann Electric makes several with AIRs of 25,000 amps (Appendix A). Some dc-rated, *UL*-Recognized supplemental circuit breakers have an AIR of only 3,000 amps. Listed, dc-rated fuses normally have an AIR of up to 20,000 amps if they are of the current-limiting variety.

Fuses or circuit breakers **shall never be** paralleled or ganged to increase current-carrying capability [240-8].

Fusing of PV Source Circuits

The *NEC* allows supplementary fuses to be used in PV source circuits [690-9c]. A supplementary fuse is one that is designed for use inside a piece of listed equipment. These fuses supplement the main branch-circuit fuse and do not have to comply with all of the requirements of branch fuses. They should, however, be dc rated and able to handle the short-circuit currents they may be subjected to. Unfortunately, many supplemental fuses are not dc rated, and if they are, the AIR (when available) is usually less than 5,000 amps. The use of ac-rated supplementary fuses **is not** recommended for the dc circuits of PV systems.

Current-Limiting Fuses—Stand-Alone Systems

A current-limiting fuse **must** be used in each ungrounded conductor from the battery to limit the current that a battery bank can supply to a short-circuit and to reduce the short-circuit currents to levels that are within the capabilities of downstream equipment [690-71c]. These fuses are available with *UL* ratings of 125, 300, and 600 volts dc, currents of 0.1 to 600 amps, and a dc AIR of 20,000 amps. They are classified as RK5 or RK1 current-limiting fuses and should be mounted in Class-R rejecting fuse holders or dc-rated, fused disconnects. Class J or T fuses with dc ratings might also be used. For reasons mentioned previously, time-delay fuses should be specified, although some designers are getting good results with Class T fast-acting fuses. One of these fuses and the associated disconnect switch should be used in **each** bank of batteries with a paralleled amp-hour capacity up to 1,000 amp-hours. Batteries with single cell amp-hour capacities higher than 1,000 amp-hours will require special design considerations, because these batteries may be able to generate short-circuit currents in excess of the 20,000 AIR rating of the current-limiting fuses. When calculating short-circuit currents, the resistances of all connections, terminals, wire, fuse holders, circuit breakers, and switches must be considered. These resistances serve to reduce the magnitude of the available short-circuit currents at any particular point. The suggestion of one fuse per 1,000 amp-hours of battery size is only a general estimate, and the calculations are site specific. The fuses shown in Figure 12 are current limiting.

For systems less than 65 volts (open circuit), Heinemann Electric 25,000 AIR circuit breakers may be used (Appendix A). These circuit breakers are not current limiting, even with the high interrupt rating, so they cannot be used to protect other fuses or circuit breakers. An appropriate use would be in the conductor between the battery bank and the inverter. This single device would minimize voltage drop and provide the necessary disconnect and overcurrent features.

Current-Limiting Fuses—Grid-Connected Systems

Normal electrical installation practice requires that service entrance equipment have fault-current protection devices that can interrupt the available short-circuit currents [230-65, 208]. This requirement applies to the utility side of any power conditioning system in a PV installation. If the service is capable of delivering fault currents in excess of the AIR rating of the overcurrent devices used to connect the inverter to the system, then current-limiting overcurrent devices **must** be used (110-9).

Dead Fuses

Whenever a fuse is used for an overcurrent device and is accessible to unqualified persons, it **must** be in a circuit where all power can be removed from both ends of the fuse for servicing. It is not sufficient to reduce the current to zero before changing the fuse there must be no voltage present on either end of the fuse prior to service. This may require the addition of switches on both sides of the fuse location—a complication that increases the voltage drop and reduces the reliability of the system [690-16, Diagram 690-1]. Because of this requirement, the use of a fusible pullout-style disconnect or circuit breaker is recommended. For the charging and dc-load circuits, it is recommended that a current-limiting fuse be used at the battery with a switch located between the battery and the current-limiting fuse. Circuit breakers can be used for all other overcurrent devices in circuits where the potential fault currents do not exceed their AIR or where they are protected by a current-limiting fuse.

DISCONNECTING MEANS

There are many considerations in configuring the disconnect switches for a PV system. The *National Electrical Code* deals with safety first and other requirements last—if at all. The PV designer must also consider equipment damage from over voltage, performance options, equipment limitations, and cost.

A photovoltaic system is a power generation system, and a specific minimum number of disconnects are necessary to deal with that power. Untrained personnel will be operating the systems; therefore, the disconnect system must be designed to provide safe, reliable, and understandable operation.

Disconnects may range from nonexistent in a self-contained PV-powered light for a sidewalk to those found in the space-shuttle-like control room in a large, multi-megawatt, utility-tied PV power station. Generally, local inspectors will not require disconnects on totally enclosed, self-contained PV systems like the sidewalk illumination system or a pre-wired attic ventilation fan. This would be particularly true if the entire assembly were *UL*-Listed as a unit and there were no external contacts or user serviceable parts. However, the situation changes as the complexity of the device increases and separate modules, batteries, and charge controllers having external contacts must be wired together and possibly operated and serviced by unqualified personnel.

Photovoltaic Array Disconnects

Article 690 requires all current-carrying conductors from the PV power source to have a <u>disconnect</u> provision. This includes the grounded conductor, if any [690-13, 14; 230 F]. <u>Ungrounded</u> conductors **must** have a switch or circuit breaker disconnect. <u>Grounded</u> conductors which normally remain connected at all times **must** have a bolted disconnect that can be used for service operations and meet the *NEC* requirements.

In an ungrounded 12- or 24-volt PV system, both positive and negative conductors **must** be switched, since both are ungrounded. Since all systems **must** have an equipment-grounding system, costs may be reduced by grounding 12- or 24-volt systems and using one-pole disconnects on the remaining ungrounded conductor.

Equipment Disconnects

Each piece of equipment in the PV system **shall** have disconnect switches to disconnect it from all sources of power. The disconnects **shall** be circuit breakers or switches and **shall** comply with all of the provisions of Article 690-17. DC-rated switches are expensive; therefore, the ready availability of moderately priced dc-rated circuit breakers with ratings up to 48 volts and 70 amps would seem to encourage their use in all 12- and 24-volt systems. When properly located and used within their approved ratings, circuit breakers can serve as both the disconnect and overcurrent device. In simple systems, one switch or circuit breaker disconnecting the PV array and another disconnecting the battery may be all that is required.

A 2,000-watt inverter on a 12-volt system can draw nearly 200 amps at full load. Disconnect switches must be rated to carry this load and have appropriate interrupt ratings. Again, a dc-rated, *UL*-Listed circuit breaker may prove less costly and more compact than a switch and fuse with the same ratings.

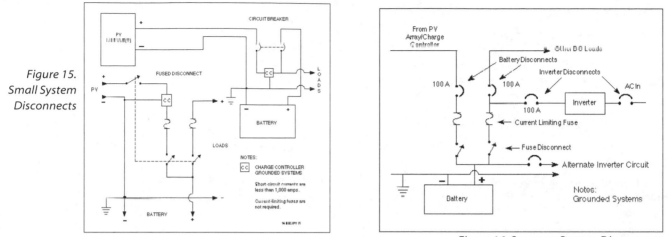

Figure 15.
Small System
Disconnects

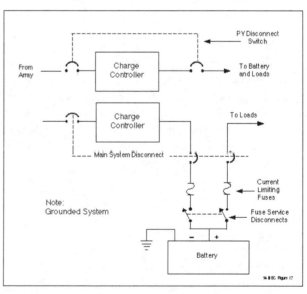

Figure 16. Separate Battery Disconnects

Battery Disconnect

When the battery is disconnected from the stand-alone system, either manually or through the action of a fuse or circuit breaker, care must be taken that the PV system not be allowed to remain connected to the load. Small loads will allow the PV array voltage to increase from the normal battery charging levels to the open-circuit voltage, which will shorten lamp life and possibly damage electronic components.

This potential problem can be avoided by using ganged multipole circuit breakers or ganged fused disconnects as shown in Figure 15. This figure shows two ways of making the connection. Separate circuits, including disconnects and fuses between the charge controller and the battery and the battery and the load, as shown in Figure 16, may be used if it is desired to operate the loads without the PV array being connected. If the design requires that the entire system be shut down with a minimum number of switch actions, the switches and circuit breakers could be ganged multipole units.

Charge Controller Disconnects

Some charge controllers are fussy about the sequence in which they are connected and disconnected from the system. Most charge controllers do not respond well to being connected to the PV array and not being connected to the battery. The sensed battery voltage (or lack thereof) would tend to rapidly cycle between the array open-circuit voltage and zero as the controller tried to regulate the nonexistent charge process. This problem will be particularly acute in self-contained charge controllers with no external battery sensing.

Again, the multipole switch or circuit breaker can be used to disconnect not only the battery from the charge controller, but the charge controller from the array. Probably the safest method for self-contained charge controllers is to have the PV disconnect switch disconnect both the input and the output of the charge controller from the system. Larger systems with separate charge control electronics and switching elements will require a case-by-case analysis—at least until the controller manufacturers standardize their products. Figure 17 shows two methods of disconnecting the charge controller.

Figure 17. Charge Controller Disconnects

Non-Grounded Systems

Systems that do not have one of the current-carrying conductors grounded **must** have disconnects <u>and</u> overcurrent devices in all of the ungrounded conductors. This means two-pole devices for the PV, battery, and inverter disconnects and overcurrent devices. The additional cost is considerable.

Multiple Power Sources

When multiple sources of power are involved, the disconnect switches **shall** be grouped and identified [230-72, 690-15]. No more than six motions of the hand will be required to operate all of the disconnect switches required to remove all power from the system [230-71]. These power sources include PV output, the battery system, any generator, and any other source of power. Multipole disconnects or handle ties should be used to keep the number of motions of the hand to six or fewer.

PANEL BOARDS, ENCLOSURES, AND BOXES

Disconnect and overcurrent devices **shall** be mounted in approved enclosures, panel boards, or boxes [240-30]. Wiring between these enclosures must use a *NEC*-approved method [690-13a]. Appropriate cable clamps, strain-relief methods, or conduit **shall** be used. All openings not used **shall** be closed with the same or similar material to that of the enclosure [370-8]. Metal enclosures **must** be bonded to the grounding conductor [370-4]. Use of wood or other flammable materials is discouraged. Conductors from different systems such as utility power, gas generator, hydro, or wind **shall not** be placed in the same enclosure, box, conduit, etc., as PV source conductors unless the enclosure is partitioned [690-4b]. This requirement stems from the need to keep "always live" PV source conductors separate from those that can be turned off.

When designing a PV distribution system or panel board, an approved NEMA style box and approved disconnect devices and overcurrent devices should be used. The requirements for the internal configuration of these devices are established by *NEC* Articles 370, 373, and 384 and **must** be followed. Dead front-panel boards with no exposed current-carrying conductors, terminals, or contacts are generally required. Underwriters Laboratories also establishes the standards for the internal construction of panel boards and enclosures.

BATTERIES

In general, *NEC* Articles 480 and 690-71, 72, 73 should be followed for installations having storage batteries. Battery storage in PV systems poses several safety hazards:

- Hydrogen gas generation from charging batteries
- High short-circuit currents
- Acid or caustic electrolyte
- Electric shock potential

Hydrogen Gas

When flooded, non-sealed, lead-acid batteries are charged at high rates, or when the terminal voltage reaches 2.3 - 2.4 volts per cell, the batteries produce hydrogen gas. Even sealed batteries may vent hydrogen gas under certain conditions. This gas, if confined and not properly vented, poses an explosive hazard. The amount of gas generated is a function of the battery temperature, the voltage, the charging current, and the battery-bank size. Hydrogen is a light, small-molecule gas that is easily dissipated. Small battery banks (i.e., up to 20, 220-amp-hour, 6-volt batteries) placed in a large room or a well-ventilated area do not pose a significant hazard. Larger numbers of batteries in smaller or tightly enclosed areas require venting. Venting manifolds may be attached to each cell and routed to an exterior location, but these manifolds are not recommended because flames in one section of the manifold may be easily transmitted to other areas in the system.

A catalytic recombiner cap (Hydrocap® Appendix A) may be attached to each cell to recombine some of the hydrogen with oxygen in the air to produce water. If these combiner caps are used, they will require occasional maintenance. If hydrogen gas is still a concern, the batteries may be installed in a tight box and the box vented top and bottom with a 3 - 4 inch ID flexible hose/pipe, which will allow air to flow through the box to the outside. It is rarely necessary to use power venting [480-8]. Flame arrestors may be installed in Hydrocap® Vents as an option.

Certain charge controllers are designed to minimize the generation of hydrogen gas, but lead-acid batteries need some overcharging to fully charge the cells. This produces gassing that should be dissipated.

In **no case** should charge controllers, switches, relays, or other devices capable of producing an electric spark be mounted in a battery enclosure or directly over a battery bank. Care must be exercised when routing conduit from a sealed battery box to a disconnect. Hydrogen gas may travel in the conduit to the arcing contacts of the switch.

Battery Rooms and Containers

Batteries are capable of generating tens of thousands of amps of current when shorted. A short circuit in a conductor not protected by overcurrent devices can melt wrenches or other tools, battery terminals and cables, and spray molten metal around the room. Exposed battery terminals and cable connections must be protected. Live parts of batteries must be guarded. This generally means that the batteries should be accessible only to a qualified person. A locked room, battery box, or other container and some method to prevent access by the untrained person should minimize the hazards from short circuits and electric shock. The danger may be reduced if insulated caps or tape are placed on each terminal and an insulated wrench is used for servicing, but in these circumstances, corrosion may go unnoticed on the terminals. The *NEC* requires certain spacings around battery enclosures and boxes to allow for unrestricted servicing—generally about three feet [110-16]. Battery voltages must be less than 50 volts in dwellings unless certain protective criteria are met [690-71]. Batteries should not be installed in living areas.

Acid or Caustic Electrolyte

A thin film of electrolyte can accumulate on the tops of the battery and on nearby surfaces. This material can cause flesh burns. It is also a conductor and, in high-voltage battery banks, poses a shock hazard. The film of electrolyte should be removed periodically with an appropriate neutralizing solution. For lead-acid batteries, a dilute solution of baking soda and water works well. Commercial neutralizers are available at auto-supply stores.

Charge controllers are available that minimize the dispersion of the electrolyte at the same time they minimize battery gassing. They do this by keeping the battery voltage from climbing into the vigorous gassing region where the high volume of gas causes electrolyte to bubble out of the cells.

Battery servicing hazards can be minimized by using protective clothing including face masks, gloves, and rubber aprons. Self-contained eyewash stations and neutralizing solution would be beneficial additions to any battery room. Water should be used to wash acid or alkaline electrolyte from the skin and eyes.

Anti-corrosion sprays and greases are available from automotive and battery supply stores which reduce the need to service the battery bank. Hydrocap® Vents also reduce the need for servicing by reducing the need for watering.

Electric Shock Potential

Storage batteries in dwellings must operate at less than 50 volts unless live parts are protected during routine servicing [690-71]. It is recommended that live parts of any battery bank should be guarded [690-71b(2)].

GENERATORS

Other electrical power generators such as wind, hydro, and gasoline/propane/diesel must comply with the requirements of the *NEC*. These requirements are specified in the following *NEC* articles:

Article 230	Services
Article 250	Grounding
Article 445	Generators
Article 700	Emergency Systems
Article 701	Legally Required Standby Systems
Article 702	Optional Standby Systems
Article 705	Interconnected Power Production Sources

When multiple sources of ac power are to be connected to the PV system, they must be connected with an appropriately rated and approved transfer switch. AC generators frequently are rated to supply larger amounts of power than that supplied by the PV/battery/inverter. The transfer switches must be able to safely accommodate either power source.

Grounding, both equipment and system, must be carefully considered when a generator is connected to an existing system. There must be no currents flowing in the equipment-grounding conductor under any operating mode of the system.

The circuit breakers or fuses that are built into the generator are not sufficient to provide *NEC* required protection for the conductors from the generator to the PV system. An external (branch circuit rated) overcurrent device (and possibly a disconnect) **must** be mounted close to the generator. The conductors from the generator to this overcurrent device must have an ampacity of 115 percent of the name plate current rating of the generator [445-5]. Figure 18 show a typical one-line diagram for a system with a backup generator

CHARGE CONTROLLERS

A charge controller or self-regulating system **shall** be used in a stand-alone system with battery storage. The mechanism for adjusting state of charge **shall** be accessible only to qualified persons [690-72].

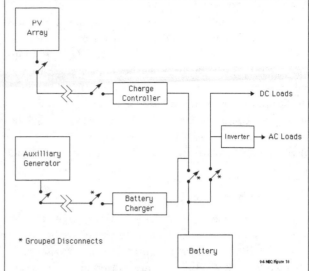

Figure 18. Disconnects for Remotely Located Power Sources

There are only two or three charge controllers on the market that have been tested by *UL* or other recognized testing organizations.

Surface mounting of devices with external terminals readily accessible to the unqualified person will not be accepted by the inspection authority. These charge controllers should be mounted in a listed enclosure with provisions for ventilation. Dead-front panels with no exposed contacts are generally required for safety. A typical charge controller such as shown in Figure 19 should be mounted in a *UL*-Listed enclosure so that none of the terminals are exposed. Enclosures containing charge controllers should have knockouts for cable entry and some method of attaching conduit where required. Internal space must be allocated to provide room for wire bending.

Electrically, charge controllers should be designed with a conductor between the negative input and output terminals. No shunts or other signal processing should be placed in that conductor. This design will allow the controller to be used in a grounded system with the grounded conductor running through the controller.

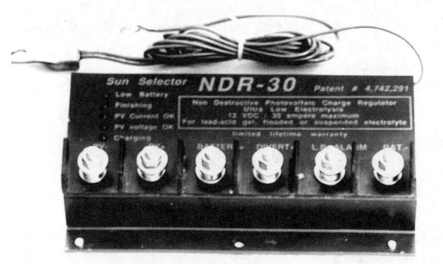

Figure 19. Typical Charge Controller

602

Distribution Systems

The *National Electrical Code* was formulated when there were abundant supplies of relatively cheap energy. As the Code was expanded to include other power systems such as PV, many sections were not modified to reflect the recent push toward efficient use of electricity in the home. Stand-alone PV systems **may** be required to have dc services with 60- to 100-amp capacities to meet the Code [230-79]. DC receptacles and lighting circuits **may** have to be as numerous as their ac counterparts [220, 422]. In a small one- to four-module system on a remote cabin or small home, these requirements are too excessive, since the power source may be able to supply only a few hundred watts of power.

The local inspection authority has the final say on what is, or is not, required and what is, or is not, safe. Reasoned conversations may result in a liberal interpretation of the Code. For a new dwelling, it seems appropriate to install a complete ac electrical system as required by the *NEC*. This will meet the requirements of the inspection authority, the mortgage company, and the insurance industry. Then the PV system and its dc distribution system can be added. If an inverter is used, it can be connected to the ac service entrance.

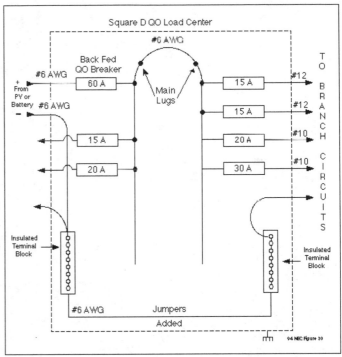

Figure 20. 12-Volt DC Load Center

DC branch circuits and outlets can be added where needed, and everyone will be happy. If or when grid power becomes available, it can be integrated into the system with minimum difficulty. If the building is sold at a later date, it will comply with the *NEC* if it has to be inspected. The use of a *UL*-Listed power center, such as the APT Power Center will facilitate the installation and the inspection (Appendix A).

Square D has received a direct current (dc), *UL* listing for its standard QO residential **branch** circuit breakers. They can be used up to 48 volts (125% PV open-circuit voltage) and 70 amps dc. The AIR is 5,000 amps, so a current-limiting fuse (RK5 or RK1 type) must be used when they are connected on a battery system. The Square D QOM **main** breakers (used at the top of the load center) **do not** have this listing, so the load center must be obtained with main lugs and no main breakers (Appendix A).

In a small PV system (less than 5000 amps of available short-circuit current), a two-pole Square D QO breaker could be used as the PV disconnect (one pole) and the battery disconnect (one pole). Also, a fused disconnect or fusible pullout could be used in this configuration. This would give a little more flexibility since the fuses can have different current ratings. Figure 15 on page 48 shows both systems with only a single branch circuit.

In a system with several branch circuits, the Square D load center can be used. A standard, off-the-shelf Square D residential load center without a main breaker can be used for a dc distribution panel in 12-volt dc systems. The main disconnect would have to be a "back fed" QO breaker, and it would have to be connected in one of the normal branch circuit locations. Back-fed circuit breakers **must** be identified for such use and **must** be clamped in place [690-64b(5), 384-16(f)]. Since the load center has two separate circuits (one for each phase), they will have to be tied together to use the entire load center. Figure 20 illustrates this use of the Square D load center.

Square D has listed one of their load centers that uses the QO breakers for DC operation. This load center is available with a *UL*-Listed, dc-rated, current-limiting fuse from a number of sources. The manufacturer is listed in Appendix A.

Another possibility is to use one of the phase circuits to combine separate PV source circuits, then go out of the load center through a breaker for the PV disconnect switch to the charge controller. Finally, the conductors would have to be routed back to the other phase circuit in the load center for branch circuit distribution. Several options exist in using one and two-pole breakers for disconnects. Figure 21 presents an example.

603

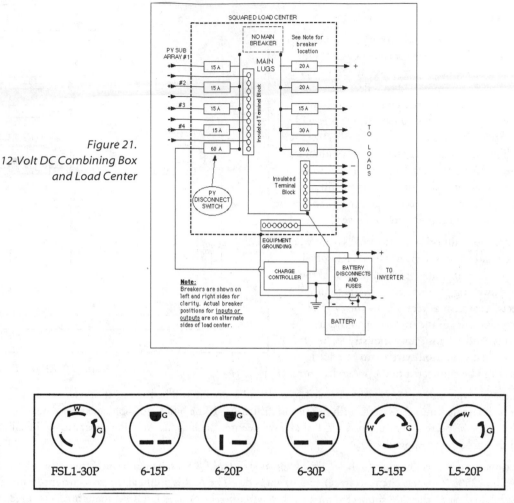

Figure 21.
12-Volt DC Combining Box
and Load Center

Figure 22. Plug Configurations

Interior Wiring and Receptacles

The interior wiring used in a PV system **must** comply with the *NEC*. Nonmetallic sheathed cable (type NM - "Romex") may be used, and it **must** be installed in the same manner as cable for ac branch circuits [300, 690-31a]. The bare grounding conductor in such a cable **must not** be used to carry current and cannot be used as a common negative conductor for combination 12/24-volt systems [336-25]. Exposed, single-conductor cables are not permitted—they **must** be installed in conduit [300-3(a)]. Wires carrying the same current (i.e., positive and negative battery currents) must be installed in the same conduit or cable to prevent increased circuit inductances that would pose additional electrical stresses on disconnect and overcurrent devices [300-3(b)]. Equipment-grounding conductors may be run apart from the current-carrying conductors [250-57(b)Ex2].

The receptacles used for dc must be different than those used for any other service in the system [210-7f, 551-20f]. The receptacles should have a rating of not less than 15 amps and must be of the three-prong grounding type [210-7a, 720-6]. Numerous different styles of approved receptacles are available that meet this requirement. These requirements can be met in most locations by using the three-conductor 15-, 20-, or 30-amp 240-volt NEMA style 6-15, 6-20, 6-30 receptacles for the 12-volt dc outlets. If 24-volt dc is also used, the NEMA 125-volt locking connectors, style L5-15 or L5-20, are commonly available. The NEMA FSL-1 is a locking 30-amp 28-volt dc connector, but its availability is limited. Figure 22 shows some of the available configurations. Cigarette lighter sockets and plugs frequently found on "PV" and "RV" appliances **do not** meet the requirements of the *National Electrical Code* and should not be used.

It is not permissible to use the third or grounding conductor of a three-conductor plug or receptacle to carry common negative return currents on a combined 12/24-volt system. This terminal must be used for equipment grounding and may not carry current except in fault conditions [210-7].

A 30-amp fuse or circuit breaker protecting a branch circuit (with No. 10 AWG conductors) **must** use receptacles rated at 30 amps. Receptacles rated at 15 and 20 amps **must not** be used on this 30-amp circuit [Table 210-21(b)(3)].

Smoke Detectors

Many building codes require that smoke and fire detectors be wired directly into the ac power wiring of the dwelling. With a system that has no inverter, two solutions might be offered to the inspector. The first is to use the 9-volt or other primary-cell, battery-powered detector. The second is to use a voltage regulator to drop the PV system voltage to the 9-volt or other level required by the detector.

The regulator must be able to withstand the PV open-circuit voltage and supply the current required by the detector alarm.

On inverter systems, the detector on some units may trigger the inverter into an "on" state, unnecessarily wasting power. In other units, the alarm may not draw enough current to turn the inverter on and thereby produce a reduced volume alarm or, in some cases, no alarm at all. Small, dedicated inverters might be used, but this would waste power and decrease reliability when dc detectors are available.

Several states now require detectors that are connected to the power line and have a battery backup. Units satisfying this requirement might also be powered by dc from the battery and by a primary cell.

Ground-Fault Circuit Interrupters

Some ac ground-fault circuit interrupters (GFCI) do not operate reliably on the output of some non-sine-wave inverters. If the GFCI does not function when tested, it should be verified that the neutral (white-grounded) conductor of the inverter output is solidly grounded and bonded to the grounding (green or bare) conductor of the inverter in the required manner. If this bond is present and does not result in the GFCI testing properly, other options are possible. Changing the brand of GFCI may rectify the solution. A direct measurement of an intentional ground fault may indicate that slightly more than the 5 milliamp internal test current is required to trip the GFCI. The inspector may accept this. Some inverters will work with a ferro-resonant transformer to produce a wave form more satisfactory for use with GFCIs, but the no-load power consumption may be high enough to warrant a manual demand switch. A sine-wave inverter could be used to power those circuits requiring GFCI protection.

Interior Switches

Switches rated for "ac only" **shall not** be used in dc circuits [380-14]. AC-DC general-use snap switches are available on special order from most electrical supply houses, and they are similar in appearance to normal "quiet switches." *UL*-Listed electronic switches with the proper dc ratings might also be used, but the nonstandard appearance may require that the *UL*-Listing specifications be provided to the inspector.

There have been some failures of dc-rated snap switches when used as PV array and battery disconnect switches. If these switches are used on 12- and 24-volt systems and are not activated frequently, they may build up internal oxidation or corrosion and not function properly. Periodically activating the switches under load will keep them clean.

SYSTEM LABELS AND WARNINGS

Photovoltaic Power Source

A permanent label **shall** be applied near the PV disconnect switch that contains the following information: [690-52]

• Operating Current (System maximum-power current)   • Operating Voltage (System maximum-power voltage)

• Open-Circuit Voltage   • Short-Circuit Current

This data will allow the inspector to verify proper conductor ampacity and overcurrent device rating. It will also allow the user to compare system performance with the specifications. This label **must** be applied by the system installer [690-52].

Multiple Power Systems

Systems with multiple sources of power such as PV, gas generator, wind, hydro, etc., **shall** have diagrams and markings showing the interconnections [705-10]. These diagrams should be placed near the system disconnects.

Switch or Circuit Breaker

If a switch or circuit breaker has all of the terminals energized when in the open position, a label should be placed near it indicating: [690-17]

• WARNING - ELECTRIC SHOCK HAZARD - DO NOT TOUCH - TERMINALS ENERGIZED IN OPEN POSITION

General

Each piece of equipment that might be opened by unqualified persons should be marked with warning signs:

• WARNING - ELECTRIC SHOCK HAZARD - DANGEROUS VOLTAGES AND CURRENTS - NO USER SERVICEABLE PARTS INSIDE - CONTACT QUALIFIED SERVICE PERSONNEL FOR ASSISTANCE

Each battery container, box, or room should also have warning signs:

• WARNING - ELECTRIC SHOCK HAZARD - DANGEROUS VOLTAGES AND CURRENTS - EXPLOSIVE GAS - NO SPARKS OR FLAMES - NO SMOKING - ACID BURNS - WEAR PROTECTIVE CLOTHING WHEN SERVICING

INSPECTIONS

Involving the inspector as early as possible in the planning stages of the system will begin a process that should provide the best chance of obtaining a safe, durable system. The following steps are suggested.

• Establish a working relationship with a local electrical contractor or electrician to determine the requirements for permits and inspections.

• Contact the inspector and review the system plans. Solicit advice and suggestions from the inspector.

• Obtain the necessary permits.

• Involve the inspector in the design and installation process. Provide information as needed. Have one-line diagrams and complete descriptions of any non-listed, non-standard equipment available.

INSURANCE

Most insurance companies are not familiar with photovoltaic power systems. They are, however, willing to add the cost of the system to the homeowner's policy if they understand that it represents no additional liability risk. A system description may be required. Evidence that the array is firmly attached to the roof or ground is usually necessary. The system must be permitted and inspected if those requirements exist for other electrical power systems in the vicinity.

Some companies will not insure homes that are not grid connected because there is no source of power for a high-volume water pump for fighting fires. In these instances, it may be necessary to install a fire-fighting system and water supply that meets their requirements. A high-volume dc pump and a pond might suffice.

As with the electrical inspector, education and a full system description emphasizing the safety features and code compliance will go a long way toward obtaining appropriate insurance.

APPENDIX A

Sources of Equipment Meeting the Requirements of
The *National Electrical Code*"

A number of PV distributors and dealers are stocking the equipment needed to meet the *NEC* requirements. These additional sources are presented as sources of specialized equipment.

CONDUCTORS

Standard multiconductor cable such as 10-2 with ground Nonmetallic Sheathed Cable (NM and NMC), Underground Feeder (UF), Service Entrance (SE), Underground Service Entrance (USE), larger sizes (8 AWG) single-conductor cable, uninsulated grounding conductors, and numerous styles of building wire such as THHN can be obtained from electrical supply distributors and building supply stores.

The highest quality USE-2 cable will be listed by *UL* and will also have XLP (or XLPE), RHW-2, and 600V markings. Flexible USE and RHW cables in large sizes (1/0 - 250 kcmil) and stranded 8-, 10-, and 12-gage USE single conductor cable can be obtained from some electrical supply houses and wire distributors, including:

Paige Electric Corp.
1071 Hudson Street
P.O. Box 368
Union, New Jersey 07083
800-327-2443

Anixter Bros.
2201 Main Street
Evanston, Illinois 60202
800-323-8166 for the nearest distributor

For grid-connected systems requiring cables with voltage rating higher than 600V, cable must be special-ordered. Rubber-insulated cables are available with up to 2,000-volt insulation. They should be marked RHW-2, XLP and be sunlight resistant when used for exposed module interconnects. The following manufacturers can supply such cable.

American Insulated Wire
36 Freeman Street
P.O. Box 880
Pawtucket, RI 02862
401-726-0700

The Okonite Company
PO Box 340
Romsey, NJ 07446
201-825-9026

MISCELLANEOUS HARDWARE

Stainless steel nuts, bolts and screws, and other hardware, insulated and uninsulated crimp-on terminals, battery terminals, copper lugs for heavy cable, battery cable, weather-resistant cable ties, heat shrink tubing and more may be obtained from the following source:

Chesapeake Marine Fasteners
10 Willow Street
P.O. Box 6521
Annapolis, Maryland 21401
800-526-0658

Dealer's price sheet is available

The company listed below makes plastic strain reliefs that fit the standard 1/2" electrical knockout (7/8" diameter). These watertight strain reliefs are needed for older ARCO modules and the current production Solarex modules as well as others. The single-conductor versions are hard to find, and the metal types are very expensive. A catalog and information on product 3224 (for AWG 10) or 3231 (for larger wire) can be requested. The company also makes UV-resistant black cable ties and copper, heavy-duty lugs, as well as other products that might be useful.

Heyco Molded Products, Inc.
Box 160
Kenilworth, New Jersey 07033
800-526-4182 or 908-245-0033
Quantity purchases only

DC-RATED FUSES

15, 20, 30 amps and higher rated fuses can be used for branch-circuit overcurrent protection depending on conductor ampacity and load. Larger sizes (100 amp and up) are used for current-limiting and overcurrent protection on battery outputs. DC rated, *UL*-Listed fuses are manufactured by the following companies, among others:

Bussmann	Gould Inc.	Littelfuse
P.O. Box 14460	374 Merrimac Street	Power Fuse Division
St. Louis, MO 63178-4460	Newburyport, MA 01950	800 E. Northwest Highway
314-527-3877	508-462-6662	Des Plaines, Illinois 60016
314-527-1270 (Technical Questions)		(708) 824-1188
		800-TEC FUSE (Technical Questions)
		800-227-0029 (Customer Service)

The following fuses may be used for battery circuit and branch circuit overcurrent protection and current limiting. If transients are anticipated in PV circuits, these fuses can also be used in those locations.

Fuse Description	Size	Manufacturer	Mfg #
125-volt dc, RK5 Time delay, current-limiting	.1-600 amp	Bussmann	FRN-R
" " Littelfuse FLNR			
300-volt dc, RK5 Time delay, current-limiting fuse.	1-600-amp	Bussmann	FRS-R
" " Gould TRS-R			
" " Littelfuse FLSR			
600-volt dc, RK5 Time delay, current-limiting fuse.	1-600 amp,	Littelfuse	IDSR
" 70-600 amp Gould TRS70R-600R			

The following fuses should be used for PV source-circuit protection if problems are not anticipated with transients. They may also be used inside control panels to protect relays and other equipment.

Fuse Description	Size	Manufacturer	Mfg #
Fast-acting, current-limiting midget fuse	.1-30 amp	Bussmann	KLM*
" "		Gould	ATM**
" "		Littelfuse	KLK-D**

* *UL*-Recognized

** *UL*-Listed

Fuse Holders (Also See Fused Disconnects)

Each fuse manufacturer makes fuse blocks matching the voltage rating and current rating of the selected fuse.

Marathon Special Projects also makes suitable fuse holders. Information and the names of distributors of Class R and Class M (midget fuse holders) should be requested. The company also makes power-distribution blocks for control panels.

Marathon Special Products
P.O. Box 468
Bowling Green, Ohio 43402
419-352-8441

Fused Disconnects (Also See Circuit Breakers)

Since fuses must not have power applied to either end when servicing, a combination switch and fuse can be mounted in a single enclosure to meet some, if not all, of the requirements.

Indoor fused switches, 250-volt dc—JN and JF series
Outdoor fused switches, 250-volt dc—JR and FR series
Siemens I-T-E
Siemens Energy & Automation, Inc.
3333 State Bridge Rd.
Alpharetta, Georgia 30202
404-751-2000

Call for nearest regional sales office that can direct you to a stocking distributor

Indoor fused switches

250-volt-dc—H22x, H32x, and H42x series

600-volt-dc—H26xx and H36xx series

Outdoor fused switches

250-volt-dc—H22xR, H32xR, and H42xR series

600-volt-dc—H26xR and H36xR series

Square D Company

800-634-2003 for the nearest

Square D electrical supply distributor

Rainshadow Solar installs a current-limiting fuse in a *UL*-Listed, dc-rated Square D load center.

Rainshadow Solar

P.O. Box 242

Guthrie Cove Road

Orcas, WA 98280

206-376-5336

Ananda Power Technologies manufactures a line of *UL*-Listed power centers (dc source circuits, charge controller, and load circuits) with numerous options.

Ananda Power Technologies, Inc

14618 Tyler Foote RD #143

Nevada City, CA 95959

916-292-3834

Boltswitch, Inc., makes pull-out fused disconnects that are dc rated for higher current applications.

Contact factory for applications.boltswitch, Inc.

6107 West Lou Avenue

Crystal Lake, IL 60014

815-459-6900

CIRCUIT BREAKERS

Square D QO circuit breakers (common ac residential breakers).

UL-Listed at 5000 AIC at 48 volts dc; 1 and 2 pole, 10-70 amps; 3 pole, 10-60 amps

Square D FA circuit breakers; 125- and 250-volt dc ratings, multiple currents

Enclosures for QO breakers

2 and 3 pole units

Indoor QO21xxBN, QO3100BN

Rainproof QO21xxBNRB, QO3100BNRB

Any of the load centers for Square D QO breakers without main breakers may be used—main lugs should be requested instead.

Square D Company

800-634-2003 for the nearest

Square D electrical distributor

Heinemann makes a full line of dc-rated, *UL*-Listed and recognized supplemental circuit breakers, but they must be mounted in custom-built enclosures. (The metal is punched by the installer).

CD-CE-CF 5000 AIC at 125-volt dc, 15-110 amp

25,000 AIC available on special order. Polyester case, spun rivets, and *UL*-Listed units should be requested.

GH 10,000 AIC at 250-volts dc, 15-100 amp

GJ 10,000 AIC at 125-volts dc, 100-250 amps

GJ 25,000 AIC at 65-volts dc, 100-250 amps

GJ1P 10,000 AIC at 160-volts, 25,000 AIC at 65-volts dc, 100-700 amps

Eaton Corporation
Heinemann Products
2300 Northwood Drive
Salisbury, Maryland 21801
410-546-9778
Call for nearest source and catalog
Applications engineering available

Philips Technology (formerly AIRPAX) also makes a full line of dc-rated, *UL*-Listed and recognized supplemental circuit breakers, but they must be mounted in custom-built enclosures.

Philips Technology
P.O. Box 520
Cambridge, Maryland
301-228-4600
Call for nearest source and catalog.
Applications engineering available.

Rainshadow Solar markets a Heinemann 250 amp circuit breaker in an enclosure that is suitable of use with 4,000-watt, 24-volt inverters.

Rainshadow Solar
P.O. Box 242
Guthrie Cove Road
Orcas, WA 98280
360-376-5336

ENCLOSURES AND JUNCTION BOXES

Indoor and outdoor (rainproof) general-purpose enclosures and junction boxes are available at most electrical supply houses. These devices usually have knockouts for cable entrances, and the distributor will stock the necessary bushings and/or cable clamps. Interior component mounting panels are available for some enclosures, as are enclosures with hinged doors. If used outdoors, all enclosures, clamps, and accessories must be listed for outdoor use. For visual access to the interior, NEMA 4x enclosures are available that are made of clear, transparent plastic.

HYDROCAPS

Hydrocap® Vents are available from Hydrocap Corp. and some PV distributors on a custom-manufactured basis. Flame arrestors are an option.

Hydrocap
975 NW 95 St
Miami, FL 33150
305-696-2504

SURGE ARRESTORS

Delta makes a full line of large, silicon-oxide surge arrestors starting at 300 volts and up that are usable on low-voltage systems to clip the tops of large surges.

Ananda Power Technologies sells a Delta unit for low-voltage systems.

Delta Lightning Arrestors Inc.
P.O. Box 1084
Big Spring, TX 79721
915-267-1000

APPENDIX B
NEC and *UL* Requirements
Too Conservative?

Introduction

As the photovoltaic (PV) power industry moves into a mainstream position in the generation of electrical power, some persons question the seemingly conservative and redundant requirements established by Underwriters Laboratories (*UL*) and the *National Electrical Code* (*NEC*) for system and installation safety. This short discourse will attempt to address those concerns and highlight the unique aspects of PV systems that dictate the requirements.

The *National Electrical Code* (*NEC*) is written with the requirement that all equipment and installations are approved for safety by the authority having jurisdiction (AHJ) to enforce the *NEC* requirements in a particular location. The AHJ readily admits to not having the resources to verify the safety of the required equipment and relies exclusively on the testing and listing of the equipment by independent testing laboratories such as Underwriters Laboratories (*UL*). The AHJ also relies on the requirements for field wiring specified in the *NEC* to ensure safe installations and use of the listed equipment.

The standards published by *UL* and the material in the *NEC* are closely harmonized by engineers and technicians throughout the electrical equipment industry, the electrical construction trades, the national laboratories, the scientific community, and the electrical inspector associations. The *UL* Standards are technical in nature with very specific requirements on the construction and testing of equipment for safety. They in turn are coordinated with the construction standards published by the National Electrical Manufacturers Association (NEMA). The *NEC* is deliberately written in a non-technical manner for easy understanding and application by electricians, electrical contractors, and electrical inspectors in the field.

The use of listed (by *UL* or other laboratory) equipment ensures that the equipment meets well-established safety standards. The application of the requirements in the *NEC* ensures that the listed equipment is connected with field wiring and is used in a manner that will result in an essentially hazard-free system. Use of listed equipment and installing that equipment according to the requirements in the *NEC* will contribute greatly to not only safety, but also the durability, performance, and longevity of the system.

Sometimes Controversial Areas

The *NEC* does not present many highly detailed technical specifications. For example, the term "rated output" is used in several cases with respect to PV equipment. The conditions under which the rating is determined are not specified. The definitions of the rating conditions (such as Standard Test Conditions (STC) for PV modules) are made in the *UL* Standards that establish the rated output. This procedure is appropriate because of the *NEC* level of writing and the lack of appropriate test equipment available to the *NEC* user.

UL Standards

UL Standard 1703 requires that the instructions for listed PV modules contain specific requirements for the installation of such modules. The rated (at Standard Test Conditions) open-circuit voltage and the rated short-circuit current of crystalline PV modules are to be multiplied by factors of 125 percent before further calculations are made for conductor and overcurrent devices.

The 125 percent factor on the open-circuit voltage (Voc) is needed because, as the operating temperature of the module decreases, Voc increases. The rated Voc is measured at a temperature of $25°C$ and while the normal operating temperature is 40-50°C when ambient temperatures are around 20°C, there is nothing to prevent sub-zero ambient temperatures from yielding operating temperatures significantly below the 25°C standard test condition.

A typical module will have a voltage coefficient of -0.38 percent / °C. A system with a rated open-circuit voltage of 595 volts at 25°C might be exposed to ambient temperatures of -30 °C. This voltage (595) could be handled by the commonly available 600-volt rated conductors and switchgear. At dawn and dusk conditions, the module will be at the ambient temperature of -30°C, will not experience any heating, but can generate open-circuit voltages of 719 volts (595 x (1 + (25 + 30) x 0.0038)). This voltage substantially exceeds the capability of 600-volt rated conductors, fuses, switchgear, and other equipment. The very real possibility of this type of condition substantiates the *UL* requirement for the 125 percent factor on the rated open-circuit voltage.

The *UL* Standard 1703 also requires that the rated (at STC) short-circuit current of the PV module be multiplied by 125 percent before any other factors are applied such as those in the *NEC*. This *UL* factor is to provide a safe margin for wire sizes and overcurrent devices when the irradiance exceeds the standard 1000 W/m^2. Depending on season, local weather conditions, and atmospheric dust and humidity, irradiance exceeds 1000 W/m^2 every day around solar noon. The time period can be as long as four hours with irradiance values approaching 1200 W/m^2, again depending on the aforementioned conditions and the type of tracking being used. These daily irradiance values can increase short-circuit currents 20 percent over the 1000 W/m^2

value.

Enhanced irradiance due to reflective surfaces such as sand, snow, or white roofs, and even nearby bodies of water can increase short-circuit currents by substantial amounts and for significant periods of time. Cumulus clouds also can increase irradiance by as much as 50 percent.

Another factor that must be addressed is that PV modules typically operate at 30-40°C above the ambient temperatures. In crystalline silicon PV modules, the short-circuit current increases as the temperature increases. A typical factor might be 0.1 percent/°C. If the module operating temperature were 60°C (35°C over the STC of 25°C), the short-circuit current would be 3.5 percent greater than the rated value. PV modules have been measured operating as high as 72°C. The combination of increased operating temperatures, irradiances over 1000 W/m^2 around solar noon, and the possibility of enhanced irradiance certainly justify the *UL* requirement of 125 percent on the rated short-circuit current.

NEC Requirements

The *NEC* requires that the short-circuit current of the module, source circuit, or array be multiplied by 125 percent before calculating the ampacity of any cable or the rating of any overcurrent device used in these circuits. This factor is in addition to the *UL* required 125 percent.

Since short-circuit currents in excess of the rated value are possible from the discussion of the *UL* requirements above, and these currents are independent of the *NEC* requirements, good engineering practice dictates that both factors should be used at the same time. This yields a multiplier on short-circuit current of 1.56 (125 percent x 125 percent).

The *NEC* also requires that the ampacity of conductors be derated for the operating temperature of the conductor. This is a requirement because the ampacity of cables is given for cables operating in an ambient temperature of 30°C. In PV systems, cables are operated in an outdoor environment and should be subjected at least to a temperature derating due to an ambient temperature of 40°C. PV modules operate at high temperatures and in some installations as high as 73°C (concentrating modules operate at even higher temperatures). The temperatures in module junction boxes approach these temperatures and conductors in free air that lie against the back of these modules are also exposed to these temperatures. Temperatures this high require that the ampacity of cables be derated by factors of 0.33 to 0.58 depending on cable type, installation method (free air or conduit), and the temperature rating of the insulation.

Cables in conduit where the conduit is exposed to the direct rays of the sun are also exposed to elevated operating temperatures.

Cables with insulation rated at 60°C have no ampacity at all when operated in environments with ambient temperatures over 55°C. This precludes their use in most PV systems.

Redundancy and Conservatism or Not?

There appears to be little question that the 125 percent *UL* factor on voltage is necessary in any location where the ambient temperatures drop below 25°C. Even though the PV system can provide little current under open-circuit voltage conditions, these high voltages can damage electronic equipment and stress conductors and other equipment by exceeding their voltage breakdown ratings.

In ambient temperatures from 25 to 40°C and above, module short-circuit currents are increased at the same time conductors are being subjected to higher operating temperatures. Enhanced irradiance can occur at any time. Therefore the *UL* and *NEC* factors for short-circuit current output and *NEC* conductor temperature deratings are not redundant.

Good engineering practice suggests that the *UL* Standard 1703 requirements and the *NEC* requirements are neither conservative nor redundant and that they should be applied to all systems.

APPENDIX C
Grid-Connected Systems

Grid-connected systems present some unique problems for the PV designer and installer in meeting the *NEC*. There are less than 500 of these systems in existence and the current economics of PV dictate that increases in this number will be limited. Although the installations sometimes do not have batteries or charge controllers, the availability of *UL*-Listed inverters and other equipment is extremely limited due to the low production volume.

Inverters

Some of the grid-tied inverters that are available do not currently meet the draft standard established for inverters by *UL*. Some of the inverters cannot have both the dc PV circuits and the ac output circuits grounded without causing parallel ground current paths. Newer versions of these inverters may have solutions for this problem.

Other inverters have the internal circuitry tied to the case and force the central grounding point to be at the inverter input terminals. In some installations, this design is not compatible with ground-fault equipment and does not provide the flexibility needed for maximum surge suppression.

PV Source-Circuit Conductors

Some grid-tied inverters operate with PV arrays that are center tapped and have open-circuit voltages of ±325 volts and above. The system voltage of 650 volts or greater exceeds the insulation rating of the commonly available 600-volt insulated conductors. Each disconnect and overcurrent device and the insulation of the wiring **must** have a voltage rating exceeding the system voltage rating. Type G and W cables are available with the higher voltage ratings, but are flexible cords and do not meet *NEC* requirements for fixed installations. Cables suitable for *NEC* installations requiring insulation greater than 600 volts are available on special order (Appendix A).

Other inverters operate on systems with open-circuit voltages exceeding ±540 volts requiring conductors with 2000-volt or higher insulation. See Appendix D for a full discussion of this area.

Overcurrent Devices

When *UL* tests and lists fuses for dc operation, the voltage rating is frequently one-half the ac voltage rating. This makes a 600-volt ac fuse into a 300-volt dc fuse. Finding fuses with high enough dc ratings for grid systems operating at ±300 volts (600-volt system voltage) and above will pose problems. There are a limited number of 600-volt fuses available. See Appendix A.

Although not *UL*-Listed, Heinemann Electric Company (Appendix A) can series connect poles of dc-rated circuit breakers to obtain 750-volt ratings. Square D and others have similar products.

Circuit breakers that are "back fed" for any application (but particularly for utility interactive inverter connection to the grid) **must** be identified (in the listing) for such use and **must** be fastened in place with a screw or other additional clamp [690-64b(5), 384-16(f)].

Disconnects

In addition to the Heinemann circuit breaker mentioned above, manufacturers such as GE, Siemens, and Square D may certify their switches for higher voltage when the poles are connected in series.

Blocking Diodes

Although blocking diodes are not overcurrent devices, they do block currents in direct-current circuits, in some cases, and help to control circulating ground-fault currents if used in both ends of high-voltage strings. Lightning induced surges are tough on diodes. If isolated case diodes are used, at least 3500 volts of insulation is provided between the active elements and the normally grounded heat sink. Choosing a peak reverse voltage as high as is available but at least twice the PV open-circuit voltage, will result in longer diode life. Substantial amounts of surge suppression will also improve diode longevity.

Surge Suppression

Surge suppression is covered only lightly in the *NEC* because it affects performance more than safety and is mainly a utility problem at the transmission line level in ac systems [280]. PV arrays mounted in the open, on the tops of buildings, act like lightning rods. The PV designer and installer must provide appropriate means to deal with lightning-induced surges coming into the system.

Array frame grounding conductors should be routed directly to ground rods located as near as possible to the arrays. Grounding conductors for array frames should not be routed parallel or adjacent to current-carrying conductors to minimize the coupling of surges into the system. The *NEC* allows this separation on dc systems in Section 250-57(b) Ex 2.

Metal conduit will add inductance to the array-to-building conductors and slow down any induced surges as well as provide some electromagnetic shielding.

Metal oxide varistors (MOV) commonly used as surge suppression devices on PV systems have several deficiencies. They draw a small amount of current continually. The clamping voltage lowers as they age and may reach the open-circuit voltage of the system. When they fail, they fail in the shorted mode, heat up, and frequently catch fire. In many installations, the MOVs are protected with fast acting fuses to prevent further damage when they fail, but this may limit their effectiveness as surge suppression devices. Other devices are available that do not have these problems.

Silicon Oxide surge arrestors do not draw current when they are off. They fail open circuited when overloaded and, while they may split open on overloads, they rarely catch fire. They are not normally protected by fuses and are rated for surge currents up to 100,000 amps. They are rated at voltages of 300 volts and higher and are available from electrical supply houses or Delta Lightning Arrestors, Inc. (Appendix A).

Several companies specialize in lightning protection equipment, but much of it is for ac systems. Electronic product directories, such as the *Electronic Engineers Master Catalog* should be consulted.

APPENDIX D

Cable and Device Ratings at High Voltages

There is a concern in designing PV systems that have system open-circuit voltages above 600 volts. The concern has two main issues—device ratings and *NEC* limitations.

Equipment Ratings

Some utility-intertie inverters operate with a grounded, bipolar (three-wire) PV array. In a bipolar PV system, where each of the monopoles is operated in the 220-235-volt peak-power range, the open-circuit voltage can be anywhere from 290 to 380 volts, and above, depending on the module characteristics such as fill factor. Such a bipolar system can be described as a 350/700-volt system (for example) in the same manner that a 120/240-volt ac system is described. This method of describing the system voltage is consistent throughout the electrical codes used not only in residential and commercial power systems, but also in utility practice.

In all systems, the voltage ratings of the cable, switchgear, and overcurrent devices are based on the higher number of the pair (i.e., 700 volts in a 350/700-volt system). That is why 250-volt switchgear and overcurrent devices are used in 120/240-volt ac systems and 600-volt switchgear is used in systems such as the 277/480-volt ac system. Note that it is not the voltage to ground, but the higher line-to-line voltage that defines the equipment voltage requirements.

The *National Electrical Code* (*NEC*) defines a nominal voltage for ac systems (120, 240, etc.) and acknowledges that some variation can be expected around that nominal voltage. Such a variation around a nominal voltage is not considered in dc PV systems, and the *NEC* requires that the open-circuit array voltage must be used. The open-circuit voltage is defined at STC because of the relationship between the *UL* Standards and the way the *NEC* is written. The *NEC* Handbook elaborates on the definition of "circuit voltage," but this definition may not apply to current-limited dc systems. Section 690-7(a) of the *NEC* requires that the voltage used for establishing dc circuit requirements in PV systems be the open-circuit voltage.

The 1996 *NEC* specifically defines the system voltage as the sum of the absolute value of the open-circuit bipolar voltages [690-7(a)].

The comparison to ac systems cannot be carried too far; there are differences. For example, the typical wall switch in a 120/240-volt ac residential or commercial system is rated at only 120 volts, but such a switch in a 120/240-volt dc PV system would have to be rated at 240 volts. The inherent differences between a dc current source (PV modules) and a voltage source (ac grid) bear on this issue. Even the definitions of circuit voltage in the *NEC* and *NEC* Handbook refer to ac and dc systems, but do not take into account the design of the balance of systems required in current-limited PV systems. In a PV system, all wiring, disconnects, and overcurrent devices have current ratings that exceed the short-circuit currents by at least 25 percent. In the case of bolted or ground faults involving currents from the PV array, the overcurrent devices do not trip because they are rated to withstand continuous operation at levels above the fault levels. In an ac system, bolted faults and ground faults generally cause the overcurrent devices to trip or blow removing the source of voltage from the fault. Therefore, the faults that pose high-voltage problems in PV, dc systems cause the voltage to be removed in ac, grid-supply systems. For these reasons, a switch rated at 120 volts can be used in an ac system with voltages up to 240 volts, but in a dc, PV system, the switch would have to be rated at 240 volts.

Underwriters Laboratories (*UL*) Standard 1703 requires that manufacturers of modules listed to the standard include, in the installation instructions, a statement that the open-circuit voltage should be multiplied by 125 percent (crystalline cells), further increasing the voltage requirement of the Balance of Systems (BOS) equipment.

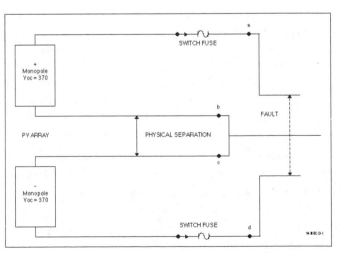

Current PV modules that are listed to the *UL* Standard 1703 are listed with a maximum system voltage of 600 volts. Engineers caution all installers, factory and otherwise, to not exceed this voltage. This restriction is not modified by the fact that the modules undergo high-pot tests at higher voltages. *UL* Standard 1703 allows modules to be listed up to 1000 volts.

Although not explicitly stated by the *NEC*, it is evident that the intent of the Code and the *UL* Standards is that all cable, switches, fuses, circuit breakers, and modules in a PV system be rated for the maximum system voltage. This is clarified in the 1996 *NEC* [690-7(a)].

While reducing the potential for line-to-line faults, the practice of wiring each monopole (one of two electrical source circuits) in a separate conduit to the inverter does not eliminate the problem. Consider the bipolar system presented in Figure D-1 with a bolted fault (or deliberate short) from the negative to the positive array conductor at the input of the inverter. With the switches closed, array short-circuit current flows, and neither fuse opens.

Now consider what happens in any of the following cases.

    1. A switch is opened

    2. A fuse opens

    3. A wire comes loose in a module junction box

    4. An intercell connection opens or develops high resistance

    5. A conductor fails at any point

In any of these cases, the entire array voltage (740 volts) stresses the device where the circuit opens. This voltage (somewhere between zero at short-circuit and the array open-circuit voltage) will appear at the device or cable. As the device starts to fail, the current through it goes from Isc to zero as the voltage across the device goes from zero to Voc. This process is very conducive to sustained arcs and heating damage.

Separating the monopoles does not avoid the high-voltage stress on any component, but it does help to minimize the potential for some faults. There are other possibilities for faults that will also place the same total voltage on various components in the system. An improperly installed grounding conductor coupled with a module ground fault could result in similar problems.

Section 690-5 of the *NEC* requires a ground-fault device on PV systems that are installed on the roofs of dwellings. This device, used for fire protection, must detect the fault, interrupt the fault current, and "disable" the array. "Disable" is not clearly defined in the *NEC*, but the *NEC* Handbook (containing supplementary guidance) says one means of disabling an array is to crowbar or short-circuit the array terminals. This crowbar creates, as a designed-in function, the fault described above. Several ground-fault devices that have been prototyped and produced have this crowbar feature.

Some large (100 kW) grid-connected PV systems like the one at Juana Diaz, Puerto Rico have inverters that, when shut down, crowbar the array. The array remains crowbarred until the ac power is shut off.

NEC Limitation

The second issue associated with this concern is that the 1993 *NEC* in Section 690-7(c) only allows PV installations up to 600 volts. Inverter and system design issues may favor higher system voltage levels.

SOLUTIONS

Near Term

System designers can select inverters with lower operating and open-circuit voltages. Utility-intertie inverters are available with voltages as low as 24 volts. They also can work with the manufacturers of higher voltage inverters to reduce the number of modules in each series string to the point where the open-circuit voltage is less than 600 volts. The peak-power voltage would also be lowered. Transformers may be needed to raise the ac voltage to the required level. At least one inverter manufacturer has pursued this option and is offering inverters which can operate with arrays that have open-circuit voltages of less than 600 volts.

Cable manufacturers can produce a *UL*-Listed, cross-linked polyethylene, single-conductor cable marked USE-2, RHW-2, and Sunlight Resistant. The cable is rated at 2000 volts. This cable could be used for exposed module interconnections and in conduit after all of the other *NEC* requirements are met for installations above 600 volts.

Several manufacturers issue factory certified rating on their three-pole disconnects to allow higher voltage, non-load break operation with series-connected poles. The *NEC* will require an acceptable method of obtaining non-load break operation.

Some OEM circuit breaker manufacturers will factory certify series-connected poles on their circuit breakers. Units have been used at 750 volts and 100 amps with 10,000 amps of interrupt rating. Higher voltages may be available.

High-voltage industrial fuses are available, but dc ratings are unknown at this time. Fuse holders with higher voltage ratings are usually the bolt-in types.

Individual 600-volt terminal blocks can be used with the proper spacing for higher voltages.

Module manufacturers can have their modules listed for higher system voltages.

Power diodes may be connected across each monopole. When a bolted line-to-line fault occurs, one of the diodes will be forward biased when a switch or fuse opens, thereby preventing the voltage from one monopole from adding to that of the other monopole. The diodes are mounted across points a-b and c-d in Figure D-1. Each diode should be rated for at least the system open-circuit voltage and the full short-circuit current from one monopole. Since diodes are not listed as over-voltage protection devices, this solution is not recognized in the *NEC*.

The 1996 *NEC* allows PV installations over 600 volts in non-residential applications, which will cover the voltage range being used in most current designs. Article 710 should be consulted for all of the numerous requirements dealing with the installation of electrical systems with voltages over 600 volts.

APPENDIX E

Example Systems

The systems described in this appendix and the calculations shown are presented as examples only. The calculations for conductor sizes and the ratings of overcurrent devices are based on the requirements of the 1993 *National Electrical Code* (*NEC*) and on *UL* Standard 1703 which provides instructions for the installation of *UL*-Listed PV modules. Local codes and site-specific variations in irradiance, temperature, and module mounting, as well as other installation particularities, dictate that these examples should not be used without further refinement. Tables 310-16 and 310-17 from the *NEC* provide the ampacity data and temperature derating factors.

EXAMPLE 1 Direct-Connected Water Pumping System

Array Size: 4, 12-volt, 60-watt modules   Isc = 3.8 amps, Voc = 21.1 volts

Load: 12-volt, 10-amp motor

Description

The modules are mounted on a tracker and connected in parallel. The modules are wired as shown in Figure E-1 with number 10 AWG USE-2 single-conductor cable. A large loop is placed in the cable to allow for tracker motion without straining the rather stiff building cable. The USE-2 cable is run to a disconnect switch in an enclosure mounted on the pole. From this disconnect enclosure, number 8 AWG XHHW-2 cable in electrical non-metallic conduit is routed to the well head. The conduit is buried 18 inches deep. The number 8 AWG cable is used to minimize voltage drop.

The *NEC* requires the disconnect switch. Because the PV modules are current limited and all conductors have an ampacity greater than the maximum output of the PV modules, no overcurrent device is required, although some inspectors might require it and it might serve to provide some degree of lightning protection. A dc-rated disconnect switch or a dc-rated fused disconnect must be used. Since the system is ungrounded, a two-pole switch must be used. All module frames, the disconnect enclosure, and the pump housing must be grounded, whether the system is grounded or not.

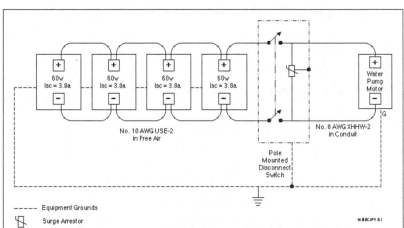

Figure E-1. Direct Connected System

617

Calculations

The array short-circuit current is 15.2 amps (4 x 3.8).

NEC 125 percent: 1.25 x 15.2 = 19 amps

UL 125 percent: 1.25 x 19 = 23.75 amps

The ampacity of 10 AWG USE-2 at 30°C is 55 amps.

The ampacity at 61-70°C is 31.9 amps (0.58 x 55) which is more than the 23.75 amp requirement.

The equipment grounding conductors should be number 10 AWG.

The minimum voltage rating of all components is 26 volts (1.25 x 21.1).

EXAMPLE 2 Water Pumping System with Current Booster

Array Size: 10, 12-volt, 53-watt modules Isc = 3.4 amps, Voc = 21.7 volts

Current Booster Output: 90 amps

Load: 12-volt, 40-amp motor

Description

This system has a current booster before the water pump and has more modules than in Example 1. Initially, number 8 AWG USE-2 cable was chosen for the array connections, but this cable has inadequate ampacity. As the calculations below show, the array was split into two subarrays. There is potential for malfunction in the current booster, but it does not seem possible that excess current can be fed back into the array wiring, since there is no other source of energy in the system. Therefore, these conductors do not need overcurrent devices if they are sized for the entire array current. If smaller conductors are used, then overcurrent devices will be needed.

Since the array is broken into two subarrays, the maximum short-circuit current available in either subarray wiring is equal to the total array short-circuit current under fault conditions. Overcurrent devices are needed to protect the subarray conductors under these conditions.

A grounded system is selected, and only one-pole disconnects are required. Equipment grounding and system grounding conductors are shown in Figure E-2

If the current booster output conductors are sized to carry the maximum current (3-hour) of the booster, then overcurrent devices are not necessary, but again, some inspectors may require them.

Calculations

The array short-circuit current is 34 amps (10 x 3.4).

NEC 125 percent: 1.25 x 34 = 42.5 amps

UL 125 percent: 1.25 x 42.5 = 53.1 amps

The ampacity of 8 AWG USE-2 cable at 30°C in free air is 80 amps.

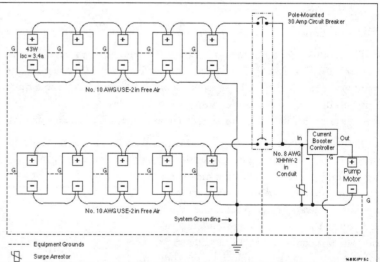

Figure E-2. Direct-Connected PV System with Current Booster

The ampacity at 61-70°C is 46.4 amps (0.58 x 80), which is less than the 53.1 amp requirement. Number 8 AWG is the largest conductor that can be connected to the modules. Therefore, the array is split into two subarrays. Each is wired with number 10 AWG USE-2 conductors.

The subarray short-circuit current is 17 amps (5 x 3.4).

NEC 125 percent: 1.25 x 17 = 21.3 amps

UL 125 percent: 1.25 x 21.25 = 26.6 amps

The ampacity of number 10 AWG USE-2 at 30°C in free air is 55 amps.

The ampacity at 61-70°C is 31.9 amps (0.58 x 55), which is more than the 26.6 amp requirement. Since this cable is to be connected to an overcurrent device with terminals rated at 75°C, the ampacity of the cable must be evaluated with 75°C insulation. Number 10 AWG 75°C cable operating at 40°C (the disconnect operating temperature) has an ampacity of 30.8 amps (0.88 x 35), which is more than the 21.3 amps requirement.

Thirty-amp circuit breakers are used to protect the number 10 AWG subarray conductors.

The current booster maximum current is 90 amps.

The current booster average long-term (3-hours or longer) current is 40 amps.

NEC 125 percent: 1.25 x 40 = 50 amps

The ampacity of number 8 AWG XHHW-2 at 30°C in conduit is 55 amps.

The ampacity at 36-40°C is 50 amps (0.91 x 55), which meets the requirements but may not meet the overcurrent device connection requirements.

The number 8 AWG conductors are connected to the output of the fuse holders, and there is a possibility that heating of the fuse may occur depending on how the holder is connected. It is therefore good practice to make the calculation for terminal overheating. The ampacity of a number 8 AWG conductor evaluated with 75°C insulation (the maximum temperature of the terminals on the overcurrent device) at 40°C is 44 amps (50 x 0.88), which is greater than the 40-amp requirement. This means that the overcurrent device will not be subjected to overheating when the number 8 AWG conductor carries 40 amps.

All equipment grounding conductors should be number 10 AWG. The grounding electrode conductor should be number 8 AWG or larger.

Minimum voltage rating of all components: 1.25 x 21.1 = 26 volts

EXAMPLE 3 Stand-Alone Lighting System

Array Size: 4, 12-volt, 64-watt modules   Isc = 4.0 amps, Voc = 21.3 volts

Batteries: 200-amp-hours at 24 volts

Load: 60 watts at 24 volts

Description

The modules are mounted at the top of a 20-foot pole with the metal-halide lamp. The modules are connected in series and parallel to achieve the 24-volt system rating. The lamp, with an electronic ballast and timer/controller, draws 60 watts at 24 volts. The batteries, disconnect switches, charge controller, and overcurrent devices are mounted in a box at the bottom of the pole. The system is grounded as shown in Figure E-3.

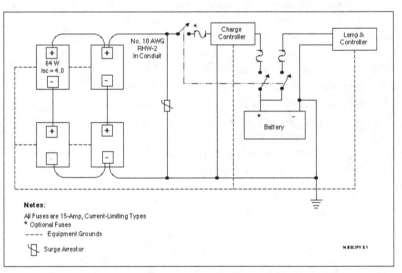

Figure E-3. Stand-Alone Lighting System

Calculations:

The array short-circuit current is 8 amps (2 x 4).

NEC 125 percent: 1.25 x 8 = 10 amps

UL 125 percent: 1.25 x 10 = 12.5 amps

Load Current: 60/24 = 2.5 amps

NEC 125 percent: 1.25 x 2.5 = 3.1 amps

Number 10 AWG USE-2 is selected for module interconnections and is placed in conduit at the modules and then run down the inside of the pole.

The modules operate at 61-70°C, which requires that the module cables be temperature derated. Number 10 AWG USE-2 cable has an ampacity of 40 amps at 30°C in conduit. The derating factor is 0.58. The temperature derated ampacity is 23.2 amps (40 x 0.58), which exceeds the 12.5-amp requirement. Checking the cable with a 75°C insulation, the ampacity at the fuse end at 40°C ambient is 30.8 amps (35 x 0.88), which exceeds the 10-amp requirement. This cable can be protected by a 15-amp fuse or circuit breaker.

The same USE-2, number 10 AWG cable is selected for all other system wiring, because it has the necessary ampacity for each circuit.

A three-pole fused disconnect is selected to provide the PV and load disconnect functions and the necessary overcurrent protection. The fuse selected is a RK-5 type, providing current-limiting in the battery circuits. A pull-out fuse holder with either Class RK-5 or Class T fuses could also be used for a more compact installation. Fifteen amp fuses are selected to provide overcurrent protection for the number 10 AWG cables. They are used in the load circuit and will not blow on any starting surges drawn by the lamp or controller. The 15 amp fuse before the charge controller could be eliminated since that circuit is protected by the battery. The disconnect switch at this location is required.

The equipment grounding conductors and the system grounding conductor to the ground rod should be number 10 AWG conductors.

The dc voltage ratings for all components used in this system should be at least 53 volts (2 x 21.3 x 1.25).

EXAMPLE 4 Remote Cabin DC-Only System

Array Size: 6, 12-volt, 75-watt modules   Isc = 4.8 amps, Voc = 22 volts

Batteries: 700 amp hours at 12 volts

Load: 75 watts peak at 12-volts dc

Description

The modules are mounted on a rack on a hill behind the house. Non-metallic conduit is used to run the cables from the module rack to the control panel. A disconnect and control panel are mounted on the back porch, and the batteries are in an insulated box under the porch. All the loads are dc with a peak combined power of 75 watts at 12 volts due, primarily, to a pressure pump on the gravity-fed water supply. The battery bank consists of four 350-amp-hour, 6-volt, deep-cycle batteries wired in series and parallel. Figure E-4 shows the system schematic.

Calculations

The array short-circuit current is 28.8 amps (6 x 4.8).

UL 125 percent: 1.25 x 28.8 = 36 amps

NEC 125 percent: 1.25 x 36.0 = 45 amps

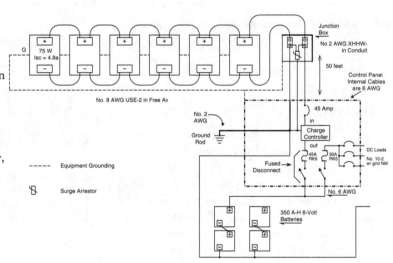

Figure E-4. *Remote Cabin DC-Only System*

The module interconnect wiring and the wiring to a rack-mounted junction box will operate at 65°C. If USE-2 cable with 90°C insulation is chosen, then the temperature derating factor will be 0.58. The required ampacity of the cable at 30°C is 77.6 amps (45/0.58), which can be handled by number 8 AWG cable with an ampacity of 80 amps in free air at 30°C. Conversely, the ampacity of the number 8 AWG cable is 46.4 amps (80 x 0.58) at 65°C which exceeds the 45 amp requirement.

From the rack-mounted junction box to the control panel, the conductors will be in conduit and exposed to 40°C temperatures. If XHHW-2 cable with a 90°C insulation is selected, the temperature derating factor is 0.91. The required ampacity of the cable at 30°C would be 45/0.91 = 49.5 amps in conduit. Number 8 AWG cable has an ampacity of 55 amps at 30°C in conduit, which exceeds the 49.5-amp requirement. Conversely, the number 8 AWG conductor has an ampacity of 50 amps (55 x 0.91) at 40°C in conduit which exceeds the 45 amp requirement at this temperature.

The number 8 AWG cable, evaluated with a 75°C insulation, has an ampacity at 40°C of 44 amps (50 x 0.88) which is greater than the 36 amps that might flow through it under enhanced irradiation conditions.

The array is mounted 200 feet from the house, and the round trip cable length is 400 feet. A calculation of the voltage drop in 400 feet of Number 8 AWG cable operating at 36 amps (125 percent Isc) is 0.778 ohms per 1000 feet x 400 / 1000 x 36 = 11.2 volts. This represents an excessive voltage drop on a 12-volt system, and the batteries cannot be effectively charged. Number 2 AWG cable (with a voltage drop of 2.8 volts) was substituted; this substitution is acceptable for this installation.

The PV conductors are protected with a 45-amp single-pole circuit breaker on this grounded system.

Number 6 AWG THHN cable is used in the control center and has an ampacity of 95 amps at 30°C when evaluated with 75°C insulation. Number 2 AWG cable from the negative dc input is used to the point where the grounding electrode conductor is attached instead of the number 6 AWG conductor used elsewhere to comply iwth grounding requirements.

The 75-watt peak load draws about 6.25 amps and number 10-2 with ground (w/gnd) nonmetallic sheathed cable was used to wire the cabin for the pump and a few lights. DC-rated circuit breakers rated at 20 amps were used to protect the load wiring, which is in excess of the peak load current of 7.8 amps (1.25 x 6.25) and less than the cable ampacity of 30 amps.

Current-limiting fuses in a fused disconnect are used to protect the dc-rated circuit breakers, which do not have an interrupt rating sufficient to withstand the short-circuit currents from the battery under fault conditions. RK-5 fuses were chosen with a 45-amp rating in the charge circuit and a 30-amp rating in the load circuit. The fused disconnect also provides a disconnect for the battery from the charge controller and the dc load center.

The equipment grounding conductors should be number 10 AWG and the grounding electrode conductor should be number 2 AWG.

All components should have a voltage rating of at least 1.25 x 22 = 27.5 volts.

EXAMPLE 5 Small Residential Stand-Alone System

Array Size: 10, 12-volt, 51-watt modules Isc = 3.25 amps, Voc = 20.7 volts

Batteries: 800 amp-hours at 12 volts

Loads: 5 amps dc and 500-watt inverter with 90 percent efficiency

Description

The PV modules are mounted on the roof. Single conductor cables are used to connect the modules to a roof-mounted junction box. UF two-conductor sheathed cable is used from the roof to the control center. Physical protection (wood barriers or conduit) for the UF cable is used where required. The control center, diagrammed in Figure E-5, contains disconnect and overcurrent devices for the PV array, the batteries, the inverter, and the charge-controller.

Calculations

The module short-circuit current is 3.25 amps.

UL 125 percent: 1.25 x 3.25 = 4.06 amps

NEC 125 percent: 1.25 x 4.06 = 5.08 amps per module

The module operating temperature is 68°C.

The derating factor for USE-2 cable is 0.58 at 61-70°C.

Number 14 cable has an ampacity at 68°C of 20.3 amps (0.58 x 35) (max fuse is 15 amps).

Number 12 cable has an ampacity at 68°C of 23.2 amps (0.58 x 40) (max fuse is 20 amps).

Number 10 cable has an ampacity at 68°C of 31.9 amps (0.58 x 55) (max fuse is 30 amps).

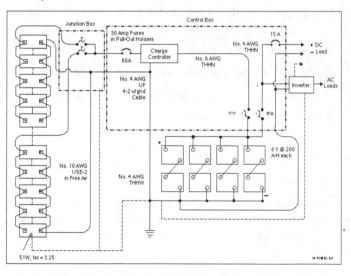

Figure E-5. *Small Residential Stand-Alone System*

621

Number 8 cable has an ampacity at 68°C of 46.4 amps (0.58 x 80).

The array is divided into two five-module subarrays. The modules in each subarray are wired from module junction box to module junction box and then to the array junction box. Number 10 AWG USE-2 is selected for this wiring, because it has an ampacity of 31.9 amps under these conditions, and the requirement for each subarray is 5 x 5.08 = 25.4 amps. Evaluated with 75°C insulation, a number 10 AWG cable has an ampacity of 30.8 amps (35 x 0.88) at 40°F, which is greater than the actual requirement of 20.3 amps (5 x 4.06). In the array junction box on the roof, two 30-amp fuses in pull-out holders are used to provide overcurrent protection for the number 10 AWG conductors.

In this junction box, the two subarrays are combined into an array output. The ampacity requirement is 50.8 amps (10 x 5.08). A number 4 AWG UF cable (4-2 w/gnd) is selected for the run to the control box. It operates in an ambient temperature of 40°C and has an ampacity of 57.4 amps (70 x 0.82). This is a 60°C cable with 90°C conductors. Care must be used when connecting to fuses that are rated for use only with 75°C conductors.

A 60-amp circuit breaker in the control box serves as the PV disconnect switch and overcurrent protection for the UF cable. The *NEC* allows the next larger size; in this case, 60 amps, which is over the 57 amps ampacity of the cable. Two single-pole, pull-out fuse holders are used for the battery disconnect. The charge circuit fuse is a 60-amp RK-5 type.

The inverter has a continuous rating of 500 watts at 10.75 volts and an efficiency of 90 percent at this power level. The ampacity requirement of the input circuit is 64.6 amps ((500 / 10.75 / 0.90) x 1.25).

The cables from the battery to the control center must meet the inverter requirements of 64.6 amps plus the dc load requirements of 6.25 amps (1.25 x 5). A number 4 AWG THHN has an ampacity of 85 amps when placed in conduit and evaluated with 75°C insulation. This exceeds the requirements of 71 amps (64.6 + 6.25). This cable can be used in the custom power center and be run from the batteries to the inverter.

The discharge-circuit fuse must be rated at least 71 amps. An 80-amp fuse should be used, which is less than the cable ampacity.

The dc-load circuit is wired with number 10 AWG NM cable (ampacity of 30 amps) and protected with a 15-amp circuit breaker.

The grounding electrode conductor is number 4 AWG and is sized to match the largest conductor in the system, which is the array-to-control center wiring.

Equipment grounding conductors for the array and the charge circuit can be number 10 AWG based on the 60-amp overcurrent devices [Table 250-95]. The equipment ground for the inverter must be a number 8 AWG conductor.

All components should have at least a dc voltage rating of 1.25 x 20.7 = 26 volts.

EXAMPLE 6 Medium Sized Residential Hybrid System

Array Size: 40, 12-volt, 53-watt modules Isc = 3.4 amps, Voc = 21.7 volts

Batteries: 1000 amp-hours at 24 volts

Generator: 6 kW, 240-volt ac

Loads: 15 amps dc and 4000-watt inverter, efficiency = .85

Description

The 40 modules (2120 watts) are mounted on the roof in subarrays consisting of eight modules mounted on a single-axis tracker. The eight modules are wired in series and parallel for this 24-volt system. Five source circuits are routed to a custom power center. Single-conductor cables are used from the modules to roof-mounted junction boxes for each source circuit. From the junction boxes, UF sheathed cable is run to the main power center.

Blocking diodes are not used to minimize voltage drops in the system.

A prototype array ground-fault detector provides experimental compliance with the requirements of *NEC* Section 690-5.

The charge controller is a relay type.

DC loads consist of a refrigerator, a freezer, several telephone devices, and two fluorescent lamps. Peak current is 15 amps.

The 4000-watt sine-wave inverter supplies the rest of the house.

The 6-kW natural gas fueled, engine-driven generator provides back-up power and battery charging through the inverter. The 240-volt output of the generator is fed through an isolation transformer to step it down to 120 volts for use in the inverter and the house. Figure E-6 presents the details.

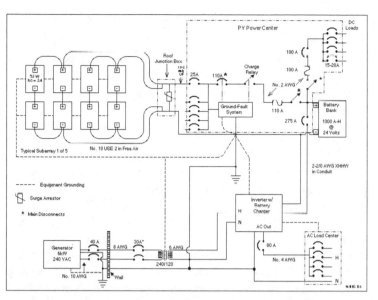

Figure E-6. *Medium Sized Residential Hybrid System*

Calculations

The subarray short-circuit current is 13.6 amps (4 x 3.4).

UL 125 percent: 1.25 x 13.6 = 17 amps

NEC 125 percent: 1.25 x 17 = 21.25 amps

The temperature derating factor for USE-2 cable at 61-70°C is 0.58.

The ampacity of number 10 AWG USE-2 cable at 70°C is 31.9 amps (55 x 0.58).

The temperature derating factor for UF cable at 36-40°C is 0.82.

The ampacity of number 10-2 w/gnd UF cable at 40°C is 24.6 amps (30 x 0.82). Since the UF cable insulation is rated at 60°C, no further temperature calculations are required when this cable is connected to circuit breakers rated for use with 75°C conductors.

The source-circuit circuit breakers are rated at 25 amps.

The PV array short-circuit current is 68 amps (5 x 13.6).

UL 125 percent: 1.25 x 68 = 85 amps

NEC 125 percent: 1.25 x 85 = 106 amps

A 110-amp circuit breaker is used for the main PV disconnect after the five source circuits are combined.

A 110-amp RK5 current-limiting fuse is used in the charge circuits of the power center, which are wired with number 2 AWG THHN (170 amps with 75°C insulation) conductors.

The dc-load circuits are wired with number 10-2 w/gnd NM cable (30 amps) and are protected with 20- or 30-amp circuit breakers. A 100-amp RK-5 fuse protects these discharge circuits from excess current from the batteries.

Inverter

The inverter can produce 4000 watts ac at 22 volts with an efficiency of 85 percent.

The inverter input current ampacity requirements are 267 amps ((4000 / 22 / 0.85) x 1.25).

Two 2/0 AWG USE-2 cables are paralleled in conduit between the inverter and the batteries. The ampacity of this cable (rated with 75°C insulation) at 30°C is 280 amps (175 x 2 x 0.80). The 0.80 derating factor is required because there are four cables in the conduit.

A 275-amp circuit breaker with a 25,000-amp interrupt rating is used between the battery and the inverter. Current-limiting fusing is not required in this circuit.

623

The output of the inverter can deliver 4000 watts ac (33 amps) in the inverting mode. It can also pass up to 60 amps through the inverter from the generator while in the battery charging mode.

Ampacity requirements, ac output: 60 x 1.25 = 75 amps. This reflects the *NEC* requirement that circuits are not to be operated continuously at more than 80% of rating.

The inverter is connected to the ac load center with number 4 AWG THHN cable in conduit, which has an ampacity of 85 amps when used at 30°C with 75°C overcurrent devices. An 80-amp circuit breaker is used near the inverter to provide a disconnect function and the overcurrent protection for this cable.

Generator

The 6-kW, 240-volt generator has internal circuit breakers rated at 27 amps (6500-watt peak rating). The *NEC* requires that the output conductors between the generator and the first field-installed overcurrent device be rated at least 115 percent of the nameplate rating ((6000 / 240) x 1.15 = 28.75 amps). Since the generator is connected through a receptacle outlet, a number 10-4 AWG SOW portable cord (30 amps) is run to a NEMA 3R exterior circuit breaker housing. This circuit breaker is rated at 40 amps and provides overcurrent protection for the number 8 AWG THHN conductors to the transformer. These conductors have an ampacity of 44 amps (50 x 0.88) at 40°C (75°C insulation rating). The circuit breaker also provides an exterior disconnect for the generator. Since the isolation transformer isolates the generator conductors from the system electrical ground, the neutral of the generator is grounded at the exterior disconnect.

A 30-amp circuit breaker is mounted near the PV Power Center in the ac line between the generator and the transformer. This circuit breaker serves as the ac disconnect for the generator and is grouped with the other disconnects in the system.

The output of the transformer is 120 volts. Using the rating of the generator, the ampacity of this cable must be 62.5 amps ((6000 / 120) x 1.25). A number 6 AWG THHN conductor was used, which has an ampacity of 65 amps at 30°C (75°C insulation rating).

Grounding

The module and dc-load equipment grounds must be number 10 AWG conductors. Additional lightning protection will be afforded if a number 6 AWG or larger conductor is run from the array frames to ground. The inverter equipment ground must be a number 4 AWG conductor based on the size of the overcurrent device for this circuit. The grounding electrode conductor must be 2-2/0 AWG or a 500 kcmil conductor, unless there are no other conductors connected to the grounding electrode; then this conductor may be reduced to number 6 AWG [250-93 exceptions].

DC Voltage Rating

All dc circuits should have a voltage rating of at least 55 volts (1.25 x 2 x 22).

EXAMPLE 7 Roof-Top Grid-Connected System

Array Size: 24, 50-volt, 240-watt modules
Isc = 5.6, Voc = 62

Inverter: 200-volt dc input, 240-volt ac output at 5000 watts with an efficiency of 0.95.

Description

The roof-top array consists of six parallel-connected strings of four modules each. A junction box is mounted at the end of each string which contains a surge arrestor, a blocking diode, and a fuse. All wiring is THHN in conduit. The inverter is located adjacent to the service entrance load center where PV power is fed to the grid through a back-fed circuit breaker. Figure E-7 shows the system diagram.

Figure E-7. *Roof-Top Grid-Connected System*

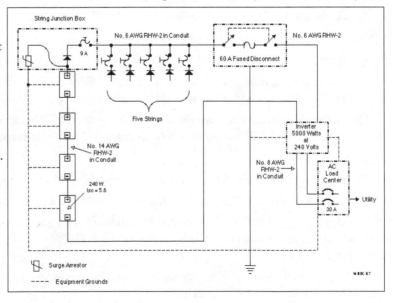

Calculations

The string short-circuit current is 5.6 amps.

UL 125 percent: 1.25 x 5.6 = 7 amps

NEC 125 percent: 1.27 x 7 = 8.75 amps

The array short-circuit current is 33.6 amps (6 x 5.6).

UL 125 percent: 1.25 x 33.6 = 42 amps

NEC 125 percent: 1.25 x 42 = 52.5 amps

The modules in each string are connected in series. The conductors operate at 63°C. The temperature derating factor for RHW-2 at this temperature is 0.58. The required 30°C ampacity for this cable is 15.1 amps (8.75 / 0.58). Number 14 AWG cable has an ampacity of 25 amps with 90°C insulation and 20 amps with 75°C insulation so there is no problem with the end of the cable connected to the fuse since the 7 amps is below either ampacity.

This cable is protected with a 9-amp fuse.

The cable from the string J-Boxes to the main PV disconnect operates at 40°C. The temperature derating factor for RHW-2 with 90°C insulation is 0.91. This yields a 30°C ampacity requirement of 58 amps (52.5 / 0.91). Number 6 AWG meets this requirement with an ampacity of 75 amps (90°C insulation), and a number 6 AWG cable with 75°C insulation has an ampacity of 65 amps, which also exceeds the 48 amp (42 / 0.88) requirement.

Overcurrent protection is provided with a 60-amp fused disconnect. Since the negative dc conductor of the array is grounded, only a single-pole disconnect is needed.

The inverter output current is 21 amps (5000 / 240).

NEC 125 percent: 1.25 x 21 = 26 amps.

The cable from the inverter to the load center operates at 30°C. Number 8 AWG XHHW-2 (evaluated with 75°C insulation) has an ampacity of 50 amps.

A back-fed 30-amp, two-pole circuit breaker provides an ac disconnect and overcurrent protection in the load center.

The equipment grounding conductors for this system should be at least number 10 AWG conductors. The system grounding electrode conductor should be a number 6 AWG conductor.

All dc circuits should have a voltage rating of at least 310 volts (1.25 x 4 x 62).

EXAMPLE 8 Integrated Roof Module System, Grid Connected

Array Size: 192, 12-volt, 22-watt thin-film modules

Isc = 1.8 amps, Vmp = 15.6 volts, Voc = 22 volts

Inverter: ±180-volt dc input, 120-volt ac output, 4000 watts, .95 efficiency

Description

The array is integrated into the roof as the roofing membrane. The modules are connected in center-tapped strings of 24 modules each. Eight strings are connected in parallel to form the array. A blocking diode is placed in series with each string. Strings are grouped in two sets of four and a series fuse protects the module and string wiring as shown in Figure E-8. The bipolar inverter has the center tap dc input and the ac neutral output grounded. The 120-volt ac output is fed to the service entrance load center (fifty feet away) through a back-fed circuit breaker.

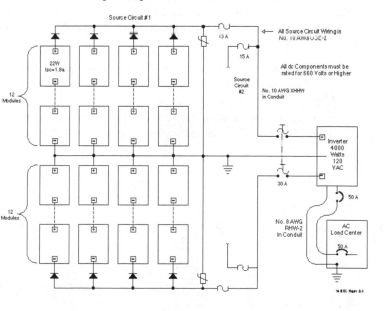

Figure E-8. Integrated Roof Module PV System

Calculations

Each string short-circuit current is 1.8 amps.

UL 125 percent (estimated for thin-film modules): 1.25 x 1.8 = 2.25 amps

NEC 125 percent: 1.25 x 2.25 = 2.8 amps

Each source circuit (4 strings) short-circuit current is 7.2 amps (4 x 1.8).

UL 125 percent: 1.25 x 7.2 = 9 amps

NEC 125 percent: 1.25 x 9 = 11.25 amps

The array (two source circuits) short-circuit current is 14.4 amps (2 x 7.2).

UL 125 percent: 1.25 x 14.4 = 18 amps

NEC 125 percent: 1.25 x 18 = 22.5 amps

USE-2 cable is used for the module cables and operates at 75°C when connected to the roof-integrated modules. The temperature derating factor in the wiring raceway is 0.41. For the strings, the 30°C ampacity requirement is 6.3 amps (2.6 / 0.41).

Each source circuit conductor is also exposed to temperatures of 75°C. The required ampacity for this cable (at 30°C) is 27.4 amps (11.25 / 0.41).

Number 10 AWG USE-2 cable is selected for moisture and heat resistance. It has an ampacity of 40 amps at 30°C (90°C insulation) and can carry 35 amps when limited to a 75°C insulation rating. This cable is used for both string and source-circuit wiring. Fifteen-amp fuses are used to protect the string and source-circuit conductors.

The array wiring is inside the building and XHHW is used in conduit. It is operated at 50°C when passing through the attic. The temperature derating factor is 0.82, which yields a 30°C ampacity requirement of 27.4 amps (22.5 / 0.82). Number 10 AWG cable has an ampacity of 40 amps (90°C insulation) or 35 amps (evaluated with 75°C insulation). Both of these ampacities exceed the 27.4-amp requirement. Thirty-amp fuses protect these cables. Since the inverter has high voltages on the dc-input terminals (charged from the ac utility connection), a pull-out fuse holder is used.

The inverter is rated at 4000 watts at 120 volts and has a 33-amp output current. The ampacity requirement for the cable between the inverter and the load center is 42 amps ((4000 / 120) x 1.25) at 30°C. Number 8 AWG XHHW-2 cable in conduit connects the inverter to the ac-load center, which is fifty feet away and, when evaluated at with 75°C insulation, has an ampacity of 50 amps at 30°C. A 50-amp circuit breaker in a small circuit-breaker enclosure is mounted next to the inverter to provide an ac disconnect for the inverter that can be grouped with the dc disconnect. Another 50-amp circuit breaker is back-fed in the service entrance load center to provide the connection to the utility.

The modules have no frames and, therefore, no equipment grounding requirements. The inverter and switchgear should have number 10 AWG equipment grounding conductors. The system grounding electrode conductor should be a number 8 AWG conductor.

All dc components in the system should have a minimum voltage rating of 660 volts (24 x 22 x 1.25). This voltage exceeds the commonly available 600-volt rated equipment. Conductors should have a 1000 or 2000-volt insulation rating, and all fuses, fuse-holders, and switches should be rated for a voltage over 660 volts. As a design alternative, the input voltage requirements of the inverter might be lowered so that only 20 series-connected modules would be required in each string. This would lower the voltage requirement to 550 volts (22 x 20 x 1.25).

AC Device	Device Watts	X	Hoursof DailyUse	X	DaysofUse per Week	÷	7	=	Average Watt-hrs per Day
		X		X		÷		=	
		X		X		÷		=	
		X		X		÷		=	
		X		X		÷		=	
		X		X		÷		=	
		X		X		÷		=	
		X		X		÷		=	
		X		X		÷		=	
		X		X		÷		=	
		X		X		÷		=	
		X		X		÷		=	
		X		X		÷		=	
		X		X		÷		=	
		X		X		÷		=	
		X		X		÷		=	
		X		X		÷		=	
		X		X		÷		=	
		X		X		÷		=	
		X		X		÷		=	
		X		X		÷		=	
		X		X		÷		=	

1. Total AC Watt-Hrs./Day

2. X 1.1 = Total Corrected DC Watt-Hrs./Day

DC Device	Device Watts	X	Hoursof DailyUse	X	DaysofUse per Week	÷	7	=	Average Watt-hrs per Day
		X		X		÷		=	
		X		X		÷		=	
		X		X		÷		=	
		X		X		÷		=	
		X		X		÷		=	
		X		X		÷		=	

3. Total DC Watt-Hrs./Day

3	(from previous page) Total DC Watt-Hrs./Day	
4	Total Corrected DC Watt-Hrs./Day from Line 2 +	
5	Total Household DC Watt-Hrs./Day =	
6	System Nominal Voltage (usually 12 or 24) ÷	
7	Total DC Amp-Hrs./Day =	
8	Battery losses, wiring losses, safety factor X 1.2	
9	Total Daily Amp-Hour Requirement =	
10	Estimated Design Insolation (hours per day of sun, see map) ÷	
11	Total PV Array Current in Amps =	
12	Select a Photovoltaic Module for Your System	
13	Module Rated Power Amps ÷	
14	Number of Modules Required in Parallel =	
15	System Nominal Voltage (from line 6 above)	
16	Module Nominal Voltage (usually 12) ÷	
17	Number of Modules Required in Series =	
18	Number of Modules Required in Parallel (from Line 14 above) X	
19	**Total Modules Required** =	

Battery Sizing Worksheet

20	Total Daily Amp-Hour Requirement (from line 9)	
21	Reserve Time in Days X	
22	Percent of Useable Battery Capacity ÷	
23	Minimum Battery Capacity in Amp-Hours =	
24	Select a Battery for Your System, Enter Amp-Hour Capacity ÷	
25	Number of Batteries in Parallel =	
26	System Nominal Voltage (from line 6)	
27	Voltage of Your Chosen Battery (6 or 12 usually) ÷	
28	Number of Batteries in Series =	
29	Number of Batteries in Parallel (from line 25 above) X	
30	**Total Number of Batteries Required**	

SOLAR INSOLATION MAP
Average Hours of Sun for the Worst Month Yearly

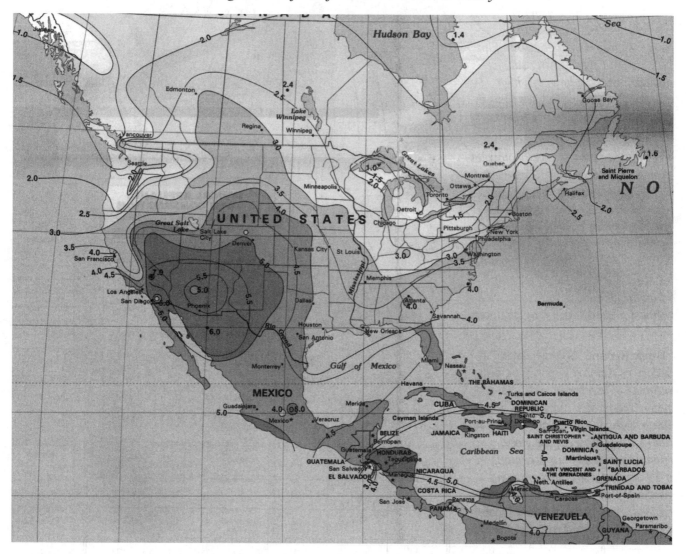

—Solarex Corporation has extended permission to Real
Goods to print the Solarex World Design Insolation Map.
Copyright© SOLAREX CORPORATION.

MAGNETIC DECLINATIONS IN THE UNITED STATES

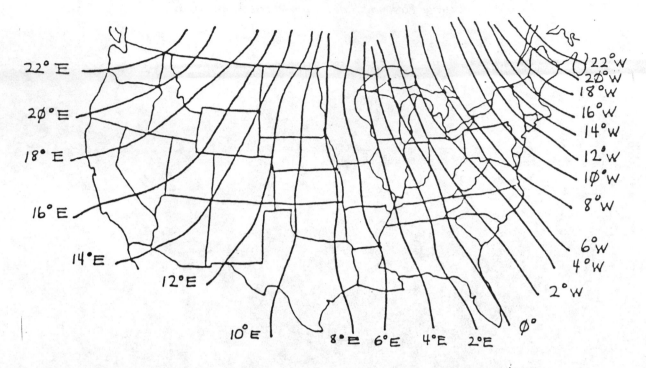

Figure indicates correction of compass reading to find true north. For example, in Washington state when your compass reads 22°E, it is pointing due north.

MAXIMUM NUMBER OF CONDUCTORS FOR A GIVEN CONDUIT SIZE

Conduit size	½"	¾"	1"	1¼"	1½"	2"
#12	10	18	29	51	70	114
#10	6	11	18	32	44	73
#8	3	5	9	16	22	36
#6	1	4	6	11	15	26
#4	1	2	4	7	9	16
#2	1	1	3	5	7	11
#1		1	1	3	5	8
#1/0		1	1	3	4	7
#2/0		1	1	2	3	6
#3/0		1	1	1	3	5
#4/0		1	1	1	2	4

Conductor size (row label, left side)

630

WATTAGE REQUIREMENTS FOR COMMON APPLIANCES

Definitions:

Volts = electric potential, similar to pressure in water systems

Amps = electrical current flow, similar to gallons/min. in water

Watts = rate of energy production or consumption = volts x amps, similar to speed (rate of movement; miles per hour)

Watt-hours = total power produced or consumed = watts x hours

Kilowatt = 1,000 watts; 1 kilowatt-hour (1 kWh) = 1,000 watt-hours

AC = Alternating current; electrical energy form as is available from utilities, generators, and inverters

DC = Direct current; electrical energy form as is stored in batteries

TYPICAL WATTAGE REQUIREMENTS FOR COMMON APPLIANCES

(figures in parentheses are additional starting wattage required when applicable)

Use the manufacturer's specs if possible, but be careful of nameplate ratings which are the highest possible electrical draw for that appliance.

Description	Watts

Refrigeration:

22 cu. ft. auto defrost (approximate run time 7-8 hours per day)	700(2200)
12 cu. ft. Sun Frost refrigerator (approximate run time 6-9 hrs. per day)	58(700)
Standard freezer (runs approximately 7-8 hrs. per day)	700(2200)
10 cu. ft. Sun Frost freezer (runs approximately 6–9 hrs. per day)	88(700)

Kitchen Appliances:

Dishwasher: cool dry	700(1400)
hot dry	1450(1400)
Trash compactor	1500(1500)
Can opener (electric)	100
Microwave (.5 cu. ft.)	1200
Microwave (.8 to 1.5 cu. ft.)	2100
Exhaust hood	144
Coffee maker	1200
Food processor	400
Toaster (2 slice)	1200
Coffee grinder	100
Blender	350
Food dehydrator	600
Mixer	120
Range, small burner	1250
Range, large burner	2100

Water Pumping:

AC Jet Pump (1/3 hp), 300 gal per hour, 20' well depth, 30 psi	750(1400)
AC Submersible Pump (1/2 hp) 40' well depth, 30 psi	1000(6000)
DC pump for house pressure system (typical use is 1-2 hrs. per day)	60
DC submersible pump (typical use is 6 hrs. per day)	50

Entertainment/Telephones:

TV (25-inch color)	170
TV (19 inch color)	80
TV (12-inch black & white)	15
Video Games (not incl. TV)	20
Satellite system, 12 ft. dish with auto orientation/remote control	45
VCR	30
Laser disk/CD player	30
AC powered Stereo (avg. volume)	55
AC Stereo, home theater	500
DC powered Stereo (avg. volume)	15
CB (receiving)	10
Cellular telephone (on standby)	20
Radio telephone (on standby)	25
Electric piano	30
Guitar amplifier (avg. volume)	40
(Jimi Hendrix)	8500

General Household:

Typical fluorescent light (60W equivalent)	15
Incandescent lights as indicated on bulb	
Electric clock	4
Clock radio	5
Electric blanket	400
Iron (electric)	1200
Clothes washer	1150
Dryer (gas)	500(1800)
Dryer (electric)	5750(1800)
Vacuum cleaner, average	900
Central vacuum	1500(1500)
Furnace fan: 1/4 hp	600(1000)
1/3 hp	700(1400)
1/2 hp	875(2350)
Garage door opener: 1/4 hp	550(1100)
Alarm/security system	6
Air conditioner	1500/ton or /10,000 BTU (2200)

Office/Den:

Computer/Modem	55
14" color monitor	100
14" monochrome monitor	25
Ink jet printer	35
Dot matrix printer	200
Laser printer	1200
Fax machine standby	10
printing	500
Electric typewriter	200
Adding machine	8
Electric pencil sharpener	100

Hygiene:

Hair dryer	1500
Waterpik	90
Whirlpool bath	750(1000)
Hair curler	750
Electric toothbrush (charging stand)	6

Shop:

Worm drive 7-1/4" saw	1800(3000)
AC table saw, 10"	1800(4500)
AC grinder, 1/2 hp	1080(2500)
Hand drill, 3/8"	400
Hand drill, 1/2"	600

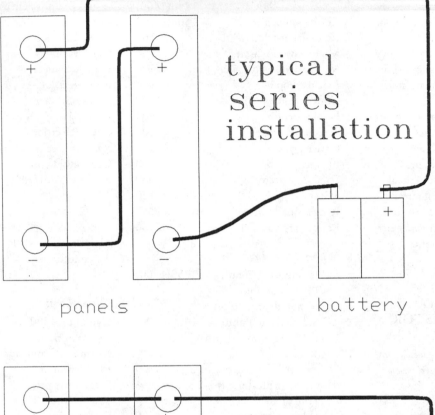

typical
series
installation

panels   battery

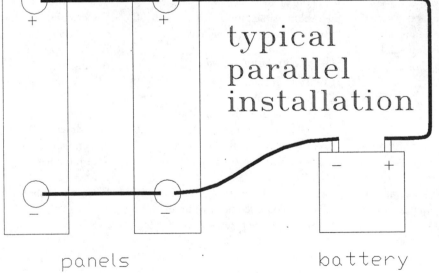

typical
parallel
installation

panels   battery

WIRE SIZING CHART/FORMULA

We could give you some incomprehensible voltage drop charts (like we've done in the past), but this all-purpose formula works better.

This chart is useful for finding the correct wire size for any voltage, length, or amperage flow in any AC or DC circuit. For most DC circuits, particularly between the PV modules and the batteries, we try to keep the voltage drop to 3% or less. There's no sense using your expensive PV wattage to heat wires. You want that power in your batteries!

Note that this formula doesn't directly yield a wire gauge size, but rather a 'VDI' number which is then compared to the nearest number in the VDI column, and then read across to the wire gauge size column.

1. Calculate the Voltage Drop Index (VDI) using the following formula:

$$VDI = AMPS \times FEET \div (\% \text{ VOLT DROP} \times VOLTAGE)$$

Amps = Watts Divided by Volts   Feet = One way wire distance
% Volt Drop = Percentage of voltage drop acceptable for this circuit (typically 2% to 5%)

2. Determine the appropriate wire size from the chart below.
    A. Take your VDI number you just calculated and find the nearest number in the VDI column, then read to the left for AWG wire gauge size.
    B. Be sure that your circuit amperage does not exceed the figure in the Ampacity column for that wire size. (This is not usually a problem in low voltage circuits.)

Wire Size AWG	Copper Wire VDI	Copper Wire Ampacity	Aluminum Wire VDI	Aluminum Wire Ampacity
0000	99	260	62	205
000	78	225	49	175
00	62	195	39	150
0	49	170	31	135
2	31	130	20	100
4	20	95	12	75
6	12	75	•	•
8	8	55	•	•
10	5	30	•	•
12	3	20	•	•
14	2	15	•	•
16	1	•	•	•

Example: Your PV array consisting of 4 Siemens PC4 modules is 60 ft. from your 12-volt battery. This is actual wiring distance, up pole mounts, around obstacles, etc. These modules are rated at 4.4 amps, times 4 modules = 17.6 amps maximum. We'll shot for a 3% voltage drop. So our formula looks like:

$$VDI = 29.3 \quad \frac{17.6 \times 60}{3[\%] \times 12 [V])}$$

Looking at our chart, a VDI of 29 means we'd better use #2 wire in copper, or #0 wire in aluminum. Hummm. Pretty big wire.

What if this system was 24-volt? The modules would be wired in series, so each <u>pair</u> of modules would produce 4.4 amps. Two pairs, times 4.4 amps = 8.8 amps max.

$$VDI = 7.3 \quad \frac{8.8 \times 60}{(3[\%] \times 24 [V])}$$

Wow! What a difference! At 24-volt input you could wire your array with little ol' #8 copper wire.

FRICTION LOSS CHARTS FOR WATER PUMPING

Friction Loss- PVC Class 160 PSI Plastic Pipe
Pressure loss from friction in psi per 100 feet of pipe.

Bold Numbers Indicate 5 Feet per Second Velocity

Flow GPM	1	1.25	1.5	2	2.5	3	4	5	6	8	10
1	0.02	0.01									
2	0.06	0.02	0.01								
3	0.14	0.04	0.02								
4	0.23	0.07	0.04	0.01							
5	0.35	0.11	0.05	0.02							
6	0.49	0.15	0.08	0.03	0.01						
7	0.66	0.20	0.10	0.03	0.01						
8	0.84	0.25	0.13	0.04	0.02						
9	1.05	0.31	0.16	0.05	0.02						
10	1.27	0.38	0.20	0.07	0.03	0.01					
11	1.52	0.45	0.23	0.08	0.03	0.01					
12	1.78	0.53	0.28	0.09	0.04	0.01					
14	2.37	0.71	0.37	0.12	0.05	0.02					
16	**3.04**	0.91	0.47	0.16	0.06	0.02					
18	3.78	1.13	0.58	0.20	0.08	0.03					
20	4.59	1.37	0.71	0.24	0.09	0.04	0.01				
22	5.48	1.64	0.85	0.29	0.11	0.04	0.01				
24	6.44	1.92	1.00	0.34	0.13	0.05	0.02				
26	7.47	2.23	1.15	0.39	0.15	0.06	0.02				
28	8.57	**2.56**	1.32	0.45	0.18	0.07	0.02				
30	9.74	2.91	1.50	0.51	0.20	0.08	0.02				
35		3.87	**2.00**	0.68	0.27	0.10	0.03				
40		4.95	2.56	0.86	0.34	0.13	0.04	0.01			
45		6.16	3.19	1.08	0.42	0.16	0.05	0.02			
50		7.49	3.88	1.31	0.52	0.20	0.06	0.02			
55		8.93	4.62	**1.56**	0.62	0.24	0.07	0.02			
60		10.49	5.43	1.83	0.72	0.28	0.08	0.03	0.01		
65			6.30	2.12	0.84	0.32	0.09	0.03	0.01		
70			7.23	2.44	0.96	0.37	0.11	0.04	0.02		
75			8.21	2.77	1.09	0.42	0.12	0.04	0.02		
80			9.25	3.12	1.23	0.47	0.14	0.05	0.02		
85			10.35	3.49	**1.38**	0.53	0.16	0.06	0.02		
90				3.88	1.53	0.59	0.17	0.06	0.03		
95				4.29	1.69	0.65	0.19	0.07	0.03		
100				4.72	1.86	**0.72**	0.21	0.08	0.03	0.01	
150				10.00	3.94	1.52	0.45	0.16	0.07	0.02	
200					6.72	2.59	**0.76**	0.27	0.12	0.03	0.01
250					10.16	3.91	1.15	0.41	0.18	0.05	0.02
300						5.49	1.61	**0.58**	0.25	0.07	0.02
350						7.30	2.15	0.77	0.33	0.09	0.03
400						9.35	2.75	0.98	0.42	0.12	0.04
450							3.42	1.22	**0.52**	0.14	0.05
500							4.15	1.48	0.63	0.18	0.06
550							4.96	1.77	0.76	0.21	0.07
600							5.82	2.08	0.89	0.25	0.08
650							6.75	2.41	1.03	0.29	0.10
700							7.75	2.77	1.18	0.33	0.11
750							8.80	3.14	1.34	**0.37**	0.13
800								3.54	1.51	0.42	0.14
850								3.96	1.69	0.47	0.16
900								4.41	1.88	0.52	0.18
950								4.87	2.08	0.58	0.20
1000								5.36	2.29	0.63	0.22
1500									4.84	1.34	0.46
2000										2.29	0.78
2500										3.46	1.18
3000											1.66

Friction Loss- Polyethylene (PE) SDR-Pressure Rated Pipe
Pressure loss from friction in psi per 100 feet of pipe.

Numbers in Bold indicate 5 Feet/Second Velocity

Flow GPM	0.5	0.75	1	1.25	1.5	2	2.5	3
1	0.49	0.12	0.04	0.01				
2	1.76	0.45	0.14	0.04	0.02			
3	3.73	0.95	0.29	0.08	0.04	0.01		
4	**6.35**	1.62	0.50	0.13	0.06	0.02		
5	9.60	2.44	0.76	0.20	0.09	0.03		
6	13.46	3.43	1.06	0.28	0.13	0.04	0.02	
7	17.91	4.56	1.41	0.37	0.18	0.05	0.02	
8	22.93	**5.84**	1.80	0.47	0.22	0.07	0.03	
9		7.26	2.24	0.59	0.28	0.08	0.03	
10		8.82	2.73	0.72	0.34	0.10	0.04	0.01
12		12.37	**3.82**	1.01	0.48	0.14	0.06	0.02
14		16.46	5.08	1.34	0.63	0.19	0.08	0.03
16			6.51	1.71	0.81	0.24	0.10	0.04
18			8.10	2.13	1.01	0.30	0.13	0.04
20			9.84	2.59	1.22	0.36	0.15	0.05
22			11.74	**3.09**	1.46	0.43	0.18	0.06
24			13.79	3.63	1.72	0.51	0.21	0.07
26			16.00	4.21	1.99	0.59	0.25	0.09
28				4.83	2.28	0.68	0.29	0.10
30				5.49	**2.59**	0.77	0.32	0.11
35				7.31	3.45	1.02	0.43	0.15
40				9.36	4.42	1.31	0.55	0.19
45				11.64	5.50	1.63	0.69	0.24
50				14.14	6.68	**1.98**	0.83	0.29
55					7.97	2.36	0.85	0.35
60					9.36	2.78	1.17	0.41
65					10.36	3.22	1.36	0.47
70					12.46	3.69	**1.56**	0.54
75					14.16	4.20	1.77	0.61
80						4.73	1.99	0.69
85						5.29	2.23	0.77
90						5.88	2.48	0.86
95						6.50	2.74	0.95
100						7.15	3.01	**1.05**
150						15.15	6.38	2.22
200							10.87	3.78
300								8.01

TEMPLATES

For best use of these templates please photocopy them at 100% (same size) on thicker paper.

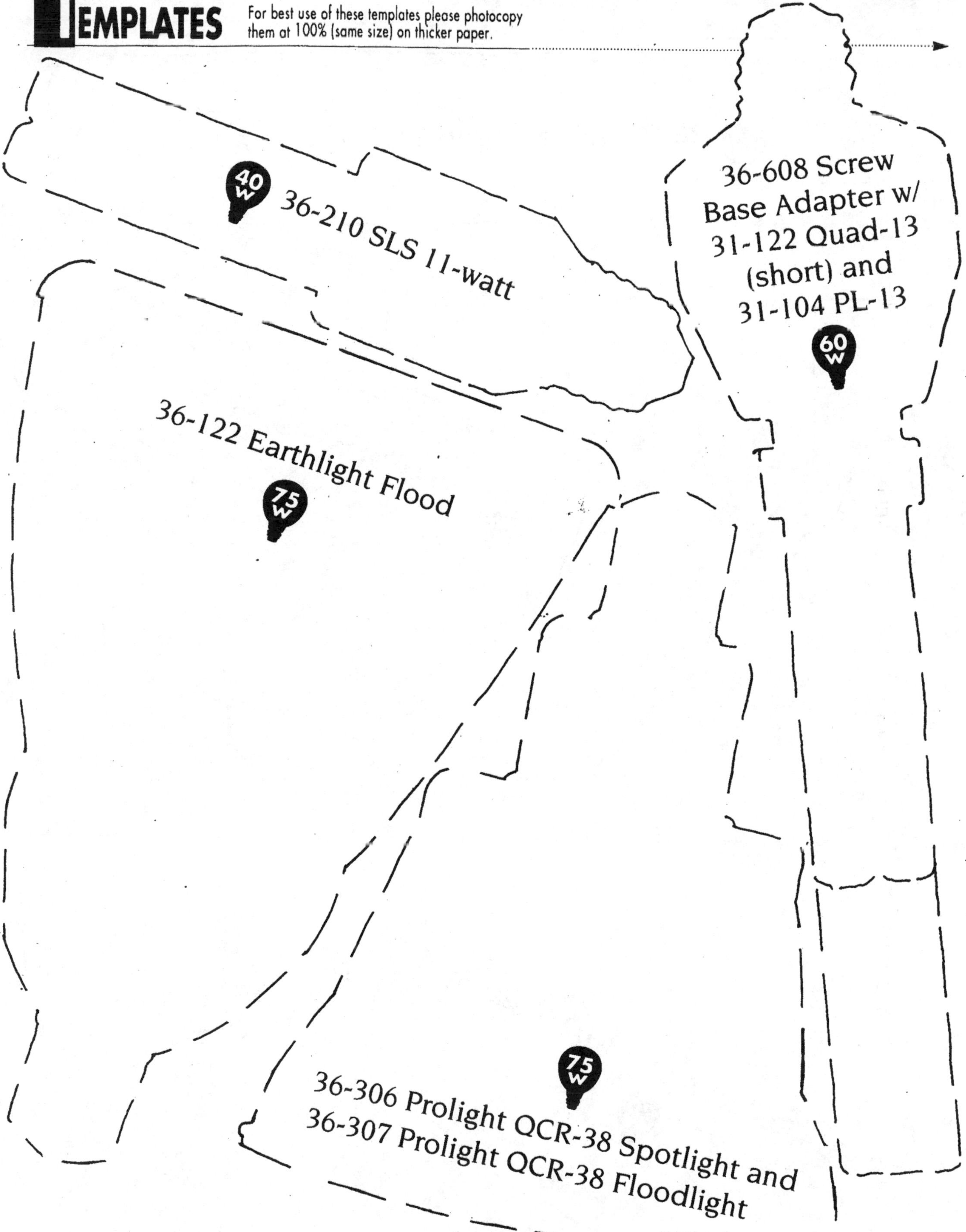

40 W 36-210 SLS 11-watt

36-608 Screw Base Adapter w/ 31-122 Quad-13 (short) and 31-104 PL-13

60 W

36-122 Earthlight Flood

75 W

36-306 Prolight QCR-38 Spotlight and
36-307 Prolight QCR-38 Floodlight

75 W

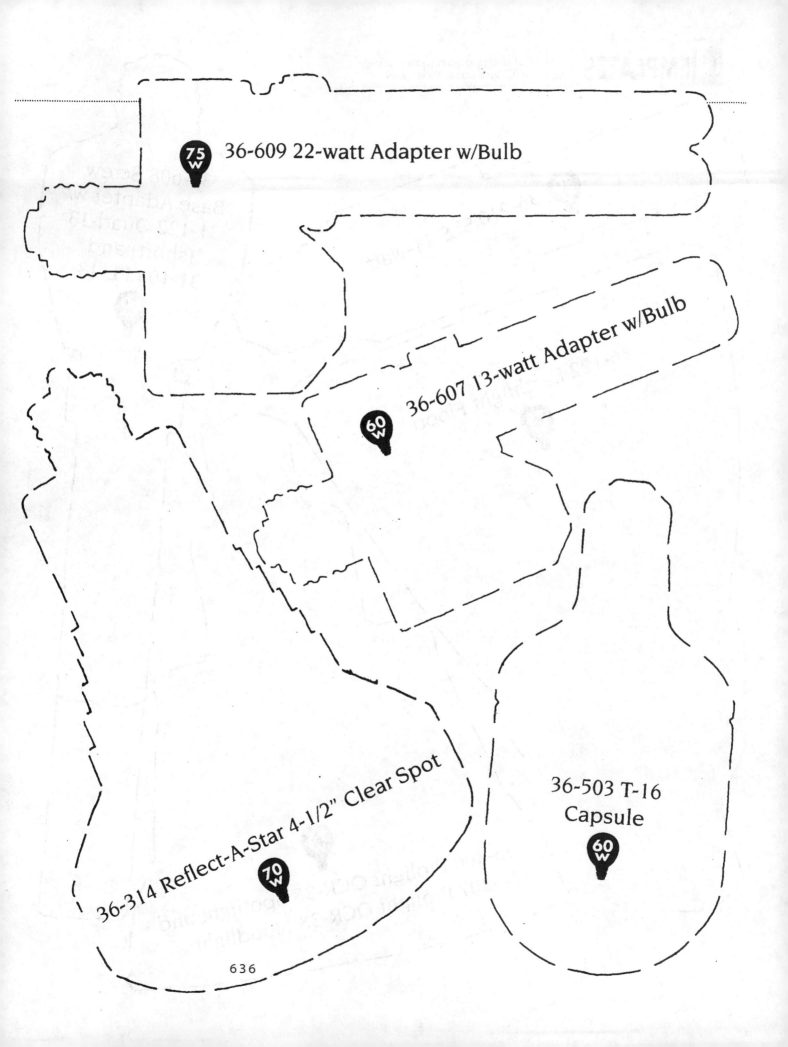

36-609 22-watt Adapter w/Bulb **75W**

36-607 13-watt Adapter w/Bulb **60W**

36-314 Reflect-A-Star 4-1/2" Clear Spot **70W**

36-503 T-16 Capsule **60W**

636

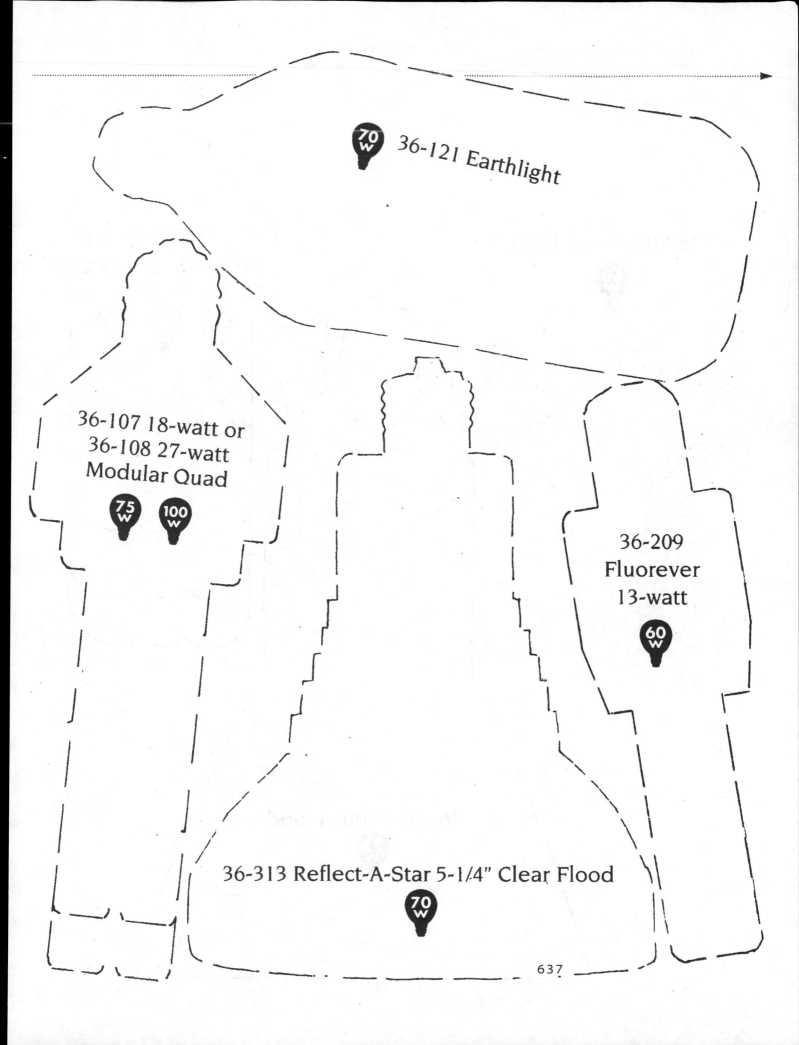

36-121 Earthlight

36-107 18-watt or
36-108 27-watt
Modular Quad

36-209
Fluorever
13-watt

36-313 Reflect-A-Star 5-1/4" Clear Flood

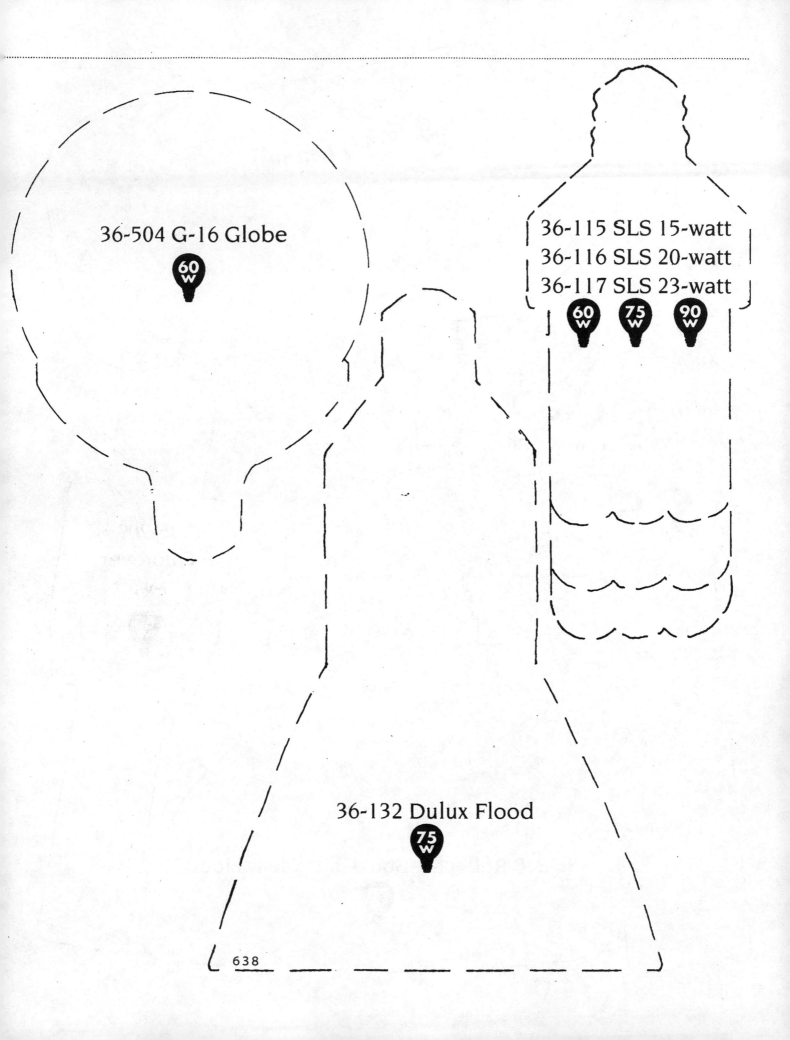

36-504 G-16 Globe
60W

36-115 SLS 15-watt
36-116 SLS 20-watt
36-117 SLS 23-watt
60W 75W 90W

36-132 Dulux Flood
75W

THE REAL GOODS ENVIRONMENTAL RESOURCE LIST

Although the *Sourcebook* is called "the most complete source of renewable energy and environmental products" by many, we can't be everywhere at once. Here is our current list of trusted outside resources, organized in categories. We do our best to pick the best organizations for the listing, and since we have worked directly with many of them, we are giving you the benefit of our experience. Do we need to say that we do not endorse all the actions of each group listed, nor are we responsible for what they say or do? Consider it said.

This list can never be complete. We apologize for obvious resources that we have forgotten, and urge you to send us names of organizations you think we should know about: a brief description of the organization to *Real Goods Eco-Desk, Attention: Resource List* will suffice.

OTHER ENVIRONMENTAL RESOURCE GUIDES

Interior Concerns Resource Guide
Box 2386
Mill Valley, CA 94942
415-389-8049 fax -388-8322 E mail-Intconc@NBN.com
Encyclopedic listing of sources for recycled and non-toxic building and decorating materials
WEB:http:/www.Numenet/intconc

RESEARCH ORGANIZATIONS

Ecology Action, John Jeavons
5798 Ridgewood Road
Willits, CA 95490
707-459-0150
Researches methods of growing food in small spaces.

The National Renewable Energies Laboratory (NREL)
1617 Cole Blvd.
Golden, CO 80401
303-275-3000
Tests products, conducts experiments, and provides information on renewable energy. E-mail-www.NREL.GOV

The Rocky Mountain Institute
1739 Snowmass Creek Road
Old Snowmass, CO 81654-9199
970-927-3851 fax: -4178 E-mail- Orders@RmI.ORG
Information and studies on energy efficiency and sustainable technology.

The Union of Concerned Scientists
2 Brattle St.
Cambridge, MA 02238
617-547-5552 E-mail-USC@IGC.APC.ORG
Organization of scientists and citizens concerned with the impact of advanced technology on society. Programs focus on energy policy and national security.

ENVIRONMENTAL CLEANUP, TESTING, ANALYZING, AND CONSULTING SERVICES

Ecology and Environment, Inc.
Buffalo Corporate Center
368 Pleasantview Dr.
Lancaster, NY 14086-1397
716-684-8060
Environmental auditing services; air, water, and groundwater monitoring; analytical laboratory services; emergency spill response; several offices covering the globe.

CALMAX
916-255-2369 or 800-553-2962
California materials exchange for re-use and recycling

ENVIRONMENTAL COMPUTER NETWORKS

Daily Planet BBS
sysop@tdp.org
808-572-4857
Fidonet, K12net, free Internet mail

Econet
18 DeBoom St.
San Francisco, CA 94107
415-442-0220 to sign up

Environet
Greenpeace
568 Howard St. 3rd fl.
San Francisco, CA 94105
415-512-9025
Access 415-512-9108

Electric Ideas Clearinghouse BBS
Western Area Power Authority
800-PWR-PLUG from 16 western states
(800-797-7584 for you numerate dialers)

Home Power Magazine BBS
707-822-8640

EDUCATIONAL ORGANIZATIONS

Environmental Careers Organization, Inc.
286 Congress St.
Boston, MA 02210
617-426-4375 E-mail@eco.ORG
Offers information about environmental careers and jobs.

The Institute For Independent Living
c/o Real Goods Trading Corporation
555 Leslie St.
Ukiah, CA 95482
800-762-7325
Offers workshops and classes on renewable energy and
sustainable living.

International Institute for Bau-Biologie & Ecology
P.O. Box 387
Clearwater, FL 34615
813-461-4371
Correspondence course on how to create healthy environ-
ments

Solar Energy International
P.O. Box 715
Carbondale, CO 81623-0715
970-963-8855 E-mailSEI@Solarenergy.ORG
Teaches classes on sustainable technology.

RENEWABLE ENERGY ORGANIZATIONS AND INFORMATION SERVICES

The American Hydrogen Association
216 S. Clark Dr. Ste. 103
Tempe, AZ 85281
602-921-0433E-mail - AHA@GETNET.COM
Promotes the use of Hydrogen for fuel and energy storage.
Publishes "Hydrogen Today."

The American Solar Energy Society
2400 Central Avenue, Unit G-1
Boulder, CO 80301
303-443-3130 E-mail-ASES@ASES.ORG
USA's member of the of the International Solar Energy
Society. Publishes "Solar Today" and sponsors a major solar
convention every year.

The American Wind Energy Association
122 C St NW 4th Fl
20001-2109
202-383-2500

Appropriate Technology Transfer for Rural Areas (ATTRA)
University of Arkansas
P.O. Box 3657
Fayetteville, AR 72702
800-346-9140

Energy Efficiency & Renewable Energy Clearinghouse
(EREC)
P.O. Box 3048
Merrifield, VA 22116
800-523-2929 E-mailhttp://www.eren.deo.gov
The renewable energy info arm of the DOE. Publications on
energy conservation and renewable energy. A terrific source
of info on a wide variety of subjects.

The Florida Solar Energy Research Center
1679 Clearlake Rd.
Cocoa, Fl. 32922
407-638-1000
Publishes up to date information about solar energy and
energy conservation.

ENVIRONMENTAL ORGANIZATIONS

The Alliance to Save Energy
1725 K St. NW, Ste. 509
Washington, DC 20006-1401
202-857-0666 E-mail@ASE.ORG
Energy efficiency through research, demonstration projects,
lobbying, and education.

The Audubon Society
700 Broadway
New York, NY 10003
212-979-3000
Rational strategies for energy development, and protects life
from pollution, radiation, and toxic substances. Publishes
"Audubon" Magazine.

CERES
711 Atlantic Ave.
Boston, MA 02111
617-451-0927
The coalition for Environmentally Responsible Economics is
a non-profit org dedicated to setting standards for
environmental responsibility within the corporate world.

Defenders of Wildlife
1101 14th St. NW, Ste. 1400
Washington, DC 20005
202-682-9400
Preservation, enhancement, and protection of wildlife.

Earth First!
P.O. Box 1415
Eugene, OR 97440
503-741-9191 E-mail-earthfirst@IGC.ADC.ORG
Eco-activist grassroots organization.

Earth Island Institute
300 Broadway, Ste. 28
San Francisco, CA 94133
415-788-3666 E-mail-WWW,earthisland.ORG/EI/
Environmental organizing and publishing network.

The Environmental Defense Fund
257 Park Ave. South
New York, NY 10010
212-505-2100 E-mail MEMBERS@EDEF.ORG
An organization committed to combining the efforts of
scientists, economists, and attorneys to devise solutions to
environmental problems.
Greenpeace USA
1436 U St. NW
Washington, DC 20009
202-462-1177
An international non-profit organization dedicated to
protecting the environment.

The Mendocino Environmental Center
106 W. Standley
Ukiah, CA 95482
707-468-1660 E-mail-http://www.Mec@Pacific.Net
Grassroots environmental resource organization.

The Natural Resources Defense Council
40 West 20th St.
New York, NY 10011
Promotes the protection of natural resources by combining
legal action, scientific research, and citizen action.

The Northeastern Sustainable Energy Association (NESEA)
50 Miles St
Greenfield, MA 01301
413-774-6051 E-mail-NESEA@NESEA.ORG
Primary focus on development of electric vehicles.

PACE (People's Action for Clean Energy)
101 Lawton Rd.
Canton, CT 06019-2209
860-693-4813 860-693-2822 Fax #
Local action group with periodic home tours

Public Citizen
315 Circle Ave. #2
Takoma Park, MD 20912
301-270-2258
Sponsors of the Sun Day Campaign

The Rainforest Action Network
450 Sansome St., Ste. 700
San Francisco, CA 94111
415-398-4404 E mail Rainforest@Ran.Org
Website http\\www.Ran.ORG/Ran/
Dedicated to saving our vital rainforests.

The Redwood Alliance, Michael Welch
761 8th. St. #4/Box 293
Arcata, CA 955518
707-822-7884 E-mail Redwood.Alliance@homepower.org

Renew America
1400 16th St.. NW, Ste. 710
Washington, DC 20036-2217
202-232-2252
Membership org that promotes public involvement in
creation of a sustainable future. They track and report on
outstanding environmental programs and policies across the
country.

Save America's Forests
4 Library Court, SE
Washington, DC 20003
202-544-9219
A nationwide coalition of grassroots organizations working
to pass laws to protect our forest ecosystems.

Sea Shepherd
3107A Washington Blvd.
Marina del Rey, CA 90292
310-394-3198 E-mail NVOTH@IGC.APC.ORG
Dedicated defenders of the ocean environment.

The Solar Energy Expo and Rally (SEER)
733 S. Main St. #234
Willits, CA 95490
707-459-1256 459-0366 Fax E-maiREDI@Pacific.Net
Sponsors solar fair every other year in Northern California.

Zero Population Growth
1400 Sixteenth St., NW
Washington, DC 20036
202-332-2200 E-mail ZBG@IGC.APC.ORG
Addresses population concerns.

SUSTAINABLE PRODUCTS AND CATALOGS

Ag Access
603 4th St.
Davis, CA 95616
916-756-7177 E-mail agaccess@Davis.Com
Web http::/www.mother.Com/Agaccess
A great catalog for books on sustainable agriculture.

641

Business Ethics Network Catalog
52 South Tenth St., Ste. 110
Minneapolis, MN 55403
612-962-4700
A toolbox for socially responsible businesses.

Chelsea Green Publishing Company
P.O. Box 428
White River Junction, VT 05001
or 10 Water St Rm 310
Lebanon, NH 03766
(warehouse)
800-639-4099
Books for sustainable living.

Dripworks
380 Maple St.
Willits, CA 95490
800-522-3747 E-mail -DRIPWRKS@Pacific.NET
Broad selection of drip and garden supplies.

Earth Care Paper Company
Ukiah, CA 95482-8507
800-347-0070
Excellent selection of recycled paper products.

Harmony Farm Supply
P.O. Box 460
Graton, CA 95444
707-823-9125
Also offers drip irrigation and gardening supplies.

Gardener's Supply Company
128 Intervale Rd.
Burlington, VT 05401
802-863-1700 E-mail Gardners@Gardners.Com
Fine gardening supplies and accessories.

Music For Little People
P.O. Box 1460
Redway, CA 95560
800-727-2233
Catalog of great children's music for peace.

Rhino Records
10635 Santa Monica Blvd.
Los Angeles, CA 90025
310-474-4778
A great source for music of all types.

Seeds of Change
PO Box 15700
Santa Fe, NM 87506
505-438-8080
Certified organic seeds and precious heirloom seeds.

ENVIRONMENTAL BUILDING

Energy Conserving Passive Solar Houses
Drawing Room Graphic Services
Box 86627
North Vancouver, BC V7L 4L2
604-689-1841 E-mail-Solplan@Cyberstore.ca
Provides house plans for passive solar homes.

Energy Rated Homes of America
5401 JFK Blvd. STEI
N Little Rock, AR 72116
501-771-2299
Nonprofit organization developing a uniform energy rating system designed to give the shelter industry a common method for rating potential energy efficiency.

Future Deck
P.O. Box 34321., Stn. D
Vancouver, BC V6J 4P3
Recycled plastic lumber products.

National Design Assistance Center
Center for Resourceful Building Technology
P.O. Box 100
Missoula MT 59806
406-728-1864 or -549-7678 (Demo home)
E-mail CRBT@Montana.Com
Non-profit information and consultation on green shelter design

Building Educational Center
812 Page St.
Berkeley, CA 94702
510-525-7619

Passive Solar Environments (Paino & Assoc.)
821 W. Main St.
Kent, OH 44240
216-673-7449
Passive solar house design and planning.

Phoenix Recycled Plastics
225 Washington St.
Conshohocken, PA 19428
610-940-1590
Lumber products made out of recycled plastic.

Yestermorrow School
RR1 Box 97-5
Warren, VT 05674
802-496-5545 E-mail YMSCHOOL@AOL.COM
Classes on building and designing your house.

Southface Energy Institute
P.O. Box 5506
Atlanta, GA 30307
404-525-7657 E-mail-Tech@Southface.org
Provides assistance for owner-builders interested in passive
solar.

ELECTRIC VEHICLE CLUBS, PRODUCTS, AND
SERVICES

Electric Auto Assoc.
2710 St.. Giles Lane
Mountain View, CA 94040
415-685-7580   Fax: 415-306-0137
The largest consumer/user organization for electric vehicles.
$35 annual dues with excellent monthly news letter.

Electro Automotive
P.O. Box 1113
Felton, CA 95018
408-429-1989
Classes and parts for electric vehicles.

GOVERNMENT AGENCIES

The United States Department of Energy
1000 Independence Ave. SW
Washington, DC 20585
202-586-6210

The United States Environmental Protection Agency
401 M St. SW
Washington, DC 20460
202-260-2090

Your Senator
The Senate Office Building
Washington, DC 20510

Your Representative
The House Office Building
Washington, DC 20515
The Capitol Switchboard
202-224-3121

Please contact the President and your Representatives if you
have something on your mind. They need to know your
opinion! And please vote!

GLOSSARY

A

AC: Alternating Current, electricity which changes voltage periodically, typically sixty times a second (or fifty in Europe). This kind of electricity is easier to move.

activated stand life: the period of time, at a specified temperature, that a battery can be left stored in the charged condition before its capacity fails

active solar: any solar scheme employing pumps and controls which use power while harvesting solar energy

A-frame: a building which looks like the capital letter A in cross-section

air lock: two doors with space between, like a mud room, to keep the weather outside

alternating current: AC electricity which changes voltage periodically, typically sixty times a second

alternative energy: "voodoo" energy not purchased from a power company, usually coming from photovoltaic, micro-hydro, or wind

ambient: the prevailing temperature, usually outdoors

amorphous silicon: a type of PV cell manufactured without a crystalline structure. Compare with single-crystal and multi- (or poly-) crystalline silicon.

ampere: an instantaneous measure of the flow of electric current; abbreviated and more commonly spoken of as an "amp"

amp-hour: a one-ampere flow of electrical current for one hour; a measure of electrical quantity; two 60-watt 120-volt bulbs burning for one hour consume one amp-hour.

angle of incidence: the angle at which a ray of light (usually sunlight) strikes a planar surface (usually of a PV module). Angles of incidence close to perpendicularity (90°) are desirable.

anode: the positive electrode in an electrochemical cell (battery) toward which current flows; the earth ground in a cathodic protection system

antifreeze: a chemical, usually liquid and often toxic, which keeps things from freezing

array: an orderly collection, usually of photovoltaic modules connected electrically and mechanically secure

array current: the amperage produced by an array in full sun

avoided cost: the amount utilities must pay for independently-produced power; in theory, this was to be the whole cost, including capital share to produce peak-demand power, but over the years supply-side weaseling redefined it to be something more like the cost of the fuel the utility avoided burning.

azimuth: horizontal angle measured clockwise from true north; the equator is at 90°.

B

back-up: a secondary source of energy to pick up the slack when the primary source is inadequate. In alternatively-powered homes, fossil fuel generators are often used as "back-ups" when extra power is required to run power tools or when the primary source — sun, wind, water — are not providing sufficient energy.

Balance Of System: (BOS) equipment that controls the flow of electricity during generation and storage

baseline: a statistical term for a starting point; the "before" in a before-and-after energy conservation analysis

baseload: the smallest amount of electricity required to keep utility customers operating at the time of lowest demand; a utility's minimum load

battery: a collection of cells which store electrical energy; each cell converts chemical energy into electricity or vice versa, and is interconnected with other cells to form a battery for storing useful quantities of electricity.

battery capacity: the total number of ampere-hours that can be withdrawn from a fully charged battery usually over a standard period

battery cycle life: the number of cycles that a battery can sustain before failing

berm: earth mounded in an artificial hill

bioregion: an area, usually fairly large, with generally homogeneous flora and fauna

biosphere: the thin layer of water, soil, and air which supports all known life on earth

black water: what gets flushed down the toilet

blocking diode: a diode which prevents loss of energy to an inactive PV array (rarely used with modern charge controllers)

boneyard: a peculiar location at Real Goods where "experienced" products may be had for ridiculously low prices

Btu: British thermal unit, the amount of heat required to raise the temperature of one pound of water one degree Fahrenheit. 3,411 Btus equals one kilowatt-hour.

bus bar: the point where all energy sources and loads connect to each other; often a metal bar with connections on it

bussbar cost: the average cost of electricity delivered to the customer's distribution point

buy-back agreement or contract: an agreement between the utility and a customer that any excess electricity generated by the customer will be bought back for an agreed-upon amount

C

cathode: the negative electrode in an electrochemical cell

cell: a unit for storing or harvesting energy. In a battery, a cell is a single chemical storage unit consisting of trodes and electrolyte, typically producing 1.5 volts; several cells are usually arranged inside a single container called a battery. Flashlight batteries are really flashlight cells. A photovoltaic cell is a single assembly of doped silicon and electrical contacts that allow it to take advantage of the photovoltaic effect, typically producing .5 volts; several PV cells are usually connected together and packaged as a module.

CF: compact fluorescent, a modern form of light bulb using an integral ballast

CFCs: Chlorinated Fluoro-Carbons, an industrial solvent and material widely used until implicated as a cause of ozone depletion in the atmosphere

charge controller: device for managing the charging rate and state of charge of a battery bank

controller terminology:

adjustable set point: allows adjustment of voltage disconnect levels

high voltage disconnect: the battery voltage at which the charge controller disconnects the batteries from the array to prevent overcharging

low voltage disconnect: the voltage at which the controller disconnects the batteries to prevent over-discharging

low voltage warning: a buzzer or light that indicates low battery voltage

maximum power tracking: a circuit which maintains array voltage for maximal current

multistage controller: a unit which allows multilevel control of battery charging or loading

reverse current protection: prevents current flow from battery to array

single-stage controller: a unit with only one level of control for charging or load control

temperature compensation: a circuit that adjusts setpoints to ambient temperature in order to optimize charge

charge rate: the rate at which a battery is recharged, expressed as a ratio of battery capacity to charging current flow

clearcutting: a forestry practice, cutting all trees in a relatively large plot

cloud enhancement: the increase in sunlight due to direct rays plus refracted or reflected sunlight from partial cloud cover

compact fluorescent: a modern form of light bulb using an integral ballast

compost: the process by which organic materials break down, or the materials in the process of being broken down

concentrator: mirror or lens-like additions to a PV array which focus sunlight on smaller cells; a very promising way to improve PV yield

conductance: a material's ability to allow electricity to flow through it; gold has very high conductance.

conversion efficiency: the ratio of energy input to energy output across the conversion boundary. For example, batteries typically are able to store and provide 90% of the charging energy applied, and are said to have a 90% energy efficiency.

cookie-cutter houses: houses all-alike and all-in-a-row. Daly City, south of San Francisco, is a particularly depressing example. Runs in direct contradiction of our third guiding principle, "encourage diversity."

core/coil-ballasted: the materials-rich device required to drive some fluorescent lights; usually contains Americium (see electronic ballast)

cost-effectiveness: an economic measure of the worthiness of an investment; if an innovative solution costs less than a conventional alternative, it is more cost-effective.

cross section: a "view" or drawing of a slice through a structure

cross-ventilation: an arrangement of openings allowing wind to pass through a structure

crystalline silicon: the material from which most photovoltaic cells are made; in a single crystal cell, the entire cell is a slice of a single crystal of silicon, while a multi-crystalline cell is cut from a block of smaller (centimeter-sized) crystals. The larger the crystal, the more exacting and expensive the manufacturing process.

cut-in: the condition at which a control connects its device

cut-out: the condition at which a control interrupts the action

cutoff voltage: of a charge controller, the voltage at which the array is disconnected from the battery to prevent over-charging

cycle: in a battery, from a state of complete charge through discharge and recharge back to a fully-charged state

D

days of autonomy: the length of time (in days) that a system's storage can supply normal requirements without replenishment from a source, also called days of storage

DC: Direct Current, the complement of AC, or alternating current, presents one unvarying voltage to a load

deep cycle: a battery type manufactured to sustain many cycles of deep discharge, in excess of 50% of total capacity

degree-days: a term used to calculate heating and cooling loads; the sum, taken over an average year, of the lowest (for heating) or highest (for cooling) ambient daily temperatures. Example, if the target is 68° and the ambient on a given day is 58°, this would account for ten degree-days.

design month: the month in which the combination of insolation and loading require the maximum array output

depth of discharge: the percent of rated capacity that has been withdrawn from the battery; also called DOD

diode: an electrical component which permits current to pass in only one direction

elecdirect current: the complement of AC, or alternating current, presents one unvarying voltage to a load

discharge: electrical term for withdrawing energy from a storage system

discharge rate: the rate at which current is withdrawn from a battery expressed as a ratio to the battery's capacity; also known as C rate

disconnect: a switch or other control used to complete or interrupt an electrical flow between components

doping, dopant: small, minutely controlled amounts of specific chemicals introduced into the semiconductor matrix to control the density and probabilistic movement of free electrons

downhole: a piece of equipment, usually a pump, that is lowered down the hole (the well or shaft) to do its work

drip irrigation: a technique which precisely delivers measured amounts of water through small tubes; an exceedingly efficient way to water plants

dry cell: a cell with captive electrolyte

duty cycle: the ratio between active time and total time; used to describe the operating regime of an appliance in an electrical system

duty rating: the amount of time that an appliance can be run at its full rated output before failure can be expected

E

earthship: a rammed-earth structure based on tires filled with tamped earth; the term was coined by Michael Reynolds.

Eco Desk: an ecology information resource maintained by many ecologically-minded companies including Real Goods

edison base: a bulb base designed by (a) Enrico Fermi, (b) John Schaeffer, (c) Thomas Edison, (d) none of the above. The familiar standard residential lightbulb base.

efficiency: a mathematical measure of actual as a percentage of the theoretical best. See conversion efficiency.

energy-efficient: one of the best ways to use energy to accomplish a task; for example, heating with electricity is never energy efficient, while lighting with compact fluorescents is.

electrolyte: the chemical medium, usually liquid, in a battery which conveys charge between the positive and negative electrodes; in a lead-acid battery, the electrodes are lead and the electrolyte is acid.

electromagnetic radiation: EMR, the invisible field around an electric device. Not much is known about the effects of EMR, but it makes many of us nervous.

electronic ballast: an improvement over core/coil ballasts, used to drive compact fluorescent lamps; contains no radioactivity

embodied: of energy, meaning literally the amount of energy required to produce an object in its present form. Example: an inflated balloon's embodied energy includes the energy required to blow it up.

EMR: electromagnetic radiation, the invisible field around an electric device. Not much is known about the effects of EMR, but it makes many of us nervous.

energy density: the ratio of stored energy to storage volume or weight

equalizing: periodic overcharging of batteries to make sure that all cells are reaching a good state of charge

externalities: considerations, often subtle or remote, which should be accounted for when evaluating a process or product, but usually are not. For example, externalities for a power plant may include down-wind particulate fallout and acid rain, damage to life-forms in the cooling water intake and effluent streams, and many other factors.

F

fail-safe: a system designed in such a way that it will always fail into a safe condition

feng shui: an Oriental system of placement which pays special attention to wind, water, and the cardinal directions

ferro-cement: a construction technique, an armature of iron contained in a cement body, often a wall, slab, or tank

fill factor: of a photovoltaic module's I-V (current/voltage) curve, this number expresses the product of the open circuit voltage and the short-circuit current, and is therefore a measure of the "squareness" of the I-V curve's shape.

firebox: the structure within which combustion takes place

fixed-tilt array: a PV array set in a fixed position

flat-plate array: a PV array consisting of nonconcentrating modules

float charge: a charge applied to a battery equal to or slightly larger than the battery's natural tendency to self-discharge

FNC: Fiber-Nickel-Cadmium, a new battery technology

frequency: of a wave, the number of peaks in a period. For example, alternating current presents sixty peaks per second, so its frequency is sixty hertz. Hertz is the standard unit for frequency when the period in question is one second.

G

gassifier: a heating device which burns so hotly that the fuel sublimes directly from its solid to its gaseous state and burns very cleanly

gassing: when a battery is charged gasses are often given off; also called out-gassing

golf cart batteries: industrial batteries tolerant of deep cycling, often used in mobile vehicles

gotcha!s: an unexpected outcome or effect, or the points at which, no matter how hard you wriggle, you can't escape

gravity-fed: water storage far enough above the point of use (usually 50 feet) so that the weight of the water provides sufficient pressure

graywater: all other household effluents besides black water (toilet water); gray water may be reused with much less processing than black water

groundwater: as distinct from water pumped up from the depths, groundwater is runoff from precipitation, agriculture, or other sources.

grid: a utility term for the network of transmission lines that distribute electricity from a variety of sources across a large area

grid-connected system: a house, office, or other electrical system that can draw its energy from the grid; although usually grid power-consumers, grid-connected systems can provide power to the grid.

H

heat exchanger: device that passes heat from one substance to another; in a solar hot water heater, for example, the heat exchanger takes heat harvested by a fluid circulating through the solar panel and transfers it to domestic hot water.

high tech glass: window constructions made of two sheets of glass, sometimes treated with a metallic deposition, sealed together hermetically, with the cavity filled by an inert gas and, often, a further plastic membrane. High Tech glass can have an R-Value as high as 10.

homeschooling: educating children at home instead of entrusting them to public or private schools; a growing trend quite often linked to alternatively-powered homes

homestead: the house and surrounding lands

homesteaders: people who consciously and intentionally develop their homestead

house current: in the United States, one hundred seventeen volts root mean square of alternating current, plus or minus seven volts; nominally one-ten volt power; what comes out of most wall outlets

hot tub: a quasi-religious object in California; a large bathtub for several people at once; an energy hog of serious proportions

HVAC: Heating, Ventilation, and Air Conditioning; space conditioning

hydro turbine: a device which converts a stream of water into rotational energy

hydrometer: tool used to measure the specific gravity of a liquid

hydronic: contraction of hydro and electronic, usually applied to radiant in-floor heating systems and their associated sensors and pumps

hysteresis: the lag between cause and effect, between stimulus and response

I

incandescent bulb: a light source that produces light by heating a filament until it emits photons — quite an energy-intensive task

incident solar radiation: or insolation, the amount of sunlight falling on a place

indigenous plantings: gardening with plants native to the bioregion

Inductive transformer / rectifier: the little transformer device that powers many household appliances; an "energy criminal" that takes an unreasonably large amount of alternating electricity (house current) and converts it into a much smaller amount of current with different properties, for example, much lower voltage direct current

infiltration: air, at ambient temperature, blowing through cracks and holes in a house wall and spoiling the space conditioning

infrared: light just outside the visible spectrum, usually associated with heat radiation

infrastructure: a buzz word for the underpinnings of civilization, roads, water mains, power and phone lines, fire suppression, ambulance, education, and governmental services are all infrastructure. "Infra," is Latin for beneath. In a more technical sense, the repair infrastructure is local existence of repair personnel and parts for a given technology.

insolation: a word coined from incident solar radiation, the amount of sunlight falling on a place

insulation: a material which keeps energy from crossing from one place to another. On electrical wire, it is the plastic or rubber that covers the conductor. In a building, insulation makes the walls, floor, and roof more resistant to the outside (ambient) temperature.

Integrated Resource Planning: an effort by the utility industry to consider all resources and requirements in order to produce electricity as efficiently as possible

interconnect: to connect two systems, often an independent power producer and the grid; see also intertie

interface: the point where two different flows or energies interact; for example, a power system's interface with the human world is manifested as meters, which show system status, and controls, with which that status can be manipulated.

internal combustion engines: gasoline engines, typically in automobiles, small standalone devices like chainsaws and lawnmowers, and generators

inverter: the electrical device that changes direct current into alternating current

intertie: the electrical connection between an independent power producer — for example, a PV-powered household — and the utility's distribution lines, in such a way that each can supply or draw from the other.

irradiance: the instantaneous solar radiation incident on a surface; usually expressed in Langleys (a small amount) or in kilowatts per square meter. The definition of "one sun" of irradiance is one kilowatt per square meter.

irreverence: the measure, difficult to quantify, of the seriousness of Real Goods techs when talking with utility suits

IRP: Integrated Resource Planning: an effort by the utility industry to consider all resources and requirements in order to produce electricity as efficiently as possible

I-V curve: a plot of current against voltage to show the operating characteristics of a photovoltaic cell, module, or array

K

kilowatt: one thousand watts, a measure of instantaneous work. Ten one-hundred watt bulbs require a kilowatt of energy to light up

kilowatt-hour: the standard measure of household electrical energy. If the ten bulbs left unfrugally burning in the preceding example are on for an hour, they consume one kilowatt-hour of electricity.

L

landfill: another word for dump

lead-acid: the standard type of battery for use in home energy systems and automobiles

life: You expect an answer here? Well, when speaking of electrical systems, this term is used to quantify the time the system can be expected to function at or above a specified performance level.

life-cycle cost: the estimated cost of owning and operating as system over its useful life

line extensions: what the power company does to bring their power lines to the consumer

line-tied system: an electrical system connected to the powerlines, usually having domestic power generating capacity and the ability to draw power from the grid or return power to the grid, depending on load and generator status

load: an electrical device, or the amount of energy consumed by such a device

load circuit: the wiring that provides the path for the current that powers the load

load current: expressed in amps, the current required by the device to operate

Langley: the unit of solar irradiance; one gram-calorie per square centimeter

low pressure: usually of water, meaning that the head, or pressurization, is relatively small

low-emissivity: applied to high tech windows, meaning that infra-red or heat energy will not pass back out through the glass

low-flush: a toilet using a smaller amount (usually about 6 quarts) of water to accomplish its function

low-voltage: usually another term for 12- or 24-volt direct current

M

maintenance free battery: a battery to which water cannot be added to maintain electrolyte volume. All batteries require routine inspection and maintenance.

maximum power point: the point at which a power conditioner continuously controls PV source voltage in order to hold it at its maximum output current

ME: Mechanical Engineer; the engineers who usually work with heating and cooling, elevators, and the other mechanical devices in a large building

meteorological: pertaining to weather; meteorology is the study of weather

micro-climate: the climate in a small area, sometimes as small as a garden or the interior of a house. Climate is distinct from weather in that it speaks for trends taken over a period of at least a year, while weather describes immediate conditions.

Micro-hydro: small hydro (falling water) generation

millennia: one thousand years

milliamp: one thousandth of an ampere

module: a manufactured panel of photovoltaic cells. A module typically houses 36 cells in an aluminum frame covered with a glass or acrylic cover, organizes their wiring, and provides a junction box for connection between itself, other modules in the array, and the system.

N

naturopathic: a form of medicine devoted to natural remedies and procedures

net metering: a desirable form of buy-back agreement in which the line-tied house's electric meter turns in the utility's favor when grid power is being drawn, and in the system owner's favor when the house generation exceeds its needs and electricity is flowing into the grid. At the end of the payment period, when the meter is read, the system owner pays (or is paid by) the utility depending on the net metering.

NEC: the National Electrical Code, guidelines for all types of electrical installations including (since 1984) PV systems

nicad: slang for nickel-cadmium, a form of chemical storage often used in rechargeable batteries

nontoxic: having no known poisonous qualities

normal operating cell temperature: defined as the standard operating temperature of a PV module at 800 W/m°, 20° C ambient, 1 meter per second wind speed; used to estimate the nominal operating temperature of a module in its working environment

nominal voltage: the terminal voltage of a cell or battery discharging at a specified rate and at a specified temperature; in normal systems, this is usually a multiple of 12.

N-type silicon: silicon doped (containing impurities) which give the lattice a net negative charge, repelling electrons

off-peak energy: electricity during the baseload period, which is usually cheaper. Utilities often must keep generators turning, and are eager to find users during these periods, and so sell off-peak energy for less.

offpeak kilowatt: a kilowatt hour of off-peak energy

off the grid: not connected to the powerlines, energy self-sufficient

ohm: the basic unit of electrical resistance; I=RV, or Current (amperes) equals Resistance (ohms) times Voltage (volts).

on-line: connected to the system, ready for work

on-the-grid: where most of America lives and works, connected to a continent-spanning web of electrical distribution lines

open circuit voltage: the maximum voltage measurable at the terminals of a photovoltaic cell, module, or array with no load applied

operating point: the current and voltage that a module or array produces under load, as determined by the I-V curve

order of magnitude: multiplied or divided by ten. One hundred is an order of magnitude smaller than one thousand, and an order of magnitude large than ten.

orientation: placement with respect to the cardinal directions North, East, South, and West; azimuth is the measure of orientation.

outgas: of any material, the production of gasses; batteries outgas during charging; new synthetic rugs outgas when struck by sunlight, or when warm, or whenever they feel like it.

overcharge: forcing current into a fully charged battery; a bad idea except during equalization

overcurrent: too much current for the wiring; overcurrent protection, in the form of fuses and circuit breakers, guards against this.

owner-builder: one of the few printable things building inspectors call people who build their own homes

panel: any flat modular structure; solar panels may collect solar energy by many means; a number of photovoltaic modules may be assembled into a panel using a mechanical frame, but this should more properly be called an array or subarray.

parallel: connecting the like poles to like in an electrical circuit, so plus connects to plus and minus to minus; this arrangement increases current without affecting voltage.

particulates: particles that are so small that they persist in suspension in air or water

passively-heated: a shelter which has its space heated by the sun without using any other energy

passive solar: a shelter which maintains a comfortable inside temperature simply by accepting, storing, and preserving the heat from sunlight

patch cutting: clear cutting (cutting all trees) on a small scale, usually less than an acre

pathetic fallacy: attributing human motivations to inanimate objects or lower animals

peak demand: the largest amount of electricity demanded by a utility's customers; typically, peak demand happens in early afternoon on the hottest weekday of the year.

peak load: the same as peak demand but on a smaller scale, the maximum load demanded of a single system

peak power current: the amperage produced by a photovoltaic module operating at the "knee" of its I-V curve

peak sun hours: the equivalent number of hours per day when solar irradiance averages one sun (1 kW/m°). "Six peak sun hours" means that the energy received during total daylight hours equals the energy that would have been received if the sun had shone for six hours at a rate of 1000 watts per square meter.

peak kilowatt: a kilowatt hour of electricity take during peak demand, usually the most expensive electricity money can buy

peak watt: the manufacturer's measure of the best possible output of a module under ideal laboratory conditions

pelton wheel: a special turbine, designed by someone named Pelton, for converting flowing water into rotational energy

periodic table of elements: a chart showing the chemical elements organized by the number of protons in their nuclei and the number of electrons in their outer, or valence, band

payback: the time it takes to recoup the cost of improved technology as compared to the conventional solution. Payback on a compact fluorescent bulb (as compared to an incandescent bulb) may take a year or two, but over the whole life of the CF the savings will probably exceed the original cost of the bulb, and payback will take place several times over.

PG&E: Pacific Gas and Electric, the local and sometimes beloved utility for much of Northern California

phantom loads: "energy criminals" that are on even when you turn them off: instant on TVs, microwaves with clocks; symptomatic of impatience and our sloppy preference for immediacy over efficiency

photon: the theoretical particle used to explain light

photophobic: fear of light (or preference for darkness), usually used of insects and animals. The opposite, phototropic, means light-seeking.

photovoltaics: PVs or modules which utilize the photovoltaic effect to generate useable amounts of electricity

photovolactics: an indicator of ignorance on the order of "nuke-ular;" makes us giggle

photovoltaic cell: the proper name for a device manufactured to pump electricity when light falls on it photovoltaic system: the modules, controls, storage, and other components which constitute a standalone solar energy system

plates: the thin pieces of metal or other material used to collect electrical energy in a battery

plug-loads: the appliances and other devices plugged into a power system

plutonium: a particularly nasty radioactive material used in nuclear generation of electricity. One atom is enough to kill you.

pn-junction: the plane within a photovoltaic cell where the positively- and negatively-doped silicon layers meet

pocket plate: a plate for a battery in which active materials are held in a perforated metal pocket on a support strip

pollution: any dumping of toxic or unpleasant materials into air or water

polyurethane: a long-chain carbon molecule, a good basis for sealants, paints, and plastics

power: kinetic, or moving energy, actually performing work; in an electrical system, power is measured in watts

power conditioning equipment: electrical devices which change electrical forms (an inverter is an example) or assure that the electricity is of the correct form and reliability for the equipment for which it provides; a surge protector is another example.

power density: ratio of a battery's rated power available to its volume (in liters) or weight (in kilograms)

PUC: Public Utilities Commission; many states call it something else, but this is the agency responsible for regulating utility rates and practices

PURPA: this 1978 legislation, the Public Utility Regulatory Policy Act, requires utilities to purchase power from anyone at the utility's avoided cost.

PVs: photovoltaic modules

R

radioactive material: a substance which, left to itself, sheds tiny highly energetic pieces which put anyone nearby at great risk. Plutonium is one of these. Radioactive materials remain active indefinitely, but the time over which they are active is measured in terms of halflife, the time it takes them to become half as active as they are now; plutonium's half life is a little over twenty two thousand years.

ram pump: a water-pumping machine that uses a water-hammer effect (based on the inertia of flowing water) to lift water

rated battery capacity: manufacturer's term indicating the maximum energy that can be withdrawn from a battery at a standard specified rate

rated module current: manufacturer's indication of module current under standard laboratory test conditions

renewable energy: an energy source that renews itself without effort; fossil fuels, once consumed, are gone forever, while solar energy is renewable in that the sun we harvest today has no effect on the sun we can harvest tomorrow.

renewables: shorthand term for renewable energy or materials sources

resistance: the ability of a substance to resist electrical flow; in electricity, resistance is measured in ohms

retrofit: install new equipment to a structure which was not prepared for it. For example, we may retrofit a lamp with a compact fluorescent bulb, but the new bulb's shape may not fit well with the lamp's design.

romex: an electrician's term for common two-conductor-with-ground wire, the kind houses are wired with

root mean square: RMS, the effective voltage of alternating current, usually about seventy percent (the square root of two over two) of the peak voltage. House current typically has an RMS of 117 volts, and a peak voltage of 167 volts.

RPM: rotations per minute

R-value: Resistance value, used specifically of materials used for insulating structures. Fiberglass insulation three inches thick has an R-value of thirteen.

S

seasonal depth of discharge: an adjustment for long-term seasonal battery discharge resulting in a smaller array and a battery bank matched to low-insolation season needs

secondary battery: a battery which can be repeatedly discharged and fully recharged

self discharge: the tendency of a battery to lose its charge through internal chemical activity

semiconductor: the chief ingredient in a photovoltaic cell, a normal insulating substance which conducts electricity under certain circumstances

series connection: wiring devices with alternating poles, plus to minus, plus to minus; this arrangement increases voltage (potential) without increasing current.

set-back thermostat: combines a clock and a thermostat so that a zone (like a bedroom) may be kept comfortable only when in use

setpoint: electrical condition, usually voltage, at which controls are adjusted to change their action

shallow-cycle battery: like an automotive battery, designed to be kept nearly fully charged; such batteries perform poorly when discharged by more than 25% of their capacity.

shelf life: the period of time a device can be expected to be stored and still perform to specifications

short circuit: an electrical path which connects opposite sides of a source without any appreciable load, thereby allowing maximum (and possibly disastrous) current flow

short circuit current: current produced by a photovoltaic cell, module, or array when its output terminals are connected to each other, or "short-circuited"

showerhead: in common usage, a device for wasting energy by using too much hot water; in the Real Goods home, low flow showerheads prevent this undesirable result.

silicon: one of the most abundant elements on the planet, commonly found as sand and used to make photovoltaic cells

single-crystal silicon: silicon carefully melted and grown in large boules, then sliced and treated to become the most efficient photovoltaic cells

slow-blow: a fuse which tolerates a degree of overcurrent momentarily; a good choice for motors and other devices which require initial power surges to get rolling

slow paced: a description of the life of a Real Goods employee . . .not!

solar aperture: the opening to the south of a site (in the northern hemisphere) across which the sun passes; trees, mountains, and buildings may narrow the aperture, which also changes with the season.

solar cells: see photovoltaic cells

solar fraction: the fraction of electricity which may be reasonable harvested from sun falling on a site. The solar fraction will be less in a foggy or cloudy site, or one with a narrower solar aperture, than an open, sunny site.

solar hot water heating: direct or indirect use of heat taken from the sun to heat domestic hot water

solar oven: simply a box with a glass front and, optionally, reflectors and reflector coated walls, which heats up in the sun sufficiently to cook food

solar panels: any kind of flat devices placed in the sun to harvest solar energy

solar resource: the amount of insolation a site receives, normally expressed in kilowatt-hours per square meter per day

specific gravity: the relative density of a substance compared to water (for liquids and solids) or air (for gases.) Water is defined as 1.0; a fully charged sulfuric acid electrolyte might be as dense as 1.30, or thirty percent denser than water. Specific gravity is measured with a hydrometer.

stand-alone: a system, sometimes a home, that requires no imported energy

stand-by: a device kept for times when the primary device is unable to perform; a stand-by generator is the same as a back-up generator.

starved electrolyte cell: a battery cell containing little or no free fluid electrolyte

state of charge: the real amount of energy stored in a battery as a percentage of its total rated capacity

state-of-the-art: a term beloved by technoids to express that this is the hottest thing since sliced bread

stratification: in a battery, when the electrolyte acid concentration varies in layers from top to bottom; seldom a problem in vehicle batteries, due to vehicular motion and vibration, this can be a problem with static batteries, and can be corrected by periodic equalization.

stepwise: a little at a time, incrementally

sub-array: part of an array, usually photovoltaic, wired to be controlled separately

sulfating: formation of lead-sulfate crystals on the plates of a lead-acid battery, which can cause permanent damage to the battery

super-insulated: using as much insulation as possible, usually R-50 and above

surge capacity: the ability of a battery or inverter to sustain a short current surge in excess of rated capacity in order to start a device which requires an initial surge current

sustainable: material or energy sources which, if managed carefully, will provide at current levels indefinitely. A theoretical example: redwood is sustainable if it is harvested sparingly (large takings and exportation to Japan not allowed) and if every tree taken is replaced with another redwood. Sustainability can be, and usually is, abused for profit by playing it like a shell game, by planting, for example, a fast-growing fir in place of the harvested redwood.

system availability: the probability or percentage of time that an energy storage system will be able to meet load demand fully

T

temperature compensation: an allowance made by a charge controller to match battery charging to battery temperature

temperature correction: applied to derive true storage capacity using battery nameplate capacity and temperature; batteries are rated at 20° C.

therm: a quantity of natural gas, 100 cubic feet, roughly 100,000 Btus of potential heat

thermal mass: solid, usually masonry volumes inside a structure which absorb heat, then radiate it slowly when the surrounding air falls below their temperature

thermoelectric: producing heat using electricity; a bad idea

thermography: photography of heat loss, usually with a special video camera sensitive to the far end of the infrared spectrum

thermosiphon: a circulation system which takes advantage of the fact that warmer substances rise. By placing the solar collector of a solar hot water system below the tank, thermosiphoning takes care of circulating the hot water, and pumping is not required.

thin-film module: an inexpensive way of manufacturing photovoltaic modules; thin-film modules typically are less efficient than single-crystal or multi-crystal devices; also called amorphous silicon modules.

tilt angle: measures the angle of a panel from the horizontal

tracker: a device that follows the sun and keeps the panel perpendicular to it transformers: a simple electrical device that changes the voltage of alternating current; most transformers are inductive, which means they set up a field around themselves, which is often a costly thing to do.

transparent energy system: a system that looks and acts like a conventional grid-connected home system, but is independent

trickle charge: a small current intended to maintain an inactive battery in a fully charged condition

troubleshoot: a form of recreation not unlike riding to hounds in which the technician attempts to find, catch, and eliminate the trouble in a system

tungsten filament: the small coil in a light bulb that glows hotly and brightly when electricity passes through it

turbine: a vaned wheel over which a rapidly moving liquid or gas is passed, causing the wheel to spin; a device for converting flow to rotational energy

turnkey: the jail warden; more commonly in our context, a system that is ready for the owner-occupant from the first time he or she turns the key in the front door lock

TV: a device for wasting time and scrambling the brain

two-by-fours: standard building members, originally two by four inches, now 1.5" by 3.5"; often referred to as "sticks" in a "stick-frame" house

U

ultrafilter: of water, to remove all particulates and impurities down to the sub-micron range, about the size of giardia and larger viruses

uninterruptible power supply: an energy system providing ultra-reliable power; essential for computers, aircraft guidance, medical, and other systems; also known as a UPS

V

Varistor: a voltage-dependent variable resistor, normally used to protect sensitive equipment from spikes (like lightning strike) by diverting the energy to ground

VCR: video-cassette recorder, a device for making TVs slightly more responsive and useful

VDTs: Video Display Terminals, like televisions and computer screens

vented cell: a battery cell designed with a vent mechanism for expelling gasses during charging

volt: measure of electrical potential: 110-volt house electricity has more potential to do work than an equal flow of 12-volt electricity.

voltage drop: lost potential due to wire resistance over distance

W

watt: the standard unit of electrical power; one ampere of current flowing with one volt of potential; 746 watts make one horsepower

watt-hours: one watt for one hour. A 15-watt compact fluorescent consumes 15 of these in 60 minutes.

waveform: the characteristic trace of voltage over time of an alternating current when viewed on an oscilloscope; typically a smooth sine wave, although primitive inverters supply square or modified square waveforms.

wet shelf life: the time that an electrolyte-filled battery can remain unused in the charged condition before dropping below its nominal performance level

wheatgrass: a singularly delicious potation made by squeezing young wheat sprouts; said to promote purity in the digestive tract

whole-life cost analysis: an economic procedure for evaluating all the costs of an activity from cradle to grave, that is, from extraction or culture through manufacture and use, then back to the natural state; a very difficult thing to accomplish with great accuracy, but a very instructive reckoning nonetheless

wind-chill: a factor calculated based on temperature and wind speed which expresses the fact that a given ambient feels colder when the wind is blowing

wind-spinners: fond name for wind machines, devices that turn wind into usable energy

ABBREVIATIONS:

ABS:	acrylonitrile butadiene styrene		LPG:	liquified propane gas
AC:	alternating current		mA:	milliamp or milliampere
AGA:	American Gas Association		mAh:	milliamp-hour
Ah:	amp-hour		MCM:	Thousandths of Circular Mils
AIC:	ampere interrupting capacity		MIPT:	male iron pipe thread
AWG:	American Wire Gauge		NEC:	National Electrical Code
Btu:	British thermal units		NSF:	National Sanitation Foundation
CFC:	chlorofluorocarbon		NTL:	National Testing Laboratories
cfm:	cubic feet per minute		OD:	outside diameter
cu. ft.:	cubic feet		oz:	ounce
DC:	direct current		psi:	pounds per square inch
DPDT:	double pole double throw		PV:	photovoltaic
EPA:	U.S. Environmental Protection Association		PVC:	Poly Vinyl Chloride
FDA:	U.S. Food and Drug Administration		RF:	radio frequency
FIPT:	female iron pipe thread		RO:	reverse osmosis
GFCI:	ground fault circuit interruption		RV:	recreational vehicle
gpm:	gallons per minute		SASE:	self addressed stamped envelope
hp:	horsepower		TDS:	total dissolved solids
hr:	hour		UL:	Underwriters Laboratories
Hz:	Hertz (formerly "cycles per second")		U.S.:	United States
ID:	inside diameter		UV:	ultraviolet
kW:	kilowatt		VAC:	volts AC
kWh:	kilowatt-hour		VDC:	volts DC
lbs:	pounds		W-hr:	watt-hour
LED:	light emitting diode			

PRODUCT INDEX

Every product carried in the Sourcebook is listed in this Index by item number. Page number on which the item is found follows the description.

13-321 — Zomeworks 2 Siemens M65-M75-M55, 117

13-323 Zomeworks 2 MSX-60, 117

13-331 — Zomeworks 3 Siemens M65-M75-M55, 117

13-341 — Zomeworks 4 Siemens M65-M75-M55, 117

13-342 — Ground/Roof Rack 2 Siemens M-series, 116

13-343 — Ground/Roof Rack 3 Siemens M-series, 116

13-344 — Ground/Roof Rack 4 Siemens M-series, 116

13-345 — Ground/Roof Rack 6 Siemens M-series, 116

13-346 — Ground/Roof Rack 8 Siemens M-series, 116

13-347 — Ground/Roof Rack 2 Siemens PC-4, 116

13-348 — Ground/Roof Rack 3 Siemens PC-4, 116

13-349 — Ground/Roof Rack 4 Siemens PC-4, 116

13-350 — Ground/Roof Rack 5 Siemens PC-4, 116

13-351 — Ground/Roof Rack 2 Solarex, 116

13-352 — Ground/Roof Rack 3 Solarex, 116

13-353 — Ground/Roof Rack 4 Solarex, 116

13-354 — Ground/Roof Rack 5 Solarex, 116

13-355 — Ground/Roof Rack 6 Solarex, 116

13-399 — Zomeworks 1 Siemens PC-4, 118

13-400 — Zomeworks 1 Siemens M-series, 117

13-401 — Zomeworks 2 Siemens M-series, 117

13-402 — Zomeworks 4 Siemens M-series, 117

13-403 — Zomeworks 6 Siemens M-series, 117

13-404 — Zomeworks 8 Siemens M-series, 117

13-405 — Zomeworks 12 Siemens M-series, 117

13-407 — Zomeworks 2 MSX-60, 117

13-408 — Zomeworks 3 Siemens M-series, 117

13-410 — Zomeworks 3 MSX-60, 117

13-412 — Zomeworks 4 MSX-60, 117

13-414 — Zomeworks 6 MSX-60, 117

13-416 — Zomeworks 8 MSX-60, 117

13-418 — Zomeworks 12 MSX-60, 117

13-419 — Zomeworks 14 Siemens M-series, 117

13-421 — Zomeworks 2 Siemens PC-4, 117

13-422 — Zomeworks 4 Siemens PC-4, 117

13-423 — Zomeworks 6 Siemens PC-4, 117

13-425 — Zomeworks 3 Siemens PC-4, 117

13-430 — Zomeworks 10 Siemens M-series, 117

13-431 — Zomeworks 8 Siemens PC-4, 117

13-432 — Zomeworks 10 Siemens PC-4, 117

13-433 — Zomeworks 12 Siemens PC-4, 117

13-434 — Zomeworks 10 MSX-60, 117

13-464 — Zomeworks 1 PC4, 117

13-465 — Zomeworks 2 PC4, 117

13-466 — Zomeworks 3 PC4, 117

13-467 — Zomeworks 4 PC4, 117

13-702 — Solar Mount, PV Modules, 118

13-909 — 1 Siemens M-series, 118

13-911 — 1 Solarex, or PC4, 118

13-912 — 2 Siemens M-series, 118

13-914 — 2 Solarex, or PC4, 118

13-915 — 1 Siemens M-series, 118

13-917 — 1 Solarex, or PC4, 118

13-918 — 2 Siemens M-series, 118

13-920 — 2 Solarex, or PC4, 118

15-101 — Deep Cycle Battery, 6V/220A-hr, 181

15-102 — Deep Cycle Battery, 6V/350A-hr, 181

15-103 — RV/Marine Battery, 12V/85A-hr, 181

15-104 — RV/Marine Battery, 12V/105A-hr, 181

15-200 — Gel Cell Battery, 12V/7A-hr, 182

15-202 — Gel Cell Battery, 12V/12A-hr, 182

15-203 — Gel Cell Battery, 12V/95A-hr, 182

15-204 — Gel Cell Battery, 12V/31A-hr, 182

15-205 — Gel Cell Battery, 12V/48A-hr, 182

15-206 — Gel Cell Battery, 12V/70A-hr, 182

15-207 — Gel Cell Battery, 12V/160A-hr, 182

15-208 — Gel Cell Battery, 12V/225A-hr, 182

15-421 — Chloride Battery, 12V/525A-hr, 183

15-422 — Chloride Battery, 12V/635A-hr, 183

15-423 — Chloride Battery, 12V/740A-hr, 183

15-424 — Chloride Battery, 12V/845A-hr, 183

15-425 — Chloride Battery, 12V/950A-hr, 183

15-426 — Chloride Battery, 12V/1055A-hr, 183

15-427 — Chloride Battery, 12V/1160A-hr, 183

15-428 — Chloride Battery, 12V/1270A-hr, 183

15-429 — Chloride Battery, 12V/1375A-hr, 183

33-103 — Quartz Lamp, 20W, 406

33-104 — Quartz Lamp, 35W, 406

33-105 — Quartz Lamp, 50W, 406

33-108 — 20-Watt Halogen Flood Lamp, 12V, 406

33-109 — Replacement Halogen Bulb, 5W, 405

33-114 — 42-Watt Halogen (set of 2), 397

33-115 — 52-Watt Halogen (set of 2), 397

33-116 — 72-Watt Halogen (set of 2), 397

33-117 — 50w Halogen Floodlight, 397

33-123 — Halogen PAR 38 Spot, 45-Watt, 397

33-124 — Halogen PAR 38 Spot, 90-Watt, 397

33-125 — Halogen PAR 38 Flood, 45-Watt, 397

33-126 — Halogen PAR 38 Flood, 90-Watt, 397

33-306 — 12" Littlite, 405

33-307 — 18" Littlite, 405

33-308 — 12" Littlite, 12V Plug, 405

33-309 — 18" Littlite, 12V Plug, 405

33-401 — Littlite Weighted Base, 405

33-402 — Littlite Snap Mount, 405

33-403 — Littlite Mounting Clip, 405

33-404 — Tail Light Bulb Adapter, 404

33-501 — Littlite 120V Transformer, 405

34-320 — Prime Light, 407

34-321 — Pathmarker Solar Light (set of 2), 406

34-323 — Pathway Light, 406

34-325 — Classic Prime Light, 407

34-332 — Walklite repl. bulb, 407

35-101 — Humphrey Propane Lamp, 433

35-102 — Propane Mantle, 433

35-320 — Kerosene Lamp, Patronne, 432

35-321 — Spare Chimney, Patronne, 432

35-323 — Spare Burner, Patronne, 432

35-324 — Replacement Ball, Patronne, 432

35-326 — Kerosene Lamp, Vintners, 432

35-327 — Lamp Parts, Vintners, 433

35-328 — Lamp Parts, Patronne, 433

35-329 — Kerosene Lamp, Aladdin, 432

35-329A — Spare Mantle for Aladdin, 432

35-329B — Spare Chimney for Aladdin, 432

35-329C — Spare Wick for Aladdin, 432

35-329D — Insect Screen for Aladdin, 432

35-337 — Adjustable Halogen Desk Lamp, 396

35-340 — Kerosene Lamp, Concierge, 433

35-341 — Spare Chimney, Concierge, 433

35-343 — Spare Burner, Concierge, 433

35-344 — Replacement Ball, Concierge, 433

35-346 — Lamp Parts, Concierge, 433

36-107 — LOA Quad 18-Watt, 390

36-108 — LOA Quad 27-Watt, 390

36-109 — LOA 18-Watt Replacement Bulb, 390

36-110 — LOA 27-Watt Replacement Bulb, 390

36-115 — Philips SLS Triple Tube 15-Watt, 390

36-116 — Philips SLS Triple Tube 20-Watt, 390

36-117 — Philips SLS Triple Tube 23-Watt, 390

36-123 — Halogen PAR 38 Spot, 45w, 397

36-124 — Halogen PAR 38 Spot, 90w, 397

36-125 — Halogen PAR 38 Flood, 45w, 397

36-123 — Halogen PAR 38 Flood, 90w, 397

36-127 — 13-w Floodlight, 395

36-145 — 30-Watt Circular, 392

36-147 — 22-Watt Circular, 392

36-181 — Naturelite Single Lamp, 407

36-182 — Naturelite 20w repl. bulb, 407

36-183 — Naturelite Dual Lamp, 407

36-184 — Naturelite 10w repl. bulbs, 407

36-190 — SLS/R-30 15-Watt Flood, 391

36-191 — SLS/R-40 20-Watt Flood, 391

36-204 — Elba Ballast for 2 F40T12 tubes, 393

36-205 — Elba Ballast for 2 F32T8 tubes, 393

36-206 — Elba Ballast for 3 F32T8 tubes, 393

36-207 — Elba Ballast for 4 F32T8 tubes, 393

36-209 — Flouorever Lamp, 390

36-210 — SLS 11-w Lamp, 389

36-306 — Prolight QCR-38 Spot, 391

36-307 — Prolight QCR-38 Flood, 391

36-338 — 26-w Ceiling Fixture, 394

36-339 — 54-w Ceiling Fixture, 394

36-402 — Sensor Socket, 395

36-403 — 10" Lampshade Harp, 396

36-404 — 12" Lampshade Harp, 396

36-405 — Socket Extender, 396

36-503 — Panasonic T-16, 389

36-504 — Panasonic G-16, 389

36-607 — 13-w Adaptor w/bulb, 388

36-608 — Screwbase Adaptor, 388

36-609 — 22-w Adaptor w/bulb, 388

37-301 — Christmas Tree Lights, 12V, 405

37-302 — Solar Lantern, 408

37-308 — Dynalite Flashlight, 408

37-310 — Dashlite Flashlight, 408

37-312 — Replacement Bulbs, 2 pack, 408

37-329 — Siemens Solar Lantern, 408

41-132 — Rust & Dirt Cartridge (2), 301

41-134 — El Sid Circulation Pump, 299

41-137 — Inline Sediment Filter, 301, 331

41-140 — Pressure Switch, 301, 310

41-148 — Easy Install Kit #505, 301

41-149 — 1/3 hp PV Pool Pump, 300

41-153 — Solar Star 1450, 24V, 301

41-155 — 1/2 hp PV Pool Pump, 300

41-156 — 1 hp PV Pool Pump, 300

41-157 — V625 Solar Fountain Pump, 299

41-158 — V1250 Solar Fountain Pump, 299

41-159 — Solar Star 1450, 12V, 301

41-301 — Bronze 12V Centrifugal Pump, 298

41-401 — Pressure Tank, 2 gal., 310

41-404 — Pressure Tank, 20 gal., 310

41-406 — Pressure Tank, 8-1/2 gal., 310

41-407 — Pressure Tank, 32 gal., 310

41-410 — Check Valve, Bronze 3/4", 311

41-411 — Check Valve, Bronze 1", 311

41-412 — Check Valve, Bronze 1-1/4", 311

41-413 — Check Valve, Bronze 1-1/2", 311

41-414 — Check Valve, Bronze 2", 311

41-420 — Foot Valve, Bronze 3/4", 311

41-421 — Foot Valve, Bronze 1", 311

41-422 — Foot Valve, Bronze 1-1/4", 311

41-423 — Foot Valve, Bronze 1-1/2", 311

41-424 — Foot Valve, Bronze 2", 311

41-436 — Taste & Odor Cartridge (2), 331

41-442 — Pressure Gauge 0 to 30 psi, 311

41-443 — Pressure Gauge 0 to 60 psi, 311

41-444 — Pressure Gauge 0 to 100 psi, 311

41-445 — Pressure Gauge 0 to 300 psi, 311

41-450 — SHURflo Low-Flow Pump, 302

41-451 — SHURflo Medium-Flow Pump, 12V, 302

41-452 — SHURflo Medium-Flow Pump, 24 volt, 302

41-453 — SHURflo High-Flow Pump, 302

41-454 — SHURflo 115 volt AC Pump, 303

41-455 — SHURflo Solar Submersible, 304

41-457 — Solar Star 1075, 12V, 300

41-458 — Solar Star 1075, 24V, 300

41-459 — Solar Star 1140, 12V, 301

41-460 — Solar Star 1140, 24V, 301

41-505 — March Circulating Pump, 1/100 hp, 298

41-526 — Hartell Circulator Pumps, 298

41-633 — Controller/Converter, 12V to 24V, 305

41-637 — Float Switch "U," 305, 312

41-638 — Float Switch "D," 305, 312

41-671 — 12V Submersible Pump, 298

41-701 — Bowjon Pump, Homesteader, 307

41-711 — Bowjon Pump, Rancher, 307

41-801 — High Lifter Pump, 4.5:1 Ratio, 309

41-802 — High Lifter Pump, 9:1 Ratio, 309

41-803 — High Lifter Ratio Conversion Kit, 309

41-804 — High Lifter Rebuild Kit, 4.5:1 ratio, 309

41-805 — High Lifter Rebuild Kit, 9:1 ratio, 309

41-811 — Ram Pump, 3/4", 308

41-812 — Ram Pump, 1", 308

41-813 — Ram Pump, 1-1/4", 308

41-814 — Ram Pump, 1-1/2", 308

42-000 — Water Test (non-city water), 323

42-003 — NTL Check w/Pesticide Option, 322

42-218 — Hydrotech Countertop R.O. Unit, 324

42-219 — Replacement TFC Membrane for 42-218, 324

42-220 — Sediment Prefilter for 42-218, 324

SUBJECT INDEX

Information about products available from Real Goods is indicated either under the sub-entry "listings", or under specific product names. Page numbers follow the description.

A

I

J

Notes

Notes

Notes

Notes

0996
555 Leslie St.
Ukiah, CA 95482-5507

For credit card orders call toll-free: **1-800-762-7325**
Monday–Friday: 6am–8pm; Saturday: 8am–6pm (PST)
Fax: 707-468-9486; Overseas orders: 707-468-9214

ORDERED BY:

A

Name _____

Address _____

City _____ State _____ Zip _____

❑ **Please do not send a catalog with this order.**

DAYTIME PHONE: _____
(in case we have questions about your order)

B | **ALTERNATIVE SHIPPING OR GIFT ADDRESS**
Name _____
Address _____
City _____ State ____ Zip _____
Message _____

C | **GIFT ADDRESS**
Name _____
Address _____
City _____ State ____ Zip _____
Message _____

ITEM #	SIZE/COLOR	QTY.	PAGE	DESCRIPTION	CIRCLE ADDRESS	PRICE	MEMBER PRICE (5% off)	TOTAL
				LIFETIME MEMBERSHIP	A B C	$50	✕	
					A B C			
					A B C			
					A B C			
					A B C			
					A B C			
					A B C			
					A B C			
					A B C			
					A B C			
					A B C			
					A B C			
					A B C			
					A B C			
					A B C			
					A B C			
					A B C			
					A B C			
					A B C			

SHIPPING AND HANDLING CHARGES			
Under $25:	$4.95	$100–$149.99:	$12.25
$25–$34.99:	$6.25	$150–$199.99:	$15.50
$35–$49.99:	$7.25	$200–$999.99:	add 8%
$50–$74.99:	$8.75	$1000 and up:	add 6%
$75–$99.99:	$9.95		

Does not include Freight Collect items. For Canada add $5.00 to the above charges. For delivery to AK, HI, PR, VI, GU, or addresses requiring parcel post, add $7 for orders over $50.

EXPRESS DELIVERY
For faster delivery we recommend Express Economy or Express Standard shipping. For only $6 additional to rates above, you can have your package delivered within 3 business days of the time you place your order. For $12 dollars additional you can have your order within 1–2 business days (if you place your order by 1pm Pacific Time). **This rate applies to all packages under 20 lbs. and to in-stock merchandise shipped within the Continental United States.** Call for rates on deliveries to Alaska, Hawaii, Virgin Islands, Puerto Rico and Guam.

PAYMENT METHOD:
❑ Money order or Check
❑ Credit Card (MasterCard, Visa, Discover, or Amex)
Account Number (please include all numbers)

Expiration Date: ☐☐ — ☐☐

Signature _____

TOTAL OF GOODS	
SALES TAX (7 1/4% CA, 5% WI) Deliveries only.	
SHIPPING (each Address) See chart at left.	
EXPRESS OR ADDITIONAL SHIPPING CHARGES	
TOTAL ENCLOSED (US Dollars only)	

THANK YOU FOR YOUR ORDER
Prices subject to change without notice.

Credit Card Orders By Telephone & Fax

For fast, efficient and friendly service, call us, toll free, at 1-800-762-7325 or fax us with your order, 707-468-9486. Please have your order form filled out completely and your credit card handy. Include your return phone number on faxed orders.

Ordering By Mail

Be sure to supply all of the information requested on the order form. Funds must be in US dollars drawn on a US bank. Where prices may have increased because a catalog is outdated, we will charge the balance to your credit card (if it is less than $10) or contact you if the difference is more.

Overseas Orders

To place an order please call us at 707/468-9214, Monday through Friday, 6am–8pm and Saturday, 8am–6pm (Pacific Time) or Fax 707/468-9486. Call or fax for shipping quote.

Shipping

We ship most merchandise via UPS within 24 hours of receiving your order. Allow 7 to 14 working days for delivery depending on your distance from our Distribution Center in Mendocino County, California. Overweight or bulky items may require additional shipping charges. For fastest delivery, we recommend our Express Delivery service.

There are a few items that we ship directly from the manufacturer (they are indicated by a (•) next to the item number in the catalog). These items are designated on your invoice and are billed at the time of order. Allow extra time for delivery of these items.

Freight Collect

Some large items (like Sun Frost refrigerators) are shipped freight collect. Shipment will arrive via truck. The trucker is responsible for curbside delivery only, not placement within your premises. Actual charges are due at the time of delivery.

Freight Damage, Returns & Adjustments

Inspect all shipments upon arrival. If you discover damage, please notify the carrier. Be sure to save the damaged cartons in their original condition until any claim is settled. If you return merchandise to us, be sure to complete the return form on the reverse of your invoice. Items must be returned in their original carton and in original condition to be eligible for replacement or refund. Ship to us via UPS ground or postal insured. We may have to charge a restocking fee on items shipped directly from manufacturers (these items are indicated by a (•) by the item code), special orders or items not in their original condition or container.

Express Delivery

For fastest delivery, we recommend Express Economy or Standard Express shipping. For only $6 additional, you can have your package delivered within 3 business days from the time you place your order. For $12 additional, you can have your order within 1–2 business days (if you order by 1pm Pacific Time).

This rate applies to all packages up to 20 lbs. and to in-stock merchandise only. Sorry, Express Delivery is not available on products shipped directly from the manufacturer. Call for rates on heavy or bulky packages, international shipments or deliveries to Alaska, Hawaii, Virgin Islands and Puerto Rico.

Toll-free Customer Service

If you have a problem with your order, our Customer Service Representatives are available between 8:30 am and 5:00 pm (Pacific Time) Monday through Friday. Call us toll-free at 1-800-994-4243.

Product Information

Our representatives can provide you with detailed information about any of our products. If you need to know more, please call us at 1-800-762-7325.

Technical Assistance

If you require technical assistance with the design of a renewable energy system or installation help, please call 1-800-919-2400, Monday through Friday, 9am–5pm. Our Technical Department can also be reached by faxing 707-462-4807.

Gift Certificates & Wedding Registry

If you are unable to decide upon a gift, we offer Real Goods Gift Certificates. We will be happy to send a gift certificate in any amount to the person of your choice along with our latest edition of the Real Goods Catalog. Many of our products make great gifts for new households.

00-001 Specify $ Amount
Call if you would like to use our Wedding Registry service.
(Lifetime Membership discount does not apply to Gift Certificates.)

Prices subject to change without notice.

0996
555 Leslie St.
Ukiah, CA 95482-5507

For credit card orders call toll-free: **1-800-762-7325**
Monday–Friday: 6am–8pm; Saturday: 8am–6pm (PST)
Fax: 707-468-9486; Overseas orders: 707-468-9214

ORDERED BY:

A

Name _____

Address _____

City _____ State _____ Zip _____

❏ **Please do not send a catalog with this order.**

DAYTIME PHONE: _____
(in case we have questions about your order)

B | **ALTERNATIVE SHIPPING OR GIFT ADDRESS**
Name _____
Address _____
City _____ State ____ Zip _____
Message _____

C | **GIFT ADDRESS**
Name _____
Address _____
City _____ State ____ Zip _____
Message _____

ITEM #	SIZE/COLOR	QTY.	PAGE	DESCRIPTION	CIRCLE ADDRESS	PRICE	MEMBER PRICE (5% off)	TOTAL
				LIFETIME MEMBERSHIP	A B C	$50	✕	
					A B C			
					A B C			
					A B C			
					A B C			
					A B C			
					A B C			
					A B C			
					A B C			
					A B C			
					A B C			
					A B C			
					A B C			
					A B C			
					A D C			
					A B C			
					A B C			
					A B C			
					A B C			

SHIPPING AND HANDLING CHARGES			
Under $25:	$4.95	$100–$149.99:	$12.25
$25–$34.99:	$6.25	$150–$199.99:	$15.50
$35–$49.99:	$7.25	$200–$999.99:	add 8%
$50–$74.99:	$8.75	$1000 and up:	add 6%
$75–$99.99:	$9.95		

Does not include Freight Collect items. For Canada add $5.00 to the above charges. For delivery to AK, HI, PR, VI, GU, or addresses requiring parcel post, add $7 for orders over $50.

EXPRESS DELIVERY
For faster delivery we recommend Express Economy or Express Standard shipping. For only $6 additional to rates above, you can have your package delivered within 3 business days of the time you place your order. For $12 dollars additional you can have your order within 1–2 business days (if you place your order by 1pm Pacific Time). **This rate applies to all packages under 20 lbs. and to in-stock merchandise shipped within the Continental United States.** Call for rates on deliveries to Alaska, Hawaii, Virgin Islands, Puerto Rico and Guam.

PAYMENT METHOD:
❏ Money order or Check
❏ Credit Card (MasterCard, Visa, Discover, or Amex)
  Account Number (please include all numbers)

[ ][ ][ ][ ][ ][ ][ ][ ][ ][ ][ ][ ][ ][ ][ ][ ]

Expiration Date: [ ][ ] — [ ][ ]

Signature _____

TOTAL OF GOODS	
SALES TAX (7 1/4% CA, 5% WI) Deliveries only.	
SHIPPING (each Address) See chart at left.	
EXPRESS OR ADDITIONAL SHIPPING CHARGES	
TOTAL ENCLOSED (US Dollars only)	

THANK YOU FOR YOUR ORDER
Prices subject to change without notice.

Credit Card Orders By Telephone & Fax

For fast, efficient and friendly service, call us, toll free, at 1-800-762-7325 or fax us with your order, 707-468-9486. Please have your order form filled out completely and your credit card handy. Include your return phone number on faxed orders.

Ordering By Mail

Be sure to supply all of the information requested on the order form. Funds must be in US dollars drawn on a US bank. Where prices may have increased because a catalog is outdated, we will charge the balance to your credit card (if it is less than $10) or contact you if the difference is more.

Overseas Orders

To place an order please call us at 707/468-9214, Monday through Friday, 6am–8pm and Saturday, 8am–6pm (Pacific Time) or Fax 707/468-9486. Call or fax for shipping quote.

Shipping

We ship most merchandise via UPS within 24 hours of receiving your order. Allow 7 to 14 working days for delivery depending on your distance from our Distribution Center in Mendocino County, California. Overweight or bulky items may require additional shipping charges. For fastest delivery, we recommend our Express Delivery service.

There are a few items that we ship directly from the manufacturer (they are indicated by a (•) next to the item number in the catalog). These items are designated on your invoice and are billed at the time of order. Allow extra time for delivery of these items.

Freight Collect

Some large items (like Sun Frost refrigerators) are shipped freight collect. Shipment will arrive via truck. The trucker is responsible for curbside delivery only, not placement within your premises. Actual charges are due at the time of delivery.

Freight Damage, Returns & Adjustments

Inspect all shipments upon arrival. If you discover damage, please notify the carrier. Be sure to save the damaged cartons in their original condition until any claim is settled. If you return merchandise to us, be sure to complete the return form on the reverse of your invoice. Items must be returned in their original carton and in original condition to be eligible for replacement or refund. Ship to us via UPS ground or postal insured. We may have to charge a restocking fee on items shipped directly from manufacturers (these items are indicated by a (•) by the item code), special orders or items not in their original condition or container.

Express Delivery

For fastest delivery, we recommend Express Economy or Standard Express shipping. For only $6 additional, you can have your package delivered within 3 business days from the time you place your order. For $12 additional, you can have your order within 1–2 business days (if you order by 1pm Pacific Time).

This rate applies to all packages up to 20 lbs. and to in-stock merchandise only. Sorry, Express Delivery is not available on products shipped directly from the manufacturer. Call for rates on heavy or bulky packages, international shipments or deliveries to Alaska, Hawaii, Virgin Islands and Puerto Rico.

Toll-free Customer Service

If you have a problem with your order, our Customer Service Representatives are available between 8:30 am and 5:00 pm (Pacific Time) Monday through Friday. Call us toll-free at 1-800-994-4243.

Product Information

Our representatives can provide you with detailed information about any of our products. If you need to know more, please call us at 1-800-762-7325.

Technical Assistance

If you require technical assistance with the design of a renewable energy system or installation help, please call 1-800-919-2400, Monday through Friday, 9am–5pm. Our Technical Department can also be reached by faxing 707-462-4807.

Gift Certificates & Wedding Registry

If you are unable to decide upon a gift, we offer Real Goods Gift Certificates. We will be happy to send a gift certificate in any amount to the person of your choice along with our latest edition of the Real Goods Catalog. Many of our products make great gifts for new households.

00-001 Specify $ Amount
Call if you would like to use our Wedding Registry service.
(Lifetime Membership discount does not apply to Gift Certificates.)

Prices subject to change without notice.

0996
555 Leslie St.
Ukiah, CA 95482-5507

For credit card orders call toll-free: **1-800-762-7325**
Monday–Friday: 6am–8pm; Saturday: 8am–6pm (PST)
Fax: 707-468-9486; Overseas orders: 707-468-9214

ORDERED BY:

A

Name _____

Address _____

City _____ State ___ Zip _____

❑ **Please do not send a catalog with this order.**

DAYTIME PHONE: _____
(in case we have questions about your order)

B | ALTERNATIVE SHIPPING OR GIFT ADDRESS
Name _____
Address _____
City _____ State ___ Zip _____
Message _____

C | GIFT ADDRESS
Name _____
Address _____
City _____ State ___ Zip _____
Message _____

ITEM #	SIZE/COLOR	QTY.	PAGE	DESCRIPTION	CIRCLE ADDRESS	PRICE	MEMBER PRICE (5% off)	TOTAL
				LIFETIME MEMBERSHIP	A B C	$50	✕	
					A B C			
					A B C			
					A B C			
					A B C			
					A B C			
					A B C			
					A B C			
					A B C			
					A B C			
					A B C			
					A B C			
					A B C			
					A B C			
					A B C			
					A B C			
					A B C			
					A B C			
					A B C			

SHIPPING AND HANDLING CHARGES			
Under $25:	$4.95	$100–$149.99:	$12.25
$25–$34.99:	$6.25	$150–$199.99:	$15.50
$35–$49.99:	$7.25	$200–$999.99:	add 8%
$50–$74.99:	$8.75	$1000 and up:	add 6%
$75–$99.99:	$9.95		

Does not include Freight Collect items. For Canada add $5.00 to the above charges. For delivery to AK, HI, PR, VI, GU, or addresses requiring parcel post, add $7 for orders over $50.

EXPRESS DELIVERY
For faster delivery we recommend Express Economy or Express Standard shipping. For only $6 additional to rates above, you can have your package delivered within 3 business days of the time you place your order. For $12 dollars additional you can have your order within 1–2 business days (if you place your order by 1pm Pacific Time). **This rate applies to all packages under 20 lbs. and to in-stock merchandise shipped within the Continental United States.** Call for rates on deliveries to Alaska, Hawaii, Virgin Islands, Puerto Rico and Guam.

PAYMENT METHOD:
❑ Money order or Check
❑ Credit Card (MasterCard, Visa, Discover, or Amex)
  Account Number (please include all numbers)

Expiration Date: ☐☐ – ☐☐

Signature _____

TOTAL OF GOODS

SALES TAX
(7 1/4% CA, 5% WI)
Deliveries only.

SHIPPING
(each Address)
See chart at left.

EXPRESS OR
ADDITIONAL
SHIPPING CHARGES

TOTAL ENCLOSED
(US Dollars only)

THANK YOU FOR YOUR ORDER
Prices subject to change without notice.

Credit Card Orders By Telephone & Fax

For fast, efficient and friendly service, call us, toll free, at 1-800-762-7325 or fax us with your order, 707-468-9486. Please have your order form filled out completely and your credit card handy. Include your return phone number on faxed orders.

Ordering By Mail

Be sure to supply all of the information requested on the order form. Funds must be in US dollars drawn on a US bank. Where prices may have increased because a catalog is outdated, we will charge the balance to your credit card (if it is less than $10) or contact you if the difference is more.

Overseas Orders

To place an order please call us at 707/468-9214, Monday through Friday, 6am–8pm and Saturday, 8am–6pm (Pacific Time) or Fax 707/468-9486. Call or fax for shipping quote.

Shipping

We ship most merchandise via UPS within 24 hours of receiving your order. Allow 7 to 14 working days for delivery depending on your distance from our Distribution Center in Mendocino County, California. Overweight or bulky items may require additional shipping charges. For fastest delivery, we recommend our Express Delivery service.

There are a few items that we ship directly from the manufacturer (they are indicated by a (•) next to the item number in the catalog). These items are designated on your invoice and are billed at the time of order. Allow extra time for delivery of these items.

Freight Collect

Some large items (like Sun Frost refrigerators) are shipped freight collect. Shipment will arrive via truck. The trucker is responsible for curbside delivery only, not placement within your premises. Actual charges are due at the time of delivery.

Freight Damage, Returns & Adjustments

Inspect all shipments upon arrival. If you discover damage, please notify the carrier. Be sure to save the damaged cartons in their original condition until any claim is settled. If you return merchandise to us, be sure to complete the return form on the reverse of your invoice. Items must be returned in their original carton and in original condition to be eligible for replacement or refund. Ship to us via UPS ground or postal insured. We may have to charge a restocking fee on items shipped directly from manufacturers (these items are indicated by a (•) by the item code), special orders or items not in their original condition or container.

Express Delivery

For fastest delivery, we recommend Express Economy or Standard Express shipping. For only $6 additional, you can have your package delivered within 3 business days from the time you place your order. For $12 additional, you can have your order within 1–2 business days (if you order by 1pm Pacific Time).

This rate applies to all packages up to 20 lbs. and to in-stock merchandise only. Sorry, Express Delivery is not available on products shipped directly from the manufacturer. Call for rates on heavy or bulky packages, international shipments or deliveries to Alaska, Hawaii, Virgin Islands and Puerto Rico.

Toll-free Customer Service

If you have a problem with your order, our Customer Service Representatives are available between 8:30 am and 5:00 pm (Pacific Time) Monday through Friday. Call us toll-free at 1-800-994-4243.

Product Information

Our representatives can provide you with detailed information about any of our products. If you need to know more, please call us at 1-800-762-7325.

Technical Assistance

If you require technical assistance with the design of a renewable energy system or installation help, please call 1-800-919-2400, Monday through Friday, 9am–5pm. Our Technical Department can also be reached by faxing 707-462-4807.

Gift Certificates & Wedding Registry

If you are unable to decide upon a gift, we offer Real Goods Gift Certificates. We will be happy to send a gift certificate in any amount to the person of your choice along with our latest edition of the Real Goods Catalog. Many of our products make great gifts for new households.

00-001 Specify $ Amount
Call if you would like to use our Wedding Registry service.
(Lifetime Membership discount does not apply to Gift Certificates.)

Prices subject to change without notice.